The Families
of Flowering Plants

Other Books by RICHARD HEADSTROM

Spiders of the United States
Frogs, Toads, and Salamanders as Pets
A Complete Field Guide to Nests in the United States
Nature in Miniature
The Origin of Man
The Story of Russia
Adventures With a Microscope
The Living Year
Garden Friends and Foes
Adventures With Insects
Birds' Nests
Your Insect Pet
Whose Track Is It?
Birds' Nests of the West
Adventures with a Hand Lens
Lizards as Pets
Adventures with Freshwater Animals
The Beetles of America

The Families of Flowering Plants

Richard Headstrom

South Brunswick and New York: A. S. Barnes and Company
London: Thomas Yoseloff Ltd

A. S. Barnes and Co., Inc.
Cranbury, New Jersey 08512

Thomas Yoseloff Ltd
Magdalen House
136–148 Tooley Street
London SE1 2TT, England

Library of Congress Cataloging in Publication Data

Headstrom, Birger Richard, 1902–
The families of flowering plants.

Bibliography: p.
Includes index.
1. Plants. 2. Plant lore. I. Title.
QK50.H4 582'.13 74-30967
ISBN 0-498-01516-5

PRINTED IN THE UNITED STATES OF AMERICA

To My Wife

Contents

Preface

To contribute another book to the vast amount of literature on botany would seem to be an exercise in futility unless such a contriibution might offer something new. Certainly the present contribution does not offer anything in the sense of new material, only the manner in which the material is presented might be said to be new. What the author has attempted to do in the present book is to provide the general reader with a readable and narrative account of the families of flowering plants with a minimum or almost complete absence of technical jargon.

Many people accept plants as natural features of the landscape and think no more about them, except that they may enjoy looking at them occasionally. Others grow them in the garden for their aesthetic value, or as a means of providing food for the table. Still others are interested in plants in order to enrich their understanding and knowledge of the world around them, and use identification manuals to accomplish their purpose. A very few pursue botany courses in college and become professional botanists.

The present book is about the flowering plants, in other words, the plants that produce flowers, such as the rose, sweet pea, pansy, and marigold. It is about the plants that many know of first hand, that grow in the fields and meadows, in the woods and along the roadside, in swamps and marshes, along the banks of streams, whether they be brooks or rivers, in the ponds and lakes, and along the seashore. It is about those that grow in the garden, and in pots that are indoors, as well as those that have been heard of and read about, or those that provide something useful, such as the wood from the mahogany tree, the drug quinine from the bark of the cinchona tree, and such fruits as the banana and pineapple.

The present book is not a field book, nor an identification manual, nor a treatise on botany, but a reading account of some of the better known of the two hundred and fifty thousand plants that have so far been described, about their virtues and vices and the legends and folk tales that are associated with many of them. In other words, it is a popular book about plants and somewhat different from the many that have already been published. At least, that is what the author hopes he has accomplished. Whether he has done so is for the reader to decide.

Introduction

There are about two hundred and fifty thousand species of flowering plants on the earth and perhaps more; at least that many have been described, and new species are added each year as the survey of the earth's vegetation continues and other materials are more critically examined. To the number of species there should also be added the varieties, races, and strains that number in the thousands, and that have been brought into being through man's efforts.

The flowering plants are the most numerous of the groups wherein the plant kingdom has been divided and are the most varied and useful as well. This is because they provide food directly or indirectly for all other living things, though many of them may seem quite worthless as weeds. They also provide hiding places for the creatures of the wild and furnish man with innumerable materials for his welfare and happiness.

The flowering plants exhibit a great variety of form; thus, they may be simple, minute disks that float on the water, or of massive size, such as some of the trees that may grow to a height of three hundred feet and, in structure, rival that of the higher animals in complexity. Some of them may only live a few weeks, others for centuries. And they may be found in every sort of habitat: in the frozen wastes of the polar regions, in hot deserts, in the equatorial rain forests, on mountaintops, and on the seashore. Able to manufacture their own food, they are independent organisms, though some have become degenerate parasites or saprophytes while others have turned carnivorous, feeding on animals that they capture by means of ingenious traps.

The outstanding feature of a flowering plant is the flower. Most people know what a flower is, and, yet, there are plants that produce flowers that most would not recognize as flowers. There are others that produce structures that would be taken as flowers, but

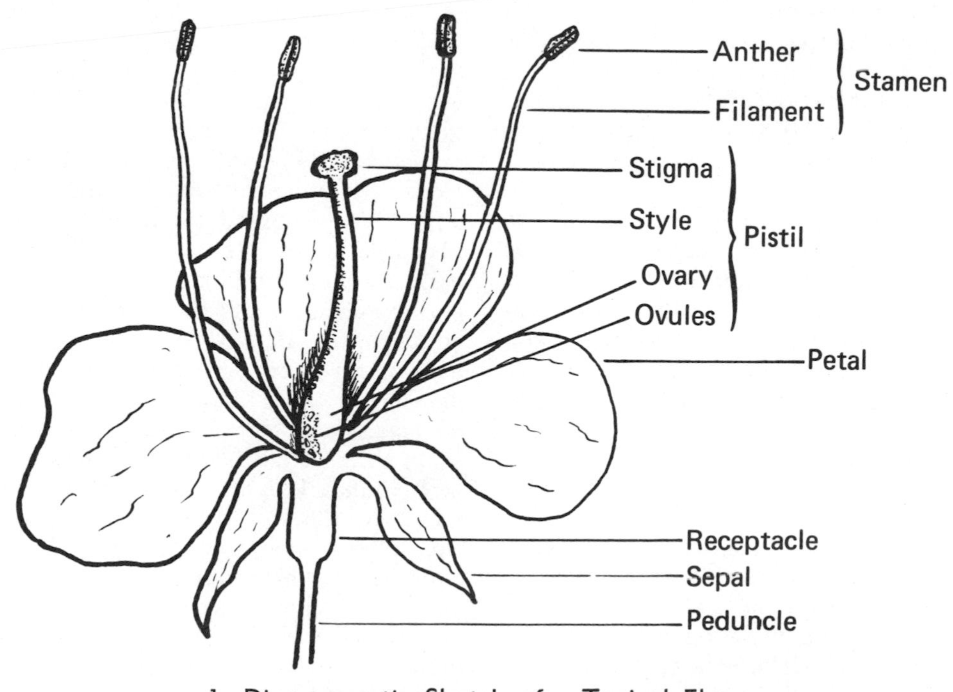

1. Diagrammatic Sketch of a Typical Flower

are not in a botanical sense, or in the economy of the plant. To define a flower is not easy, because a flower is too complex and variable in structure and behavior to allow an easy definition. Succinctly put, it can be said that a flower is a sex organ or combinations of sex organs, or that part of the plant concerned with reproduction and the maintenance of the species. It is designed to produce male and female sex cells and is so constructed that a union of the two may take place; that is, the fertilization of the egg with the resulting production of fruits and seeds. And though flowers may vary greatly in detail and in form, color, odor, size, etc., they are all basically the same.

Thus, the typical flower is borne on a stem called the *peduncle* and the enlarged end of this stem where the parts of a flower are attached is known as the *receptacle,* the parts or floral organs being the *sepals, petals, stamens,* and *pistils.* The outer or lower of these floral organs are the sepals. They are usually small, green, leaflike structures, though they may be of almost any color, that enclose and protect the other floral organs in the bud before they are fully developed. Collectively they are known as the *calyx.* Brightly colored petals, usually conspicuous above and inside the sepals, are collectively known as the *corolla.* The number of petals is usually the same as the sepals, or sometimes it is a multiple of the sepal number. Their purpose is to attract the insects whose visits are important in the reproductive functions of flowers. Flowers draw insects to them in several ways: by bright colors, by a sweet liquid, called nectar, secreted in glands called nectaries that are desired by bees and other kinds of flower-visiting insects; and by odors from essential oils and other substances produced by special modified cells. Together the calyx and corolla constitute the floral envelope or the *perianth,* important to the flower because it protects the delicate sex organs that lie within.

Inside and above the petals are located the male (stamens) and female (pistil) reproductive organs. The stamens may number a few or many in a flower. Each stamen consists of a slender stalk or filament, surmounted at its apex by a single enlarged, often cylindrical or ovoid *anther* that produces the pollen grains, later leading to the formation of the male reproductive cells or sperms. The anthers are commonly yellow and each may produce a great many pollen grains, thousands or even millions in some flowers. These grains are liberated when slits or pores are formed in the anthers.

Situated in the center of the flower is the pistil, around which are located the stamens, petals, and sepals. It consists of three fairly distinct parts: an enlarged globular or bulbous base called the *ovary;* an elongated stalk or *style* that extends upward from the ovary; and at the top of the style a slightly enlarged *stigma* where the pollen grains fall, or are brought by the wind or insects. The stigma is often expanded into a bulb or disk, or cut into two, three, or more slender segments. It frequently is very rough or bristly and sometimes is covered with a sticky fluid so that the pollen grains will adhere to it. The ovary has a well-developed cavity where the immature seeds or *ovules* are produced. The ovules, usually attached to the central axis of the ovary, are where the female reproductive cells or eggs are formed.

The ovary may be simple or compound. A simple ovary is composed of a single pistil or *carpel,* a carpel merely being a floral organ that bears and encloses ovules. A compound ovary is made up of two or more pistils or carpels, firmly grown together. In compound pistils the styles are also usually united, sometimes even to the very top of the stigmas. A simple pistil or carpel (cell) of a compound pistil may contain a single ovule or many ovules. The pistil (ovary) normally develops into a fruit and, in turn, each ovule becomes a seed contained in the fruit. The sepals and petals are usually referred to as accessory organs, the stamens and pistils as essential organs—the former because they are not directly concerned with the reproductive processes, whereas the latter are directly involved in the formation of seeds.

A flower wherein all the floral structures are present is said to be *complete;* a flower lacking one or more is, of course, incomplete. Flowers that have both stamens and pistils are *perfect* flowers; those that have only stamens or pistils are imperfect. Imperfect flowers with stamens are called *staminate* flowers; imperfect flowers with pistils are described as *pistillate.* Where the stamens and pistils occur in separate flowers on the same plant, such plants are said to be *monoecious;* where the two kinds of flowers occur on separate plants, such plants are said to be *dioecious.* A few species of flowering plants that produce their stamens and pistils in different flowers also produce perfect flowers. They are said to be *polygamous.* In such species where the staminate and pistillate flowers occur on separate plants, the staminate plants obviously cannot bear fruit.

In many flowers the sepals and petals are separate from each other whereas in other flowers, the sepals appear to be united to produce a continuous girdle about the flower. The petals also appear united basically forming a tubular or rotate corolla. The united part is the tube, and the spreading parts or the lobed border is the limb. Most flowers are beautifully symmetrical and are said to be *regular* while others are unsymmetrical or lopsided because some of the petals on one side of the corolla are larger, or differently shaped than other parts of the flower, and are said to be irregular. Finally, the size of the flowers, as represented by the spread of the open perianth, may vary from almost microscopic in certain grasses and water plants to an enormous structure four or five feet in diameter, weighing as much as fifteen pounds. Such flowers occur in certain species of *rafflesia,* which are parasitic on the roots of grapes in the tropical forests of the Malay archipelago.

In many species of flowering plants the individual flowers are borne at the end of a long stalk or stem known as a *peduncle*. Such flowers are usually fairly conspicuous because of their relatively large size and bright colors. In most species, however, the flowers, whether they are small or large, brilliantly colored or not, occur in groups or clusters, and the stalks or stems of the individual flowers are known as *pedicels,* the main stem or axis of the cluster being the peduncle. Flower clusters are generally known as *inflorescences.*

There are a number of different kinds of inflorescences such as head, spike, catkin, spadix, raceme, corymb, umbel and cyme. The *head* is a dense cluster of sessile or nearly sessile flowers on a short axis or receptacle, such as the dandelion. A *spike* is an inflorescence wherein the sessile flowers are arranged on a basically elongated common axis, such as plantain. Much like a spike, a *catkin* is really a spikelike inflorescence that bears only staminate or pistillate apetalous (without petals) flowers as in hickory. A *spadix* is a spike with a fleshy axis bearing sessile flowers, for example the jack-in-the-pulpit. The spadix is commonly surrounded and partly enclosed by the spathe, which is a large bract or pair of bracts. A

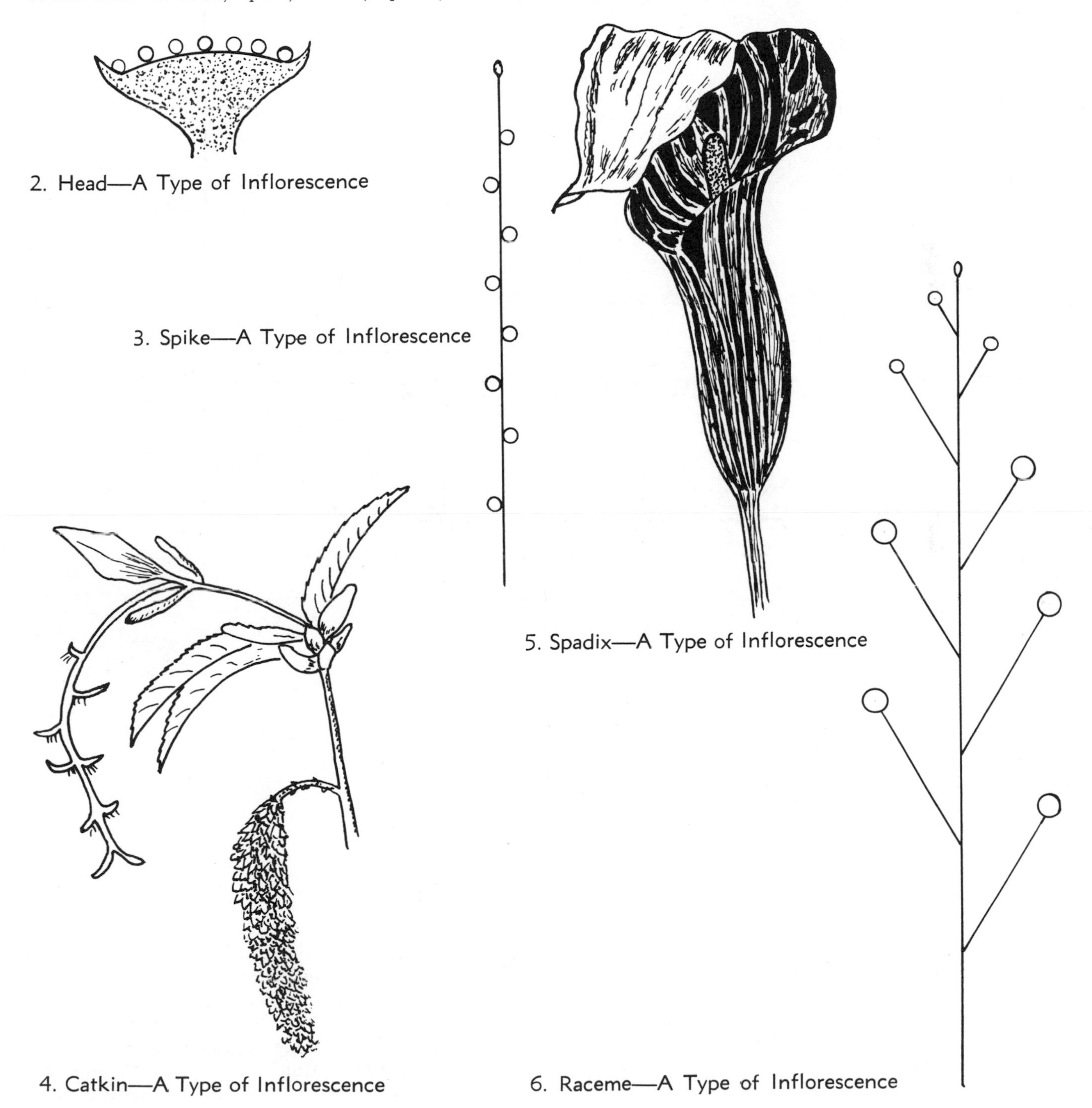

2. Head—A Type of Inflorescence

3. Spike—A Type of Inflorescence

4. Catkin—A Type of Inflorescence

5. Spadix—A Type of Inflorescence

6. Raceme—A Type of Inflorescence

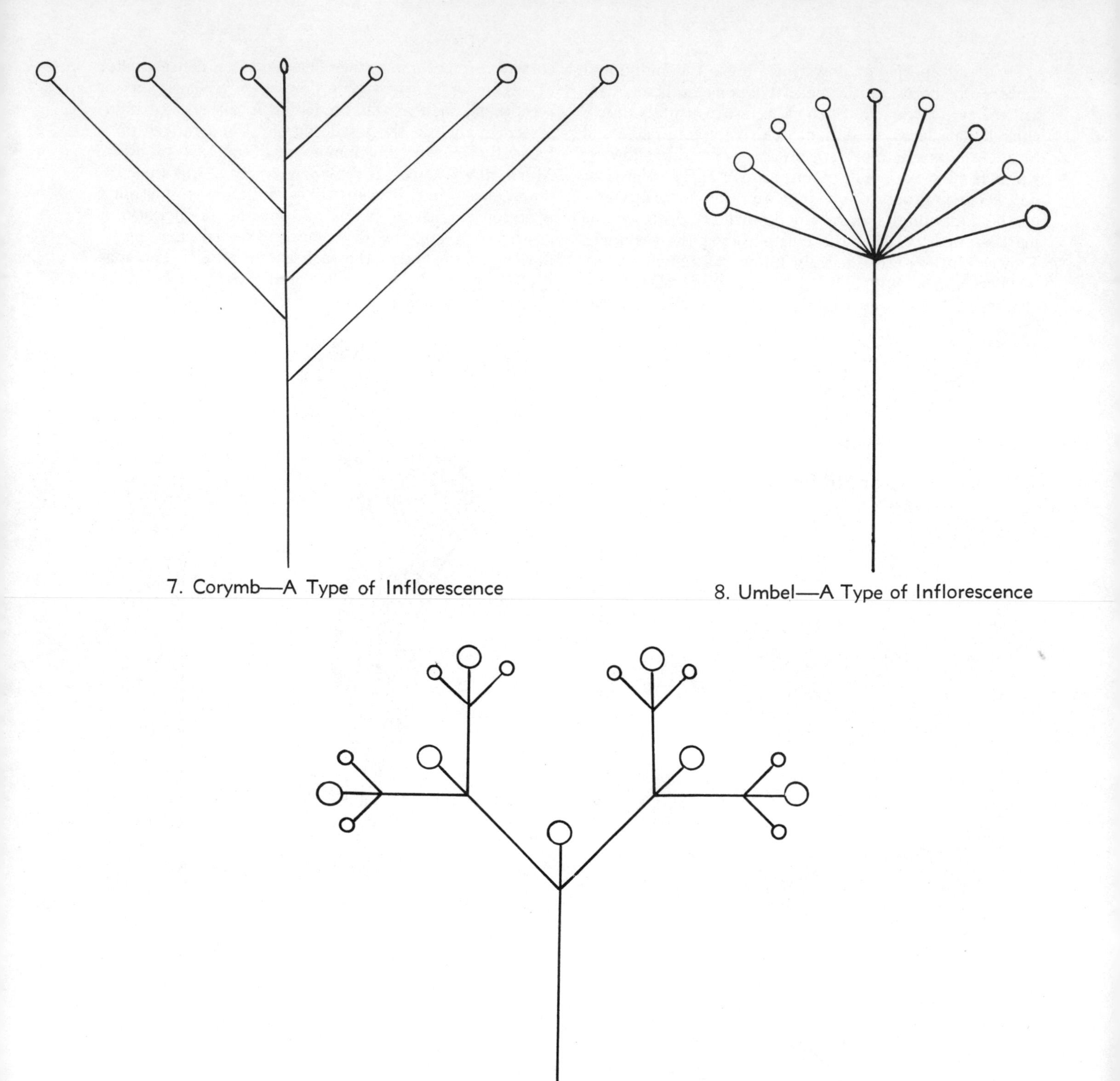

7. Corymb—A Type of Inflorescence

8. Umbel—A Type of Inflorescence

9. Cyme—A Type of Inflorescence

bract is a modified leaf associated with a flower or an inflorescence. A *raceme* is a simple inflorescence wherein the flowers, each with their own pedicel, are spaced along a common, more or less elongated, axis such as the lily-of-the-valley. A *corymb* is a simple inflorescence wherein the pedicels, growing along the peduncle, are of unequal length, those of the lowest flowers being the longest, those of the upper flowers shortest. This can be evidenced in the candytuft or cherry flower cluster. An *umbel,* as in the wild carrot, is an inflorescence wherein the stems of the flowers are approximately the same length and grow from the same point. A *cyme* is an inflorescence wherein the apex ceases growth early, and all its meristematic (growing) tissues are used up in the formation of an apical flower. Other flowers develop farther down the axis, and the younger they are the farther they are from the apex. This is seen in the crabapple and forget-me-not.

There are also many kinds of branched or compound flower clusters such as a panicle, compound

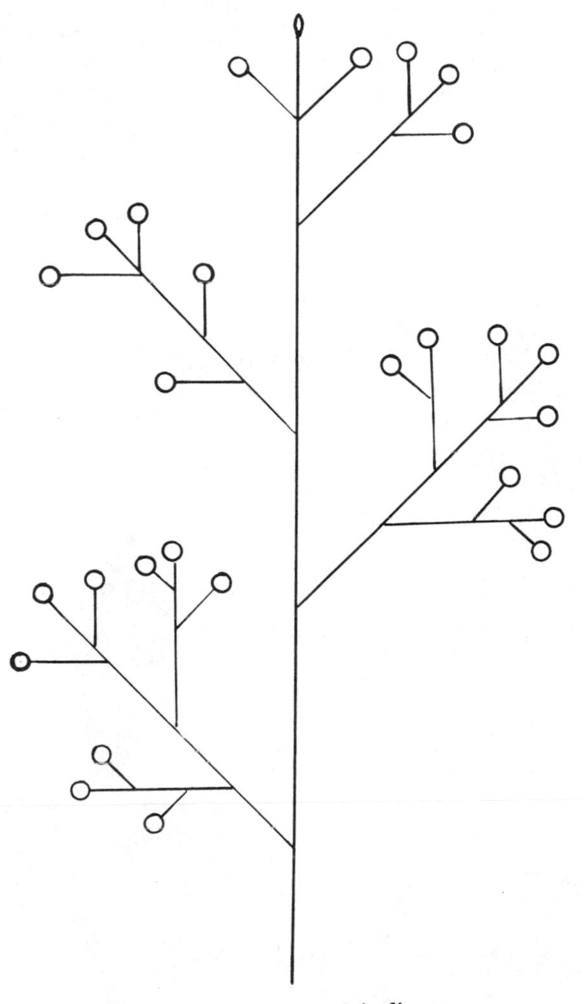

10. Panicle—A Type of Inflorescence

spike, compound umbel and the thyrse. A *panicle* is a compound inflorescence that has several main branches with pedicellate flowers arranged along its axis such as the Spanish bayonet. In a *compound* spike the branches of the axis are spikes, as in wheat, and in a compound umbel, the carrot, the axis is much shorter and its branches arranged like an umbel, each branch itself bearing an umbel. A *thyrse* is a form of mixed inflorescence wherein the main axis is racemose and the secondary or later axes are cymose. Such a flower cluster is illustrated by that of the lilac.

A botanist will say that a fruit is a matured ovary of a flower, including one or more seeds and any part of the flower that may be closely associated with the ovary. Look at an apple, for example, and the remnants of the sepals and the stamens will be seen. This

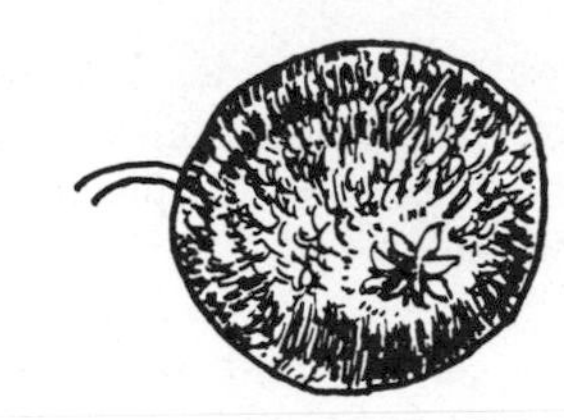

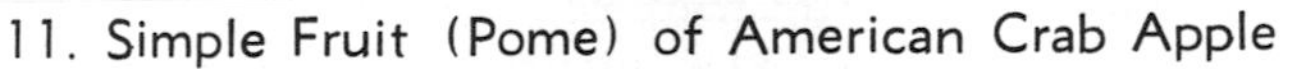

11. Simple Fruit (Pome) of American Crab Apple

tells only part of the story, however, for there are simple fruits, aggregate fruits, and multiple fruits. A *simple* fruit is developed from a single ovary; an *aggregate* fruit is a cluster of fruits derived from a single flower having many pistils or ovaries; and a *multiple* fruit is a cluster of fruits formed from the ovaries of many separate, yet closely clustered flowers such as the pineapple. Both the strawberry and raspberry are aggregate fruits, the former consisting of a number of

12. Aggregate Fruit of Raspberry

13. Multiple Fruit of Red Mulberry

one-seeded fruits called *achenes,* and the latter consisting of a number of *drupelets,* or small stony fruits.

There are many kinds of simple fruits that differ from one another in certain structural features and in the number of ovaries. They are generally divided into fleshy fruits and dry fruits. The fleshy fruits are the berry, pome, pepo, hesperidium, and drupe. A berry is a thin-skinned, fleshy, or pulpy fruit containing a number of seeds, the fleshy portion representing the ovary wall, as in the smilax. Such seemingly di-

14. Berry of Smilax

15. Drupe of Buckthorn

verse fruits as the grape, banana, tomato and the blueberry are all berries. A pome, or an apple, is also a fleshy fruit containing seeds, but differs from a berry in that the fleshy portion represents the receptacle. Both the pepo and hesperidium are a type of berry having a hard and leathery rind. They are represented by the squash, cucumber, and watermelon, and by the orange, lemon, and grapefruit. A drupe is a one-seeded or stony fruit, as in buckthorn.

The dry fruits are also divided into two groups depending upon the number of seeds produced and whether the fruit opens naturally at maturity (*dehiscent*), or remains closed (*indehiscent*). The dehiscent fruits are the legume, follicle, capsule, and silique; the indehiscent fruits are the achene, caryopsis, samara, schizocarp, and nut. The *legume* (bean or pea) is a simple, dry, dehiscent fruit formed of a single pistil and splitting along two sutures. The *follicle,* milkweed for example, is a simple, dry, dehiscent fruit, producing many seeds and composed of one carpel that splits along one seam. The *capsule* is a dry, dehiscent fruit developed from a compound ovary wherein each carpel opens longitudinally along the carpellary septa (the azalea), or in the middle of the carpels (the lily), or by the pores toward the top of each carpel (the poppy.) The *silique* is a simple, dry, dehiscent fruit developed from two fused carpels that separate at maturity, leaving a persistent partition between them. The silique is a fruit characteristic of the members of the mustard family. The achene is a small, dry, hard, one-celled, one-seeded indehiscent fruit characteristic of the buttercup. The *caryopsis,*

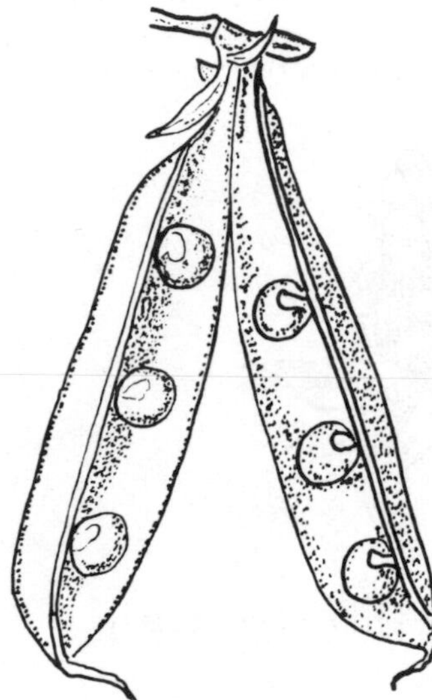

16. Pod of Pea

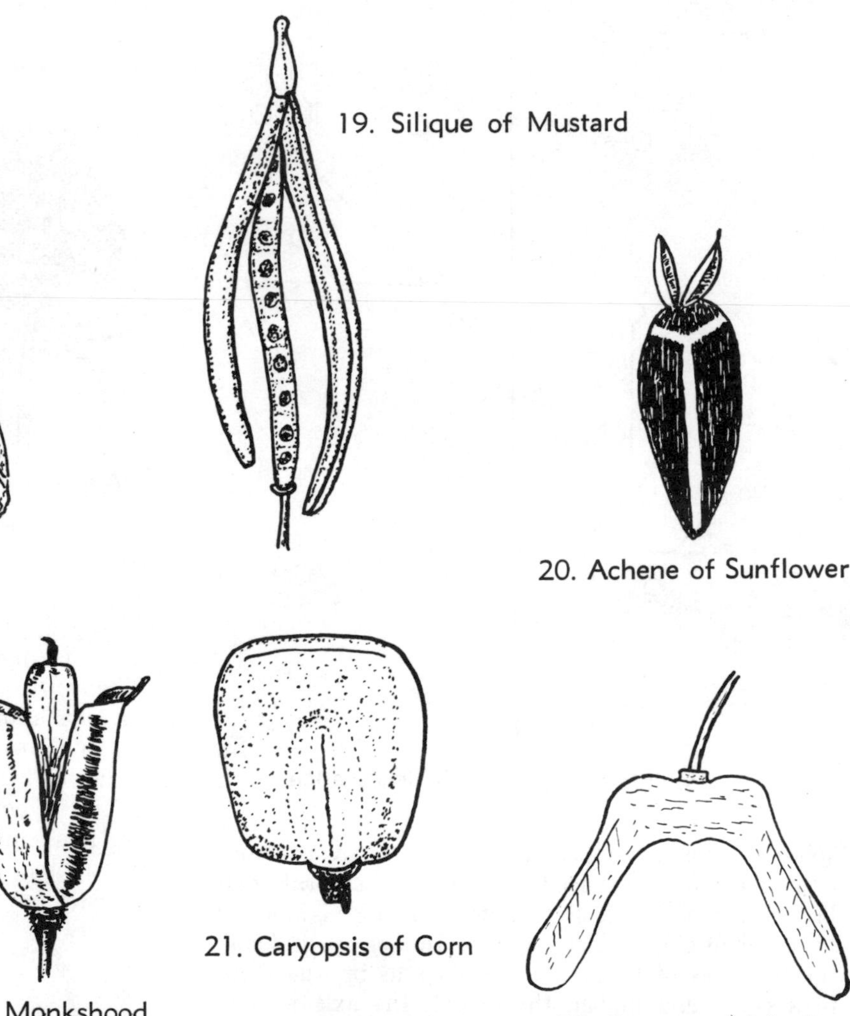

19. Silique of Mustard

20. Achene of Sunflower

21. Caryopsis of Corn

22. Samara of Maple

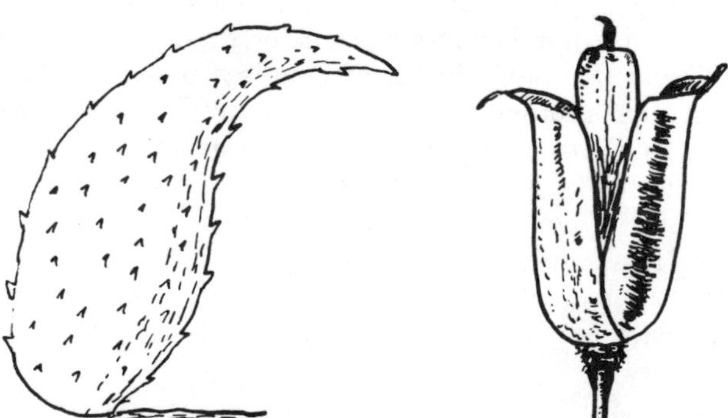

17. Follicle of **A,** Milkweed and **B,** Monkshood

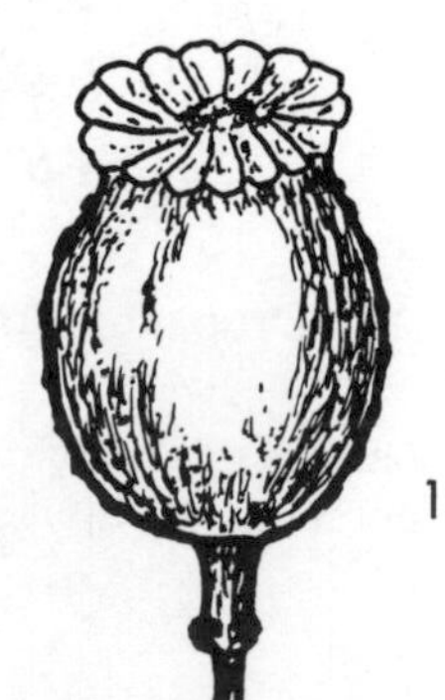

18. Capsule of Poppy

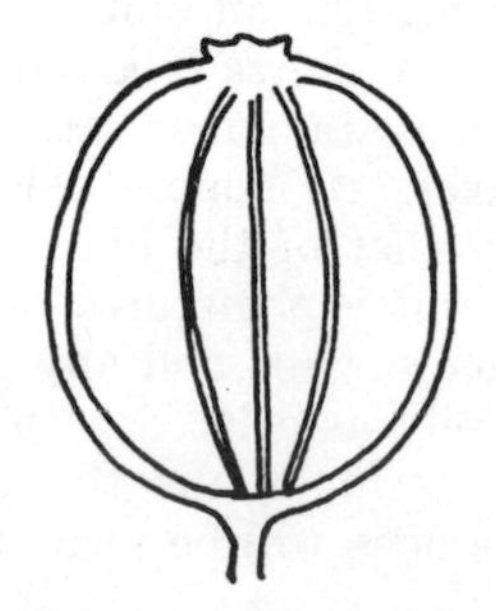

23. Schizocarp of Carrot

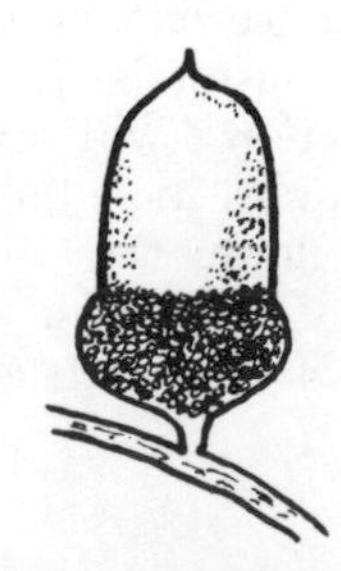

24. Nut of Oak

example corn, is a dry, one-seeded, indehiscent fruit with the seed coat and *pericarp* (fruit wall) completely united. The *samara,* of the maple, is a dry, indehiscent, one-seeded winged fruit, while the *schizocarp,* of the carrot, is a simple, dry, indehiscent fruit composed of two fused carpels that split apart at maturity, each part usually with one seed. The nut, or acorn for example, is an indehiscent, dry, one-seeded, hard-walled fruit, produced from a compound ovary.

As important as the flowers are to a flowering plant, the leaves are equally as important in carrying on the functions of photosynthesis (food making) and transpiration (water outgo)—necessities if a flowering plant is to survive. The leaves also serve other secondary functions.

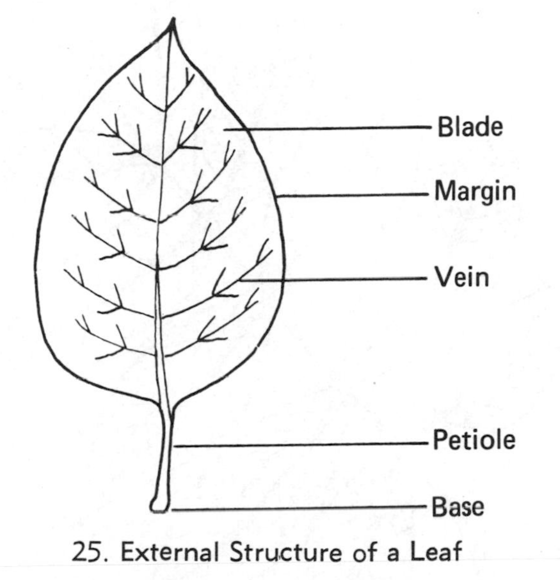

25. External Structure of a Leaf

A typical leaf consists of three parts: the expanded *leaf blade,* the slender stalk or *petiole,* and the broadened attachment to the stem called the *leaf base.* Some leaves have no petioles and are said to be *sessile.* Frequently there are outgrowths, sometimes leaflike, known as *stipules,* from either side of the leaf base. When present, these structures serve various functions. They may protect the parts within the bud or, when large, function as leaves, or they may form spines and tendrils.

The form of the leaf blade may be oval, oblong, linear, elliptic, lance shaped, or heart shaped, and is designed to provide a large surface for absorption of the light energy needed in photosynthesis and for water loss through transpiration. The petiole that attaches the blade to the stem conducts materials to and from the leaf blade and also maintains the blade in the position most favorable for the performance of its functions. The leaf base provides a firm attachment of the leaf blade to the stem, which, in some plants, is partly or completely encircled by the leaf base. In many plants the leaf base forms a long sheath that surrounds the stem, and in some of them, the sheaths protect the stem at the base of the internodes where its tissue remains soft and capable of growth. In young grasses and in the banana, these sheathing leaf bases are responsible for the mechanical strength of the stem.

One needs only to look at a leaf to observe that it is veined. These *veins* are conducting vessels. The veins are arranged in two principal types, parallel-veining and net-veining. When the veins run parallel to each other, the leaf is said to be parallel-veined. Parallel veins, as in grasses, may run lengthwise down the leaf and converge somewhat at the base and at the tip. They may also be directed outward from the midrib to the margin of the leaf as in the banana. When the veins branch frequently and join again so that they form a network, the leaf is said to be net-veined. Some net-vcincd leaves have a single primary vein or midrib where the smaller veins branch off like the divisions of a feather, and these leaves are said to be pinnately veined, as in the willow. Others

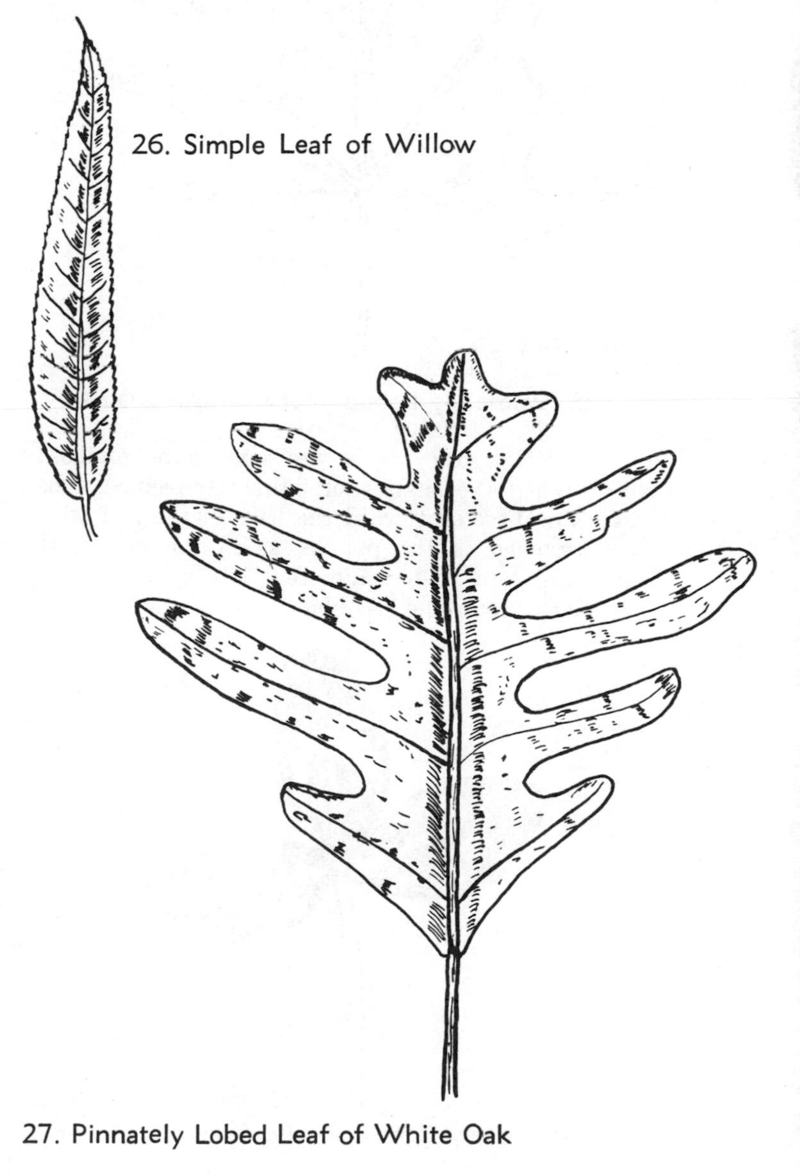

26. Simple Leaf of Willow

27. Pinnately Lobed Leaf of White Oak

have several principal veins spreading out from the upper end of the petiole and these leaves, like those of the maple, are said to be palmately veined.

When the blade is in one piece, as it is in most plants, the leaf is said to be simple, but in many species the simple leaves have the blade consisting of a number of lobes separated by deep clefts or sinuses. Such leaves, as those of the oak and maple, are said to be lobed. When the sinuses of pinnately veined, lobed leaves are directed toward the midrib,

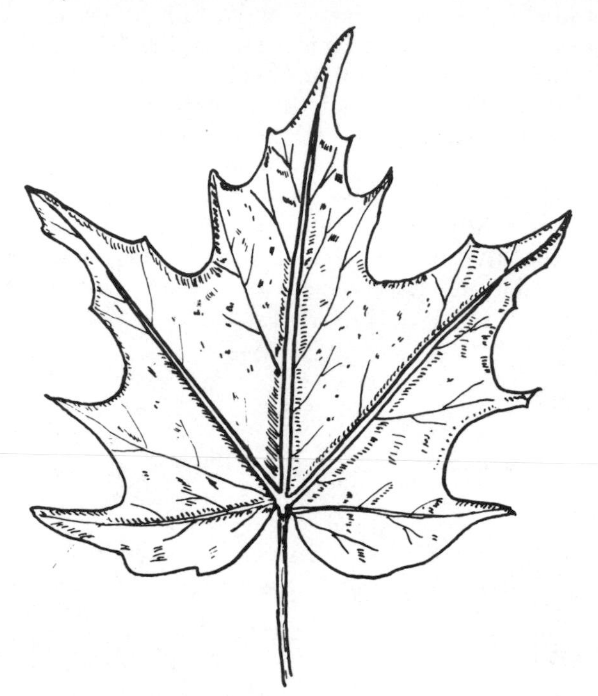

28. Palmately Lobed Leaf of Sugar Maple

as in the oak, the leaves are said to be pinnately lobed. When the sinuses of palmately veined, lobed leaves are directed toward the base of the leaf where it is attached to the petiole, as in the maple, the leaves are said to be palmately lobed.

In many plants the leaf blade is divided into a number of separate segments called leaflets. Such a leaf is said to be compound. If, for instance on the horse chestnut and clover, the leaflets are attached directly to the end of the petiole the leaves are palmately compound and the leaflets are said to be arranged finger-fashion. If the petiole is extended into a long slender structure, called the rachis, corresponding to the midrib of an entire leaf, and if the leaflets are arranged at intervals along the rachis, the leaf is pinnately compound and the leaflets are said to be arranged feather-fashion. This is typical of both the pea and the butternut.

29. Palmately Compound Leaf of Red Clover

30. Pinnately Compound Leaf of Butternut

If the margin of a leaf is smooth and uninterrupted, the leaf is said to be *entire*, but if it is cut into small divisions it is said to be *toothed*. An irregularly waved margin is said to be *scalloped*. The teeth may vary greatly in size and form.

If the leaves are attached to a twig not exactly opposite to each other the leaves are said to be *alternate;* if they are attached precisely opposite to each other they are said to be *opposite;* and if they are attached in a circle they are said to be *whorled*.

Although the members of the pine family do not bear the conventional type of flowers—their flowers lack sepals, petals, and an ovary, the staminate consists merely of naked, pollen-bearing organs in various sorts of clusters and the pistillate bears naked ovules between the scales of a cone—it is, nevertheless, included in this book. The reason is that the pines and their relatives are an integral feature of the landscape and are familiar to us all. Excluding them would make the book rather incomplete. For the same reason several other families that do not bear conventional flowers have also been included. It might also be added that a number of small, uninteresting, mostly tropical families have been excluded. This leads us to the question: What is a family?

It is the inherent nature of man to classify things. Thus, we classify stamps, coins, and a variety of other man-made objects into various categories, so that we can refer to any one of such objects readily and establish a sort of relationship between them. It is for the same reason that we classify plants and animals. One of the cardinal virtues of science is to systematize knowledge. One way of doing so is to attempt to arrange plants and animals into groups and subgroups. Obviously, the placing of plants and animals into such groups and subgroups is an artificial, man-made device. And as botanists and zoologists learn more about plants and animals and as new ones are discovered, such a system must necessarily be in a constant state of flux and so revisions are constantly taking place.

The attempt to arrange plants into some sort of a system of groups and subgroups is commonly spoken of as "classification." Over the years, or should it be said, centuries, certain practices, rules or principles have been developed gradually into a branch of the science of biology that has come to be known as "taxonomy," a word derived from two Greek words meaning arrangement and law. Hence, taxonomy treats of the laws governing the arrangement or classification of plants.

The classification of plants may be based on a great variety of characteristics. Thus, such features as environmental conditions, methods of obtaining food, anatomy, size, color of flowers may be used, and we may distinguish various groups as land and water plants, shade plants and sun plants, and so on. Such a classification is still, however, an artificial one.

It is an accepted fact by botanists that all plants are related through time as well as space; in other words, if plants are traced back far enough they will be found to have common ancestors. What this means is that the plant world may be regarded as a vast assemblage of organisms in many diverse groups, but all linked closely together by similarities of structure and development. It is true that such similarities may be close or distant, or their relationships near or remote. Yet such relationships form a convenient basis for classifying the thousands of plants found on the earth, a system of classification that is called natural classification. Thus, natural classification is based on the principle of evolution and is an effort to show the true relationship of plants in contrast with "artificial" classification that is based upon any other characteristics, even to the point of ignoring the fact of descent.

One needs only to look about to observe that there are many different and various kinds of plants. Botanists designate these different kinds as species and give each species a particular name. Thus, one may easily recognize different species of oak, maple, or asters, each having its own name. Thus, there are red oaks, black oaks, red maples, silver maples, and so on. The naming of plants and the principles involved in naming plants are included in that phase of botany known as "nomenclature."

There have been described some two hundred and fifty thousand different species of flowering plants and each is represented by many individuals, no two of which are exactly alike, though the differences may be within narrow limits. Thus, a species may be regarded as a group containing all of the individuals of a particular plant, or as a group of individual plants more like each other than anything else. All the different species may further be grouped together to form a genus (plural genera), with somewhat wider limits of variation. Thus a genus is a group of species linked together by usually obvious, but sometimes rather puzzling, botanical characteristics, and includes those broader characteristics that are common to all the species. A genus, then, such as *Quercus* includes all the species of oaks, the genus *Acer* all the species of maples, and the genus *Aster* all the species of asters. Some genera have only a single species, such as *Sanguinaria* that contains only the plant we know as the bloodroot. Others have several or many species, and some are enormous, as the genus *Solanum,* which contains over twelve hundred species.

In the same manner that species are grouped into genera, the genera are in turn grouped into categories known as families. A plant family, then is a group of related genera united together because they all have a family resemblance, although quite distinct from one another. As an example we might cite the poppy family, known to science as the Papaveraceae, a family name composed of *Papaver* (the poppy) and *aceae* (belonging to). The poppy family comprises about twenty-five genera all having similar flowers or fruit structures that are also the basis for the identity and scope of all the other plant families. A few plant families have such a strong or distinct likeness that even to the uninitiated, plants belonging to them can easily be recognized, as for example, the grass family (Gramineae), the daisy family (Compositae), the pea family (Leguminosae), the mint family (Labiatae), and the lily family (Liliaceae). The families are further classified into groups called orders, and orders into classes, but such groups are beyond the present scope of this book.

Scientific names seem to be a source of dread to the general reader and seem to frighten him with a sense of the text being too scientific; hence, this author has left out the scientific names in the text, but has included both the scientific name and common name of the plants mentioned in the book in the appendix for anyone wishing to know what they are, and for reference.

Most people prefer common names for plants. They are easier to pronounce and make the plants seem more alive and meaningful. The trouble with using common names, however, is that they so often differ from locality to locality so that a plant called by one name in one place may be known by another elsewhere. This leads, of course, to confusion. A plant may even have a number of different names in the same place. Thus, two people may be talking about the same plant, but not know they are doing so since they both call it by a different name. Hence, the need for stability in plant nomenclature, and so, the need for universal scientific names. Two botanists then, or even two amateur plant enthusiasts, living in different parts of the country, can correspond with each other and not confuse themselves by using a common name for a plant. Take, for instance, the white oak. It may apply to two different plants in various places, but when addressed by the Latin name, *Quercus alba,* there is then no question as to the particular white oak in mind.

Latin is the language used in botanical nomenclature, for it is universal and precise. In naming plants it has become the custom to use both the genus name and specific name as the scientific name of a plant, written together as a binomial. Thus, for the white oak we have *Quercus alba,* for the red maple *Acer rubrum,* and so on, the first name being the generic name, the second the specific name, the first letter of the generic name being a capital.

It often happens that a third name may be necessary to further designate a plant that already has a generic and specific name. This third name, also in Latin, designates a particular variety of a species. For instance, the scientific name of the maiden pink is *Dianthus deltoides,* which has green leaves, but there is also a variety of this plant that has bluish gray leaves and so it is designated as *Dianthus deltoides glaucus*—*Dianthus* being the generic name, *deltoides* the specific name, and *glaucus* the varietal name. Species names are usually, but not always, supposed to indicate some feature of the particular plant, as *alba* for white, *rubrum* for red, *canadensis* from Canada, and so on. It must be added, however, that some species names are misleading because the person who first described a particular plant may have been mistaken as to its origin or confused as to its true identity. Once a plant has been named, the species name cannot be changed. Finally, it is the accepted practice to place after the scientific name of a plant an abbreviation of the man who first described it, or who first used the combination of genus and specific names, as, for instance, *Beta vulgaris* L., the L. being the abbreviation for Linnaeus, perhaps the greatest botanist of all time and the father of our modern system of naming plants.

The Families of Flowering Plants

1. Acanthus (Acanthaceae)

The Acanthus family consists of about one hundred and ninety genera and about two thousand species of plants, most being herbs, though there are some shrubs and vines. They are widely distributed in the tropical and subtropical regions of the world, but a few occur in Temperate Zones.

They have opposite, simple leaves, perfect, nearly always irregular flowers that are generally crowded in a leafy cluster of some form. The flowers have a four to five-lobed calyx, a four to five-lobed corolla, two or four stamens, if four the stamens are in two pairs, and a two-celled ovary with one to many ovules in each cell. The fruit is a capsule.

Many species and varieties of this family are grown in the greenhouse for ornamental purposes, others are for the garden such as the bear's breech of the genus *Acanthus, Acanthus* being Greek for thorn, introduced into England from Italy four hundred years ago. It is a striking, upright plant with mostly basal leaves that are about two feet long, and white, lilac, or rose-colored flowers in a spike from a foot and a half to two feet long.

One of the plants introduced by Thunberg, the great collector of Japanese plants in the first half of the nineteenth century, is the species known as thunbergia or black-eyed susan. It belongs to the genus *Thunbergia,* named after Carl Peter Thunberg, a Swedish botanical author and traveller, and a student of Linnaeus. It is an herbaceous vine with oval or triangular-toothed leaves and white, purple-throated, long-stalked, solitary flowers. The best known of the genus, however, and a handsome, evergreen, woody vine is the sky-flower, with oval, toothed leaves and blue flowers in drooping racemes, which are very showy. It is a native of India and is planted in warm climates such as Florida or California.

A few members of the family occur in the United States. Some of them belong to the genus *Ruellia,* named for Jean de la Ruelle, a French botanist. They are perennial herbs with simple, opposite leaves and flowers that may either be solitary or in clusters. The smooth ruellia, found in dry woods, has an erect, slender stem that may either be simple or branched, oblong or oval leaves, and blue flowers that may either be solitary or in clusters of several in the axils. The hairy ruellia is similar to the preceding, but is hairy or pubescent.

Other members, known as water willows, are of the genus *Dianthera,* Greek for double anthers. They are mostly perennial herbs with entire or sometimes toothed leaves, and small or large irregular flowers, variously clustered or solitary in the axils. The dense-flowered water willow is found in water and moist places, and has an erect, grooved and angled, slender, simple stem, lance-shaped or linear leaves, and violet or nearly white flowers at the ends of slender axillary stems. Another species is the loose-flowered water willow that also grows in wet soil. It has a simple, sparingly branched stem, ovate-oblong or oblong leaves, and flowers in loose axillary spikes.

2. African Violet (Gesneriaceae)

Probably the best-known members in a group of mostly tropical herbs called the African violet family are the plants of the same name. They are hairy plants of the genus *Saintpaulia,* named for Baron Walter von Saint Paul who discovered the first species, with long-stalked basal leaves that form an open rosette. Their flowers are generally violet, and are very showy, in long-stalked, few-flowered clusters called cymes. The corolla consists of a short tube and its lobes are beautifully two-lipped. There are two fertile stamens, and two infertile ones, and the fruit is a two-valved, oblong capsule.

The genus contains four species, one being the African or Usambara violet, mostly grown as a pot plant and sometimes as a window-box plant, though it does not like too much sun or wind. Its leaves are round or oval, and form a wide rosette. Its flowers are nearly an inch wide and occur in clusters of one to six. There are two varieties; one has larger, deeper violet flowers, while the other has yellow and white markings on its leaves.

The African violet family is large with over eighty genera and some five hundred species, about a dozen of them being cultivated in greenhouses. They chiefly have simple, opposite leaves and irregular flowers. The flowers have a tubular, five-lobed calyx, a five-lobed, tubular corolla with the lobes usually unequal, invariably four stamens with two shorter than the others, and a two-carpelled ovary. The fruit may be dry or fleshy.

One of the better-known members of the genus *Sinningia* is the gloxinia, having long been cultivated for its showy flowers. It is a tuberous, rooted, nearly stemless herb with oblong or oblong-oval, toothed leaves, and flowers either solitary or a few in a cluster. The corolla is tubular or bell-shaped, usually violet or purple, but sometimes is reddish and white-spotted. The fruit is a two-valved capsule.

Some members of the family are popular plants for hanging baskets, such as those of the genus *Episcia*. Two genera, *Haberlea* and *Ramonda*, are European and can be grown outdoors in some parts of the United States. Among the more important genera for the greenhouse are *Achimenes, Corytholoma, Kohleria, Naegelia,* and *Streptocarpus.*

3. Akebia (Lardizabalaceae)

The akebia is a dainty Japanese and Chinese vine introduced to the United States in 1884 by Robert Fortune, an English botanist. There are four species in the genus *Akebia,* a popular Japanese name for these plants, but the one most commonly planted is the fiveleaf akebia. It is a handsome, hardy vine, cultivated for shade and for covering walls and arbors. Its

31. Compound Leaf of Akebia

leaves are compound of five, entire, small, green, oval leaflets, which are inconspicuous. Fragrant rosy purple, or violet brown flowers grow in loose clusters, the staminate and pistillate are separate, and the three

32. Flower of Akebia

sepals of each flower appear to be petals and beautifully cupped. The fruit is a dark purple or indigo berry, containing numerous black seeds, and is used as a food in Japan.

The akebia family is a small group of mostly woody vines comprising about seven genera and twelve species, chiefly Asiatic, with alternate, compound leaves, the leaflets arranged finger-fashion. The flowers are unisexual or polygamous, not very showy, and are borne in racemes. The individual flowers have three or six sepals, no petals, six stamens, and an ovary usually containing many ovules. The fruit is either a berry or a pulpy pod.

4. Amaranth (Amaranthaceae)

The amaranth family is rather large, consisting of forty genera and perhaps five hundred species, mostly annual and perennial herbs, and a few woody plants. They are widely distributed, but are far more abundant in the tropics.

They have alternate or opposite, simple, leaves, which are sometimes fleshy and small, and inconspicuous flowers. The close-packed clusters in which they occur, however, are often very brightly colored and showy. They have no petals, but have three to five sepals that are free or slightly united, one to five stamens that are opposite the calyx segments, and a one-celled ovary. The fruit is capsulelike, as in the utricle, and opens by a lid, though sometimes it is more of an achene or even berrylike. The family includes many weeds and a number of garden flowers.

Most weeds of the family belong to the genus *Amaranthus,* from the Greek meaning not fading, because of the everlasting character of some species. It is a group of coarse, annual weeds, with alternate leaves and small green or purplish flowers in axillary or terminal spiked clusters. A common species is the green amaranth or rough pigweed, an introduced species from the Old World, and an annoying weed in cultivated ground and in gardens, with light green, roughish, long-stemmed leaves and dull green flowers in a stiff, bristly spike.

33. Utricle—A Capsulelike Fruit

Another weed of the same genus is the species known as tumbleweed, common in cultivated ground and in waste places. It is a low, smooth, greenish, white-stemmed, spreading plant with pale green, finger-shaped leaves and small green flowers crowded in small close axillary clusters. When mature the leaves fall off, the withered plant is uprooted, and it tumbles over the ground in the wind—hence, its popular name.

Several amaranths are garden flowers, such as love-lies-bleeding, a very popular annual with one variety having blood red foliage, green or yellow spikes, and another with headlike, red flower clusters. A second amaranth is the prince's-feather, a showy plant three to four feet high, with reddish foliage and dense flower clusters, multi-branched and red or brownish red. A third is the Joseph's coat, a variable annual, with leaves sometimes blotched and colored, and flowers in stalkless, headlike clusters in the leaf axils, or sometimes spikelike.

Other garden representatives of the amaranth family belong to the genera of *Telanthera, Iresine, Celosia,* and *Gomphrena.* The *Telanthera* are low-growing Brazilian plants used essentially for carpet bedding. They are perennials with opposite, narrow, small leaves, often colored, and minute, chaffy flowers in dense clusters in the leaf axils. The flowers are rarely produced, however, because of the shearing necessary to keep them low enough for carpet bedding.

The *Iresines* are a group of South American plants with brilliant foliage and usually grown for bedding plants. They have opposite, stalked, generally oval leaves, and small, whitish flowers crowded in dense spikes that are gathered in branched panicles. The flowers, however, are rarely produced. The *Iresines,* Greek for a woolly harvest garland, an allusion to the woolly flowers, are excellent, showy, summer-bedding, foliage plants, but extremely sensitive to cold and can take no frost.

The *Gomphrena* are Old World herbs, the best known being the globe amaranth, which was first introduced into England from India in 1714, and ever since has been a deservedly popular garden annual. It is grown for its terminal, globular flower heads, whose persistent, colored bracts are retained when dried so that the plant is one of the more decorative everlastings. It is eight to twenty inches high with opposite, oblong or elliptic, downy leaves, and red, pink, white, or yellow chaffy flowers in dense, long-stalked, cloverlike heads. The globe amaranth is found in many horticultural forms, and is excellent for late summer bloom, and for cutting.

The *Celosia* is a large genus of tropical herbs and shrubs whose best-known member is the celosia or cockscomb. It is one of the oldest plants in cultivation, having arrived in England from Asia in 1570 according to an old English record. The primitive form had erect, pyramidal, plumy panicles, but the present garden form is something of a monstrosity in that the stems and plumes have become joined laterally with a partial stoppage of their growth upwards. There are a number of species and forms, but all have alternate leaves and minute, chaffy flowers, crowded into dense spikes, crested, or otherwise modified, as the cockscombs. They are often brilliantly colored in shades of red, yellow, green, purple or white, the coloring due to a vast number of small bracts intermingled with the flowers. The name celosia is from the Greek for burned, in allusion to the brilliant scarlet inflorescence.

There are a few other wild flowers of the family that warrant mention. These are the water hemp, Juba's bush, and froelichia. The water hemp of wet meadows, swamps, marshes, and ditches looks like a large, succulent amaranth. It has a smooth, erect stem, three to six feet tall, but sometimes as much as ten feet, with many slender branches. It has entire, smooth, lance-shaped leaves with prominent veins and pointed at both ends, and small, greenish, dioecious flowers, each blossom guarded by one to three awllike bracts, in dense terminal or axillary spikes.

The Juba's bush, a close relative of the genus *Iresine,* is a conspicuous plant because of its white flowers and the sometimes red color of its leaves. A native of tropical America it seems to prefer sterile, sandy, or gravelly soil and, thus, is found in dry fields and dry banks. It has an erect, furrowed stem, swollen at the nodes, two to five feet high, with opposite, narrowly ovate, long-pointed, entire leaves. The flowers are in large terminal, branching, nearly leafless panicles, though the blossoms themselves are small, with a silvery, white five-parted calyx, subtended by three dry, white, papery bracts.

Like the green amaranth or rough pigweed, the froelichia is an equally annoying weed, intruding in all sorts of crops, and by its long flowering season, from July to November, requiring late tillage. It is an unpleasant, woolly haired plant, with a rather slender, erect, and somewhat branching stem. The leaves are opposite, rather thick, narrowly lance-shaped, and sessile, the lower ones somewhat spatulate and downy on the lower surface. The very small flowers are densely crowded on spikes, the latter disposed oppositely in branching panicles, the calyx densely woolly, tubular, and five-toothed with broad wings. The froelichia is common in fields, waste places, and in cultivated crops.

Although many members of the amaranth family are troublesome weeds, they are not entirely without some value. The seeds of the green amaranth or rough pigweed were once used by the Indians as food. These black seeds were parched or ground into a meal that was baked in cakes or used for porridge. Even the leaves were used as food. As a matter of fact, if taken when young the plant can be cooked like spinach and served with butter and vinegar. In India the prince's-feather is cultivated for its seeds because they are eaten. The seeds also serve as food to the Indians of the Southwest. The seeds of various amaranths are important to many species of game birds and songbirds, and other wildlife animals such as the pocket mouse and kangaroo rat. Moreover, as the seeds of some species persist in the densely clustered spikes well into the winter, they become available when other foods become scarce.

5. Amaryllis (Amaryllidaceae)

The yellow star grass is not really a grass, though if one looks at its leaves it becomes quite evident how it got its name, for they are deep green, linear, and grasslike, somewhat covered with hairs, and all from an egg-shaped corm, an underground stem. It is a small plant of fields and meadows and one would think that its yellow flowers, at the summit of a rough, hairy scape, from two to six inches high, would be lost among the buttercups and dandelions. Somehow it manages to get its starlike blossoms high enough to call itself to our attention. More importantly, the smaller bees and butterflies find it readily and visit it often, for the flowers produce an abundant supply of pollen as if to make sure of their continued visits.

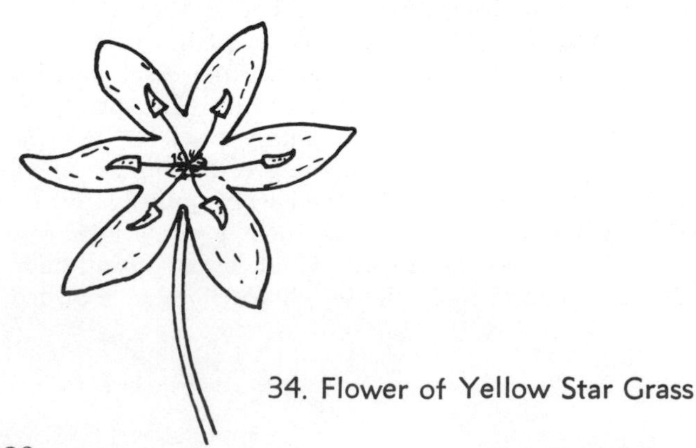

34. Flower of Yellow Star Grass

Like the yellow star grass, the atamasco lily is not a lily, although it may look like one. It is a plant of moist situations in the South, with narrow, long and straight, bright green leaves, forming a basal cluster to the flowering stalk, which has a single, white, funnel-shaped flower with six elliptical sepals, six stamens, and a pistil. Bees are the chief agents of cross-fertilization.

Again called lilies, but not lilies, are the swamp lily and the spider lily. The swamp lily is one of about a hundred mostly tropical plants found in swampy woods and meadows from Florida to Texas, with long narrow leaves that are chiefly basal. A slender flowering stalk with two or more fragrant, white blossoms with long, tubular perianths ends in narrow, lobed segments. Some species of swamp lilies are cultured for their showy flowers, and are grown outdoors in the South.

Like the swamp lily, the spider lily is also a plant of wet places and is found in swamps and on the marshy banks of streams from southern Illinois and Missouri to northern Georgia and Alabama. It also has long narrow leaves and white flowers with an elongated, tubular perianth. There are forty known species of spider lilies, most being tropical plants, though there are some wild native species in the South—from South Carolina to Florida. Only two species seem to be of much cultural interest, and these are grown outdoors in the South. In the North they would need greenhouse care.

Nearly everyone has heard of the century plant, but few probably know that there are some one hundred and fifty species that are most abundant in tropical America. They belong to the genus *Agave,* Greek for noble or illustrious. The most cultivated century plant is a native of Mexico and is often found growing wild from Florida to Texas. Like many other species it blooms once and then dies; however, there are species that bloom periodically.

The members of the genus *Agave,* generally known as agaves, have leaves in a basal rosette and that are usually spiny-margined. Often they have a strong, terminal prickle, and persist for years. The flowers, greenish yellow, funnel-shaped, with six segments, are set in a long, terminal cluster at the end of a very long stalk. The fruit is a capsule.

In the mature, cultivated century plant, the leaves are large—up to five or six feet long and six to eight inches wide—decidedly grayish, and their tips are usually recurved with the marginal spines stout and also recurved. The stalk of the flower cluster may grow from twenty-five to forty feet high and the numerous flowers are two and a half inches long.

In the southeast, there are two native species of agaves: the false sisal and the wild century plant. The former has a short, stout trunk ending in stiff leaves, and greenish yellow flowers, while the latter is similar, but has smaller, gracefully recurved leaves. From Virginia, southward and westward to Ohio and Indiana, another native agave, known as the false aloe, grows on dry and rocky banks. It has entire or toothed leaves and greenish yellow, fragrant flowers in a loose, wandlike spike. And on the rocky mountain slopes of California one may find the desert agave, with a basal rosette of relatively short, thick, gray green leaves and a flowering stalk, six to nine feet high, ending in a cluster of bright yellow flowers.

There are other western species of agaves from which the Apache Indians prepare their famous mescal, an alcoholic drink made from the fermented juice of the plants and said to be palatable and wholesome.

The yellow star grass, the atamasco, swamp, and spider lilies, as well as the agaves, belong to the amaryllis family, a large family consisting mostly of perennial, bulbous herbs, or with stems and leaves arising from rootstocks. It includes some seventy genera and eight hundred or more species that are found in the warm and temperate parts of the world. The members of the family are of very diverse habit and often resemble lilies; in fact, they are frequently confused with them. They may be distinguished, however, from the lilies by the fact that the ovary of the flower is attached to the calyx, and the stamens are attached to the inside of the tube of the corolla. In the lilies, the ovary is free from the calyx; that is the ovary is above the point of insertion of the calyx and it is said to be superior. Thus, an inferior ovary is beneath the point of insertion of the calyx.

In the amaryllis family the leaves are alternate or basal, narrow and without marginal teeth, though sometimes prickly. The flowers are perfect, regular or nearly so, usually one or more on a leafless scape subtended by two or more bracts, of three sepals and three petals that are inserted on a tubular receptacle, with six stamens and one pistil; the fruit is a capsule or berrylike.

To most people, the family is of interest because it includes many highly prized ornamentals that are grown either as pot plants indoors, or planted in lawns and gardens outdoors. Some of these, in addition to the swamp and spider lilies, are the belladonna lily, daffodil, narcissus, jonquil, and snowdrop.

The belladonna lily, a native of South Africa, belongs to the genus *Amaryllis,* which Linnaeus named for one of Virgil's nymphs. It has strap-shaped leaves, usually appearing after the plant blooms, and typically rose red, sweet-scented flowers, about three and a half inches long in dense clusters at the end of a stout, naked stalk that is frequently eighteen inches long.

Both the daffodil and narcissus are such familiar plants that it almost seems unnecessary to say anything about them. They both belong to the genus *Narcissus,* named for the mythological youth who was so fond of his reflection in the water of a fountain that he was changed into the flower that bears his name. It is an important group of the family, comprising hardy, bulbous plants, most of them European, very widely grown for ornament or fragrance. All have bulbs, with generally rushlike leaves, more or less round in cross-section in the jonquil and its relative, but flat or nearly so in the common daffodil, basal in all kinds, and usually about the same length as the flowering stalk. The flowers are prevailingly white or yellow, frequently nodding, with the calyx and corolla not separate, but modified in two ways: (1) the flower with a central crown or corona that is long and tubular; or (2) the central crown or corona is reduced to a shallow, ringlike cup. Outside are the six segments that compose the petals and sepals. There are six stamens, usually hidden in the crown, and the fruit is a three-lobed capsule containing many seeds.

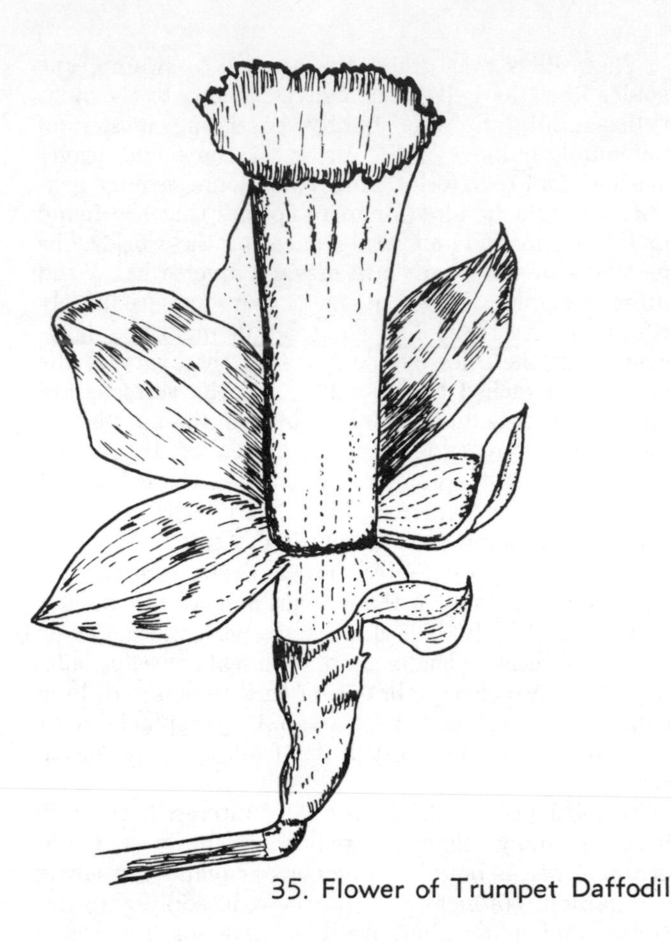

35. Flower of Trumpet Daffodil

The best-known daffodil of our gardens, the trumpet daffodil, is a native of southern Europe, naturalized in England centuries ago where it has been a favorite from Anglo-Saxon times. I do not doubt that the children still chant, as they did three hundred or more years ago: "Daffy-down-dilly, just come to town With a yellow petticoat and a green gown."

Now of many forms and in many-named garden types, the daffodil has flat leaves, twelve to eighteen inches long, and a corolla about two inches long, typically pale yellow, the segments and corona usually of a slightly different shade, the latter very long, deeply wavy or even slightly fringed. Not without reason has the daffodil become such a popular member of our gardens, for it has an exquisite grace and something of a spiritual beauty.

A very popular narcissus is the species known as the paper white narcissus. This is the one that is commonly grown in bowls of pebbles and water in the house. As its name indicates, the flowers are pure white, fragrant, and usually in clusters of four to eight. The tube is about an inch long, while the corona is much shorter than the segments and usually pale yellow.

Of the various species of narcissi, the poet's narcissus is doubtlessly the most vigorous. It naturalizes readily and may be grown either in garden beds or scattered through the grass. The leaves are flat and grass-like, the flowers very fragrant and white, the tube is about an inch long, and the corona is very shallow, considerably shorter than the segments, with wavy edges and red-margined.

36. Flower, Leaves, and Bulb of Poets' Narcissus

There is often considerable confusion between the jonquil and the daffodil. Bearing in mind that the leaves of the jonquil are rushlike, whereas those of the daffodil are flat, however, there should be no need for mistaking one for the other. Moreover, in the jonquil the central crown or corona is a shallow, ringlike cup, whereas in the daffodil it is long and tubular.

One of the earliest-blooming bulbs, appearing in midwinter if the weather is warm enough to thaw the surface of the ground so that its leaves may push through, the snowdrop is a pretty little flower, so white that it almost seems to have been formed from the snow that may still lie on the ground. It has small bulbs, two or three leaves six to nine inches high and bright green, and a solitary, white, nodding flower at the end of the scape. The flower appears as if it were made up of three concave, white petals surrounding a greenish tube, but a closer look will prove that the tube itself is made up of three separate parts that are white, marked, and blotched with green.

There are three native species of snowflakes that are commonly grown in the garden; one blossoms in early spring, one in late spring, and one in the fall. They are all native to Europe, with small bulbs and a hollow flower stalk, basal, narrow leaves, and nodding flowers that are not tubular, but with the inner and outer segments alike, though often colored differently. The leaves appear with the spring-flowering species, but after the flowers in the fall-flowering species. In the early spring species, the flowers are white, green-tipped, and solitary; in the late spring species they are also white and green-tipped, but occur in clusters of two to eight, each flower on a slender, individual stalk; in the fall species the flowers are white, but red-tinged, one to three on individual stalks.

The common English name of the plant tuberose provides us with an example of how a word can be changed from its real meaning to something quite different. The name tuberose conveys the idea of a tubular blossom with an odor somewhat roselike. Neither the tube nor the odor, however, had anything to do with the selection of its name. Originally the word was an adjective, tu-ber-ose, meaning tuberlike in reference to its thickened rootstock. And so the plant was called *tuberose polianthes,* the latter the name of the genus to which it belongs, but in some manner the word *polianthes* was dropped and the plant became simply tuberose.

The tuberose is the only cultivated species of the genus *Polianthes,* comprising about twelve species—all native to Mexico. It grows to a height of three and a half feet, with the stem leaves clasping the stem. They are somewhat smaller than the basal leaves that are one, to one and a half feet long, and one-half inch wide, bright green, but reddish at the base. The flowers are waxy white and fragrant in short terminal racemes. The tuberose is not a hardy plant and must be taken up before heavy frost, stored in a warm place, and set out again after the middle of May in the North, or April in the South. It is a delightfully fragrant plant, though sometimes rather overpowering, and best for the back of the summer border.

6. Annatto (Bixaceae)

The annatto family consists of a single genus *Bixa,* South American vernacular for tree, and a single species, a tropical American tree called the annatto or achiote. It is cultivated in Florida as an ornamental tree, where it is not over twenty-five feet high, but in the tropics it is grown as the source of annatto or achiote dye, used for coloring butter and cheese. It has alternate, simple, entire leaves and rather showy red or pink regular flowers having five sepals, five petals, many stamens, and a one-celled ovary. The fruit is a reddish, two-valved, nearly egg-shaped pod (capsule) that is covered with soft, reddish, weak prickles. The pulp around the seeds furnishes the red or yellow dye.

7. Arrowgrass (Juncaginaceae)

The arrowgrass family is a small group of four genera and twelve species of marsh herbs, often growing in standing saline water or freshwater bogs, widely distributed in temperate and subarctic regions.

They have mostly basal, rushlike leaves, broadly sheathing at the base, small, perfect flowers in terminal spikes or in few-flowered, loose racemes, the flowers with three sepals and three petals, three to six stamens, three to six carpels with one to two ovules in each carpel, and a folliclelike fruit.

There are only about four species known to occur in North America. The bog arrowgrass, genus *Scheuchzeria,* named in honor of Jacob Scheuchzer, a Swiss botanist, is a plant of mostly cold northern bogs, with leaves four to eighteen inches long and a few white flowers in a loose cluster. Growing in salt marshes along the Atlantic seaboard, as well as in fresh and saline marshes across the continent, is the seaside arrowgrass, genus *Triglochin,* from the Greek in allusion to the three-pointed fruit. Its leaves are less than an eighth of an inch wide, often nearly round in cross-section, and its flowers are small and whitish in a slender cluster about fifteen inches long, individually on short stalks. A related species, the marsh arrowgrass, has shorter leaves and grows in bogs, and another related species, the three-ribbed arrowgrass, is found in the saline marshes of the South, from Maryland to Florida and Louisiana.

8. Arrowroot (Marantaceae)

The arrowroot family is of interest primarily because it includes the arrowroot whose root furnishes the commercial arrowroot, a nutritious and easily digested starch, and a number of handsome foliage plants of greenhouse culture. The family comprises some thirty genera and about two hundred and sixty species of herbs almost wholly tropical. Most have tuberous rootstalks, and the leaves are entire, mostly basal, usually narrowed into a sheathing leafstalk and often, in the horticultural varieties, variegated with a metallic sheen. The perfect flowers, with three sepals and three petals, one stamen, and a three-celled ovary, are irregular, not showy, and occur in panicles, racemes, or spikes—the latter surrounded by sheathing bracts. The fruit is capsular or berrylike.

The arrowroot belongs to the genus *Maranta,* named for a Venetian botanist, a group of tropical American foliage plants, many grown in the greenhouse for ornament. Although the arrowroot, like other members of the genus, is cultivated in the greenhouse for ornament, it is grown in tropical regions for its starchy root. It is a slender-branched shrub, two to six feet high, with mostly basal leaves and white flowers. The plant has become naturalized in Florida.

Another cultivated genus is *Calathea,* Greek for basket, in allusion to the basketlike flower cluster. It contains over a hundred species, most being tropical American, but a few occur in tropical Africa. They are handsome foliage plants, the cultivated species suggesting, in their finely marked and colored leaves, the fancy-leaved caladium. Perhaps the most common species in cultivation is the zebra plant, of compact habit and not over three feet high, usually much less. The upper side of the leaves, which are elliptic and twelve to twenty inches long, is generally velvety green, but from the midrib the side veins are alternately barred with pale yellow green and much darker green. The lower surface is purplish red. The plant is a native of Brazil.

There is a native species of the family, called the powdery thalia, or water canna, found in ponds and swamps from South Carolina to Louisiana. It is a white, powdery, stemless herb, with ovate-lanceolate, toothless leaves, sixteen to twenty inches long, and heart-shaped at the base. The leafstalks are slender and nearly two feet long. The dull violet or purple flowers occur in spikes at the end of a hollow stalk, four to five feet high. The corolla is irregular and some of the sterile stamens are petallike. The fruit is a globose or ovoid capsule. The water canna is one of about a dozen species of the genus *Thalia* (named for Johann Thalius, a German naturalist), a group of cannalike, swamp or aquatic perennial herbs from the warmer parts of America.

9. Arum (Araceae)

Although its name may remind us of a mephitic odor and prompt many to regard it with a degree of revulsion, the author has always had a special feeling for the skunk cabbage because it is the true harbinger of spring. When it begins to appear above the ground, winter, though still lingering perhaps, is nevertheless, getting ready to leave until another year. To those who look for a more sophisticated kind of beauty, the skunk cabbage is hardly a floral delight and, yet, it is not without its redeeming qualities. If its somewhat fetid odor can be ignored, examine with a magnifying glass the minute flesh-colored flowers profusely scattered over the stout and fleshy spadix, and it will be found they are not unpleasing to the eye. Note that

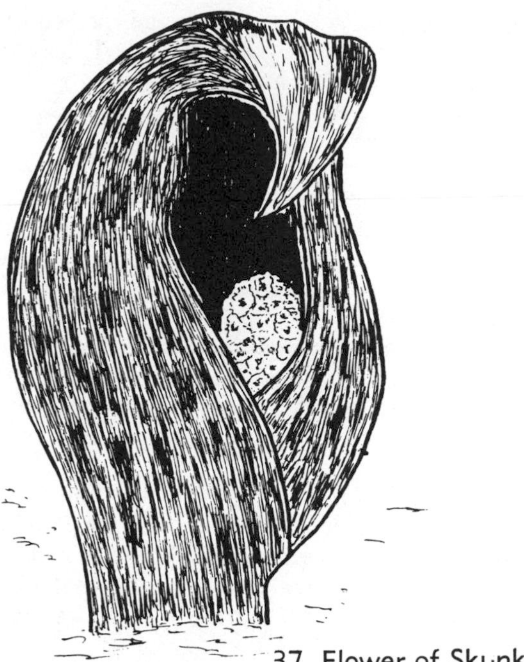

37. Flower of Skunk Cabbage

the flowers have both stamens and pistils and that the grayish straw-colored anthers are fairly conspicuous. Observe, too, that the hive bees are among the first of the insects to visit it, but being ungrateful creatures, they soon leave it alone when other and more attractive flowers appear. Perhaps it is just as well, for the hive bees never entered into the skunk cabbage's scheme of things, and many a bee that manages to gain entrance into the horn, manifestly designed for smaller insects, finds the going too slippery on its way out, falls back, and perishes miserably. Small flies and gnats that have lived under the fallen leaves during the winter and have gradually warmed into active life, however, will soon swarm about the spathes.

The skunk cabbage is a plant of the swamps, brooksides, and wet glades. It appears above the ground as early as January and blossoms in March even while snow still lies on the ground. After the flowering, the vivid green crowns of leaves begin to appear and when they have become completely unfolded, they are one to two feet long and cabbagelike. The young leaves, it is said, are good to eat if boiled to tenderness and then seasoned with butter, pepper, and salt. In September, the fruit is ripe. It is the enlarged and fleshy spadix with bulletlike seeds embedded in its surface.

One would not expect this malodorous plant to bear any relationship to the stately calla lily so widely used for floral decoration, but they are distant cousins. Compare the two and many points of similarity will be observed, notably the presence of a spadix and spathe in both. It is the showy, solitary, beautifully colored spathe for which the calla lily is cultivated, but the spathe of the skunk cabbage, purplish brown to greenish yellow and mottled, is not entirely without an aesthetic interest as there is a certain charm in the way the colors are blended.

A mere glance at the Indian turnip or Jack-in-the-pulpit shows its relationship to the skunk cabbage and the calla lily, too. The minute, greenish yellow flowers are clustered on the lower part of a smooth, club-shaped, slender spadix within a spathe that curves in a broad, pointed flap above it. The flowers are usually staminate on one plant, pistillate on another; hence, the plant must depend on insects for cross-pollination. These are small insects, the fungus gnats of the genus *Mycetophila,* which may frequently be found imprisoned in the narrow confines between the bases of the spadix and the spathe. Sometimes both kinds of flowers may be found on one plant. This is the exception rather than the rule, but it suggests the possibility that the plant's dependence on insects for fertilization may be a recent development. The curiously shaped, green and maroon or whitish-striped spathe is somewhat variable in color. In the open and exposed to sunlight, it is paler than when found in the deep woods where it is dark purple.

The Indian turnip is at home in moist woodlands and thickets where it grows to a height of one to two and a half feet. It is not an unattractive plant with its two long-stemmed, three-parted green leaves that overshadow the hooded flower below at the junction of the leaf stems. The handsome clustering berrylike fruit is at first green, but later becomes a brilliant scarlet.

The Indians utilized the farinaceous root (corm), which is shaped somewhat like a turnip, as an article of food. The author would advise against trying it, for unless properly prepared, it can produce blisters in the mouth. Both the skunk cabbage and the Indian turnip belong to the arum family, as do such water-loving plants as the dragon arum, arrow arum, water arum, golden club and sweet flag.

The arum family is a large one numbering about one hundred genera and fifteen hundred species of perennial herbs, although the stems of some tropical species may become woody and even treelike. They have a tuberous rootstock or corm, are erect, prostrate, or climbing, and occur in damp or wet places. All contain a bitter, often poisonous juice that is sometimes milky. The leaves are extremely varied, mostly basal, with long stems, simple or compound, without teeth, but often deeply lobed, and very showy in some genera. The flowers, almost microscopic in size are densely aggregated in a fleshy spike, the spadix, which is surrounded by a leaflike and funnel-shaped bract, called the spathe, which is often conspicuously colored, both the spadix and spathe forming the "flower" in a popular sense. The real flowers are perfect or of one sex only, without sepals and petals, though these may be replaced by scalelike substitutes. There are one to many stamens and one ovary that develops into a berry. The fruiting spadix is densely clustered with the berries.

Many members of the arum family, often referred to as aroids, are of value as ornamentals, probably the best known being the calla lily, a native of South Africa and grown throughout the world. Others are the elephant ear and the ceriman or "delicious monster," with its large, perforated leaves. A few of the aroids are of food value, such as the dasheen or taro whose starchy root is cultivated in the Pacific Islands and other tropical regions.

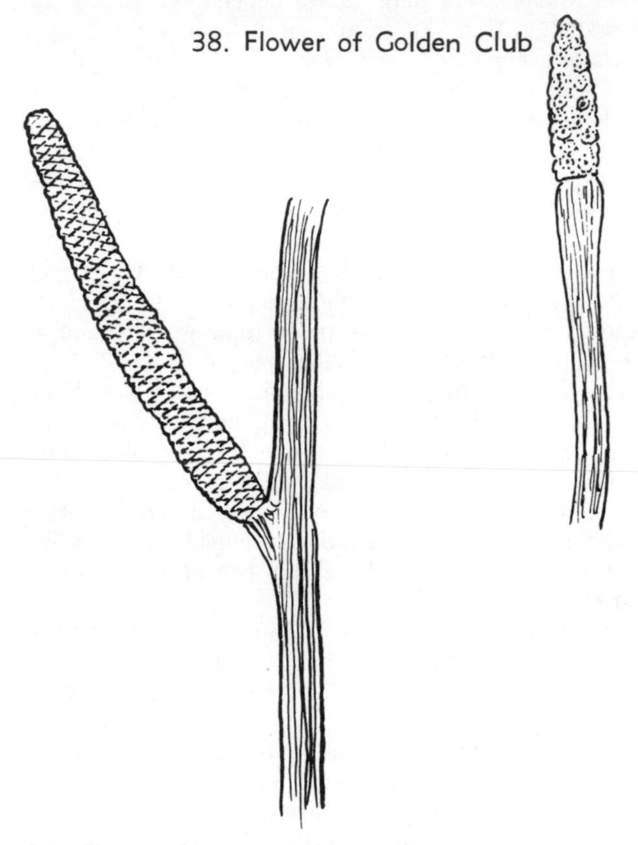

38. Flower of Golden Club

39. Flower Cluster of Sweet Flag

The dragon arum, arrow arum, water arum, golden club and sweet flag, which have been mentioned as belonging to the arum family, are all native wild flowers. The dragon arum, or dragon-root, as it is also called, grows in wet woods or low grounds where it sends up a narrow greenish or whitish, tapering spathe, one or two inches long, which envelopes a slender, pointed spadix, the latter sometimes projecting seven inches beyond its tip. The entire flower is beneath a large, solitary long-stemmed, green leaf, divided into five to seventeen segments, spreading like an umbrella. The reddish orange berries are the dragon arum's most conspicuous feature.

According to Captain John Smith the starchy root of the arrow arum was in great favor with the Indians of Virginia, who roasted it in pits for a day or two. Then after cooking, the roots were dried and ground into a meal. They also seemed to have boiled the spadix and green berries and eaten them as a luxury. The arrow arum occurs in shallow water where it arrests our attention by its stately growth and the beauty of its bright, lustrous green, arrow-shaped leaves, sometimes as much as thirty inches long.

One need only glance at the water arum to detect a resemblance to the calla lily of greenhouses, for its flower is a close copy of the potted plant. The true flowers—staminate above, perfect below—appear on a fleshy receptacle, the spadix, which is short, thick, and lies against a broad, flattish, multi-pointed, pure white spathe. The leaves are deep green and heart-shaped, and on long stems come up from the creeping rootstock. The berries, red and distinct, in a head like those of the Indian turnip, are ripe in August. Bees and pond snails assist in fertilization. The water arum is at home in cold bogs. Thoreau in writing of it says—

> **that having found it in one place, I now find it in another. Many an object is not seen, though it falls within the range of our visual ray, because it does not come within the range of our intellectual ray. So, in the largest sense, we find only what we look for.**

The water arum is a common plant in Lapland, where bread is made from its caustic and acrid roots, and other Northern countries where it grows in such abundance as to cover entire marshes.

Both the Indians and the white settlers ate the dried seeds of the golden club. They resemble peas. The bulbous rootstalks are also edible when cooked. The golden club is an aquatic perennial whose prominent golden yellow spadix (club), hardly much thicker than its sinuous stem, is densely clustered with minute, bright yellow, perfect flowers. The spathe is undeveloped, is removed from the spadix, and is like a mere leaflet on the flower stem. The leaves, long-stemmed, oblong, and dark green, float upon the waters of pond shallows.

A native of Europe and Asia, as well as America, the sweet flag or calamus resembles the iris. The stiff, swordlike leaves, that give the plant a rigid character are yellow green and glossy, whereas those of the iris are bluish green and dull. They are one to three feet long, about an inch wide, and have sharp edges. The inconspicuous flowers, yellowish green, are borne on a spadix that emerges about halfway up from one side of a leaflike scape and stands out at an angle, two or three inches long. The fruit is a small berry, at first gelatinous, then finally dry, but the plant is largely propagated by its fleshy rootstock. What boy of yesterday did not know that it was good to eat, especially when boiled, cut in slices, and dried in sugar. Indeed, it is used by confectioners and sold as a candy, and it is not unlike candied ginger. The pungent, aromatic rootstock also furnishes the drug calamus and is a familiar commodity of the apothecary. The sweet flag grows to a height of one to four feet beside small streams and in wet ground. Why the name of sweet flag? Simply because the inside of the leaflike scape is sweet.

10. Australian Oak (Proteaceae)

The Australian oak family is a large group of over fifty genera and approximately one thousand species of trees and shrubs found mostly in Australia, but also in South Africa, Asia, and South America. They have alternate or scattered, tiny and awl-shaped or needlelike leaves without marginal teeth, and flowers in dense clusters. The flowers are without petals, but have four sepals that are united or tubular, four stamens, and a fruit that may be a nut, capsule, follicle, or drupe.

A popular street tree in southern California, but also planted elsewhere in the South, is the Australian silk oak of the genus *Grevillea.* The silk oak or silver oak grows to one hundred or to one hundred and fifty feet high, but in the juvenile, greenhouse state grows only two to three feet high. It has leaves that are twice-divided into graceful, feathery, fernlike segments, and orange flowers in one-sided racemes that are up to four inches long. The silvery green foliage of this stately tree from Queensland is so attractive and lacelike that small seedlings are used as decorative pot plants. It has the floral characteristics of the family and its fruit is a woody follicle.

Another member of the family native of South Africa, widely planted in California and similar climates for its beautiful silvery and silky foliage is the silver tree of the genus *Leucadendron.* It has scattered leaves, without marginal teeth and nearly stalkless, and the staminate and pistillate flowers are on separate plants. The staminate flowers are numerous in dense, stalkless, headlike clusters, while the pistillate flowers are in conelike heads, beneath which is a series of woody bracts. The fruit is a nut.

One of the genera (Banksia, named for Sir Joseph Banks, a noted British botanist) of the family contains over forty species of shrubs and trees that are known as the Australian honeysuckles. They are semidesert plants with alternate leaves and yellow flowers in dense, mostly terminal spikes. These spikes are between rather showy bracts, and are followed by a woody, conelike fruit with winged seeds.

Still another Australian shrub rather popular in California is the sea urchin, of the genus *Hakea,* also known as the pincushion flower. It is a tall shrub or small tree, sometimes up to twenty-five feet high, with elliptic, alternate leaves that are narrowed at the base into an obvious stalk. The crimson flowers are in dense, stalkless, nearly globe-shaped clusters. The fruit is an egg-shaped, hard, woody, capsule.

In both California and Florida one will find the Queensland nut of the genus *Macadamia,* which is chiefly grown for its edible seeds. It is a tree up to fifty feet high, with somewhat oblong, bright green leaves in whorls of three or four, and slightly spiny-toothed, suggesting a long holly leaf, and small white flowers that are borne in pairs and grouped in racemes nearly twelve inches long. They are followed by hard drupes with edible, globe-shaped seeds.

11. Banana (Musaceae)

The banana is a berry, botanically similar to the blueberry, the remnants of the seeds (rarely fertile) being the minute, black specks in the center of the pulp. Actually, the only difference between the banana and the blueberry is the shape.

The banana plant is an herb. It grows to the size of a small tree, fifteen to thirty feet high, with a subterranean stem. What appears to be the trunk is merely the concentric bases of the enormous feather-veined leaves that are a foot in width and five feet in length. Subject to tropical winds, they are easily torn and, as a result, are merely fragments. The yellowish white flowers, each with three sepals and three petals, are borne on a large, nodding spike, three or more feet in length, with purple bracts, and develop into the bunch of fruit (bananas) that we see in the stores. Each plant produces only a single bunch, whereupon it dies, but a new plant then grows up from the base.

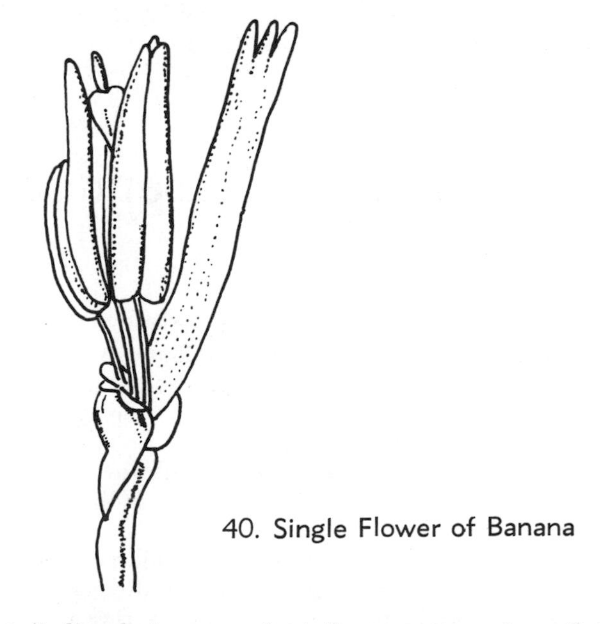

40. Single Flower of Banana

The banana plant was cultivated early in India and was well known to the Arabs and the Greeks, but for centuries remained only a curiosity, until rapid means of transportation was developed that could bring the perishable fruit quickly to market. It belongs to the genus *Musa* (named for Antonio Musa, physician to the first Emperor of Rome), a group of giant herbs chiefly from the Indo-Malayan region with huge, fleshy, treelike stems formed of the tightly packed sheaths of the leaf bases. They have exceptionally large rhizomes from which develops the single, trunklike stem that flowers only once and is then replaced by suckers from the base. Their leaves are simple, entire, and often very large with a single stout midrib. From this midrib diverge many parallel, transverse veins along which the leaves split in the wind. The flower clusters of the banana plant are terminal, appearing among the crown of leaves. Each cluster consists of a long, usually drooping spike of colored, tightly overlapping bracts enclosing the flowers. The flowers are highly irregular, perfect or monoecious, with a tubular calyx of three sepals, three petals, and six stamens—one of them sterile and petallike—and a three-celled ovule with one or many ovules. The fruit is the familiar banana, regularly seedless and sterile in the common banana, but producing seeds in some others.

The common banana that furnishes most of the fruit of commerce is now found in innumerable varieties, two of the best known being the small, red-skinned kind; the other, a small-fruited type known as the ladyfinger banana. The plantain, sometimes obtainable in stores, is the fruit of a plant of the same name. The plantain closely resembles the common banana, but its fruit is larger, green when ripe, and good only when cooked. It is a staple food of millions of poor people in the tropics. The dwarf, or Chinese banana, not over six feet high, is cultivated chiefly for ornament; and the Abyssinian banana, a huge, treelike herb, twenty to forty feet high, produces a fruit with a few large black seeds. Another species of the same genus is the abaca that yields the Manila hemp. Its fruit is inedible, but its leafstalks contain the finest cordage fiber in the world. It is a native to the Philippine Islands and is grown there as a major industry.

All banana plants are members of the banana family, a small group of only six genera and some seventy species, but containing the largest herbs in the world. They are widely distributed throughout the tropics. The family characteristics are essentially those of the genus *Musa*.

Only three other members of the family need to be mentioned: the curious traveler's tree of Madagascar, the bird-of-paradise, and the wild plantain. The famous traveler's tree is an extraordinary bananalike plant often cultivated for its striking habit. It has a stout, palmlike trunk, twenty to forty feet high, crowned with a tuft of immense bananalike leaves so arranged that they appear to be a gigantic fan. The fruit of this plant is quite unlike that of the banana, being a three-celled woody capsule. The bird-of-paradise is a tender South African perennial cultivated outdoors in Florida, California, and other warm states. It grows to three feet high and has underground, woody rootstocks and bananalike leaves, among which a half dozen orange and blue flowers are borne between purplish boat-shaped bracts. The wild plantain is a large herb of tropical America ten to fifteen feet high, with bananalike leaves. The leafstalks are very long and arise from the ground, hence the wild plantain differs from the banana in that it does not have a true stem. The flowers borne on long stalks are showy, but more striking are the scarlet, boat-shaped bracts bearing the flowers. The fruit is a bluish capsule that eventually separates into berrylike segments.

12. Barberry (Berberidaceae)

Some people like the fruit of the mayapple, finding it sweetish and having a strawberry flavor, while others consider it as having a mawkish taste. The Indians valued it for medicinal purposes. The mayapple, or mandrake, is a common, handsome, woodland plant, also known as the wild lemon in allusion to its peculiar lemonlike fruit. It is a large, yellowish, egg-shaped, many-seeded berry.

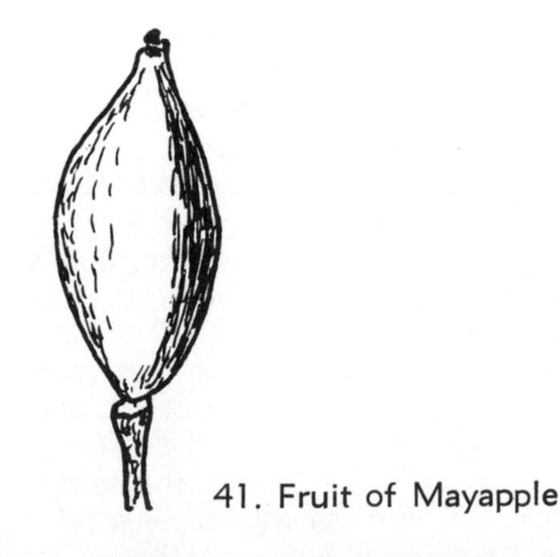

41. Fruit of Mayapple

The mayapple is remarkable for its large, poisonous leaves, frequently measuring a foot in diameter. When they first unfold in the spring, like a tiny umbrella, children of another generation used to say, "The umbrellas are out." The curious leaves betray the hiding place of the ill-smelling but handsome white flower, nearly two inches broad.

42. Leaf and Flower of Mayapple

The mayapple may be found not only in the open woods, but also along fences and roadsides. Its long, horizontal rootstalks are poisonous and remain in the ground year after year, each spring sending up a one or two-leaved plant twelve to eighteen inches high. The one-leaved plant is flowerless and has only a single leaf with seven to nine lobes. It is shield-shaped and, like an umbrella, is supported by the stem in the center. The two-leaved plant has two symmetrical leaves that are somewhat smaller and are attached to the two forks of the stem by their inner edges. From between the two forks of the stem droops the flower that usually has six petals and twice as many stamens. Though it has no nectar, it is nevertheless cross-fertilized by early bees and bumblebees. Despite its poisonous character, the mayapple is included in the dietary of the variegated fritillary, the leaves being eaten by the caterpillars.

Also poisonous and a relative, the early flowering blue cohosh is a common perennial herb of moist rich woodlands and woodland pastures. It has a short, knotty rootstock and a simple, erect stem bearing a large compound leaf at the top. The leaf is three-divided, the divisions three-lobed. The blue cohosh has a raceme or panicle of yellow green or yellowish flowers, one-half inch across, with six sepals, six petals, six stamens, and a solitary pistil. The pistil bursts while small, exposing the two ovules that develop into dark, blue, naked seeds about as large as peas, and that children should be warned against eating. The plant is very irritating to mucous surfaces and people handling it have often suffered from dermatitis. Also known as papoose-root and squaw-root, the blue cohosh was used in the Indian's *materia medica*.

As its name suggests, the twinleaf has twin leaves appearing as if a single leaf had been cut into two equal leaflets. The twinleaf is a little plant, not much more than eight inches tall, but grows taller later in the season when in fruit. The two leaflets at the end of a long petiole come from the root, and a single white flower, about an inch broad, has eight petals and half that number of sepals which fall early. It is an inhabitant of rich, shady woods.

Doubtless the plants giving their name to the barberry family are better known to most than those just mentioned, for the barberries are of considerable horticultural importance and widely planted in our gardens. The barberry family comprises ten or more genera and about two hundred species of herbs and shrubs widely scattered over the north Temperate Zone. They have alternate, simple, or compound leaves and flowers that are either solitary, or in racemes or panicles. They are perfect with six sepals and six petals, the latter often replaced with nectaries, with about as many stamens as petals, and a one-celled ovary, with few to many ovules. The fruit is a berry or capsule.

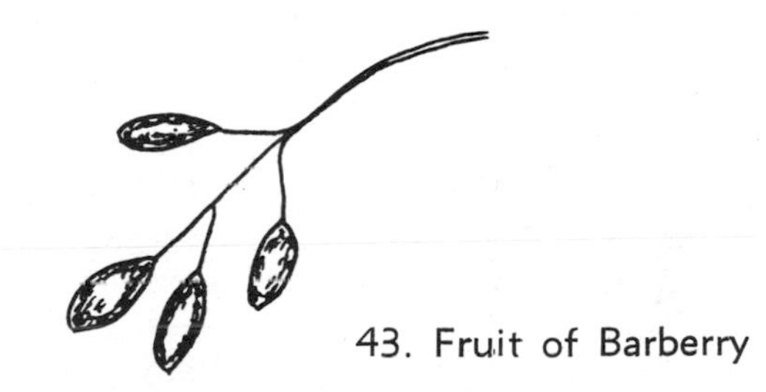
43. Fruit of Barberry

The barberries belong to the genus *Barberis*, a Latinized form of an Arabic word for the fruit. They are all more or less spiny, with yellow wood and inner bark (cambium), simple leaves in small clusters at the ends of short spurs, and yellow flowers in clusters blooming in the spring and developing into berries.

There are about one hundred and seventy-five species of barberries coming mostly from Asia, but a few do come from North Africa and South America. The European barberry, imported into this country and planted for hedges, has become naturalized and is now found running wild in the woods and thickets of New England and the Middle States. It has now been largely supplanted by the Japanese barberry, possibly more widely cultivated than any other shrub. The European barberry is an arching shrub, five to nine feet high, with finely toothed, pale green, elliptic leaves that are reduced to thorns. The small, yellow flowers occur in drooping clusters (racemes) from the leaf axils. It is a graceful bush, less decorative in the spring, perhaps because its flowers are small, but considerably more beautiful in the fall when its branches arch with the weight of clusters of red or scarlet berries. This barberry has earned a somewhat evil reputation by being the alternate host for the destructive wheat rust. A yellow dye is obtained from the bark of the root.

Our native barberry, known as the American barberry or Allegheny barberry, is a woodland shrub of the Allegheny Mountains and is similar to the European barberry except that it is smaller and lower-growing, with slender, terra-cotta brown branchlets. Its leaves are less finely toothed, and its flowers and berries are in smaller clusters. The latter are also more oval than those of the European species, but both are used in making jelly and sometimes pies.

There are fifty or more species of barberries that are cultivated in America. Some of them, like the European barberry, serve as the alternate host for the wheat rust and are excluded from the cereal-growing states. A few of them are evergreen such as the Magellan barberry and the wintergreen barberry.

In the western states there are a number of species that were formerly placed in the same genus as the barberries. Collectively known as Oregon grapes they now have a genus of their own, *Mahonia,* named for Bernard McMahon, an American horticulturist. They have alternate, compound, evergreen leaves, and the leaflets are pinnate, rarely in threes, spiny, often turning purplish in autumn. The Oregon grapes have yellow, fragrant flowers, with six petals and nine sepals in terminal racemes or panicles, and dark blue berries that are usually covered with a bloom.

A very beautiful, low shrub is the trailing mahonia, also known as the Oregon grape, of the Rocky Mountains and the northwestern coast. It has compound leaves of about five ovate, hollylike leaflets that are dark, lustrous, green to bronzy green on top, paler and smoother underneath, and changing to a dull red or yellow in autumn. The flowers are waxy gold in erect, terminal clusters, and the berries are purple and globelike. It is found in open woodlands.

A second species, known as the Oregon grape or the water holly, is a shrub about two feet high, with eleven to nineteen leaflets that are oval to narrow, leathery, shiny above, and paler beneath with spiny teeth on the margin. The flowers, in erect terminal clusters, are bright yellow and fragrant, and the berries are oval, dark blue with a bloom. A third species, called the California barberry, is also a shrub up to twelve feet high, with seven to thirteen leaflets, pale yellow flowers, and round purplish black berries. The creeping barberry, although a shrub, is not more than a foot high. It has underground rooting stems, three to seven roundish, oval leaflets that are dull, bluish green above with spiny marginal teeth, flowers in terminal clusters, and small black berries with a bloom. Its foliage is relished by the mule deer.

A native of China and Japan, widely cultivated in the South, is the nandina, which is sometimes called sacred bamboo, although it has nothing to do with bamboo. It is an attractive shrub, six to eight feet high, with alternate leaves twice or thrice-compound, which turn a beautiful red in the fall. The flowers are small and white, in handsome clusters (panicles) a foot long, and the berries are red, two-seeded and when ripe are the plant's chief attraction.

Of somewhat unusual interest because the sepals of the flowers assume petaloid forms and colors, the petals becoming spurs for the production of nectar, the large-flowered barrenwort is brilliantly colored and is a native of Japan, widely cultivated for its fine flowers. It is an erect herb, six to nine inches high with thrice-compound leaves, the leaflets sharply toothed. The flowers occur in terminal panicles. The eight sepals are in two sets, with the outer set a bright red, the inner set violet. The four petals in the form of spurs are an inch long, white and nectar-bearing, and the fruit is a few-seeded capsule. The plant is used chiefly in the rock garden, but is also planted as ground cover.

Often planted in the wild garden is a perennial herb of the southeastern woodlands called the umbrella leaf. It is an erect, stout species, with a single, basal, long-stalked leaf twelve to twenty inches in diameter and two stem leaves that have shorter stalks. The leaves are deeply cut into two lobes, while the lobes themselves are also cut or lobed, with the petioles arising from or near the middle of the blade. The flowers are white in a small terminal cluster, and develop into blue berries.

13. Batis (Batidaceae)

Along the seashore of California and along the seashore from North Carolina to Florida, there may be found growing a low shrubby bush that is the lone species of its family. It is known as batis (genus of the same name), and grows one to three feet high, with simple, opposite, fleshy, sessile, entire leaves, and dioecious flowers in catkins that are crowded in a terminal spike. The staminate flowers have four or five stamens, and petallike staminodia; the pistillate flowers lack a perianth and have a four-celled ovary with a single ovule in each cell. The fruit is a fleshy egg-shaped mass.

14. Bayberry (Myricaceae)

> At the mouths of their rivers, and all along upon the sea, and near many of their creeks and swamps, the myrtle grows, bearing a berry of which they make a hard, brittle wax of a curious green color, which by refining becomes almost transparent. Of this they make candles, which are never greasy to the touch and do not melt with lying in the hottest weather; neither does the stuff of these ever offend the sense like that of a tallow candle, but instead of being disagreeable if an accident puts the candle out, it yields a pleasant fragrance to all who are in the room, insomuch that nice people often put them out on purpose to have the incense of the expiring snuff.

So wrote Robert Beverly in his *History of Virginia* of the bayberry, a stiff, somewhat crooked shrub found growing in a variety of situations in the coastal plain from Nova Scotia southward to Florida and Alabama.

The bayberry is a shrub three to eight feet high, with a brownish gray bark, elliptical, leathery leaves that are broadest at the tip. They are colored a pale green, tinged with red that darkens to a bronze purple in autumn, with little change in color, and the flowers are in catkins. The fruit is a dry, waxy, gray white, drupelike nut and is used in making the bayberry

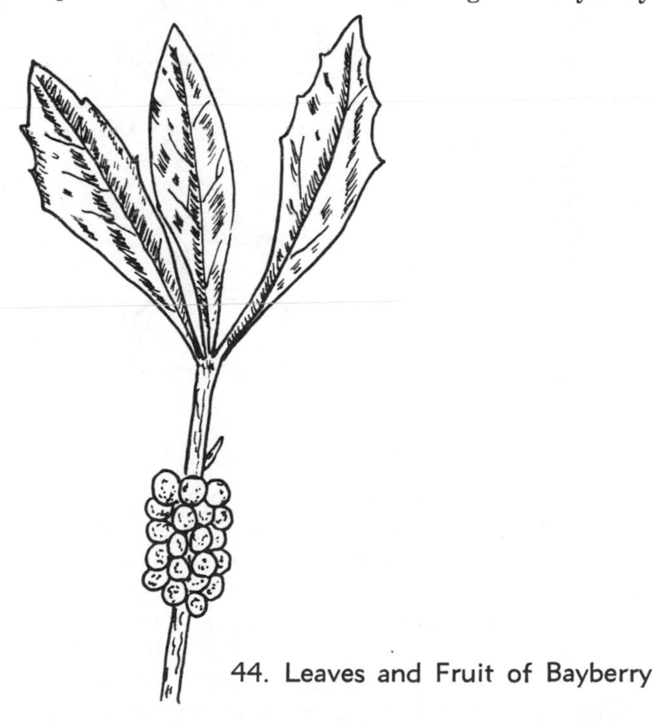

44. Leaves and Fruit of Bayberry

candle. The fruit, as well as the leaves and recent shoots, are aromatic with a balsamic odor coming from the minute, transparent, resinous dots profusely covering the recent shoots and the undersurface of the leaves. It should be mentioned that our native bayberry is not the true bayberry, the latter being a tropical American tree of another family.

The bayberry family is a small group of forty species of shrubs and small trees contained in only two genera found usually in swamps or on dry sandy soils of the temperate and warmer parts of the world. They have alternate, aromatic, sweetly scented leaves that are almost leathery in texture. The small flowers are in catkins, the staminate and pistillate on the same plant or on different plants. They lack both sepals and petals, but have two to sixteen stamens and a one-celled ovary. The fruit is a small, often waxy nut or berrylike drupe.

The bayberry belongs to the genus *Myrica* consisting of thirty-five species having the family characteristics. It includes such plants as the sweet gale, wax myrtle, and California wax myrtle. The sweet gale, also called moor or bog myrtle, is a shrub not usually over four feet high. It is found in swamps and pond margins, and has dark brown, ascending stems, and wedge-shaped leaves with a few teeth at the apex. These leaves are dark green above, pale and usually downy beneath. The staminate flowers are in terminal catkins, and the pistillate occur at the base of the leaves. The fruit is resinously waxy, berrylike, dotted, and crowded in a cluster of two to six nutlets. The leaves have a pungent, spicy odor and are said to give a pleasant flavor to roasts.

The wax myrtle is a small, slender, southern tree, never more than thirty-five feet high and is found in sandy swamps. It has brown gray bark, small, lance-shaped leaves that are a lustrous green, very fragrant when crushed, and almost evergreen. The flowers are like those of the preceding species, and the fruit is grayish, waxy, aromatic, and berrylike. The California wax myrtle is a beautiful western species of low hills, sand dunes, and riverbanks with deep green, willow-like leaves and purple black berries covered with white wax in crowded clusters.

The second genus of the family, *Comptonia,* named for Henry Compton, Bishop of Oxford, contains only

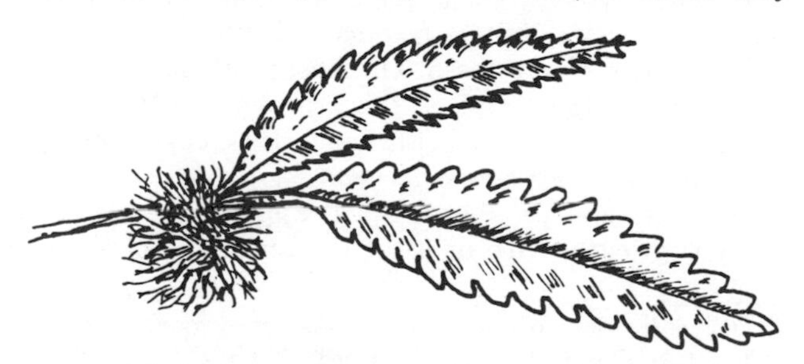

45. Leaves and Fruit of Sweet Fern

one species—the sweet fern, a plant that looks like a fern and grows like a bush and one familiar to everyone who knows the outdoors. It is a fragrant, round-headed shrub about two feet high, of hillsides, stony uplands, and dry pastures, with commonly a dull red or dark brown, scragged, woody stem. The leaves are alternate, very narrow, almost linear, the margins cut into many rounded lobes, dark green, stalked, and very fragrant when crushed. The leaves suggest a fern and so the common name is not inappropriate. It is said that during the Revolutionary War the leaves were dried and used for tea. The flowers are small, inconspicuous and green, the staminate in drooping catkins, the pistillate in globular and burlike catkins, both sometimes on the same plant, but more commonly on separate plants. The fruit is a small nutlet that is enveloped in eight linear bractlets, thus, the fruit is burlike, pale, or rusty green, and very aromatic when crushed.

15. Beech (Fagaceae)

Probably no group of trees appeals to our imagination more than the oaks, largely because of their grandeur, sturdiness, and durability—indeed, qualities that men have admired since ancient times. They are long-lived trees, too. There are oaks in England believed to have been old trees at the time of William the Conqueror. Pliny even mentions an oak that was an old tree when Rome was founded and that still stood during his time. And they are of an ancient lineage; their remains are found in the rocks of the Miocene and Eocene Periods. Many historic trees are oaks.

The oaks, numbering several hundred species, are the finest hardwood timber trees of the north Temperate Zone and are found in North America, Europe, and Asia. Besides their value as timber, some species produce cork,* others tannin, and still others produce ink from leaf galls. Their nuts are a very important source of food for wildlife; the sweet acorns of several western species were used extensively as food by the Indians. Many species are planted as ornamentals on lawns, parks, streets, or for the home woodlot.

Most of the oaks are evergreen, especially the Asiatic species, and in North America the group as a whole just misses being evergreen; many species have leaves, usually withered, that persist through most of the winter. They have a strong, deep tap-root system from which many horizontal roots spread outward near the surface of the ground. The leaves are alternate, stalked, variously lobed, toothed, or divided in most species, but a few have leaves that are unlobed and without teeth. Of those that have lobed or toothed leaves, about a third are set with bristles, while the rest do not, although this character does hold in the evergreen species.

The flowers are monoecious, appearing on the old or new growth, with or after the unfolding of the new leaves. The staminate flowers are in drooping, clustered catkins, the individual flowers are without petals, but have a four to seven-lobed calyx that usually encloses six stamens. The pistillate flowers, also without petals, are either solitary, or in few to many-flowered spikes from the axils of the new leaves. The individual flowers have a six-lobed calyx surrounding a usually three-celled ovary, the whole partly enclosed in an involucre. The fruit is a true nut (the acorn), maturing in one or two seasons, and is set in a cuplike involucre that may partly or completely cover the nut only at the base. Sometimes the cup is fringed.

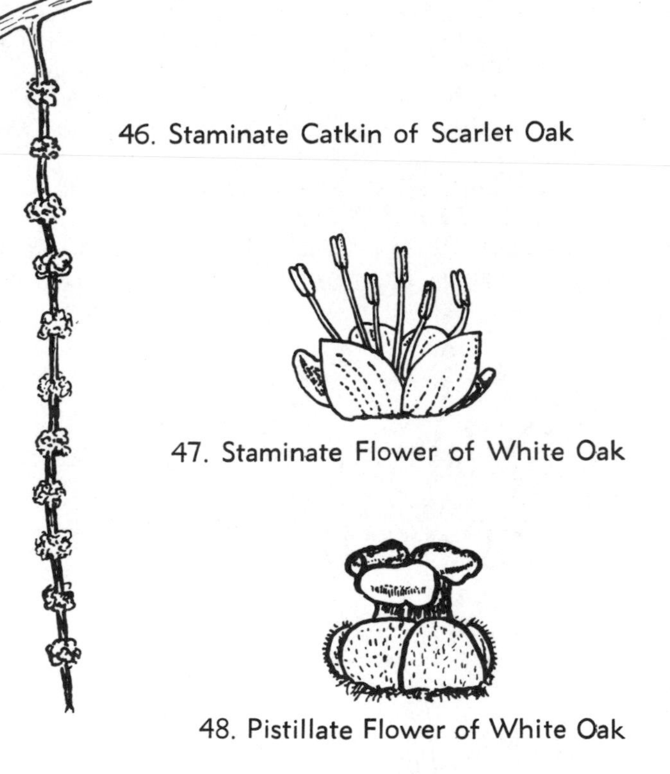

46. Staminate Catkin of Scarlet Oak

47. Staminate Flower of White Oak

48. Pistillate Flower of White Oak

* The cork oak of the Mediterranean region, sometimes planted as an ornament in the United States, is the source of commercial cork.

The oaks are generally divided into two groups: those that have leaves without bristles on their lobes or teeth, blooming in the spring and maturing their acorns the same year; and those that have bristles on their lobes or teeth also blooming in the spring, but not having their acorns mature until the second year.

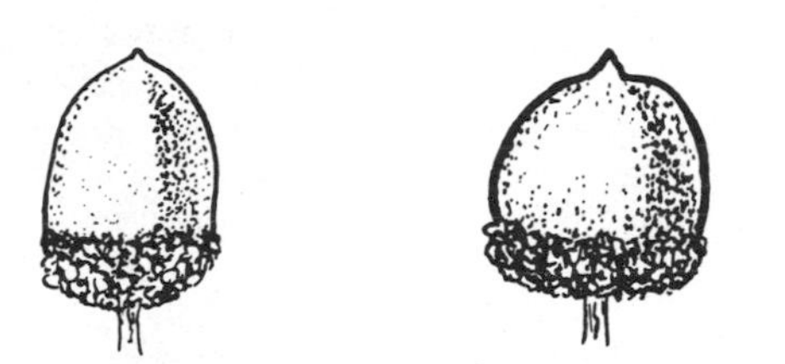

49. Nut or Acorn of White Oak and Red Oak

In the first group belong such species as the white, post, overcup, bur, swamp white, chestnut, yellow, and dwarf chinquapin oaks; in the second group belong such species as the red, scarlet, black, Spanish, pin, bear, black Jack, shingle and willow oaks, the last two species having no bristles on their leaves, but maturing their acorns the second year.

The white oak, supposedly so-called from the color of its bark that is, however, more of an ashen gray, is a magnificent round-headed tree (in the open), fifty to seventy feet tall, perhaps the most stately and beautiful of our oaks and certainly one of the most valuable. In the spring, the leaves unfold a bright red, fade to a soft pink, then become a silvery white, and finally a yellow green. As the young leaves are covered with a soft down, the entire tree acquires a frosty, misty appearance quite pleasing to the eye, though unfortunately this effect lasts only a few days. All summer the leaves are a bright green, but then in autumn the red comes back with blue or purple tones and the tree glows in the woods with a splendor that softens the dreary November days. The white oak is tolerant of many soils and is found in dry, upland woods, on sandy plains and on gravelly ridges; but it reaches its best development in coves or on the higher bottomlands.

Julia Rogers in her *Tree Book* says that the post oak looks like a tree with its trunk buried in the ground. It is a round-headed tree of dry, sandy soil and gravelly uplands; its branches characteristically are gnarled or somewhat twisted in appearance, and though often a shrub, it usually reaches a height of fifty to sixty feet. Its leaves are thick, leathery, and somewhat cruciform in appearance, and in the fall they turn yellow or brown. They remain on the twigs until the opening shoots push them out of the way. The post oak is characteristically leafy all winter.

A large tree, fifty to seventy feet high, of swampy lands, often with a crooked or twisted bole and a large open crown of crooked branches, the overcup oak derives its name from the fact that the acorn is nearly covered by the rough, scaly, thin, fringed cup. The tree is found on coastal flats or in alluvial sediments of rivers from New Jersey along to the lower Mississippi and the valleys of its tributaries, where it is one of the most common of the swampland trees.

One of the most distinctive of our eastern oaks, the bur oak extends farther to the northwest than any other oak of the east and varies in size from a shrub in Manitoba to as much as one hundred and sixty feet, being one of the tallest oaks in the eastern states. It is a rugged-looking tree of rich bottomlands with three marked characteristics. Its leaves are a lustrous, deep, olive green turning a dull, rusty or maroon red in the fall, and have a peculiar though variable outline. Rarely, if ever, will one find two that are alike and, yet, they all resemble each other so that there is no difficulty in distinguishing them. Each leaf, moreover, is cut, usually in the middle, by two opposite sinuses nearly to the midrib. The terminal lobe is toothed or repand, while the lower divisions may either be lobed or entire. Yet, with all the variations, the leaves still retain a general similarity. The second characteristic is the acorn that is often two inches long and set in a thick, hairy cup that, with its frayed and ragged edge (the upper scales being long-awned), resembles a miniature bird's nest. Moreover, the mossy cup—hence the tree is known as the mossy-cup oak—fairly covers the nut, covering two-thirds to three-quarters of its surface, sometimes almost entirely covering it. The third distinctive characteristic of the tree is its corky twigs or ridges, which begin to form the third or fourth year and remain for several years, eventually disappearing as the branches get old.

Unlike the leaves of the white oak, which unfold a beautiful red color, those of the swamp white oak appear bronze green; and, again, unlike those of the white oak that turn red in the fall, those of the swamp white oak turn dull yellow, or pale yellow brown. The swamp white oak, fifty to sixty feet high and sometimes higher, is a medium-sized tree of streambanks, moist flats, and swamps, with flaky bark and large leaves six to eight inches long. Looking at the tree standing in the open as outlined against the winter sky, one will see that it has a round-topped open head, sometimes wider than high, with the upper limbs ascending and the lower limbs rather small, horizontal or declined even to the ground, with numerous tufted, small scraggly, lateral, pendant branchlets. The lower branches spread out below to form a fringe that resembles an old-fashioned hoopskirt, a characteristic whereby all can recognize the tree from a distance.

The books state that the chestnut oak is a tree of dry hillsides and rocky woods, but it is also found in better soil. It is a vigorous, medium-sized tree, forty to fifty feet high, bearing deep, yellow, olive green leaves with about ten shallow lobes on each side, narrowly elliptical or inversely ovate in general outline, similar to that of the chestnut leaf. Called the basket oak because its wood is used in making baskets, crates etc., the basket oak is a large southern species of riverbanks and swamps, seventy to ninety feet high. It is a well-formed tree with a straight massive trunk and narrow crown, with shining, dark green leaves that turn crimson in the fall.

Like the chestnut oak, the yellow oak is also found on dry hillsides as well as on limestone ridges. It is a tall, handsome, light-colored tree, forty to one hundred and sixty feet tall, with bright yellow green leaves, hence its popular name of yellow oak.

Unlike the preceding species the dwarf chinquapin oak is a shrub, three to nine feet high, generally growing in clumps on rocky hillside pastures and spreading by underground stems. The opening leaves are silvery below and orange red above; later they become a deep yellowish olive green, and in the fall they turn a beautiful rusty red. Its leaves look much like a chestnut leaf and from its leaves and habit of growth, the shrub is also called the scrub chestnut oak. The kernel of the acorn is sweet and edible and is produced abundantly.

A handsome tree and the tallest of our northern oaks, fifty to one hundred and forty feet high, the red oak is a common species in rich or rocky woods. It has a columnar trunk and a round, symmetrical head of stout, spreading branches with a nearly smooth, grayish brown bark on young trees, while dark brown on old ones. The leaves are wavy-edged, with many pointed lobes, pink and covered with soft silky down when they unfold, later a dark green, turning a very dark, rich, maroon red in autumn. The acorn is chestnut red or light brown, with a saucerlike shallow cup. "What delicate fans are the great Red Oak leaves," writes Thoreau, "now just developed, so thin and so tender a green! They hang loosely flaccidly down at the mercy of the wind, like a newborn butterfly or dragon fly. A strong cold wind would blacken and tear them. They have not yet been hardened by exposure, these raw and tender lungs of the tree."

The scarlet oak derives its name from the color of its leaves in the fall when the tree blazes like a torch against the duller reds and browns of the woodlands. The color is not scarlet, however, but more of a bright cardinal red. The scarlet oak is a medium-sized tree, forty to fifty feet high, found growing on dry, sandy or gravelly soil, with a slender, tapering trunk, graceful curving branches, and round head. The bark is dark brown, with irregular, shallow furrows, and the thin, bright olive green leaves are deeply cut into about seven lobes. The acorn has a thin, saucer-shaped or top-shaped cup.

The name of the black oak is derived from the color of its bark, though it is not black, but a dark gray or even brown. The inner bark is deep orange yellow and is a means of identifying the tree if it is cut into. The black oak is one of the finest and largest of the eastern oaks, fifty to seventy feet tall, but sometimes a great deal taller, with slender branches and a narrow open head. The leaves are a lustrous green on the upper surface, dull, whitish, olive green on the lower surface, and turn a russet brown or a dull red in autumn. The acorn is light brown with a top-shaped cup covering about half the nut. The bark of the black oak, commonly found on dry, gravelly uplands, is rich in tannin, and formerly a yellow dye, known commercially as quercitron, was obtained from it.

A southern species of oak, one of the commonest of the upland southern oaks as a matter of fact, and found in dry upland or sandy lowland soil, being particularly characteristic of the dried, poorer soils of the Piedmont plateau, is the oak known as the Spanish oak or southern red oak. It is a medium-sized tree, fifty to eighty feet high, with a sepia brown, often blackish, bark, variable dull, dark, green leaves, and a globular acorn with a saucer-shaped cup.

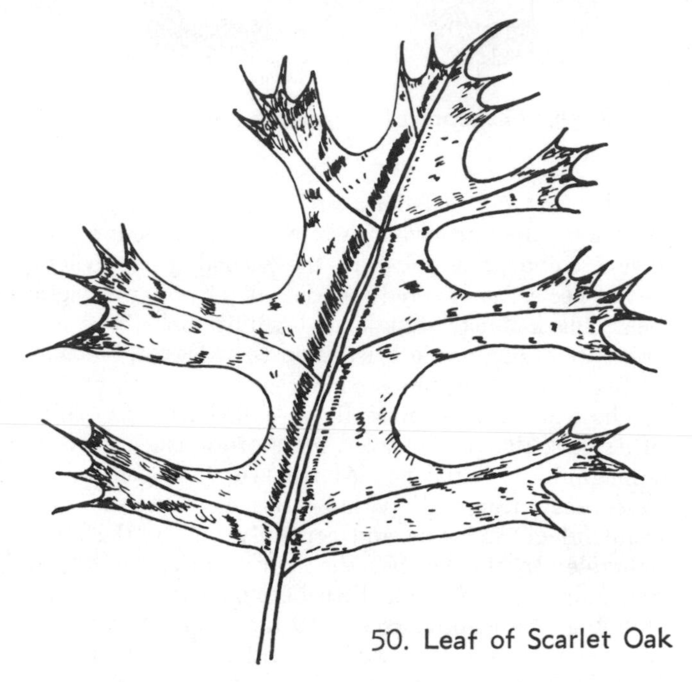

50. Leaf of Scarlet Oak

Of the pin oak the French botanist and traveller Michaux writes:

> **Its secondary branches are much more slender and numerous than is common on so large a tree and are so intermingled as to give it at a distance the appearance of being full of pins. This singular disposition renders it distinguishable at first sight in winter and is perhaps the cause of being called the Pin Oak.**

The pin oak is a medium-sized tree, fifty to eighty feet high, with a root system shallower than in many of the other oaks. When young, the pin oak, which loves a moist, rich soil and is found on the borders of swamps and in river bottoms, is a graceful tree, with an unbroken trunk and a pyramidal head with somewhat pendulous branches, the latter set with tiny branchlets. The leaves are small, dark, shining green and deeply cut, with two or three narrow lobes on each side. The acorn is small, somewhat spherical, often striped, with a shallow, saucerlike cup. With age the pin oak loses some of its symmetry and beauty, but retains much of its dignity even in its decline.

The bear oak, named by the early settlers because the bears delighted in its little bitter acorns, is commonly a shrub, three to ten feet high, with numerous intertwined branches. The little, straggling, shrubby oak is fond of rocky hillsides and dry, sandy, barrens, and often forms dense thickets on mountainsides. The bark is dark grayish brown, and the leaves are dark yellow green above, grayish white and downy beneath, and turn rusty yellow or russet brown in autumn. The acorns are light brown with top-shaped cups, and are produced in great numbers, making a fruiting branch a very picturesque object.

The blackjack oak is so-called because the early American settlers used the word *jack* to signify something worthless or disgraceful. The blackjack oak is a pariah of its kind because its wood is not valuable when compared to all the other oaks. Anyway, it is a small, shrubby tree, twenty-five to forty feet high, with a small trunk, spreading and contorted branches, growing on barren sandy, or clay soils. It is not common in the North, but very abundant in the South, and in the Southwest it makes up a large part of the forest growth in the poorest of soils. The bark is rough, sepia brown, often nearly black. The leaves are dark, lustrous, olive green above, paler and rusty-haired beneath, quite variable in form, and turn brown or yellow russet in autumn. The acorn is small, nearly spherical, with a top-shaped, coarse-scaled cup. What the tree lacks in beauty it is more than offset by a ruggedness of character.

Michaux, already quoted, named the tree the shingle oak because he found that the early pioneers used the wood to make split shingles or shakes. He describes the tree in part as follows:

> **The Shingle Oak has a smooth bark and for three-fourths of its height is laden with branches. It has an uncouth form when bare in winter, but is beautiful in summer when clad in its thick tufted foliage. The leaves are long, lanceolate, entire, and of a shining green.**

No one would take the tree to be an oak from the appearance of the leaves because they are more like willow or peach leaves than oak leaves. Yet, upon seeing its acorns that are nearly globular and set shallowly in saucerlike cups, no one can deny that it is a true oak.

The shingle oak is a medium-sized tree, forty to sixty feet high, with a broad, pyramidal head when young. It resembles the pin oak with its horizontal and drooping branches, but without the "pins," becoming in old age broad-topped and open. It is a tree of rich woods, rare in the East, but abundant in the lower Ohio Valley, where it reaches its best development on moist soils along the streams. It is also known as the laurel oak.

The willow oak is an anomaly among the oaks, for in its shape and general appearance it is like a willow. Its shoots are straight and slender like those of the willow, and its leaves are lance-shaped like those of the willow. It is, indeed, appropriately named.

The willow oak is a graceful, quick-growing tree, forty to sixty feet high (in favorable situations, eighty feet), with slender branches that form a conical, round-topped head. Its bark is a deep, ruddy brown; its leaves are bright, light olive green above, paler beneath; and its acorn is yellowish brown set in a shallow, saucerlike cup. The willow oak is a southern bottomland tree, and is commonly found on poorly drained, loamy or clay flats. As an ornamental, the tree has few if any superiors throughout the South, and in southern cities is widely planted along the streets.

A small southern tree, up to forty feet high, but in favorable situations as much as eighty feet high, the water oak is a plant of wet flats and swamps, with slender branches that form a conical, round-topped head. The bark is rather rough, reddish brown, and the twigs are slender and dull red. The leaves are thick, often evergreen, but usually deciduous, and are dull olive green above, paler beneath, with three, blunt, rounded lobes at the tip or without lobes altogether. The acorn is hemispherical set in a shallow, saucerlike cup.

The live oak, so-named because its leaves are evergreen, is a southern tree, forty to fifty feet high, with a thick trunk and horizontal branches of great length that form a low-spreading dome, resembling an old apple tree. The leaves are not as showy and beautiful as the leaves of many of our northern oaks, as they are small and without lobes. They are lustrous, dark green above, pale and somewhat hoary beneath, but they last all winter until the new leaves appear in

51. Leaf of Live Oak

the spring. The acorns are dainty, dark brown, and are set in long-stemmed, hoary, top-shaped cups.

Writing three centuries ago, the English naturalist Catesby relates that the Indians gathered the acorns "to thicken their venison-soop," and also cooked them in several other ways. "They likewise draw an Oil, very pleasant and wholesome, little inferior to that of almonds."

The live oak is commonly found in dry soil, and in the more southern part of its range is often draped in Spanish moss, giving it an unkempt and rather eerie look, though some people say that it adds to its charm. The live oak is planted as an ornamental and the streets of many southern cities are also lined with it.

The oaks that have been mentioned thus far are oaks found in the eastern half of the country. There are also western species—such as the California live oak, found in the coast foothills and valley flats; the interior live oak; the mesa oak, with conspicuous blue green foliage; the canyon oak, a shrubby species of mountain canyons; the blue oak, a small tree of the rocky foothills of California; the valley oak, a large graceful tree; the Oregon oak; and the California black oak found on the lower mountain slopes of Oregon and California.

The oaks are of considerable importance to our wildlife. Their acorns are eaten by many game birds and songbirds, as well as by several species of mammals. Their twigs, foliage, and bark are also consumed by various mammals and by such hoofed browsers as the white-tailed and mule deer and the peccary. In addition to their food value, the oaks also provide useful cover and their leaves and twigs are used by many birds as nesting material. The caterpillars of a number of butterflies include the oaks in their dietary: gorgone checkerspot; white m hairstreak; southern hairstreak; Edward's hairstreak; banded hairstreak; striped hairstreak; dreamy dusky wing; sleepy dusky wing; and Juvenal's dusky wing.

The oaks belong to the genus *Quercus,* the classical Latin name for the oak, though the name is said to have originated from the Celtic *quer,* meaning fine, and *cruz,* meaning tree. The beech, from which the family derives its name, belongs to the genus *Fagus,* the classical Latin name for the beech. The beech family, also called the oak or chestnut family, comprises six genera and about six hundred species of trees and shrubs scattered throughout both hemispheres, but mostly of the north Temperate Zone. Their leaves are alternate, usually deciduous, but evergreen in some species, simple, entire, lobed, or cleft. Their flowers are moneocious and without petals, the staminate in clusters or catkins, the individual flowers usually with four to six sepals and a few to many stamens. The pistillate are solitary or in few-flowered clusters with four to six sepals, and there is a three to six-celled ovary, one or two ovules in each cell, with only one maturing. Their fruit is a one-seeded nut, partly enclosed in a cup (oaks), or completely surrounded by a bur (chestnut), and by bracts in some other genera.

Beneath the shade which beechen boughs diffuse,
You, Tityrus, entertain your sylvan muse,
Virgil

I ran to meet you as a traveller
Gets from the sun under a shady beech.
Theocritus

The ancients valued the beech as a shade tree, but they also knew the wood absorbed little water and hence made excellent bowls.

No wars did men molest
When only beechen bowls were in request.
Virgil

In beechen goblets let their beverage shine,
Cool from the crystal spring their sober wine.
Milton

That the beech seems to have had some significance for lovers may be gathered from the following:

Or shall I rather the sad verse repeat
Which on the beech bark I lately writ?
Virgil

On the smooth beechen rind the pensive dame
Carves in a thousand forms her Tancred's name
Tasso

The name beech has a very ancient origin and may mean *book,* since both can be traced to Anglo-Saxon words of similar meaning, a conclusion further enhanced by the fact that the ancient Saxons and Germans both wrote runes on pieces of beechen board. Anyway, the beech, about which the above was written, is the European beech, or copper beech, a tree whose leaves, buds, and bark are darker than those of our native species. The European beech is an important timber tree of Europe, where its fruit, known as "beech mast," has long been used for fattening hogs and also as a source of vegetable oil.

There is only one native beech in North America, the American beech, or simply beech as it will be called here. It is a clannish tree, and should one find a single specimen in the woods, there is sure to be a miniature forest of beeches around it as soon as it begins to bear. Not only do the blue jays and squirrels, as well as the wind, carry the nuts, but the nuts have considerable vitality and the seedlings do well even in the densest shade. Thus, they have an advantage over the young of other trees.

The beech is both a distinctive tree and a common one. It is found on rich uplands and on moist, rocky ground, with a continuous trunk and almost horizontal, slender branches forming a symmetrical, round-topped

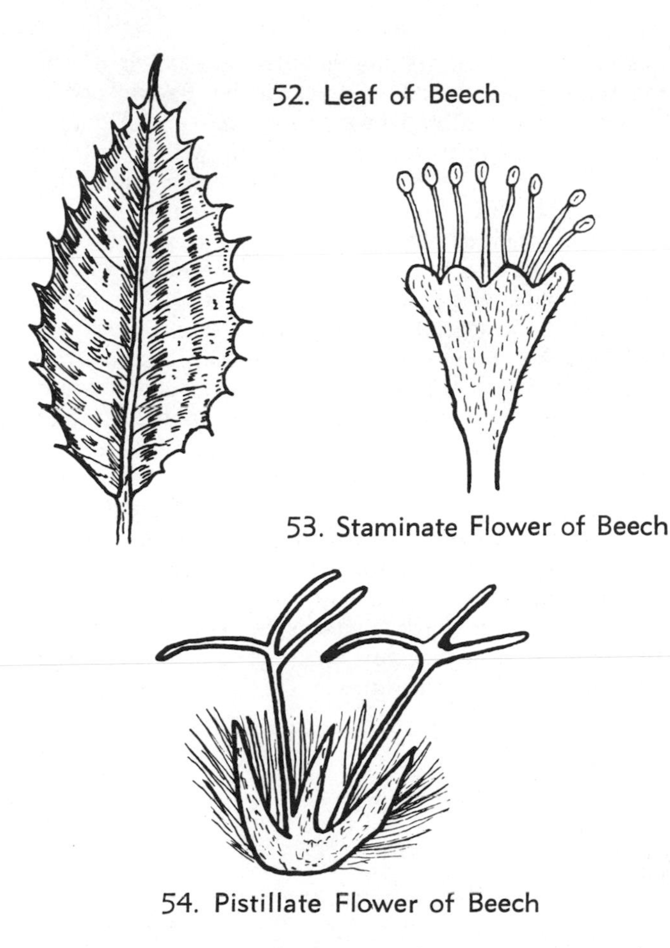

52. Leaf of Beech

53. Staminate Flower of Beech

54. Pistillate Flower of Beech

or conical head. It grows forty-five feet to fifty high, occasionally one hundred feet, and has a shallow, but extensive, root system, and a smooth, pale gray or blue gray, often mottled bark that is easily recognized even at a distance. The leaves are papery thin, sharp-pointed, ovate, a light green that turns a light, gold yellow in the fall. The flowers appear after the unfolding of the leaves in the spring. The staminate are in globose heads, and each flower has a four to eight-lobed calyx and eight to ten stamens. The pistillate are in two to four-flowered spikes, surrounded by an involucre, and each flower has a four to five-lobed calyx attached to a three-celled ovary. It has been said that beech trees have to be at least forty years

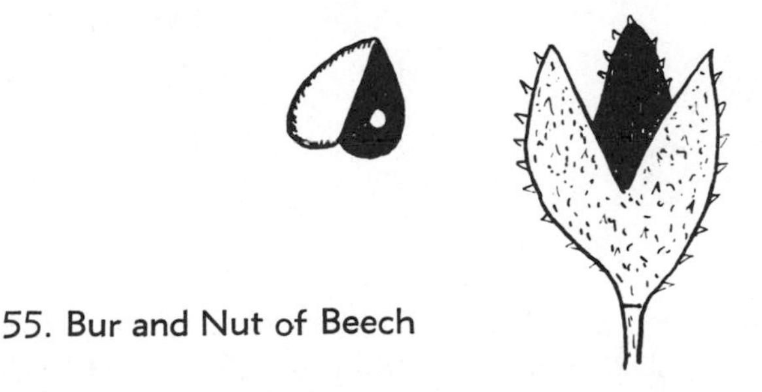

55. Bur and Nut of Beech

old before they can begin to bear fruit. This author does not know how true this is, but since the trees grow to a venerable age—three to four hundred years—it might be so. The tree is not consistent in bearing fruit, but produces large crops at irregular intervals. The fruit is a tiny bur, with soft, spreading, recurved prickles. It opens in four sections and remains until after the nuts have fallen. The nuts—brown, shining, and triangular—are sweet and edible. There are two, sometimes three, to a bur. Needless to say, they are eaten by various birds as well as mammals.

No longer does the chestnut grace our woodlands as it once did. No longer is there a bountiful crop of nuts for either man or beast. The chestnut was always a dependable source of mast for the woodland animals,

56. Nut of Chestnut

as it produced a crop of nuts every year—something that the beech and oaks did not always do since their blossoms were often killed by a late freeze. Those of the chestnut, however, did not appear until June or even July when all danger of frost had passed. So:

> . . . when the chestnut burr
> Broke open with the frosts, its store
>   disclosing,
> The squirrel, wild with joy.
> Disputed with the boy
> His right of thus his winter food
>   disposing.
>
> Walter Cooper

In some places we fynd chestnuts, whose wild fruiet I maie well saie equalize the best in France, Spaine, Germany, Italy or those so commended in the Black Sea by Constantinople, all of which I have eaten.

Historie of Travaile into Virginia Britannia

The chestnut was a majestic tree, highly valued for its wood and nuts around the turn of the century, but then in 1904 it was attacked by a fungus disease, the chestnut blight, and was virtually exterminated throughout its entire range, although a few old trees remain. Because of its ability to sucker freely from the bottom and to send up sprouts the chestnut still exists, though the fungus persists in the living stumps and eventually kills the larger sprouts. Experiments in hybridizing the native species with the Japanese and Chinese chestnuts hold promise that a disease-resistant, fast-growing hybrid may eventually be developed.

The chestnut belongs to the genus *Castanea,* the classical Latin name for the chestnut, a small group of about ten species of important nut and timber trees found in southern Europe, northern Africa, southwestern and eastern Asia, and the eastern United States. The Chinese chestnut, the Japanese chestnut, and the Spanish chestnut are all planted as ornamentals, as well as for their fruits, though the Japanese chestnut is possibly more useful for possibly breeding blight free hybrids than for its fruit. The fruit of the Spanish chestnut is the finest of all chestnuts.

A chestnut in miniature, the chinquapin is usually a shrub, though sometimes a small tree, eight to fifteen feet high, which forms thickets on hillsides, on bare ridges, or on the margins of swamps. The leaves are a deep green, densely white and downy beneath, somewhat oval in outline, acutely-pointed, and sharply, almost spinily toothed. The bur is one to one and a half inches in diameter, and the nut is solitary.

Another genus of the beech family, *Castanopsis,* Greek for resembling the chestnut, is represented in our flora by two species. The first is the golden chinquapin, a tree up to one hundred feet high of the mountainslopes that face toward the Pacific ocean. It has oblong, entire leaves, green above and golden beneath, and the fruit is a spiny bur with a solitary nut. The second is a small, little-known timberline shrub of the high Sierra Nevada of California.

Still another genus, *Lithocarpus,* Greek for stone fruit, referring to the hard shell of the acorn, is represented by a single species—an evergreen tree abundant along the coast from southern Oregon to northern California and known generally as the tanbark oak or simply tanoak. Under favorable conditions it grows seventy-five to one hundred feet high, and has alternate, leathery, toothed, oblong leaves that are covered on the lower surface with matted hairs that eventually disappear. The flowers are in upright catkins, with the male flowers toward the top of the catkin and the few female ones toward the base. The fruit is a bitter acorn in a shallow cup, lined with lustrous red pubescence. The bark contains appreciable amounts of tannin, and at one time was an important commercial source of the substance.

16. Beefwood (Casuarinaceae)

The beefwood family is a curious group of chiefly Australian trees with only one genus, *Casuarina,* in reference to the branches that resemble the feathers of the cassowary, and consisting of some twenty species. They are slender-branched, open-topped trees with apparently leafless twigs that are covered with minute scalelike leaves. True foliage leaves are absent. The twigs with their scalelike leaves suggest the horsetail, an allusion further enhanced by the twigs being joined and ridged, as in some horsetails. The flowers are extremely simple, and lack true sepals and petals. The flower structures are scalelike in club-shaped or globose catkins. The fruit is a winged nutlet. The family gets its name from the reddish color of the wood.

Three species have been introduced to California and Florida, where they have successfully made themselves at home. One of them, known as the beefwood or horsetail tree, is a tall tree with drooping branches and leafless twigs that sway wildly in the wind. It is much planted in Florida, where it has become naturalized, and makes a splendid windbreak for the citrus groves.

17. Begonia (Begoniaceae)

The begonia family, distinguished by its asymmetrical leaves and monoecious flowers, comprises five genera and about four hundred species of widely distributed tropical, mostly succulent, herbs or undershrubs.

They have alternate, simple, often oblique or asymmetrical leaves that are stalked and succulent, often large and beautifully variegated, with the margin entire, lobed or divided, and showy flowers, mostly in axillary clusters. The staminate flowers have two opposite sepals, two to five or no petals, and many stamens. The pistillate flowers have a perianth similar to that of the staminate flowers, with a two to four-celled ovary and numerous ovules. The fruit is a winged or angular capsule, or berry with many fine seeds.

All but a very few of the species, are included in the genus *Begonia,* named for Michel Begon, a French antiquary and patron of botany, and by far the largest of the genera. Numerous species and varieties are grown as glasshouse ornamentals and houseplants, and many are cultivated outdoors in warm countries. Probably the most familiar begonia is the common rex begonia, sometimes called the beefsteak geranium, found in many homes and greenhouses. It has creeping rootstalks, large and long-stalked leaves that are marbled, blotched, or banded with metallic markings above, reddish beneath, with rose pink flowers. Other species or varieties have red, pink, yellow, or white flowers.

18. Bellflower (Campanulaceae)

I saw the harebell for the first time along the grassy edge of a quiet pond and at the time I did not know that this seemingly delicate plant is as much at home in some inaccessible crevice of a precipice, among the rocks sprayed by the waters of a swiftly flowing mountain stream, in a windswept upland meadow as in a sun-kissed pondside.

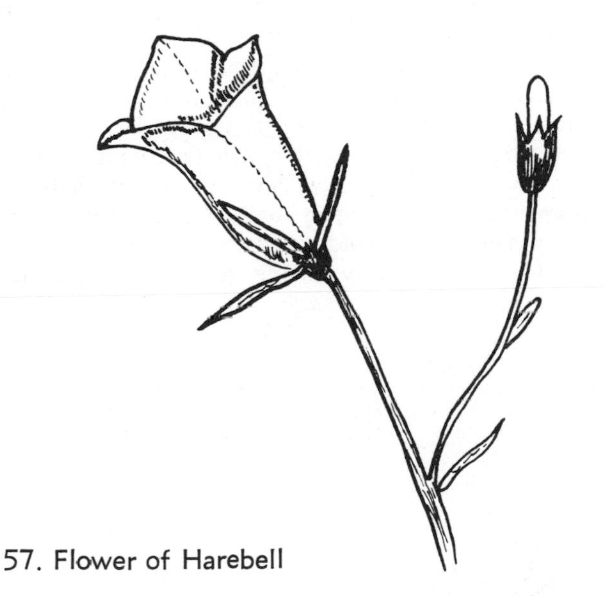

57. Flower of Harebell

The harebell is the bluebell of Scotland, the bluebell of literature, a cosmopolite growing in Europe, Asia, and America. Despite its frail appearance, it is a hardy plant capable of surviving the cold and storms of mountaintops over five thousand feet above sea level. In the spring it shows a tuft of round leaves that soon wither and disappear before the flowers appear, and which are succeeded by a tall, wiry stem, from six inches to three feet high, which is supple and bends easily without damage before the strongest wind. It has narrow, pointed, pale olive green leaves that offer little resistance to moving air currents however strong, and bright blue or violet blue nodding bell-shaped flowers depending from threadlike pedicels. The calyx has five awl-shaped lobes, the corolla is five-lobed, the stamens are five in number, the anthers have a delicate lavender tint, and the prominent pistil is tipped with a three-lobed stigma that at first is green, but later white. The chief benefactor is the bumblebee, but mining bees and the flowerflies are also visitors.

The harebell is a bellflower. It belongs to the genus of bellflowers known as the *Campanula,* which is Latin for little bell. There are some two hundred and fifty known species of bellflowers, the name of the family of which they are members. The bellflower family includes fifty genera and from one thousand to fifteen hundred species of annual and perennial herbs. It is of considerable garden interest because it contains not only the bellflowers, but also the balloon flower and both are widely cultivated for their showy flowers. All the species are cosmopolitan in range. They are very common in temperate and subtropical regions, though

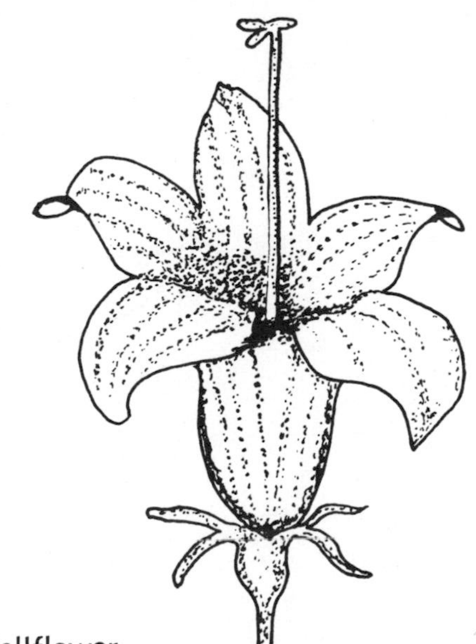

58. Flower of Bellflower

they frequently extend into cold climates in the north and into high mountains. They have simple, alternate leaves, perfect flowers that are often large and showy with a calyx commonly of five sepals, a bell-shaped corolla of five lobes, often blue or white, five stamens, and a two to five-celled ovary with many ovules. The flowers are solitary or in clusters, and the fruit is a capsule with many seeds.

It can be said that no garden flowers offer such a variety of blue color as do the bellflowers. A species long in cultivation and still much-planted is the Canterbury bells, a biennial with hairy, long oblong, toothed leaves and violet blue flowers that may either be solitary or in loose clusters (racemes). The name of the plant can be traced back to the time of the Canterbury pilgrims who wore, on their return from the shrine of St. Thomas à Becket, small images of the saints as evidence of their pilgrimage. Their horses were also decorated with small bells as evidence of the pilgrimage performed and as a charm against any accidents that otherwise might befall them. A garden variety, known as the cup-and-saucer, has a calyx of the same color as the petals and is shaped like a saucer, below the cuplike corolla.

A second garden species, also long in cultivation and mentioned by Gerarde in 1595, is the peach bells, a perennial with narrow, finely toothed leaves and blue or white flowers in showy terminal racemes. Other garden bellflowers are the Coventry bells, an exceedingly hardy and robust perennial that is apt to run wild; the Carpathian bellflower, a tufted plant about six inches high with erect solitary blue flowers that are quite handsome; the chimney bellflower, so-called because it was at one time grown in pots to adorn unused fireplaces in summer; and the creeping bellflower, named because of its inveterate habit of spreading. It was one of the first of the bellflowers to be brought to America, and long ago it took to the road, and so today it is found in fields and along the waysides.

One of the few native campanulus is the tall bellflower. One will find it in moist thickets and woods where its long, slender wands, studded with blue or sometimes whitish flowers, wave high above the ground. Looking closely at the flat, round, corolla, which is deeply cleft into five pointed petals, one receives the effect of a miniature pinwheel in motion. The honeybee, the bumblebee, and the yellow jacket are among its most frequent visitors. Another native species is the marsh bellflower. It is common in swamps, but it takes a little looking to find it. Quite often it may be found around a mossy, wet stump. It has a weak, triangular, reclining stem about twenty inches long, with pale lavender or white flowers scattered on slender pedicels. By means of tiny hooks it is able to lift itself above the surrounding vegetation, and, thus, avoid being smothered by it.

The Venus's looking-glass is not a campanula, but a bellflower, nevertheless with all the characteristics of the family. It differs from the campanulas by its rotate corolla and narrowly oblong ovary. Its rather picturesque name stems from the fact that Venus's looking-glass was the medieval name of the campanula in England. At any rate, the Venus's looking-glass is common in sterile, open ground, thin meadows, and upland pastures. It has a long, slender, weak and usually prostrate stem with small, curved, shell-shaped leaves that are scalloped-toothed, light green, and that clasp the stem. There are two kinds of flowers. One kind is rudimentary, being without a corolla, and

59. Flower and Leaves of Platycodon

never opens, and is self-fertilized in the bud, producing much seed that is often ripened and sown before the second kind of flowers appear. These are violet blue, are set at the hollows of the leaves, and are wheel-shaped with a deeply five-lobed corolla, five stamens, and a three-lobed pistil. They are also fertile and are visited by insects that cross-pollinate them as a safeguard against the plant becoming weakened through self-fertilization.

The balloon flower, mentioned above as a member of the bellflower family, is a Eurasian, showy, perennial herb, also known as the Platycodon, and widely cultivated for ornament. It is an erect plant, eighteen to thirty inches high, with leaves several times longer than wide, broadest at the base, and a usually solitary flower on a long stem. The flower is broadly bell-shaped or deeply saucer-shaped, two to three inches long, and dark blue, though pale or white in some horticultural forms. The name *platycodon* is from the Greek for broad bell in allusion to the shape of the corolla. There are dwarf forms and one with semi-double flowers; that is, with an inner bell whose lobes alternate with the outer, thus making it appear like a ten-pointed star.

19. Bignonia (Bignoniaceae)

The best-known member of the bignonia family is doubtless the catalpa tree, although the trumpet creeper and cross vine are also well known.

The catalpa is a low spreading tree, twenty-five to fifty feet high, with a broad irregular head of coarse twigs. It prospers on riverbanks and in shady woods, is much planted for ornament, and is a valuable lawn and street tree because of its profuse flowering. Only the horse chestnut can rival the catalpa in showy flower clusters. Those of the catalpa are often ten inches high, loosely conical, and bloom from the base upwards. It is a late bloomer in the North, the flowers appearing in late June or early July and covering the tree so thickly that it almost conceals the full-grown leaves.

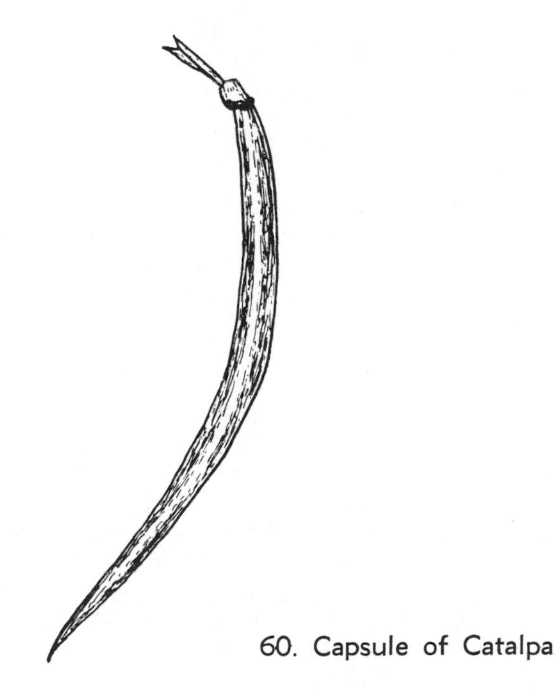

60. Capsule of Catalpa

A glance at an individual flower may well be in order. The calyx is globular, a green or light purple with two lobes. The corolla is tubular and irregular with five lobes, the two above are smaller than the three below. It is also white, marked on the inner surface with two rows of yellow blotches and in the throat on the lower lobes with purple spots. These markings are an invitation to bumblebees to explore the nectaries, the fragrance being a further reassurance. The stamens number two and are mature before the stigma that rises between them, hence, a bee that brushes against them receives pollen that it will carry to the stigmas of other blossoms, thus ensuring cross-pollination. The stamens of the other blossoms in the meantime wilt, and, thus, the plant avoids self-pollination. The ovary is two-celled with many ovules.

The bark of the catalpa is light brown tinged with red, smooth, and contains tannin. The leaves are opposite or in threes, heart-shaped, with a prolonged, sharp point, toothless, a bright green above, paler beneath, and have a disagreeable odor when bruised. The fruit is a long, cylindrical, somewhat curved, beanlike or cigarlike pod, light brown, eight to eighteen inches long, and contains numerous ragged winged seeds, fringed with whitish hairs. They hang in loose clusters through the winter and swing and rattle in the wind.

The catalpa is a southern tree and appears to have been first observed growing in the fields of the Cherokee Indians by the white man. The Cherokees called it catalpa, which is also the genus name to which it belongs. The genus consists of seven to ten species, most being North American, but a few Asiatic. There are two native species—the common catalpa that has been described and also known as the Indian bean from the shape of its fruit; and the hardy catalpa, also called the cigar tree from the shape of its fruit, and the catawba tree. The hardy catalpa is a native of the Central United States where it is found on rich bottomlands and along the margins of streams. It is similar to the common catalpa, but is hardier and

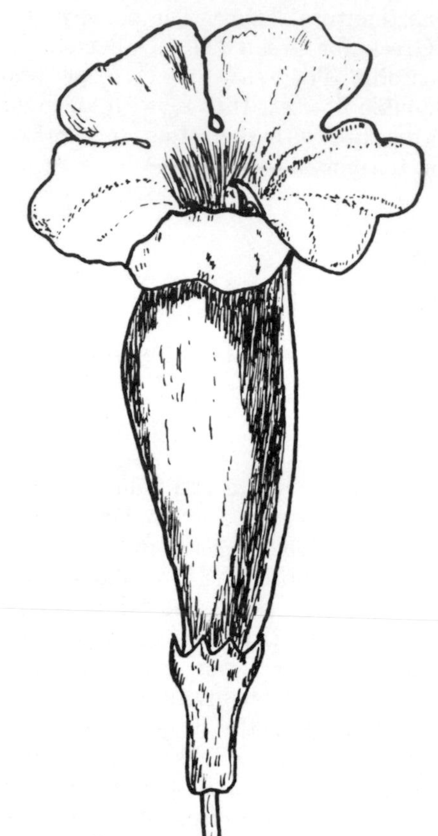

61. Flower of Trumpet Creeper

grows in more upright form. It has stout, thick-walled fruits, more pointed leaves, and fewer flowers in a cluster. They are also less gaily spotted than the flowers of the catalpa.

The genus *Catalpa* is one of about a hundred genera of the bignonia family, a group that comprises some seven hundred and fifty species of trees, shrubs, and woody vines (and a very few species of herbs), most being tropical in distribution. They have opposite, rarely alternate or whorled, simple or commonly compound leaves, and perfect flowers that are often very showy in terminal or axillary racemes or panicles. The individual flowers have a bell-shaped or tubular, five-lobed or no-lobed calyx, a usually more or less irregular corolla of five unequal lobes, sometimes two-lipped, four stamens, and a one or two-celled ovary with many ovules. The fruit is usually a capsule, and its seeds are generally winged.

Besides the two species of catalpas, we have only two other native arborescent species—the desert willow and the black calabash tree. The desert willow of the genus *Chilopsis,* from the Greek for liplike, is a small tree or shrub, not more than fifteen feet high, of southwestern United States. It has narrow, willowlike leaves, three to five inches to a foot long, white flowers shaded into pale purple or lilac with a pair of yellow stripes inside in a short, terminal raceme, and a cylindric, many-seeded pod, nearly twelve inches long. It grows in the banks of streams and depressions in the deserts, usually in dry, gravelly porous soil.

The black calabash tree is found on rich hummocks only near the shores of Bay Biscayne in southeastern Florida, and is a tree about twenty feet high, with obovate, leathery, dark green leaves, a solitary flower having a purplish yellow tube with a flaring border, and a fruit that is a berry, three or four inches long, and shaped like a peach or plum. It has a hard, shiny shell that encloses many flattish seeds. A related, but much larger species, the West Indian calabash tree is somewhat cultivated in southern Florida. The fruit is a hard-rinded berry, more or less oval, from six inches in diameter to twenty inches long, and is made into drinking cups and a variety of culinary utensils.

The bignonia family is sometimes called the trumpet-creeper family, after a vine of the same name. It is a stout, woody vine climbing by means of aerial rootlets up to thirty feet high and found in moist woods, thickets, or along fencerows. It has opposite, pinnately compound leaves, with nine to eleven almost stalkless leaflets that are elliptic or ovate, pointed at both ends, coarsely and sharply toothed. The flowers are orange scarlet, trumpet-shaped, and five-lobed, and the fruit is a cylindrical, somewhat flattened, two-ridged capsule containing many winged and flattened seeds.

The trumpet creeper is a handsome and general utility vine cultivated for ornament and used to adorn a fence, cover a porch, a chicken house, or any outbuilding. The blossoms contain a great deal of nectar and are attractive to hummingbirds. However, some people may suffer a case of dermatitis from handling the leaves or flowers. A Chinese species, the Chinese trumpet creeper, is also widely cultivated for ornament. It does not climb as high as our native species and has fewer aerial rootlets.

Another native, high-climbing vine is the beautiful woody vine, usually called the cross vine, though sometimes it is known as the trumpet flower, growing in moist woods and swamps. It has evergreen, compound leaves with two stalked, oblong to egg-shaped leaflets that are toothless and pointed at the tip, and a terminal tendril. The flowers, in axillary clusters of two to five are funnel-shaped, somewhat irregular, five-lobed, reddish orange and yellow within. The fruit is a long, narrow, cylindrical, slightly flattened capsule, containing many winged and flattened seeds.

Many tropical species of the family are highly valued as ornamentals and many of them are planted in Florida and California. Those that should be mentioned are the jacaranda, or Jack tree of Brazil, with its wonderful masses of blue flowers; the sausage tree of Africa with its sausagelike fruits that swing on long, cordlike stalks; the candle tree of Panama, with its candlelike, hanging, yellowish white fruit; the flame vine of Brazil, considered to be the finest vine cultivated in Florida next to the Bougainvillaea; the yellow elder, a very popular ornamental shrub or tree because of its profusion of late-blooming flowers; and the Cape honeysuckle of South Africa, which can be grown as a vine or pruned as a scrambling shrub.

The bignonia family has only a single herbaceous genus, *Incarvillea,* named for Father Incarville, a French Jesuit in China. The species of the genus are tender perennials, native to Tibet and China. They have alternate, pinnately compound leaves, trumpet-shaped red to yellow flowers in terminal clusters, and a capsule as fruit. Several species are cultivated in southern gardens.

20. Birch (Betulaceae)

How many species of birches occur in North America has never been fully established, but there are certainly no more than fifteen. Of this number, most are arborescent, while the remainder are simply shrubs.

The birch trees, genus *Betula,* are beautiful, short-lived, medium or tall trees, with a singular grace and distinction of character. They have fibrous roots, a smooth bark that has a decided color, with conspicuous, horizontal slits (lenticels), that usually curls back in thin layers or plates. The leaves are simple, alternate, deciduous, toothed, and feather-veined. The monoecious flowers lack petals, the staminate occur in early blooming catkins or spikes, the pistillate in small conelike, but leafy-bracted clusters (strobiles). The bracts are three-pointed and drop at maturity with the fruit, a winged and scalelike nutlet.

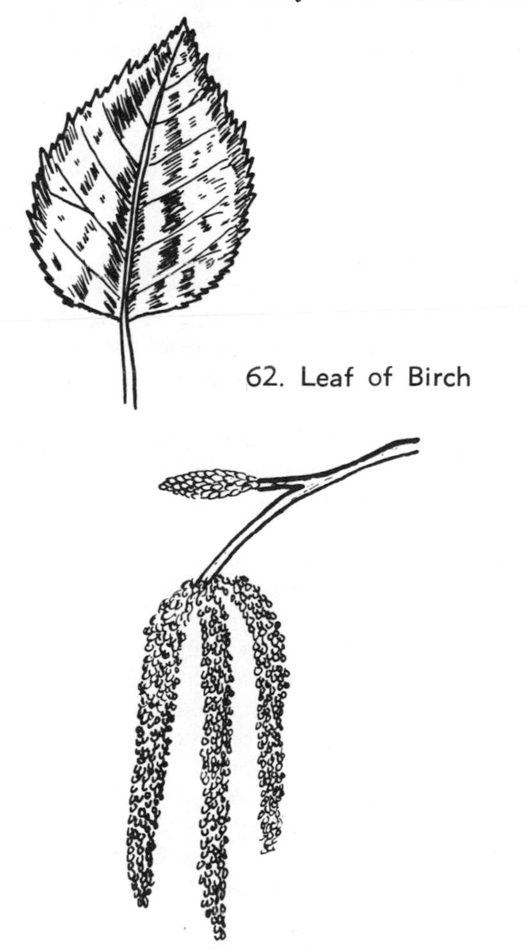

62. Leaf of Birch

63. Three Staminate and One Pistillate Catkin of Black Birch

The Indians knew the paper birch well and made use of its papery bark to frame their tents and make their canoes. Their canoes were light, graceful, and durable, and the Indians handled them with skill and dexterity. An early letter writer in describing them says that these "delicate canowes so light that two men will transport one of them overland whether they list, and one of them will transporte tenne or twelve Salvages by water at a time." Many other articles were also made of the bark, such as dishes, baskets, buckets, and the like.

The paper birch, also known as the canoe birch, is a large handsome tree, fifty to seventy feet high, although some are one hundred feet, and grow in northern woods and rocky uplands. Its white bark separates into papery layers and may be distinguished from the white bark of other birches by its chalky whiteness that rubs off on clothing. The leaves are deep green, ovate, sometimes heart-shaped at the base, and the flowers appear before the leaves in April. The staminate catkins are three to four inches long, pendulous, and clustered or paired, the pistillate with bright red styles in catkins one to one and a half inches long. Thoreau relates that the Indians of Maine made a tea from the leaves and it is also said that the people of the extreme north mixed the inner bark with their food for the starch that it contains.

Said to be the least common of the birches, the gray birch may easily be recognized by the triangular dark markings at the base of each branch. It is a small tree, twenty to thirty feet high, of wasteland, rocky slopes, dry, gravelly, barren margins of swamps and ponds, old fields and abandoned farms. It has a dull white bark and bright, shining green leaves that flutter in the slightest breeze like those of the aspen, because their petioles, though not laterally compressed, are long, slender, and slightly twisted.

The gray birch closely resembles the European white birch, which is the commonest white birch in cultivation. It is found in many attractive forms. There is one with finely divided leaves, another with drooping branches almost a weeping form, and still another has a compact, bushy crown that has no drooping branches.

In days past, the white birch of Europe has been found useful in many ways. It has been used as a fuel and made into charcoal, and has served in the manufacturing of wooden spoons, shoes, and furniture. Its tiny, winged seeds have served as a food for birds and doubtless still do, its sap has been made into birch mead, and its bark yields tannin, a yellow dye that has been used for paper. Plutarch says that the famous books of Numa Pompilius were written on birch bark. Parkinson relates that

> many other civill uses the Birch is put unto, as first to decke up Houses and arbours, both for the fresh greennesse and good sent it casteth; it serveth to make hoopes to bind caskes withall; the young branches being fresh are writhed, and serve for bands unto faggots: of the young twiggs are made broomes to sweep our houses, as also rods to correct children at schoole, or at home, and was an ensigne borne in bundles by the Lictors or Sargeants before the Consulls in the old *Romans* times, with which, and with axes borne in the like manner, they declared the punishment for lesser, and greater offences, to their people.

One needs only look at its bark to recognize the red birch, for it is reddish or reddish brown from twig to root. Every genus of trees, or so it seems, has at least one species that delights to grow in water, and the birches have the red. This tall, graceful tree, a veritable fountain of leafy sprays, crowds to the water's edge and is found on the banks of streams, ponds, and in low marshy woodlands from Massachusetts to Florida and Louisiana where, in the swamps and bayous, it reaches its greatest size.

The red birch is a beautiful, medium-sized tree, thirty to eighty feet high. Its trunk is often divided into two or three slightly diverging limbs, forming a round-topped, picturesque head. The bark peels easily into translucent, curling, papery layers, and the bright green leaves are broadly ovate, somewhat wedge-shaped at the base, double-toothed, whitish downy beneath and in autumn turn a pale, dull yellow.

As the red birch is recognized by the color of its bark, in a like manner the yellow birch can be identified, although its bark is not a pure yellow, but a silvery, pale yellow gray peeling horizontally into very thin ribbonlike layers that give the trunk of the tree a lacerated appearance. The yellow birch is a tall tree, fifty to sixty-five feet high, sometimes ninety-five feet, with an irregular, perpendicular trunk, a broad, round head, and drooping branches. Its leaves are deep, dull green above, yellow green below, ovate with a slight point, double-toothed, and somewhat heart-shaped at the base. The flowers appear in early spring before the leaves, the staminate catkins long and pendant, the scales pale yellow green below the middle, dark brown above, the pistillate erect, stemless, reddish green, and hairy. The yellow birch is a tree of rich moist woodlands.

As an historical sidelight, it is said that the bark of the black birch saved the lives of many of Garnett's Confederate soldiers during their retreat after the Battle of Carricks Ford in 1861. It is not known how true this is, but the bark is edible, for in Kamchatka the natives used to strip the inner bark into long shreds, drying them for use as food during the winter, boiling them with caviar and with fish.

The white, red, and yellow birches were named after the color of their bark, and, thus, it would be assumed that the bark of the black birch is black. But it is a dark, slate gray, brown on young trees, not peeling, strongly marked with long horizontal lines (lenticels), on old trees, broken into irregular gray brown plates. The bark is reminiscent of that of the common cherry tree, hence the black birch is also known as the cherry birch. The inner bark is very fragrant and has a pleasant spicy taste, hence, the black birch is further known as the sweet birch. The flavor is due to an essential oil similar to that obtained from the true wintergreen, as a matter of fact, the oil of wintergreen has been distilled from the bark and twigs. It should also be mentioned that birch beer is made from the sap; that the sap may be used in making sugar about half as sweet as that obtained from the sugar maple; and that the rapidly growing young twigs make a most delightful tea.

The black birch is a handsome, round-headed woodland or forest tree, forty-five to fifty feet tall and in favorable situations may grow as high as eighty feet, with slender, often tortuous, but graceful branches, the lower ones drooping. The twigs are slender, smooth, glossy, red brown and have a strong wintergreen flavor, as have the leaves that are bright green above, lighter beneath, ovate, double-toothed, and sharp-pointed. The staminate catkins are pendulous, the scales are a bright red brown above the middle, pale brown below, the pistillate catkins are erect, and the scales are a pale green.

In the West, there is a western species of birch, a graceful little tree with a smooth, shining orange brown or copper red bark that often has a plum-colored bloom and that, unlike the black birch, sheds its bark in thin, papery layers. There are several other birch trees of local distribution.

The genus *Betula* includes a number of shrublike species, such as the swamp birch, a low, upright often mat-forming plant of northern bogs and swamps. It has soft hairy twigs and egg-shaped, coarsely toothed leaves, dull above, and pale green to white beneath. A second species is the northern birch, a spreading or erect shrub with densely, white hairy twigs and elliptic to broadly egg-shaped, double-toothed leaves, and occurring in poor soil. A third is the Newfoundland dwarf birch, a very small shrub creeping by means of underground stems, with erect branches, brownish woolly twigs and leathery leaves. A fourth is the dwarf birch, an erect but often low and mat-forming shrub of the tundras and Alpine mountain summits, having twigs with wartlike dots and leathery, oval or nearly circular leaves. Finally there is the minor birch, a spreading or erect shrub of Arctic and Alpine areas, with warty twigs and egg-shaped, double-toothed leaves.

The birch family is a small group of trees and shrubs found mainly in the cooler regions of the northern hemisphere and comprising about six genera and some one hundred species. Their characteristics are essentially those of the birches, with alternate, simple, mostly toothed leaves, and the male and female flowers in separate catkins on the same plant. The staminate flowers are small, numerous, in long, pendulous terminal or lateral catkins, with two to four or no stamens and no petals. The pistillate flowers are in short, cylindric, conelike spikes, with three-lobed bracts and four sepals or none, and the ovary is two-celled, each cell with one ovule. The fruit is a small, indehiscent, one-seeded nut, with or without membranous wings.

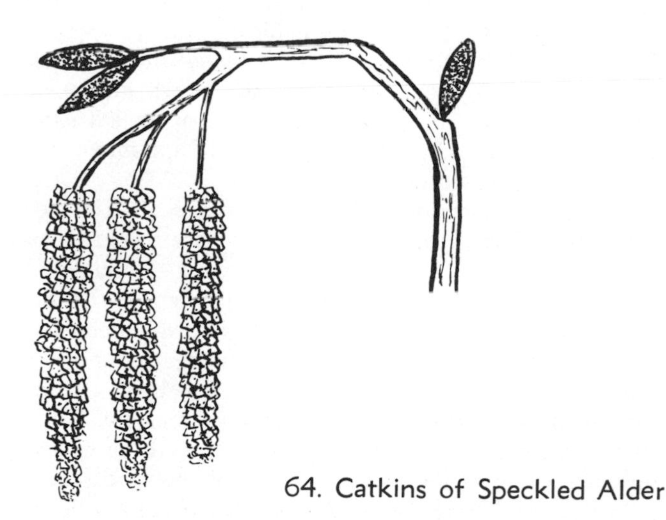

64. Catkins of Speckled Alder

One of the genera of the family is a large group of shrubs and trees known as the alders. The genus is *Alnus,* from the Latin name for alder. They have rather handsome, often somewhat burnished or sticky leaves, flowers in catkins, the staminate and pistillate on the same plant, however, the pistillate are in smaller catkins than the staminate. The pistillate develop into small, slightly winged nutlets that are borne on the axils of woody scales of conelike structures peculiar to the alders.

To those who roam through the northern woodlands and wander along the country roadsides, the stiff uncompromising catkins of the speckled alder, conspicuously outlined against the winter sky, are a familiar sight. Then when March comes, if March is mild and springlike, the scales open and the long, plumed, pendent tassels emerge to wave like pennants in every transient breeze. Sometimes it is not until April that they open.

If examined with a magnifying glass or hand lens, the catkin, so-called from its resemblance to a cat's tail, is found to consist of brown and purple scales surmounting a central axis. The scales are set on short stalks, and beneath each scale are three flowers, each having a three to five-lobed calyx cup and three to five stamens whose anthers are covered with yellow pollen. Pistils seem to be lacking, but if one of the shorter, erect, deep purple catkins is examined, every fleshy scale will be seen to enclose two flowers, each having a pistil with a scarlet style.

Because the pollen grains are transferred from one flower to another by the wind, the alder long ago dispensed with petals, as did many other plants that depend upon the wind as an agent of cross-fertilization. The petals would only prevent the wind from picking up the pollen grains and transferring them to the stigmas of other flowers, an adaptation of distinct advantage. Once pollination has been carried out and fertilization completed, the pistils develop into small woody structures that resemble pinecones. The scales protect the seeds formed beneath them, and when the seeds mature the scales open and release them to be scattered by the wind.

The speckled alder, so named because its stems are sprinkled with numerous conspicuous light gray spots that are actually breathing pores (lenticels), is a shrub or small tree, eight to twelve feet high. It crowds the edge of water courses and also grows in swamps and moist thickets.

The speckled alder essentially is a plant of the northern states, but a related species, the smooth alder, is widely distributed throughout the South. The two, often found growing together where their ranges overlap, are very much alike in habit and have similar catkins. The smooth alder does not have, however, as many lenticels as its relative; moreover, its leaves are finely toothed, whereas those of the speckled alder are irregularly double-toothed.

Also a northern species, the mountain or green alder is a shrub, two to ten feet high, growing on rocky shores, slopes, and southward on mountain balds. It has roundish to egg-shaped or somewhat heart-shaped finely toothed, somewhat sticky leaves that are green and smooth on both sides though sometimes downy or velvety beneath. Unlike the speckled, smooth, and mountain alders, the seaside alder does not flower until August or September. It is a shrub or small tree of restricted coastal distribution, found about ponds and on streambanks, with oval to egg-shaped leaves, lustrous above, and pale green below. An introduced European species, commonly planted and often an escape in eastern North America, is the black alder, a tree up to seventy feet, with sticky twigs and nearly round or ovalish, coarsely toothed leaves.

In the West there are several species—such as the mountain alder, a shrub or small tree of the alpine meadows with a dark grayish brown bark and ovate leaves; the red alder, a tree of the stream banks and shore flats with a smooth, ashen gray bark and coarsely toothed, ovate leaves that are rusty-haired beneath; and the white alder, a tree found growing in gravelly soil along mountain streams with a dark brown bark and light yellow green leaves that are paler below.

The hazelnut responds to the warmth of spring in much the same manner as the alder. During the winter the staminate catkins hang stiff and rigid, but under the influence of the spring sun they relax, develop their pollen, and fade away. The flowers consist of four stamens with two-cleft filaments, each fork

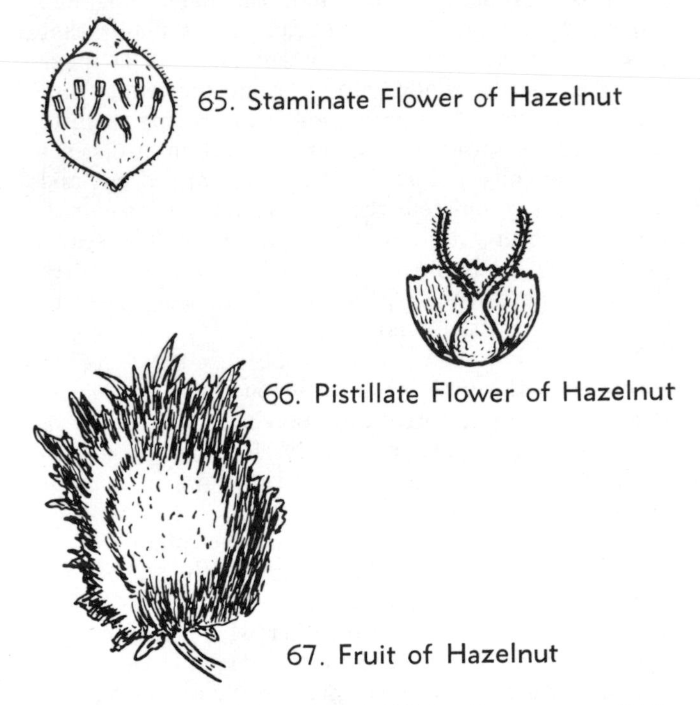

65. Staminate Flower of Hazelnut

66. Pistillate Flower of Hazelnut

67. Fruit of Hazelnut

bearing an anther. The pistillate flowers are little starlike tufts of crimson stigmas of surprising beauty when viewed through a magnifying glass or hand lens. The nuts, enclosed in a leafy husk, are sweet and rival the filberts in quality. The chipmunk is extremely fond of them because they are not only good eating, but fairly accessible since the hazel is a shrub of rather low growth. This little animal is not a good climber, so the nuts of the tall trees are not available until they fall.

Everyone knows the hazelnut as a nut but not as the plant itself, and, yet, it is quite common and can be found growing in moist to dry thickets, fencerows, roadsides, and woodland borders. It reaches a height of three to ten feet, has egg-shaped to roundish leaves that are somewhat heart-shaped at the base, abruptly pointed at the tip, double-toothed, dark yellow green above, and pale green and finely haired beneath. The fruit is an almost globular, chestnut brown nut enclosed within a pair of leafy, cut-toothed bracts that are covered with down or with glandular bristles at

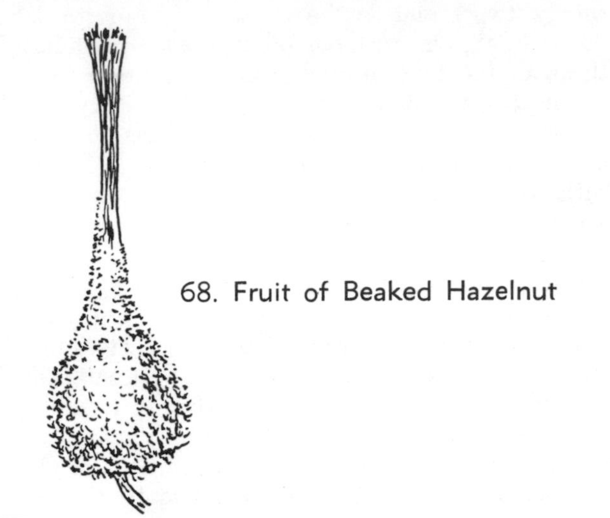

68. Fruit of Beaked Hazelnut

the base. There is a related species, the beaked hazelnut, found in similar situations and differing only in the marked character of the involucre that surrounds the nut. The two bracts are united and form a prolonged, tubelike bristly haired beak that is reminiscent of a long-necked bottle with the nut inside. As the bracts mature they develop a color scheme that varies from yellows to browns to reds, the dense hairs giving them a velvety look. The nuts are worth gathering for they are easy to crack and there is little waste. They are sweet and pleasing to the taste and much like those of the common hazelnut.

The hop hornbeam is a lonely forest tree, of a rather shy and retiring disposition. It is never found in masses, but is scattered usually in cool, fertile, and shaded situations. A mere glance at it shows its relationship to the birches. Its leaves are much like those of the birches, and, in winter, its slender branches bear three catkins in a drooping cluster that wait for spring like the birch catkins do. Then when spring comes, they swing out, like those of the birches. The pistillate catkins are not easily discernible, but by looking carefully for them they will be found erect at the ends of leafy side shoots. Eventually they will develop into a hoplike fruit, formed of several papery-leafed inflated bags, each containing a hard nutlet.

The hop hornbeam is a small tree, twenty-five to forty feet high, with a gray brown bark, furrowed and broken into narrow oblong strips that in color and stripping resemble the shagbark hickory, though on a smaller scale. It is a rather pretty tree with yellow green leaves that turn clear yellow in the fall, but whatever claim it has to individuality lies in its exceedingly hard and strong wood.

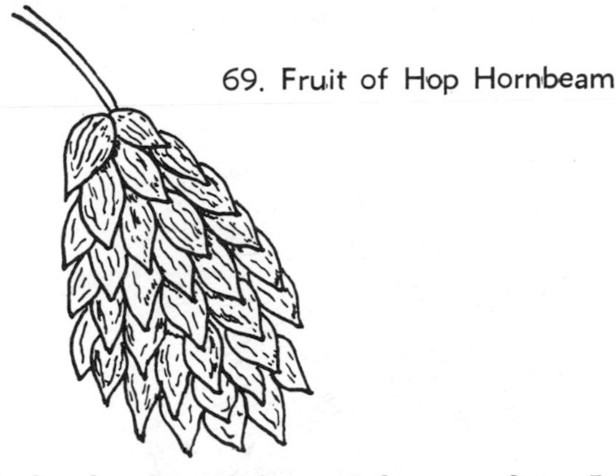
69. Fruit of Hop Hornbeam

The hop hornbeam belongs to the genus *Ostra,* Latin for a tree of hard wood; the American hornbeam to the genus *Carpinus,* the ancient name of the hornbeam. Though the two trees belong to different genera the American hornbeam resembles the hop hornbeam in many ways. Its leaves, its flowers and its twigs are similar but not its bark, which is like that of a young beech, smooth, and blue gray. It is therefore known also as the blue beech, with flutings that course up the trunk and out on the larger limbs.

The American hornbeam is a small, somewhat shapeless tree with irregular branches, often pendulous, and slender wiry twigs. It is an attractive tree with blue green leaves that turn to orange and scarlet in the fall. It is quite common in the woods and along streams where it grows in the shade of other trees. The staminate flowers appear in long, loose, pendulous catkins, the pistillate in loose, half-erect catkins. The fruit is not hoplike. The hard, paired nutlets are instead each supported by a large, leaflike, three-lobed bract. Its wood is also hard like that of the hop hornbeam as the early colonists discovered, for the *New England Prospects* reads: "The Horne bound tree is a tough kind of wood that requires so much paines in riving as is almost incredible, being the best for to make bolles and dishes, not being sunject to cracke or leake."

The value of the birches to wildlife is considerable, but is confined largely to the North and to northern animals such as the grouse, the redpoll, and pine siskin that feed on the seeds. Such browsing and wood-eating

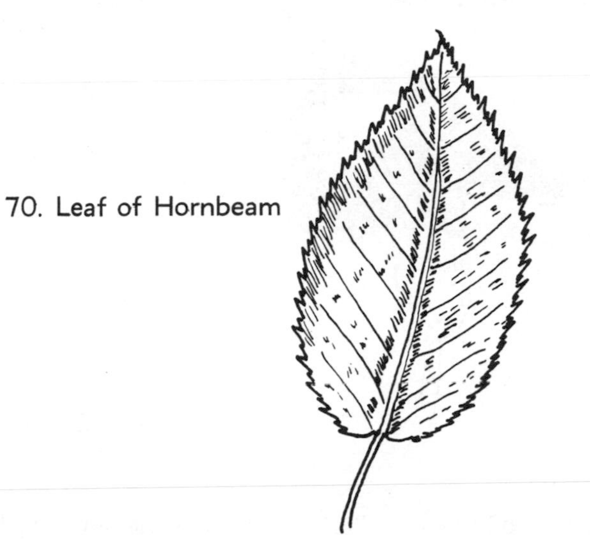
70. Leaf of Hornbeam

mammals as the moose, snowshoe rabbit, porcupine and beaver feed on the twigs, foliage, and bark. The wildlife food value of the alders is of less value, though their seeds are eaten by redpolls, pine siskins, and goldfinches, while their leaves and twigs form part of the diet of some game birds and the browsers. The same can be said of the hornbeams, although their nuts are eaten by several kinds of birds and squirrels. The hazelnuts, on the other hand, are of more importance. Squirrels, chipmunks, and other rodents feed on the nuts, grouse commonly eat the catkins, while rabbits, deer, and moose browse on the twigs and leaves. Moreover, the low, dense growth of these spreading shrubs provides useful cover and nesting sites.

Various members of the birch family serve as food plants for the caterpillars of several species of butterflies: the green comma, the Compton tortoise shell, and the white admiral or banded purple, feed on birch; while the green comma feeds on alder, and the red spotted purple on hornbeam.

21. Birthwort (Aristolochiaceae)

Unless one knows the wild ginger or recognizes the leaves, which are broadly heart-shaped or kidney-shaped and three to six inches wide, there may be some trouble finding it. Even though it is common in our woodlands, it is a rather low-growing plant and, when in flower, the blossom is close to the ground and partially concealed by the leaves rising above it. Moreover, its sober color not infrequently resembles the leaf mold just beneath it, hence, increases the difficulty of finding it.

The wild ginger is a somewhat curious plant with an aromatic rootstalk, once used dried and pulverized as a remedy for whooping cough and as a substitute for ginger, and two long-stemmed, deep green, veiny,

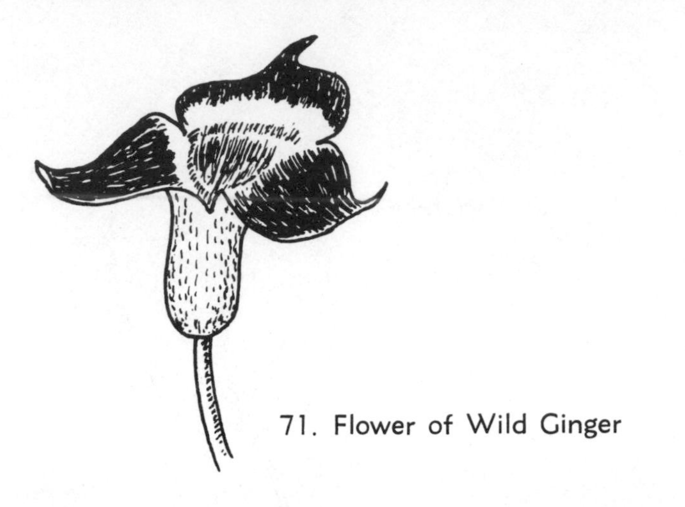

71. Flower of Wild Ginger

soft, woolly leaves. There is but a single flower to the plant. It is on a short nodding stem from between the bases of the leafstalks and near the surface of the ground. It has no petals, and the calyx is bell-shaped with three pointed, brownish or madder purple lobes closely united to the ovary. It persists until the seeds, contained within a fleshy, roundish capsule, have ripened, when the capsule bursts irregularly. The low position of the flower and the frequent visits of fungus gnats and early flesh flies suggests that these insects are the ones most actively engaged in cross-fertilization. There are several species of the wild ginger, differing primarily in the shape of the calyx lobes. The caterpillars of the pipe vine swallowtail feed on the leaves, as well as those of two related species—the Virginia snakeroot, and Dutchman's pipe.

The wild ginger, the Virginia snakeroot, and the Dutchman's pipe are all members of the birthwort family, a fairly small group of some two hundred species of widely distributed herbs and vines or twining shrubs. The leaves are alternate, stalked, and often heart-shaped, and the flowers are perfect without petals, the calyx is curiously tubular or bent, and has a six-celled ovary that forms a many-seeded capsule or berry in fruit.

The Virginia snakeroot is a woolly stemmed woodland plant with an aromatic root used in medicine. It is a low-growing herb with long, heart-shaped leaves that are narrow and tapered to a point and green on both sides. The flowers are dull greenish, and solitary at the tips of curving, crooked stems, the calyx curiously curved like the letter s. They are at or near the ground level, and sometimes are fertilized in the bud without opening.

A tall, twining shrub or vine of rich woodlands, the Dutchman's pipe, or pipe vine as it is also known, is widely cultivated for its fine foliage and is often seen trailing over arbors, porches, and verandas. Its leaves are roundish, kidney-shaped, and light green, and its

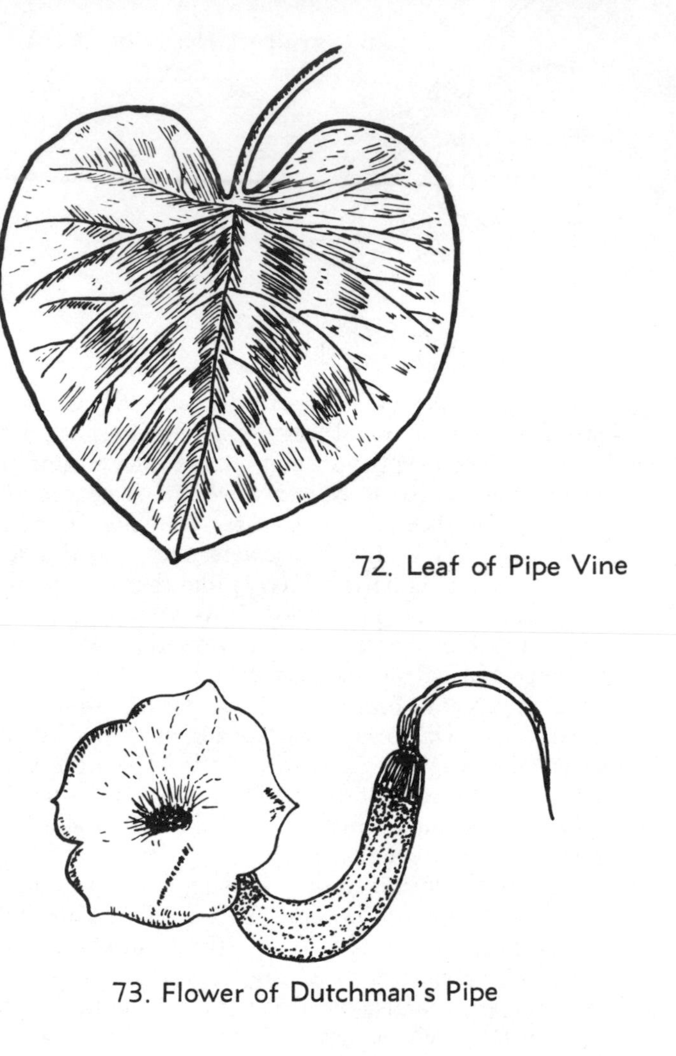

72. Leaf of Pipe Vine

73. Flower of Dutchman's Pipe

flowers are hooked-shaped, the yellowish green, veiny tube (calyx) with a flat, three-lobed, purple brown throat that looks like a Dutchman's pipe. In May when the Dutchman's pipe blooms, small flies and gnats seek the nectar within the curving tube and enter easily enough, but, once inside, become imprisoned because hairs at the entrance, approached from within, bar their escape. However, when the danger of self-fertilization has passed, the tube begins to wither, the hairs at the exit become limp, and the entrapped insects are able to escape, none the worse for their experience. Meanwhile they have deposited their load of pollen and acquired a fresh load that they will carry to other young flowers.

22. Bladdernut (Staphyleaceae)

One needs only glance at the extraordinary inflated seed pods of the bladdernut to suspect that they are the distinguishing characteristic of the shrub. A little imagination and they will appear like tiny balloons. Anyway, when the seeds are ripe they break away from their attachments within the pod and rattle about inside when the pod is shaken.

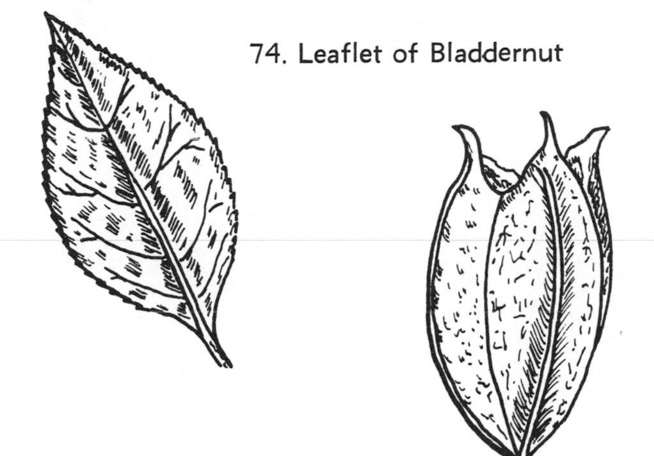

74. Leaflet of Bladdernut

75. Fruit of Bladdernut

The bladdernut is a slender shrub, six to twelve feet high, with light gray stems, conspicuously striped with green, spreading branches. This shrub is found in moist thickets, hillsides, and stream banks. The leaves are opposite, compound, on long petioles, with three egg-shaped or elliptic, pointed, coarsely toothed, dark green leaflets. The flowers are small and white in drooping clusters and though the individual flowers are not particularly attractive the clusters cover the bush in spring and produce a pleasing effect. The fruit is a three-sided, inflated, baglike capsule, one to three inches long, and contains three to five light brown, bony seeds.

The bladdernut is one of twenty-five species of the family of the same name, all from the north Temperate Zone, and contained in about six genera. They are trees or shrubs with mostly opposite, compound leaves of three leaflets, regular, perfect flowers, white or greenish white in terminal or axillary clusters. The flowers have five sepals, petals, and stamens and mostly three carpels. The fruit is a bladdery capsule. They are chiefly Asiatic, one of them being cultivated in the United States; so, too, is our own native bladdernut.

23. Bladderwort (Lentibulariaceae)

Most people know the sundew, the pitcher plant, and the Venus flytrap with their traps for capturing insects, but the bladderwort is not so well known and, yet, has traps more ingeniously contrived than those of the others. There are one hundred and eighty species of bladderworts contained in about four genera that comprise the bladderwort family. They are a group of small herbs growing in water and in wet places, and widely distributed in warm and temperate regions of the world.

They are rather delicate and vinelike in appearance, as they float beneath the surface among the water lilies and pondweeds, with their slender stems and leaves that are finely divided into threadlike segments. The leaves are divided so close to the base that each leaf appears like two leaves growing opposite to each other. The flowers are perfect, regular, with a two-lobed calyx and a two-lipped corolla, two stamens, and a one-celled ovary. They are borne on tall slender stems that rise above the water. The fruit is a capsule. The bladderworts have no true roots, but, instead, have twiglike rhizoids that serve the same purpose.

The traps of the bladderworts are small bladders or utricles that occur on the leaves. When the plant is in flower they become filled with water and then act as pontoons to keep the blossoms out of the water where the insects can visit them, but at other times they function as traps.

The bladder, or trap, is a slightly compressed sac with a slitlike opening. Armed with teeth and bristling hairs, this opening is guarded by a valve or trap door that only opens inwardly. When an unwary animal swims into the opening, the valve is stimulated and opens. As it does, water flows into the bladder so that its walls, which are ordinarily compressed inward, are now pushed outward under the pressure of the water. This outward movement creates a suction that draws more water into the bladder, and the water carries with it not only the animal that triggered the trap, but others that may have been nearby. The action

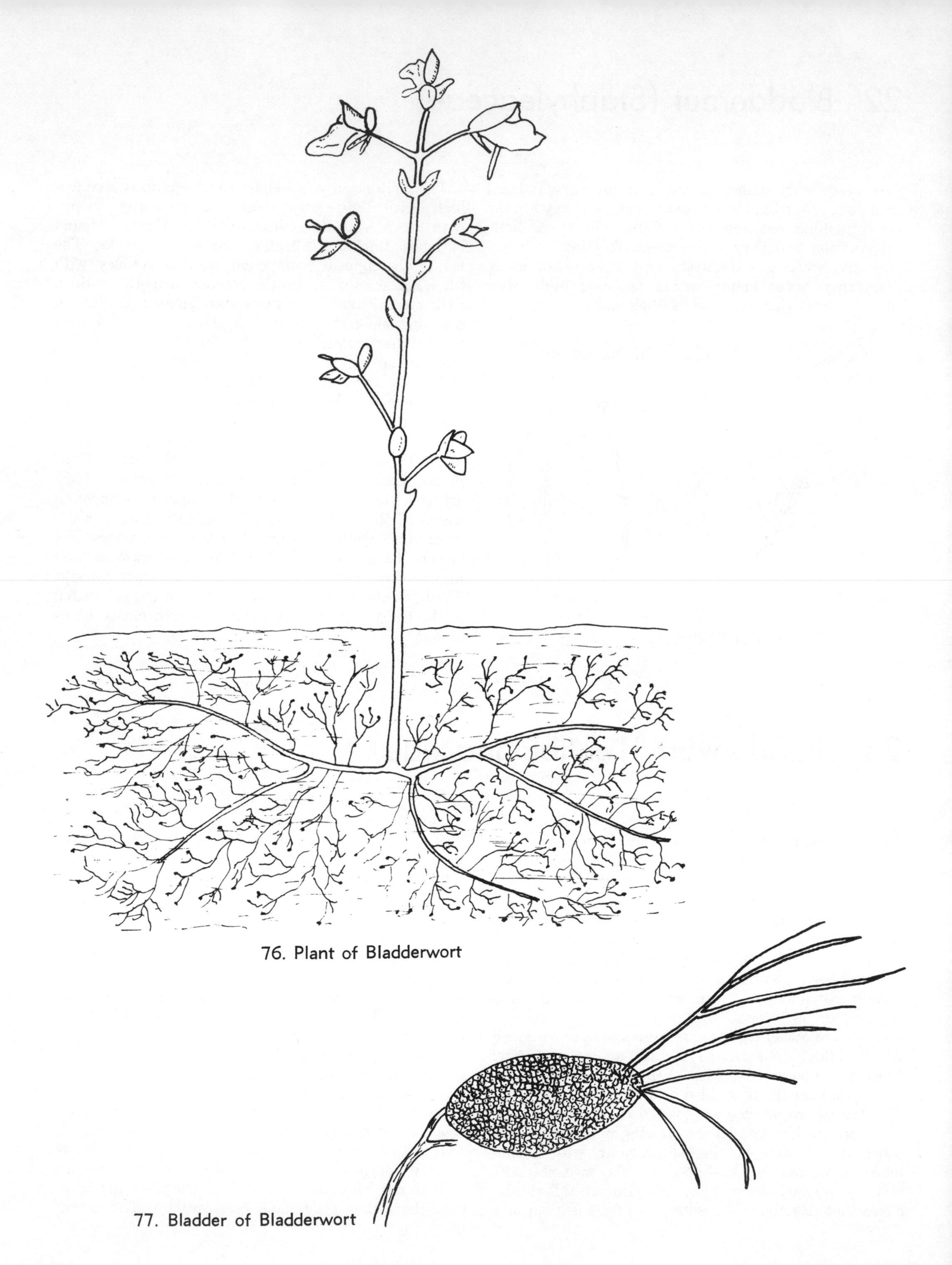

76. Plant of Bladderwort

77. Bladder of Bladderwort

is much like that of a sinking ship that pulls down anything within the area effected by its suction. Once within the bladder there is no escape because the valve is strongly elastic and shuts tightly once the animals are inside. What is the fate of the imprisoned animals? They are digested by fluids secreted by the bladder walls and are used by the plant as food. It has been estimated that all the bladders on a large plant when filled could contain as many as one hundred and fifty thousand living organisms, including protozoans, rotifers, nematode worms, minute insects, and crustaceans.

The bladderworts belong to the genus *Utricularia*, from the Latin utriculus meaning a little bladder or bag. A species common in brooks and ponds is the greater bladderwort. It has three to twenty yellow flowers in a raceme and numerous bladders. Another species, the purple bladderwort, found in ponds, has violet purple flowers and many large bladders. A third species, the closed bladderwort, grows in sandy or muddy shores, and has flowers that may either be purplish or whitish.

A second genus of the family, *Pinguicula*, from the Latin word pinguis, or fat, in allusion to the leaves appearing greasy to the touch, contains thirty species of herbs known as the butterworts. They are of wide geographical distribution, chiefly in temperate and cold regions. They are more terrestrial plants than the bladderworts, growing on moist soil in the pinelands and in shallow-water pools, with a basal rosette of long, narrow leaves from the center of which extend the flowering stalks bearing yellow, purple, or violet flowers. The flowers have a five-lobed calyx, a two-lipped spurred corolla, two stamens, and a one-celled ovary.

The slender tapering leaves are slightly hollowed like a trough, and the surface is covered with a large number of hairlike, stalked glands. The top of the glands are moist with a sticky substance, fatal to any small insect that should happen to alight on the leaf. As the insect comes in contact with the glands, they are stimulated to secrete more of the sticky substance and as it struggles to free itself it becomes more securely trapped on what might be likened to a living fly paper. At times the leaf margin will curl over the insect and, thus, aid in its capture. As with the bladderworts, a digestive fluid, secreted by other leaf glands, converts the insect into soluble nitrogeneous substances that are absorbed by the leaves. In Lapland, the digestive action of the butterworts has been used for centuries in making a junketlike food from milk by pouring it over the leaves.

About six species of butterworts occur in the United States. The common butterwort, also known as the bog or marsh violet, grows on wet calcareous rocks and in gravelly riverbeds in northern New York, Michigan, Minnesota, and further north. It has spatulate or elliptical leaves, and violet purple flowers on a somewhat pubescent flower stalk.

24. Bloodwort (Haemodoraceae)

There are only about nine genera and thirty-five species comprising the bloodwort family. Most are native to South Africa and Australia, a few in tropical America, and only one in North America. They are perennial herbs with erect stems, narrowly linear leaves, and regular or somewhat irregular, small, perfect flowers in terminal cymose panicles. The flowers have a six-parted or six-lobed perianth, three stamens, and a three-celled ovary. The fruit is a three-valved capsule.

Our only species is a stout herb with fibrous red roots and known as the redroot. It is the only one of its genus *Lachnanthes*, Greek for wool and flower in allusion to the woolly flower clusters, and occurs in bogs from Massachusetts to Florida, mostly near the coast. It has mostly basal, narrow leaves with a few on the stem, which is gray-haired at the top. The flowers are a dingy yellow and are borne in woolly, terminal clusters. The fruit is a small, globe-shaped capsule. Its red roots once furnished a dye.

25. Boldo (Monimiaceae)

Although the boldo is cultivated in Caliifornia only for its fragrant, evergreen foliage, it is a valuable economic tree in Chile where it furnishes wood, charcoal, fruit, dyes, and medicine. It is about twenty feet high, with leathery, warty, rough, oval leaves, white flowers grouped in small clusters, (panicles), the staminate and pistillate on separate trees, and a fruit that is a collection of two to five small, stalked, sweet, edible drupes.

The boldo is the only species of the genus *Peumus*. It is one of about thirty genera of the boldo family, a group of aromatic tropical trees and shrubs with opposite leaves, rather small, inconspicuous flowers with many stamens, and a fruit consisting of small one-seeded drupes inclosed in the enlarged calyx. Several species provide wood for building purposes, and the leaves of others possess tonic properties.

26. Borage (Boraginaceae)

Who does not know the little forget-me-nots that grow in the wet, grassy banks of brooks: "The sweet forget-me-nots/ That grow for happy lovers," as Tennyson would have it, but he seems to ignore the melancholy legend that tells of the lover who, when gathering some of the blossoms for his sweetheart, fell into a deep pool and, as he was about to disappear from her sight forever, threw a bunch of them on the bank calling out at the same time "forget me not."

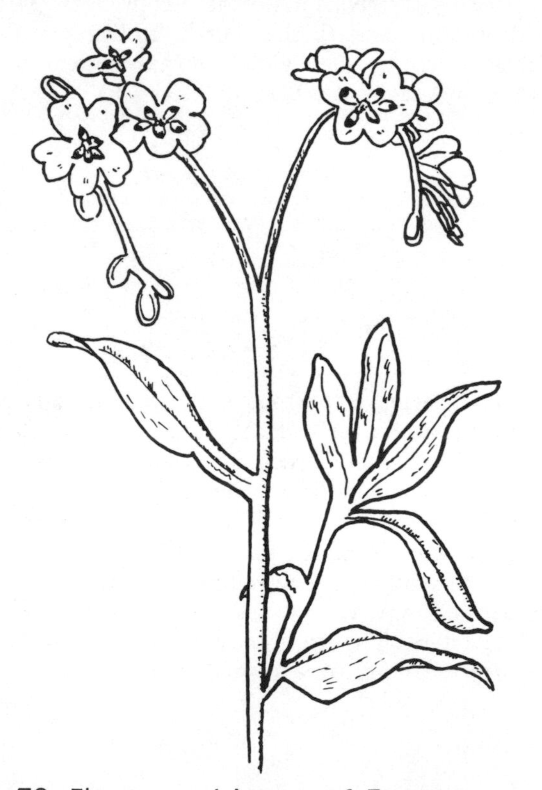

78. Flowers and Leaves of Forget-me-not

The true forget-me-not is a native of Eurasia, long cultivated in the garden and an escape to wet places. It is a perennial with a slender, sprawling stem, somewhat hairy, gray green, lance-shaped leaves that are stemless or nearly so, and small light blue flowers with a yellow eye. There are our own forgot-me-nots: the smaller forget-me-not is a somewhat similar species also found in wet ground; and the spring forget-me-not is a species with very bristly haired stems and leaves and small white flowers, found on dry banks and in rocky woods.

One of the favorite, early spring flowers of the Midwest, the Virginia cowslip is a beautiful plant with a smooth, erect stem, deep green, toothless, obovate, veiny leaves, and light violet blue flowers nearly an inch long. The blossoms are visited by a great variety of insects and, in early spring, are especially attractive because their color varies from red purple to brilliant blue. Since the plants naturally grow in clumps, and the flowers in clusters, the massed effect is delightful to the eye.

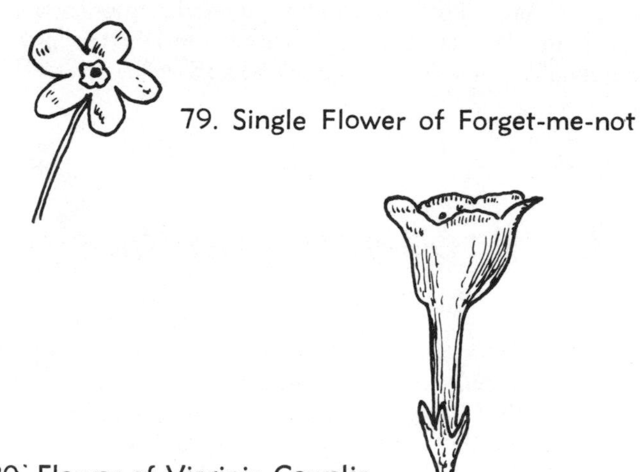

79. Single Flower of Forget-me-not

80. Flower of Virginia Cowslip

Both the forget-me-nots and the cowslip belong to the borage family, a group of eighty-five genera and over fifteen hundred species that are mostly herbs. Many are well-known garden plants, but a few are also shrubs and trees. They are widely scattered throughout the temperate and tropical regions. As for family characteristics, the leaves are alternate, simple, mostly entire, while both the stems and leaves generally are rough and hairy. The flowers are usually small, perfect, often blue or white, in attractive one-sided clusters, both the calyx and corolla are five-cleft, and the latter is often with an appendage. There are five stamens, and one pistil, and the fruit is usually four bony nutlets, each with one seed, or the fruit is sometimes two nutlets each with two seeds.

The plant generally known as viper's bugloss, though it is also called viper's herb or grass, or snake flower is often called blueweed because the flower is sufficiently of a blue tone, although the flower does range in color between lilac, purple, and violet. It is a coarse and bristly plant growing about two feet high along the wayside and in dry fields where, often covered with dust, it has a most plebian appearance. The flowers, however, are rather showy and tend to offset its weedlike character, for, indeed, it is a rather pernicious weed and is likely to cause dermatitis if contact is made with the bristly hairs of its leaves and stems. At one time, simple people believed that the plant would cure snakebites; hence, its generic name is from the Greek for viper.

81. Flowers and Leaves of Hound's Tongue

Few flowers have as many adaptations for survival as the hound's tongue. It has a rank odor and the foliage tastes even worse than it smells; hence, grazing animals such as cattle leave it alone. It has hairs on the stems that serve as a protection against pilfering ants, and humps on the petals that hide the nectar from winged creatures that would trespass on what the bees and butterflies regard as their own domain. Its seeds too are armed with barbs that catch in the fur of any passing mammal and, thus, are dispersed to distant places. The hound's tongue with its stout branching stem, lance-shaped leaves, and small, dull purplish red flowers can be found in dry fields and waste places.

A related species, the wild comfrey is a plant of thin woods. It usually has a simple stem with deep green, basal leaves and pale violet flowers. Both plants serve as food for the caterpillars of certain tiger moths.

Plants of the genus *Lappula,* the name a diminutive of *lappa* for bur, are known as stickseeds because the nutlets are armed with barbed prickles that catch in the fur of mammals. They are rough-haired and grayish herbs, with small blue to whitish flowers in racemes or spikes. The Virginia stickseed, with lavender white flowers on slender flower spikes, is common on the borders of dry woods, thickets, and upland brushy pastures. The European stickseed, with many small, gray green linear leaves and tiny, light violet flowers scattered on slender branches, grows in dry fields, roadsides, and waste places. The multi-flowered stickseed, with a stem five feet tall, roughly haired leaves, and racemes of many blue flowers, is a plant of the western plains.

Two garden members of the borage family are the common borage and the common heliotrope. Pliny speaks of the former and, as a matter of fact, seems to have held it in fairly high esteem. It was supposed to exhilarate the spirits and drive away melancholy. It is a hardy annual not over two feet high, with bristly leaves that are slightly oblong, and blue flowers, more or less wheel-shaped, in a loose, leafy cluster. The heliotrope, a native of Peru, is actually a perennial, but normally is treated as a tender annual. It has oval or oblong leaves and small purple or white flowers that are vanilla-scented.

27. Box (Buxaceae)

It is generally agreed that our most valuable broad-leaved evergreen is perhaps the box or, as it is more commonly known, the boxwood. There are some thirty species of boxwoods of the genus *Buxus,* the classical Latin name for box, but only two are commonly cultivated. Yet these two and their many horticultural varieties have been widely planted and have given an atmosphere of grace and charm to many historic gardens. Ever since Roman times the box or boxwood has been the best of all plants for hedges and topiary work.

The boxwoods are evergreen plants with entire, opposite leaves, small and inconspicuous flowers without petals, the staminate with four sepals and the pistillate with six on the same plant, with the latter developing into a three-horned capsule. The common box, a native of southern Europe, northern Africa, and western Asia, ranges from a dwarf, globular shrub to a tree up to twenty-five feet high, but it grows into a tree only in the most favorable sites. It is not hardy in the North, but is often grown there in locally favorable places, or with winter protection. The second cultivated boxwood, a Japanese species, is hardier, especially in two varieties. It is rarely over three feet high and resembles the common boxwood, but has smaller leaves and four-angled or winged branchlets.

The boxwoods are members of the box family, a small group of about six genera and some forty species of wide distribution, mostly in subtropical and tropical regions. They are trees, shrubs, or perennial herbs with alternate or opposite simple, mostly evergreen leaves, and small greenish, inconspicuous monoecious or dioecious flowers, with or without a four to twelve-parted calyx, but with no petals. There are four to seven stamens, a mostly three-celled ovary, and a fruit that is a capsule or drupe.

There are only two native species of the family: the Allegheny or mountain spurge in the southeastern states and a California species. The Allegheny spurge (genus *Pachysandra,* from the Greek for thick and men, in allusion to the stamens) is a low-growing, perennial, woodland herb with a matted rootstalk, stems that trail at first and then become erect, and alternate, ovate or oval leaves that are a dingy green. The flowers, white or purplish, are borne in spikes produced from the leaf-bearing stems, the upper staminate flowers having four sepals and four stamens, the lower pistillate having four sepals and a pistil. The latter develops into a capsule of three two-seeded carpels. The plant is often cultivated for its early spring flowers. A related species, the Japanese spurge, with thick, glossy foliage forming a dense mat, is one of the best evergreen ground covers for partly shaded places.

28. Brazil Nut (Lecythidaceae)

The Brazil nuts of commerce are the seeds of the Brazil nut tree, a tall tree wiith leathery, oblong leaves nearly two feet long, cream-white flowers in spikelike clusters (racemes), the flowers with no petals, but with colored sepals, and a hard, woody, brown structure about four inches in diameter, containing eighteen to twenty-four of the familiar nuts that are often called niggertoes. It is one of two species comprising the genus *Bertholletia,* named for C. L. Berthollet, a French chemist.

The genus is one of eighteen genera with perhaps two hundred and fifty species of exclusively tropical trees forming the Brazil nut family. They have alternate leaves that are often crowded towards the ends of the branches, white flowers in clusters (racemes), the flowers regular with four or six sepals, sometimes replacing lacking petals, four or six petals when present, many stamens, and a two or more-celled ovary. The fruit may either be a hard, woody pod or a large berry.

The only other genus of the family of any interest is *Barringtonia,* named for Daines Barrington, an English naturalist. It contains thirty species of very showy trees, one being found in southern California and extreme southern Florida. It has evergreen, entire leaves that are mostly crowded at the ends of the twigs, few flowers in a loose cluster, with white petals and purple stamens, and a four-sided, boxlike, woody pod.

29. Broomrape (Orobanchaceae)

Ages ago, some plants, preferring to live at the expense of others, discarded their food factories and took up a parasitic existence. Such a plant is the one-flowered broomrape, a beautiful little plant despite its habits, with a few brownish ovate bracts near the root, and one to four erect, slender, brownish, scapelike peduncles, each with a curved, tubular, five-lobed, delicately fragrant purplish or light-violet flower. It is found in damp woodlands.

There are other broomrapes, all living directly on

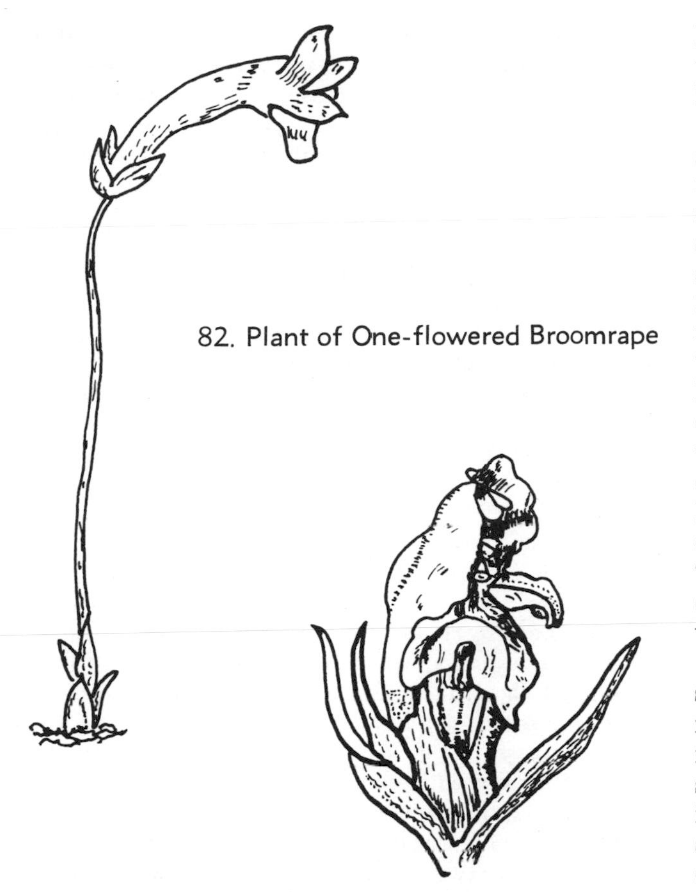

82. Plant of One-flowered Broomrape

83. Single Flower of Broomrape

their neighbors by attaching strong haustoria, or suckers, to their roots, penetrating the tissues, and absorbing the food materials made by the host plants for their own growth and development. They are essentially alike except for certain minor structural differences, having yellowish scales instead of leaves, the plants brownish, purplish, or whitish, and the flowers urn-shaped, in terminal spikes, the corolla with a two-lobed upper lip, the lower lip three-lobed and spreading. The branched broomrape is found in hemp and tobacco fields and is parasitic on tomatoes. The clover broomrape occurs in clover fields where it is parasitic on clover, and the Louisiana broomrape is a parasite on several wild plants and occasionally has attacked tobacco. Its thick underground stems, white and tender, are eaten, it is said, by the Paiute Indians.

The broomrape family, of the same name as the genus of broomrapes; *Orobanche,* is from the Latin for broomrape, and the Greek for bitter vetch. It consists of eleven genera and more than two hundred species of erect, simple or branched, yellowish or almost colorless parasitic plants of wide distribution in the northern hemisphere. The plants are often nearly leafless or with stems scaly only at the base. The leaves are generally reduced to alternate scales lacking chlorophyll. The flowers are usually perfect, often crowded in terminal, bracted spikes, but sometimes are solitary, with a four to five-toothed, lobed, or deeply cleft calyx, an irregular five-lobed, two-lipped corolla, and four stamens in two sets on the tube of the corolla and alternate with the corolla lobes. They also have a one-celled ovary with numerous ovules and a two lobed stigma, and the fruit is a two-valved capsule with many small seeds.

The family has no economic value except that some species parasitize certain cultivated plants. It has no representatives in the garden except those that might find their way into it as parasites, and only a few species growing in the wild are of any particular interest. In addition to the broomrapes that have already been mentioned, there are the beechdrops, or cancerroot, and squawroot. The beechdrops, as its name might suggest, is parasitic on beech trees and generally occurs in beech woods or woodlands where the beech is found in companionship with other trees. It has a tough brownish or reddish-tinged, slender, stem branching above, six inches to two feet tall, and two kinds of flowers: small, dull purple and white, tawny or brownish-striped, scattered along tiny bracted, ascending branches; below them are located the second kind, cleistogamous flowers, which never open and which are self-fertilized. To prevent degeneracy some of the upper flowers are cross-fertilized by bees.

Should one want to find the squawroot, look among fallen oak leaves, for the plant is generally parasitic on oaks, though it will also select other trees as hosts. It is a stout, yellowish or brownish plant with a stem densely covered with scales and pale, yellow flowers on a scaly spike that, with their attendant bracts, look like fir cones, making the squawroot easily identifiable.

30. Buckeye (Hippocastanaceae)

Under a spreading chestnut tree
The village smithy stands. . .

These familiar words of Longfellow are known to most, but the poet's chestnut tree is not our native chestnut, but the horse chestnut of southern Europe and Asia, much planted in this country as an ornamental.

Few trees can match the horse chestnut when in

flower, for then it is a "pyramid of green supporting a thousand pyramids of white." The flower cluster is what the botanists call a *thyrse,* a mixed sort of inflorescence wherein the main stem is racemose and the secondary ones cymose. It is very complicated. These are certainly rare clusters; the only other familiar examples that come to mind are the lilac, syringa, and privet. The corolla of each flower in the horse chestnut is white, spotted with red and yellow, and the curving, yellow stamens extend beyond the ruffled border. In view of the vast number of blossoms there would be expected a large harvest of nuts; but actually a cluster forms only two or three burs, and some clusters none. The reason is that only a few of the flowers are fertile; most of them have only stamens, and an aborted pistil.

The horse chestnut is a symmetrical and beautiful tree, forty to eighty feet high, with an erect trunk and ascending branches that form a superb conelike head. The bark is dark brown roughened with flat scales or divided by shallow fissures. The leaves are compound, dark green, usually with seven wedge-shaped, round-tipped, coarsely toothed leaflets. The flowers are white and showy, as was mentioned above, and the fruit is a leathery pod, with scattered soft spines, at first green, but finally rusty brown, containing a single, large, shining, chestnut red nut that is not edible.

The horse chestnut began to be cultivated in Europe in the seventeenth century, where it has been a favorite tree for avenues and parks. In America it grows better than in the Old World and, because it casts the densest shade of almost any cultivated tree, it has been a welcome street tree. It always seems to be dropping something, however—first the bud scales, then the flowers, followed by the leaves of the crowded interior that turn rusty yellow and fall all summer, and the unripe fruits that drop in all stages, and finally in the fall by both the leaves and fruits—and many people object to the litter that is continuously strewn on the ground.

The horse chestnut may easily be recognized in all seasons by its clustered showy flowers, its compound leaves of seven leaflets that are arranged like the fingers of the hand, its characteristic pod, and, when all these have gone in the winter, by its large, gummy, resinous brown, terminal buds. The name horse-chestnut has been accounted for in many ways. One reasonable explanation is the shape of the leaf scar, the scar left on the twig when the leaf falls off, is singularly like the print of a horse's hoof. Or could it have been as Gerarde suggests "for that the people of the east countries do with the fruit thereof cure their horses of the cough, shortness of breath, and such like diseases?" There is a red-flowered horse chestnut that is believed to be a hybrid between the horse chestnut and the red buckeye, a shrub or small tree native to our southern states.

The horse chestnut is one of twenty species of the buckeye family, also known as the horse chestnut family, which are contained in two genera—one being Chinese, the other *Aesculus,* being represented in North America by the horse chestnut and the buckeyes. They are very ornamental trees and shrubs with compound leaves, showy flowers in profuse clusters, and large seeds in a more or less prickly pod.

The species of the genus *Aesculus,* the classical name of an oak that produces edible acorns and given by Linnaeus to the horse chestnut, though its seeds are of no value, have very scaly, gummy-coated buds, and large, compound leaves with long stalks and five to nine digitate, toothed leaflets. The flowers are very showy in large multi-flowered clusters. The individual flowers have a bell-shaped or tubular four to five-toothed calyx, four to five petals, five to nine stamens, and a three-celled ovary with two ovules in each cell. The fruit is a large three-valved, often spiny capsule containing one or two very large seeds.

There are very few native trees whose leaves are placed opposite on the twigs. They are the buckeyes, ashes, maples, dogwoods, catalpas, viburnums, and elders. Of these, only the buckeyes, ashes, and elders have compound leaves, and of these the ashes and elders have pinnately compound leaves—the buckeyes have palmately compound leaves. So any tree having opposite, palmately compound leaves is sure to be a buckeye, or horse chestnut. All the buckeyes have the characteristics of the genus: large winter buds, showy flowers in thyrselike clusters, large, rather handsome leaves, and large seeds or nuts in three-valved husks. The name buckeye is said to be due to the brown nut marked with white, leading someone to remark at one time that it looked like the eye of a deer.

Of our six native species of buckeyes three of them

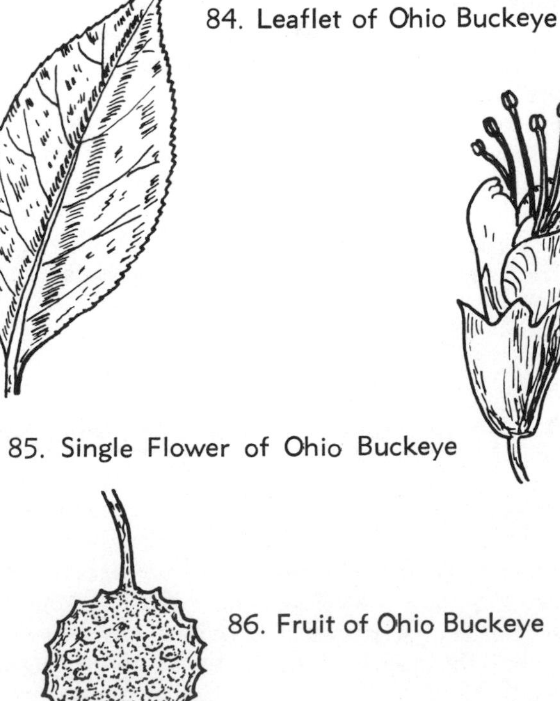

84. Leaflet of Ohio Buckeye

85. Single Flower of Ohio Buckeye

86. Fruit of Ohio Buckeye

are trees, and three are shrubs. The trees are the Ohio buckeye, the sweet buckeye, and the California buckeye, while the shrubs are the red buckeye, the painted buckeye, and the bottlebrush, or dwarf buckeye. The Ohio buckeye, also known as the fetid buckeye due to the strong, disagreeable odor exhaled by its bark, is a small tree of woodlands and river bottoms in the Midwest, with five leaflets that are elliptic or broadest towards the tip, pale yellowish green flowers, and an inverted, egg-shaped prickly pod. The French explorer and botanist Michaux wrote in his *Sylva of North America* that he principally found this tree in Ohio, especially abundant on the Ohio river, and so he called it the Ohio buckeye. Hence, Ohio's nickname, the Buckeye State. Actually the distribution of the tree is far wider than the Frenchman supposed.

The sweet buckeye, of rich woods is a more eastern species. It is also a larger tree reaching a height of sixty feet or more, with leaves similar to those of the Ohio buckeye, yellow flowers, and a nearly globe-shaped prickless pod. It is a beautiful tree and called the sweet buckeye simply because it lacks the disagreeable odor of its relative. The nuts are said to have been eaten by cattle.

The California buckeye, growing on the dry hills and canyons of the California mountains, is a small tree, not more than twenty feet high, with four to seven elliptical leaflets, white or pink flowers in large cylindrical clusters, and smooth, pear-shaped fruits.

A southern species, the red buckeye is a shrub four to ten feet high, occasionally a small tree twenty feet high, growing in hammocks and pinelands near the coast. The leaflets are five in number, the flowers are a bright red, and the fruit is roundish or egg-shaped. Also a southern species growing in rich woods or along streams, the painted buckeye is a small shrub two to ten feet high, with five leaflets, pale yellow or greenish yellow flowers, often tinged with red, and a smooth fruit. Another southern species and a handsome ornamental widely planted as a lawn shrub, the bottlebrush or dwarf buckeye is a widely spreading shrub eight to twelve feet high, of rich woods, with five to seven leaflets, white flowers having pink protruding and showy stamens, and an inverted, egg-shaped fruit. It is the last of the genus to flower and becomes an effective shrub in midsummer when it blossoms profusely in July and August when most of the woody plants have passed into the fruiting stage. It has the habit of spreading broadly, as the outermost and lowest of its many stems often become horizontal and rest on the ground. Those on the ground readily form roots, and, thus give rise to new plants that blend with the old, increasing the circumference of the original.

31. Buckthorn (Rhamnaceae)

It is said that when the Oriental tea brought to the American colonies in English vessels, just prior to the Revolution, became unpopular and its use considered unpatriotic, Colonial housewives used the leaves of the plant known as the New Jersey tea as a substitute. A tea brewed from the dried leaves is not as pleasant as the beverage made of real tea, but it is palatable and doubtless many an American soldier gladly found it refreshing.

The New Jersey tea, common in open woods and thickets throughout the eastern half of the country, is a shrubby species with coarse, woody, brown green or bronzy upright stems, one to three feet high, and alternate ovate leaves, pointed at the apex and often heart-shaped at the base with saw-toothed edges and conspicuously three-ribbed. The flowers appear in May or about the first of June, and are small, white, and lightly odorous and are arranged in cone-shaped, plumy clusters on long stems from the leaf angles. The shrub is quite conspicuous when in blossom, is attractive and decorative, and may well have a place on our lawns. The fruits are small three-lobed cap-

87. Leaves and Flower Cluster of New Jersey Tea

sules containing three pale brown seeds. The New Jersey tea is sometimes called the redroot because of the color of its roots, from which a red dye has been obtained, but the name redroot is better applied to a similar species with narrower, almost willowlike elliptical leaves.

Both the New Jersey tea and redroot belong to the buckthorn family, which is composed of five hundred species of shrubs, trees, and woody vines of very wide distribution in tropical and temperate climates.

Their leaves are commonly alternate, simple, and toothed. The flowers are small, generally perfect, with five sepals, four to five petals or sometimes none at all, with the same number of stamens, a two to three-celled ovary, and mostly in cymelike clusters. The fruit is either a drupe or capsule.

The genus, *Ceanothus,* to which the New Jersey tea and redroot belong, also includes a number of other native plants that are essentially found in the west. They are the deer brush, a diffusely branching shrub of hillsides and mountainslopes with tough leathery leaves and a strong resinous or cinnamon odor; the snow brush of the prairie states, a taller shrub with reddish branches and more rounded leaves; and the California lilac, a colorful plant of the hillsides and mountain canyons with dense clusters of white, pale, or lavender blossoms that are sometimes produced in such profusion that they tint the hillsides with a bluish haze.

In the eastern part of the United States the most familiar plants of the buckthorn family, aside from the New Jersey tea, are the various species of buckthorns: the alder-leaved, the lance-leaved, the Carolina, the alder, and the species simply known as the

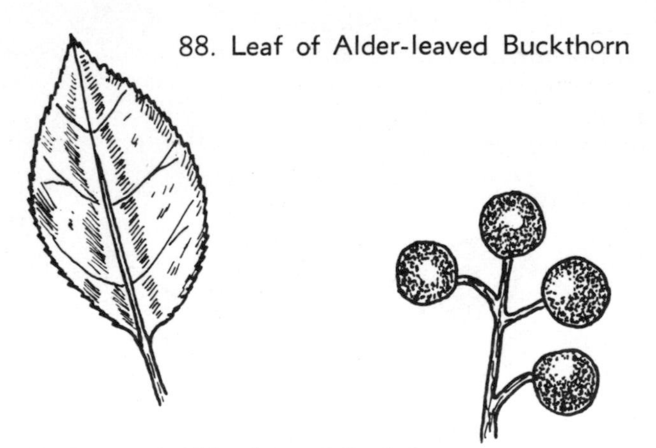
88. Leaf of Alder-leaved Buckthorn

89. Fruit of Alder-leaved Buckthorn

buckthorn. The alder-leaved buckthorn is a low shrub of swamps, two to three feet high, with gray brown or gray stems, and deep green leaves that are elliptical and pointed at either end. The flowers are small, green, without petals, and usually solitary at the base of the leaf-stems; the fruit is a spherical, purple black berrylike drupe, with three grooved seeds. The lance-leaved buckthorn is a slightly taller shrub found in damp, swampy ground or on gravelly riverbanks. The light olive green leaves are rather lance-shaped, downy on the veins beneath, and the flowers are generally yellow green, and the staminate and pistillate flowers are on separate plants. They are followed by purple black, berrylike drupes, one-quarter of an inch in diameter, with two seeds.

A considerably taller shrub than the preceding two species, at times a small tree from ten to thirty-four feet high, the Carolina buckthorn, also known as Indian cherry, is essentially southern in its range, occurring in New Jersey, Kentucky, Missouri, Kansas, and southward. It grows on moist, shaded banks and in swampy ground, and has large, olive green elliptical leaves, inconspicuous green flowers, and a fruit that is at first red, then finally purple black. Unlike the fruits of our northern buckthorns that are disagreeable and astringent, those of the Carolina buckthorn are sweet and edible.

Both the alder buckthorn and the species known as buckthorn, though this plant is also known as the common buckthorn, are introduced species from Europe. The alder buckthorn is a handsome lawn shrub that has escaped locally in some of the Eastern States. It is well formed, often ten feet high, with inconspicuous flowers, but the shining foliage and purple black berries give it an air of distinction.

Because its twigs are often armed with formidable thorns, the common buckthorn is often planted as a hedge. It is eminently suitable for hedges as its roots extend only a short distance and, because they are fibrous, the roots do not interfere with other plants. It is also not subject to injury by insects, and its foliage is a rich dark green and persists well into the fall.

The common buckthorn has escaped from cultivation and one may find it growing along roadsides and in the vicinity of dwellings. It is a tall upright shrub with brownish gray stems, deep green leaves, and inconspicuous yellowish green or whitish green flowers that are staminate and pistillate on separate plants. They are succeeded by black, shining, berrylike drupes that were once used medicinally, but have long since been discarded for less-violent remedies. A dye, known as "Chinese green," was at one time obtained from them, but in this synthetic age it is probably no longer used for this purpose.

In the western part of the United States, there occur several native related species. One is the red berry, a common shrub of the California hillsides, with spiny branches, leathery, oval leaves, and clusters of red berries. Another is the coffee berry, an evergreen shrub found on hillsides and desert slopes. Its berries are red at first, but then turn black. Still another is the Cascara sagrada, also called bearberry, Chittamwood, and shittimwood. It is a tall shrub, or occasionally a tree to forty feet high, with elliptic to oblong leaves, flowers in small hairy clusters (umbels), and purplish black fruit. The bark contains a cathartic substance sold as cascara, thus making this tree one of the few important medicinal plants that grow in the United States.

32. Buckwheat (Polygonaceae)

In the northeastern states, during the summer months, the Ameriican copper, a small, bright metallic coppery butterfly, may be seen flying about with careless abandon in city parks and village yards, as well as in the more open fields and woodlands. The caterpillars of this species feed on the field sorrel, and the butterfly is often seen stopping to lay her eggs on the leaves and stems. The coppery red color of the butterfly blends so well with the rusty red of the blossoms that sometimes the butterfly is not noticeable on the plant until it moves its wings or suddenly takes to flight.

The field sorrel is a common plant, often a troublesome weed, from Europe. It is found in fields, meadows, pastures, roadsides, and waste places, with an extensively creeping, branched, yellowish rootstalk, halberd-shaped leaves on long petioles, acid to the taste, and inconspicuous flowers in branching spikes. They are green at first, then later turn a brown red. As a matter of fact, the entire plant often turns ruddy, especially in dry, sterile fields. The staminate and pistillate flowers occur on separate plants and, hence, are fertilized by the insects, especially the bumblebees, honeybees, and the smaller butterflies.

The field sorrel belongs to the genus *Rumex,* the Latin for sorrel, a group of perennial herbs, generally known as the docks or sorrels, with strong roots, simple, basal, or stem leaves, and small flowers, usually greenish white in long, branching clusters. Sometimes the flowers on the same plant are of two kinds, some with stamens only, others with pistils only. They lack a corolla, but have a calyx of six sepals and develop into a three-sided capsule that is often winged.

90. Leaves and Flowers of Field Sorrel

The docks, most coming from the Old World, are rather uninteresting plants that encumber fertile ground and decorate waste places. One species, the patience dock, has long been cultivated in Europe for early greens. It is sometimes cultivated in the United States as a potherb. Also known as spinach dock, it is a strong-growing perennial six feet high, with basal leaves tapering at both ends and having wavy margins, and tiny green flowers in branching clusters. It is a common species of farmyards, waste places, and roadsides.

Two other species of dock also used as potherbs are the curled dock and bitter dock. The curled dock is a common plant of waste places, pastures, and cultivated fields, and is a smooth, dark green plant, one to three feet tall, with a deep yellow root, oblong or lance-shaped leaves, wavy on the margins, and greenish flowers in simple or compound racemes often a foot in length. The bitter dock, in some localities the most common of the docks, is a large, robust plant in many respects similar to the curled dock, but with broader leaves and a more bitter taste. When using the leaves of the docks as greens, the leaves should be gathered in the spring when young and tender and parboiled to remove the bitter taste. They are generally cooked with a little bacon, ham, or salt pork and a small amount of vinegar. Some housewives often prefer to mix the leaves with those of dandelions, and the tender tops of mustard.

A wild dock, native from California to Texas and whose roots are used in tanning leather, is the species known as canaigre or wild rhubarb. The stems of the leaves and stalk are crisp and tart, and are often used as a substitute for rhubarb, which it resembles. Two species, the great water dock and the swamp dock, occur in wet places. The former is a tall stout species with flowers in circling clusters, while the latter is a similar species with a grooved stem and flowers also in circles. A third species, the golden dock, grows in sand along the seashore, and is more or less woolly.

The buckwheat family, to which the docks belong, is important chiefly because of two plants, the buckwheat and rhubarb, rather than for its genera of showy flowers, only a few being grown for ornament. It is a large family of some forty genera and eight hundred species of herbs, vines, shrubs, and trees that are widely distributed throughout both warm and cold regions. Most of the plants have jointed stems that are often swollen at the nodes. They have simple, alternate, rarely opposite or whorled leaves, the stipules often forming a membranous sheath about the stem, and small, perfect, and regular flowers, often in spikes or racemes. The flowers have three to six sepals that are separate or slightly united, frequently becoming greatly enlarged, winged, spiny, and brightly colored in fruit, and no petals, six to nine stamens, the ovary with two to four carpels, and one-celled. The fruit is an indehiscent, three-angled or lenticular nut or achene that is sometimes enclosed by an enlarged calyx.

Aside from the docks, most of our wild members of the family belong to the genus *Polygonum,* from the Greek for many-jointed in reference to the nature of the stems. The genus comprises about two hundred species of erect, trailing, or climbing herbs of diverse habits that are found throughout the world. Our species are known as the smartweeds or knotweeds.

A species common everywhere is the common persicaria, also known as the Pennsylvania persicaria, pink knotweed, or plain smartweed. It is a somewhat red, jointed species with erect pink spikes that brighten roadsides, fields, waste places, streamsides, and ditches from midsummer to frost. The flowers open anywhere on the spike and attract many insects, especially the smaller bees. This species, like many other smartweeds, has glandular hairs on the upper part of the stems to discourage unwanted pilferers. Shortly after fertilization the black, smooth, lens-shaped achenes begin to form within the persistent calyces so that the flower spikes often contain more pink buds and shining achenes than flowers.

Another common species and similar to the preceding, with crimson pink or deep magenta flowers, is the lady's thumb. It is a species easily recognized because the leaves have a dark triangle spot in the middle. The story has it that Joseph once injured his hand while working in his carpenter's shop. Mary wanted to make a healing poultice with this plant, but "she could not find it at her need/ and so she pinched it for a weed," and ever since the leaves have borne the imprint of her thumb—hence, its name.

91. Leaves and Flowers of Lady's Thumb

An insignificant little plant found everywhere in waste places, roadsides, and farmyards is the knotgrass or doorweed. It is a prostrate plant trailing its leafy, jointed stems over the ground, but at other times erect to display its tiny greenish or white pink edged flowers. Few insects visit it, for it secretes little, if any, nectar. The erect knotweed is a stouter, erect plant, two feet or less tall, with a yellowish green stem, nearly oval leaves, and greenish yellow flowers, common along waysides.

A somewhat slender, pale edition of the common persicaria, the mild water pepper is a water plant, growing in shallow water, swamps, and moist places. It rises three feet or less, has narrow, lance-shaped leaves, and pale pink flowers in slender, erect spikes two inches long. It may easily be recognized by the cylindrical sheaths around the swollen joints of the stem that are fringed with long bristles. Another water pepper or smartweed is so-called because its leaves are very acrid and pungent. They are fringed with tiny bristles. This species is also found in wet places. A third species, the water smartweed, is taller than the two preceding species. It also has larger and longer leaves with white or purplish flowers in stiff, upright spikes.

Run your finger down the stem of the halberd-leaved tearthumb and it will feel as smooth as satin, but run your finger upward and it will be scratched by a thousand vicious prickles—hence, its name, though the halberd part comes from the shape of its leaves. By means of these prickles the plant climbs all over its neighbors. Another tearthumb, the arrow-leaved tearthumb, has arrowhead-shaped leaves and prickles only at the angles of the four-angled stem. Both species have insignificant pink flowers, and both occur in wet places.

In woods, thickets, and along the wayside, the climbing false buckwheat trails its way over neighboring shrubs and rocks to a height of twelve feet or more. It is a smooth species with a slender reddish stem, arrowhead-shaped leaves, and tiny, green white or pink flowers in leafy flower spikes that are practically hidden from view in August and September when the more showy composites are in their glory.

A number of polygonums have found their way into the garden where they are grown for their foliage and flowers. One of them is the mountain fleece, a hardy perennial with a spreading rootstalk, oval leaves with wavy margins, and rose or white flowers in terminal spikes. A second is the silver lace vine, a hardy, twining, woody perennial with slender stems, growing to twenty-five feet, broadly lance-shaped leaves, and greenish white, fragrant flowers in long erect or drooping clusters. A third is the Mexican bamboo, which is neither Mexican nor a bamboo, a strong-growing hardy perennial, with roundish leaves, and many small, greenish white flowers in loose clusters. And a fourth is the prince's feather, a hairy, multibranched annual growing to six feet high, with oval leaves, and pink or rose flowers in branching spikes. All of these garden ornamentals are importations.

Many of our native polygonums serve as food plants for some of our butterflies, such as the gray hairstreak, the bronze copper, the purplish copper, and the pipe vine swallowtail.

The common buckwheat, from which the family derives its name, is not a wheat at all, as true wheat is a grass. The name buckwheat is, in all probability, a corruption or modification of the word "beechwheat," from the resemblance of the fruits to beechnuts, and the fact that a flour is prepared from the seeds. Whatever the origin of its name, the common buckwheat is a tender, annual grain plant, a native of Siberia, cultivated as a crop plant, but also an escape everywhere. It is an erect, branching, slightly hairy plant with alternate, halberd-shaped or arrow-shaped leaves, and greenish white flowers, sometimes pinkish, the calyx five-divided, and with eight honey glands alternating with the stamens. The flowers are fertilized mainly by honeybees. The honey made by the honeybees is of a peculiarly fragrant character, but dark in color. The seeds of the buckwheat are richer in starch than those of the cereal grains.

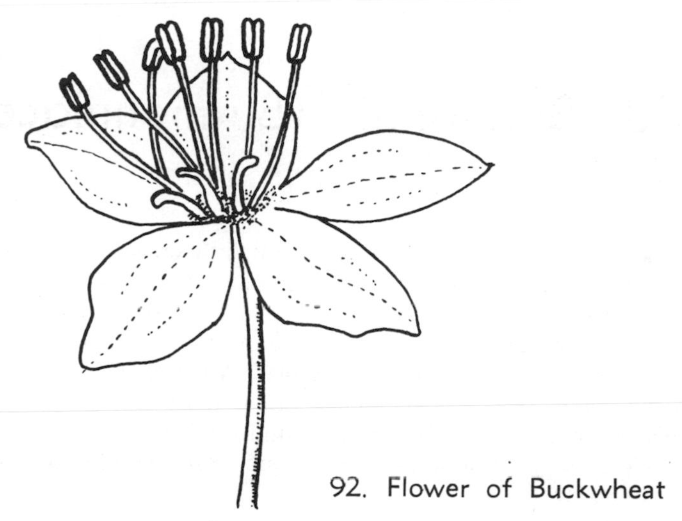

92. Flower of Buckwheat

Like the celery, the rhubarb is cultivated for its leaf stalks, both the leaf and root being without value. The rhubarb, a native of Siberia, is a strong perennial with thick, clustered roots, succulent, pleasantly acid, wholesome leaf stalks, large roundish leaves, and small whitish flowers borne in tall, leafy, densely flowered panicles. The rhubarb has been in cultivation since prehistoric times, so long that actually no one knows of its ancestral home, though it is generally believed to be Siberia. There are a number of species of rhubarb, and all are Asiatic herbs. One species is grown in the garden for its striking foliage; another species is grown in China for its large underground stem that contains medicinal substances and is used by pharmacists.

There are a number of other plants of the buckwheat family that deserve mention, such as the woolly herbs of the deserts, plains, mesas, and mountainslopes of the western states. They chiefly have basal leaves without teeth—the stem leaves when present are alternate or whorled—and small inconspicuous flowers in various sorts of clusters. One of them, known as the wild buckwheat or flattop, has white flowers; another, the sulphur flower, has golden yellow flowers.

A beautiful plant, with threadlike stems and leaves, is the coast jointweed. It is a plant of sandy places with small, white or rose-colored flowers in small racemes. And chiefly along the coastal plain from South Carolina to Florida and west to Texas, the brunnichia is found. It is a vine with grooved stems and climbing by means of tendrils at the tips of the branches. It grows in swamps and along the banks of streams, has alternate, egg-shaped, leaves pointed at the tip, and greenish flowers two to five in a cluster.

Only two trees of the buckwheat family grow in

the United States and both occur in Florida. One of them, the sea grape, is a stoutly branched tree of the sandy seashore with large rounded leaves that have prominent red midribs, small white flowers in terminal and axillary racemes that develop into purple or greenish white, translucent berrylike fruits with a thin, juicy flesh, and a thin-walled, light red nutlet. The other, the pigeon plum, is somewhat similar, but grows to be a taller tree with small leaves.

33. Burmannia (Burmanniaceae)

The burmannia family is a small, chiefly tropical group of low herbs with threadlike stems and fibrous roots. The family comprises about ten genera and some sixty species that are widely distributed in tropical regions. The group is represented in North America by one genus and two species. They have basal grasslike leaves, but often with minute and scalelike leaves, perfect, regular flowers, with a six-cleft, corollalike perianth, three or six stamens, and a one or three-celled ovary. The fruit is a multiseeded capsule.

One of our two species is known as the northern burmannia of the genus *Burmannia,* named for Johann Burmann, a Dutch botanist of the eighteenth century, and is paradoxically a plant of southern peaty bogs, with simple stems and small, alternate scalelike or bractlike leaves and one or several terminal flowers.

34. Bur Reed (Sparganiaceae)

The bur reed family is a small group of thirty to forty species—all of a single genus, *Sparganium,* most of them being found in the temperate and cold regions of the northern hemisphere. They are reedlike, aquatic, or marsh herbs with perennial or fibrous roots. Their leaves are alternate, linear, flat, grasslike and sheathing at the base, in general appearance much like those of an iris. The flowers lack both sepals and petals, but have a perianth of a few small, chaffy scales, and are crowded in globular heads at the nodes of the upper zigzag parts of the stems and branches. The globular heads and zigzag stems make them easily recognizable among other marsh plants. The flowers are unisexual, the upper heads being the staminate flowers, the lower heads the pistillate. There are several stamens, usually five, the ovary is one or two-celled, the style is threadlike, and the fruit is nutlike or drupelike.

There are about two dozen species of bur reeds in the United States. A common species is the great bur reed, a plant three to seven feet high. It grows on the borders of ponds and streams, has deep green leaves, downy, brown white flowers in dense round heads, and a green fruit that is a burlike sphere made up of nutlets that are wedge-shaped below, and flattened above with an abrupt point in the center so that the entire surface looks like that of a pineapple.

The smaller bur reed is a smaller edition, with narrower leaves, and only one to two feet high. Some-

93. Leaves and Flower Cluster of Bur Reed

times it is found growing in the water, at other times in the mud in the shores of ponds and rivers. A third species, the branching bur reed, may be distinguished from the other bur reeds by its branching and somewhat angular flower stem that grows out at the point where the leaf is joined with the plant stem.

35. Buttercup (Ranunculaceae)

"The flowery May" wrote Milton of the fifth month of the year. No one can dispute the poet, for during this month many colorful and different-sized flowers and blossoms appear on the landscape. They may be seen in field and meadow, in the thicket and woodland, along the roadside and brookside, by the water's edge of pond and stream, and even in the water itself. Now, if there is any one flower that one can truly associate with May, it would probably be the buttercup. This association can be traced back to childhood days, when the bright yellow blossoms were a special delight. There is an early buttercup that blossoms in April, but the one known so well appears in May. The scientific name of the flower is *Ranunculas,* also the genus name, and is from the Latin for little frog in allusion to the wet places where buttercups tend to grow. Although there are some species that prefer moist habitats others are common in the woods, fields, and roadside banks.

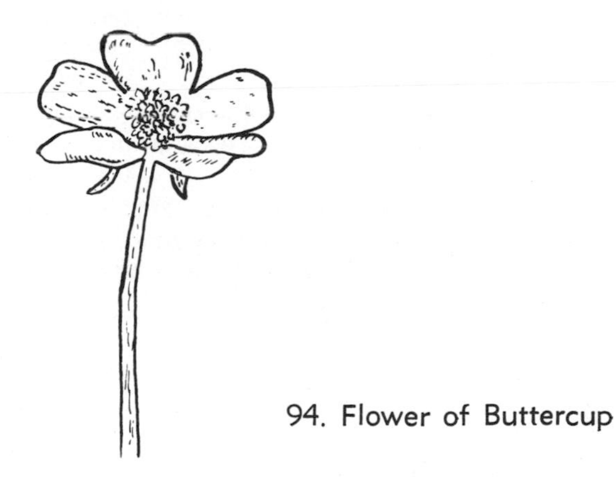

94. Flower of Buttercup

There is nothing complex about a buttercup's blossom. It has five, pale yellow sepals with brownish tips, though they are green in the bud, five petals that are pale beneath but a bright yellow above, and shining as if they were varnished, numerous stamens, and several pistils. Each petal is wedge-shaped with its broad outer edge curved to form a cuplike flower. If a petal were removed, and examined under a magnifying glass, a small scale would be seen at its base. It covers the nectariferous pit.

A newly opened buttercup reveals the anthers huddled in the center. Later they form a fringy ring about the pale green pistils, each pistil having a short, yellowish stigma. It is interesting to note that the anthers open away from the pistils to prevent self-fertilization, and also that they shed much of their pollen before the stigmas are ready to receive it. The smaller bees, butterflies, and flowerflies, wasps and beetles, too, visit the flowers, and serve as agents of cross-fertilization. Later the flowers become a tiny cluster of dry achenes.

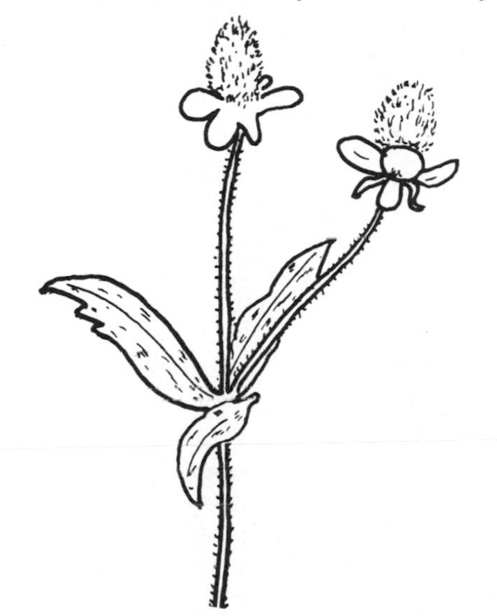

95. Leaves and Flowers of Bristly Crowfoot

There are several species of buttercups. They have tuberous or fibrous roots and simple or compound leaves that are often cut, lobed, or divided, and flowers that are prevailingly yellow. The tall buttercup is the common buttercup of fields and meadows and is the one that has already been mentioned. It is an immigrant from Europe with a hairy, branched stem two to three feet tall and deep green leaves with three to seven stemless divisions that are cleft into several narrow, pointed lobes. Both the stem and leaves contain a peculiarly acrid juice that will cause blisters if applied to the skin and cause a poison if eaten. Grazing cattle are aware of its toxic character and shun it.

Like the tall buttercup, the bulbous buttercup is also an immigrant from Europe, but now is at home in fields and roadsides. It resembles the tall species in most respects, but it is a low and generally more hairy plant and also one of the most acrid of its tribe. Much less common is the creeping buttercup that spreads by runners to form large patches in the same situations that the two preceding species are found. Its leaves are frequently white-spotted or blotched.

A woodland or hillside species, the early buttercup is a rather low plant, with fine, silky hairs on the stem and leaf and deep yellow flowers almost an inch broad. It is the first buttercup to blossom, as it appears in April. The next to appear is the swamp buttercup, confined to swamps and low, wet grounds. Its hollow stem is generally smooth, but at times may be finely haired. Its deep green leaves are divided into three leaflets—each with a distinct stem and three-lobed—and its deep yellow flowers are fully an inch broad. The species, however, is quite variable in both size and foliage.

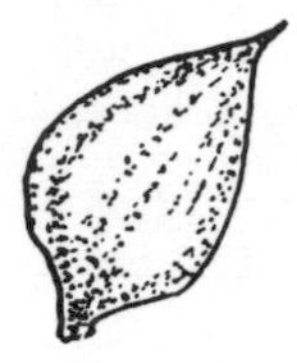

96. Achene of Buttercup

Some species of the genus *Ranunculus* are known as crowfoots, such as the bristly crowfoot, for instance, which is often and incorrectly called a buttercup. It is common in wet situations, and has a hollow, coarse, light green stem that is unusually beset with stiff hairs. The leaves are three-divided, each division three-lobed, and the flowers are thimble-shaped green heads formed of the pistils with insignificant, round, yellow petals and quite unlike the cup-shaped flowers of the buttercups.

Another species also occurring in wet places is the cursed crowfoot. It has a thick, hollow stem, sometimes two feet tall, with rather thick leaves, and small yellow flowers. Its juice is very acrid and blistering. A rather attractive species is the small-flowered crowfoot commonly found beside the woodland brook, but also growing in shady, moist ground. Its lower leaves are somewhat kidney-shaped, but the upper ones are slashed like those of the buttercup, and they are bright green and smooth. The small flowers have globular heads and reflexed or drooping, yellow petals. Also a woodland species, the hooked crowfoot is distinguished by its hooked seed vessels. The flowers are light yellow with the calyx curved backward, and the stem and leaves are olive green, the latter usually three-lobed and toothed.

In such a galaxy of flowers as the hepatica, wood anemone, early meadow rue, marsh marigold, columbine, and buttercup, only the marsh marigold and buttercup would seem to have anything in common. Yet, all are of the same family—the buttercup family, sometimes called the crowfoot family, and a family considered by botanists to be a primitive representative of the angiosperms. As that may be, the family is an important one, especially horticulturally, and the flowers mentioned above are among the most delightful flowers of the wildwood, bringing cheer and joy to those who travel the byways of the nature world.

The buttercup family is a large one, of about thirty genera and some twelve hundred species of essentially herbaceous plants, though some are low shrubs and a few are woody climbers. Most of them come from the cooler parts of the north Temperate Zone, and many of them are highly prized ornamentals.

Their leaves are alternate or opposite, though basal in many genera, usually divided or even compound, but sometimes also undivided. The flowers are commonly perfect and complete, though irregular in two genera, and spurred. They have from three to many sepals that are sometimes petallike and in some genera are hooded and irregular, three to five petals that are absent in many species, many stamens, and few or many pistils, all distinct. The fruit is a dry pod, an achene, or a berry.

It is difficult to describe one's feelings upon finding the exquisite and seemingly little hepatica among the winter snows, for normally flowers are associated with the warmth of summer.

> Blue as the heaven it gazes at,
> Startling the loiterer in the naked groves
> With unexpected beauty; for the time
> Of blossoms and green leaves is yet afar.

Yet, looking at it a bit more closely, it does not seem so delicate after all, for its leathery, green leaves wrapped in fuzzy furs appear ample protection against the rigors of winter. Delicate or not, it is a capricious plant, for its flowers vary in color—blue, lavender, purple, lilac, pink, or white—and in fragrance as well. Sometimes it is the purple flowers that are sweet-scented; other times the white; then sometimes the pink. Only by sniffing them can one find those that are fragrant. The odor is faint and is reminiscent of violets.

97. Leaves and Flowers of Hepatica

Unless one knows where to look, the hepatica will only be found by chance, blooming beneath the decaying leaves of the woodland floor, in some hidden nook, or beneath the lingering snows that have withstood the warm rays of the spring sun. When it is found, if the flowers are examined with a magnifying glass, it will be seen that what appears to be petals are, instead, colored sepals, leaflike structures that enclose and protect the other floral organs within the bud before they are fully developed. Three small, sessile leaves forming an involucre directly beneath the flowers may also be seen. They simulate a calyx and might easily be mistaken for one. The three-lobed, olive green leaves last throughout the winter. Sometimes the hepatica is also called the liverwort or liverleaf from the shape of the leaves.

Perhaps no plant is more typical of early spring than the wood anemone or windflower, whose tremulous, starlike blossoms quiver in the slightest breeze along the woodland border and on shaded hillsides. It has a slender, though tough and pliable, stem and a horizontal rootstock that firmly anchors it in the ground. Its solitary white flower with many stamens and carpels, and with many petallike sepals, is set off by a background of whorled leaves, the better to advertise its wares to the insects that visit it. The flower also nods in cloudy weather—a refinement to ensure fertilization by wind-carried pollen in case insects fail to do so. Thus, the dainty anemone is as well equipped to survive in the endless struggle for existence as the more familiar and highly successful dandelion.

The name anemone comes from the Greek word meaning wind. According to one poetic Greek tradition,

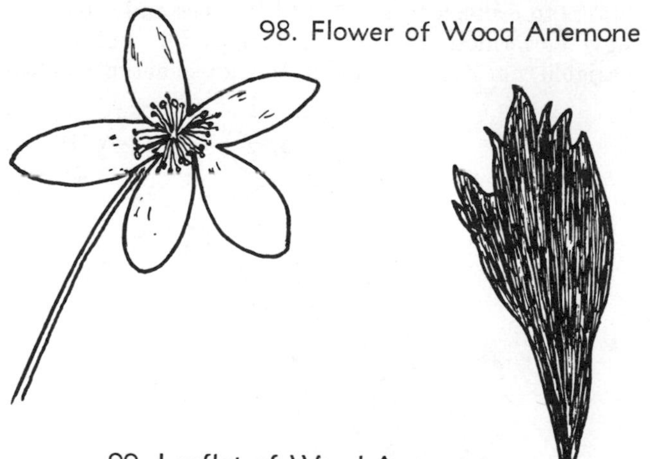

98. Flower of Wood Anemone

99. Leaflet of Wood Anemone

Anemos, the wind, used the little starlike namesakes to announce his coming in early spring, and Pliny declares that only the wind could open anemones. Bion, the Greek bucolic poet, had this to say:

> As many drops as from Adonis bled,
> So many tears the sorrowing Venus shed:
> For every drop on earth a flower there grows:
> Anemones for tears; for blood the rose.

He refers, of course, to Venus being grief-stricken over the death of her youthful lover.

It is the European anemone that is extolled by the poets. There are quite a few species of native anemones, all belonging to the genus *Anemone*—a group of perennial herbs with compound leaves or if simple, divided or dissected and mostly basal, and with usually showy flowers without petals that develop into short-beaked achenes.

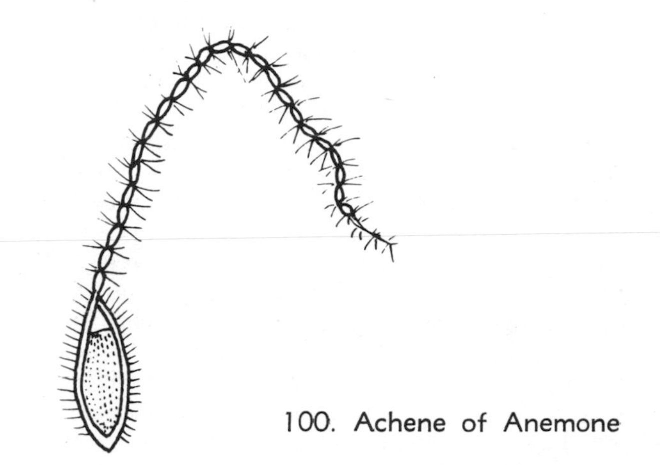

100. Achene of Anemone

In upland meadows and pastures, along the borders of woods, by fencerows, and in waste places the tall anemone is generally found. It is a tall, hairy plant, with a stem two to three feet high, deep olive green leaves, with flowers generally having five inconspicuous sepals that are white or greenish white inside and greener outside. The flower head is usually one inch or less across and is followed by an enlarged fruit head that resembles a good-sized thimble; hence, the plant is also known as the thimbleweed. Also a tall species, the long-fruited anemone is common in dry woods and by wooded roadsides. Both the stem and leaves are silky haired. The leaves are dark green, veiny, and ornamentally cut into three to five parts. The flowers are solitary, greenish white, and on a tall stem, while the fruit is a narrow, cylindrical, burlike head.

A somewhat northern species, the Canada anemone has a rather coarse stem, multibranched, with broad, sharply toothed, three-divided leaves, and a flower with five white sepals; the fruit head is globular. It is generally found in low, moist ground. The mountain anemone, which grows chiefly in the mountains of the South, is similar to the wood anemone, but with a stouter stem and trifoliate leaves.

Often found growing together with the wood anemone is a frail and delicate spring flower called the rue anemone and sometimes confused with it. It may easily be distinguished from it, however, as it bears two or three flowers in a cluster, the wood anemone having a solitary blossom. The flowers are usually white, though sometimes rosy-hued, and its deep olive green leaves in groups of three are much like those of the meadow rue. It is common in thin woodlands. Incidentally, it belongs to a different genus, but is closely related to *Anemone*.

The meadow rue that has just been mentioned is a lovely plant of soft, feathery flowers and delicate foliage, growing from three to ten feet high in wet meadows. Its stem is stout and light green, its leaves are compound, with lustreless, blue olive green leaflets, and its decorative white flower clusters are often a foot long. The flowers may be staminate, pistillate, and perfect on the same or on different plants. The staminate flowers are a decided tone of greenish white, and prettier than the pistillate flowers and delicate-scented. Bees, the smaller butterflies, and moths visit the flowers and aid in their cross-fertilization.

There are several species of meadow rues; the early and the purple are two for instance. The early meadow rue is a much smaller plant, only one to two feet high and not as showy, yet, it is a beautiful species with staminate and pistillate flowers on separate plants. The staminate flowers generally have four small, green sepals and fairly long stamens tipped with terra-cotta, and the pistillate ones are inconspicuously pale green. It is not the flowers that give this plant its charm, but its exquisite foliage that is reminiscent of the maidenhair fern. It is a plant of the woodlands. The purple meadow rue grows on the borders of wooded hills and in copses. Its flowers are white, tinged with purple, and its stem is stained with madder purple that serves to identify it.

During the summer months of July and August the fleecy white clusters of the wild clematis or virgin's bower will be found everywhere throughout the East. It is a handsome trailing vine that drapes itself over the roadside shrubbery, over the vegetation of the woodland border, or even over a fence or wall. The wild clematis, named for the Greek word meaning a climbing or slender vine, does not climb by means of tendrils, but by the leaf petioles or leaf stalks that fasten themselves about twigs and branches in a kind of sailor's knot. Its leaves are a dark green with three, coarsely toothed leaflets, and its flowers are of two kinds, staminate and pistillate, usually on different plants, but sometimes on the same one. The staminate flowers have white plumy stamens, those in the very center being pale yellow, and four greenish white sepals. The pistillate flowers have a group of carpels giving them a green center. They are visited by the bees, the beelike flies, and the brilliantly colored flowerflies. After the flowers have been fertilized, the styles begin to grow and become long, hairy tails attached to the seed vessels so that by autumn the flower clusters have become gray feathery masses that are more noticeable than they were when in blossom. They look so much like an old man's beard that the vine is also called the old man's beard. As for virgin's bower, Gerarde so named the plant because he thought it was "fit for the bower of a virgin."

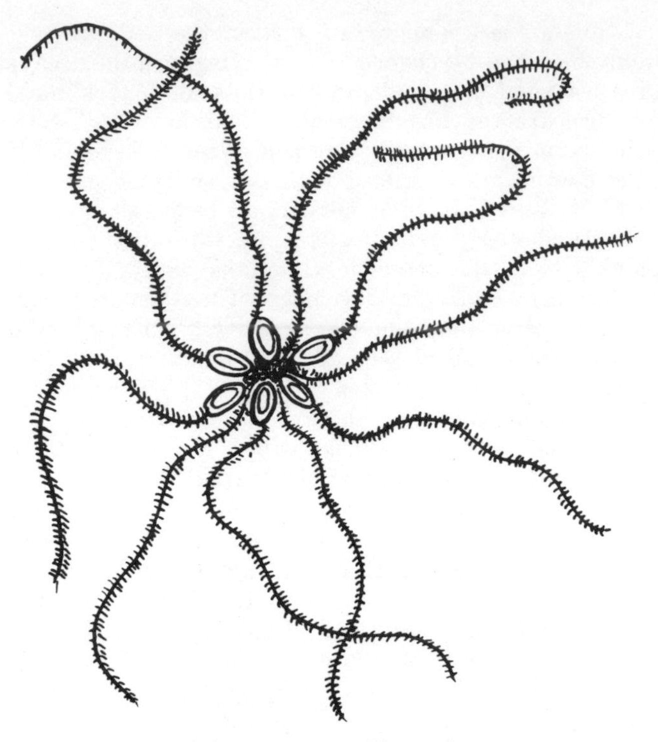

101. Fruit of Clematis

There are some three hundred species of *Clematis,* many being garden ornamentals such as the leatherleaf clematis, the Japanese clematis, the scarlet clematis, the sugarloaf, the golden clematis, and traveler's-joy.

A visit to a woodland in April, where a brook whose rushing waters from melted snows courses its way, will usually bring into view the marsh marigolds huddling on little islands and opening their golden flowers to prove a festive board for bees and the handsome flowerflies. The marsh marigold is not a true marigold, nor even a cowslip, though sometimes called

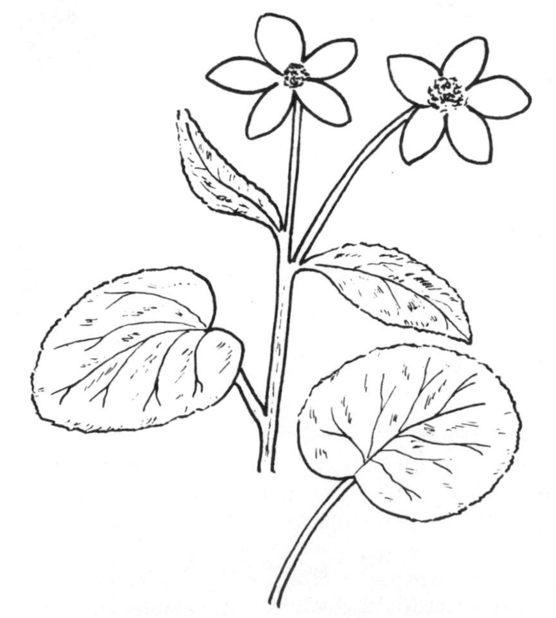

102. Leaves and Flowers of Marsh Marigold

a cowslip, but is more of a buttercup. Its scientific name, *Caltha palustris* means marsh cup, from *Caltha* meaning cup, and *palus* meaning marsh. The Indians called it Onondaga—"it blooms in the swamps." The marsh marigold is a marshy plant, at home in swamps, low meadows, riverbanks, and even ditches. Its bright yellow flowers are a welcome relief from the dull hues of the early spring landscape.

If a flower is examined closely it will be found that there are no petals, but instead, petallike sepals. It will also be seen that the anthers and stigmas mature simultaneously, and that the anthers open outwardly, and the outer ones, or those farthest from the stigmas, opening first—a nice refinement to ensure cross-pollination.

The marsh marigold is a thick and hollow-stemmed, stocky plant with round or kidney-shaped, deep green leaves that are much used as a potherb, especially in the spring or near the flowering season. They are boiled and served in the same way as spinach and many say that the marsh marigold is equal to if not superior to it. In some parts of the country the tender buds are pickled and used as a substitute for capers, but do not have the same piquancy.

When one comes upon the goldthread, a tiny woodland plant with lustrous, dark green, evergreen leaves that are three-lobed and scalloped, and a solitary white flower terminating a long slender stem, one may

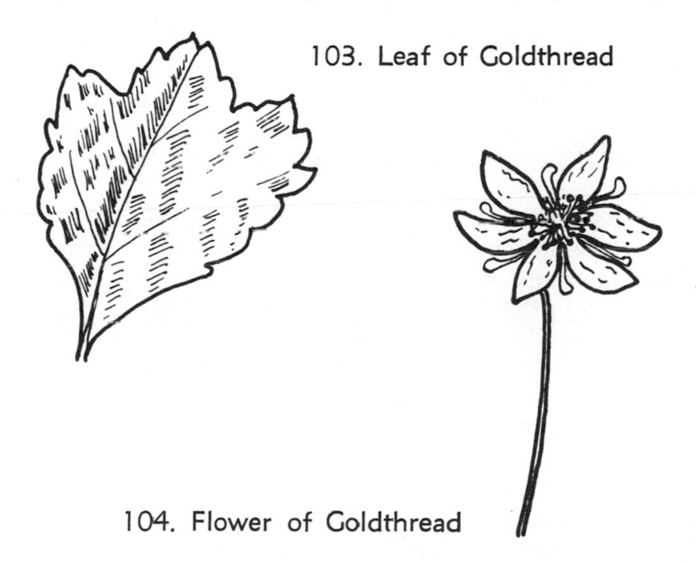

103. Leaf of Goldthread

104. Flower of Goldthread

well wonder how it got its name. If the plant is dug up, and the golden yellow threadlike roots are revealed, the reason at once becomes apparent. In the old days these extremely bitter roots furnished a tea greatly valued as a spring tonic and as a remedy for sore throat. A close look at the flower shows that it has five to seven white sepals and as many club-shaped petals that are hollow at the apex and which are actually nectaries that provide refreshment for thirsty insects. But the plant's chief benefactors are a fungus gnat and a small elongated beetle that belongs to a family known as the tumbling flower beetles.

105. Leaves and Flowers of Columbine

The cultivated columbines of gardens, though considerably larger and more showy than the wild varieties, somehow lack the appeal of those found in the woods, where their scarlet and yellow cornucopias dance in the breeze with elfin charm. Paradoxically, the columbine is a delicate though hardy plant, common on rocky hillsides and in the woodlands. It has a branching stem, one to two feet high, and long-stemmed, compound leaves, with three-lobed, light olive green leaflets. The flower is unique and interesting. The petals are five in number, red on the outside and lined with yellow within, funnel-shaped, and narrowing into long, erect, very slender hollow spurs that are rounded at the tip and united below by the five ruddy yellow sepals. Between these sepals the straight spurs ascend, and the knobs at the end are filled with nectar. There are many stamens and five pistils, the latter developing into as many long, erect pods, tipped with the slender styles. Various insects visit the flowers, but the long-tongued bees and the hummingbird are the plant's real benefactors. Its leaves incidentally serve as food for the caterpillars of the columbine dusky wing.

There are forty known species of columbines, many of them garden ornamentals. The columbine bloomed in Plymouth gardens. It was the wild columbine of English fields, long since transferred to English gardens and whose seeds were carried to America by the early settlers. Its flowers are usually blue, but sometimes purple or white. There are, however, many color forms and hybrids—some dwarf, white and double-flowered, some lilac, and one with yellow-lined leaves. This garden columbine, a heavier and less-graceful flower than our native columbine, has found a home in the woods and fields and has become a charming addition to our flora.

Few plants have contributed so much charm to gardens, woodlands, and roadsides as the larkspurs. There are over two hundred and fifty species, and all belong to the genus *Delphinium,* Latin for dolphin in allusion to the shape of the flowers. They are annual or perennial herbs with alternate leaves that are lobed or divided finger-fashion and showy flowers that generally are in a long, terminal cluster (raceme or spike) on stalks that are sometimes several feet high. The flowers are prevailingly blue, but in some horticultural forms they may be of other colors. They are irregular with five, petallike sepals, one being produced into a long spur, and with five petals. Of the latter, two are short-spurred, the others (sometimes absent) are short, small, and usually clawed. The fruit is a collection of small follicles. The larkspurs are mostly fertilized by the beelike flies, honeybees, and bumblebees. They all have a poisonous juice and with the exception of the locoweeds they cause greater losses among the cattle of the western plains than any other poisonous plant. Sheep and horses are also sometimes poisoned by them.

A woodland plant south and west of Pennsylvania, the tall larkspur is a slender and smooth species with deep green leaves and light purple or blue violet flowers. Throughout much of the same range there is also to be found the dwarf larkspur, a lower growing plant, but sometimes three feet high. Its slightly ascending spur, its three widely spreading seed vessels, and the deeply cut leaves of from five to seven divisions are the identifiable characteristics. A very beautiful species and often cultivated in gardens, the sky blue larkspur is a slender, downy species with finely cleft leaves and blue flowers occasionally varying to white. It occurs in the South and on the prairies in the North and West. An immigrant from Europe and a garden ornamental, but an escape to fields, roadsides, and waste places, the field larkspur is a lovely plant, both in leaf and flower. Its deep green leaves are deeply dissected and its flowers are one inch broad, long spurred, and vary in color from pale magenta, lilac, and purple to ultramarine blue. A more common larkspur of the garden, however, is the species known as the rocket larkspur. It has rather finely divided leaves and flowers of violet, rose, pink, blue, or white.

There are over a hundred species of monkshoods. A few are of garden interest, several species are native wildings. One species, the common monkshood, yields the drug aconite that is used as a heart sedative. All are highly poisonous, and the toxic substances are found in all parts of the plants, but especially in the roots and seeds. They are not dangerous to the touch, however; only when eaten.

The monkshoods have thick or even tuberous roots, cleft leaves, and irregular flowers that are mostly in terminal clusters (panicles or racemes). They are prevailingly blue or purple, but in some species white or yellow. They have five petallike sepals, one of them being large and hood-shaped, hence their name, and two to five petals, two of them spurlike and contained in the hood, while the others are small or absent. The fruit is a follicle.

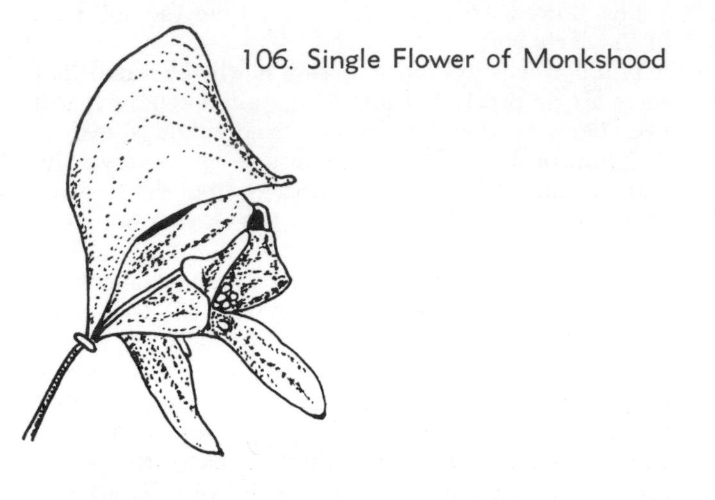
106. Single Flower of Monkshood

The common monkshood often cultivated in the garden is a beautiful and stately plant not over four feet high, with deeply cut leaves and blue flowers. In the general appearance of the leaf and flower stalk, it resembles the larkspur—tall, rather cylindrical, with a crowded raceme of flowers that seems to have wandered far from the buttercup type. The upper sepal has become a hood instead of a spur, and the two petals beneath the hood appear more like two hammers with long handles, but to the people of another day, appeared to have suggested two hidden monks or two monks' hoods.

The species generally referred to as the wild monkshood is a native perennial found in rich, moist woods, especially on the banks of small streams, from Pennsylvania to Georgia. It is a handsome wild flower with rather slender, weak stems, its delicate character not unlike that of the columbine. It has deep green leaves with three to five lobes and toothed, and blue flowers whose regular visitors are the bumblebees.

The word *cohosh* is said to be of Indian origin, though what it means does not seem quite clear. It is a word, however, that has been applied to several plants having medicinal qualities, such as the black cohosh, also known as the black snakeroot. The black cohosh is a tall spreading, slender-stemmed woodland plant with compound light green leaves, the leaflets small, cut, and toothed, and white flowers that occur in a wandlike cluster (raceme), nine to twenty-four inches long. The plant has a disagreeable odor and was once supposed to be poisonous to insects. However, it is one of the food plants of the spring azure butterfly.

The dictionary defines the word *bane* as being harmful or injurious, hence, one would suspect that the baneberry would be a plant that should be regarded with care. There are two native species: the red baneberry whose berries are considered to be poisonous, and the white baneberry. The red baneberry is a handsome woodland plant, especially when in fruit. It is rather bushy with compound three to five-parted leaves, the leaflets toothed, and lobed, and tiny, white, perfect flowers with four to ten petals, numerous stamens, and four or five petallike sepals that fall when the flowers open. The flowers are in a short, thick cluster (raceme) at the end of a two foot-high stem and are followed by coral red, oval berries.

The white baneberry is a somewhat similar plant of the woodlands with more deeply cut and pointed leaves and white berries instead of red, usually on red stalks. The berries have a conspicuous purple black spot on the end that has led to them being called "doll's eyes." Both baneberries are visited by the smaller bees, and when in fruit both are most conspicuous in the woods.

An interesting little plant of early spring, the orangeroot has a short, knotty rootstock and yellow roots that send up a single green, round rootleaf, and a simple, hairy stem that is two-leaved near the summit. From the uppermost of these two leaves there springs a single, greenish white flower with numerous stamens, and about a dozen pistils that develop into a small head of tiny red berries. It is a native of the rich woodlands of eastern United States and is cultivated in the Pacific Northwest. The rootstock and, to some extent, the leaves are used in medicine. If eaten they will produce ulcerations and catarrhal inflammation of mucous surfaces.

Writing about A.D. 70 Pliny the Elder, after having mentioned the virtues of moly, a fabulous herb of occult power and said to have been given by Hermes to Odysseus to counteract the spells of Circe, and those of dodecatheon, sacred to the Olympian gods, goes on to say:

> The plant known as Paeonia is most ancient of them all. It still retains the name of him who was the first to discover it. This plant is a preservative against the illusions practised by the Fauni in sleep. It is generally recommended to take it up at night; for if the woodpecker of Mars should perceive a person doing so it will immediately attack his eyes in defence of the plant.

The plant Pliny mentions, that is, *paeonia,* is the Latin for peony and is believed to have been named for Paean, the physician of the gods or the god of healing.

That the peony had healing qualities was a belief that endured for a long time, and during the Middle Ages it was held to have the power to drive away evil spirits, to avert tempests, and, in general, to protect houses and the people living in them from all dangers. Indeed, children in England not too long ago wore necklaces of beads, which were made of the dried portions of the roots, in the hope that they would help in dentition and prevent convulsions.

The peonies of our gardens, magnificent flowering perennials, have come from primitive forms that have been in cultivation in Europe and Asia for thousands of years, and are the result of the gardener's art. They are all erect herbs with tuberous or thickened roots, large compound leaves, and large showy flowers, usually solitary and terminal, and variously colored. Most of the garden peonies of today are limited to five species and their varieties and derivative hybrids, though other species may be found in the collections of fanciers. These are used chiefly in making new crossbreeds.

Several other members of the buttercup family that have found their way into our gardens are the globeflower, the Christmas rose, and the love-in-a-mist. The globeflower is a perennial native to northern Europe where it grows naturally in marshy places. It is a strong-growing species, up to two feet high, multibranched, with dark green leaves that are five-lobed, cut, and coarsely toothed. One to two lemon yellow, globular flowers develop at the end of the branches, which are actually globe-shaped, hence the plant's name. The globeflower looks like a big unopened buttercup, but, as a matter of fact, differs markedly from the buttercup. What appear to be petals are really sepals, the true petals being small, narrow, nectar-bearing bodies surrounding the stamens.

The Christmas rose, so named because of its very late bloom often blossoming in the open air in some parts of the country at Christmas time, is a stemless, evergreen perennial, native to the rocky and wooded mountains of various parts of Europe. It has a thick, but fibrous root, chiefly basal, long-stalked, compound leaves, and a solitary flower nearly two and a half inches wide, white or pinkish green. The beauty of the blossom lies in its enlarged sepals, the petals being curiously turned into two-lipped tubes producing nectar.

Love-in-a-mist is a hardy annual from the Mediterranean region. It is a multibranched plant, twelve to eighteen inches high, with lacelike, bright green leaves and light blue or white flowers one and a half inches across. The flower has a rather unusual appearance and looks much like a radiant star enveloped in a green mist. A related species, the fennelflower, was a prized member of the herb gardens in the sixteenth century, its seeds being used as a spice or seasoning—they still are for that matter. It is a branching plant, growing to a height of a foot, with lance-shaped leaves and a solitary blue flower.

36. Cacao (Sterculiaceae)

The Aztecs and Mayans called it "chocolatl." It was a cold, frothy drink seasoned with peppers and spices. The Spaniards added sugar to it and brought it to Europe. Later, the English added milk, and about the beginning of the seventeenth century exclusive chocolate houses, many eventually becoming famous clubs, opened in London and Amsterdam. But chocolate was an expensive beverage and only the wealthy could afford it.

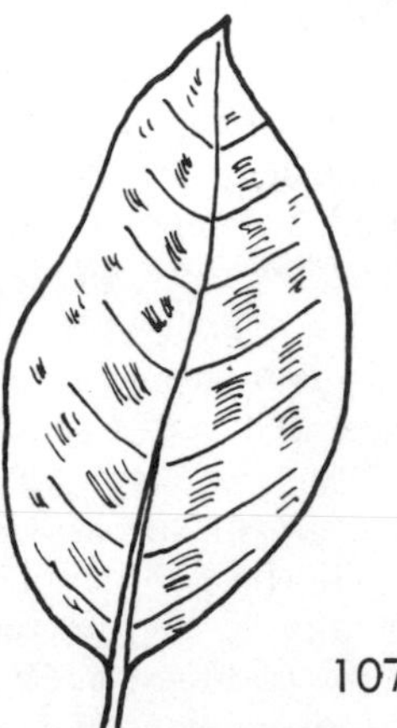

107. Leaf of Cacao Plant

Chocolate is made from the seeds of the cacao tree. The pods containing the seeds are cut off from the tree, and the seeds are removed from the pulp and placed in boxes or in piles covered with sand and allowed to ferment, bringing about the decomposition of the mucilaginous seed coats. After fermentation and a subsequent washing, the seeds are dried and roasted, then shelled and broken into pieces called "nibs." If the nibs are sent through presses and the oily cocoa butter is removed, the ground and dried product becomes cocoa. If the cocoa butter is left in and sugar is added the product is chocolate. Both chocolate and cocoa contain volatile flavoring substances as well as a drug, theobromine, which is related to caffeine. Cocoa butter is widely used in the manufacturing of confectionaries, cosmetics, ointments, and soaps. In 1876 a Swiss manufacturer introduced milk chocolate in the form of cakes for eating.

The cacao or chocolate tree (genus *Theobroma,* Greek for food of the gods, as the Aztecs and Mayans called it) is an evergreen tree, twenty to twenty-five feet high, native to tropical America, and the source of chocolate and cocoa. It has alternate, leathery, oblong leaves, small yellowish flowers borne mostly on the bark of the larger branches and the trunk, and a large, woody, ribbed, reddish brown capsule, filled with a whitish, rather evil-smelling paste embedded in which are the cacao seeds of commerce. In the tropics, the tree is widely cultivated for its seeds.

The cacao tree is one of about seven hundred and fifty species, contained in fifty genera, comprising the cacao or chocolate family, a group of almost exclusively tropical trees and shrubs. They have alternate, simple, or digitately compound leaves, and perfect or unisexual flowers that are nearly always clustered. The flowers have a five-parted tubular or bell-shaped calyx, five petals or none, five or more stamens, and a two-celled ovary. The fruit is a dry capsule, rarely berrylike.

108. Fruit (Capsule) of Cacao Plant

Another important member of the family is the cola tree that produces the kolanut or Gooranut, widely used as a stimulant in soft drinks. A native of Africa, it is a tree up to forty feet high, with alternate, simple leaves, yellowish green flowers that lack petals, and a fruit that is a collection of four to five, woody or leathery pods (follicles). It is one of about a hundred species of the genus *Cola,* the African vernacular for these trees. The inner bark of several other species furnishes tough fibers that are used for heavy mats and cordage.

The cacao family is represented in our country by two western shrubs. One is a low-growing plant of the southern California mountains, with brownish flowers, while the other is an evergreen shrub, variously called flannel bush, leatherwood, and, in California, slippery elm, although it is not an elm. It is a shrub, six to ten feet high, with alternate, lobed leaves, grayish white on the lower surface from felty hairs, showy, yellow flowers, nearly two inches wide, and solitary in the leaf axils. It is a member of the genus *Fremontia,* named after John C. Fremont who discovered it.

37. Cactus (Cactaceae)

Whether one knows the cacti only from a few specimens planted in a window terrarium garden, or knows them more intimately growing in their native home on a gravelly mesa, on a sandy desert, or in a rocky canyon of the Southwest, one has to admit that they are strikingly different from any other group of plants in the plant kingdom. Such adjectives as bizarre, fantastic, weird, and grotesque have been applied to them, but somehow do not quite do the cacti justice. They are all that the adjectives imply, but they are more. Perhaps it may be said that they are of a heroic stature, for they have elected to live on the frontier of the habitable world, where entrance is a precarious venture, in a habitat where the merciless heat of a broiling sun and the scarcity of water are inimical to any living thing, and where survival is a long shot gamble.

The cacti, however, have become adapted to such an environment. They have reduced their leaves to a minimum size or have dispensed with them altogether because the leaves with their many tiny pores would have given off precious water necessary to sustain them. So they have evolved stems to carry on the process of photosynthesis and to store the meager supply of water obtainable during the few rainy days of the year. Thus, the stems are thick and succulent, ninety-eight percent of the weight being water. Their surface, too, has been reduced to a minimum, most cacti being unbranched cylindrical or hemispherical plants. Furthermore, to offset the danger from animals that would find the thick and succulent and juicy stems good eating, they have developed modified stems in the form of spines, and have covered themselves with such an armament that would be an effective deterrent. Today, all the cacti are spiny, bristly, or thorny plants. Luther Burbank once attempted to breed out the spines that took Nature thousands of years to produce, but met with little success, for spineless cacti would be a boon to cattle men on the more arid prairie lands where only cacti can grow. Many species of cacti, we should also add, have enlarged tuberous roots for the storage of water.

The spines are interesting structures. Needle-sharp and resinous they are long lasting, in some cases, as on the trunk of the giant Sahuaros, they have remained unchanged for centuries. Moreover, they remain in the ground long after other parts of a cactus have rotted and gone back into the soil. Often they are beautifully colored and variegated in shades of yellow, orange, brown, purple, red, pink and white.

That such unnatural appearing plants should have beautiful flowers seems something of an anomaly, and yet, many of them bring forth exquisite blooms, some of the blossoms six inches in diameter and twelve inches or more in length. Cup-shaped or funnel-shaped, most of them have numerous stamens—as many as three thousand in the Sahuaro flower—that grade in color into the scarlet, purple, yellow and white of the petals. And sensitive to light, some flower only for a few hours or at most for a day, some only at night, as in the case of the night blooming cereus.

The cactus family is a large group of succulent, mostly spiny desert plants native to the dry or desert regions of the New World, especially southwestern United States where they grow in rocky mesas and mountain canyons. There are some one hundred genera and over thirteen hundred species, in habit being often fantastic, sometimes erect and treelike, sometimes climbing vines, and sometimes small and round growths in the ground. Many are cultivated for their grotesque form, their showy flowers, their sometimes edible fruits, or simply because they make desirable, if somewhat odd-looking, house plants.

They have a ridged or tubercled, continuous or jointed stem that is circular in transverse sections and that functions in the manner of leaves. True leaves, when present are small and succulent, and they soon fall except in the two genera Pereskia and Pereskiopsis. The flowers are perfect, regular and are mostly solitary, sessile, terminal or lateral, often showy, with many sepals, petals, and stamens, and a one-celled ovary. The fruit is berrylike. The cacti have little economic value except for a few that provide forage. At one time the cochineal plant had some commercial value as the cochineal insect, the source of the famous dye, fed upon it, but the dye has now been largely replaced by synthetics.

The cacti are usually thought of as plants of hot, dry deserts, but they are not exclusively so, as there are species native to Texas, New Mexico, and the prairie states that can withstand temperatures fifteen to twenty degrees below freezing. Moreover, some species grow in the high altitudes of the South American Andes, others in the grasslands of the central states where they become covered with snow in the winter, and still others in British Columbia, Alberta, and Massachusetts. There is also a number that grow in the swamps and keys of Florida, a habitat quite different from their ancestral home in the Southwest.

The old adage that there is an exception to every rule can be slightly paraphrased and applied to the cactus family, for a glance at the Barbados gooseberry, or lemon vine as it is sometimes called, would hardly lead one to take it as a member of the family, its appearance being so uncactuslike. Native to tropical America, it is a shrub or vine, sometimes a small tree, that has become naturalised in the hammocks of southern Florida, with woody, branching, spiny stems, and true, flat, broad, elliptical leaves. The flowers are fragrant, white, yellowish, or red, and occur mostly in corymbs or panicles. The fruit is a yellow berry,

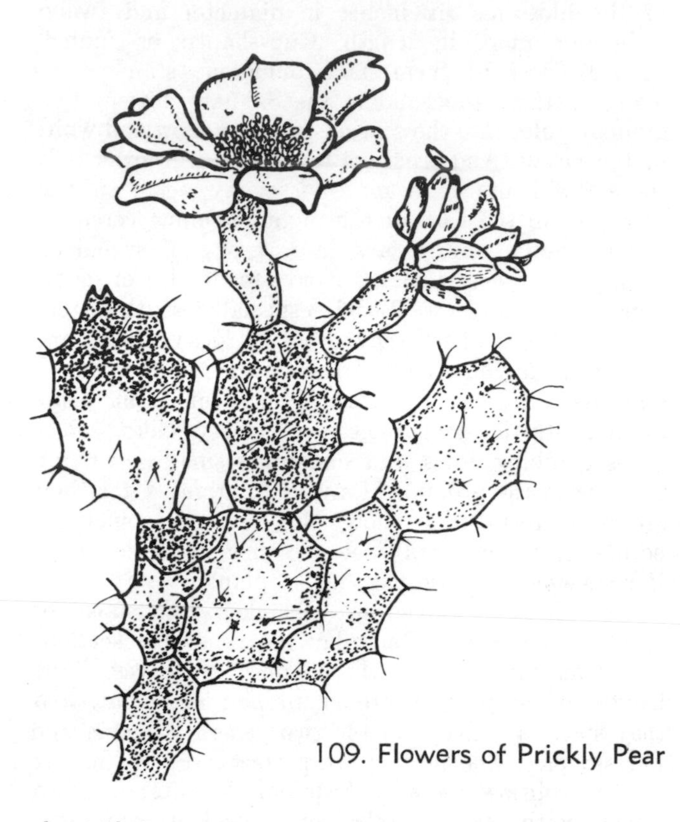

109. Flowers of Prickly Pear

about three-quarters of an inch in diameter. It looks like any other woody deciduous plant and is generally considered to be the one member of the family most closely allied to the other flowering plants. It may well be representative of what the first cacti looked like. The Barbados gooseberry belongs to the genus *Persekia,* named for N. Claude Fabry de Peiresc, a French scientist, and contains some twenty species. The Barbados gooseberry is one of two species that have long been grown for their fruit or for ornament in the tropics.

That the berries of the prickly pear are good to eat may seem surprising to anyone finding this prickly plant growing in the rocks and in sand, and yet they have a pleasant taste and may be eaten fresh or stewed. Touching the plant, however, can be a hazard, for the spiny bristles can become embedded in the skin and be quite annoying.

The common or eastern prickly pear, that is found along the coast from Massachusetts to Florida and Alabama, is a thickened, jointed, branching usually prostrate, though sometimes ascending, plant often a foot long. The joints are flat, oval or oblong, from two to five inches long, while the leaves are tiny, awl-shaped and lie appressed against the stem. The bristles are short and greenish yellow, and the spines, sharp-pointed and solitary, are grayish and nearly an inch long. Sometimes they are lacking. The flowers are yellow, two to three inches wide, with eight to ten petals and many stamens. The berries are red and pear-shaped, an inch to an inch and a half long, and are pulpy and juicy.

There are about two hundred and fifty species of other prickly pears, some of them also called cholla and tuna, of the genus *Opuntia,* probably named from a town in Greece where some species have been introduced. They are distributed from New England to the Argentine or Tierra del Fuego. They have fleshy, jointed, green stems, that take the place of leaves and that consist of either cylindrical or flattened joints known as "pads." The leaves are rudimentary and scalelike and develop in the axils of the spines. The leaves and spines, as well as the flower buds of the cacti, all appear in special areas of the stem known as "areoles," and in the prickly pears these areoles also bear delicate hairlike bristles. The flowers are usually solitary and conspicuous, beautiful in many species. They are often yellow, but are also red and white. The fruit is usually a juicy berry. The Opuntias are grouped loosely in two groups: those with flat or broad joints, some being called tuna, and those with cylindrical or roundish joints, some being known as cholla. In habit the many species vary widely. Some are prostrate or climbing plants without a trunk, others are treelike, but do not reach the height that other treelike cacti do, and these latter are mostly tropical. Besides the spines, the opuntias generally have barbed bristles on the joints and fruits that must be removed from the edible fruits or they may choke whoever eats them. As a rule, the opuntias are not cultivated in cactus gardens because they grow too rapidly and form large, ungainly shapes, although in the Southwest they are often used as hedges.

A western species of the prickly pear is the western prickly pear or devil's tongue. It is similar to the eastern species in many respects, but the green joints are more oval or nearly round, the bristles are reddish brown, and the spines are three to four in a group, one being larger than the others. The flowers are yellow, sometimes with a reddish center, and the berries are club-shaped with a purplish edible pulp. The western prickly pear occurs in dry sandy or rocky soil from Ohio, Michigan and Minnesota, south to Texas.

In the Southwest there are many species that bear edible fruit, such as the Indian fig and tuna, both of which, it is said, were in cultivation long before the discovery of America. The Indian fig is a flat-jointed cactus, bushy or treelike, and sometimes fifteen feet high. The joints are oblong and usually spineless, the flowers yellow, and the fruit pear-shaped, red, and juicy. The tuna is a treelike cactus, ten to fifteen feet high, with an obvious trunk, flat and oblong joints, one to five spines in a cluster, yellow or orange flowers, and pear-shaped red fruit. It is the chief tuna in the market.

A most valuable species is the Nopal, also called cacanapa, that ranges from Louisiana to New Mexico. It is a treelike, flat-jointed cactus, eight to twelve feet high, with a definite trunk and a branching treelike maze of bluish green joints, one to six spines in a cluster, yellow or red flowers and a purple, pear-shaped fruit. The Indians, besides eating the raw berries, make a cheese, a syrup, and a tea from them, and also cook and eat the tender young joints.

Some other members of the genus include the cane cactus or tasajillo, a bushy or treelike cactus, with a trunk two to three inches in diameter, cylindric joints, slender solitary spines, greenish yellow flowers and scarlet fruits; the tree cane cactus, a shrubby cactus with many tubercled points and purple flowers; and the jumping cholla, a deceptively beautiful plant with stout, cylindrical joints that are completely covered with a mass of spines. A blow at the base of the plant sends them off in a shower, hence the name of the plant. It has yellowish or purplish green flowers and yellow green berries. Finally there is the beavertail, one of the smallest species and an innocent-looking spineless cactus with attractive blue gray joints that, when half-grown and tender, are eaten by the Indians. Its flowers are bright rose or purple. The brownish bristles of the beavertail appear to be soft hairs, but anyone careless enough to brush against them will find that they are sharp as spines, and worse, they are able to work their way deep into the skin.

The cacti seen most often in rock gardens and indoor cactus plantings are those known as the pincushion cacti of the genus *Mamillaria*, an allusion to the nipplelike tubercles, a group of some two hundred species chiefly from the deserts of North America. Most of them are only a few inches high and even in their native habitat do not grow higher than a foot. They are spherical or hemispherical in form and have little swellings known as tubercles, that are arranged in loose spirals. There is usually a small tuft of hair, wool, or bristles in pits between them and at the tip of most tubercles there is a collection of spines. The flowers are somewhat bell-shaped, not large, but often brilliantly colored and day-blooming. The fruit is berrylike.

Some of the better-known representatives of the genus are the following: the foxtail cactus, a small spherical or cylindrical plant with distinctively marked spines, each white with a brown or red tip, and straw-colored, usually solitary flowers; the fishhook cactus, also a spherical or cylindrical plant with eight rows of tubercles, the tubercles forming symmetrical spirals, from each of which grows a compact cluster of forty or more white thorns, each cluster with one or more large, hooked, reddish brown spines, and purplish flowers; the devil's pincushion, the largest of the genus, much like a pineapple in color and appearance, with red spines and bright yellow flowers; the sunset cactus, with grayish white spines and pink or rose-colored blossoms; and the Arizona pincushion, the smallest of the genus, hardly an inch in diameter or height, with reddish brown thorns and dainty flowers, that grows among the rocks of the Grand Canyon.

The members of the genus *Echinocereus,* from the Greek for spine plus *cereus,* from the Latin for wax candle, are known as the hedgehog cacti because their fruits are covered with spines, and because the bright red fruits look like strawberries. The latter can be eaten like strawberries, hence, they are also known as the strawberry cacti. There are sixty species, all from southwestern United States and Mexico. They have ovoid, oval, or cylindrical, strongly ribbed or tubercled, sometimes jointed stems, the ribs or tubercles plentifully beset with spines and solitary flowers that are borne on the sides, but towards the top of the branches and more or less funnel-shaped or bell-shaped, the tube spiny, and day-blooming. The fruit is a pulpy, thin-skinned berry.

Common throughout the entire Southwest the Indian strawberry cactus may be found in clumps of twenty or more clinging to the rocky slopes of desert canyons as well as on the sandy lowlands. It is a cylindrical plant, rarely more than a foot high with deep pink purple flowers that bloom for a single day. The strawberrylike fruit is regarded as a delicacy by the Pima and Papago Indians. Unlike most of the cacti the flowers of the claret cup cactus remain open continuously night and day during their blooming period. The plant, native to western Texas, New Mexico, and Colorado, is low-growing with five to eight ribs, each with tubercles and radiating circles of spines, and scarlet goblet-shaped flowers that suggest its name. The rainbow cactus has greenish yellow or greenish brown flowers that are borne around the middle of the plant, which is noticeably red in color because of the reddish spines. The lady finger cactus is a prostrate plant of southern Texas, with large pink or reddish purple flowers that open at noon and close by sundown for several consecutive days. The lace cactus is a delicately spined and highly ornamental plant and one of the most common of the Texas cacti, growing on rocky, well-drained hillsides. The green-flowered pitaya, the common cactus of the plains, found from Texas north to Wyoming and Kansas, is only four to eight inches high and is usually hidden among the grasses. Its cylindrical stems are covered with vari-colored spines and its yellow green flowers appear all on one side of the stems.

Except for several species of *Opuntia* and the group known as the torch cacti, the cacti known as the barrel cacti are the largest of our native cacti. They were formerly members of the genus *Echinocactus,* a confused group of generally globular or columnar-ribbed cacti occurring from southwestern United States to Mexico. The genus once contained about a thousand species, but botanists have divided it into several genera and only nine species now remain in it. One of these barrel cacti is the niggerhead cactus of the Mojave Desert and Utah and Arizona, conspicuous for its armor of silvery gray spines banded with pink. A second species, the California barrel cactus, sometimes growing to a height of eight feet, though the average is usually a foot or two in height. It has eighteen to twenty-eight ribs that are armed with hooked spines several inches long and white, pink, red or yellow, bell-shaped flowers that form a circle at the top of the stem. And a third species, the bisnaga of central Mexico, grows six to nine feet tall with a diameter of three to four feet. It weighs well over four thousand pounds, and it is said to attain an age of a thousand years.

An interesting feature of the barrel cacti is that they are a source of drinking water, one specimen in particular, found in southern and western Arizona, being known as the traveller's friend. One needs only to cut off the top of the cylindrical stem and to crush the pulpy mass to get a refreshing drink. This particular cactus also serves as a sort of compass for it always leans to the southwest.

One of the most extraordinary members of this odd and remarkable family is the saguaro or giant cactus, also known as the suwarro and sahuaro, and sometimes known as the sage of the desert. The largest cactus in the world, it is a massive, columnar plant, twenty to sixty feet high and often two feet thick with twenty to twenty-five lengthwise ridges or ribs and corresponding furrows, and stout, strong spines. Old specimens have three to four candelabralike, huge branches that curve upward. The flowers are waxy white, about four inches long and half as wide, with thousands of stamens surrounding the pistil, and are formed only at the tips of the stem or branches. Each flower blooms only for a night, then closes the following afternoon. A glance at this giant plant makes one wonder what keeps it erect, but within it there is a woody skeleton of bamboolike rods running lengthwise beneath the surface and that form a hollow cylinder with the watery tissues within and outside.

There are about two hundred of these treelike cacti, mostly in tropical America, but a few are found in the United States, most abundantly on the rocky slopes and mesas of the Arizona mountains where they form weird forests. They grow slowly, taking fifty years to reach a height of fifteen feet. Plants thirty to forty feet high must be at least two hundred years old. A full-grown plant may weigh six tons and be two hundred years or more old.

One of the rarest of our native cacti is that known as the pipe organ cactus. It might be likened to a slender saguaro, but it branches near the ground to form a dense mass of erect cylindrical stems, ten to twenty feet high, and from six to eight inches in diameter. Each stem has a dozen or more lengthwise ribs and with their clusters of spines give the plant a sort of pleated look. The pipe organ cactus occurs in the arid rocky mesas of southern Arizona and Mexico where the flowers open to the night air. The olive green fruits are used by the Indians and Mexicans to make jellies, candies, and wines.

Also rare and usually seen only in cactus gardens as young and small plants, is the old man's beard, a columnar, unbranched plant of inaccessible rocky hillsides of northern Mexico. It often grows to a height of thirty feet and its long spines are intermingled with a tawny, gray wool that gives the plant a venerable and hoary appearance.

Most people have heard of, if not seen, the night-blooming cereus. There are many species, all formerly belonging to the genus *Cereus,* which has now been divided into a number of genera. They are greatly prized for their rare and fragrant flowers. One species, known as "La Reina de Noche" of Mexico, forms a prostrate mass of tangled stems, each with its lengthwise ribs and clusters of spines. It is a climber, covering walls, hedges, and shrubs, and is rather repulsive in appearance. One night a year, however, it makes up for its ugliness when its buds open into large creamy white or pink flowers, seven inches in diameter and twelve inches long, its stamens forming a corona that is a "symphony of pale yellow and white." One blossom will perfume the night air with its fragrance for a considerable distance.

Another species of the Southwest, known simply as the night-blooming cereus, grows in clumps, and has dull, gray green slender stems covered with very short spines. It looks very much like a mass of dead sticks. Once a year it brings forth a white flower, three inches in width and six inches in length, its fragrance falling on the desert night air and serving as a guide to its presence. Of all the many cultivated species of night-blooming cerei is the species with three angled, creeping stems that climbs over trees and walls. This night-blooming cereus has been cultivated for such a long time that its wild progenitor is unknown, though it undoubtedly came from tropical America. It was introduced into Florida during the Seminole War and is now naturalized in the hammocks and on the Keys where its awkward, angular stems twist about tree trunks like gray brown serpents, and are a sharp contrast to the waxy white flowers with their fragrant ethereal beauty.

A somewhat curious cactus is the mistletoe cactus, so-called because it resembles mistletoe. It belongs to the genus *Rhipsalis,* from the Greek for wicker-work, in allusion to the branches that usually grow mixed together. The genus contains some sixty epiphytic species scattered from Florida to the Argentine. They are leafless cacti with slender, usually ropelike, spineless stems that are provided with hairs, bristles or wool, small flowers, and a one-seeded berry. The common mistletoe cactus is a many-branched plant, with cylindric, pencil-thick stems, usually three to four feet long, but as much as ten feet in Brazil, and flowers hardly a quarter of an inch wide.

Another cactus of more than passing interest is the

Christmas cactus, a tree-perching, spineless species of Brazil. It is a many-branched, hanging plant of the genus *Zygocactus,* that is from the Greek for yoke and cactus, in allusion to the irregular flowers, with the broad, leaflike stems or joints sharply cut off at the tip, and the margins with coarse, blunt teeth. The irregular flowers are red and very showy, while the fruit is also red and pear-shaped. The Christmas cactus is a favorite house plant and can easily be grown in a potting mixture. Hence, it is a fine plant for the window sill or for a hanging basket.

38. Caltrop (Zygophyllaceae)

The caltrop family is a group of twenty genera and about one hundred and fifty species of herbs, shrubs, and trees widely distributed in warm and tropical regions. They have mostly opposite, compound leaves, the leaflets entire, perfect flowers with usually five sepals, the same number of petals or none at all, as many stamens as the petals or two to three times as many, and a four to twelve-celled ovary, and a variable kind of fruit.

A fugitive from Europe the ground burnut or land caltrop is an annual pubescent herb with a prostrate or ascending stem found in waste places around the eastern seaports. Its compound leaves contain four to eight pairs of oblong leaflets, its flowers are solitary and yellow, and its fruit is five-angled, spiny, and splitting into five segments. It belongs to the genus *Tribulus,* from the Latin meaning a kind of caltrop, a caltrop being a plant with stout spines on the fruit or flower heads.

Another species, the great caltrop of the genus *Kallstroemia,* is an annual herb of dry soil with slender, prostrate branches, compound leaves with three to five pairs of oval leaflets, yellow flowers, and a ten to twelve-lobed fruit splitting into as many seeded segments.

A characteristic feature of the sparse vegetation of the desert plains and slopes from Texas to Southern California is the creosote bush or greasewood, an evergreen, balsam-scented, many-branched and resinous shrub, five to eight feet high. It has compound leaves of two leaflets, that are oblique and ovalish, yellow, solitary, and terminal flowers, with five sepals and five petals, eight to ten stamens, and a globe-shaped, white, felty, small capsule.

The lignum vitae will be found in the Florida Keys, and is of more than passing interest, for its wood is one of the toughest and hardest known to commerce, very well adaptable to many uses. It is very close-

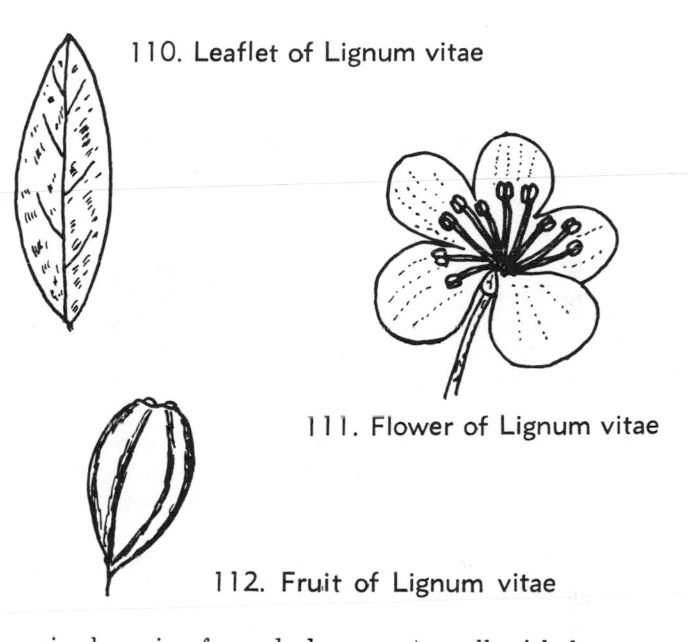

110. Leaflet of Lignum vitae

111. Flower of Lignum vitae

112. Fruit of Lignum vitae

grained, varies from dark green to yellowish brown, and has a resin content of about thirty percent of its wood weight making it suitable for many purposes where the use of other lubricants would not be practical. For this reason, the wood is ideal for ship propeller shaft bearings and for machinery used in preparing various kinds of food. It is also used for pulleys, bowling balls, casters, mallets, bushings and similar articles.

The lignum vitae is a small tree, sometimes twenty-five to thirty feet high, with a small, globelike crown of slender pendulous branches. It has opposite, pinnately compound, persistent leaves, the leaflets mostly obovate and lustrous dark green, perfect, regular, blue flowers borne in terminal clusters of two to four or rarely solitary, and a lustrous orange colored, five-angled, obovoid fleshy capsule with many black seeds, the seeds with a fleshy, scarlet outer coat.

39. Canna (Cannaceae)

The garden canna and its many varieties are known as tall, large-leaved plants with showy flowers and tuberous rootstalks. The leaves are stately, broad, often colored, toothless, but prominently veined, the flowers are not only showy, but sometimes gorgeous, in a terminal cluster. They have three greenish sepals, three petals resembling the sepals with nearly all the color coming from the much enlarged, colored, and petallike, sterile stamens (staminodes). The fruit is a three-angled, roughish capsule, surrounded by the withered calyx.

The garden canna is one of some forty species of the genus *Canna,* an old name for some reedlike plant, the only genus of the canna family, a group of some very useful and handsome tropical herbs. Most of them are nonhorticultural species.

Until recently, cannas were mostly grown for their handsome foliage, but now are chiefly grown for their gorgeous flowers that have been developed by plant breeders. Our modern cannas are all of hybrid origin and it is doubtful if any one of them can be traced directly to its wild ancestors. Even the parentage of many is entirely unknown.

113. Flower of Canna

There is only one native representative of the family—the golden canna, that is rather frequent in the marshes and swamps of the Coastal Plain from South Carolina to Florida. It has yellow flowers in showy spikes of two to four feet long.

40. Caper (Capparidaceae)

Although there are several native members of the caper family and several of garden interest, doubtless the most important member is the caper plant that is used in seasoning food. It is a spiny, somewhat straggling shrub, three to six feet high, with simple, roundish or oval leaves, and white, solitary flowers that are borne in the leaf axils. The unopened, pickled flower buds are the capers of commerce. The caper plant of the genus *Capparis,* is often grown in southern gardens (sometimes in the North where it is treated as a tender annual), but it is chiefly cultivated in southern Europe, where it is native, and where it is grown for the capers.

The caper family consists of thirty genera and four hundred and fifty species of herbs, shrubs, and trees with a watery sap and that are largely confined to the tropics. They have alternate, rarely opposite, simple or digitately compound leaves, the leaflets entire, mostly perfect flowers that are more or less irregular, solitary and axillary or in terminal racemes, with four to eight sepals, commonly four, four to eight petals that are also commonly four though sometimes none, six to many stamens, and a one-celled ovary with few to many ovules. The fruit is a capsule or berry, being a berry in the caper plant.

The caper family is represented in the United States by several species, but only one, the Rocky Mountain bee plant or pink cleome, is of any interest. The bee plant of the genus *Cleome,* is an annual herb, two to three feet high, with compound leaves of three leaflets and pink or white showy flowers. It grows on the western prairies where it is an important bee plant, especially in California where it is grown for that purpose.

Also found growing wild, but a native of tropical America and an escape from gardens is the spider flower, also of the same genus. It is an annual, bushy herb, four to five feet high, with compound leaves of five to seven leaflets, though near the top of the stem the leaves may be simple and smaller, and rose purple or white flowers two to three inches wide. The flowers have six long purple, antennaelike stamens and four clawed petals that give them a certain spidery look. The spider flower is a very popular garden annual.

41. Carpetweed (Aizoaceae)

The carpetweed, like purslane and chickweed, notorious pests of the garden, can be said to be almost domesticated because of its liking for cultivated fields and gardens. It is found not only in gardens and on lawns, but in fields, in waste places, and along the roadside. It also springs up in the crevices of city pavements and sidewalks.

It is a prostrate plant, with stems three inches to a foot long and smooth, branching in all directions from a slender root and forming circular mats. It has spatulate, entire leaves in whorls of five or six and very small axillary flowers without petals, but having a five-parted calyx, white inside and green outside, five stamens and a three-celled ovary. The fruit is an ovoid capsule containing many fine, brown, kidney-shaped seeds.

The carpetweed is one of twelve species of the genus *Mollugo*, an old Latin name for some soft plant, most being tropical excluding itself, and another species found in the southwestern states. The genus is one of some twenty genera of the carpetweed family that contains about five hundred species of widely distributed herbs most occurring in warm regions. They have alternate, opposite, or whorled leaves that are often succulent or at least thick. The flowers are small, perfect, and regular with a four to five-cleft or four to five-parted calyx, small or no petals, stamens equivalent in number to the sepals, and a three to five-celled ovary. The fruit is a capsule.

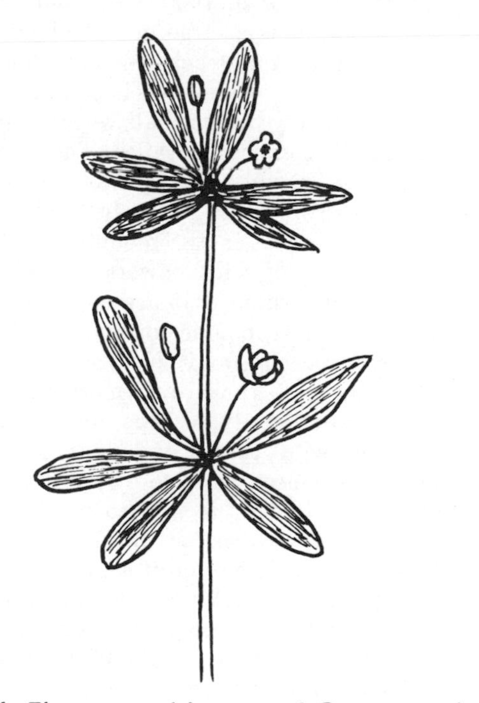

114. Flowers and Leaves of Carpetweed

Another low-growing plant is the sea purslane, of the genus *Sesuvium,* and found on the sands of the seashore from Long Island to Florida. It has obovate or spatulate entire leaves, and solitary flowers in the axils of the leaves, each of them with five sepals that are tinted purple on the inner surface. There are no petals.

The best known genus *Mesembryanthemum,* Greek for midday flower in allusion to the flowers opening in sunshine and closing at night or during cloudy weather, contains one thousand species, all but a handful, being South African. They have fleshy, three-angled or round, sometimes flattened leaves and large, showy white, red or yellow flowers that often have a superficial resemblance to a daisy. One native species may be found on the sand dunes of southern California. It is called the sea fig and has purplish red flowers. Several other species have escaped from cultivation and are common along the seacoast.

A commonly cultivated species, grown for its thick foliage that has a glistening appearance because of many small glands on its surface, is the ice plant, also called sea fig and sea marigold, especially in California. It is a prostrate plant with alternate, flat, fleshy oval leaves and white or pale pink flowers. Another cultivated species is the hottentot fig, a common border plant. It is a sprawling or prostrate perennial, with three-sided, fleshy leaves and very showy yellow flowers. In Africa its fleshy fruit is edible.

42. Carrot (Umbelliferae)

One would have to examine the lacy umbels of the wild carrot with a magnifying glass or hand lens to appreciate their delicate structure and perfection of detail—the naked eye cannot do it. What appears to the casual observer as a single cluster, is actually a number of small clusters. Notice that the small white flowers are disposed in a radiating pattern like a handmade piece of lace. In the very center of the cluster there is a tiny purple floret that is not part of any of the smaller clusters, but is set upon its own isolated stalk. No one has yet been able to account for it. Over sixty different kinds of insects may be taken from the flat-topped clusters of the wild carrot, because the flowers secrete an abundant amount of nectar that is easy to reach. Even the shortest-tongued insects can sip it in less time than it takes them to sip from the tubular florets of the Compositae, hence the plant's wide distribution since it may be found growing in fields and waste places almost everywhere. When the flowers have served their purpose, the entire cluster dries and curls up to resemble a bird's nest.

115. Flower Cluster of Wild Carrot

The characteristic flower cluster (umbel) of the wild carrot sets the tone for the carrot family, also variously called the parsnip, celery, or parsley family, and anyone may easily recognize any member by its distinctive form. In the flower cluster the stems of all the individual small flowers arise from the same place, thus presenting a somewhat flattened or perhaps slightly convex mass of flowers with gracefully arching stalks. Each flower is regular and perfect with five sepals that are sometimes reduced to small teeth or absent altogether, five petals that are separate and usually bent inwards over the center of the flower, five stamens that are alternate with the petals, and a two-celled ovary with two styles. The fruit consists

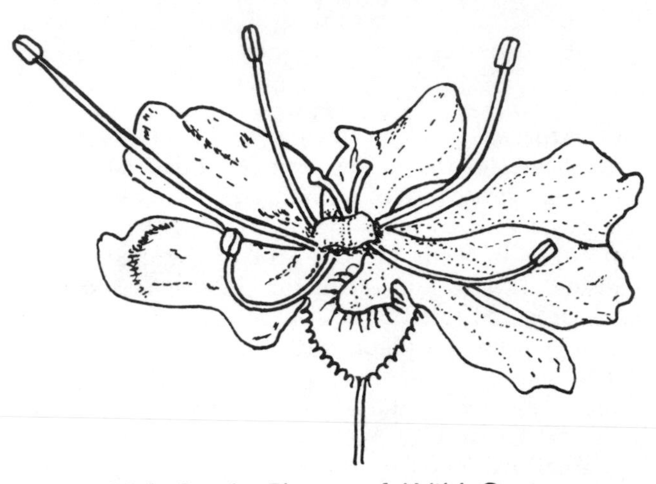

116. Single Flower of Wild Carrot

of two often ribbed or winged, one-seeded carpels that frequently contain longitudinal oil canals or tubes that give the fruit its characteristic odor and flavor.

The carrot family is a large one of about two hundred and fifty genera and between two thousand and twenty-five hundred species. They are essentially herbaceous, or rarely, somewhat woody plants with alternate, mostly compound leaves. They are of very wide distribution mainly throughout the north temperate and subtropical regions, but largely absent from the tropics except in the mountains. The family includes a number of wild flowers as well as many plants of economic value that are cultivated as herbs and vegetables. These include coriander, caraway, fennel and carrot, parsnip, and celery.

Doubtless, the best known and most familiar member of the carrot family, except for the cultivated species, is the wild carrot that has already been mentioned. It is a near relative of the garden or cultivated carrot, indeed, it is said to be the progenitor of the plant whose root we eat, but having none of its succulent sweetness. It has been said to be poisonous, but it is not, though if cows should eat it in quantity their milk is likely to become tainted with a bitter flavor and some people are likely to come down with a dermatitis if they come in contact with its leaves, especially when they are wet.

The wild carrot is an immigrant from Europe, but has long since been an escape to fields, meadows, pastures, roadsides, waste places and the like. It is a coarse, and hairy-stemmed biennial with exceedingly finely cut leaves and dull white flowers in clusters of a radiating pattern as fine as lace, hence, it is also known as Queen Anne's lace. As the fruits mature the entire flower cluster curls up and resembles a bird's nest, by which characteristic it is also known. The wild carrot is a weed, troublesome at times, but if it were less common it might find favor in the garden because of its finely cut leaves and the soft, fine appearance of the flower clusters.

The garden carrot is not dissimilar to the wild carrot, indeed it is considered a variety of the wild carrot and likely an improved form with a larger root development. It is a native of Europe and Asia where it has been cultivated for four hundred years. The view has been advanced that the cultivated carrot was brought to America, and eventually escaped from cultivation to become the wild carrot. Hence, there appears to be some doubt what carrot was the original stock.

If botanists are unable to agree as to what was the original carrot, they seem to agree that the cultivated parsnip and the wild parsnip are the same species, in other words that one is not the type species and the other merely a variety. However, the root of the wild parsnip is aromatic, mucilaginous, sweet, and slightly acrid, whereas cultivation has greatly modified the acridity and increased both the size and fleshiness of the root.

The wild parsnip is a common plant of waysides and field borders, with a tough, strongly grooved, smooth stem, dull deep green, compound leaves, pinnately dissected, and light gold, yellow flowers in small clusters on slender stems. Like the wild carrot, the wild parsnip is an immigrant from Europe, now escaped from gardens and found wild everywhere.

Pliny says that parsnips, that were cultivated beyond the Rhine in the days of Tiberius, were brought annually to Rome for the emperor's table. Just when the Romans began cultivating the plant for its thick, white, esculent root is not known, however, it has been under cultivation in Europe for a long time. Tradition has it that the wild parsnip is poisonous though this is not ordinarily so. Nevertheless, some people that come in contact with its leaves or flowers develop a skin eruption. Cattle, it is said, leave the plant alone, but the root has been fed to them in Europe for years. Caterpillars of the black swallowtail butterfly feed on the leaves of both the garden and wild parsnip, as a matter of fact also on the carrot, parsley, dill, caraway and celery. Sometimes the butterfly is called the parsnip butterfly.

The word parsnip has been applied to a number of other plants—early meadow parsnip, meadow parsnip, water parsnip, and cow parsnip—although none are parsnips, and all belong to different genera. The early meadow parsnip is a common species, its yellow umbels appearing in moist fields, meadows, and swamps in April. The meadow parsnip, sometimes called golden Alexanders, is a western species found in thickets and woodlands. It has a smooth stem, one to two feet high, with few branches. The root leaves are heart-shaped, sharply toothed with long petioles, those on the stem have three leaflets, and the golden yellow flowers are set in sparse flat-topped clusters. A variety has purple flowers. The water parsnip, as its name suggests, is a plant of shallow water and muddy banks. Its stout, grooved, branching stem, from two to six feet tall, its extremely variable pinnate leaves that may be divided into from three to six pairs of narrow and sharply toothed leaflets, and its compound umbels of small white flowers are its distinguishing features. It has been suspected of being poisonous, but its toxic principle has not been described as yet.

In contrast to some of the more dainty members of the carrot family, the cow parsnip is a coarse, vigorous species with a stem sometimes eight feet tall, thus, making it one of the largest of the umbellifers. The stem is stout, hollow, deeply ridged, sometimes two inches thick at the base and sometimes stained lightly with dull brown red. The leaves are large, dark green, in three divisions, toothed and deeply lobed, densely covered on the lower surface with a network of white, woolly hair, and the small flowers are dull white with five petals, each deeply notched and of unequal proportions. It grows in wet or moist ground and on the sides of ponds, ditches, and streams and has the reputation of being poisonous to cattle though there is little evidence that it is harmful to cattle under ordinary conditions. Its scientific name of *Heracleum lanatum* is from the Greek for Hercules, the giant, a name given to it ostensibly because of its size.

An old book on gardening, written about 1440, describes the parsley as being "much used in all sortes of meates, both boyled, roasted and fryed, stewed, etc, and being green it serveth to lay upon sundry meates. It is also shred and stopped into powdered beefe." The seeds of the plant were also added to cheese to give it a flavor. Charlemagne, according to an old story, is supposed to have eaten some cheese containing parsley seeds and liked it so well that henceforth he had two cases of the cheese sent annually to Aix-la-Chapelle.

The ancient Greeks used parsley as a decoration at funerals, laying it upon graves. This led to the saying "to be in need of parsley," in other words, to be at death's door. This seems to have given the plant a somewhat bad reputation, for Plutarch offers the story of a Greek force, when marching against the enemy, suddenly panicking upon encountering some mules laden with parsley that they regarded as an evil omen.

The parsley that we cultivate for its leaves and that is grown in market gardens is a multi-branched herb, ten to fifteen inches high, with pinnately compound leaves, the leaflets cut and cleft, and small, greenish yellow flowers in compound umbels though the plant is rarely allowed to bloom. It is occasionally found as an escape.

The hemlock parsley is not a parsley and neither is the fool's parsley, also called the false parsley. The former is a smooth species somewhat similar in appearance to the wild carrot, but with a flower cluster of far less showy flowers. It grows in cool swamps. The latter, an immigrant from Europe and found in fields and waste places, is an erect plant with a spindle-shaped root, a hollow, striated stem, dark green, finely divided, rather glossy leaves, and white flowers in compound umbels. The flowers are unpleasantly scented. The fool's parsley is poisonous. Cows have died from eating the leaves and people who have mistaken the leaves and roots for those of the true parsley have also been stricken with fatal results.

That a family, which has given us the carrot, parsnip, parsley, celery, caraway, fennel, and coriander should also contain some poisonous species, might seem rather strange. And yet, some members such as the cowbane, water hemlock, and poison hemlock in addition to the fool's parsley are toxic. The cowbane is a tall and slender species of the swamps, with large tubiferous roots, long-stemmed, deep green leaves of three to nine lance-shaped leaflets that are, however, more or less variable. Tiny, dull white flowers occur in slender clusters. Also a plant of wet places such as the borders of swamps and moist meadows, the water hemlock, also known as the spotted cowbane, may be recognized by the brown and purple streaks on the stem. It is an erect, slender, usually many-branched species with deep green, smooth leaves often tinged ruddy, twice or thrice divided, the leaflets lance-shaped and rather coarsely toothed, inconspicuous, dull white flowers in a flat, somewhat straggling cluster, and with thick, fleshy, tuberous roots, two to four inches long and bunched in a cluster at the swollen base of the stem. The plant as a whole is poisonous, but especially the roots. As their taste is pleasantly aromatic, somewhat like that of its harmless relative, the sweet cicely, they may be mistaken for those of the latter, and generally with fatal results. They are also mistaken for parsnips.

Even more dangerous is the poison hemlock, for all parts of the plant are exceedingly poisonous. Domestic animals usually die from eating the leaves in spring, though sometimes they recover. Children have been killed from mistaking its seeds for fennel or caraway, and adults have been fatally stricken from having eaten the leaves or roots thinking the leaves were those of parsley, and the roots to be parsnips.

The poison hemlock is a native of Europe and Asia, but has become naturalized in the United States, being rather common on waysides, in waste places, and about farm buildings. It is a smooth, purple-spotted, hollow-stemmed plant, with compound leaves resembling those of the parsley, tiny white flowers in large, open, compound umbels, and a white, parsniplike root. Bees and wasps appear to be the most numerous of the plant's visitors, indeed, the poison hemlock seems to attract more of these insects than any other member of the carrot family. It is believed that a decoction made from the roots of this plant was used by the ancient Athenians to put to death state prisoners, and to have furnished the "cup of death" given to Socrates.

A delicate woodland plant with fernlike leaves and white blossoms that are far more noticeable in a dim light than colored ones, the sweet cicely might easily be mistaken for the poison hemlock though the latter is not a woodland plant. When young, it might be recognized by its fine hairiness, but this characteristic is not so evident as the plant grows older. Unlike the root of the poison hemlock that it resembles in general appearance, the anise-flavored root of the sweet cicely, is edible and often eaten by children, though they should be cautioned in gathering it so that they do not dig up that of the poison hemlock instead. The seeds are linear, compressed, with bristles on the ribs to catch in clothing and in the fur of animals.

Insignificant though they may be, the tiny, pale greenish yellow flowers of the sanicle or black snakeroot are, however, of more than passing interest. At first the five petals of each floret are curiously tucked into the center of each little flower, beneath them the five stamens are imprisoned and so restrained while there is still danger of self-fertilizing the stigma. The few perfect flowers have their styles protruding from the beginning so that any incoming insect will leave its pollen, obtained from staminate florets, on the early maturing stigmas. Once cross-fertilization has been effected, the styles curve backward so that the withering stigmas are safely out of the way, thus not allowing a grain of pollen to reach them. Meanwhile, the petals unfold and release the stamens when the anthers are ready to shed their pollen. So that the plant may not be fertilized with its own pollen the anthers are sometimes held closely between the petals until all the pollen has been shed.

The sanicle or black snakeroot is a plant of rich woodlands with a smooth, light green, slightly grooved stem, hollow like most members of the carrot family, with bluish green leaves, palmately divided into from five to seven oblong, or palm-shaped, irregularly toothed leaflets, the upper leaves stemless, the lower ones with long petioles, a feature that serves in its identification. Hooked bristles and slender curved styles that extend from the tiny ovoid bur and that catch in the fur of animals are effective in bringing the sanicle to new colonizing grounds.

To look at the water pennywort is almost to doubt that it is a member of the carrot family, for in its outward appearance it fails to resemble the conventional member of this large group. It is a small, creeping marsh plant with threadlike stems, kidney-shaped

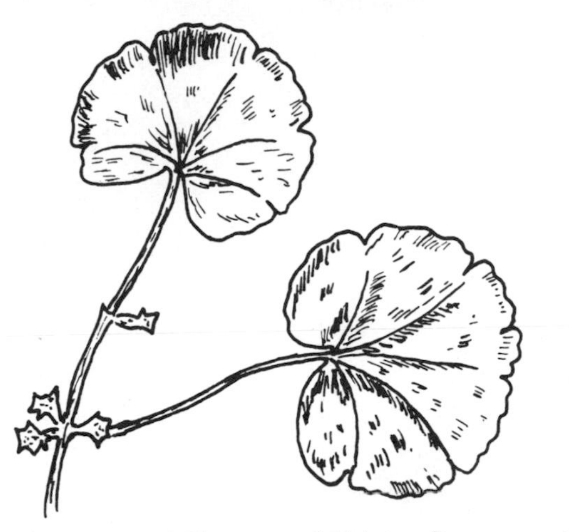

117. Leaves and Flowers of Water Pennywort

or heart-shaped, light green, smooth and shining leaves that are doubly scalloped, and tiny white flowers, one to five, clustered in the leaf axils.

In brackish marshes along the Atlantic coast may be found the mock bishop weed a dainty plant with finely dissected, fringy leaves and compound umbels two to four inches in diameter of tiny white florets.

Probably one of the least known members of the carrot family is the plant known as the harbinger-of-spring. It is a small, delicate, species, hardly eight inches high, and an inhabitant of rich woods, with thrice-compound leaves, very small white flowers in a terminal, leafy umbel, and a collection of kidney-shaped, five-ribbed carpels as fruit. It hardly lives up to its name as it blooms late in April or early May.

Widely cultivated for the border or rock garden, the sea hollies are very striking and handsome, perennial, spiny-leaved herbs. Unlike most members of the carrot family, they have simple leaves that are generally cut or lobed with usually spiny margins. The flowers are prevailingly blue in the horticultural forms, but sometimes white or green in dense, bracted, head-like clusters.

The species longest in cultivation is very likely the one called the sea holly, the blue thistle, or the star thistle. It is native to the southern Alps and has the likeness of a thistle. A rather stiff, rigid plant, it is blue in stem, leaf, and flower. A native species, known as the button snakeroot, is a plant of low ground and marshes. It has spiny, toothed, and lobed leaves and white or green flowers clustered in heads and mixed with bracts that are sometimes also colored.

Everyone is familiar with caraway seeds that are used to flavor bread, cookies, buns and other bakery products, but it is doubtful if there are many who are acquainted with the plant that produces them. Four native species of caraway are found in the United States, but the garden species wherefrom the seeds are obtained is a European plant, one to two feet high, with leaves so finely cut that they appear threadlike, and small white flowers grouped like those of the wild carrot, but far less showy. The seeds are oblong, slightly curved, plainly ribbed, and very aromatic.

What has been said about the caraway, might be repeated for the dill. Dill pickles are familiar to most people, but how many know the plant whose seeds are used in making this particular brand of cucumber pickles. It is a robust, weedy plant of a strong aromatic taste and odor, of perhaps two and a half feet high, with finely dissected, usually thrice-compound leaves, small yellow flowers in a large umbel, and broad oval, flattened seeds with sharp, threadlike dorsal ridges. At one time the dill was believed to be a defense against witchcraft, and also, having the power of working spells of the blackest magic. Such virtues have long since disappeared in favor of a more humble use. Some three hundred years ago it was said that dill, if "added to pickled cucumbers, gave the cold fruit a pretty spicie taste." The diarist John Evelyn speaks highly of "gerckens muriated with the seeds of dill," and the essayist Joseph Addison said "I am always pleased with that particular time of the year which is proper for the pickling of dill and cucumbers."

Coriander is another plant grown for its seeds that are used in seasoning food and in medicine. It is native to the Mediterranean regions and in general appearance resembles the caraway.

Of fennel that is grown for its aromatic seeds and leaves, Parkinson had this to say:

> Fennel is of great use to trim up and strowe upon fish, as also to boyl or put among fish of divers sorts, cow-cucumbers pickled and other fruits. The roots are used with Parsley rootes to be boyled in broths. The seed is much used to put in Pippin pies and divers other such baked fruits, as also into bread to give it the better relish.

Among the ancients, fennel was believed to be an antidote against all poisons and had the power of restoring sight to the blind and strength to those who were weak and infirm. Although fennel is essentially a perennial it is usually grown as an annual or biennial. It is a plant three to five feet high, its stem bluish green, with compound leaves, the leaflets many and threadlike, and small yellow flowers in compound umbels, the umbels large with fifteen to twenty-five flowers in each umbel. Its generic name of *Foeniculum* is a Latin diminutive for hay in allusion to the odor of its foliage.

Like so many other members of the carrot family, celery, long cultivated as food, is a native of Europe. It is a strong-smelling herb with compound leaves of many leaflets, the stalks (celery) channeled and sheathing, and with small white flowers in a compound umbel. For its best development, celery requires cool weather and plenty of moisture such as prevails in the commercially important growing areas along the Great Lakes. When grown in the warmer states, it is a late fall or early spring crop. A variety of the celery, known as celeriac, is a somewhat similar plant, but cultivated for its edible, thickened, turniplike root.

43. Cattail (Typhaceae)

Who is there not acquainted with the cattail? Especially in late summer when the picturesque red brown wands become conspicuous on the landscape. *It is then, my wife being partial to bouquets of all kinds, that I have to pay a visit to the nearest cattail marsh and tempt an uncertain fate for the cattails are usually not too accessible and it generally means venturing into the muck in which they grow. The wands are rather decorative and add a touch of the wild to our porch, though when they begin to break up they become rather messy.*

Like other aquatic plants, such as the reeds and bulrushes, the cattail is admirably fitted to live in a watery environment. Its roots are fine and fibrous and especially fitted to thread the mud of marshy ground. Its leaves are long, thin and strong and, being flexible, yield to the wind rather than defy it. In June and July, the tip of the cylindrical flower stalk will be covered with a fine drooping fringe of olive yellow that, when examined closely as with a magnifying glass, is seen to be a mass of crowded anthers packed with pollen. This pollen falls on the pistillate flowers below on the same flower stalk, or with every passing breeze, is showered to neighboring flowers leeward. It is virtually useless to look for the pistillate flowers for, lacking both petals and sepals and covered with down, they are hidden from view. Even with the aid of a magnifying glass they are difficult to find. It is the down, of course, that forms the familiar cattail of late summer and early fall.

The cattail family is a small one with but a single genus and a dozen species. They are reedlike plants with long, narrow erect leaves that taper to a slender tip, and minute flowers, lacking both sepals and petals, but with a perianth of small bristles that are crowded together in a dense terminal, brownish spike. Fluffy hairs attached to the tiny nutlike fruits aid in wind dispersal. They have been used in stuffing pillows and the leaves used in the manufacture of matting.

The common cattail, that grows to a height of six to eight feet, and that is found throughout the world, has light green leaves five to nine feet long and a cylindrical, brownish flower spike, about six to eight inches long, the upper half consisting of the stamens, the lower half of the pistils. The narrow-leaved cattail is a more slender species with obviously narrower leaves. It is not too different from the common cattail, from which it may generally be distinguished by its interrupted flower spike, in other words, there is a distinct and somewhat considerable separation between the staminate and pistillate flowers on the stem.

The flower head of the cattail does not appear to be a likely place for an insect wherein to dwell, and yet, females of the little cattail moth lay their eggs in the lower pistillate half of the flower spike, and when the eggs hatch the caterpillars spin silken threads to form a protective covering that also serves as a snug retreat from the hazards of winter. The silken threads, moreover, fasten the seeds in place and thus, assure the caterpillars of an adequate food supply until the time comes for them to pupate. As many as seventy-six cocoons have been taken from a single cattail spike.

118. Staminate and Pistillate Flowers of Cattail

Other insects also find the cattail an equally hospitable food plant. During the summer, plant lice suck the juices of the leaves, two species of leaf-mining moths tunnel in them, and the larvae of a snout beetle feed on the starchy core of the rootstalk. The Indians also used the rootstalk for food, grinding it into a meal, and it is said that the early settlers of Virginia were fond of the roots. Both the roots and the lower part of the stem have been used in salads and it is said that the young fruiting spikes are edible when roasted.

The cattail marsh is the home of the red-winged blackbird where we can see him perched on a swaying stem, his sable plumage shimmering in the golden sunshine, or flashing red as he streaks across the open water, and listen to his familiar song that somehow seems to strike an optimistic note in the pervading gloom of the marsh. Indeed, he is so much a part and so inseparable from the marsh that he is not often seen elsewhere. The marsh wren is also a bird of the cattail marsh, but will not often be seen unless there are those willing to brave the treacherous ooze and swarming mosquitoes. The best way is to float silently in a canoe along the marshy shore of a sluggish river.

44. Coca (Erythroxylaceae)

The coca family is a group of shrubs or trees contained in three genera with the species occurring from Mexico and Cuba to the American tropics and sparingly in Africa. They have simple, entire or toothed, alternate leaves, relatively small flowers with five petals, each with a two-lobed appendage or projections on the upper surface, ten stamens in two series, three carpels, and a fruit that is a berry or drupelike.

The only species of any interest is the coca plant, genus *Erthyroxylon* from the Greek for red wood, in allusion to the color of the wood of some species, whose leaves are the source of cocaine, a valuable anesthetic, but also a habit-forming drug. It is a shrub eight to twelve feet high with rusty brown branches, leathery leaves and small, regular, yellowish flowers that develop into a reddish drupe. The Andean Indians chew the leaves to reduce fatigue. The coca plant is grown commercially in South America, Java, and Ceylon.

45. Composite (Compositae)

It is human nature to view the common and familiar with a certain amount of indifference, and sometimes this indifference may border on the contemptuous. It seems that no one speaks well of the dandelion, probably the best known of all flowers, simply because it is a weed and the bane of every homeowner or so it seems. Were it a rare exotic everyone would extol its virtues. Yet, it is those very virtues that make it a weed, if a weed can be defined as a plant that grows where it is not wanted to grow, for the dandelion seems able to grow almost everywhere and to survive in the face of stiff competition offered by thousands of other plants. It is one of the most successful of plants, if success can be measured in terms of survival, and if examined carefully the reason for such success becomes fairly obvious.

Note how deeply the stocky root penetrates into the ground, far below where heat and drought can affect it or where nibbling rabbits, moles, and grubs can break through and feast. Watch the winds buffet and bend the stem, and though it is a hollow tube, how invincibly strong it must be, since no harm befalls it. As any engineer will confirm, a hollow tube is stronger than a solid one. Why are grazing cattle not tempted by it even though they devour other succulent plants indiscriminately? Is it because the rosettes of leaves secrete bitter juices?

Examine the golden yellow flower head with a magnifiyng glass and what is revealed? Not one flower, but often three hundred minute, perfect florets, all cooperating to ensure cross-pollination from small

119. Flower Cluster of Dandelion

bees, wasps, flies, and other insects that come seeking the nectar, secreted in each little tube, and the abundant pollen, both greatly appreciated in early spring when food is scarce. And after flowering, the golden head is transformed into a globular white, airy mass of tiny parachutes, each one a seed, and each one ready to sail away on the slightest breeze, to be carried, perhaps, untold distances before finding a resting place.

The dandelion represents a type of inflorescence known as a head. Such an inflorescence consists of a number of usually small flowers that are sessile (with-

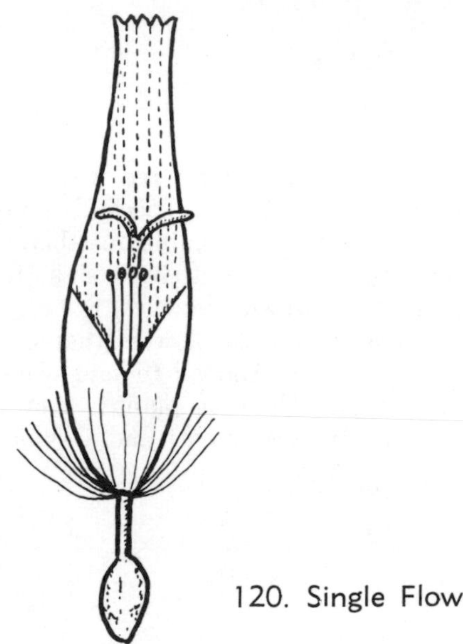

120. Single Flower of Dandelion

out a pedicel or stem), and crowded together on a flattened or convex receptacle. The bunching together of many small flowers intensifies the color and scent attractive to insects, and thus, lures them more efficiently than a number of small individual flowers would be able to do. Many plants accomplish this objective by grouping their flowers closely together in the form of a raceme, spike, umbel, corymb, cyme, and so on, but the most effective kind is the head, as found in the dandelion, sunflower, daisy, aster, chicory, thistle, and goldenrod. In many of these species there are two kinds of flowers: tubular and strap-shaped. In the dandelion they are all strap-shaped. Once upon a time, the flowers may have been five-petaled blossoms, for the five teeth at the top of the corolla and the five lines descending from them would seem to indicate that once distinct parts had been fused together to form a more showy and suitable corolla. Observe the five anthers that create a tube wherefrom the pistil extends with its two-lobed stigma. Also note that the florets may be at various stages of development. Those in the outer row of the dandelion head blossom first. After a corolla has opened, the anther tube first appears and then later the pistil that gradually rises out of the anther tube and extends above it, when the stigma lobes curl back.

"Dear common flower that grow'st beside the way
Fringing the dusty road with harmless gold."

So wrote Lowell. The common dandelion, using the word common to distinguish it from other dandelions, but of different genera, is an immigrant from Europe now common everywhere and generally regarded as a weed, but there are some who have spoken of it as

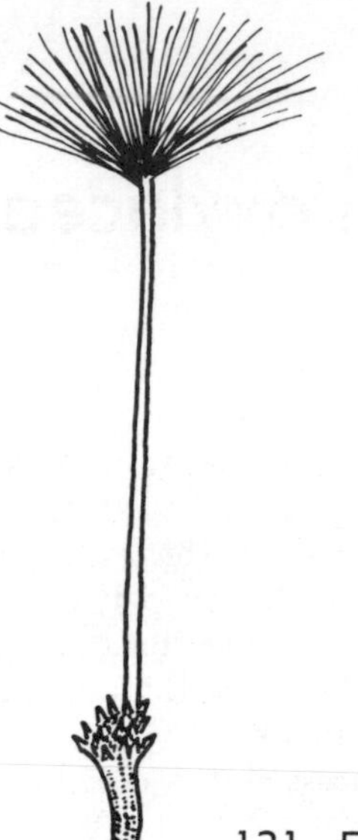

121. Fruit (Achene) of Dandelion

the "tramp with the golden crown." It has a deep taproot, a golden yellow flower head, sometimes two inches in diameter, supported on a pale green, hollow stem, the individual flowers later developing into minute, plumed achenes, and basal leaves, highly prized as a spring green, that are irregularly and angularly broad-toothed, the jagged edge suggesting the row of teeth in a lion's jaw, hence its common name that is a corruption of the French *dent-de-lion.*

The common dandelion is one of the six known species of the genus *Taraxacum,* from the Greek for disquiet or disorder, in reference to the medicinal qualities of the leaves. Another species is the red-seeded dandelion a similar, but smaller plant with flower heads barely an inch in diameter and bright terracotta seeds.

When one sees the fall dandelion for the first time, one may wonder if the common dandelion of spring has not suddenly become aberrant. But a second and closer look shows that it is not the same species. Yet, in a way, it can be looked upon as a smaller edition of the larger and more robust dandelion, for both have certain features in common, such as a yellow golden flower head and the same backward-turned, sharp-pointed lobes, or "lion's teeth." The fall dandelion, too, is an immigrant from Europe and is found in fields and along roadsides where it begins to blossom in July, in spite of its name.

Also called a dandelion, but again not a true dandelion, for it, too, like the fall dandelion belongs to a different genus, is the dwarf dandelion. It is a small plant, two to twelve inches tall, with many long, slender flower stalks, rising from a circle of small, irregularly lobed leaves, each with a golden yellow flower head. It will be found in moist meadows, woodlands, and shady rocky places where it appears in May.

The typical flower head of the dandelions gives its character to the composite family, *Compositae,* from the Latin compositus, meaning made up of parts, and that has variously been called the thistle, aster, daisy, or goldenrod family. It is the largest of all the plant families, so large as a matter of fact, that it has been divided into several families, but the author has included all such families in one. The *Compositae* consists of one thousand genera and between fifteen to twenty-three thousand species of mostly annual and perennial herbs, with a few shrubs and low trees, of worldwide distribution. They occur in the deserts and in water and in every habitat gradation between these two extremes. They are equally at home in the tropics as in the frigid Arctic alpine regions. Included in this family are such familiar plants as the dandelion, aster, daisy, thistle, sunflower, goldenrod, marigold, cosmos, zinnia, chrysanthemum, dahlia and a host of others, many familiar, many unfamiliar.

The leaves of the many members of this large family vary considerably. Though they are generally alternate, they are often opposite or whorled, frequently crowded in basal rosettes. The flowers (florets) are small, in few to many-flowered heads, and are closely grouped on a common disklike, dome-shaped, conical or cylindrical axis. The axis is in turn surrounded by an involucre composed of few to many, more or less, leafy, or scaly bracts. The end result is a flower cluster that resembles a single flower wherein the involucre is the calyx and the massed flowers the corolla, a condition or structure that is popularly regarded as a single flower. Each flower may have a scalelike or bristly structure that may be regarded as the floral bract, in which case the receptacle is said to be "scaly," "chaffy," "bristly," etc., or such a structure may be wanting when the receptacle is said to be "naked."

The composite flower cluster or head may be one of three types. In one form there is a fringe of several to many, radiating, irregular flowers, called ray flowers, on the margin of the common receptacle just inside the involucral bracts, the remaining flowers are regular and are called disk flowers. In the second form, all of the flowers are irregular and there is no distinction between ray flowers and disk flowers. In the third form, all the flowers are regular, the ray flowers being either female or sterile, the disk flowers usually being perfect though sometimes staminate or pistillate. The calyx is a ring, or crown, or a fringe of scaly, bristly, or hairy appendages on the rim of the floral axis that often persist on the ripe fruit. The corolla may be either united five-lobed or five-divided, regular in the disk flowers, or irregular, two-lipped, or strap-shaped in the ray flowers. The stamens are five in number inserted on the corolla tube, the ovary has two carpels, the style is slender, mostly two-lobed or two-parted in the perfect flowers, and the fruit is a one-seeded achene varying greatly in form, with the persistent calyx as a crown or tuft of pappus scales, bristles, or hairs that greatly aid in the dispersal of the species.

In the dandelion, all the florets that compose the flower head are perfect and strap-shaped, in the aster, the disk flowers are perfect and the outer ray flowers

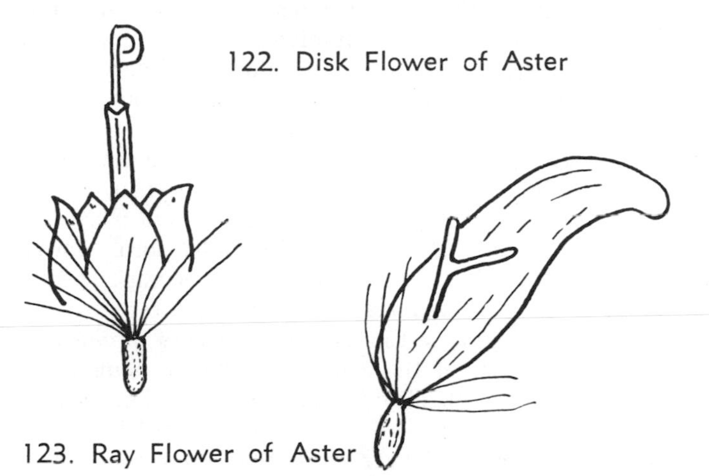

122. Disk Flower of Aster

123. Ray Flower of Aster

are pistillate. There is said to be two hundred and fifty species of asters, but few know them all. They are a common feature of the autumn landscape—indeed, they are more common in North America than in Europe or Asia—with a variety of color in their ray flowers. They are usually bushy plants with alternate leaves and flower heads that may either be solitary or in clusters. The latter consist of a hemispherical or bell-shaped involucre that is covered with bracts, a marginal circle of ray flowers, that may be blue, purple, pink, or white, and a central area of disk flowers that have tubular corollas with five short teeth. The disk flowers are usually yellow, but often darken in color with age to a purple or reddish brown. The asters belong to the genus of the same name, that is *Aster,* Latin for star, hence, the asters are often known as starworts.

Of the many species of asters that are found growing in the wild, the following are probably the ones that come mostly to our attention and the ones we are best acquainted with. The large-leaved aster, so-called because of its broad, heart-shaped leaves is a stout, stiff, purplish-stemmed species of open woods and thickets with pale lavender or violet flower heads of about sixteen rays each. A somewhat bushy species is the heart-leaved aster, with, as its name implies, heart-shaped leaves but that, however, are somewhat variable. It is a familiar small-flowered aster, whose masses of lavender or lilac flower heads appear as a mist suspended above the ground in woods and shady roadsides.

An aster that is easy to recognize is the wavy-leaved aster whose leaves have broad, winged stems. They are heart-shaped where they clasp the stem and have wavy margins. The flowers, that are pale blue to violet, often grow along one side of the axis as well as in the usual raceme. This aster may be found in dry places and on shaded roadsides. One of the earlier asters, coming into bloom before the middle of August, the spreading aster is a common species on dry ground and in open places. It has a hairy stem about two feet high, thick, somewhat rigid, oblong leaves, and flowers with twenty to thirty light violet purple rays. A handsome species, the smooth or blue aster has light green, lance-shaped leaves and sky blue or violet flower heads that are common along woodland borders and roadsides during September and October.

Probably the loveliest of all the asters is the showy aster. It has a stiff, usually unbranched stem, one to two feet high, olive green bladelike or narrowly oblong leaves, and handsome, bright, violet purple flower heads as broad as a half dollar. Unfortunately this aster is found only along the coast. Despite its rather local name the New England aster occurs throughout the eastern half of the country. It has a stout, rigid stem, bristly with stiff hairs, long, bladelike, softly haired, toothless leaves, and numerous beautiful violet or magenta purple flower heads that shine with royal splendor as much as six feet above the ground in swamps, moist fields, and roadsides.

One of the tallest of all the asters is the white-flowered panicled white aster. It is bushy and coarse-stemmed, with dark green, lance-shaped leaves, and flowers, slightly larger than a "nickel" in loose or scattered clusters. This aster is rather common on low moist ground and in thickets. One aster everyone seems to know is the bushy, little, white heath aster or Michaelmas daisy that is found throughout the eastern United States, growing in fields and on roadsides. It has a smooth, many-branched stem, tiny, heathlike, linear, light green leaves, and tiny white flowers with yellow disks that appear like miniature daisies. Another tiny-flowered species, with hairy, often brownish stems, and small, linear leaves, is the many-flowered aster whose dense flower clusters are crowded with white or lilac white flowers. It is interesting to note that the wavy-leaved aster serves as a food plant for the caterpillars of the tawny crescent butterfly, the New England aster as a food plant for the caterpillars of the pearl crescent butterfly, and various asters as food plants for the caterpillars of the silvery checkerspot. Also, some birds eat the seeds and some mammals feed on the leaves.

Although called asters, but not true asters that are never yellow, the golden asters are low herbs with woolly or hairy leaves, and rather large heads of yellow ray and disk flowers that are generally in small clusters at the ends of the branches. A common species, known as the golden star, and found on dry, sandy roadsides near the coast, grows twelve to twenty inches high and has gray green, lance-shaped, stemless leaves and golden yellow flower heads that are nearly an inch in diameter. From New Jersey southward and westward in the same kind of situations may be found the grass-leaved golden aster, so named because its leaves are linear and grasslike, and in sandy soil along the coast from Massachusetts to New Jersey one may find the ground gold flower, a much lower species than the other two, with woolly stems and small linear leaves, gray green, and crowded together.

When the first goldenrod blossoms in July, one may be sure that autumn is on its way. Though it is still summer and there are a few more weeks to go, the goldenrod, like the aster, is typically a flower of the autumn landscape. There are not as many species of goldenrods as there are of asters, some hundred and forty species or so. They are coarse, somewhat weedy plants, often branched or arching, with alternate, usually toothed leaves and small, yellow flower heads that are very numerous in crowded or bushy, often elongated and plumelike, clusters. Detach a flowering stem and look at it with a magnifying or reading glass and one will be surprised at what one sees—a row of tiny goblets. It is said that no flower attracts as many insects as the goldenrod. The author does not know how true this is, but countless insects do visit the blossoms. Among the more frequent are the locust borer, a beautiful black beetle with numerous wavy yellow bands, and the blister beetle, a black beetle that is frequently found in such numbers that the golden plumes appear as if sprinkled with soot. All goldenrods, by the way, belong to the genus *Solidago*, Latin meaning to strengthen or draw together in reference to the imputed medicinal properties that they are supposed to have.

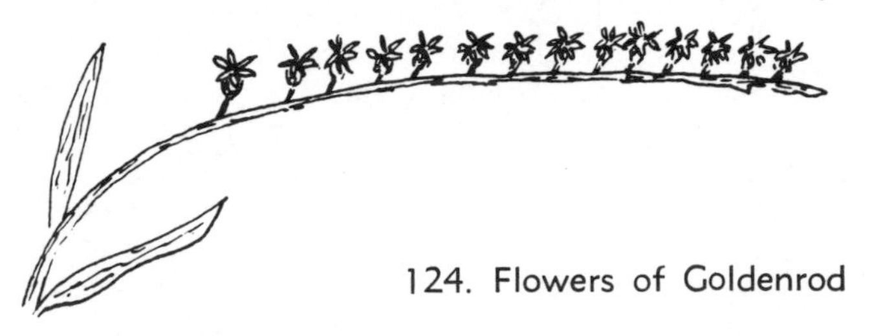

124. Flowers of Goldenrod

To the untrained eye the goldenrods may all look alike, but many of the more common species have perceptible differences that serve to identify them. Thus, for instance, the blue-stemmed goldenrod has a distinct bluish or purplish, plumlike bloom on the stem. It is a late blooming woodland species, common on woodland borders and shaded banks, lance-shaped, feather-veined leaves and small, oblong clusters of light golden flowers set in the leaf axils. In somewhat the same manner the broad-leaved goldenrod may be recognized by its zig-zag stem that grows as if waveringly not sure what direction to take. This species, of woodlands and thicket borders, has broad, ovate, saw-edged leaves pointed at both ends and small clusters of light golden yellow flowers.

One of the easiest goldenrods to identify, especially when in flower, is the white goldenrod or silverrod, for it is our only goldenrod with white flowers. They are more properly cream white, in mignonettelike clusters from the upper axils of a usually simple and hairy gray stem six inches to four feet high. It is a common species growing on dry barren ground. A species of peat bogs and swamps, the bog goldenrod has a wand-like stem with short appressed branches, lance-shaped leaves that gradually increase in size, the lowest as much as nine inches long, and light golden yellow flowers crowded on the stem.

One of the more attractive of our goldenrods is the showy goldenrod. It is a handsome stocky plant with a stout stem that rises from three to seven feet high and that often assumes reddish tints. The saw-toothed leaves are smooth, firm and broadly oval, and the golden flower heads occur in dense, somewhat pyramidal clusters. It grows in dry open woods and thickets. Crush the dotted bright green, toothless leaves of the sweet goldenrod and you will recognize the plant as they give off a pleasant anise scent. It is a goldenrod of dry sandy soil with a slender, simple, smooth stem crowned with a graceful panicle, whose branches have the florets all on one side. The rough-stemmed goldenrod has an exceptionally hairy stem and leaves. It is a perversely variable species of many forms, often branched like an elm at the top, sometimes only a foot high, at other times seven feet tall. A common species, it grows on wooded roadsides and on the margins of fields.

Doubtless, the most common of all the goldenrods, the Canada goldenrod is the familiar goldenrod of thickets, roadsides, and copse borders where its large, spreading, densely flowered, plumelike panicle crowns a rough, hairy stem sometimes as much as eight feet tall, but at other times not much more than two feet high. And if the Canada goldenrod is the most common of the goldenrod tribe, the gray goldenrod is without a doubt the most brilliant. A rather low, late-flowering species it is remarkable for its rich, deep golden yellow flowers that admirably contrast with the ashy gray stem, often covered with cottony hairs, and hoary, grayish green leaves. It is common everywhere, except at the seaside, in dry pastures and beside sandy roads, but rarely in the woods.

Moore's famous lines:

"As the sunflower turns on her god, when he sets,
The same look which she turn'd when he rose,"

have been seriously questioned, and generally have been regarded as a poet's fancy. Yet, it has been said that the head of the common sunflower does, to a certain extent, change its direction from east to west with the sun. The common sunflower is the sunflower of our gardens, where it sometimes grows as high as ten feet with flower heads a foot or so in diameter, though in the wild it commonly grows from three to six feet high with flower heads ranging from three to six inches in diameter, with the disk florets being brown, purplish brown, or purple, the ray flowers bright yellow.

The common sunflower, the state flower of Kansas, is said to be a native of the western states, from Minnesota to Idaho and south to Texas and California, and an escape in the East. It is also said, however, to be a native of South America, of Mexico and Peru, because the Spanish conquistadors found it used there as a mystic and sacred symbol, in much the same manner as the Egyptians used the lotus in their sculpture. However that may be, when Champlain and Segur visited the Indians of Lake Huron's eastern shore about three hundred years ago, they found them cultivating the plant—its stem provided them with a textile fiber, its leaves fodder, its flowers a yellow dye, and its seeds food and a hair oil. Lewis and Clark relate that when they were along the Missouri River in western Montana:

> along the bottoms, which have a covering of high grass, we observe the sunflower blooming in great abundance. The Indians of the Missouri, more especially those who do not cultivate maize, make great use of the seed of this plant for bread, or in thickening their soup. They first parch and then pound it between two stones, until it is reduced to a fine meal. Sometimes they add a portion of water, and drink it thus diluted; at other times they add a sufficient proportion of marrow-grease to reduce it to the consistency of common dough and eat it in that manner. This last composition we preferred to all the rest, and thought it at that time a very palatable dish.

The sunflower seems to be more appreciated in some countries of Europe than in our own, where it is cultivated for its flowers yielding a fine yellow dye and for its seeds yielding an oil used for cooking, burning, or for soapmaking. The oil cake makes an excellent food for cattle. In some places the seeds are roasted and used as a substitute for coffee, and in Russia, the seeds are not only used for feeding poultry, but are also ground into meal that is used in baking bread and cakes. In addition the various species of sunflowers serve as food plants for the caterpillars of the silvery checkerspot, Scudder's patched butterfly, and the painted lady.

There are about one hundred American species of sunflowers that are rather coarse, hardy, annual or perennial herbs belonging to the genus *Helianthus*, from the Greek for sun and flower. They vary in size and character, as they readily hybridize in their natural surroundings. The perennial species have thick, woody rootstocks, that may either be compact or spreading. Their leaves are, for the most part, alternate, though sometimes opposite, with the margins coarsely toothed, and their flowers occur in terminal heads, the disk florets yellow, brown, or purple, the ray florets yellow.

One of the more common species of sunflowers is the tall sunflower or giant sunflower, a species that grows in swamps and on the borders of wet meadows.

It has a bristly haired, often branching stem that reaches a height of twelve feet, often reddish, rough, firm, lance-shaped, saw-toothed leaves, and flower heads two inches broad, light yellow, the disk florets perfect, the ray flowers neutral, that is, without stamens or pistil. Also liking a somewhat moist habitat, the ten-petaled sunflower is found on the borders of copses and low, damp woods. It is a somewhat showy species with yellow flower heads two to three inches in diameter, usually with ten rays, but actually any number from eight to fifteen, a tall slender stem, and lance-shaped, deep green leaves, the lower ones opposite on the stem, those towards the summit, alternate.

In dry woodlands and in roadside thickets one will find the woodland sunflower, a slender, smooth-stemmed species with opposite, sessile, lance-shaped, toothed leaves, and flower heads about two inches across, with eight to fifteen rays around a yellow disk. In an old book, printed in 1649, and called *A Perfect Description of Virginia,* one reads that the early settlers had "rootes of several kinds, Potatoes, Sparagus, Carrets and Hartichokes," the last named having long been in cultivation by the Indians who doubtless introduced what is now known as the Jerusalem artichoke to them. As early as 1617 the artichoke appeared in Europe and some twelve years later Parkinson writes that the roots had become very plentiful and cheap in London. It appears to have been cultivated in Italy also, where it was known as *Girasole Articocco* (sunflower artichoke), *girasole* eventually becoming corrupted into Jerusalem.

Doubtless it was the Jerusalem artichoke that Lewis and Clark describe in their journal under date of 9 April 1805 as they travelled in what is now North Dakota:

> When we stopped for dinner the squaw (Sacajawea) went out, and after penetrating with a sharp stick the holes of the mice near some driftwood, brought to us a quantity of wild artichokes, which the mice collect (it seems likely that the *mice* were prairie dogs) and hoard in large numbers. The root is white, of an ovate form, from one to three inches long, and generally the size of a man's finger, and two, four and sometimes six roots are attached to a single stalk. Its flavor as well as the stalk which issues from it resemble those of the Jerusalem artichoke, except that the latter is much larger.

The Jerusalem artichoke was once extensively cultivated in this country until the potato came into favor when its use declined, though it is still widely grown. The true artichoke, a native of Southern Europe, is quite different from the Jerusalem artichoke.

The Jerusalem artichoke is found growing wild in meadows, swamps, fields, roadsides, waste places and along fencerows, and when found growing along fencerows, it can safely be assumed that the spot was once a colonial farm. It is a strong-growing perennial, with a stout and roughly haired stem, ovate, lance-shaped leaves covered with stiff hairs above, soft hairs below, and yellow flower heads, often as much as three inches or more across, with twelve to twenty rays.

There was once an Indian in New England named Joe Pye, who is said to have performed many miraculous cures, including typhus fever, with a decoction made from a plant now known as Joe-Pye Weed. It is a rank-growing herb, seven to nine feet high, with a stout stem whereupon are grouped, in circles at intervals, oblong or oval, coarsely toothed, pointed, light green leaves. The flower heads are small in rather long, round, multi-branched and showy terminal clusters, pinkish purple, and the florets are all tubular and perfect. The slight fragrance of the flowers appears to be especially attractive to butterflies, for they visit the flowers in numbers, though the long-tongued bees and flies are also visitors. The Joe-Pye weed is a familiar plant on the borders of swamps, damp meadows, moist woods, sides of streams and ditches where it may be seen flowering during August and September.

Growing in the same situations, the boneset spreads its soft leaden white bloom about the same time as the Joe-Pye weed. Butterflies do not seem to favor this plant, perhaps preferring deeper colors, and generally leave it alone. Beetles, who do not seem to care much for the flowers of the Joe-Pye weed, find the dull, somewhat odorous flowers to their liking, however, and crowd about the clusters. Flies, bees, and wasps, too, seek the nectar in the tiny florets whereof there may be ten to sixteen in a single head.

The boneset, a near relative of the Joe-Pye weed, is a coarse, stout herb, three to six feet high, with light green, pointed leaves that are so closely joined that they appear as one perforated by the plant stem that is remarkably hairy. The flowers are white in terminal clusters. An old-fashioned illness, called break-bone fever, probably the grippe of today, was often treated with a decoction known as boneset tea made from the leaves. A hillside species, known as the upland boneset and found in woods and on wooded banks, is a somewhat similar species. Unlike either of these two species, the climbing boneset is a trailing vine that straggles over bushes in swamps, brookside thickets, and moist, shady roadsides. It shows its kinship to the boneset when it comes into flower, though its opposite leaves, pointed at the tip and heart-shaped at the base, are quite different from its relative.

The Joe-Pye weed and the bonesets belong to the genus *Eupatorium,* named for Eupator, King Mithridates VI of Pontus, who is supposed to have used some member of the genus for healing, hence, the plants are often called eupatoriums. A most attractive and graceful member of this group of generally coarse plants is the white snakeroot, also called the Indian sanicle. It is a branching herb, two to four feet high, with deep green leaves that are smooth or nearly so, long-stemmed, and nearly heart-shaped, and white flower heads in a loosely branched cluster. It grows in rich, moist woods, thickets along streams, and by shady roadsides.

The tall blazing star, that is found blooming in August and September in fields and by the roadsides, may reach a height of six feet, but usually does not grow more than half that height. It is a hairy-stemmed plant with deep green, hoary leaves and flower heads an inch wide set in a showy spike, the florets magenta purple to pale violet, tubular, and perfect. A lower species, appearing in June, is also a hairy-stemmed plant, with narrow leaves and flower heads an inch and a half long, the florets tubular and a bright purple or rose purple. Unlike these two species, a third species chooses moist, low ground wherein to grow. It is perhaps the most common of the three and often blooms into October. It resembles its relative in general manner of growth, with a smooth, or somewhat hairy stem, linear leaves, and a closely set flower spike, sometimes fourteen inches long, the florets bright rose purple, but occasionally becoming white. All three species are often called button snakeroot and all attract the long-tongued bees, flies, and butterflies.

Doubtless called ironweed because of its coarse character that makes it an undesirable constituent of hay, the tall ironweed is a showy plant that vies with the Joe-Pye weed in making the low meadows bright with autumnal color. Seeing it for the first time it might be mistaken for an aster, but once having seen the yellow disk in the center of the aster it will hardly be confused with the ironweed's thistlelike head of ray florets. The tall ironweed, a common plant of fields and meadows, is a smooth-stemmed species with lance-shaped, toothed, deep green leaves, and a terminal cluster of reddish purple, thistlelike flowerheads that somehow reminds one of bachelor's buttons without petals. Bees and flies find refreshment in each tiny, tubular floret, but butterflies appear to be the most abundant guests. The New York ironweed is a similar species, but differing in its slightly rough stem, longer leaves, and acute, bristle-tipped, brown purple scales of the flower heads.

Upon seeing an aster, usually thought of as an autumn flower, blooming in May, one may wonder if nature has suddenly become capricious and has moved the calendar ahead. But, the Robin's plantain is not an aster, though it looks like one. It is one of the first in the long list of composites to appear. It is a rather large-flowered plant, with fingerlike leaves in a tuft about the root and showy flowers that are daisylike, the disk florets greenish yellow, the ray florets bluish violet or lilac. The Robin's plantain is often communistic, and when the flowers begin to open in field and meadow and along the roadside, they tint their surroundings with a delicate hue. And bees and butterflies appear to gather the pollen and nectar in exchange for providing a worthwhile service. A similar, but somewhat taller plant, known as the common fleabane, has a soft, hairy stem and reddish purple or pink florets. Both species are common everywhere and both belong to the genus *Erigeron,* from the Greek meaning old man and spring, and probably suggested by the hoariness of some spring species.

125. Flower Cluster of New York Ironweed

The members of the genus are known as fleabanes, presumably because they are supposed to drive away fleas when dried and reduced to a powder. Another common species is the daisy fleabane, found in fields, meadows, roadsides, and waste places. It has an erect stem, two to five feet tall, sparsely covered with spreading hairs, coarsely toothed, lance-shaped leaves,

126. Flower Cluster of Daisy Fleabane

and white or pale lilac flower heads with a green yellow disk. There are several other species that are all much alike including the Canada fleabane wherefrom the volatile oil of fleabane was once distilled and used in making a mosquito repellent.

On visiting a field or rocky pasture in early spring, one will see the pussytoes unfold their little clustered heads, tufts of silver white silk on stems rising from charming rosettes that can be found throughout the winter. There are many species of pussytoes that differ only in minor details. They are all perennial herbs

127. Flower Cluster of Pussytoes

with mostly basal leaves and dirty white, minute, tubular flowers in small heads in clusters. The pussytoes, also known as everlastings, belong to the genus *Antennaria,* derived from antennae because the pappus resembles the antennae of some insects.

At first glance, the pearly everlasting may not seem a composite, for the flowers look like miniature pond lilies, the tiny, petallike scales surrounding the central staminate flowers being arranged not unlike the petals of the water plants, but a composite it is, as may be seen by a study of the flowers.

The pearly everlasting is a densely woolly plant that is found in old fields and along the roadsides. The stems, as well as the leaves, are profusely covered with hairs, and particularly the leaves have such a dense layer that all the veins except the midrib are hidden from view. Nonetheless, it is a rather beautiful plant with leaves sage green above and white beneath, its flowers crowded in small terminal heads, the miniature petallike white scales surrounding the central yellow florets. It is often dried and used in winter bouquets.

128. Flower Clusters of Pearly Everlasting

A much less beautiful species, but having an aromatic odor, the sweet or common everlasting is abundant in dry open places and stony pastures throughout our range. Its stem, many-branched at the top and one to three feet tall, as well as its linear leaves, are velvety-haired and sage green in color. The flower heads are many in panicled clusters, fertile and sterile flowers in the same heads, all surrounded by dry, white scales. Dried and dyed red and blue, the plant was once a favorite for winter decorations. Although all are called everlastings, the pussytoes, pearly everlasting, and the sweet or common everlasting belong to three different genera. Though they are similar in many respects, they do differ, however, in certain technical details. Thus, for instance, the pearly everlasting differs from the sweet everlasting in that each little flower head has separate staminate and pistillate flowers, that is, they grow on separate plants. Moths and butterflies are the chief agents of cross-pollination, but other insects visit the flowers too. The everlastings serve as food plants for the caterpillars of the painted lady and the American painted lady.

For about two thousand years the home doctors of Europe and Asia have used the thick, mucilaginous roots of the elecampane as a horse medicine. When the early settlers came to our shores they brought the plant with them, believing they could not live here without it, and so, in colonial days and for many years thereafter, every country person had a patch of elecampane for the relief of asthmatic horses, "to help the heaves." Like many other immigrants it was not slow in naturalizing itself. Today it is found in old fields, pastures, roadsides, waste places, often lifting its fringy, yellow disks above lichen-covered stone walls in New England, rail fences in Virginia, and barbed wire barriers in the west.

The elecampane is one of the tall picturesque plants so characteristic of the composite family. It has a stem two to six feet high that is woolly and sometimes toned with purple gray, large, olive, yellow green leaves, rough above, woolly beneath, white-veined, toothed, and the upper partly clasping the stem. The flowers are large and yellow, the ray florets a deep, lemon yellow, narrow and curving and pistillate, the disk florets are tubular and perfect, at first yellow and later tan.

Of the chicory or succory, Virgil, only rarely in a practical mood, wrote: "And spreading succ'ry chokes the rising field," but Emerson in his poem "The Humble Bee," speaks of it as matching the sky. The flowers, an inch to an inch and a half across, are a beautiful blue color, but as the plant grows along roadsides, in waste places, and in similar situations, it is regarded as a weed and hence most people remain indifferent to its distinctive beauty. The flowers, similar in form to the dandelion, close in rainy or cloudy weather and open only in the sunshine. There are few florets in a single head, but these are highly developed, with gracefully, curving branching styles. Its stem is stout, tough, and stiff and its leaves are generally lance-shaped, coarsely toothed, and a dark gray green. The plant is sometimes associated with coffee, for the deep taproot has been used as a substitute or adulterant for this beverage. Chicory roots have also been boiled and served in place of carrots, but they never became a popular substitute. In the spring, the young leaves are often gathered and boiled as a potherb, and in France they were at one time, and perhaps are still, forced and blanched in a warm, dark place and used as a salad. Several cultivated forms of the plant have been developed and in some places have become a farm crop.

129. Flower Cluster of Chicory

Of the tansy Gerarde wrote: "In the spring time, are made with the leaves thereof newly sprung up, and with eggs, cakes or Tansies which be pleasant in taste and goode for the Stomache." Tansies were popular in the seventeenth century for Pepys speaks of them in his diary. In describing a dinner for some guests he says that, "it consisted of a brace of stewed carps, six roasted chickens, and a jowl of salmon, hot, for the first course; two neat's tongues, cheese, and a tansy for the second." The tansy was a sort of cake or fritter made from the leaves of the plant. In Cole's *Art of Simpling* it is told that maidens were assured that tansy leaves laid to soak in buttermilk for nine days "maketh the complexion very fair." A tea made from tansy leaves was, according to the faith of medieval herbalists, a cure for every human ill. At any rate, until recently, tansy tea was a favorite beverage for colds and similar ailments in this country.

The tansy is one of the more common plants of summer, its rather flat, clustered, dull orange yellow flower heads being a familiar sight along roadsides, in fields, and waste places. The flowers look like those of the daisy minus the petals, and as a matter of fact they are very buttonlike, hence, the tansy is sometimes known as bitter buttons. An immigrant from Europe and an escape from garden cultivation, the tansy is a rather rank-growing herb two to three feet high, with alternate leaves, ornamentally toothed and cut. They are strongly aromatic.

In May and June, myriads of daisies whiten the fields and meadows of New England and the Middle States, as if a belated snowstorm had covered them with a white blanket, to the dismay of every farmer and a joy to every nature lover. The daisy, or oxeye daisy to give it its complete name, is perhaps the most common plant of the field and wayside, and like many others is an immigrant from Europe. One of its centers of distribution in America is said to have been Saratoga and the route of Burgoyne's army, because his horses were fed upon fodder that came from central Germany and wherein the daisy was mixed, its seeds germinating in the wake of his army.

Looking closely at a daisy, one will likely see the thrips. They are minute, black insects, threading their way in and out of the tiny florets of the daisy head. The daisy, like the dandelion and other composites, is not a single flower, but a cluster of hundreds of tiny, tubular, yellow florets, each a perfect flower, packed tightly together within a green cup. Note how the stamens form a ring around the pistil, how the pistil rises through their midst, and how the two little hair brushes at its tip sweep the pollen from the anthers. As the pistil continues to rise, the pollen is lifted high up where any insect crawling over the florets will be dusted with it. Then, with its pollen gone, the pistil spreads its sticky, stigmatic arms, hitherto tightly closed so that self-fertilization could not occur, but now receptive to pollen from another flower.

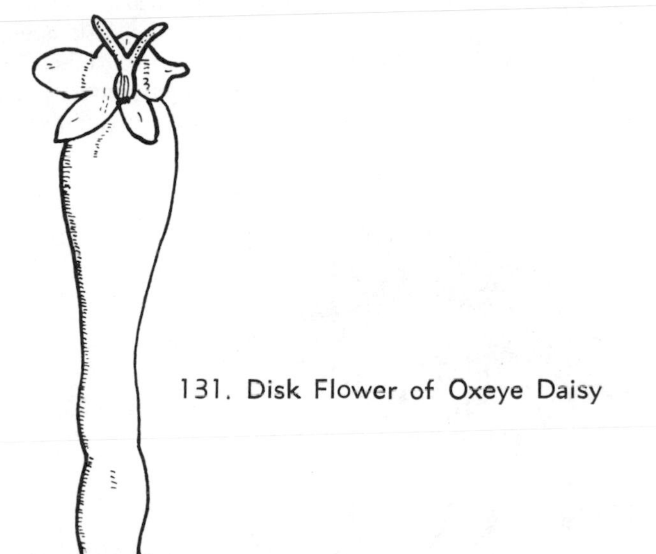

131. Disk Flower of Oxeye Daisy

130. Flower Cluster of Oxeye Daisy

The ray florets are female and have only the pistil, but this pistil does not have the hair brushes, for it does not need them. Because daisies are so conspicuous and offer so much in the way of liquid refreshment, almost every winged insect eventually finds its way to at least one of them, and with every insect visitor, seed production is enhanced.

Unlike the pistillate ray florets of the daisy, the orange yellow ray florets of the black-eyed Susan are neutral, that is, they have neither stamens nor pistils. However, the purplish brown florets that form

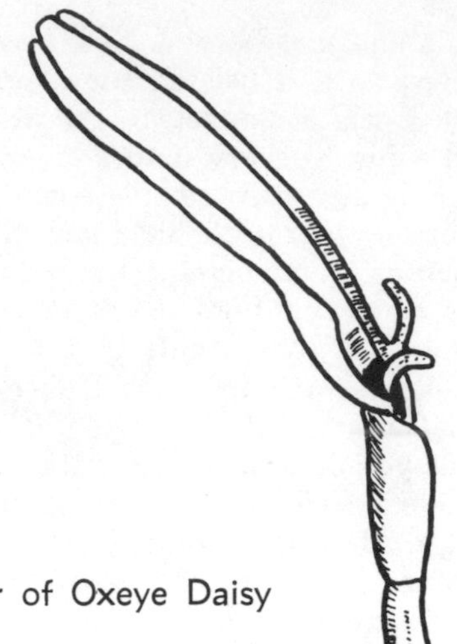

132. Ray Flower of Oxeye Daisy

the conical disk are perfect. The florets at the base of the disk open first and their pollen forms a yellow circle. Then the next higher florets on the disk open, forming another yellow circle, and this continues as blossoming circles climb toward the apex. Sometimes small caterpillars are found in the heads of the black-

133. Purple Coneflower

eyed Susan, and have a habit of attaching small pieces of the flowers to their backs, keeping them in place with silk. No doubt this bit of camouflage helps in concealment.

Unlike so many of our wild plants that landed on eastern shores and marched westward, the black-eyed Susan is a western species that travelled eastward, in bundles of hay it is said. It is a plant one to three feet tall, with stem and leaves very rough and bristly, and grows in dry fields and in open sunny places. Its close relative, the tall coneflower, however, prefers moist thickets such as border swamps and meadow runnels. It is a smooth, many-branched plant, sometimes growing twelve feet high, but usually only half that height, with smooth green leaves, the lowest compound, the intermediate irregularly three to five-divided, the uppermost small and elliptical, and numerous showy heads, from two and a half to four inches across. They have from six to ten bright yellow, drooping rays around a dull greenish yellow conical disk that gradually lengthens to twice its size as the seeds mature. The purple coneflower is a western species.

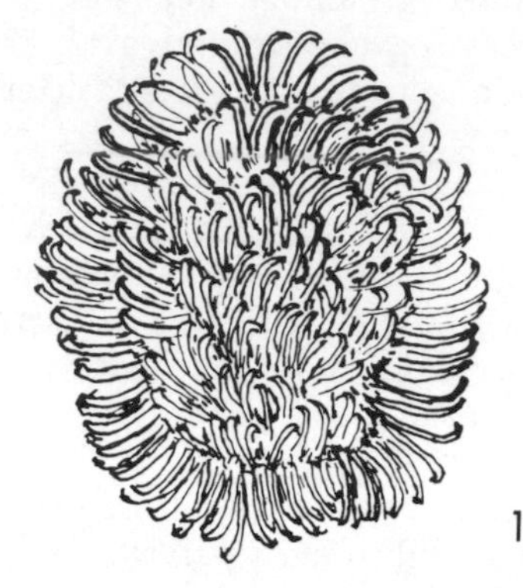

134. Bur of Burdock

When I was a boy we took great delight in tossing the burs of the burdock at each other for the burs would cling to our clothing and at times we would be covered with them. It was a mild form of sadism, I suppose. If you are not too familiar with the burs look at one closely, as with a hand lens, and you will see why they cling so tenaciously. Each bur consists of a number of achenes that are oblong, three-angled, and ribbed, with one end truncate and the other in the shape of a hook.

The burdock is a familiar plant of roadsides, waste places, and fencerows. It has an enormous root, often three inches thick, penetrating the ground straight down for a foot or more, and then branching in all directions. The leaves are large, dull green, and veiny, the lower heart-shaped, often a foot long, the upper ovate, and woolly beneath. The globular flower heads, hook-bristling green burs with magenta or nearly white perfect tubular florets, are a standing invitation to butterflies, that delight in magenta, and to bees of various kinds. To appreciate its depth of coloring one should look at a flower head with a magnifying glass or hand lens. The root and seeds of the burdock were once used in medicine and perhaps still are, and the leaves are said to be cultivated as vegetables in Japan, or were at one time. The large, tender leafstalks are peeled and eaten raw, used as a salad, or cooked as asparagus.

The burs of the burdock have one feature in their favor; they remain intact and can be removed in one piece from clothing. Brush against the beggar-ticks, and your skirt or trousers will be covered with hundreds of barbed seed vessels; it takes an interminable time to get them all off.

The beggar-ticks, also called stick-tight, devil's bootjack, pitchforks, stick-seed, popular names that suggest the plant's character, is one of a large number of weedy herbs of the genus *Bidens,* from the Latin

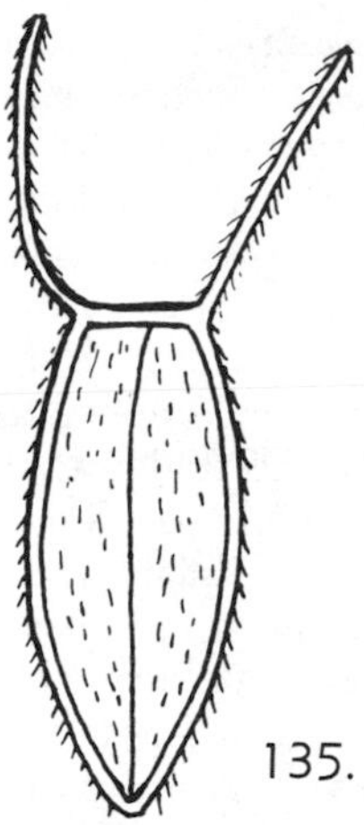

135. Fruit (Achene) of Beggar-ticks

bidens meaning two-pronged in reference to the fruit, that are commonly known as the bur marigolds. They have usually divided or dissected leaves with toothed segments, yellow flowers, and a flat, two-pronged barbed achene. The beggar-ticks has a purplish branching stem, two to five feet high, with rayless, bristly flower heads, the disk florets tubular, orange yellow, perfect, and fertile. As in all the bur marigolds, the calyx of each floret is converted into a barbed implement for grappling the fur of an animal or our clothing.

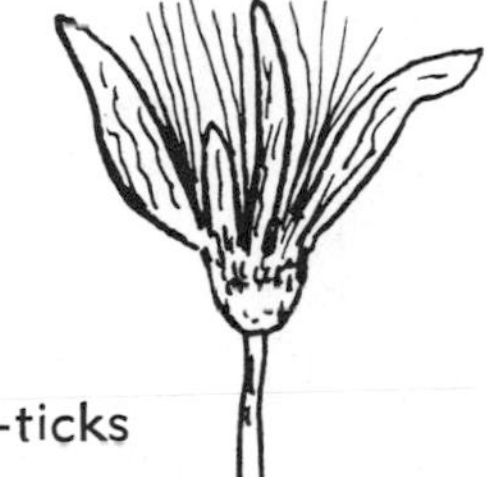

136. Flower Cluster of Beggar-ticks

The beggar-ticks grows in moist soil, swamps, ditches, meadows, and in the same situation the larger bur marigold and the smaller bur marigold are found. The larger bur marigold is a more attractive species than the smaller bur marigold that often grows in company with it, because of its light golden yellow rays, and its flower heads that are sometimes over two inches broad. Another member of the genus is the species commonly called Spanish needles. It is a slender-stemmed plant with many spreading branches, once to thrice-pinnately cut leaves, flower heads with short yellow rays and dull yellow disks, and achenes with three or four awns.

Few plants have such a worldwide distribution as the yarrow, and few appear so often in mythology, folklore, and literature. Chiron the Centaur is said to have named the yarrow after Achilles (its scientific name is *Achillea Millefolium*), who is supposed to have used it to heal his wounded soldiers at the siege of Troy. Old books mention its virtues as a love charm, a cure-all for diverse ailments, and an ingredient of an intoxicating drink. In England it was once believed

137. Fruit (Achene) of Spanish Needles

that if a spray of yarrow were placed beneath a pillow on Midsummer's Eve, the sleeper would dream of his or her future wife or husband. As an old rhyme had it:

Thou pretty herb of Venus tree
Thy true name it is Yarrow;
Now who my dearest friend shall be
Pray tell thou me to-morrow.

Aesthetically, the plant has mixed qualities. The feathery masses of finely dissected leaves have a lacelike character appealing to the eye, but its dusty-looking flower clusters give it a sort of unkempt appearance. When viewed with a hand lens, the individual flowers belie their collective appearance and are not unattractive. The perfect disk florets are at first yellow, but later turn brown. The pistillate ray flowers are white or a grayish white, or in some rare cases crimson pink.

An immigrant from Europe the yarrow is found in dry fields, along the roadside, in waste places, on banks, and if viewed closely, one is apt to find two dissimilar animals lying in ambush in the flower clusters. One is a small, white, crablike spider and the other a greenish yellow insect with a broad, black band across the expanded part of the abdomen. The spider is of interest because later in the summer it moves to a yellow flower and turns yellow, for spiders as a rule do not change color. The insect is called the ambush bug and is well named, for both the rather grotesque form of the body and its peculiar coloring simulate the blossoms whereamong it hides and help to conceal it.

Goldfinches have long been associated with the thistles, but they visit the flowers only after the seeds have been formed. It is the butterflies and bumblebees that visit the flowers when they are newly opened, for the densely clustered florets are rich in nectar. The dense, matted, woollike hairs that cover the stems and leaves discourage the climbing pilfering ants, and the spiny leaves are equally as effective against grazing animals.

There are a number of different species of thistles, some two hundred as a matter of fact, that are scattered over the north Temperate Zone. They are prickly herbs with alternate or basal leaves that are nearly always cut or lobed and horribly spiny-margined. The flowers are tiny, tubular, crowded in a dense, usually spiny, bracted head. An adventive from Europe and now found in pastures, roadsides, and waste places, the common thistle is a biennial species, the first year producing only a deep taproot crowned by a large, tufted, spreading rosette of leaves three to six inches or more long. The second year it becomes a plant with a stout stem, two to four feet high, branching and leafy to the heads, the leaves dark green, narrow, white-spined, the upper surface prickly haired, the lower webby-wooled with light brownish, fine hairs. The tubular perfect flowers, densely clustered, vary from crimson magenta to light magenta, and are remarkably sweet-scented and rich in nectar. Bumblebees that visit the flowers often become intoxicated, but the butterflies apparently do not. This species and other thistles serve as a food plant for the painted lady butterfly.

The largest flowered thistle of them all is the pasture thistle, with solitary heads two to three inches broad, the florets exceedingly fragrant and rich in nectar, light magenta lilac or nearly white. Gray called the Canada thistle "a vile pest", and in 1896 all but three of the states had laws making it an offense for anyone to mature and scatter its seeds. It is a common plant, however, in pastures, fields, and on roadsides. By horizontal rootstalks it creeps and forms patches that defy all attempts to get rid of it.

138. Flower Cluster of Canada Thistle

The small lilac, pale magenta, or rarely white flower heads contain about a hundred florets, that are abundant in nectar, attracting not only bees and butterflies, but also flies, wasps, and beetles. The name Canada thistle is a libel on our northern neighbor for it is not a native to Canada, but to the Old World. This only shows how names are sometimes applied to plants.

In Europe the sow thistle or hare's lettuce is used as a potherb and is said to be "exceedingly wholesome." It springs from a white taproot and has an angled, branching stem from one to six feet tall and filled with a milky juice. The leaves are large and decorative, usually lobed, irregularly toothed and armed with soft spines. The light yellow, thistle-shaped flower heads are grouped in a somewhat loosely spreading flat cluster. The sow thistle is not a thistle though it looks like one, and is an immigrant from Europe. It is found in fields, roadsides, and waste places. In the Old World it has also been known as hare's palace. One reads in the "Grete Herbale" that "if the hare come under it, he is sure no beast can touch hym." Another early writer says that "when hares are overcome with heat they eat of an herb called hare's lettuce, hare's house, hare's palace; and there is no disease in this beast the cure whereof she does not seek for in this herb." There are several other species of sow thistles, such as the field sow thistle and the spiny-leaved sow thistle, differing only in details.

Anyone would think that the genus of such unwelcome plants as ragweeds that cause hay fever, though there are other plants equally as guilty, would hardly have been given the name *Ambrosia*. This was in classical mythology the substance that, with nectar, formed the food and drink of the gods. Today, one thinks of the word as meaning something pleasing to the taste and smell. This can hardly be said of the common ragweed, one of our most common weeds, intruding everywhere, troublesome in dooryards and gardens, and a pest in meadows and pastures where cattle will sometimes eat it when better forage is scarce, and yielding, as a result, bitter milk with a bad odor.

There are several species of ragweeds that may be distinguished from other groups of composites by the united involucre of the staminate heads of flowers, and by the single row of spines on the involucre of the pistillate heads. The common ragweed, also known as Roman wormwood or hogweed, has a many-branched, finely haired stem and remarkably ornamental, cut leaves, deep branching roots, and flowers

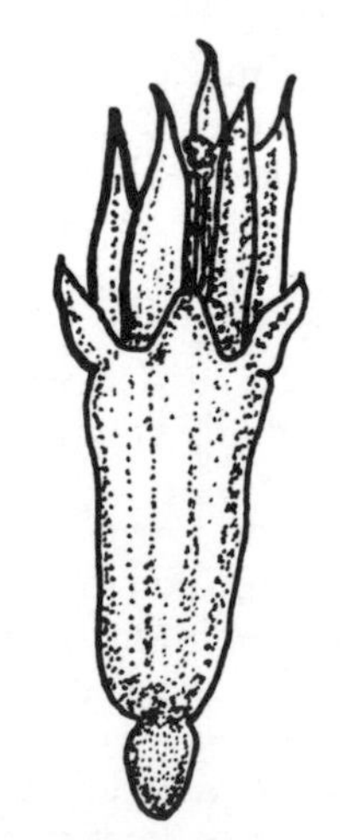

139. Single Staminate Flower and Single Pistillate Flower of Ragweed

of two kinds. The staminate heads are of six to twenty small, greenish flowers in crowded spikelike racemes at the summit of the plant. The pistillate flowers occur in the axils of the leaves, and form hard, achene-like fruits with a beaked crown, surrounded by four to six spiny points. The great or giant ragweed, one of the tallest members of the composites, is a huge, coarse plant with a stout, tough, woody stem set with bristly hairs, from four to ten and sometimes fifteen feet high, rough-haired leaves varying in shape, often more than a foot long, the staminate flowers in terminal racemes six inches to a foot in length, and the pistillate clustered in the axils of the leaves. It is a plant of fields and waste places.

At one time it was believed by the people of the Old World that hawks improved their eyesight with leaves of a certain plant that came to be known as the hawkweed, and the genus wherein it was placed is *Hieracium,* from *hierax,* a hawk. The genus is a large one, but only a dozen or less are common in the eastern and central states and even fewer in the west. One of them, painting many an old field a fiery red in June, is known as the devil's paintbrush, or somewhat more prosaically, the orange or tawny hawkweed. It is an odd, but attractive plant with a stout, unbranched stem and a flower cup covered with sepia brown hairs that gave the plant the common name in England of Grim the Collier. The leaves are coarse, blunt, lance-shaped, covered with short gray hairs, and the flowers, strap-rayed and finely fringed at the edges, are tawny orange with light golden pistils, and are grouped in small terminal clusters. They are slightly fragrant and colorful, and attract bees, flies, and butterflies.

A second species of hawkweed, known as king devil, has yellow flower heads, about three quarters or an inch broad, on glandular, hairy pedicels. It blooms during June in fields, meadows, roadsides, and waste places. A third species, the Canada hawkweed, is generally a smooth species of open woods and thickets, with a stem from one to five feet high, lance-shaped leaves, the midrib prominent and the margins sparingly but sharply toothed, and very small yellow flowers in loosely clustered heads. An earlier flowering species, the rattlesnake weed, with deeper yellow dandelionlike flowers and a generally leafless stem, is common in woodlands and thickets. Because its leaves are purplish-veined, suggesting the marks on a snake's body, a credulous generation believed it was a cure for snake bite, hence its name.

According to Herodotus, lettuce was eaten as a salad in 550 B.C. It was cultivated in Pliny's time and even blanched and stored so the Romans would have it year-round. And Parkinson, in describing the "vertues of the lettice," said "They all cool a hot and fainting stomache." Despite such references to the lettuce, the origin of our garden varieties, or what they were, is not known.

Everyone, of course, is familiar with lettuce as the leafy heads that are bought in the store, but few have seen it in flower. Lettuce gardeners growing it as head lettuce, do not permit the flowering stalk to develop. If one should want to know what the flowers of the cultivated lettuce would look like if allowed to develop, one might look at those of the wild lettuce, a common plant of fields, meadows, roadsides, and waste places, for both are much alike.

Both the wild lettuce and the cultivated lettuce belong to the genus *Lactuca,* Latin for milk in allusion to the milky juice characteristic of the genus. It is a large group of herbs mostly from the north Temperate Zone with alternate leaves (except in the cultivated lettuce), usually cut or divided and often with soft, prickly margins, and small flower heads generally in large terminal clusters (panicles). The yellow, pink, blue or white florets are all ray flowers, and develop into flat achenes (the lettuce "seed").

The wild lettuce is a tall biennial species sometimes six feet high with a smooth, stout stem, sometimes of a purplish tinge, but most commonly deep green, and with leaves that are somewhat like those of the dandelion. The insignificant pale yellow flowers occur in flower heads in a thin, scattered flower spike. They are followed by an abundant, soft, white pappus. The hairy or red wild lettuce is a similar species, but often with a dark reddish stem, peduncles, and tiny flower cups, the ray florets varying from yellow to pale reddish or purplish. The leaves are longer and deeply cut, or lobed almost to the wide midrib.

A plant of damp, shady places, the blue lettuce is the tallest of the species, growing to a height of fifteen feet. It has a stout, straight, smooth stem, deeply lobed leaves irregularly wavy-toothed, and green flower heads tipped with inconspicuous dull purplish or whitish rays. A noxious plant of wide range, the prickly lettuce is indifferent to the kind of soil wherein it lives, and thus, is found almost everywhere. It has an erect stem, with short lateral branches, sometimes attaining a height of seven feet, but usually two to five feet tall, leaves with prickly toothed edges and with spines on the lower surface. The flower heads are numerous in a large panicle at the summit of the stalk, and pale yellow. The leaves of plants growing in the open have a vertical twist at the base, causing their edges to point north and south, hence the species is also known as the compass plant.

A plant of the woods, rich, moist thickets, and roadsides, the rattlesnake root is a somewhat interesting species, with a smooth green or dark purplish red stem, rising from a bitter, tuberous root two to five feet high. The leaves are variable in form, large below, small above, and there are numerous cream-colored flower heads, occasionally tinged with lilac, fragrant, and nodding in loose, open, narrow terminal clusters. Each bell-like flower head consists of eight to fifteen ray flowers, drooping from a cuplike involucre of eight colored bracts. A similar species, but somewhat lower growing and with leaves more cut, is the lion's foot, fairly common in thickets and dry, sandy ground.

There are a number of other plants in the large composite family that grow wild and that deserve mention. The oxeye is a plant of copses, rich, low ground, and streamsides with numerous golden yellow, daisylike flower heads, easily mistaken for a sunflower and visited by a horde of insects. These include bees, flies, wasps, hornets, beetles, and small butterflies such as the pearl crescent, the common hair streak, and the tiny sooty wing. The sneezeweed, of low-lying meadows, has handsome bright yellow flower heads, an inch or two in diameter, on stems as much as six feet tall, rich in both nectar and pollen and consequently besieged by bees, wasps, flies, beetles, and butterflies. The mayweed or chamomile, is a daisylike plant of sandy roads with white flower heads and small leaves cut and slashed into complete formlessness, and remarkable for their unpleasant odor that repels the fastidious bees and butterflies. Some flies, however, seem to be attracted by it. The mugwort is a familiar plant found in waste places and near old houses, with deeply cut leaves, and inconspicuous green yellow flowers. Common in fire-scarred areas the fireweed is a tall, coarse plant with a disagreeable odor. A familiar plant of swamps and meadows, the golden ragwort blossoms earlier than many of the composites. It has a perennial root, rootleaves on long petioles, variable and sessile stemleaves, and handsome, deep golden yellow, daisylike flower heads in terminal clusters.

The composite family is well represented in the garden. Many of the ornamental composites are wildings, either native or escapes, such as the common sunflower, oxeye, sneeze weed, aster, goldenrod, chicory, and the common or oxeye daisy. The last named is a chrysanthemum, one of many grown in the garden. The chrysanthemums (genus Chrysanthemum, from the Greek for golden flower) are erect herbs, some being in cultivation in China and Japan for three thousand years, often with strong-smelling foliage, generally multi-branched, with alternate divided leaves, and flowers in heads of all colors except blue and purple, the heads in some species large and showy, in others small and buttonlike.

One of the better-known garden chrysanthemums is the summer chrysanthemum. It is a half-hardy annual, two to three feet high, with leaves cut into narrow segments, and flower heads white, red, or yellow, easily distinguished by the keeled or ridged scales of the involucre. The pyrethrum is a very popular, summer-blooming species, with fernlike leaves and large flower heads that are red, pink, lilac, or white. The costmary, also called the mint geranium and erroneously lavender, is a hardy perennial, two to three feet high, with sweet-scented leaves and numerous flower heads, the white rays very few or absent altogether. A multi-branched annual, the corn marigold grows one to two feet high. It has notched or slightly cut leaves and daisylike flower heads, solitary at the ends of the branches, and white or yellowish. The marguerite or Paris daisy is a tender, multi-branched, beautiful species, native to the Canary Islands, with coarsely divided leaves and daisylike flower heads white or pale yellow. The giant daisy is a hardy perennial, four to seven feet tall, many-branched, with narrow and sharply toothed leaves and solitary flower heads that look like particularly fine oxeye daisies. A fine perennial hybrid, one to two feet high, the shasta daisy has long, narrow, toothed leaves and flower heads two to four inches wide, and white and daisylike. An old favorite of pilgrim gardens, the feverfew is a bushy hardy perennial, two to three and a half inches high, with yellow green leaves more or less cut, and many flower heads, scarcely three-quarters of an inch wide, the rays white, short, or altogether lacking. The name marigold may be applied to several different kinds of plants, but is used here as a member of the genus *Tagetes,* supposedly named for Tages an Etruscan god. The genus is a group of tender annual herbs all native from New Mexico to Argentina. It includes the French and African marigolds, both misleading names and misappropriately applied since marigolds are neither French nor African in origin. Perennial garden favorites, the marigolds have mostly opposite and usually finely dissected, strong-scented leaves and showy flower heads that either may be solitary or clustered. Beneath them is a series of involucral bracts united into a cuplike base. There are several species of marigolds, but the two best known are doubtless the African and French marigolds. The African marigold is an erect rather bushy species, eighteen to twenty-four inches high. more or less branched, with finely divided leaves, the segments narrow and toothed, and yellow or orange flower heads two to four inches wide, the rays with a long claw. The French marigold is a many-branched species, rarely over a foot high (there is a dwarf variety), with leaves like those of the African marigold and flower heads about an inch and a half wide, and the rays yellow with red markings. There are a number of horticultural forms ranging in color from pure yellow to nearly pure red, some being double-flowered.

The garden zinnias, named in honor of Johann Gottfried Zinn, a professor of medicine at Göttingen in 1750, though of comparatively recent introduction, are among the most popular of garden plants, the tall forms being introduced about 1886, the small forms of single flowers about 1861, and of double flowers in 1871. There are about fifteen species of zinnias, chiefly found in Mexico, but also from Colorado, Texas, and Chile. They are rather stiff, erect herbs, the stems covered with short bristly hairs and somewhat woody at the base, with opposite oval or lance-shaped leaves that usually clasp the stem. The flower heads are showy, somewhat flat or cone-shaped, and solitary, each floret growing in the axil of a scale-like bract, the disk florets yellow or purplish brown, the ray florets of every shade except blue according to the variety. In some horticultural varieties the disk flowers are absent.

The dahlia, a highly prized garden ornamental and named after Andreas Dahl, a Swedish pupil of Linnaeus, is said to have been discovered sometime during the eighteenth century by the director of the Mexican Botanic Gardens whose name was Cervantes. There are about a dozen or more species of dahlias, stout and somewhat woody plants with tuberous roots, most being from the uplands of Mexico, but a few come from Central America and northern South America. Originally, the ray flowers were in a single series and mostly of some shade of red or purple, the disk flowers yellow, but by breeding, the color and shape of the ray flowers have been changed considerably, while many of the disk flowers have been changed into ray flowers. Most of our older garden varieties have been derived from a single ancestral species and these are the single and pompon types, while many of our most recently popular types, such as the peony-flowered and chrysanthemumlike forms, have been produced from the cactus dahlia, perhaps of hybrid origin and derived from a wild species of Mexico and Guatemala. Some hundred or more years ago there were sixty varieties of the dahlia, today there are several thousand.

Our common garden dahlia, is usually four to eight feet high, with opposite, compound leaves, the segments oval and more or less blunt-toothed, and usually grayish beneath. The flower heads are at least four inches wide and more or less nodding, however, in some of the forms the flower heads are wider, in others smaller. The cactus dahlia is similar, but the rays have recurved margins, are of unequal length, and overlap.

Like the marigold, the cosmos is also a native of Mexico, and like the marigold also a garden favorite. It is an annual, growing seven to ten feet high, with a green, somewhat slender stem, leaves cut into fine almost threadlike segments, and flower heads one to two inches wide or even wider in some horticultural forms. The disk florets are yellow, the ray florets white, pink, or red. A perennial species, the black cosmos, has solitary flower heads, the disk florets are red, the ray florets are velvety, dark red or purplish. The name cosmos is from a greek word meaning orderliness, hence, an ornament or something beautiful.

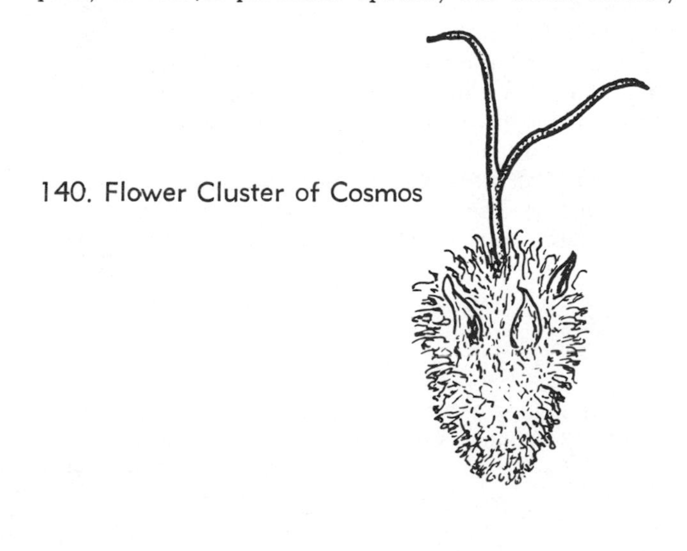

140. Flower Cluster of Cosmos

> Open fresh your round of starry folds,
> Ye ardent marigolds!
> Dry up the moisture from your golden lids,
> For great Apollo bids
> That in these days your praises should be sung.

So wrote Keats of the calendula or pot marigold, a native of southern Europe and cultivated as a popular annual for centuries. At one time it was esteemed as the "Herb-General of all pottage", the dried florets being used to add flavor to soups, and the fresh ones being made into syrups and conserves. Such uses have long since been discarded, the calendula (from the Latin *calende,* meaning throughout the months in allusion to its long blooming period) today planted in our gardens because it blooms from midsummer to frost.

The calendula or pot marigold is a plant from twelve to twenty inches high, with oblong leaves, more or less clasping the stem, and solitary, stalked flower heads orange to yellow. There are many horticultural forms.

The Europeans call it the cornflower but it is known in America as the bachelor's button. It is a widely cultivated annual, quite frequently an escape to fields, roadsides, and waste places, of somewhat sprawling habit, one to two feet high, woolly when young, green later, with narrow toothless leaves nearly five inches long. The solitary flower heads are on slender stalks, typically blue (but purple, pink, or white in horticultural forms), and consist of tubular florets only, the marginal florets expanded and raylike. A bed of bachelor's buttons, the starry heads fairly twinkling in the slightest breeze, makes an attractive display when in flower, but by midsummer, the blooming season is past and the bed becomes a scene of desolation.

The bachelor's button belongs to the genus *Centaurea* from Chiron the Centaur, for according to the legend, Chiron used the plant to heal the wound of Hercules. This genus also includes such well-known garden plants as the dusty miller, a handsome foliage plant, white-felty, with small rose purple heads in clusters and almost hidden by the leaves; the sweet sultan, the most popular of the group next to the bachelor's button, a smooth plant, one to two feet high, with leaves cut feather-fashion and solitary, usually fragrant heads, yellow, red, purple, pink, or white; and the mountain bluet, a perennial from Austria, with cottony leaves and large, blue flower heads, often three inches wide.

Perhaps not as well known as some of the preceding species, the Stokes' aster or stokesia, named for Jonathan Stokes an English botanist, is nevertheless, a good garden plant. It is a perennial herb with a purplish stem covered with white matted hairs, lance-shaped leaves to ten inches long, and lavender blue flower heads two to four inches in diameter, the disk florets tubular becoming smaller toward the center, and the ray florets large, flattening into five lobes.

The stokesia is not an aster despite its name, but belongs to the genus *Stoksia,* and as a matter of fact, is the only species of the genus. The China aster is not an aster either. It belongs to the genus *Callistephus,* Greek for beautiful crown in allusion to the double pappus that surrounds the seeds, and like the stokesia, is the only member of its genus. It is not closely related to the true genus *Aster,* and yet, to most gardeners it is "the aster."

The name China can be historically authenticated, for in 1731 seeds of the plant were sent to the Jardins des Plantes by a missionary named Father d'Incarville. The China aster, a native of China and Japan where it had probably been a garden plant long before Father d'Incarville's time, is a hairy species, nine to twenty-four inches high, with coarsely toothed, oval leaves and solitary, terminal flower heads of nearly every color, but yellow, though predominantly blue or violet.

Among the best of our garden plants, the gaillardia is a favorite with many because of its masses of beautiful and effective flowers and long blooming season, from midsummer to autumn. It is, moreover, a hardy plant, liking the sun and being able to endure considerable cold. It is often found still blooming after a frost has killed other plants.

There are several species native to North America. They are leafy, erect, branching herbs with alternate or basal, gray green leaves and handsome flower heads, two to three inches across, the disk florets purple, the ray florets yellow, orange, orange red, and white in a horticultural form. The gaillardias, named for Gaillard de Marentonneau, a French botanist, are often called blanketflowers. A popular annual species has the ray florets yellow at the tip, rose purple at the base, while another popular perennial species has yellow ray florets.

A remarkably handsome annual from southwestern Africa and frequently planted in the garden is the arctotis. It is a many-branched, bushy plant with thick, grayish green leaves that are covered with white, woolly hairs and usually toothed or deeply cut, and flower heads about three inches wide, blue and yellow, though occasionally white or violet. The arctotis, the name being Greek for bear's head in allusion to the silky, shaggy achene, is of the daisy type, and hence, often called the African daisy. It is a desirable plant for the garden, not only because of its showy flowers, but also because of its long blooming season that extends from early summer to late autumn.

A group of wild flowers native to our western and southwestern states and widely cultivated for their showy blooms are the plants commonly called tick-seeds, but better known perhaps to the gardener as the coreopsis. There are some seventy species, all members of the genus *Coreopsis,* Greek for bug in allusion to the shape of the achene. They are annual or perennial herbs, the annuals being the most popular, with either entire or lobed or cut leaves and solitary or clustered flower heads. The disk florets are usually yellow, the ray florets prevailing yellow, but white, pink, or sometimes variegated in certain horticultural varieties. A full bed of coreopsis makes a beautiful display, the flowers seemingly glow in the sun and on their long stalks dance on the wind.

141. Flower Cluster of Coreopsis

The best-known and most popular of the annual species is the golden coreopsis, an erect tall plant, twenty to thirty-six inches high, the flower heads red brown in the center, the ray florets notched and pure yellow. Another annual is the golden wave that is not quite so tall, the ray florets yellow at the ends, brownish purple at the base, so the head thus having a dark center. A third species, the tall coreopsis, is an anise-scented perennial with yellow flower heads clustered in a large corymb.

Because of its cheerful, early bloom the English daisy has become a favorite bedding plant and, also, one widely used for window boxes. This is the true daisy of history and literature. The English poets from Chaucer to Tennyson have all felt its charm and sung its praises.

Of all the flowers in the mede,
Than love I most these floures white and rede,
Soch that men callen daisies in our town.

Chaucer

When daisies pied and violets blue
And lady-smocks all silver white
And cuckoo buds of yellow hue
Do paint the meadows with delight.

Shakespeare

Meadows trim with daisies pied.

Milton

Wee, modest, crimson-tilled flower.

Burns

The English daisy is a dwarf perennial herb not over six inches high with leaves in basal tufts and flower heads solitary on stiff, erect stalks, the disk florets yellow, and the ray florets white or pink.

Equally a favorite bedding plant and perhaps the most popular of all bedding plants is the ageratum, a tropical American annual herb. There are some thirty species of ageratums, but the common or blue ageratum is the one most generally found in gardens. The word, ageratum, incidentally comes from the Greek for not growing old, however, it is not applicable to these plants. The blue ageratum is a somewhat bushy species, between one and two feet high, with leaves somewhat heart-shaped at the base, and flower heads about one-half inch across and blue, but white or pink in some horticultural forms. It is an easily grown plant and flowers most of the summer.

A plant widely grown in gardens, not only for its bloom, but also for its use as a dry winter bouquet, is the strawflower, an annual herb from Australia. It is usually branched, reaches a height of twenty-four to thirty-six inches, has oblong or lance-shaped leaves and one to two inch wide flower heads, the bracts of the involucre petallike, red, yellow, orange or white in some varieties. The disk florets are yellow.

A plant of great dignity is the way the globe thistle has been described. It is a decidedly handsome, thistlelike herb of the Old World and a popular garden flower. There are a number of species of globe thistles, but one of the most commonly planted is the tall globe thistle, a species three to four feet high, with a whitish, woolly stem, prickly leaves, and large blue flower heads. A perennial species and perhaps the most widely planted has white felty leaves and does not grow quite as tall.

A long standing garden favorite is the perennial golden glow, a variety of the tall coneflower, a native species found growing in the woods and thickets, and also a garden flower. It is a smooth, coarse, multibranched herb, three to eight feet high with deeply lobed, toothed leaves and flower heads, two to four inches wide, the center of the head cone-shaped, the disk florets greenish yellow, and the ray florets drooping and yellow. It belongs to the same genus as the black-eyed susan, *Rudbeckia,* named for two professors Rudbeck, father and son, predecessors of Linnaeus at Upsala.

The composite family is sparingly represented in our flora by woody plants and these are shrubs. The groundsel bush or groundsel tree is a multi-branched shrub, six to ten feet high, with light, gray brown stems, oblong, short-stalked, sage green leaves, wedge-shaped at the base and coarsely toothed, and small flower heads in dense clusters, white or yellowish, the staminate and pistillate flowers on separate plants. The shrub is a conspicuous object in autumn by reason of its profuse white-haired flowers. The fruit is a single, small-ribbed achene in clusters. The groundsel bush is a plant of brackish marshes and is common along the seashore. It is, however, also found inland in swampy thickets, sandy, open woods, and inland fields. A similar, but southern species, has brighter green leaves and larger flower heads, and a western species occurs from western Kansas and eastern Colorado to Texas.

142. Fruiting Cluster of Groundsel Tree

A partly herbaceous and shrubby species, woody at the base, and from three to eleven feet high, the marsh elder is a plant of the coastal salt marshes and seashores. Its leaves are narrowly elliptical, pointed, coarsely toothed, and deep green, while its flowers are small, greenish, in drooping heads, the latter in terminal, spirelike clusters. A much lower-growing shrub, one to two and a half feet high, the sea oxeye occurs on the coastal plain from southeastern Virginia to Florida, and west to Texas. It has grayish green leaves with short, silky hairs on both surfaces, rather narrow and pointed at both ends with some wavy teeth on the margins, and flowers in daisylike heads, the disk florets brownish, the ray florets bright yellow.

Also known as the lavender cotton, the santolina is an evergreen, aromatic, silvery gray undershrub with leaves finely divided, feather-fashion and solitary, yellow, globe-shaped flower heads about three-quarters of an inch thick. It is a useful border or low edging plant. The name santolina is of uncertain origin, but the plant may have been named for the Santoni, a people of Aquitania.

46. Coriaria (Coriariaceae)

The coriaria family is a small group with a single genus *Coriaria,* from *coriacea,* leathery, in reference to one species used for tanning. It consists of shrubs or half-shrubs that are widely distributed. They have simple, entire, opposite, or whorled leaves, perfect or sometimes unisexual flowers, solitary or in racemes, with five sepals, five petals, ten stamens in two series, five to ten carpels, and a fruit that is an achene, but surrounded by the persistent fleshy petals and appearing drupelike. The fruits, leaves, and roots are said to be strongly poisonous.

A European species is cultivated for the black dye that it yields, a second species, a native of New Zealand, is known as the wineberry. The tutu is also a native of New Zealand. It has angled branches and drooping branches of handsome flowers. It is said to be harmful to cattle.

A Japanese species is cultivated for ornament. It usually grows to three, rarely to seven to eight feet high, and has nearly stalkless, opposite, oval leaves. The flowers are small, regular, greenish, with five sepals and five petals, the latter eventually becoming fleshy and enclosing the fruit that is bright red at first, but later becomes black, and is about the size of a pea.

47. Corkwood (Leitneriaceae)

The corkwood family has a single genus, *Leitneria,* named after a German naturalist, Dr. E. T. Leitner, who travelled and was killed in Florida, and a single species, the corkwood, a southern shrub or tree, three to nine feet high, though sometimes twenty feet high, of swamps, with a brown gray, smooth bark, and alternate, bright green leaves that are narrowly or broadly elliptical, pointed at either end, and very finely downed. The flowers, without either sepals and petals, but with three to twelve stamens and a one-celled ovary, occur in catkins, the staminate and pistillate on separate trees, and appear in March. The fruit is an obovoid, somewhat compressed, brown leathery drupe. The corkwood is of no commercial value, but is of interest because it produces the lightest wood grown in the United States, indeed, it is one of the lightest of all woods. The wood has a specific gravity of 0.21 compared to that of cork, 0.24, and of balsa wood, 0.12.

143. Leaf of Corkwood Tree

48. Crowberry (Empetraceae)

A plant living in the Arctic stirs the imagination, for most people think of plants as living where it is warm. A plant that can survive near the limit of everlasting snow must have a great deal of adaptability and be able to adjust to the harshest climate found anywhere on the earth. It must also have taken eons of variation and natural selection for it to be able to do so.

These characteristics aptly describe the black crowberry, a low, spreading, evergreen shrub that forms dense masses wherever it elects to grow. Its habitats are the barren mountain summits and cold, sandy, rocky situations from the region of the Arctic south to the American-Canadian border. The black crowberry grows only two to three inches high, has rough, scraggy, dark brown stems and somewhat gray-powdered branches, and tiny, linear, alternate or sometimes whorled, bright green leaves with margins so rolled that they almost meet at the back. The flowers are small and inconspicuous, solitary in the axils of the upper leaves, dioecious, and purplish, the staminate with three sepals, petals, and stamens, the pistillate with three sepals and petals and a spherical six to nine-celled ovary, the latter forming a globular, black, berrylike drupe that is eaten by several species of arctic birds. A similar shrub, the purple crowberry, has dark red or purplish berries.

The crowberry family contains only three genera and only a few species of low, evergreen, heathlike shrubs with small, crowded leaves, small, usually dioecious flowers in axillary or terminal heads, the flowers with two or three petals or none and three sepals, the staminate with mostly three stamens, the pistillate with a two to several-celled ovary, and the fruit a berrylike drupe.

Aside from the black crowberry of the genus *Empetrum,* Greek for *upon* and a *rock* in allusion to the growth of these plants in rocky places, the only other interesting species of the family are the broom crowberry and the sandhill rosemary. The broom crowberry of the genus *Corema,* Greek for a broom, in allusion to the bushy habit is a small-leaved, low shrub barely eighteen inches high, of sandy and rocky places and native from New Jersey to Newfoundland. The leaves are tiny, deep green, and the flowers, without petals, are small and in heads at the tips of the branches. The fruit is a tiny, dry, madder brown, berrylike drupe, scarcely larger than a pinhead.

The sandhill rosemary of the genus *Ceratiola,* is a multi-branched, evergreen, aromatic shrub, one to five feet high, growing in dry and sandy coastal plain pinelands and on sandhills, with slender and erect branchlets, alternate, closely crowded leaves with their margins so enrolled that they appear needlelike. The red or yellowish flowers are small and inconspicuous, the staminate and pistillate usually on separate plants. The fruit is a roundish, yellow or red berrylike drupe.

49. Custard Apple (Annonaceae)

To find a tree in our northern forests with a fruit that looks like a banana is to wonder if nature has not been in one of her tantalizing moods and is trying to deceive us with a bit of the tropics, for even the leaves themselves give the tree a tropical look. And yet, the tree is real enough and one must conclude that it is perhaps a fugitive member of a family that belongs in a land where there is no winter.

The early settlers called it the papaw because its fruit resembles the fruit of the tropical papaw. It is oblong in shape, somewhat cylindrical, rounded, often pointed at both ends, rather curved, irregular in outline, three to five inches long, green at first and finally a dull sepia brown, with flat and wrinkled seeds. Once called the "poor man's banana" the fruit is edible, but with a taste or flavor that might be disagreeable to some, albeit pleasing to others. Like many tropical fruits it is a matter of getting to like it. One may dislike it at first, but after a while, may become accustomed to it and really begin to enjoy it.

The "wild banana tree" of the North, as it sometimes affectionately referred to, is a somewhat unusual tree, at least to those who are more accustomed to the conventional oaks, maples, and hickories. In order for a forest tree to survive, it must either reach the top where it can get air and sunshine, or become habituated to grow in the shade. So the papaw, known incidentally also as the custard apple, is a small tree

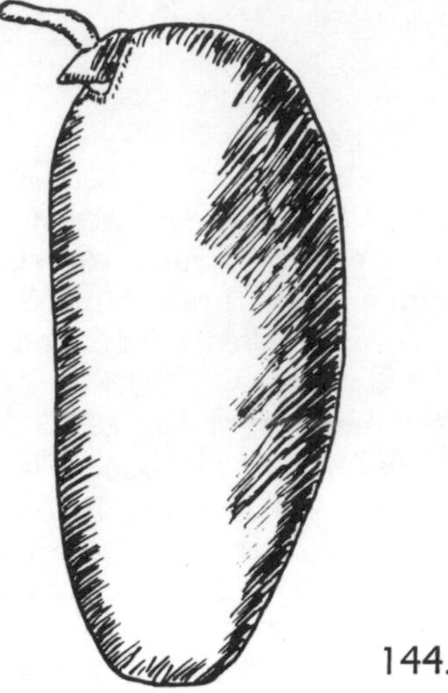

144. Fruit of Papaw

or more often a shrub, from ten to forty feet high, growing in swampy or rich alluvial soil where it often forms a dense undergrowth in the forest.

Being a member of a tropical family, one would expect the flowers to be large and showy, but they are more interesting than beautiful. They appear with the leaves in April and at first are green, but gradually darken in color until they become a deep, dull magenta red. Borne singly they are nearly one and three-quarters inches broad, with three large sepals and six petals, the latter in two sets, the three inner ones being smaller and more erect than the three forming the outer circle. Most people find their scent unpleasant, but the bees find them worth visiting.

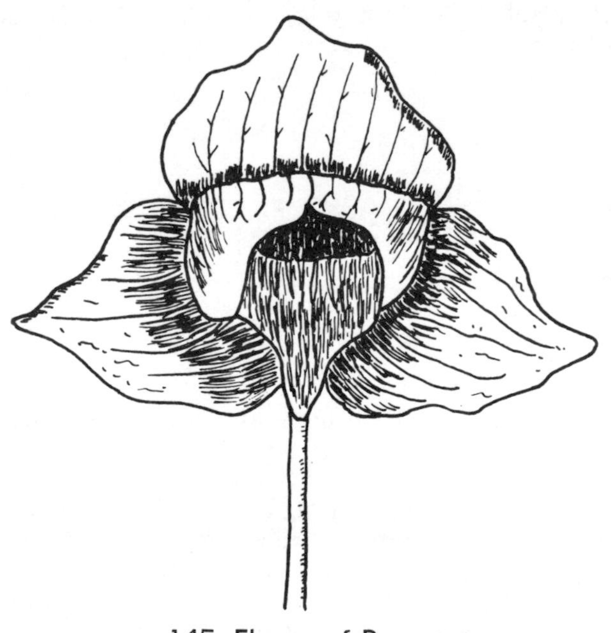

145. Flower of Papaw

The custard-apple family, also known as the papaw family, contains about forty-five genera and some eight hundred species of mostly tropical trees, shrubs and vines, but a few species are found in temperate climates. They have alternate, simple, entire leaves, mostly perfect flowers with commonly three sepals, separate or somewhat united, usually six petals, often in two series, many stamens and many carpels that may be separate or sometimes united in the fleshy fruit. The flowers develop into an elongated, fleshy, aggregate fruit or a large berry. The better known and more widely distributed species are tropical, such as the "soursop," the "sweetsop," and the "custard apple," all producing large edible fruits. The soursop is extensively planted and the sap of the large fruit is used as a beverage and for the preparation of jellies, and the juicy pulp for conserves.

The family is represented in the United States by two genera, *Asimina,* a Latinized version of a French and Indian vernacular name for the papaw; and *Annona,* a Latinized version of some name for the trees of the genus. The papaw, mentioned above, is one of four species of *Asimina,* the others being the dwarf papaw, a small shrub, two to five feet high, growing in dry oak and pinewoods of the South and also called the possum-simmon; the narrow-leaf papaw, also a small shrub of dry sandy, southern pinelands; and the showy papaw, also a small shrub of southern pinelands. The dwarf papaw has blunt, short-pointed or notched-tip leaves that are wedge-shaped at the base, reddish brown to purplish brown flowers, and brown fruits. The narrow-leaf papaw has narrow leaves, somewhat leathery in texture, and pointed at both ends, creamy white flowers, the inner petals often reddish purple at the base, and brown fruit. The showy papaw has leaves broadly rounded at the tip, narrowed at the base, white or creamy white flowers and a fruit the same color as the others.

Our only representative of the genus *Annona* is the pond apple, a southern species somewhat similar to the papaw. It is an evergreen tree, thirty to forty feet high, with wide-spreading, crooked branches, dark, ruddy brown bark, broadly elliptical, pointed, leathery, bright green leaves, pale dull yellowish white flowers, and an inverted pear-shaped fruit that is yellow with brown blotches. The fruit is edible, but rather insipid and is used for making jellies. The Seminole Indians used the fruit rather extensively. The pond apple grows in the swamps of southern Florida.

50. Cyrilla (Cyrillaceae)

The cyrilla family is a small group of shrubs or small trees comprising three genera and about six species, exclusively American. The leaves are alternate, simple, entire, thick, evergreen or nearly so, while the flowers are small, regular, perfect with four to eight sepals and petals, four to ten stamens, and a two to five-celled ovary, each cell with one to four ovules. The fruit is a small, corky drupe or pod. Only two of the genera are found in the United States with the following two species of any interest: the leatherwood or swamp cyrilla, and the buckwheat tree, the latter also called the black titi.

The leatherwood or swamp cyrilla is a southern shrub or small tree, fifteen to thirty-five feet high, common along streams and in swamps. It has a pale gray or whitish bark, silvery gray smooth twigs, and many spreading branches. The leaves are leathery, evergreen or nearly so, lustrous, deep green above, paler beneath, narrowly elliptic or broadest above the middle, and wedge-shaped at the base. The flowers are white, small, numerous, in long, slender, clusters, and the fruit is a tiny, corky, yellowish brown capsule.

The buckwheat tree is also a small southern shrub or small tree, five to fifteen feet high, occasionally higher, of swamp borders and streamsides where it often forms thickets. The leaves are much like those of the preceding species as are the flowers. The fruit is a two to four-winged, dry, light brown capsule.

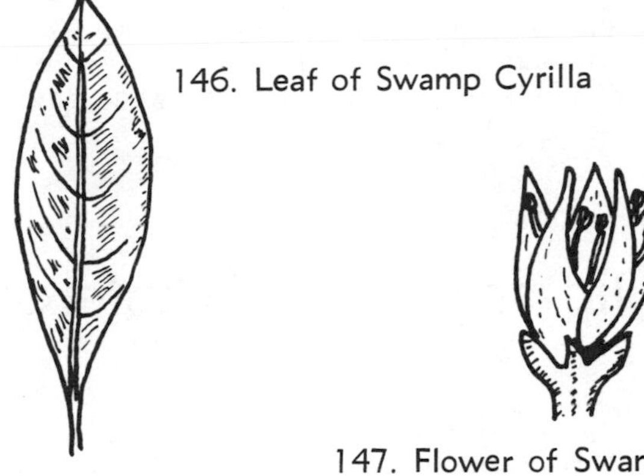

146. Leaf of Swamp Cyrilla

147. Flower of Swamp Cyrilla

148. Fruit of Swamp Cyrilla

51. Diapensia (Diapensiaceae)

Only in the pine barrens of New Jersey to North Carolina will one find the pyxie or flowering moss, a creeping, tufted, evergreen, mosslike, delightful little plant, with very small, scalelike, pointed, somewhat hairy leaves, and white or pale pink flowers that are bell-shaped and numerous, but not clustered, and rather solitary. They are followed by a tiny globe-shaped capsule. Upon seeing the dense, mossy cushions, sprinkled rather profusely with pink buds and white flowers, they are so beautiful that one can hardly resist the temptation to take a tuffet or two for a rock garden, but it is unsuited to the ordinary garden and requires a specialized site.

149. Flower of Pyxie Moss

The diapensia family consists of six genera and about eight species of low, perennial herbs or tufted shrubs, all of the north Temperate Zone. They have alternate or basal, simple leaves and small white pink or purple perfect, regular flowers, the latter solitary in the axils or in racemes at the end of the scapes. The stamens are five in number and are borne on the corolla, and the ovary is three-celled, the ovules few or numerous. The fruit is a dry capsule.

In addition to the pyxie our other native members of the family are the diapensia, the galax, and the shortia. The diapensia, found on alpine summits, is a low dense, cushionlike plant with short woody stems, stalkless, small evergreen leaves crowded all along the stems, and solitary waxy, white flowers. The galax, of open woods, is an evergreen herb, with a thick matted tuft of scaly creeping rootstalks that send up round, heart-shaped, veiny, shining leaves with rounded teeth and a slender naked scape bearing a wandlike spike or raceme of small and minutely bracted white flowers. The shortia, a plant of the mountains of North and South Carolina, is a low-growing evergreen herb, with creeping underground stems, basal, roundish, or heart-shaped, shining green, stalked leaves, and solitary white flowers. The shortia, sometimes called Oconee bells, is adaptable for the rock garden, but must be grown in shade and in soil composed of sandy peat and leaf mold. It is, however, difficult to establish.

52. Dipterocarpus (Dipterocarpaceae)

The dipterocarpus family consists of some ten genera and about one hundred species of trees, natives chiefly of tropical Asia, known for their aromatic oils and resins, and distinguished by their two-winged fruit. They have simple, alternate leaves, perfect flowers in axillary panicles, with five sepals that are enlarged and winglike at fruiting time, five petals, numerous stamens, three carpels, with two ovules in each carpel, and a one-seeded, indehiscent fruit.

The family includes the sal tree, an East Indian timber tree whose light brown, close-grained hard wood is second only to teak in value; the piny resin of southern India, the source of Indian copal; and the Sumatra camphor tree that produces borneol, a kind of camphor.

53. Dogbane (Apocynaceae)

During the month of July it seems that the spreading dogbane is in evidence everywhere. It is not, of course, but it can be found in fields and thickets, beside roads, lanes, and stone walls, its delicate and beautiful white pink and rose-veined bells suggesting pink lilies of the valley. The little veins show where the nectar is, and butterflies, bees, flies, and beetles come to feast. Some of these, however, pay for having trespassed on what is, in a sense, the butterflies's preserve, since butterflies are best fitted to serve the dogbane. Many bees, flies, and other insects remain imprisoned in the flower's trap of horny teeth until death from starvation at last releases them.

The spreading dogbane, also called honeybloom, bitterroot, and wild ipecac, is an attractive and graceful plant, growing usually in patches or colonies because of its extensive creeping, horizontal rootstalks, wherefrom new plants are sent up at intervals. It has a somewhat shrubby stem, one to three feet tall, generally reddish on the side exposed to the sun, and opposite light blue green, ovate, toothless leaves. The flowers are in small clusters at the ends of the branches and develop into twin follicles, long, slender, curved, four inches long and stuffed with many, thin, flat, brown

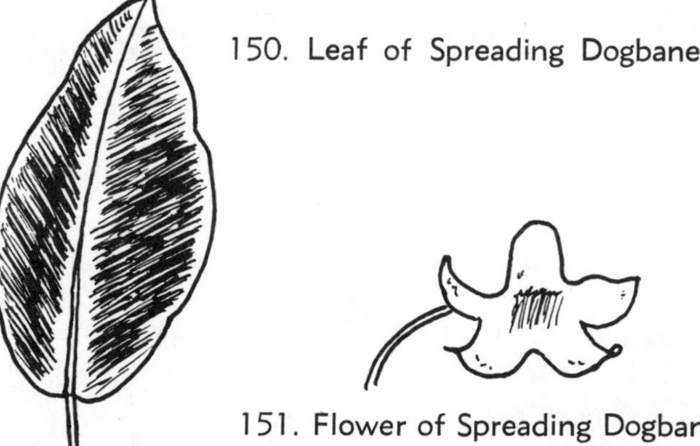

150. Leaf of Spreading Dogbane

151. Flower of Spreading Dogbane

seeds with a tuft of long, silky hairs that are reminiscent of the milkweed seeds. As well they may, for the spreading dogbane once belonged to the milkweed tribe since it has many of the same structural details of the milkweeds including the milky juice. For further proof that the dogbane once belonged to the milkweed tribe, the caterpillars of the monarch butterfly, though they feed almost exclusively on milkweeds, will at times include the dogbane in their dietary.

In July the dogbane beetle adopts the plant and remains inseparable from it. The insect has never been found on any other plant by this author. Resplendent in metallic green and red of incomparable luster, the beetle is a beautiful insect, and when it is numerous on the foliage the dogbane appears as if studded with jewels. To prevent capture, the beetle draws its legs up beneath its body and drops to the ground, where it becomes lost to view in the grass as its colors blend so well with the herbage.

The dogbane derived its name from the fact that it was considered poisonous to dogs. How much truth there is to it or that dogs have ever been poisoned from eating it is not known, but the plant is poisonous and horses, cattle, and sheep have died from eating its leaves.

The dogbane belongs to the family of the same name, a family of some one hundred and thirty-five genera and over eleven hundred species of herbs, shrubs, trees, and vines scattered throughout the world, but most abundant in the tropics and subtropics. They have a milky juice, alternate, opposite, or whorled simple, entire leaves. The flowers solitary or in clusters (cymes or panicles), are regular and perfect, with a five-lobed calyx, a usually bell-shaped or funnel-shaped corolla, five stamens inserted on the tube of the corolla, and mostly two ovaries. The fruit may be berrylike or a follicle, the seeds often hairy or winged.

Closely related to the spreading dogbane, as a matter of fact a member of the same genus, *Apocynum*, the classical name for dogbane, the Indian hemp is a far less attractive species, though somewhat similar but less spreading and more upright, with greenish white, tiny flowers erectly five-pointed. Unlike its relative that has traps to catch unwanted visitors the Indian hemp solicits the aid of insects to insure cross-fertilization by providing its nectar in shallow receptacles so that even short-tongued bees may feast without harm. And so, one may see about the clusters honeybees and mining bees, wasps, various flies and beetles, and such butterflies as the spangled fritillary and the banded hair streak. The Indian hemp is found in fields and thickets, in meadows and waste places, and on sandy riverbanks. The Indians used to substitute its tough fibers for hemp in making fish nets, mats, baskets, and clothing, hence its name. Its root furnishes an emetic and a cathartic.

The dogbane family is poorly represented in our flora. Beside the two preceding species our only other native species are the climbing dogbane, a twining plant with small, pale yellow flowers found in damp grounds; the Texas star, a bushy species with pale blue flowers; and the blue dogbane a species of the dry slopes of the southwestern deserts.

The dogbane family is also poorly represented among our cultivated plants with only two species, the periwinkle and the oleander, being popular in the North though several tropical and subtropical species may be found in southern gardens. The periwinkle or vinca or creeping myrtle, as it is variously called, is a trailing, hardy evergreen with thin and wiry stems, opposite, broadly lance-shaped, shiny, dark green leaves and light blue flowers three-quarters of an inch across. The periwinkle, a native of Europe and known to Pliny as pervinca, the corruption being the vinca, is used as a ground cover under trees, and as a covering for shady banks. It has escaped to roadsides and similar places. Two other species of periwinkle are the larger periwinkle, a larger edition of the periwinkle and not so hardy, especially in the North, and, of which, a variegated form is much used in window boxes, and the rose or Madagascar periwinkle, a native of Madagascar and a tender, erect, ever-blooming perennial, growing to two feet, with opposite, lance-shaped, prominently veined leaves and showy pink or white flowers with a red eye. Though a perennial, it is mostly grown as a tender annual and chiefly used for summer bedding, as the plants bloom continuously from the time they are set out until frost.

The oleander is not a hardy shrub and in the North is used as a tubbed plant for the outdoors in summer. It also makes a good pot plant for indoors, but as its juice is poisonous children should be warned against it. The common oleander, the most cultivated species, is a shrub or small tree, eight to twenty-five feet high, with narrow, oblong, dark green leaves that are somewhat paler beneath, and white, red, pink, or purple flowers that are often double in some horticultural forms. The other cultivated species is the sweet oleander, a smaller plant, usually not over eight feet high, with narrow, lance-shaped leaves and fragrant white or pink flowers. In both species the corolla is funnel-shaped, its limb bell-shaped, and with five fringed or broad teeth, slightly twisted to the right. The fruit is a cluster of two long, cylindrical follicles.

54. Dogwood (Cornaceae)

It is doubtful that there is a more decorative ornament in the wildwood than the flowering dogwood when its naked branches become ornamented with a multitude of white or pink-flushed blossoms. Strictly speak-

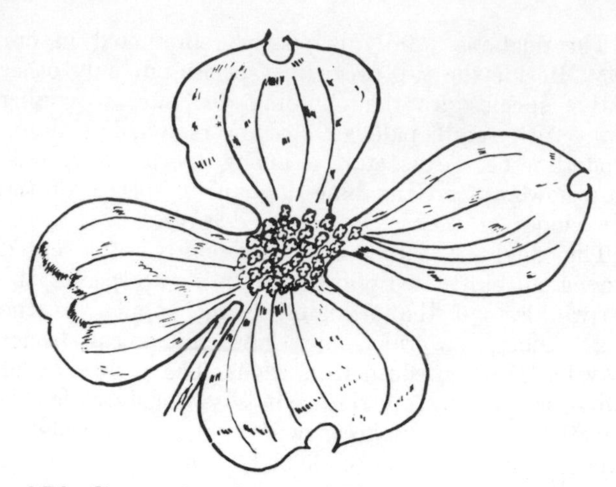

152. Showy Involucre of Flowering Dogwood

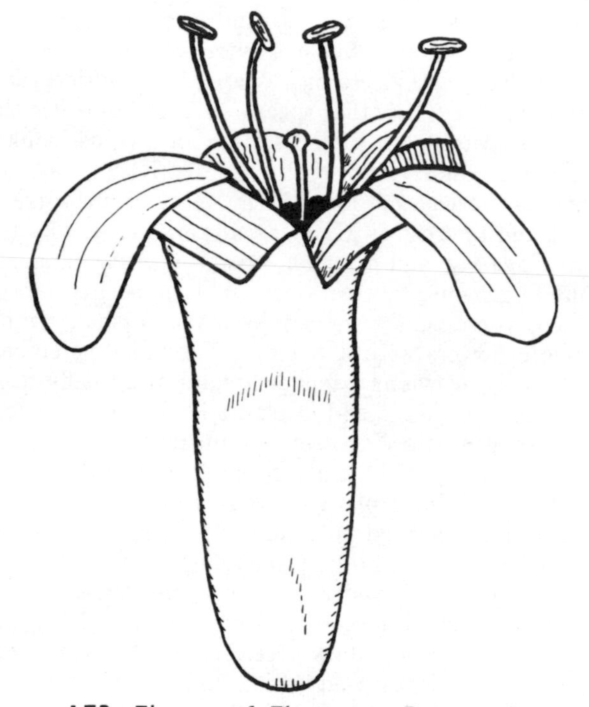

153. Flower of Flowering Dogwood

ing, they should not be called blossoms, for the actual or true blossoms are small and greenish and somewhat tubular. What, then, are the four large white petallike structures that are so eyecatching along a woodland trail or along a country lane? They are merely involucral bracts or scales, reduced leaves designed to attract the mining bees and other insects with similar appetites to the nectar in the floral tubes. A close look at the flowers, in clusters of ten to thirty, and they will appear to have been made on a plan of four; four lobes to the calyx, four divisions of the corolla, and four stamens. There is only a single pistil. Later it will become an egg-shaped scarlet drupe.

The floral pattern of the flowering dogwood is the same for all the dogwoods. There are about forty species native in the north Temperate Zone. They are mostly trees and shrubs with generally opposite, entire leaves and in many species the flowers are grouped in flat-topped or round clusters. These resemble the viburnums, but the latter have five stamens and a united corolla. All the dogwoods belong to the genus *Cornus,* from Latin for cornu, a horn, in allusion to the hardness of the wood. Many of the dogwoods are grown for their handsome flowers, often brightly colored fruits, and in some species for the winter effect of their colored twigs.

The flowering dogwood is a bushy tree, from fifteen to thirty feet high, with a short trunk and spreading branches making a flat-topped head. Its leaves are dark green above, pale and downy beneath, and somewhat clustered at the ends of the branches. It is an attractive, even handsome tree, and during all seasons. In early spring before the leaves have appeared the "white blossoms" transform the tree into one large, glorious bouquet of the landscape. In summer its branching habit and the beauty and grace of the leaves appeal to our aesthetic senses. In the fall its red leaves and shining red berries signal its presence in the splendor of the autumn woods, while in winter its curious turban-shaped, gray buds and the alligator-skinned appearance of its gray, checkered bark are so distinctive that it is easily seen amidst other trees.

The fleshy fruits of the flowering dogwood, as well as those of other dogwoods, are very valuable to wildlife, especially in the Northeast. They ripen in late summer and not only are they available in the fall, but in some species they persist on the plants into the winter months. Both songbirds and game birds feed on the drupes as do many mammals, some of the latter also feeding on the leaves. The flowering dogwood, and probably other dogwoods as well, serve

154. Fruit (Drupe) of Flowering Dogwood

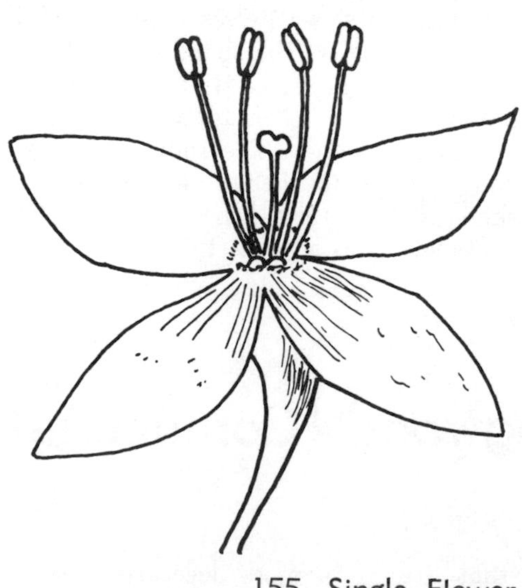

155. Single Flower of Dogwood

as food plants for the caterpillars of the spring azure butterfly.

There seems to be no rhyme or reason for the names of some plants and the dogwoods are a case in point. It is not known how these plants came to be called dogwoods. One suggestion that has been offered is that the name is a corruption of dag-wood or dagger-wood in allusion to the use of the wood for skewers to hold meat together in cooking. Another suggestion is that the leaves and bark were used in making a decoction for washing mangy dogs. But as Parkinson says, in speaking of the European dogwood, "We for the most part call it the dogge-berry tree because the berries are not fit to be given to a dogge."

No matter how or why the dogwoods received their name they are a most attractive group of plants, and not the least, is the red osier dogwood whose purplish stems give a bit of warmth to the winter landscape. It is a low, straggling shrub of wet places and is easily recognizable, not only by the color of its stems and twigs, but also by its opposite, slender, petioled leaves somewhat pointed at the apex and roughish on both sides. Its small, flat-topped flower clusters, and its white or lead-colored fruit are other identifiable features. In a somewhat similar manner the round-leaved dogwood may be recognized by its greenish, warty twigs, its broadly ovate, abruptly pointed leaves that are light green above and woolly beneath, its small, flat white, flower clusters, and light gray drupes. Both dogwoods may be found on shady hillsides, in open woodlands, and in roadside thickets.

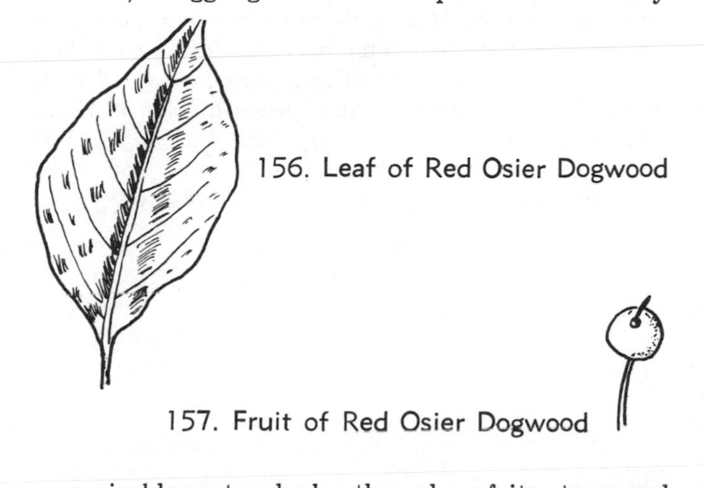

156. Leaf of Red Osier Dogwood

157. Fruit of Red Osier Dogwood

The latest of the dogwoods to blossom, the silky dogwood or kinnikinnick, is a shrub six to ten feet high of low woods and streamsides. Its dull reddish twigs, oval or oblong leaves, round at the base and tapering to a point at the apex, and usually rusty-downed on the lower surface, its compact, flat clusters of white flowers, and its bluish fruit are the features that serve to identify it. A tall shrub, often ten or twelve feet high, and found along the borders of streams and on the margins of lowland woods, the panicled dogwood is a beautiful plant when in bloom with its many convex, loose clusters of white flowers at the ends of the twigs. This species has comparatively long and narrow leaves and white fruit on bright red pedicels. Unlike the other dogwoods, the alternate-leaved dogwood has alternate leaves instead of opposite ones. It is sometimes a shrub, at other times a tree, but in either case, the stem and twigs are green often streaked with white, the leaves have slender petioles, are oval or ovate, dark green above, paler green and pubescent beneath, the flowers are white in broad, flat clusters, and the fruit is dark blue and somewhat less bitter than some of the other species.

The genus *Cornus* has only one herb, the bunchberry, a dainty and attractive little plant common in cool, rich woods. It is rarely more than six inches high, with the short stem ending in a whorl of oval and pointed leaves. In the center of these leaves is a cluster of small yellowish or greenish flowers usually surrounded by four large white bracts having the semblance of petals and giving on the whole the appearance of a single blossom about an inch broad. The flowers are visited by mining bees and woodland flies and are succeeded in late August by a compact bunch of exceedingly beautiful scarlet berries of the most vivid hue. They are edible, but somewhat insipid and tasteless, however, the Indians are said to have eaten them.

The dogwood family includes some ten genera and about one hundred species of mostly trees and shrubs with widely scattered distribution, but the only genera of interest to us are *Cornus*, the dogwoods that have already been described; *Aucuba*, a small genus of Asiatic evergreen shrubs, very popular as foliage plants with opposite leaves, small greenish flowers, the male and female on separate plants, and a usually orange or scarlet berry; and *Nyssa* that some botanists consider as forming a separate family. The genus *Nyssa*, said to be the name of a nymph and "so-called because it (the original species) grows in water," is a small genus of North American and Asiatic trees, two of our native species being known as tupelo, sour gum, or pepperidge, and tupelo gum, bay poplar, or cotton gum.

To see the tupelo at its best is to see it in the fall when its foliage becomes a glowing mass of scarlet, a veritable pillar of fire among the yellowing ashes and hickories, doubtless the most brilliant and fiery of a magnificent group that includes the maple, sumach, dogwood, sassafras, and liquidambar. The tupelo is commonly a small tree about twenty to fifty feet high, but in the South, occasionally reaches a height of one hundred feet or more. It has slender, nearly horizontal branches that gradually and gracefully droop near the ends. The leaves are oval, entire, and glossy dark green, the flowers are also green and are inconspicuous, the staminate and pistillate on different trees in very small clusters, and there are purple black drupes on long stalks. The fruit is rather sour, but it is eaten by a number of birds and is relished by the black bear

158. Leaves and Fruit of Tupelo

and the fox. It is said that it has been used for preserves.

The tupelo is found throughout the East where it grows in lowlands, swamps, along water courses, and borders of swamps. The name tupelo is an Indian name and this is the name whereby the tree is generally known in the Northeast. In the south it is usually called the sour gum, probably because at one time its scarlet foliage led people to think it was a relative of the sweet gum. As for its name of pepperidge, it is not known by this author how it received this name, though this is the name whereby it seems to be called in the Midwest.

In the marshes of the South there is a variety of the tupelo with smaller, narrower leaves. It is also smaller than the parent species and has a much swollen base with large roots that extend out of the water. The flowers are twin-clustered. The important point of difference is the flat and furrowed stone of the commonly twin-clustered fruit. The stone of the type species is ovoid and slightly ridged.

A tree similar to the tupelo and found in the river swamps of the South, the tupelo gum, or sour gum, or bay poplar, or cotton gum, as it is also called, has an unusually broad base, numerous corky roots, and a pyramidal crown. It may often have a trunk diameter of three to four feet, and may reach a height of a hundred feet. The leaves are large, oval, slightly toothed, green above, paler beneath, the pistillate flower, rather large, is solitary, and the single, dark violet blue, berrylike drupe is ellipsoidal and about an inch long. It is used in making a preserve.

55. Duckweed (Lemnaceae)

In midsummer one will often see what appears to be a green blanket on the surface of a pond, along the quiet edges of a lake or a slow-moving stream, or in

159. Duckweed

a ditch. This green blanket consists of innumerable small plants called duckweeds. They belong to the family of the same name, a small group of some three genera and a few dozen species that are widely distributed, and all being floating aquatics without leaves or stems. They are all small, either globular or flattish in form, and consist merely of a plant body called a thallus, generally known as a leaf since it looks very much like one. They may have one to several roots, or none at all, and their flowers are

160. Greater Duckweed

very simple. They are minute and grow directly out of the edge or upper surface of the plant body. The male flowers consist of only one stamen, the female of only one pistil. Actually both flowers are rare, for the duckweeds usually reproduce by "budding" off, or the thallus merely divides into two parts, each producing roots and becoming a new plant. The duckweeds are among the smallest of the flowering plants, though not quite the smallest, and thousands of them may often cover the entire surface of a pond or slow stream. Ducks are very fond of them, hence their name. They are eaten by fish, too.

In many places the greater duckweed is the common species. It belongs to the genus *Spirodela,* Greek in allusion to the cluster of rootlets, and on the surface of ponds, rivers, pools, and shallow lakes it forms an extensive green blanket of thousands of individual plants, each with five to fifteen rootlets. The plant body, or thallus, is circular and about a quarter of an inch in diameter, green above and pinkish purple beneath, with five to fifteen simple veins.

Of the several species of duckweeds belonging to the genus *Lemna,* Greek in allusion to these small plants growing in swamps, the lesser duckweed is the commonest. The thallus is a fraction of an inch long and is obovate or subcircular with only one root and one to five simple veins, but in spite of its small size, it forms green blankets on the surface of ponds, lakes, and stagnant waters. A characteristic of the ivy-leaved duckweed of the same genus, is that it forms zigzag chains or lattices floating just below the surface of the water. The thallus is oval or lanceolate and is often without rootlets. It occurs in ponds and ditches and especially in cattail marshes.

Even smaller than the duckweeds are the wolffias. They are the smallest of all flowering plants. The wolffias look like green grains floating just below the surface film, and are often wedged in between the duckweed plants. The Columbia wolffia has a spherical thallus, the Brazil wolffia has an oblong one. The former occurs in stagnant ponds, pools, and shallow lakes, and is common in the eastern and central states, while the latter occurs in stagnant waters of the central and southern states. In June and July the wolffias of the genus *Wolffia,* named in honor of Johann F. Wolff, a German physician and naturalist, become very abundant and then disappear, moving to the bottom where they spend the winter, rising again to the surface in the following spring.

161. Wolffia

Besides serving as food for ducks and fishes, the duckweeds also serve as food plants for a certain fly and beetle, the larvae of the fly boring into the thallus and the larvae of the beetle mining into it. A moth and the larva of a caddis fly both make cases out of the thalli, and such animals as hydras and planarians browse over their surface.

56. Ebony (Ebenaceae)

They have a plumb which they call pessemmins, like to a medlar, in England, but of a deeper tawnie cullour; they grow on a most high tree. When they are not fully ripe, they are harsh and choakie, and furre in a man's mouth like allam, howbeit being taken fully ripe, yt is a reasonable pleasant fruict, somewhat lushious. I have seene our people put them into their baked and sodden puddings; there be whose tast allows them to be as pretious as the English apricock; I confess it is a good kind of horse plumb. (*The Historie of Travaille into Virginia Brittania*)

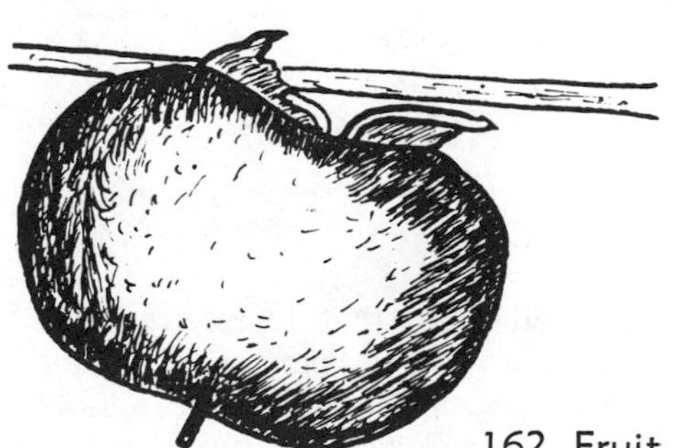

162. Fruit of Persimmon

If one has never tasted a persimmon and does so before it ripens, one will probably hesitate to taste it when it is ripe because it is very astringent and puckers the mouth when still in the ripening stage. The fruit is a large globular berry, about an inch and a half in diameter, at first green, but orange-colored or reddish yellow when ripe, with several large flat seeds. It is sweet and edible when ripe, especially after a frost, although this partially destroys the flavor, but before it is ripe it is very astringent due to the presence of tannic acid that later disappears. The early settlers made a bread from the fruit, and it has been said that it was superior to gingerbread, a custom very likely learned from the Indians. The fruit is also valuable to wildlife. It hangs on trees well into winter, and is an important food for foxes, raccoons, opossums, skunks, and ringtailed cats as well as birds, such as the catbird, robin, and cedar waxwing.

Although the persimmon is essentially a southern tree, common in fields and open woods, it has a northern range that extends into southern New England. It is a slender, tall tree, forty to sixty feet high, occasionally higher, with a handsome round head. It has a sepia brown or grayish brown, somewhat corky, bark broken into thick, scaly plates, angular twigs, and ascending branches. The leaves are oval to ovate, a lustrous green, but paler and a trifle downy beneath, the flowers are small, greenish yellow, bell-shaped,

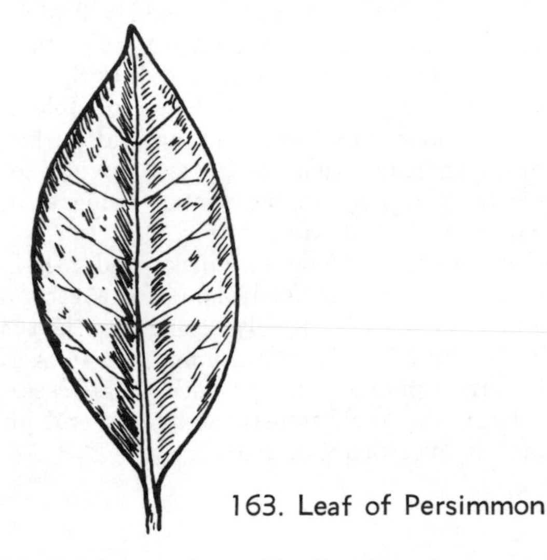

163. Leaf of Persimmon

with four lobes, and usually dioecious, but sometimes perfect flowers occur, the staminate in small clusters, and the pistillate solitary. The tree is inclined to vary in the character and quality of its fruit, in size from that of a small cherry to a small apple, and also in flavor. A tree, for instance, may bear fruit that is never palatable, even after a frost, or a tree may bear delicious fruit even before the first frost.

The persimmon belongs to the genus *Diospyros*, Greek for Jove's grain, in allusion to the edible fruit, a genus of some two hundred species of tropical and subtropical plants. There are two native species, the common persimmon just described and the black persimmon, a scrubby tree of Texas with dark, leathery leaves and a black insipid fruit. Its juice is used as a black dye. Widely cultivated in the South and in California, the Japanese persimmon bears large luscious fruits that are superior to those of our native species. It is a deciduous tree, reaching a height of forty feet or more, with oval leaves that are glossy above, somewhat downy beneath, yellowish flowers, and an orange-colored fruit.

The persimmon is a member of the ebony family, a group of largely tropical usually hard-wooded trees and shrubs classified into six genera and about three hundred species. The leaves are alternate, entire, often leathery, while the flowers are regular, sometimes perfect, solitary or in cymose clusters, the staminate and pistillate flowers often on separate trees, but always separate if on the same tree. The calyx is three to seven-lobed, the corolla is bell-shaped or tubular, also three to seven-lobed, the stamens number two to four times the number of corolla lobes, and the ovary is three to many-celled with one to two ovules in each cell. The fruit is a berry.

Other than the persimmons the only other members in the family of any interest are the various species, of the same genus as the persimmons, that produce ebony wood, a hard, heavy, and durable wood that exhibits great variation in color, the most highly prized being black because it is susceptible of a fine polish.

57. Elm (Ulmaceae)

It is the law of the wildwood that forest trees shall bring forth their flowers before their leaves. Poplars and aspens, the American elm, and silver and red maples do so in early spring, while birches, oaks, and hickories do so somewhat later. The elm, tracing its lineage to the Miocene epoch, is one of the first trees to accept the challenge of March with its meteorological vagaries, and even when the snow is still on the ground the flower buds begin to swell and shine. Before April is well on its way, these buds shake off their brown scales and the tree appears as in a coppery mist. A close look at a twig reveals eight to twenty tiny flowers in umbellike clusters. Viewed through a magnifying glass or hand lens, the flowers are most attractive with their bell-shaped, reddish green calyx cups, bright red anthers, and pale green pistils. Just before the lopsided leaves appear a few weeks later, the blossoms will have given way to little flat, oval, green samaras winged all around for flight on the air currents. They can be found on the ground, but for the most part they lie there unnoticed.

There are about eighteen species of elms of the genus *Ulmus*, the classical Latin name of the elm, all from the north Temperate Zone. They are mostly tall trees with alternate, simple, short-stalked, doubly toothed leaves that are somewhat oblique. The flowers, without petals, the calyx bell-shaped and inconspicuous, appearing before the leaves unfold, or in some species after, occur on slender pedicels in fascicles or cymes, and have an ovary with a two-lobed style. The fruit is a flattened oblong or somewhat circular samara.

Almost everyone knows the American elm, for its tall, vaselike form is a conspicuous feature on the landscape and is easily recognized at a distance. The American elm comes in various forms, but the most common is the Etruscan form, a base gradually flaring into a round dome, especially when the tree grows out in the open. Then its trunk usually divides near the ground into several divergent, erect limbs that arch and end in many slender, drooping branchlets forming a vaselike crown of considerable beauty and symmetry. There are narrower forms whose tall trunks end in limbs that form a brush at the top, much like a feather duster, and then there is the "oak tree form" that is wider and broader than the vase form and somewhat suggestive of the ample crown of an oak.

The American elm, whose root system is extensive, but shallow, is a tree fifty to seventy feet high, sometimes as much as one hundred and twenty feet, of rich alluvial soil, especially on intervales and along the margins of streams, with a light or dark gray, rough bark, that is divided by many short, perpendicular furrows into flat-topped ridges, and slender, brownish twigs. Its leaves are oval to oblong, unequal at the base, smooth or roughish above, sometimes hairy beneath, and dark green. The flowers are small, in fascicles, and the fruit is notched at the tip, while hairy on the margin.

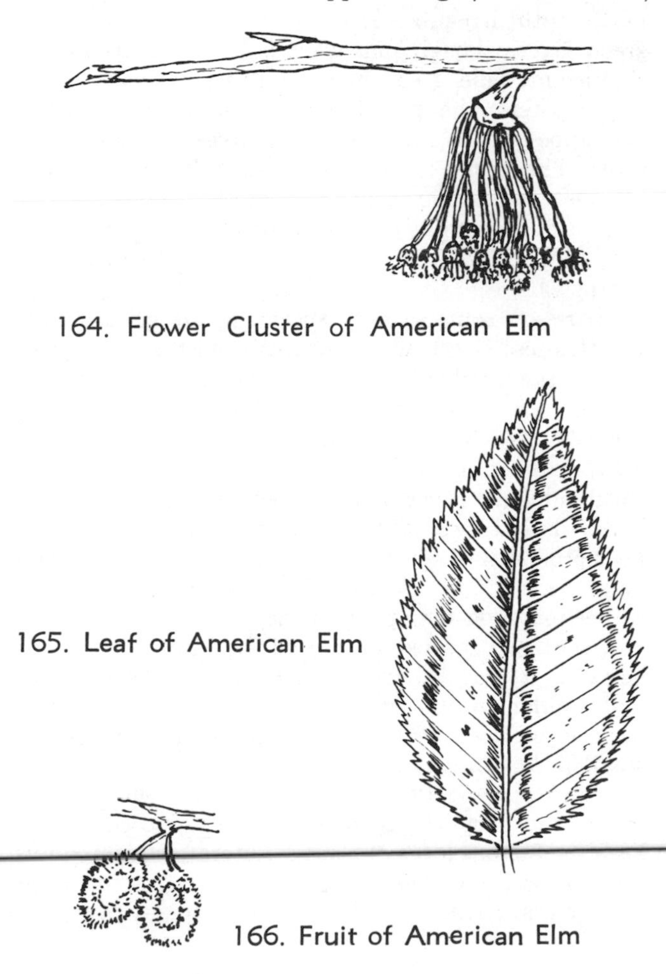

164. Flower Cluster of American Elm

165. Leaf of American Elm

166. Fruit of American Elm

In the past, the American elm has been highly valued as an ornamental and street tree, but it does not thrive in the smoke-laden atmosphere of industrial cities. In recent years the Dutch elm disease has threatened the tree with extinction and many an old and majestic elm has had to be cut down. The American elm is a favorite nesting site of the Baltimore oriole and in the winter, when the leaves have fallen, deserted nests can be seen swinging in the wind from the high outer limbs where they were effectively concealed by the foliage in the summer.

Several American elms have had historic associations. Washington is said to have taken command of the American Army beneath one of them in Cambridge, Massachusetts, and under the branches of another, William Penn made his famous treaty with the Indians. It stood in one of the suburbs of Philadelphia, and its site is now marked by a marble column.

The Indians of the Missouri River valley once cooked the bark of the slippery elm with buffalo fat in rendering the tallow, to give the latter a pleasing taste. In the early days, the pioneer woodsmen chewed its inner mucilaginous bark to quench their thirst. It was also used as a remedy for inflammation of the throat after having been steeped in water, and poultices were made of it to relieve chest troubles. It has been said that a flour prepared from the dried and ground bark and mixed with milk forms a wholesome and nutritious food for infants and invalids. The Indians of New York called the tree Oo-hoosh-ah—"It slips."

An elm can invariably be distinguished by its lopsided leaves, that is, they are uneven or oblique at the base. Those of the slippery elm are no exception. The slippery elm is a commonly slim and rough tree, forty to fifty feet high, of rich alluvial soil and rocky hillsides. Everything about the tree is rough and coarser than its relatives. Its bark, a brownish gray, is rough both on the trunk and branches, and its leaves are rough to the touch, one of the striking characteristics of the tree. They are covered with harsh, tubercular hairs and when a leaf is crumbled by hand, it grates unpleasantly on the ear. The twigs, too, are rough, grayish, and hairy. The buds are also hairy, the leaf buds are a dark chestnut brown to almost black, the flower buds often have an orange tip.

The cork elm can be recognized by the corky ridges on the twigs. It is a medium-sized tree, fifty to seventy feet tall, occasionally taller, with a straight trunk that tends to remain unbranched for some distance into the crown, but then, with short spreading branches, forms a round-topped head. Its bark is gray, tinged with red, deeply furrowed, and its leaves are similar to those of the white elm. The cork elm, also called the rock elm because of its extremely hard, tough wood, is a tree of riverbanks, gravelly or rich soil, and calcareous ridges. Apart from its corky ridges, it may be distinguished in the spring by its racemes of drooping blossoms and later by its samaras that mature when the leaves are half-grown and that are winged all around. They are somewhat oval in shape and pubescent with a ciliated margin.

The smallest of the elms with the smallest samaras is the wahoo elm, a southern species thirty to forty feet high, of dry, gravelly uplands, though it is also found in moist soil. It, too, has corky outgrowths on the twigs, but they are in the form of two opposite thin wings that are abruptly interrupted at the nodes. Another southern species is the red elm, so named because of its red brown wood. It occurs in limestone regions, on hillsides, or in valleys between streams, and its twigs, also, have two or three corky wings. An unusual feature of the tree is that it flowers very late in the summer. Blooming in late summer, also, is the cedar elm, a small-leaved species common in the bottomlands of the Mississippi valley. There appears to be no reason for its common name other than it grows with cedars on the dry limestone hills of Texas.

An introduced species, the English elm, though not essentially English, is a large tree, frequently reaching a height of ninety feet or more with an oblong, round-topped figure, often massed in two sections with the upper one being larger. It has a dark gray brown bark and its leaves are similar to those of the American elm, but very rough on the upper surface. It is a handsome tree with many varieties, some with variegated foliage, some purple-tinged, and some with more erect branches than the typical plant.

According to classical literature, the elms of ancient times were important trees. The Romans fed their leaves to cattle and they were planted in vineyards to support the vines. Virgil writes that oak branches were successfully grafted on elm trees and that swine ate the acorns that dropped from the fruiting branches of what must have been an extraordinary tree.

The genus *Ulmus* is one of about fifteen genera of the elm family, a group of about one hundred and forty species of deciduous trees and shrubs, widely distributed in the north Temperate Zone, with a few species in the tropics. The family includes several important ornamental and timber-producing species.

The leaves are alternate, simple, pinnately veined, often very unsymmetrical at the base, and coarsely toothed. The flowers are small or inconspicuous, perfect or imperfect, without petals, the staminate and pistillate separate, but on the same plant. The calyx consists of four to nine united sepals, the stamens are four to five in number, and the ovary is usually one-celled with a single ovule. The fruit is a samara, drupe, or nut.

To find an elm with berries is to wonder if nature has gone astray, for the hackberry at a casual glance looks much like an elm. But a more careful look will show the finer spray of the hackberry twigs, its more horizontal, and less drooping branches, and though the leaves are superficially like those of the elm, they have three conspicuous ribs diverging from the base, totally unlike the straight-ribbed leaves of the elm. Of course, the little axillary sugar berries are quite unlike the elm samaras. There are few months in the year when, green or ocher yellow or purple black, they are not to be found on the tree.

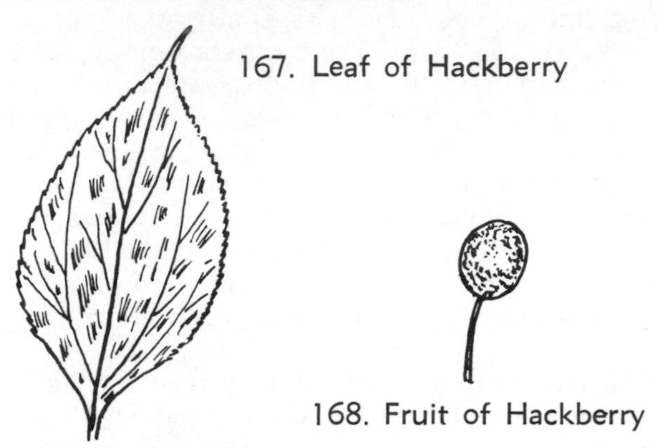
167. Leaf of Hackberry

168. Fruit of Hackberry

The hackberry is usually a small tree, thirty to forty feet tall, but actually varying in size, of rich woods and riverbanks and widespread in the United States east of the Rocky Mountains. It has a light brown bark, with irregular shallow ridges, lopsided olive green leaves though somewhat paler beneath, and insignificant, pale green flowers. The staminate are in tiny clusters, the pistillate are solitary or in pairs, and appear with the unfolding leaves or soon thereafter. The berry is cherrylike, slightly obovoid, and is sweet and edible. The twigs are so frequently disfigured by insect galls, the so-called "witches' brooms," that they could almost be given as a distinguishing character.

There are some seventy species in the genus *Celtis*, the classical Greek name for a tree with a sweet fruit, though not applicable to the hackberries, but only a few species occur in the United States. The hackberry occurs in two varieties, one a dwarf form a few feet high, with irregular straggling branches, actually a shrub, and the other a tree sometimes as much as one hundred and twenty-five feet high that may often be distinguished by its finely downed branchlets and large leaves that are heart-shaped at the base and rather rough on the upper surface.

A southern species, known as the Mississippi hackberry, is a graceful tree of rich, wooded bottomlands and similar to the preceding species except that its bark is covered with warty outgrowths, and is also smaller. An Eurasian species, known as the European nettle is a tree forty to seventy feet high, with ovalish leaves and purplish green fruit. It is supposed to have been the famous *lotus* of classical literature. Herodotus, Dioscorides, and Theophrastus all speak of its sweet and pleasant fruit, and Homer has Ulysses say:

I sent explorers forth—two chosen men,
A herald was the third—to learn what race
Of mortals nourished by the fruits of earth
Possessed the land. They went and found themselves
Among the Lotus-eaters soon, who used
No violence against their lives, but gave
Into their hands the lotus plant to taste.
Whoever tasted once of that sweet food
Wished not to see his native country more
Nor give his friends the knowledge of his fate;
And then my messengers desired to dwell
Among the Lotus-eaters, and to feed
Upon the lotus, never to return.

The elms are not as useful to wildlife as various other trees, but the buds and seeds are eaten by both songbirds and game birds as well as by squirrels. The fruits of the hackberries, however, are very popular with many winter birds especially with the cedar waxwing, yellow-bellied sapsucker, mockingbird, and robin. They are also eaten by the gray fox, opossum, and flying squirrel. Both the elm and hackberry are used as food plants by the caterpillars of such species of butterflies as the question mark, the hop merchant or comma, the mourning cloak, the tawny emperor, and the hackberry butterfly.

A single species of an elmlike tree is the planer tree or water elm, a southern species common in the swamps of the coastal plain from North Carolina to Florida and westward to Texas. It is of interest chiefly as a botanical remnant of its family. At one time, several species of its genus once grew in Alaska and in the Rocky Mountains. The planer tree, twenty-five to forty feet high, has a nearly smooth, dark brown-gray bark, small elmlike leaves, and staminate and pistillate flowers in small clusters at the base of the leaves. The fruit is a small drupe, with fleshy, irregularly crested ribs.

58. Evening Primrose (Onagraceae)

By day, the evening primrose has little to commend it, for when it is seen by the roadside, or in a thicket, or fence corner, its wilted, faded flowers and hairy capsules, crowded among the willowlike leaves at the top, give it a rather bedraggled appearance. When the sun has set in the western sky and twilight begins to creep over the landscape, then developing buds begin to open, and pure yellow, lemon-scented flowers appear gleaming like miniature moons when other flowers have melted into the deepening darkness. Bumblebees and honeybees may visit the flowers in the morning before they have had a chance to close, but it is the night-flying moths that the plant depends upon for cross-pollination. It is not without reason that

169. Leaves and Flowers of Evening Primrose

the flowers are yellow, that their fragrance becomes stronger with the night, that the nectar wells are located in tubes so deep that none but the moths' long tongues can drain the last drop, or that the golden pollen is loosely connected by cobwebby threads on eight prominent and spreading stamens.

> A tuft of evening primroses,
> O'er which the wind may hover till it dozes,
> But that it is ever started by the leap
> Of buds into ripe flowers.

The Isabella tiger moth is perhaps the plant's chief benefactor, but there are others, among them the sphinx moths. When morning comes one may find a little rose pink moth, its wings bordered with yellow, asleep in a wilted blossom, and since the flowers turn pink when faded, the moth is safe from prying eyes. Later in the summer when enough seed has been set, the primrose changes in habit, and the flowers remain open all day. If one of the brown seed capsules is cut open one will see that the small brown seeds are neatly arranged in eight parallel rows. The evening primrose can be recognized by its long calyx tube, at the end of which is the flower. The evening primrose was introduced into Europe in as early as 1614, and was at one time cultivated in the English gardens for its edible roots that, when boiled, were said to be wholesome and nutritious. The roots are also medicinally valuable.

There are several species of evening primroses that differ from one another in minor details. Both they and their day-blooming relatives, the sundrops, belong to a genus called *Oenothera.* They have alternate leaves and showy flowers that are prevailingly yellow, with a tubular calyx that is usually four-sided, its four lobes often being bent backward, and four petals. The fruit is a four-angled capsule.

As with the evening primrose, there are several species of sundrops, but the species known simply as sundrops is perhaps the most common and the one more generally cultivated. It is a somewhat variable species with an erect, branching stem, oblong or lance-shaped leaves that are short-stalked or sessile, and brilliant yellow flowers, nearly two inches wide, borne in a loose cluster or spike. The sundrops is found in fields and on roadsides where its flowers, from May to July, blatantly advertise their wares to bumblebees and butterflies, who drain the deeply hidden nectar. Mining and leaf-cutter bees, wasps, flowerflies and beetles also visit them for their abundant pollen.

170. Leaves and Flowers of Sundrops

The evening primrose family is a somewhat conspicuous group of annual and perennial plants that are mostly herbaceous and rarely woody. Most of them are native to the temperate parts of the New World, but a few interesting species occur in Europe and Africa. The family includes over thirty genera and about five hundred species. Their leaves are opposite or alternate, and usually simple. Their flowers are usually showy, perfect and regular, but sometimes irregular, occasionally solitary, but more often in handsome clusters. The sepals are four to five in number, typically four, forming a tube that is joined to the ovary. The petals are four to five, alternate with the sepals on the rim of the receptacle. The stamens are as many or twice as many as the petals, and are inserted on the throat of the calyx tube. The ovary is two to four-celled. Their fruit is a dry pod (capsule) in some genera, fleshy and berrylike in others. The members of the family are often easily recognized because the floral structures are arranged in sets of four or multiples of four.

Many of the members of the family are grown for ornamental purposes and many wild species are noteworthy for their beautiful flowers. One species, the fireweed, has the unusual distinction of appearing in a place recently scorched by a fire. In such places the spikes of the beautiful, brilliant magenta flowers, on stems as much as eight feet tall, soon appear to cover the ugliness of the blackened ground as if nature abhorred the unsightly devastation. Beginning at the bottom of the long spike, the flowers open in slow succession throughout the summer, leaving behind slender, velvety, gracefully curved, purple-tinged seed pods that are as effective as the flowers in covering the ravages of fire.

The fireweed is also known as the willow herb and sometimes as the *Epilobium,* its generic name. There are some two hundred species of *Epilobiums* with usually willowlike leaves and flowers in showy terminal clusters. Probably the most familiar species is the great or spiked willow herb that is common in fields, roadsides, and burned areas. It is a tall plant with a ruddy stem, dark olive green, lance-shaped leaves, and light magenta flowers. In general structure it is not unlike the evening primrose, but bees, not moths, are its chief benefactors. The young shoots of the plant are often used as a substitute for asparagus and in Canada and northern Europe both the stems and leaves when boiled are used as a potherb.

The fruit of the willow herb is a many-seeded, slender capsule, the seeds each with a tuft of long hairs at the end, so that when the capsule opens in the fall and liberates the seeds they give the plant a rather wild and dishevelled appearance. There are a number of willow herbs or *Epilobiums,* (the name Epilobium coming from the Greek and meaning "upon the pod" in reference to the flower being placed upon a long ovary), such as the hairy willow herb an immigrant from Europe, whose stem is densely covered with rather long, fine, straight spreading hairs, and that grows in waste places and about dwellings; the linear-leaved willow herb, a very small branching species found in swamps with very narrow acute leaves, somewhat hairy throughout, the flowers magenta pink or whitish, and the seeds with dingy threads; the downy willow herb, somewhat taller than the latter with minute, whitish hairs and sessile leaves, growing in bogs or swamps; and the purple-leaved willow herb, also a plant of wet situations, with a minutely haired branching stem, often ruddy, lance-shaped, yellow green leaves, and tiny lilac flowers, the seeds with a tuft of brown hairs.

Common in swamps the seedbox is a smooth, rather tall plant with toothless, lance-shaped opposite leaves and conspicuous, solitary yellow flowers whose sepals are nearly as long as the petals. When ripe, the seeds become loose and rattle about in the seed pod when the plant is shaken, hence, its name. Sometimes growing wholly in the water, at other times its stems lying on the mud, the water purslane is a common, but somewhat uninteresting aquatic plant of marshes and ditches. When growing in water, the inconspicuous flowers lack petals, but when growing out of water they have very small reddish ones.

Originally, the name of enchanter's nightshade was the name of a truly noxious plant of Europe, but how it came to be applied to a harmless and inconspicuous plant of damp and shady woodlands is something of a mystery. Our enchanter's nightshade is a rather low-growing species with thin, frail, opposite, deep green leaves and very small, white flowers in terminal and lateral racemes. The flowers have only two petals, but they are so deeply cut that they appear as four. Later, the flowers form little green burlike, white-haired seed pods, that are somewhat pear-shaped. A smaller species, whose stem is watery and translucent and ruddy, has club-shaped seed pods.

Several members of the evening primrose family are garden plants such as clarkia and gaura, but the best known is the fuchsia, an essentially greenhouse plant. It was first discovered by a missionary in South America, who named it in honor of Fuchs, a German botanist. It is a somewhat woody shrub with opposite, toothed leaves and flowers hanging on slender peduncles in terminal racemes and in the axils of the leaves. They are usually very showy and may be red, purple, blue, or white, or sometimes all four. There are some eighty species of fuchsias and not only are they popular as greenhouse plants, but they also make good house plants. In the summer they may be used for bedding and for window boxes, but in the South they may be grown outdoors all year.

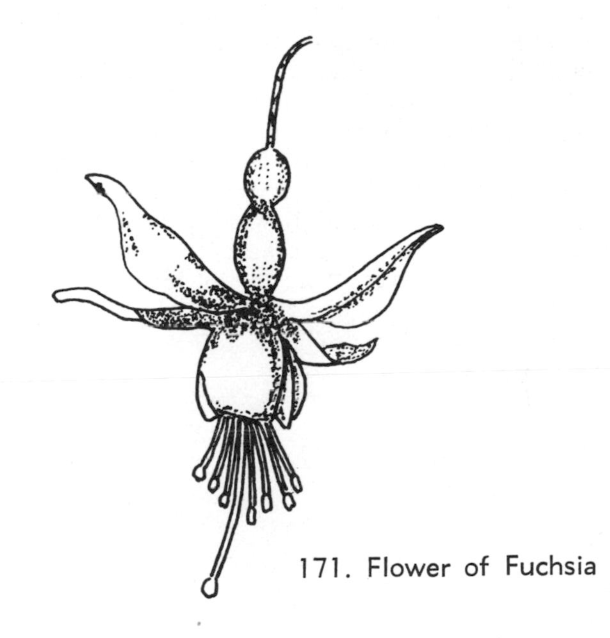

171. Flower of Fuchsia

59. False Mermaid (Limnanthaceae)

A little-known plant of marshes and riverbanks, the false mermaid is a slender, weak annual herb of the genus *Floerkea,* named from Gustav Heinrich Flörke, a German botanist, with deeply divided leaves of three or five narrow, toothless segments, and solitary white flowers at the ends of long stalks from each of the upper leaf joints.

The false mermaid is one of five species of the family

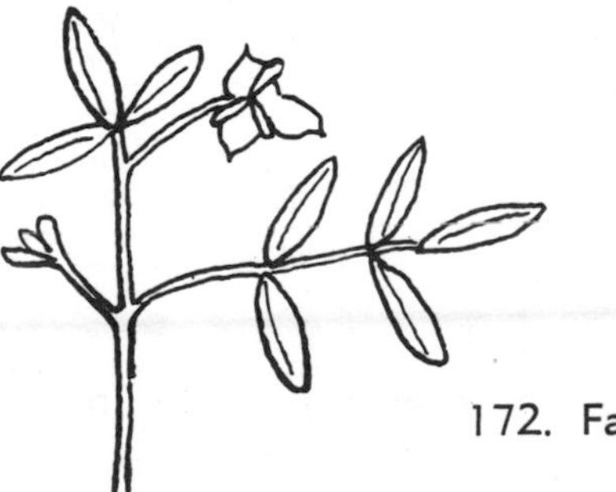

172. False Mermaid

of the same name, a small group of annual herbs comprising only two genera both being American. They have alternate, pinnately divided leaves, perfect, regular white, pink or red flowers solitary on long stalks, with three to five sepals, three to five petals, twice as many stamens, as many carpels as sepals, and one-ovuled, and a dry fruit.

The second genus *Limnanthes* Greek for marsh flowers in allusion to the habitat of the plants, is a group of four western North American herbs with the characteristics of the family. One of them, the meadow foam or marsh flower, is occasionally cultivated in the flower garden. It is a spreading or sprawling plant, four to eight inches high, and usually branches from the base.

60. Figwort (Scrophulariaceae)

If one should watch a bumblebee when it visits the butter-and-eggs one will see how the insect, when guided by the gold orange palate to the place where the curious flower opens, depresses the lower lip by its weight and makes an entrance through the gaping mouth large enough for it to pass into the throat of the corolla. One will also be able to see how its back brushes off the pollen from the stamens overhead as its tongue seeks the nectar deep within the spur, and how the gaping mouth of the flower springs tightly shut after the bee has taken its departure. A weed it may be, but a more beautiful weed is hard to find, with its canary yellow and orange cornucopias and gray green linear leaves. Were this plebian perennial less satisfied to grow everywhere without cultivation, it might well have become a garden ornamental instead of decorating a city lot or some equally unattractive place.

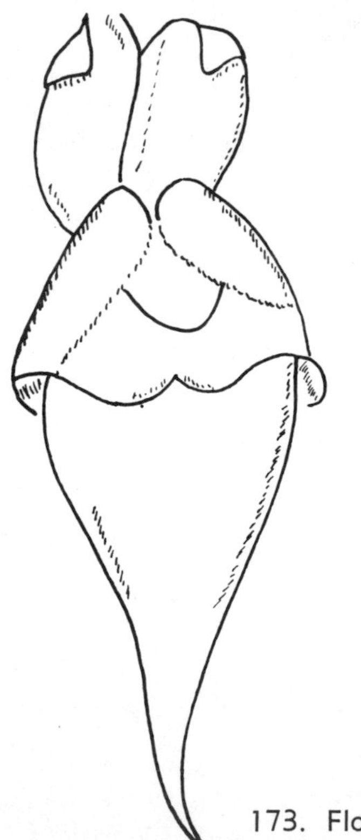

173. Flower of Butter-and-eggs

The butter-and-eggs has a number of other names: toadflax, eggs-and-bacon, flaxweed, brideweed, and wild snapdragon. It is a perennial from Europe, but long since naturalized, growing in fields, pastures, and city lots everywhere. Cattle dislike its taste and odor and leave it alone, allowing it to reproduce itself unmolested. Upon examining one of its flowers, it will be found that it is irregular with the corolla spurred at the base, two-lipped, the upper lip erect, two-lobed, the lower lip spreading, three-lobed, its base an orange-colored palate closing the throat, with four stamens in pairs within, and with one pistil.

There are other plants with a somewhat similar kind of flower, some three thousand of them in fact, in a hundred and eighty genera, and all members of the figwort family, also known as the snapdragon or foxglove family. The scientific name of the family is *Scrophulariaceae,* from the genus *Scrophularia,* so-called because one of its species was believed to be a remedy for scrofula, a form of tuberculosis that affects the lymph glands. Most of the plants of the family are herbs that are generally found in temperate regions, but there are also many shrubs and trees in the tropics, and all are widely distributed. The leaves are alternate, opposite or whorled. The flowers are irregular, perfect, with the calyx more or less tubular, four to five toothed or divided, the corolla united, its four to five lobes nearly equal or two-lipped, with four stamens, and the style entire or two-lobed. The fruit is usually a capsule. The family does not contain any fruits or vegetables all of them being either garden ornamentals or wildings. Many are poisonous, while a few are of value in medicine, notably the foxglove (*Digitalis*).

The butter-and-eggs belongs to the genus *Linaria,* from the Latin *Linum* for flax that the leaves of some species resemble. Also of the same genus is the blue toad flax, a seemingly weak and delicate plant with few small, linear, light green leaves and small pale violet or lavender flowers. It is hardy enough, however, to spread throughout the country, and now is common in dry, sandy soil everywhere. Unlike that

174. Flower of Blue Toadflax

of its relative, the butter-and-eggs, the corolla of the blue toadflax is so contracted that bees are unable to reach the nectar. Flies also visit the flowers, but it is the butterflies that are especially attracted to the slender-spurred flowers, and attend them in the greatest numbers. As these butterflies cannot secure the nectar without touching both the anthers and the stigma, they are the chief agents in cross-pollination.

Throughout the summer the great mullein, sometimes as tall as seven feet and at home in old fields, pastures, and waste places, is a conspicuous feature of the landscape with its yellow flowers an inch or so across, and its velvety-haired leaves. Merely a glance at the plant will show that in addition to the few small pointed leaves that are seated on the stem there is also a rosette of basal leaves, thick pale green, and densely woolly with branched and interlacing hairs. Grazing animals will not touch these feltlike leaves, but hibernating insects find them a safe winter shelter. Why such woolly leaves? Simply to help the exquisite rosettes formed by the year old plants, to endure through the winter so that they can send up a flower stalk the second spring. The flowers open for only a day, but are succeeded by others on an ever-lengthening spike, the older ones quickly producing capsules filled with many brown seeds that are eagerly sought by the goldfinch.

Quite different from its heavy, somewhat sluggish looking relative, the moth mullein is a sprightly and

175. Flower of Moth Mullein

slender species with light yellow or sometimes white flowers, marked with lavender or brown, and found along the wayside and in fields and pastures. It is pleasing to the eye at all times, even in the winter when its slender stem, beset with many seed vessels, rises above the snow to advertise a banquet table for hungry birds.

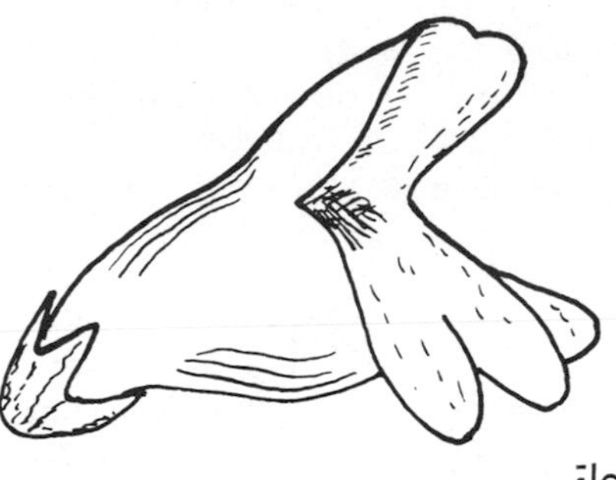

Flower of Turtlehead

It takes little imagination to invest the flower of the turtlehead with reptilianlike features. This smooth-stemmed plant, with deep green, toothed, lance-shaped leaves and white flowers, delicately tinged with magenta pink or crimson pink, and not unlike a turtle's head, sometimes grows three feet tall in swamps, along a brookside, or by the pond's edge where it may be seen mirrored in the quiet waters. The bumblebee is a regular visitor, but the Baltimore butterfly is often found hovering near, indeed, this butterfly occurs only where the turtlehead grows and rarely wanders farther than a hundred yards from it. Not that the butterfly visits the blossoms, it does only upon a rare occasion, but the reason is because the caterpillars feed on the leaves, and do so almost exclusively. They feed on the plant in a communal web until autumn, then hibernate in the web, and in the spring continue their feeding, though now more solitary, until it is time to pupate. The emerging butterflies mate, the females lay their eggs on the leaves, and the cycle is repeated.

If a little imagination can see the reptilian features of a turtlehead flower, then a little imagination can

177. Flower of Monkey flower

also see the grinning face of a monkey in the corolla of the monkey flower. A smooth perennial with an upright square stem, light green lustreless leaves, rich clear purple or violet flowers, the monkey flower is a plant of swamps, brooksides, and meadows. In similar situations the American brooklime is also found. It is a perennial species with a hollow smooth stem that creeps over the ground and then becomes erect and branching. Its tiny flowers, lavender blue, striped

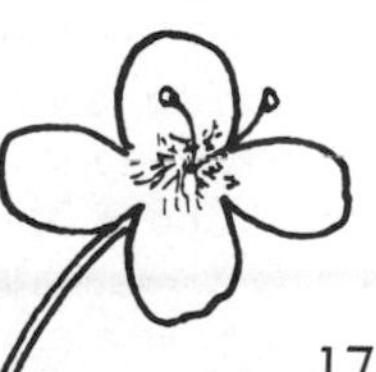

178. Flower of American Brooklime

with violet, and with a white center, might easily be mistaken for forget-me-nots by one in haste, but a more careful look will show the difference.

The American brooklime is a speedwell and probably the most beautiful of all our native speedwells, a group of plants of the genus *Veronica*, named after St. Veronica. They generally have opposite leaves and small, pink, white, blue or purple flowers mostly in terminal clusters (spikes or racemes), with a four-lobed rotate corolla, two stamens, and a compressed capsule.

There are several speedwells: the marsh speedwell, a delicate, tender plant, with racemes of small, pale blue flowers, single or in pairs, on rather zigzag stems found in swamps; the thyme-leaved speedwell, a small species, growing flat and matlike on the ground with the opposite flower branches standing three or four inches high, the flowers nearly white or light blue, with dark blue lines, in loose racemes, found in fields and thickets; and the common speedwell, a woolly species with prostrate stems that finally become erect, and pale blue flowers crowded on slender spikes, an immigrant from Europe and common in dry fields and wooded uplands. This is the speedwell whereof Tennyson said, "The little speedwell's daring blue," and it is also the *Veronica* that was named from *vera icon*, the true image. People in medieval times thought

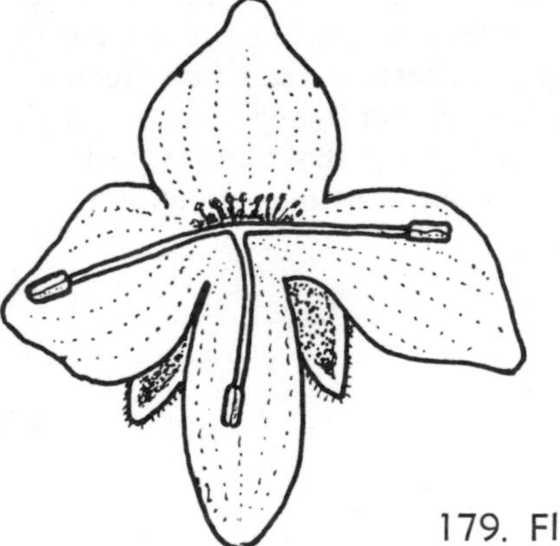

179. Flower of Speedwell

they saw in the flower a resemblance to Christ's feature in the napkin that He wiped His face with on His way to Calvary, and later becoming one of the most precious relics at St. Peter's.

Also belonging to the same genus are the Culver's root or Culver's physic and the various garden species of Veronica. Culver's root or Culver's physic, the last name doubtless in allusion to the violent emetic or cathartic that is contained in the root, is a very tall, smooth perennial species, more common in the west, with lance-shaped or oblong leaves in whorls about the stem, and small, white or pale lavender flowers in dense terminal spikes. In woods, thickets, and meadows, the flowers on stems seven feet tall, tower above the surrounding vegetation and serve as advertisements to woo the insect trade. And who was Culver, one may ask? Probably some quack doctor who discovered healing virtues in the root of the plant for his name does not appear in either botanical or medical literature.

The garden species of *Veronica* are familiar plants to those who have a garden. They are attractive plants for the border or rock garden and are of easy culture under most ordinary garden conditions. Most of them have erect stems and blue flowers of various shades or tints.

John Gerarde, the famous English herbalist, never knew he was to be honored by Linnaeus who gave his name to a group of charming American plants.* He died before the Swedish botanist was born, and likely, Gerarde never saw any of them. Some of them are called false foxgloves to distinguish them, probably, from the true foxgloves of another genus. The downy false foxglove is one of them, a handsome plant with a simple stem, yellow green, somewhat lance-shaped leaves, both the stem and leaves velvety-downed with soft hairs, and showy, pure yellow, or light lemon yellow flowers. The flowers, about an inch and a half long, trumpet-shaped with five lobes, are set in a close terminal cluster that is rather one-sided. It is a plant of the woodlands and thickets, is cross-fertilized mostly by butterflies and bumblebees, and serves as food for the caterpillars of the buckeye or peacock butterfly.

Another handsome species is the fern-leaved false foxglove, a very leafy, somewhat sticky, plant found in dry woods and thickets. Its light green leaves are fernlike, deeply cut into many toothed lobes, and are stemless or nearly so. The flowers are yellow, bell-shaped, and are visited by the bumblebee, and at times, by the ruby-throated hummingbird. The smooth false foxglove is much like the preceding two species with the same yellow flowers, though they are perhaps slightly larger, that are borne on a smooth stem somewhat taller than the other two. The foxgloves would be worthy of the garden were they not parasitic on the roots of other plants.

An immigrant from Europe and an escape to meadows, pastures, roadsides, and waste places, the foxglove, purple foxglove, or purple gerardia is a dainty plant and much cultivated in the garden. It has a generally smooth, slim, straight, rigid stem with widely spreading branches, yellowish green, small, linear and pointed leaves, and cup-shaped or bell-shaped, bright purplish pink, deep magenta, or pale to whitish flowers with five wide, flaring lobes, and four stamens with deep golden anthers. Bees attend the blossoms as do the common yellow and buckeye butterflies. When carefully grown in the garden, the foxglove, when in full flower, is a plant of dignity and beauty. It is planted in several forms, one with white flowers, and another, the leopard foxglove, being showy and spotted. The name foxglove is probably a corruption of folkglove, that meant fairy glove, the

* Genus *Gerardia*.

latter in allusion to the thimblelike shape of the flower. The leaves yield the drug digitalis extensively used in medicine as well as several other toxic substances such as digitoxin and digitonin. For this reason, the plant has become of somewhat grave concern, as horses and cattle have been poisoned by eating the leaves.

In the salt marshes along the Atlantic coast one may find the seaside gerardia, and in dry woods and thickets, the slender gerardia. The first is a small plant rarely exceeding a foot in height and sometimes barely rising four inches above the ground. The second is a very slender species growing somewhat higher. Both have linear leaves and light magenta flowers.

> Scarlet tufts
> Are glowing in the green like flakes of fire;
> The wanderers of the prairie know them well,
> And call that brilliant flower the "painted cup."

The painted cup is a rather odd species with a ruddy, soft-haired, slender stem and light green leaves, the lower ones oblong, clustered, and undivided, the upper ones three to five-lobed and all stemless, and each appearing as if stained on the tip with scarlet. The flowers are inconspicuous and greenish yellow, and are enclosed by broad, red, three-cleft floral bracts or leaves. They are completely eclipsed by these red floral leaves, and with them form a dense terminal cluster. The plant is at home in meadows, prairies, and thickets where it is somewhat parasitic on the roots of other plants.

Somewhat in contrast to the painted cup, the wood betony decorates roadside banks and open woods and copses with thick, short spikes of blossoms that rise above rosettes of coarse, hairy, fernlike leaves. At first the flowers are greenish yellow, but as the spike increases in length with added bloom, the arched upper lip becomes a dark purplish red, the lower remaining a greenish yellow, and the throat turning red. The stem is rather hairy and is often stained with dull magenta as are the feather-shaped lower leaves. Bumblebees and mining bees are frequent visitors. At one time, farmers believed that once their sheep had eaten the leaves they would suffer a skin disease, actually produced by a tiny louse, and hence, the innocent betony is also known by the repellent name of lousewort.

One of the most common plants of the woods is a low, shrubby, delicate species with a multi-branched, gray green stem, yellow green, lance-shaped leaves, the lower ones toothless, the upper ones with bristly teeth at the base, and greenish white cylindrical flowers, the corolla two-lipped, the upper arched and covering the stamens, the lower three-lobed and tinged straw yellow. It is called cow wheat, its scientific name meaning "black wheat" from the color of its seeds. The common sulphur, the buckeye, and the white cabbage butterflies visit the flowers. So, too, do various bees.

Along the rocky coast of New England the yellow rattle will be found, its name partly derived from the yellow color of its flowers and partly from its fruit that is an inflated pod with many seeds that rattle when ripe. Also in the same place there occurs a small growing plant with the name of eyebright. It is a tiny annual with inconspicuous flowers that are whitish and deeply purple-veined. Common everywhere in low, wet ground is the false pimpernel, a branching and spreading little annual with smooth, branching, leafy stems and pale, dull lilac flowers in terminal racemes, or from the leaf axils, on long, slender pedicels.

There are about one hundred and fifty species of pentstemons or beardtongues, all being North American perennial herbs with one exception, that being Asiatic. They have opposite or whorled leaves and showy, two-lipped tubular flowers, mostly in terminal clusters that bloom in the summer. The species commonly found in the eastern states is the hairy pentstemon, a sticky-haired plant with light green oblong to lance-shaped leaves, with dull violet or lilac and white flowers. There are four stamens and a sterile one that is hairy or bearded a little more than half its length, and hence, the plant's name from *pente,* five, and *stemon,* stamen. It grows in dry or rocky fields, thickets, and open woods, and is cross-fertilized mostly by butterflies.

180. Hairy Pentstemon

The large-flowered pentstemon is a species of the prairies with lavender blue flowers about two inches long. The smooth pentstemon, also of the western plains has smaller blue or purple flowers in densely crowded spikes. Another, the foxglove beardtongue, is found in thickets and moist fields, and is the species most often cultivated in the garden. It is a nearly smooth herb, three to five feet high, with ovallike to lance-shaped, toothed leaves, and white flowers about an inch long.

A very popular garden member of the foxglove family is the snapdragon

> "that has been improved and petted and coaxed and bullied until it has become virtually one great flowering stalk thickly set with blossoms, whose dropsical bodies and swollen throats have little charm separately, but the groups in mass give fine banks of color; reds, yellow, and whites.*

There are about thirty species. The common snap-

* *Our Garden Flowers* by Harriet L. Keeler.

182. Leaf of Paulownia

183. Flower of Paulownia

181. Flowering Stem of Snapdragon

dragon is a bright green perennial, reaching a height of three feet, with lance-shaped leaves and reddish purple, sometimes white flowers in long terminal racemes. There are, however, many horticultural color forms in this species. It is often found as an escape, originally being an immigrant from Europe. A small species, also an adventive from Europe and also an escape, may be seen in fields and waste places near dwellings. It is generally known as the small snapdragon, and has light green, linear leaves and light purple or white, showy and solitary flowers. There seems to be some doubt as to the origin of the name snapdragon. Some think the name was given to the plant because its flower looks like the face of a dragon, others because the seed pods resemble the snout of a dragon. But then, who has ever seen a dragon.

Another garden member of the family that deserves mention is the torenia, perhaps not as well known generally, though it has been under cultivation for over a hundred years. It is a low, midsummer blooming annual, with a many-branched, four-angled stem, six to twelve inches high, ovalish, toothed leaves, and blue, or blue violet and yellow flowers in stalked clusters in the leaf axils or terminal. Looking directly into the corolla, one will see against the glowing golden center, two pale blue stems that hold the stigma between them. The torenia is a beautiful little flower and, tolerant of many conditions, is useful for the border, for the rock garden, for edging, and for hanging baskets.

The only woody member of the family found within the North American range is the paulownia or empress tree, a native of China introduced into this country and cultivated as a lawn tree, but an escape in the eastern states from New York to Georgia. It was named for Anna Paulovna, a Russian princess, daughter of Tsar Paul I, and queen of William II of the Netherlands.

It is a medium-sized, beautiful tree, thirty to fifty feet high, with thick, stiff branches, rather open in habit, and round-topped. The bark is a light brownish gray, scored into shallow, confluent ridges. The leaves are hairy, more or less ovate, toothless, light green above, and paler beneath. The flowers are large, handsome, fragrant, light violet, in terminal, erect, pyramidal clusters, and the fruit is an ovoid, leathery capsule resembling a bishop's mitre containing tiny winged seeds.

In leaf, flower, and in general habit the paulownia is similar to the catalpa. It is a vigorous, hardy tree, its blossoms appearing before the leaves. Blue is an unusual color among tree blossoms and no tree flower has a more delicate color. When in flower, with its blossoms in clusters at the ends of the twigs, the tree is a striking object, but the lack of a leafy background to show them off scores against it, at least to a critical eye. The unsightly seed capsules, hanging on the naked branches in the winter, are hardly ornamental.

61. Flax (Linaceae)

"Blue were her eyes as the fairy flax," says Longfellow of a charming little plant that has provided man with its linen fiber and seed oil since prehistoric times. Its native land no one knows, but it has clothed the world since the dawn of civilization. It was in use before the founding of Babylon, the lake dwellers of Switzerland used it, and both Herodotus and Pliny speak of it.

The common flax is an annual herb, a rather delicate and pretty plant with a smooth, slender, erect stem growing to a height of two to three feet, with alternate, lance-shaped, sharp leaves, and bright blue flowers borne at the summit of the stem on slender pedicels in a loose cluster. It is an adventive from Europe, but now an escape to roadsides, cultivated fields, and waste places.

Unlike the common flax, the wild yellow flax is a smooth perennial that is found growing in dry woodlands, and shady places. It has small yellow, starlike flowers, about a fifth of an inch long, at the end of slender branches. A somewhat similar species has long and narrow leaves and yellow flowers about half an inch broad. There is still another species, the yellow flax, with small yellow flowers somewhat crowded on stiffish, spreading, ascending, angulate branches and found in wet woods and sandy shores.

The species of flax belong to the genus *Linum*, Latin for flax, but in Anglo-Saxon the word means to plait or weave. The genus is one of fourteen genera of the flax family, a group of some one hundred and fifty species of herbs or shrubs widely distributed throughout the temperate regions. They usually have simple alternate leaves, though sometimes opposite. The flowers are perfect, regular, usually with five sepals, free or united at the base and persistent, usually five petals, five stamens, sometimes with five additional sterile ones, and a two to five-celled ovary with as many styles. The fruit is usually a several-seeded capsule.

The flax family is an important one from the viewpoint of the utilitarian value of its products. The flax plant has been cultivated for its fiber since long before recorded history, and its use for linen cloth, cordage etc., is as old as the Egyptian mummies, many being wrapped in fine linen.

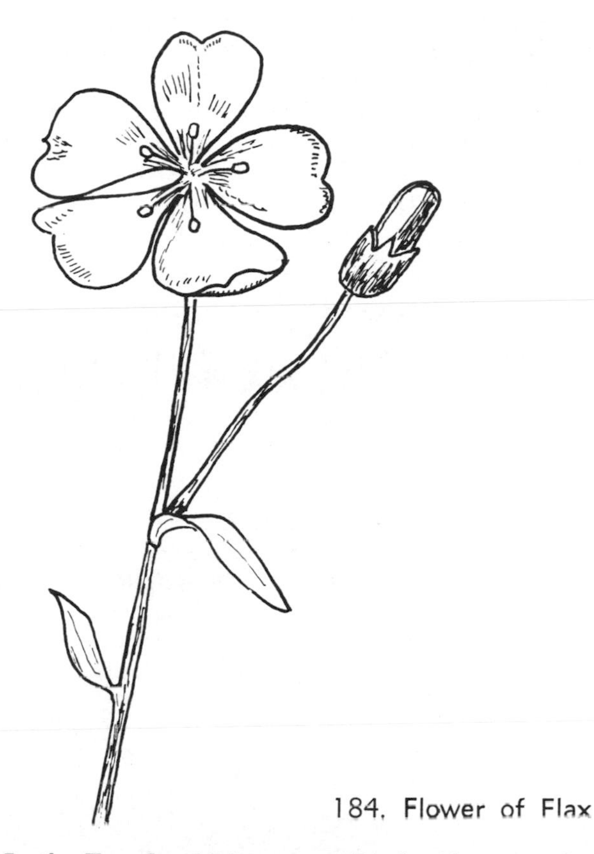

184. Flower of Flax

In the Temple of Minerva at Lindus there was kept a linen corselet of fine workmanship, which had been worn by Amasis, an Egyptian king who reigned six hundred years before Christ, each thread of which was composed of three hundred and sixty filaments.

Besides its fiber, various useful products, such as linseed oil, are manufactured from its seeds. A number of species are grown for ornamental purposes, such as flowering flax, an annual of erect, branching leafy habit about two feet high, with flowers nearly an inch and a half wide, in red or pink, or in shades of either, and the perennial flax with clear sky blue flowers.

As a final note on the family, or more accurately on one of the species, the Indians used the seeds of the blue flax in their cooking because of their high nutritional value and agreeable flavor, that they added to whatever they cooked.

62. Flowering Rush (Butomaceae)

The flowering rush family is a small group of two South American genera and two Old World genera and about ten species of aquatic or marsh plants. The plant wherefrom the family receives its name is a Eurasian aquatic or pond edge herb with long, narrow leaves, and small, rose or pinkish flowers with three sepals and three petals in a multi-flowered umbel on a naked, rushlike stalk. The fruit is a small, dry achene. The flowering rush, the only species of the genus *Butomus*, Greek for ox and to cut, in allusion to the leaves being too sharp for fodder, is hardy in all except the coldest regions, and is easy to grow in pools or along their edges.

Another member of the family, known as the water poppy, is often cultivated for the aquarium, for greenhouse pools, and outdoors in the South. A native of Brazil, it has creeping stems that root in the mud and long-stalked, oval or narrow floating leaves. The light yellow flowers, with three sepals and three petals, are borne on long stalks arising from the joints of the stem. The fruit is a collection of five to seven carpels that are not united.

63. Four-O'clock (Nyctaginaceae)

In the days of daylight savings time the four-o'clock may not open its blossoms at four o'clock in the afternoon, but it can be recalled by the author, that they did open fairly close to that time when he was a boy. At any rate, they do open late in the afternoon, remain open all night, and close when the sun is high in the sky in the morning. On cloudy days they remain open much longer. The blossoms open only for a single night, and when they close they remain closed.

185. Flower of Four-o'-clock

The four-o'clock belongs to the genus *Mirabilis*, Latin for wonderful, a group of some twelve species of tropical American perennial herbs that have thickened or tuberous roots, opposite, generally stalked leaves, and usually solitary flowers from a calyxlike involucre. The true calyx is corollalike, tubular, and variously colored, with no petals, and five to six stamens. The fruit is a leathery-ribbed achene.

Our common four-o'clock is a perennial in the South, but is grown as a tender annual in the North. It is fourteen to thirty inches high, with an erect bushy stem, ovalish, smooth leaves, and tubular or funnel-shaped, usually solitary flowers in terminal clusters.

The four-o'clock family includes about twenty-five genera and over three hundred and fifty species of herbs, shrubs, and trees of wide distribution, but predominantly tropical American. They have alternate or opposite simple, entire leaves, perfect and regular flowers, without petals, but usually very showy from the profusion of colored bracts that may be separate or united. The calyx is often tubular, with four or five wide-spreading lobes, the stamens are one to many, and the ovary is one-celled and one-ovuled. The fruit is a dry, indehiscent, often ribbed or grooved achene and sometimes enclosed in the persistent calyx.

Doubtless, the most showy vine in cultivation is the *Bougainvillaea*, named after De Bougainville, a French navigator. A South American plant, it is perhaps the most widely planted ornamental vine of the tropics and a great favorite in the South and in California. It is a tall-growing woody vine with alternate, stalked leaves, small and unpretentious flowers, all the color coming from the showy bracts. The flower characteristics are those of the family.

There are ten known species of these vines, and for arbors, porches, or for covering the corner of houses, are among the finest creepers known. One species has a spiny stem, oblong leaves, the bracts magenta or purple, while another has a hairy stem with larger and more showy reddish bracts.

A native relative of the garden four-o'clock, the heart-leaved umbrella wort is a very persistent weed, with a large, fleshy, deep-boring taproot, a smooth, or nearly so, angled stem one to three feet tall, opposite, smooth, entire broadly ovate or heart-shaped leaves, and flowers in forking, terminal clusters, the peduncles and pedicels all somewhat hairy. A saucer-shaped or umbrellalike involucre subtends each cluster of three to five flowers. It is five-lobed, persistent, and enlarges as the flowers mature until it becomes nearly an inch broad, thin and net-veined, and acting as a parachute in the dispersal of the seeds. Each small blossom has a bell-shaped, red perianth, with three to five stamens, a one-celled ovary and one style. The fruit is a small, hard, achenelike, narrowly obovoid, ribbed and hairy nutlet.

There are several species of umbrella worts as well as a number of western herbs, also relatives of the garden four-o'clock, with opposite, stalked leaves, usually sticky-haired, entire leaves, tubular flowers crowded in a loose head. Below the head there are five or more colored bracts. The fruit is one-seeded and leathery. The sand verbena, common along the California coast, is of some horticultural importance. It is a prostrate, vinelike herb, often rooting at the joints, with long-stalked, nearly oval leaves, and pink flowers, usually ten to fifteen in an umbellike cluster. Like the four-o'clock, the flowers have no corolla though the calyx looks like one.

64. Frog's Bit (Hydrocharitaceae)

Those who have tropical fish know the common aquarium plants such as elodea, or anacharis, and vallisneria, both members of the frog's bit family also known as the tapegrass family. It is a small family of fourteen or fifteen genera and about fifty species of floating or submerged herbs widely distributed in both fresh and saltwater in warm and temperate regions. Their leaves are various and are simple, linear, lanceolate, or rounded kidney-shaped, floating or submerged, and their flowers are regular, mostly dioecious, enclosed by a spathe of one to three bracts or leaflike scales. They have three sepals, usually three petals, three to twelve stamens that may be separate or united, and a one-celled or three-celled ovary with several ovules. The fruit, ripening under water, is leathery or fleshy. Some species grow so luxuriantly that they choke streams and canals, and thus, become pests.

The elodea, or anacharis, or waterweed, is a beautiful and common plant of slow streams and ponds. It has an elongated, floating stem crowded with dark green, translucent leaves in whorls of three or more. The stem is quite brittle and whenever broken, the fragments continue to grow independently, thus, rapidly increasing the number of plants. The flowers are small and inconspicuous. The male flowers form underwater, but rise to the surface where they come in contact with the female flowers that are borne on long stems that reach the surface. Snails find good forage on the leaves and one can often find a six inch branch harboring a number of them.

The vallisneria or tapegrass, or eel grass, as it is

186. Vallisneria

also known, is common in quiet waters and is a favorite food of the canvasback and other water birds. It is an excellent aquarium plant, for it grows easily and provides abundant oxygen for the fishes or other animals. It is rooted in the bottom mud of the pond or lake, and its slender ribbonlike leaves may be three feet or more long though hardly a quarter of an inch wide. The female flowers are small, greenish and are borne singly on the ends of long spiral stems. The male flowers, on short stems, develop near the base of the plant in clusters of several hundred. When mature, they become detached and are carried upward by a bubble of air, each separately. When they reach the surface they open and then float about until one meets a female flower that receives its sticky pollen. After the latter has become pollinated, the spiral stem contracts and pulls the maturing fruit down into the water.

Unlike either elodea or vallisneris, the frog's bit is found in stagnant water. It has heart-shaped or rounded leaves, often colored purplish green on the lower surface, and proliferates by runners. The staminate flowers, about three in number, are borne in a long, peduncled spathe, while the pistillate flowers are sessile or borne on short stems. When the pools or ponds dry up, the plant often remains to cover the muddy bottom with a bright green growth.

65. Fumitory (Fumariaceae)

In spite of its rather inelegant name, the Dutchman's breeches has a certain air of refinement. The finely dissected leaves and dainty, heart-shaped blossoms, pendent from a trembling stem as a gem hangs from a lady's ear, suggest the feminine rather than the masculine.

The Dutchman's breeches is a dainty plant of early

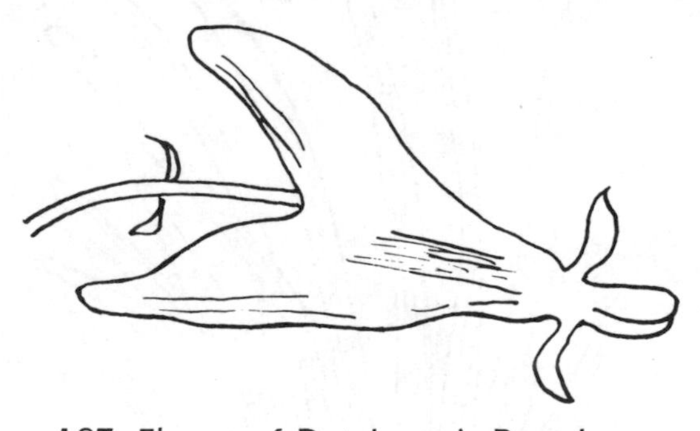

187. Flower of Dutchman's Breeches

spring found in rich woodlands. The plant has a feathery compound leaf on a long stem that comes from the root. It is grayish green in color, somewhat blue and paler beneath, its leaflets finely cut and in three parts. Like the leaf petiole, the flower scape also proceeds from the root and bears four to eight (usually) nodding white flowers. The flowers consist of two scalelike sepals and four petals that are in two pairs, two of them forming a double, two-spurred somewhat heart-shaped sack, the other two within the sack are small, narrow and united protectively over the slightly protruding stamens. The spurs are tipped with pale yellow.

The tongue of the honeybee is too short to reach the nectar secreted in the two deep spurs. Butterflies have a long enough tongue, but it is difficult for them to cling to the pendulous blossoms. Hence, cross-fertilization is effected mainly by the early bumblebees.

The two Dromios of the wildwood are the Dutchman's breeches and the squirrel corn. As a rule they are found growing together, though one may be more abundant in one locality than the other, while in another locality the reverse may be true. And like the twins of Shakespeare they are much alike. There are differences however, especially in the flowers. In those of the Dutchman's breeches, the spurs are longer than the pedicel and the crest of the inner petals is minute. In those of the squirrel corn the spurs are very short and rounded, and the crest of the inner petals projects rather conspicuously. The roots, too, are different. They are a collection of small, solid tubers enclosed in a common scaly sheath in the Dutchman's breeches, whereas in the squirrel corn the roots are round, scattered yellow tubers resembling grains of corn, hence its name. Both species belong to the genus *Dicentra,*

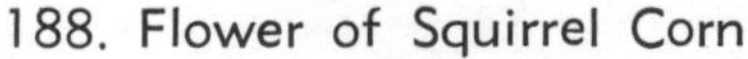

188. Flower of Squirrel Corn

Greek for two-spurred in allusion to the spurred corolla, and both have a slender, two-valved pod for a fruit, as do the remaining members of the genus.

Two other members of the genus are the bleeding heart of our gardens and the wild bleeding heart, the latter a native herb found in the woods and suited for the rock garden. Its leaves are all basal, and its rose pink flowers nodding in a branched cluster (raceme).

Also belonging to the same genus is the garden bleeding heart, a native of the Orient, a leafy-stemmed herb, from one to two feet high, with long drooping sprays of pendant rose-colored flowers, the latter heart-shaped and rather unique among floral forms.

The genus *Dicentra* is one of only five genera of the Fumitory family, a small group of only about one hundred and seventy species, all being relatively weak herbs with generally glistening foliage and highly irregular flowers, sometimes showy, with a conspicuous spur. The leaves are simple, compound, and much cut or fernlike. The flowers have two small, scalelike sepals and four petals in two pairs. One or both of the outer pairs are prolonged into a spur or simply swollen, the inner pair narrower with their tips united over the stigma. The stamens are in two sets of three each, and the ovary is above the point of insertion of the calyx. The fruit is a dry pod (capsule) except in the genus *Fumaria* that has a small, nutlike fruit.

By its pale foliage and attenuated saclike blossoms the corydalis at once shows its kinship to the Dutchman's breeches and the bleeding heart. The dainty, little pink sacs, golden yellow at the mouth, hang upside down along the graceful stems and sway with every passing breeze. The corydalis, too, favors rocky woodlands where great boulders are covered with little forest gardens. The pale corydalis, a conspicuously delicate wild flower of early spring, is probably the best representative of the genus *Corydalis,* Greek for lark, in allusion to the spur of the flower that suggests a lark's spur. This genus consists of ninety species in north Temperate Zones. The golden corydalis, common in the west, is a woodland species, with compound pale leaves beautifully cut and bright yellow flowers about half an inch long.

Climbing by means of prehensile leafstalks, the climbing fumitory or mountain fringe is a beautiful and delicate vine that trails its way over the shrubbery of woods and thickets. Its foliage is misty, pale green, and extremely delicate, its flowers attenuate, sack-shaped white and greenish tinted or pale pink. Often planted in the wild garden, it is also known as the Allegheny vine or adlumia, the latter in honor of John Adlum, an early American botanist and the first horticulturist in our country.

An old garden plant and a native of Europe now found growing wild in waste places and about dwellings, the fumitory, wherefrom the family gets its name, is a small delicate plant with finely cut light green leaves and small crimson pink or magenta pink flowers in a dense, long, narrow spike. The name fumitory comes from two Latin words, *fumus* meaning smoke, and *terre* meaning earth, and was presumably given to the plant because of the nitrous odor its roots have when first pulled from the ground. It is the only cultivated species of some forty herbs of the genus *Fumaria,* and was once widely grown as a remedy for scurvy.

66. Gentian (Gentianaceae)

> Thou waitest late and com'st alone,
> When woods are bare and birds have flown,
> And frosts and shortening days portend
> The aged year is near his end.

So wrote Bryant in his poem "To the Fringed Gentian." Strictly speaking, he was not quite accurate, for the flower begins to appear in September. However, it is still a delight to find the fringed gentian in blossom, for it is becoming much too rare. This author has never forgotten the day when he chanced, for the first time, upon this gay and lovely flower in a shady copse that fringed a woodland pool.

The fringed gentian is undoubtedly one of the most beautiful of our native plants, and has the added charm of elusiveness for it does not always reappear in the same place year after year. It may be found in one place one year and then the next year it may have disappeared from that same spot. It is a plant of low, moist ground with a leafy, perpendicular, branched stem, each branch erect and with a single terminal flower. The flower is deep, vase-shaped, with four lobes deeply fringed, and is remarkable for the misty quality of its blue color though as Schuyler Mathews says in his *Fieldbook of American Wildflowers*:

> the color varies from pale to deep violet blue, with occasionally a ruddy tinge, but never with a suspicion of true blue, though lines of a deeper blue violet appear on the outer surface of the corolla.

There are perhaps some three hundred species of

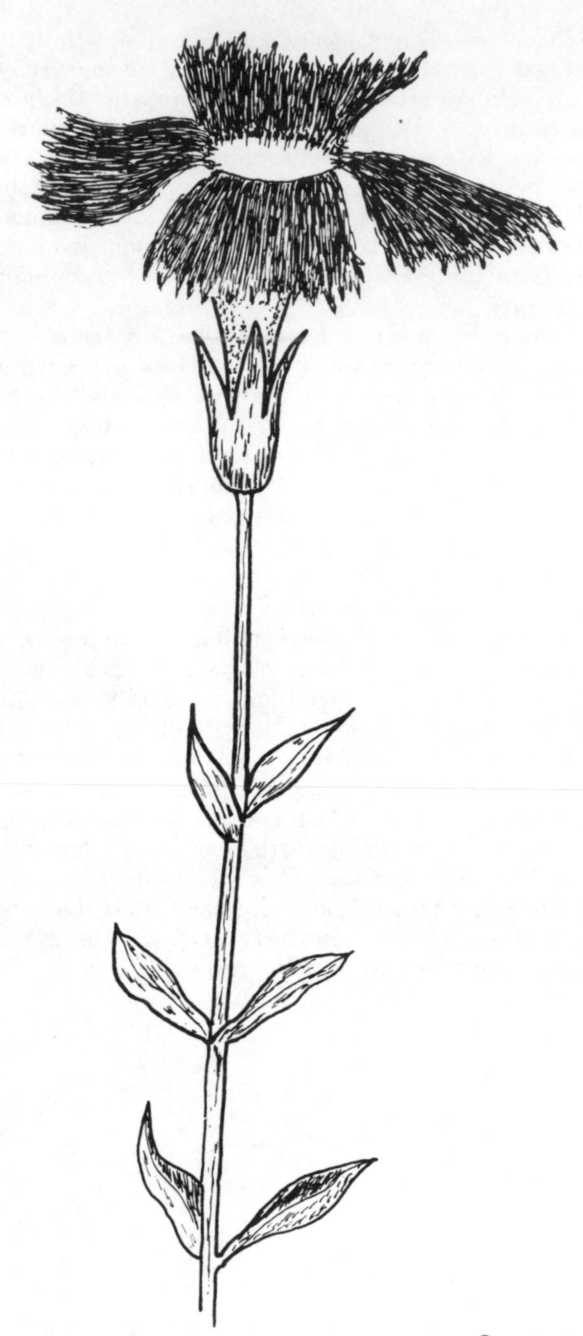

189. Leaves and Flower of Fringed Gentian

gentians belonging to the genus *Gentiana*. Most of them are perennials, some are annuals, and a few, like the fringed gentian, are biennials. They are chiefly plants of cool, moist regions, especially mountain meadows and alpine summits, with opposite leaves and showy flowers, often solitary or in few-flowered clusters, usually blue, but occasionally purple, yellow or white. A few of them are choice plants for the rock garden, border, or wild garden.

The downy gentian, a handsome perennial species of the prairies, usually has a single stem, rather minutely hairy and rough, with lance-shaped, light green leaves, and blue violet, bell-shaped flowers in terminal clusters or at the bases of the leaves. The bottle or closed gentian is also a perennial, and in the east, where it is found in rich woodland borders, is the most common of all the gentians. It has a tightly closed, bottle-shaped corolla of a violet blue, though sometimes the color is an intense deep blue, something like the color of the back of the male bluebird. The flowers are usually crowded in a terminal cluster, but some grow from the leaf bases. Bumblebees occasionally manage to get into the corolla.

A plant of the mountains, but also of moist hillsides, the five-flowered gentian has its light violet blue or lilac flowers clustered at the apex of the branches usually in groups of five, but occasionally from two to seven. Its slender, branching, ridged stem, with leaves sharply pointed at the tip and slightly clasping at the base, may rise only about two inches in dry soil, or perhaps two feet in rich, moist ground.

The gentians belong to the gentian family. The name gentian is said to have been derived from *Gentius,* an Illyrian king who was supposed to have discovered the medicinal properties of a European species, the yellow gentian or bitterwort, that has long been cultivated in Europe for its bitter rootstock used as a tonic and stomachic.

The gentian family is a large group of annual or perennial herbs (a few are woody) arranged in about sixty-five genera and over seven hundred species that are widely distributed, but more abundant in temperate and subtropical regions. Most of them have opposite leaves without marginal teeth. The showy flowers are usually perfect and regular, in axillary or terminal cymes. The calyx is tubular, four to twelve-parted, the corolla is four- to twelve-lobed, but usually five-lobed, and the ovary is one-celled, or rarely two-celled with numerous ovules and one pistil. The fruit is a capsule.

Another genus of the gentian family is the *Sabbatia.* It comprises about fifteen species of annual or biennial herbs found in eastern North America and includes the rose pink, the sea pink, the large marsh pink, and the lance-leaved sabbatia. The rose pink is a plant found in meadows and thickets, where it blooms during the months of July and August. It has delicate, fragrant flowers, an inch or more broad, pale crimson pink or sometimes white and marked in the center with a yellow green star. The calyx lobes are about one third as long as the corolla and the style is cleft at the tip, that is, into two stigmas. The plant has a stem that is sharply four-sided, rather thick and many-branched, and its light green leaves are five-ribbed, acute at the tip, and somewhat clasping at the base.

As its name suggests, the sea pink is to be found along the coast from Maine to Florida, where it is common in the salt meadows. It is a rather pretty plant, with a graceful, alternate branching stem, light green leaves, and crimson pink flowers, as large or larger than a nickel, their yellow eyes bordered with

190. Leaves and Flower of Sea Pink

carmine. A similar plant, but more graceful is the slender marsh pink with upper leaves that are almost threadlike. There is no danger in confusing the large marsh pink with its smaller relatives, for its flowers measure two and a half inches across and have from nine to twelve pink or crimson pink petals. This sabbatia usually chooses the sandy margins of brackish ponds wherein to grow, and is most frequently visited by bees and flower flies.

Without leaves, but with small, awl-shaped, opposite-growing scales instead, the yellow bartonia is something of an odd plant and is usually found in grassy places. It has a slender, stiff stem with yellow, bell-shaped flowers of a greenish tone arranged oppositely on it. There are several other species of bartonia all with threadlike stems and scales in place of leaves.

The lesser centaury and the spiked centaury are adventive from Europe, but a few native species occur in the Southeast. They are rather low-growing plants with ovate leaves and clusters of white, pink or purple flowers. The flowers are rather weak in color and the plants are more delicate than beautiful.

Several other members of the gentian family might briefly be mentioned. They are the American columbo, a tall showy plant with four whorled leaves and light greenish yellow flowers marked with small purple brown dots in pyramidal panicles; the pennywort, a low and very smooth, purplish green perennial, with a simple or sparingly branched stem, opposite, wedge-obovate leaves, and dull white or purplish flowers solitary or in clusters of three, terminal and axillary, found in moist woods and thickets; the buckbean, a perennial herb with a thickish, creeping rootstalk sheathed by the membranous bases of long petioles, with three oval or oblong leaflets, and white or slightly reddish flowers in racemes on the naked scape; and the floating heart, a perennial aquatic, with floating leaves on very long petioles that usually bear near the summit the umbel of white flowers.

67. Geranium (Geraniaceae)

The common garden geranium or house plant is so familiar that it hardly needs description. It is not the only geranium as there are many species, a few being of horticultural interest such as the show geranium, the ivy geranium, the rose geranium, the nutmeg geranium, and the fish geranium. The geranium family comprises eleven genera and some six hundred and fifty species of large annual or perennial herbs, or undershrubs, widely distributed in the north Temperate Zone and in the subtropics, with a somewhat peculiar concentration in South Africa. They have alternate or opposite, lobed, simple, dissected, or compound leaves. The flowers are perfect and usually regular, commonly conspicuous, with five sepals and five petals, the number of stamens two to three times the number of petals, but usually five or ten, some of them sterile, three to five-lobed pistils, commonly of five carpels, each prolonged into a style. The fruit is a collection of one-seeded pods, each splitting from the base upward.

The true geraniums belong to the genus *Geranium*, Greek for crane, in allusion to the beaked fruits that resemble a crane's bill, hence, the true geraniums are often known as cranesbills. It is a genus of some two hundred and fifty species that are generally low, often half prostrate herbs, with forking stems and dissected or lobed, roundish, palmately lobed leaves, regular flowers, and a fruit that is a collection of elastically splitting, beaked carpels that persist for some time. Some of the species may be found in the garden, others are wildings such as the wild geranium or cranesbill, a familiar plant of our woodlands and wooded roadsides. It has an erect, branching, hairy stem, with deeply cut, five-lobed leaves that are rough-haired, and delicate, pale or deep magenta pink flowers in

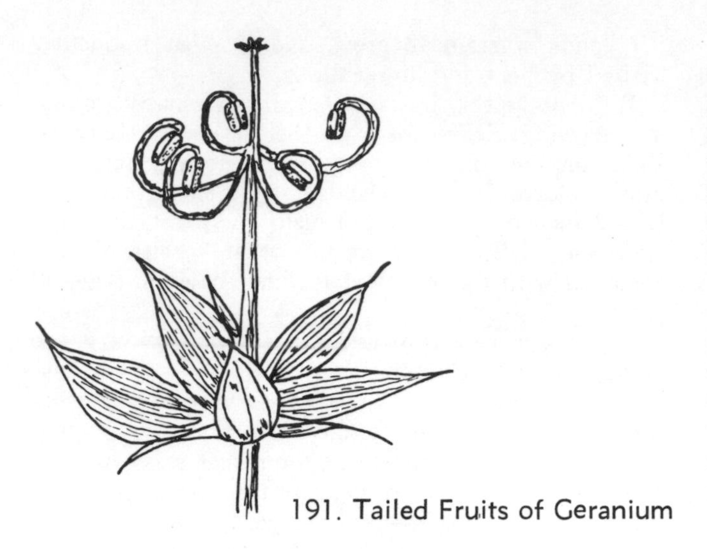

191. Tailed Fruits of Geranium

terminal or axillary loose clusters. Perhaps no flower has devised such an ingenious way to prevent self-fertilization as the wild geranium, since the pollen becomes mature and the anthers fall away before the stigma becomes receptive. Though the flower is perfect, the two reproductive structures will not be present at the same time. During cold, rainy, or cloudy weather, the flower may retain its anthers for several days before the stigma becomes receptive, while on a warm, sunny day, when insects are flying, the change may take place within a few hours. The blossoms are cross-fertilized mostly by honeybees, the smaller bees of the Halictus tribe, and the flowerflies.

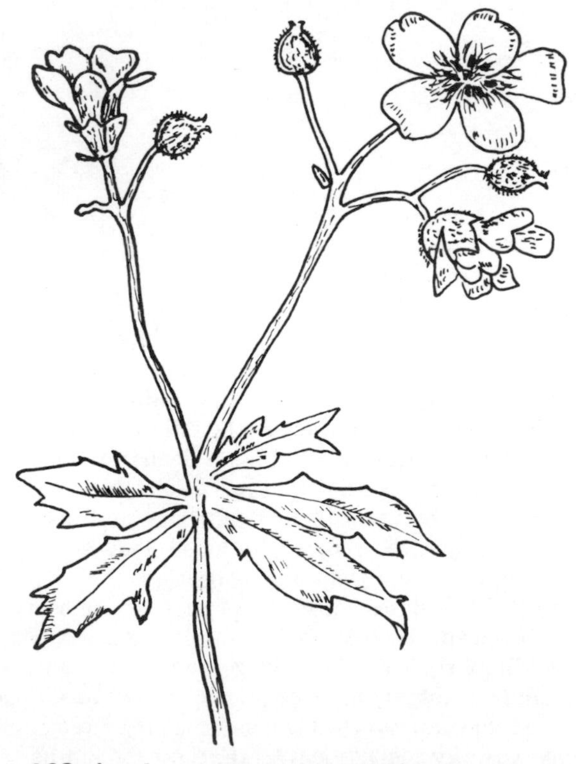

192. Leaf and Flowers of Wild Geranium

Resembling the wild geranium in general features, the Carolina cranesbill has more compact clusters of pale rose or whitish flowers. The beak of the seed vessel is nearly an inch long and is short-pointed, whereas that of the wild geranium is long-pointed. Although more commonly distributed throughout the South, the Carolina cranesbill occurs from Maine south to Florida, and grows in rocky places or in poor soil.

An adventive from Europe and found on the borders of rocky woods, the herb Robert is a handsome and decorative plant with a ruddy stem and a generally disagreeable odor, but especially so when the leaves or stem are crushed or bruised. The ornamental leaves, with three to five divisions, are deep green and the flowers are deep or pale magenta or rose pink. The seed pod of this species appears to be somewhat more elastic than that of other species, for when mature, it opens rather more strenuously and sends forth its seeds quite some distance, often several feet.

Who was the Robert for whom this plant was named. Some say that it was St. Robert, a Benedictine monk, to whom the twenty-ninth of April—the day the plant blossoms in Europe—was dedicated. Others say that it was named after Robert, the Duke of Normandy. In his *Romance of Wild Flowers,* Edward Step says the name Robert is the result of an interchange of meaning. He explains that the name Robwort, that is Redwort, was originally given the plant because of its red or ruddy character, and that in the course of time, the *w* disappeared, leaving it Rob'ort that eventually became Robert. However, as that may be, Linnaeus regarded this Robert as a personal name and gave the plant a Latin termination: *Geranium Robertianum.*

The common garden geranium, as well as the others mentioned above, that is, the show geranium, ivy geranium etc, are not true geraniums, but belong to the genus *Pelargonum,* Greek for stork, in allusion to the shape of the fruit. The essential botanical difference between the two genera is that, in most cases, the flowers of *Geranium* are regular, whereas those of *Pelargonum* are irregular, the two upper petals differing from the under in size, shape, and often in coloring. There is also in *Pelargonum* a spur that extends from the base of the upper sepal, not usually noticed by the casual observer. Indeed, in many cultivated forms it seems to have virtually disappeared.

The genus *Pelargonum* is a large one, of South African tender, perennial herbs and shrubs, with habits varying according to the species. Their stems are strong-growing or trailing, and herbaceous or woody. Their leaves are alternate, stalked, simple, without teeth and roundish or much cut and often fernlike. Some are deeply marked on the upper surface, smooth or hairy, and at times fragrant. Their irregular flowers are in umbellike clusters on a leafless stalk from the axils of the leaves, and have a calyx of five sepals, a corolla of five petals, ten stamens, and a five-celled ovary. The individual flowers are showy and vary in color from pure white to pink, crimson, and bright scarlet.

The common cultivated forms of *Pelargoniums* are all descendants of a South African plant that seems to have been introduced into England about 1690, the English and the Dutch being the first to popularize the flower. Among the cultivated forms are the ones that have already been mentioned. The show geranium is of rather straggling habit, growing to two feet high with a soft stem, roundish leaves, slightly lobed and toothed, and large white, pink, or red flowers in umbels few to many-flowered. The ivy geranium has a trailing stem, four feet long, five pointed, ivy-shaped, bright glossy green leaves, and white to deep rose flowers in umbels of five to seven flowers. The rose geranium grows to a height of three feet, is woody with roundish, fragrant leaves, five to seven-lobed, the lobes toothed, and rose pink flowers in umbels five to ten-flowered. And the fish geranium, one to six feet high, has strong and fleshy stems, roundish leaves, generally with a deep horseshoe-shaped marking or zone on the upper surface and scalloped margins, and red, salmon-pink, or white flowers in many-flowered umbels. This species has a faint fishlike odor.

A third genus of the geranium family that bears mention is *Erodium,* containing some sixty species of widely distributed herbs, a few grown as ornamentals, a few planted for forage in dry regions, and one or two of importance as bee plants in California. They are rather closely related to the true geraniums wherefrom they differ in having the outer stamens without anthers, whereas all of the stamens in the true geraniums are anther-bearing. The plants are commonly known as storksbill or heronsbill. The best known is alfilaria that has become naturalized in the United States. It has a low-spreading habit with pinnately compound leaves and pink flowers. It is widely planted for forage, and in California, is cultivated for bees. Our own native species include the desert storksbill of the California deserts, and the white storksbill of the coastal ranges and valleys of the same state.

68. Ginger (Zingiberaceae)

The ginger that is used as a spice, candied, or preserved, is obtained from the common ginger plant, one of four hundred species of the family of the same name and contained in forty genera. They are all aromatic, perennial, tropical herbs with creeping or tuberous rootstalks, short, simple stems, and simple, entire leaves that are mostly with sheathing bases, or leafstalks, and often very large. Their flowers are perfect, irregular, usually in bracted clusters, but often from between sheathlike bracts (a spathe), with a tubular or spathelike calyx of three segments, a tubular corolla, unequally three-lobed, six stamens, but only one is functional, the others sterile and sometimes broad and petallike, and a one-celled or three-celled ovary with many ovules. The fruit is a capsule.

The common ginger is one of fifty or more species of the genus *Zingiber,* the Latinized form of the Sanskrit name for ginger, all from moist, tropical forests in southeastern Asia. It has a stout canelike stem, three to four feet high, rising from a thick rootstalk wherefrom ginger is derived, oblong or narrow leaves, sheathing at the base, and yellowish green flowers, the irregular lip purple, but spotted with yellow, in bracted spikes, the bracts usually persistent and with one flower at each. The common ginger is grown as an ornamental in frostfree regions of Florida. Also grown in the South for its ornamental foliage and fragrant flowers is the ginger lily of the genus *Hedychium,* Greek for sweet snow in allusion to the often white, fragrant flowers. It has stems four to seven feet high, leaves fifteen to thirty inches long, and two to three white, fragrant flowers borne between bracts in a terminal cluster, the bracts being whitish green.

Another valuable member of the ginger family is the cardamom, the source of the spicy seeds used in seasoning and in medicine. The only species of the genus *Elettaria,* a Latinized version of a native Malabar name, it is a stout herb, five to nine feet high, with large leaves that are hairy on the lower side, flowers in loose spikes or racemes, and the fruit a capsule whose seeds are the commercial cardamoms. A related plant of the genus *Amomium,* Greek meaning a poison antidote, produces inferior seeds that are often substituted for the true cardamoms.

69. Ginkgo (Ginkgoaceae)

The ginkgo is a most remarkable and unusual tree. First of all, it is the only species and the only genus of the family of the same name, the last surviving member of a dwindling family that was once a widely distributed group extending back to the Carboniferous. Until recently it was unknown as a wild tree, all the ginkgos being cultivated specimens derived from trees preserved in the temple gardens of China where it has been cultivated for centuries. A study of the tree shows that it seems to have botanical affinities with the ferns on one hand and with the conifers on the other. There is, for instance, the singular character of the leaves. In the arborescent foliage of either America or Europe there is nothing like them.

The leaves are fan-shaped and are usually cleft with

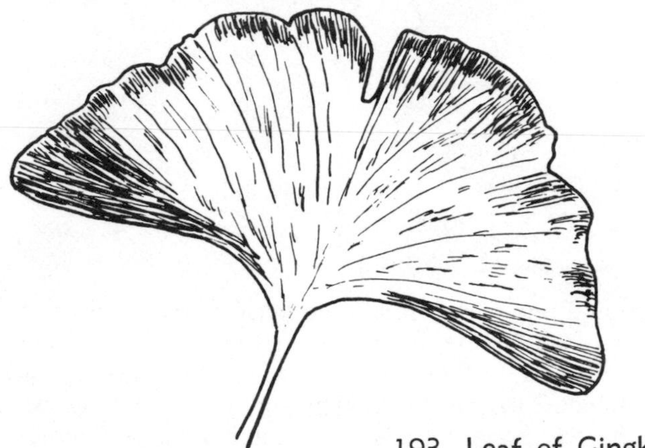

193. Leaf of Gingko

one deep suture that extends almost to the petiole with an odd sort of venation, the unbranched veins radiating to the upper border of the leaves as in the leaves of the maidenhair fern. The leaves are somewhat feathery, dull green or bright yellow green in alternating clusters, and turn a light yellow or gold in the fall before they drop.

The ginkgo is a narrow, tapering tree when young, but gradually assumes a pyramidal form as it grows older. The branches are set at an angle of approximately forty-five degrees with the trunk. The bark is very dark brown, or blackish gray, smooth on young trees, rough on old trees, the twigs stout, smooth, yellow brown, the foliage and almost upright branches producing an effect that is reminiscent of the Lombardy poplar.

The flowers appear in late spring, the staminate and pistillate on separate trees, both without sepals and petals, the staminate consisting of naked pairs of anthers in catkinlike clusters, the pistillate only of a naked ovule that is characteristic of the conifers. It is interesting to note that the ovule, unlike that of

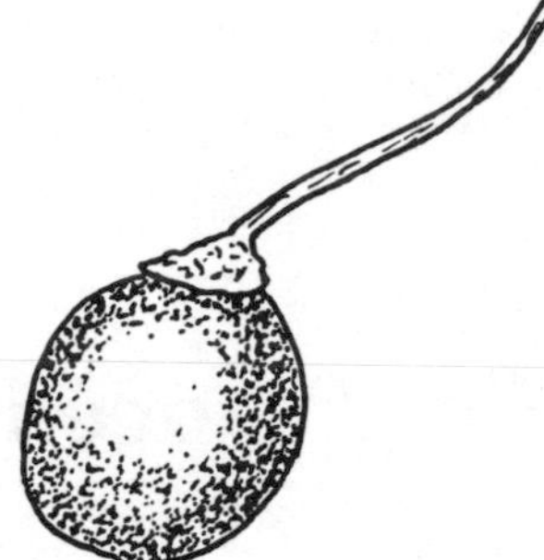

194. Fruit of Gingko

all other trees, is fertilized by motile sperm cells as in the ferns. The fruit resembles a plum in size and shape, with an almond-shaped stone, and when ripe, is pale, golden yellow. The sweet flesh has a very disagreeable odor and flavor, and yet in China, it is made into a preserve, or is baked and eaten at meals between courses to aid digestion. The kernel of the stone is also prized as a food in both China and Japan.

The ginkgo was introduced into Holland sometime between 1727 to 1737, then into England in 1754, and finally into America in 1784. The tree, also known as the maidenhair tree, has become one of the finest ornamentals in America with such important virtues as appearing not to have any insect or fungus enemies, and being resistant to smoke and drought. It does very well under modern street conditions.

70. Ginseng (Araliaceae)

The ginseng is one of the most difficult plants to find, for it is a rare plant. It has not always been so, however, only since the Jesuit missionaries discovered that it was similar to the ginseng of China and began exporting it to the Chinese. For centuries, Chinese physicians had ascribed miraculous virtues to the Chinese ginseng, regarding it as a panacea for all ills. It was naturally in great demand, indeed, in such demand that it was feared it would become extinct, and so an imperial edict was promulgated henceforth, prohibiting the Chinese from digging up their native plant. Thus, when they found that they could obtain the American ginseng, not dissimilar to their own native ginseng, they began importing as much of it as they could get. With the result that our ginseng is now difficult to find.

Our ginseng, a corruption of the Chinese Jen-shen or Jin-chen, is a plant of rich, cool woods, with a smooth, green stem a foot or a little higher, compound leaves of three in a whorl, the leaves each of five deep green and obovate, acute pointed leaflets and resembling a horse chestnut leaf, and yellowish green flowers, crowded in a single hemispherical cluster, the staminate and pistillate flowers on separate plants. The fruit is a cluster of red berrylike drupes. It is the root, that has been in such demand by the Chinese who believed it had considerable medicine value, that has brought the plant into prominence. It is aromatic, of pleasing flavor, and rather spindle-shaped and forked, giving it the appearance of having two legs, hence, the Chinese name mentioned above that means manlike.

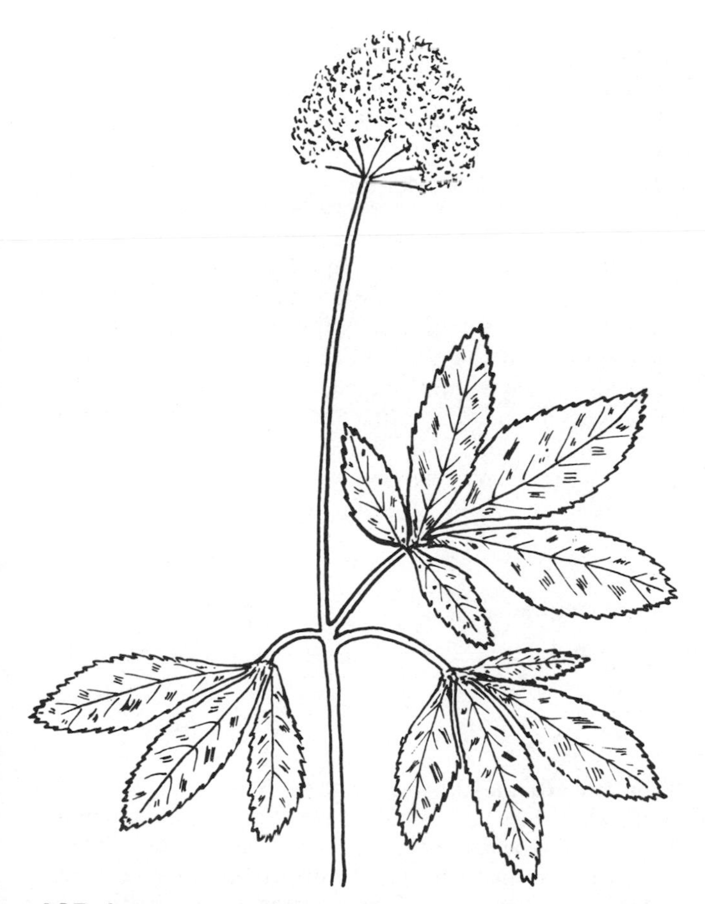

195. Leaves and Flower Cluster of Dwarf Ginseng

A second species, the dwarf ginseng, is a lovely little plant with white flowers in feathery, fluffy balls just above a whorl of three compound leaves, each divided into three to five sessile leaflets. An early spring flower, appearing in April in the woodlands, the blossoms eventually develop into yellowish berry-like fruits. The root is round, pungent, sweet, nutlike and is said to be edible and of little value. It is sunk deep in the ground and only after patient, careful digging is one likely to find it.

Although there are only a few native species, the family whereto the ginseng belongs and whereto it has given its name, is fairly large with some sixty genera and over seven hundred species of trees, shrubs, and herbs, widely distributed in temperate and tropical regions, but more abundant in the tropics. The leaves are alternate, simple or compound, pinnate or digitate. The flowers are small, greenish or whitish, regular, perfect, in umbels, racemes, or headlike clusters, the calyx small and attached to the ovary, the petals five in number, their margins touching or overlapping or even slightly united, the stamens five, and the ovary one to many-celled. The fruit is either a berry or a drupe. The family includes a number of drug plants and several important ornamentals.

A rather common plant of the woodlands whose roots are esteemed for their spicy and aromatic flavor, the spikenard is a tall, branching, smooth plant with a round, blackish stem and large compound leaves of usually fifteen to twenty-one ovate, double-toothed leaflets of a deep green, and greenish white flowers in small umbels that are arranged in a large terminal spike or perhaps several smaller spikes. They are visited by the mining bees and the flowerflies and later form round, dull brown crimson berries in compact clusters. The spikenard was once greatly sought after for its medicinal values, but whether they were real or fancied the author does not know.

Although the wild sarsaparilla is not the sarsaparilla of commerce, its aromatic root can be used as a substitute for the true sarsaparilla, and as a matter of fact, has been used to flavor a once popular summer drink. It is a low-growing herb of the woodlands, thickets, and hillsides whose true plant stem barely rises above the ground, the leaf stem and the flower stem apparently separating near the root. A single, long-stalked leaf rises to a height of a foot and has three branching divisions with three to five oval, toothed leaflets on each division. The flowerstalk is leafless with three to seven rather flat, hemispherical clusters of greenish white flowers that have their petals turned backward to make the nectar more accessible to the flies that are chiefly responsible for the clusters of purple black berries whereon the migrating birds gorge themselves in early autumn.

A related species, the bristly sarsaparilla, is a finely-haired plant of rocky woods. Its compound leaves are divided into oblong-ovate, acute, toothed leaflets,

196. Fruit of Bristly Sarsaparilla

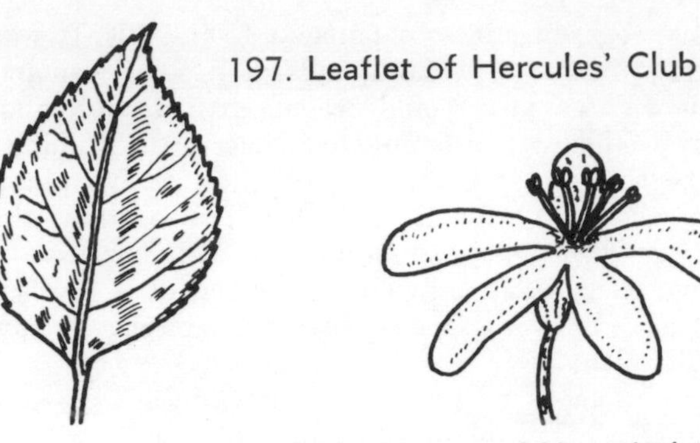
197. Leaflet of Hercules' Club

198. Flower of Hercules' Club

and terminating the stem are several umbels of tiny dull white flowers. The fruit is a dull brown crimson when ripe.

Our only native tree or shrub of the family is the Hercules' club or angelica tree, also called the devil's walking stick. It grows to a height of about forty feet and both the stem and branches are covered with hard reflexed spines. The leaves are large, doubly compound, dark green above, paler beneath, the leaflets broadly ovate, abruptly pointed, and toothed. The flowers are white in large clusters (panicles) that are sometimes twenty inches or more long. The fruit is a fleshy black berry that is eaten by thrushes and other songbirds as well as by a few mammals.

The Hercules' club, so-called because its stem when denuded of its leaves looks like a club, grows in the woods and along streams. It is often planted as an ornamental for in habit of growth and in general aspect it is a unique tree. It is also a most decorative tree for its large leaves, the largest in our flora, give it a sort of tropical palmlike appearance, and when the clouds of minute, white flowers in clusters that match the leaves in dimensions settle above the crown of leaves they provide a most effective display. Later when the berries ripen in the fall, they add their own distinctive charm.

Why the name of devil's walking stick? Presumably because a naked branch looks like a walking stick and it was once the habit to name any thorny or spiny plant for the devil.

Two introduced members of the ginseng family are

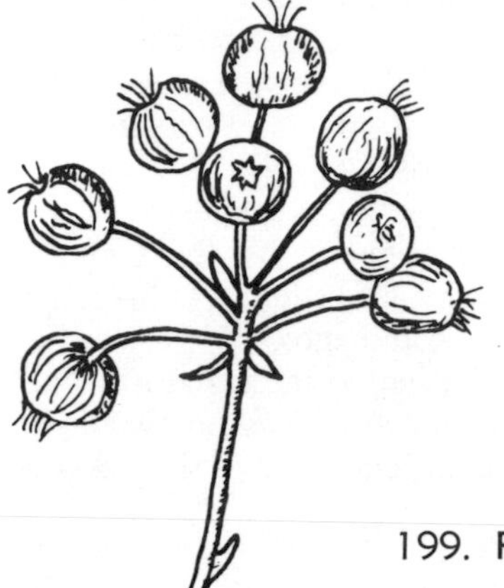
199. Fruit of Hercules' Club

the English ivy and the rice paper plant. The English ivy is well known as the vine that covers many of our older American buildings with a raiment of glossy green. It is a woody climber, with stems having aerial roots, that cling tenaciously to wood or stone surfaces. The leaves are three to seven-lobed, and the flowers occur in small, greenish clusters that appear in the fall. Its black berries ripen the following spring.

The rice paper plant, a Formosan species where it is grown as a source of rice paper, is planted in the warmer parts of the United States for its large, palmately lobed leaves and luxuriant shrubby growth. Its large leaves and spreading habit, however, require considerable space, and yet, it is a handsome plant if it has enough room wherein to grow.

71. Globe Daisy (Globulariaceae)

The globe daisy family is a group of herbs or small shrubs of Europe and Africa and consists of only three genera, the largest being *Globularia,* derived from the arrangement of the flowers in globular heads. The species of the genus are often cultivated in the border or rock garden and are known as globe daisies. They are all mostly low or prostrate herbs, with alternate leaves and small blue flowers in globular heads and crowded between small bracts, the flowers with a two-lipped corolla, the upper lip two-lobed, the lower one three-lobed. The fruit is a tiny nutlet included within the persistent calyx.

72. Goodenia (Goodeniaceae)

The goodenia family is a group of mainly Australian herbs or shrubs with flowers having an irregular two-lipped corolla, five stamens, and a one to two-celled ovary that becomes a capsule or an edible drupe. The family is represented in the United States by one species, the scaevola, a curious sort of plant found growing on the coastal dunes of southern Florida where its sprawling stems often spread beneath the sand and form extensive colonies. It has relatively large pink or white flowers, the corolla woolly within, and split on one side so that the pistil and stamens protrude conspicuously. The leaves are essentially wedge-shaped with entire margins, and the fruit is a black, juicy berry.

73. Goosefoot (Chenopodiaceae)

The goosefoot family is a rather peculiar family in that it includes uninteresting herbs on the one hand, while on the other, a few garden ornamentals of wide cultivation as well as such food plants as the beet and spinach. It is a fairly large family of seventy-five genera and about six hundred species of mostly succulent, annual or perennial herbs of homely aspect, though there are numerous species that are distinctly shrubby. They are widely distributed throughout the world, more particularly along seashores and in saline areas in the interior of the continents. Many of them are known as "saltbushes" because they are associated with saline and alkaline sites.

They have alternate, simple leaves that are often mealy or scurfy, as are the stems sometimes, inconspicuous flowers, prevailingly greenish or whitish,

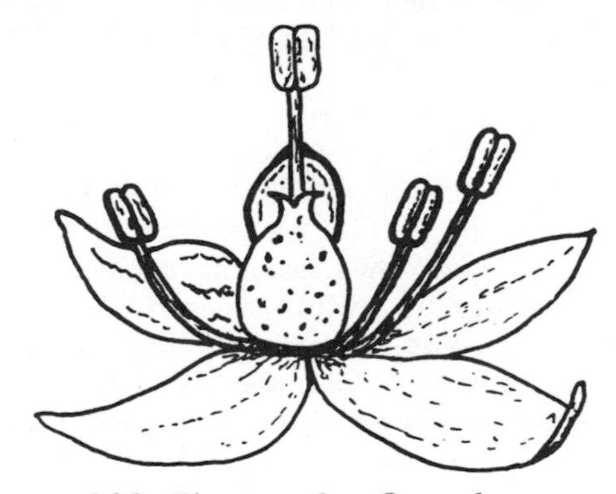

200. Flower of a Goosefoot

without petals, often unisexual, and sometimes with male and female flowers on different plants. The calyx is two to five-parted, or merely of one sepal or none, the ovary is one-celled and one-ovuled with one to three styles, and the fruit is an achene or utricle (a small, bladdery, one-seeded fruit).

As pigs and sheep are very fond of it, the pigweed or lamb's quarters are apt names for a succulent, fast-growing weed of cultivated fields, gardens and waste places. A native of Europe and Asia, but introduced early into our country, it is an annual that grows two to seven feet tall. It has a slightly grooved stem, leaves mealy white beneath, varying from rhombic-oval to lance-shaped or narrower, the lower ones with coarse teeth, and small green flowers in spiked panicles.

The Indians of the west and southwest gather and cook the plant as a spinach, and as well they may, for it is related to the garden spinach. Sometimes they eat it uncooked. They also gather the seeds that are ground into a meal for baking cakes, or served as a gruel. It has long been used as a food in Europe, though not cultivated. Indeed, the lamb's quarters is perhaps the most common wild plant or weed that might be used for human consumption, for when small, six to ten inches high, the plant is succulent and tender and highly desirable as a potherb.

The lamb's quarters belongs to the genus *Chenopodium,* Greek for goosefoot in allusion to the shape of the leaves of some species. The genus contains some sixty species with leaves that are usually mealy and the fruit a dry achene, often enclosed by the persistent calyx, that is developed from small and inconspicuous flowers. A native of Europe and Asia, but an escape and found along roadsides and in waste places, the feather geranium or Jerusalem oak is a rank-smelling, multi-branched annual, twelve to twenty-four inches high, occasionally found in the garden, though at an earlier period, it was much more popular. The leaves are wavy-margined, or somewhat toothed or cut, and the flowers, though not showy, are numerous in spreading cymelike racemes. The Mexican tea is a similarly introduced species with lanceolate leaves and a densely flowered leafy spike. It, too, occurs in waste places.

Closely resembling the lamb's quarters, but a lower-growing species and with leaves that are not mealy, the Good-King-Henry is a rank perennial herb with a deep rootstock wherefrom arise several stems one to two feet tall. Its leaves are triangular with entire or slightly wavy margins, the upper leaves with short petioles, the lower leaves with long petioles, and small green flowers that are like those of the lamb's quarters in appearance and arrangement. Like the preceding species of *Chenopodium,* it is also an immigrant, originally introduced into American gardens, but now an escape, though never abundant anywhere. Its succulent leaves were much used at one time as a potherb, and the young shoots were once a substitute for asparagus. Another native of Europe and now found in cultivated grounds and waste places, the strawberry blite is a plant with a stem six inches to two feet tall, and slender, pale green, often striped with purple, triangular or halberd-shaped leaves. The flowers occur in densely crowded, sessile clusters.

A relative of the pigweed and ranking with it as a weed, the spreading orach has a stem one to three feet in length that may be either prostrate or ascending or sometimes erect, diffusely branching and grooved, often with reddish stripes. The leaves are lance-shaped, entire or sparingly toothed, and the flowers are greenish and grouped in round clusters, axillary, in the upper leaves, and forming, along the top of the stem, leafless spikes. They are of two kinds; the staminate with a three to five-parted calyx, and the pistillate with two

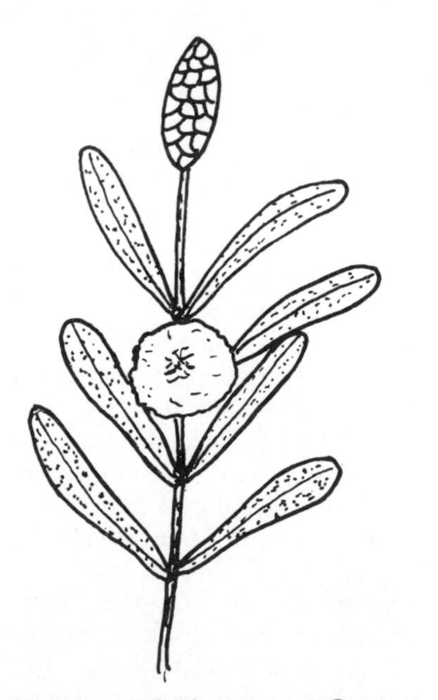

201. Leaves and Flowers of Greasewood

large bractlets underneath, united at their bases. There are several species of orachs, as for instance, the garden orach, a stout, annual herb, one to three feet or more high, whose leaves are used like spinach. They all belong to the genus *Atriplex,* Greek for orach, a large group of herbs and salt-tolerant shrubs. In the West and Southwest the wild shrubs are commonly called greasewood or saltbush, some of them being useful in the revegetation of arid regions. There are some twenty or thirty species known variously as spearscale, redscale, arrowscale, and wedgescale. The saltbush is one of the strange flora that can live on the pure gypsum sands of New Mexico.

A western species of the family, known as the bugseed, is common in grainfields and grasslands. When young, its stems are pale green, succulent, and finely haired, but become smooth, hard, and faintly ridged with age. They are often strongly zigzagged, very freely branched, six inches to two feet in length, the longer branches spreading and usually decumbent, but the shorter ones erect. The lower leaves are rather thick, sessile, tipped with a hard, rigid point, the upper floral leaves or bracts are thinner, ovate, pointed, in the axils whereof are the solitary flowers, hardly an eighth of an inch long, with the calyx consisting of one delicate sepal.

Another western species, but also found along the eastern seaboard, is a most pernicious weed that was brought to America in impure flax seed from Russia many years ago. The Dakota farmers, who first became acquainted with it, called it thistle because of its excessive prickliness, and hence, it became known as Russian thistle. It is not a thistle, but a saltwort, that is, it belongs to a genus of herbs or slightly shrubby, branching plants with fleshy and rather awl-shaped leaves and sessile axillary flowers, known as the saltworts. Such a prickly plant as the Russian thistle would seem to be unfit for food, but western housewives often collect the young, tender plants when only a few inches high, carefully cut and wash them, and then boil them until they are tender. They are served with butter, or if cream sauce is added, they may be served on toast.

Also used as a food by western housewives is the greasewood, a plant of the alkaline, clay soil of desert valleys. It is a species that usually grows three to four feet high, but sometimes reaching a height of six or eight feet with grayish white twigs and narrow fleshy leaves. The staminate flowers occur in terminal spikes, while the pistillate flowers are solitary. The tender twigs are carefully washed, cut into short pieces, and boiled until soft, then served with butter or cream sauce.

Along the seashore there are several species of the goosefoot family that have little beauty in form or color, lacking a corolla, that are known as the glassworts. One of them, the samphire, found in the salt marshes, is a low, fleshy plant, with no leaves, but

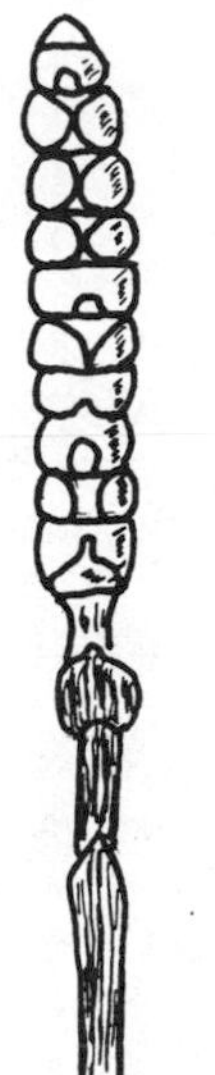

202. Glasswort

instead, fleshy, opposite, pointed scales, and with a thick spike of green flowers in groups of threes, sunk in hollows in the axils of the upper scales. With age the plant becomes reddish. A second species, the slender or marsh samphire, turns a vivid red in the fall, and a third species, the woody glasswort, has broadly ovate scales and flowers in a short spike. All of them belong to the genus *Salicornia,* from the Latin *sal* for salt and *cornu* for horn, a small group of fleshy, maritime herbs having thick, jointed, leafless stems bearing minute flowers immersed in the upper nodes, forming a spike, the fruit being a utricle containing a single seed.

Another group of plants of the goosefoot family widely distributed along the sea coasts and in alkaline desert regions is the genus *Suaeda* or *Dondia,* the latter after an Italian herbalist *Dondi.* They are herbs and shrubs, generally known as sea blites, with fleshy, terete, linear leaves and small flowers with a persistent five-lobed perianth free from the enclosed utricle. A common species, the tall sea blite, found along the sandy coast from Maine to Texas, is a fleshy plant, with long, narrow, rushlike leaves. It is prostrate on the ground or stands erect, one to two feet high. The flowers are stemless and grow in the axils of the leafy bracts.

Frequently cultivated for its bright autumnal color and locally established as a weed, the mock cypress or Mexican fire plant usually grows about two feet high. It is an erect, multi-branched, bushy annual herb, sometimes globe-shaped, with many, alternate, often hairy leaves without teeth, very narrow, and often round in cross-section. They may be red, green, or yellow, but turn a purple red in the fall. The flowers are greenish, inconspicuous, and are few in number in the leaf axils. The fruit is a utricle. The mock cypress is a showy plant usually planted for the border, and holds its leaf color all summer. The name Mexican fire plant is a misnomer, for it is not Mexican nor American, but a native to southern Europe and western Asia.

The garden beet, one of the two important members of the goosefoot family, prized for its large red root, has been in cultivation from Roman times. It belongs to the genus *Beta,* Latin for beet, a group of Old World herbs having leaves without hairs, greenish flowers, in spikes or panicles, without petals, and a fruit that is an aggregate of two or more flowers joined together at the base and forming a dry, corky utricle.

The cultivated beet is a biennial, the stem produced from the root the second year, with ovate-oblong, large greenish purple leaves, the flowers with an urn-shaped, three-bracted, five-parted calyx, five stamens, and two stigmas. As a vegetable the beet is boiled, pickled, and used as a salad, and the tops are cooked for greens in the same way as spinach is cooked. When grown for the use of the leaves as a vegetable, the variety is known as Swiss chard or simply chard. By selective breeding the leaves, instead of the roots, have become storehouses of food. An agricultural variety of the beet, with enlarged roots used for stock food, or cattle fodder is known as the mangel, or mangel-wurzel. The sugar beet has a white root instead of a red one. The culture of the sugar beet seems to have begun in France and Germany about 1830 and today has become an important crop in the plateau states of Colorado, Utah, and Montana where the late summers are cool enough to favor the formation of sugar in the roots.

The second of the two important members is the spinach, a plant grown for its densely crowded leaves that are used as a vegetable. The garden spinach, whose ancestral home is southwestern Asia, is one of only three or four species of Asiatic annual herbs of the genus *Spinacia,* Latin for spine, in allusion to the husk of the fruit. They have alternate, mostly basal or nearly so, ovalish, or in some cultural varieties, roundish leaves, usually with crinkly margins, small, unisexual flowers, and a fruit that is a utricle, surrounded by a small, prickly, capsulelike body, often, but incorrectly, called the seed. The spinach, the common potherb of the garden, is a cool season plant, because the summer leaves are tough.

74. Gourd (Cucurbitaceae)

The word gourd suggests the ornamental gourds, often of grotesque shapes, that are grown chiefly for show and that can be bought in the market or stores, often varnished and highly polished, in late summer or early fall. However, the cucumber, squash, pumpkin, cantaloup, and watermelon are also gourds.

The family whereof all belong is generally called the gourd family, but in various botany books, it will also be called the cucumber, squash, or melon family. It is a large family of some one hundred genera and perhaps eight hundred species of mostly tropical annual or perennial, generally fleshy and weak-stemmed climbing herbs, all usually bearing tendrils. They have alternate, broad, usually simple, but often deeply lobed or divided, sometimes compound leaves, regular mostly monoecious flowers, that are often large and showy, but in some genera, small, greenish, and inconspicuous. They are funnel-shaped or with five separate petals and have a calyx joined to the five-lobed ovary and five stamens, but with two pairs united so that apparently there are only three stamens. The fruit is fleshy, berrylike, and usually with a rind, or a bladdery pod.

A native of the Himalayan foothills in northwestern India, the cucumber, in spite of it having little taste and being mostly water, has been cultivated for at

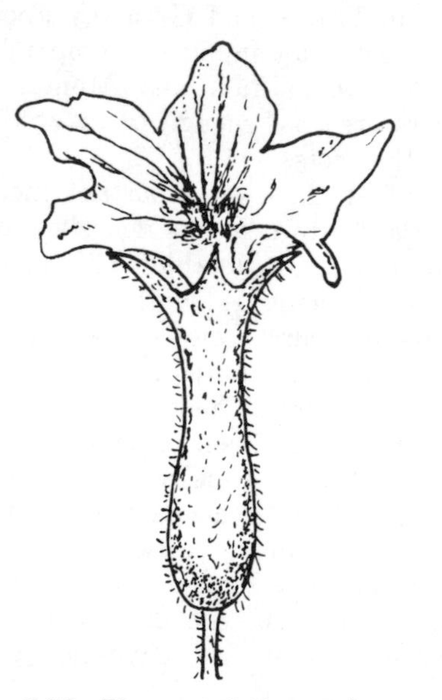

203. Flower of Cucumber

least three thousand years, and its use has been spread all over the civilized world. It is a trailing, rarely climbing roughly haired, annual vine with triangular, oval leaves, more or less angled or three-lobed, the middle lobe the longest and pointed, and yellow flowers, cup-shaped, borne in the axils of the leaves, with the staminate flowers clustered, and the pistillate solitary. The fruit is the familiar cucumber that is more or less cylindrical. When young, the surface is covered with tubercles, and armed with sharp, rigid bristles that later fall off.

The melon, that is the muskmelon or cantaloupe, honey dew, casaba, and related varieties, is also a native of India, but has been found wild in Africa, along the banks of the Nile, and in Guinea. Cultivated wherever the climate will permit, the melon is a rough, hairy, trailing vine with roundish-oval leaves, more or less angled on the margin, but hardly lobed, and yellow bell-shaped or cup-shaped flowers, with the staminate, pistillate, and perfect flowers all being found on the same plant. The fruit is globose, cylindrical, or ovate, and is the familiar melon.

Found from southern United States to Brazil, the bur gherkin is a trailing vine with angled, rough haired stems, three-lobed, rough leaves, each lobe in turn lobed, the margins waxy and toothed, and yellow, bell-shaped flowers. The fruit is oval or somewhat oblong, about two inches long, with crooked stalks and a furrowed and prickly skin. It is usually called gherkin, but the gherkins of most pickle mixtures are young cucumbers.

The familiar and well-known Hubbard or winter squash, whose origin is unknown, is an annual, prostrate vine, not very rough or prickly, with blunt, roundish or heart-shaped leaves, usually not lobed, and yellow, bell-shaped flowers, with the lobes recurved or drooping. The fruit is a berry, and variable, spherical or oblong, sometimes of great size, with white seeds. The turban squash, that is a variety of the Hubbard squash, has a turban-shaped fruit.

It is believed that the pumpkin originally came from America, as it has been found growing wild in Mexico and was cultivated by the aborigines in Florida, Mexico, and the West Indies at the time the Spaniards arrived. As that may be, it is an annual, prostrate vine, with prickly stems and leafstalks, triangular oval, usually prominently lobed leaves, and large yellow flowers, with the corolla lobes pointed, erect, or spreading. The fruit is a large berry, furrowed, usually orange, with the stalk enlarged at the point of attachment. A variety of the pumpkin produces the summer-maturing fruit we know as scallop and summer crookneck squash. A second variety includes the running vines that produce the inedible, but showy, hard-shelled, ornamental yellow gourds. A separate and distinct species is a musky-scented, quick-growing vine with a sticky-haired stem, oval or

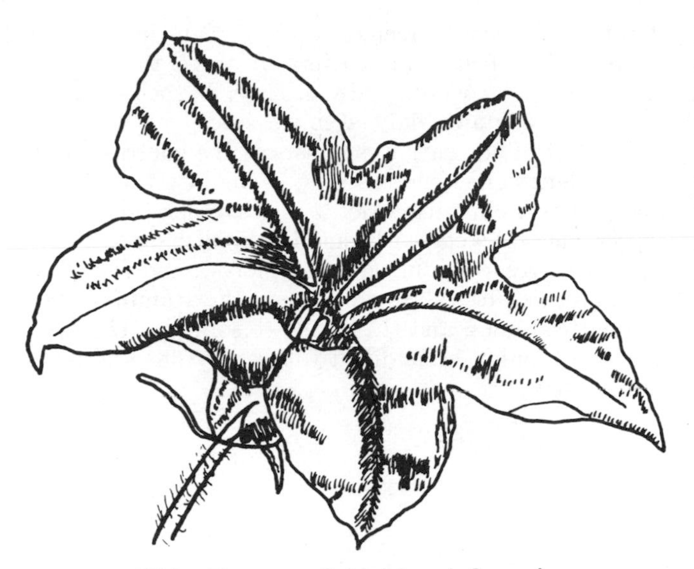

204. Flower of Hubbard Squash

kidney-shaped leaves, and white, showy flowers that develop into the ornamental gourds that are round or flattish, crook-necked, bottle-shaped, dipper-shaped, club-shaped, or shaped like a dumbbell.

It is believed that the ancient Egyptians cultivated the watermelon, as well as the ancient Hebrews and Arabs, but it does not seem to have come into use by the Romans until the beginning of the Christian era, or by the Chinese until the tenth century. There are four or five species of the genus whereto the watermelon belongs, all from tropical Africa, although one of them is also found in Asia, but the common watermelon is the only cultivated species. It is a long-running, prostrate annual vine with alternate, three to five-lobed, pale or bluish green leaves, and light yellow, solitary flowers, about an inch and a half wide, borne in the leaf axils. The fruit is the familiar watermelon, an oblong spheroid berry, greenish brown, mottled or striped, with a hard rind and a red or pink, juicy watery core wherein the seeds are embedded. A variety of the watermelon is the citron, sometimes called the preserving melon. The fruit of this watermelon is much smaller and the core is hard, white, and useful only when cooked.

There are several native species of the gourd family, as for instance, the plant known as the Missouri gourd or wild pumpkin. It is a perennial, prostrate and long-running vine with triangular heart-shaped leaves, whitish beneath, long-stalked, and toothed, and yellow flowers two and a half to four inches long. The fruit is a smooth, orange-shaped, green and yellow-splashed berry and is inedible. It grows in dry and sandy soil from Missouri and Kansas, to Texas and California.

A second native species is the climbing wild cucumber, also called the wild balsam apple. It is a beautiful, rapid-growing, and luxuriant annual climber, generally with tendrils, found growing along rivers and in waste places, but often in cultivation climbing over arbors and on fences, often reaching twenty feet. It has slender, somewhat angled stems and alternate, light green, thin leaves that are ovalish, heart-shaped, with three to seven lobes, the lobes with a minute, soft prickle. The small, six-petaled staminate, greenish white flowers are borne in many loose clusters, while the small green pistillate flowers occur singly or in

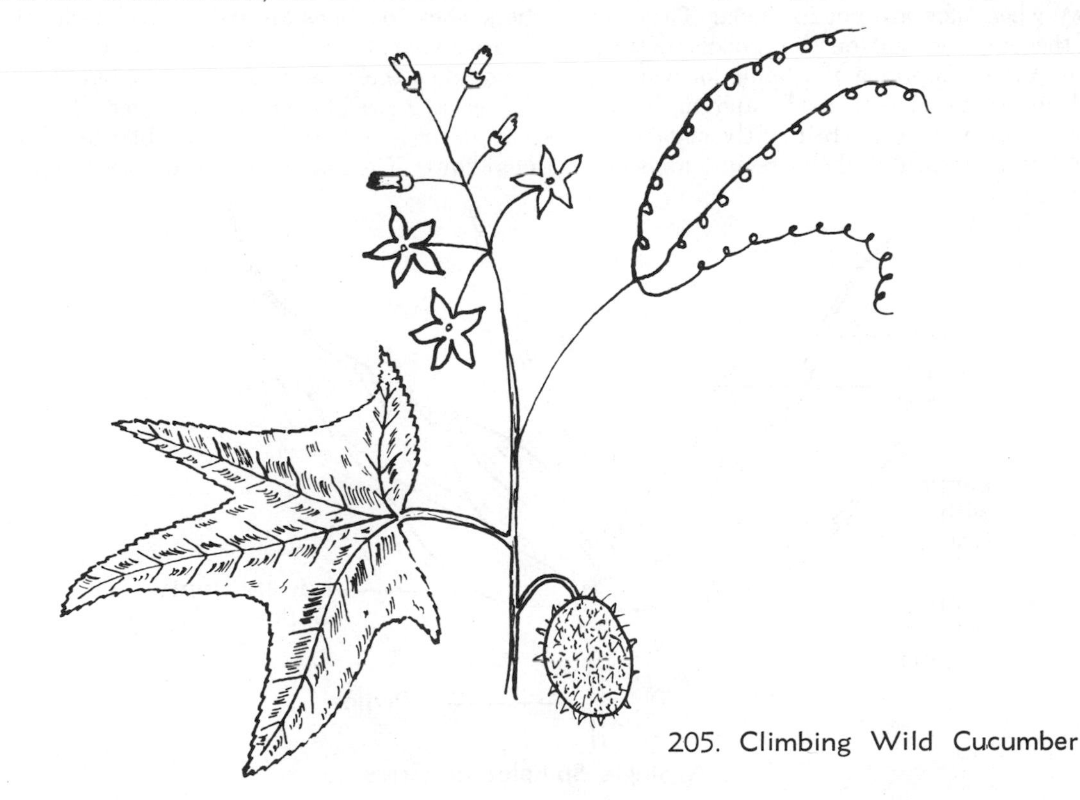

205. Climbing Wild Cucumber

twos at the angles of the leaves. The fruit is cucumber-like, two inches long or less, green, ovoid, and thickly covered with slender, weak prickles. When mature, it often bursts open so forcibly, that it discharges the large, dark seeds to a distance of several feet.

Also an annual climber, the one-seeded bur cucumber is a plant of moist places and riverbanks, but occurs also along fencerows, in thickets, and in waste places. It is an amazingly fast grower with a pale green, slender, but very tough and fibrous, angled stem, more or less sticky-haired, and with very large leaves, five-lobed, thin, rough on both sides and finely toothed. The flowers are small and greenish white, the staminate in small racemes, the pistillate in rounded clusters. The fruits, in clusters of three to ten, are yellowish and covered with prickly, barbed bristles. Each fruit contains only a single seed.

The last of our native members of the gourd family, is the plant called the creeping cucumber, though actually it is a climbing vine. It is a southern species of copses and thickets with small, roundish and heart-shaped leaves, five-angled or five-lobed, and roughish, tiny, greenish or yellowish flowers, the staminate few in small racemes, and the pistillate solitary. The fruit is a small, pulpy berry filled with many flat and horizontal seeds.

75. Grass (Gramineae)

It is a lamentable fact that of all our common plants, the most common are the ones that most people are least familiar with. Yet, everyone comes in almost daily contact with these plants that are economically the most important members of the plant kingdom. From the moment the March sun begins to warm the earth, the grasses, in green tenderness, give us the first intimation of spring. They tinge the brown hillsides even before the snows have ceased, and from then until the frosts of autumn take their toll, there is never a day when they are not in bloom. They are found along the wayside and on the woodland trail, in gardens and orchards, along the banks of winding streams, and in waste places, fields and meadows. Their blossoms may not be as brilliantly colored as the more familiar blooms, but they are just as beautiful in their rose and lavender, purple or green tints.

If one examines the tiny blossoms through a magnifying glass or hand lens, one will be amazed at their seemingly delicate and fragile quality. How they survive the buffetings of wind and falling raindrops, or withstand the merciless rays of the hot summer sun is something of a mystery. There are grasses so tall that they rise above our heads, and others that barely extend above the earth. There are grasses whose flowering spikes are hardly noticeable, and others whose panicles are half a yard in length. There are grasses that are stout and robust, and others so slender that their stems are like golden threads.

When a grass blossom is examined closely, it may seem at first to bear little resemblance to a conventional flower. Compared with a lily, the two apparently

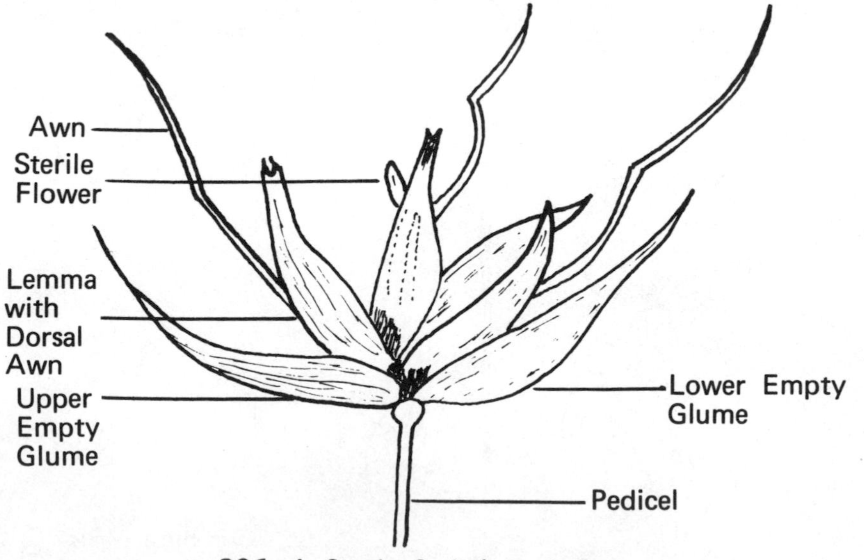

206. A Single Spikelet of Grass

do not seem to have much in common. Yet, if one were to select a lily that blooms in a spike, and were to imagine that the lily suddenly crowded the flowers and reduced the petals to mere scales, one would have a lily with a reasonably grasslike appearance. The flowers of grasses occur in clusters called spikelets. Spikelets vary in size and may be composed of one, several, or many flowers. The short stem whereupon the flowers of a spikelet are placed is called the rachilla. Sometimes the rachilla is prolonged as a tiny thread lying outside the uppermost flower.

207. A Spike of Foxtail Grass

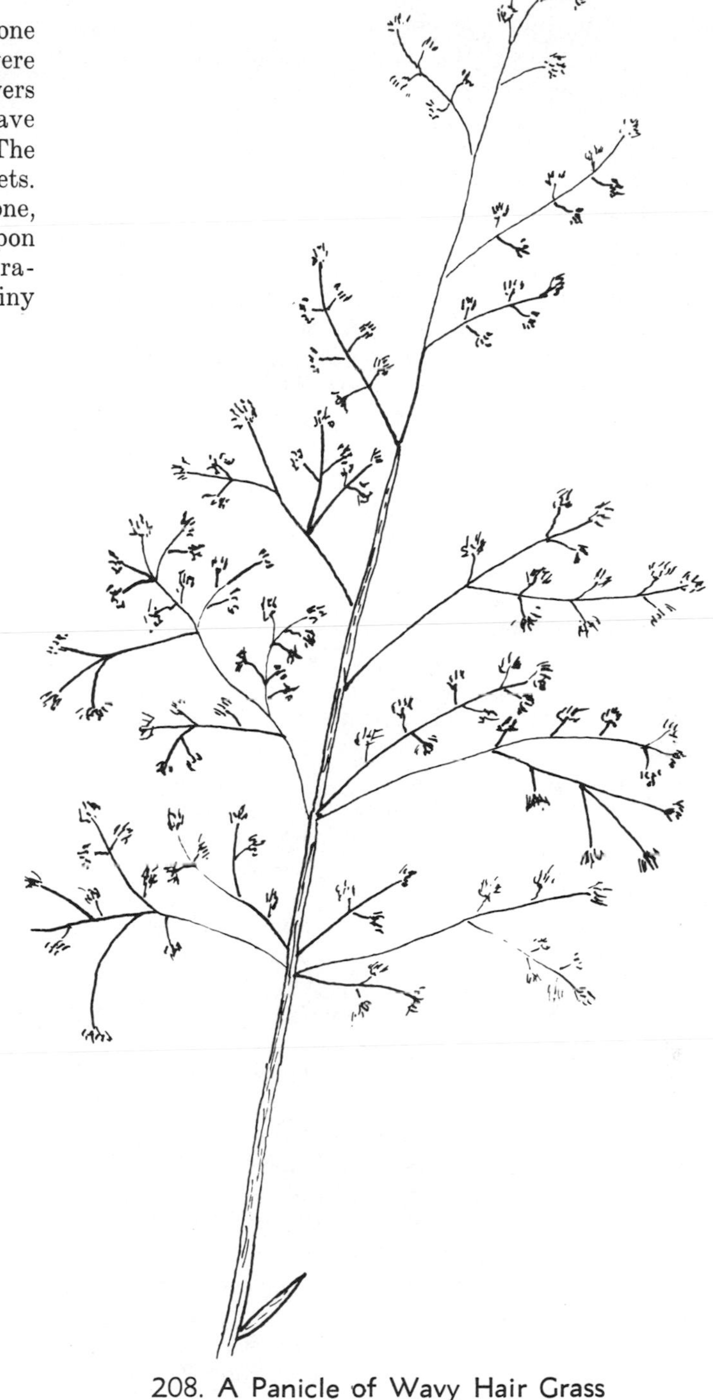

208. A Panicle of Wavy Hair Grass

A grassflower stalk usually has many spikelets, and collectively they form an inflorescence that may be either a spike or a panicle. If the flowerstalk is unbranched and the spikelets are attached directly to it, the inflorescence is called a spike. If, however, the flowerstalk is branched and the spikelets are attached to the branches, then the inflorescence is known as a panicle. In a spike, the lower spikelets bloom first, or from below upward, but in a panicle, the uppermost spikelets are the first to blossom, followed successively by those below.

In a grassflower the sepals and petals of a conventional flower appear in the form of modified leaves called bracts (sepals and petals are also modified leaves), and the rachilla bears a number of these chaffy overlapping bracts. The two at the base of the spikelets (glumes) are larger than the others and enclose the rest of the spikelets. Each flower of the spikelet is enclosed between two bracts, that are usually similar to the glumes, but smaller. The lower of these bracts is known as the lemma, the upper one as the palea or palet.

Lemmas often have a bristlelike appendage called

an awn. Awns are not always present, but when they do occur, they may be straight, bent, or twisted. In some lemmas, they are attached at one end and are called terminal awns, while in others they are attached to the back and are called dorsal awns. Many lemmas are rather flattened and folded so that the two edges are brought closely together. Then the midvein becomes prominent as a ridge on the back. Such lemmas are said to be keeled. When the veins are conspicuous, the lemma is said to be three-nerved, five-nerved, or nine-nerved, according to the number.

Unlike the lemma, the palea is awnless, and is usually two-nerved with two keels. At the base of the ovary and within the lemma and palea, are commonly two (in rare cases three) minute, thin, and translucent scalelike structures, or lodicules. They probably represent two reduced perianth segments. They are rarely seen except at the time of flowering when, for a short time, they become swollen with sap and by forcing the lemma apart, allow the flower to open.

Most grassflowers are perfect, with one to six (usually three) stamens, and usually a one-celled, one-ovuled ovary. The anthers are lightly attached to the slender filaments, and when they tremble in the wind, they release the pollen grain. As grasses depend on the wind for pollination, many pollen grains fall to the ground and are wasted, hence, to ensure sufficient seed, they are produced in vast numbers. A single anther of rye, for instance, contains no less than twenty thousand pollen grains.

The grasses are mostly erect annual or perennial herbs with fibrous roots, often with underground stems that extend horizontally and give off new tufts of leaves, but some of them, as the bamboos, are distinctly woody and shrublike or even treelike. They usually have conspicuously jointed, hollow stems, with parallel-veined, alternate leaves that are often greatly elongated and that are attached to the node by means of a sheathing base. The sheath usually is open on the side opposite the leaf and furnished with a membranous or hairy projection, called the ligule. The fruit is a single-seeded, dry fruit known as a caryopsis (grain), the ripened wall of the ovary forms a dry fruit coat that adheres to the seed as a thin husk. The seed is oval, cylindrical, or tooth-shaped with the tiny embryo in one end, the rest being stored food. The grains are usually yellow, brown, red, or purple, those of the cereal grasses having been prized foods of mankind since earliest times. The cereal grasses, wheat, corn, rye, oats, are able tc grow in a wide variety of soils and under a great diversity of climatic conditions, and thus, have become universally valuable as food crops. Moreover, the grains contain all the essential nutrients and are almost the perfect food and, being dry and not fleshy, can be stored and shipped with comparative ease and success.

The grasses exhibit a great variety of form and size, and some species are sure to be seen in every condition of soil and climate from the equator to the frozen wastelands of the Arctic and Antarctic, and from sea level to the tops of the highest mountains. Some of them are aquatic, while others are found in extremely arid and desert places. They seldom grow singly, but instead, form communities that range in size from that of our lawn to our western prairies, the pampas of South America, and the steppes of Russia. There are between forty-five hundred and seven thousand species (authorities differ as to the number), contained in about one hundred and forty genera and that constitute the grass family. The family is a confusing one to the layman, and so the author shall not attempt to describe completely the more familiar species, but just enough so that they may be recognized. The reader who wants to know more about the grasses is referred to the various books that have been written about them.

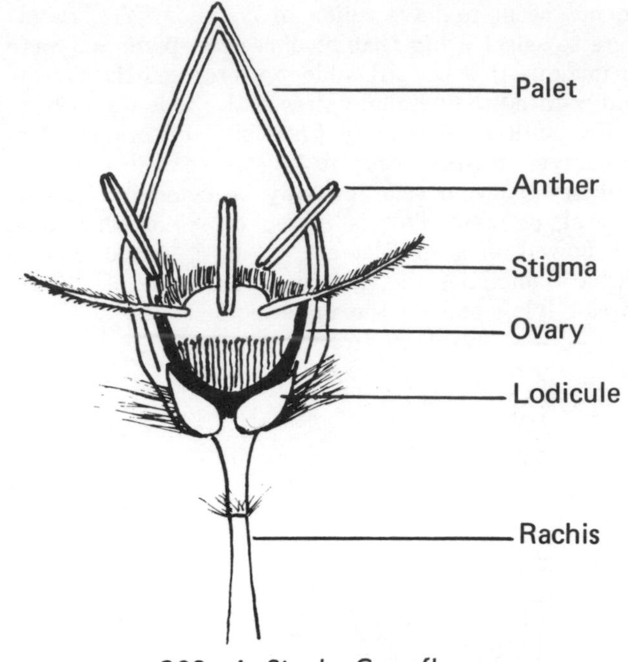

209. A Single Grassflower

Unlike most grasses, corn or maize of the genus *Zeà*, Greek for some cereal, has a solid stem composed of pithy material wherein are embedded the vascular bundles. It is a tall, annual grass with numerous, relatively broad, sword-shaped leaves, that in many varieties have a cutting edge. Ordinarily, its fibrous roots could not support ten to fifteen feet of leafy stalk, and so, special prop roots grow out of the base of the stem, arch over, and function as buttress supports. The staminate flowers are borne in a terminal cluster (spike) to form the plumelike tassel at the top of the corn plant, and that produce abundant pollen. The pistillate flowers occur in small clusters (the cob) below the staminate flowers, and it is the long, threadlike styles of these flowers that constitute corn silk. From these, develops the familiar, heavily sheathed ear of corn, the kernels being the seeds.

Maize is a subtropical and tropical American plant

whose wild ancestors are unknown, but that are generally believed to be closely related to teosinte, a maizelike plant native to Mexico. Maize was cultivated by the Mayas a thousand years ago and by various Indian tribes of Central and South America before that time. Today, maize is of considerable economic importance and is the chief feed grain of the American livestock region known as the Corn Belt. There is only one species, but many varieties, such as sweet corn, the garden variety of corn; pop corn, that resembles sweet corn, but has small ears that have very hard, pointed seeds exploding when heated; flint corn, a medium-sized corn with dark yellow, hard, smooth seeds; and dent corn, a tall variety with yellow or white seeds that are indented at the top.

Maize has many uses besides a food for stock. It is canned and made into cornmeal, and is used as a cereal breakfast food. It is a source of oil that is extracted for use in cooking and for making soaps and paints, and starch, wherefrom dextrins and gums are made for use on envelopes, postage stamps, and in the textile industry, as well as corn and glucose syrups, and it is even a source of alcohol. The stems, too, are used, the pith for making explosives and the stalks themselves for making a coarse grade of paper. Even the cobs have a use, as they are made into pipes and charcoal.

Like the grasses, the bamboo has a hollow stem, but unlike the grasses, it is woody. There are a number of species of these woody grasses of the Old World tropics and all belonging to the genus *Bambusa*, a Latinized version of the Malayan vernacular for these plants. They usually have a polished, hollow stem, interrupted by a partition at the joints, that conspicuously rings the stem, short-stalked leaves, and flowers in a few or multi-flowered spikes. They are noted for the delicate massed effect of their foliage, and the closeness with which the stems grow, as well as for their height that may sometimes reach a hundred feet. They are exceptionally rapidly growing plants, growing, in favorable places, a foot or more a day. In the regions where they thrive, they have many uses, as for construction and for making paper and textiles. The young shoots are edible and are eaten like asparagus.

Without a doubt, the wheat plant is the most important cereal grass in the world. It is one of the oldest cultivated plants, having been cultivated since antiquity, and today is unknown as a wild plant. Even its ancestral home is in doubt, though it is believed to have been somewhere in western Asia. The Spaniards introduced wheat into Mexico during the sixteenth century, and the English brought it to America in the seventeenth, but the English colonists soon abandoned it as a cereal crop in favor of the better adapted maize.

The modern wheat plant of the genus *Triticum*, is an annual grass about four feet high, with flat grasslike leaves, twelve to sixteen inches long, flowers in terminal clusters composed of fifteen to twenty spikelets, and a grain that is ovoidal in shape and cream, red, or purple in color. What is known as bran is the ovary wall and seed coat, that is rich in proteins. There are now several hundred named varieties, most being developed from a few original species.

In the United States the first wheat belt extended from Delaware and Maryland to New York, and in this area the pastry flour varieties are still grown, as well as in California and the Pacific Northwest. Later, wheat was grown in the states of the Ohio valley that became the center for soft winter wheats that are very starchy. Following the Civil War, the great plains, an ideal wheat country, became available, and it is here that the hard wheats have become the favorite varieties, the hard, red, winter wheat being grown in the southern prairie states, the hard red spring wheat in the northern. Bread flour is a blend of both. The variety known as durum wheat, grown chiefly in Minnesota and the Dakotas, is used in making spaghetti and macaroni. The different types of flour are made from different parts of the kernel, thus, graham flour is made from the complete wheat kernel, whole wheat is made from about ninety percent of the kernel, part of the bran being removed, and standard white flour is made from about three-quarters of the kernel, the bran, protein layers, and embryo all being removed.

Next to wheat, barley is probably the most widely distributed cereal crop. It has been used as a cultivated plant since prehistoric times and was used by the Romans in baking a heavy bread eaten by the soldiers and peasants. Barley of the genus *Hordeum*, the classical Latin name of barley, is a native of southwestern Asia and is an annual cereal grass usually about thirty inches high, with flat, grasslike leaves, nearly twelve inches long, and terminal, cylindrical flower clusters, mostly in dense spikes with conspicuous awns, three spikelets at each node. Barley contains little of the gluten found in wheat, and thus, cannot be used for a light bread. Most of the barley grown in the United States is used as a food for livestock and in preparing malt, which is the sprouted and dried barley grains, malt being used in making malted beverages and in brewing beer. A member of the same genus is squirreltail grass, also known as wild barley, a perennial grass of some beauty, and for this reason, has been grown for ornament, but is more often a troublesome weed. The long, barbed, reddish-golden awns become very brittle when ripe and break into small pieces that work their way into the teeth and jaws of animals that eat the grass, causing ulcerations and swellings. Injury to horses, cattle and sheep can often be considerable.

Although rye was known to the Greeks and Romans, it was not cultivated by them and, as a matter of fact, did not come into extensive use until after the Middle Ages. In Russia and Germany rye is an important food grain, but is of less importance in the United States where it is used in making rye bread and whiskey. A native of the Mediterranean region, rye of the genus *Secale,* an old Roman name for some cereal, is an annual grass, three to five feet tall, with erect, slender, bluish green stems, grasslike leaves, and flowers in a closely-set terminal spike, long and multi-awned, the spikelets on a small zigzag stalk. The fruit, the familiar grain of rye, is oblong and minutely grooved. The common rye is unknown as a wild plant.

As in the case of rye, the oat was not grown by the Greeks and Romans, though it had been cultivated by the barbarian tribes of Europe prior to the time of Christ and did not come into general use until after the Middle Ages. The common oats, genus *Avena,* Latin for oat, is an Old World annual plant with flat grasslike leaves and a terminal flower cluster (panicle), its flattish, long-awned spikelets spreading on all sides. It is grown chiefly as food for stock, only a small amount being used for breakfast cereals. The common oat is often a weed in the garden. Winter oats are hardy strains for autumn planting, maturing the following spring. All the others are spring planted annuals. A related species, the wild oat, a native of Europe and Asia, has become naturalized on the Pacific Coast.

Next to wheat, the most important cereal in the world, and without a doubt, the staple food of more people than is wheat, (it being to the Orient what wheat is to the Occident), the rice, genus *Oryza,* the Latinized version of an Arabian name for rice, is a swamp-dwelling grass, three to four feet high, with smooth, angled stems, that are mostly hidden in the long leafsheaths and grasslike leaves. The flowers are borne in a narrow, branching, terminal cluster (panicle), that is usually curved to one side, with the spikelets flat and ribbed. The fruit is the familiar rice grain that before milling has a brown fibrous covering and outer layers of protein food material. These are removed by milling, the white grain then being further scoured and polished, and finally coated with glucose and talc before being marketed. Besides its use as a food, rice is made into a beer by the Japanese, known as saké, and the straw is used as a food for stock and in making paper, hats, and strawboard. Our native wild rice of the genus *Zizania,* a Greek name for some wild grain, is a tall, annual marsh grass greatly cultivated for its nutritious grain and for ornament, it being a handsome grass seven to ten feet high. Also called water, Indian, or Canada rice, it is widely grown by sportsmen as a food for wild fowl.

210. A Grain of Oat

211. A Panicle of Oat

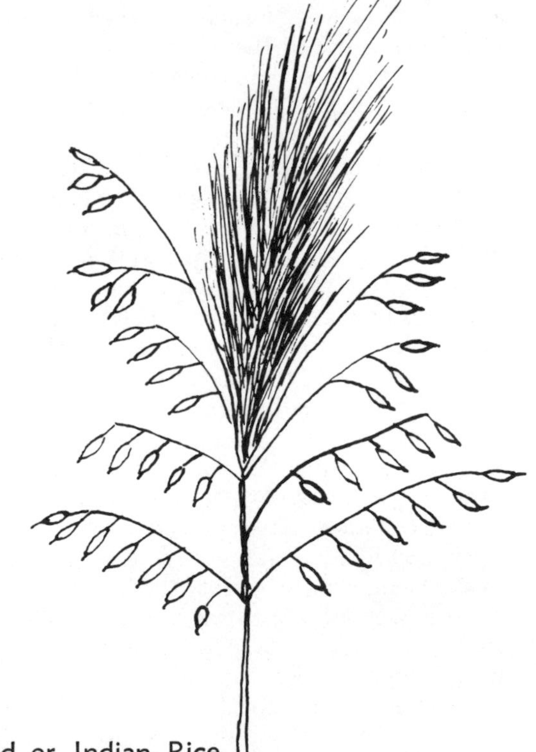

212. Wild or Indian Rice

The sugarcane, that few people think of as a grass, is a native of India, having been cultivated in that country and in China since ancient times. It was introduced into Europe after the Crusades, and later into

America and the West Indies. Before the Civil War, a considerable amount of sugarcane was grown in the southern states, but today, practically all cane sugar produced in the United States is from extreme southern Louisiana, the reason being the higher sugar yield per acre in Cuba, Java, and the Philippines.

The sugarcane of the genus *Saccharum,* an old Greek name for sugar and also used as a specific name for the sugar maple, is a solid-stemmed plant ten to fifteen feet high, the stem green or purplish, conspicuously ringed, and containing a juice that is the source of sugar. The leaves are similar to those of the common corn, but somewhat longer, and with rough or cutting margins. The terminal flower cluster is usually a branching panicle, fifteen to thirty inches long, the branches plumelike, but drooping, and rarely produced except in the tropics where the pollen is rarely fertile and mature fruits seldom formed. Hence, sugarcane is propagated by stem cuttings. In making sugar, the canes or stems are shredded and then crushed between rollers, the extracted juice is strained and evaporated, and the sugar crystallized out. If done at high temperatures, a hard-grained sugar is produced, at lower temperatures a soft sugar. The liquid molasses is separated from the crystallized sugar by centrifugal action, and the poorer grades are utilized as stock feed and in making alcoholic beverages.

Maizelike in appearance, sorghum or broom corn, a native of tropical Asia and Africa, is a stout, annual grass up to twelve feet high, with broad leaves two feet long with a large terminal, branching, flowering cluster of spikelets. There are many varieties of sorghum of the genus *Holcus,* Latin for some grain, and they are grown for grain, syrup, forage, and for the manufacture of brooms. In China they also furnish a fermented drink and provide material for making mats, baskets, roofing material, window shades, and for fuel. Some of the varieties include the Kafir, grown for its edible grain whereof there are white, red, and black forms; the chicken corn, grown in the South for chicken feed; the Sorgho, whose rich, sweet sap is a commercial source of syrup; and the broomcorn, that is commercially grown for the making of brooms. Also a species of the same genus, Johnson grass is a stout, perennial forage grass of much value and extensively grown in the South.

In Asia and Africa, the millet is grown for food, but in Europe and the United States it is grown for fodder. It is an annual, three to four feet high, with flower clusters of drooping habit. Millet belongs to the genus *Panicum,* an old Latin name for Italian millet, a group of four hundred species that includes such common grasses as the old witch grass, a beautiful grass with delicate soft and silky flowering heads; the forked panic grass, with short, spreading leaves on slender, wiry stems; the round-fruited panic grass with many flowered, purple panicles; the hispid panic grass, common in wayside thickets; and the seabeach panic grass, a characteristic plant of the sea beach.

The familiar lawn grass, the Kentucky bluegrass belongs to the genus *Poa,* an old Greek name for grass, a group of some one hundred species that includes the low spear grass, one of the earliest plants in the North to change the brown hillsides to living green; the wood spear grass, a slender grass of wooded hillsides; the Canada bluegrass, the bluest of the poas, whose leafy shoots in early spring are noticeable on sandy hills and in thickets that border deep woods; and the false redtop, the tallest of the poas, that blooms in swampy places and in wet meadows, its green spikelets showing each a tawny orange tip.

213. Low Spear Grass

When the early grasses have passed their flowering

stage, the fescues appear, and for a short time, they are the primary grasses of the fields. There are some hundred species of the genus *Festuca,* an old Latin name for some grass, of annual or perennial, usually tufted grasses and generally of small habit. The first to bloom is the sheep's fescue, but following closely on the opening of its flowers, the graceful stems of the meadow fescue may be seen by the wayside, where the red fescue, its flower cluster, green, reddish, or bluish green, is also common.

The canary grass may usually be recognized by the two sterile, or staminate florets between the empty glumes, and the one perfect floret in the spikelet. One species, the reed canary grass of the genus *Phalaris,* an old Greek name for some of the species, is a perennial grass of wet grounds, often planted as a border plant, though a variety with white and yellow-striped leaves known as ribbon grass, is more widely grown. The second species, canary grass, a native of Europe, but naturalized here, is an annual, and is cultivated for its shining, straw-colored seeds that are a favorite feed for birds.

In June the graceful manna grasses bloom in tones of dull green and purple. They are typical of spring without that suggestion of midsummer heat that the bent grasses bring to the July fields. The first to blossom is usually the nerved manna grass of genus *Glyceria,* from the Greek for sweet in allusion to the taste of the grain. It grows in the borderland between pasture and marsh, its drooping panicles easily recognized by their spreading and drooping branches and by their tiny purple and green spikelets. A stout handsome species, the tall manna grass, often seen in wet grounds, is like others of its genus a species that cattle like. During the fall migration the water fowl find a resting place along the streams where the grass grows abundantly, and feed by the thousands on its seeds. The most beautiful of the genus is the rattlesnake grass with its pendant, inflated spikelets of pale green and purple that retain their beauty until fall. It is a grass of brooks and ditches and damp waysides.

Few grasses so monopolize the field and wayside as the species of the genus *Agrostis,* Greek for field and Latin for some grass. It is a large genus of widely distributed grasses that are known as the bent grasses. They are as typical of midsummer as the goldenrod is of autumn, beginning to bloom by the waysides in June. Brown bent, common in some lawn mixtures, is also common near the coastwise marshes of New England and New Jersey where its wide open panicles of brown or brownish purple, flecked by white by the small anthers, remain untouched by the hot summer's sun. Common among the wayside grasses is the rough, hair grass, with its shining stems and delicate panicles that glisten in the sunlight "like purple cobwebs." Among the chief grasses of the July fields is the redtop and its varieties, giving color to acre upon acre with varying tones of reddish purple. The redtop, a perennial European pasture grass and naturalized in the United States, is much planted for hay and is an ingredient of lawn mixtures. In fields, meadows, and waysides the unopened panicles, in narrow spikes of green and purple, rise above the leaves in early summer and the grass may be thus recognized weeks before it blooms.

Native grasses, and found in many localities, are the Muhlenbergias. With one exception, they are unattractive grasses, with inconspicuous flowering heads of little beauty. They grow in open woods, dry fields, and on the moist banks of streams. One will find the wood Muhlenbergia and nimble Will in the woodland border, the rock Muhlenbergia in rocky situations, the marsh Muhlenbergia in wet grounds, and the meadow Muhlenbergia in fields and waste grounds. Distinct in appearance from the other Muhlenbergias, the long-awned hair grass is a delicate and beautiful species with long and glistening, gossamerlike, purple panicles. The solitary flowers, borne on such widely spreading

214. Meadow Muhlenbergia

pedicels that the flowering heads are often a foot across, are, in late summer, gauzelike and graceful, and remain so until the first frost lays its blight upon them. The Muhlenbergias belong to the genus of the same name, from Dr. Henry Muhlenberg, a distinguished American botanist.

Usually occurring as weeds in cultivated land and also found near waysides, both the green and yellow foxtails are common near gardens. They are stout grasses, with smooth stems, red-tinged at the base, and with flattened sheaths and many leaves, characteristics that enable one to recognize the grasses before the blossoming spikes appear. In flower, the two species of the genus *Setaria,* from the Latin for bristle, may be distinguished from one another by the color of the clustered bristles that cover the spikes, those of the yellow foxtail being yellow or yellowish brown, those of the green foxtail being green. Less common than either, the bristly foxtail is a more slender species with smaller spikes. Also members of the same genus are the foxtail or Italian millet, an annual grass cultivated for forage and hay and often found in waste places and by roadsides, and the palm grass, a native of the East Indies, often grown in greenhouses for ornament.

The Indians of Arizona and Southern California use the seeds of the cockspur grass for food, or did at one time. It is a stout, coarse annual grass of the genus *Echinochloa,* Greek for hedgehog grass, also known as barn or barnyard grass. It is a leafy, usually branched species, a native of Europe, but naturalized in the United States, and found growing in fields, waste places, and in cultivated lands. It is often grown for forage.

Perhaps the most obnoxious grass, at least from the farmer's viewpoint, the couch grass or quack grass is, however, a good grass for hay and makes two crops a year. Cattle eat it greedily, and also important, is that it is an excellent soil binder, its matted "couch" of interlacing rootstocks holding the soil in place in steep gullies or on road embankments where the ground must be guarded against "washouts." But it is this very quality that makes it such a pest when it gets in cultivated ground, where its stout leafy stems and flattened, two-headed spikes shoot up out of the ground with all the energy of the fabled hydra. Cut one stem and half a dozen will rise in its place.

The couch grass belongs to the genus *Agropyron,* Greek for field and wheat. The genus also includes the bearded wheat grass, a naturalized species from Europe often found in cultivated ground and in meadows, and the purple wheat grass, an alpine species.

The bane of every gardener, the crab grass is so well known that it hardly needs a description. In some parts of the world it is grown for pasturage, while in other parts, for its seeds that are used in porridge. In our country it is a weed, however, growing about our dooryards, on our lawns, and in cultivated land generally, where it spreads and roots at the joints, and sends up a long stalk with fingerlike spreading stalks. There is a large crab grass and a small crab grass, of the genus *Digitaria,* from digitus

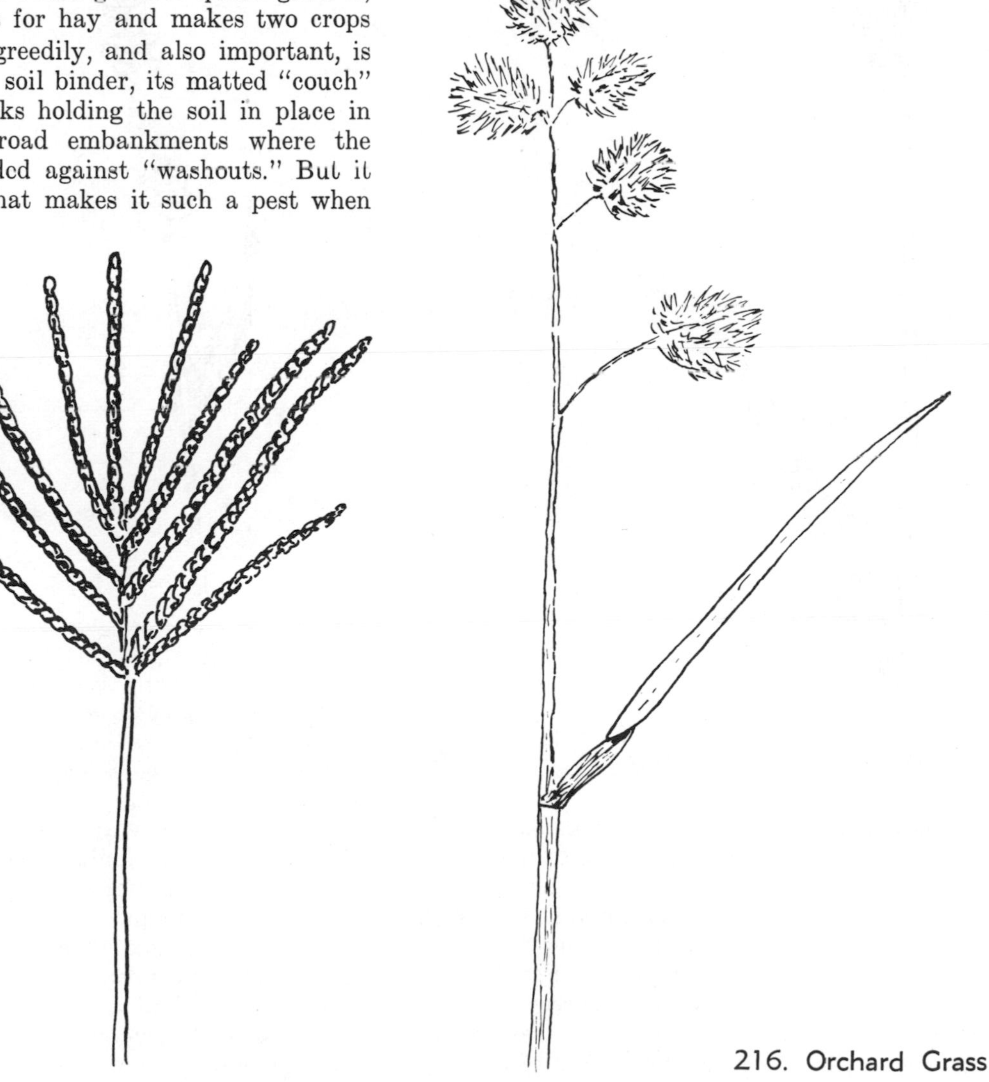

215. Crab Grass

216. Orchard Grass

a finger, and both are much alike, differing chiefly in size and in the number of fingers.

An immigrant from Europe, but naturalized here and common almost everywhere, indeed, in most states one of the most common species, the orchard grass, with its spreading tufts of blue green leaves, is very noticeable by the roadside in early spring before the grasses of summer have appeared in every lane and byway. The only species of its genus, *Dactylis,* from the Greek for finger, it is a stout perennial with coarse panicles painted with large anthers of purple and yellow, terra-cotta and pink.

"I have just made out my first grass, hurrah, hurrah!" wrote Darwin of the sweet vernal grass, one of the first grasses to attract our attention in early spring

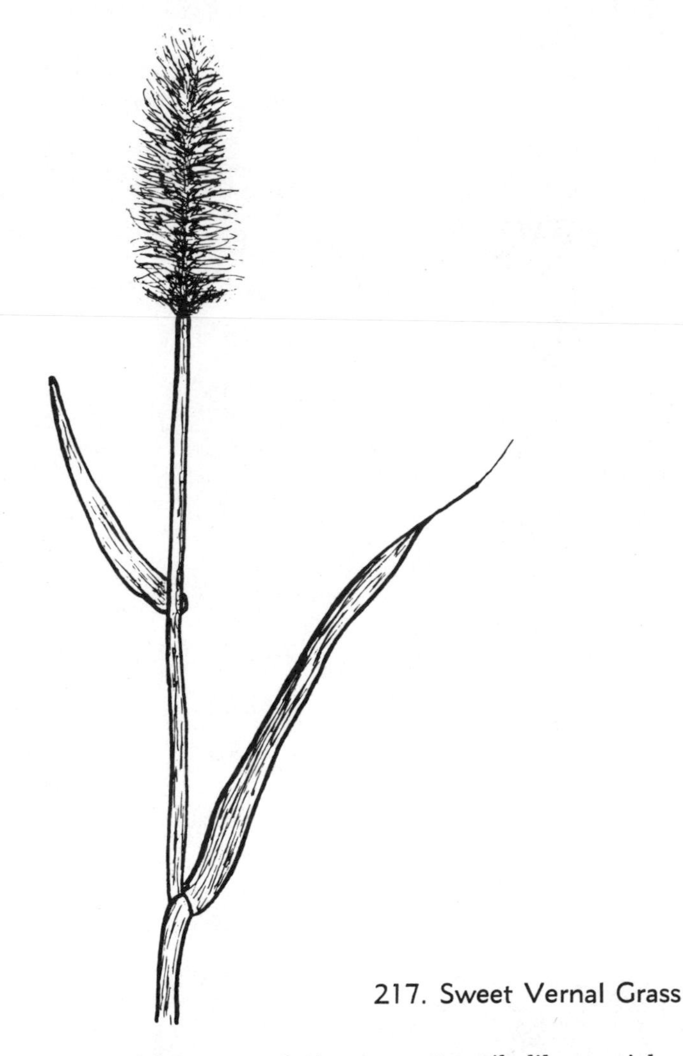

217. Sweet Vernal Grass

as it pushes upward its compact, spikelike panicles that soon expand into open blossoms with large violet anthers. The sweet vernal grass of the genus *Anthoxanthum,* probably from the Greek for yellow flower, in allusion to the pollen, is an Eurasian perennial grass naturalized throughout the whole of North America, with flat, very narrow leaves and narrow spikelike flower clusters, and is one of our only two strongly fragrant species, its fragrance being due to the principle coumarin.

The second fragrant species is the vanilla grass of the genus *Hierochloe,* Greek for sacred and grass, a perennial grass with flat leaves and a terminal panicle, the latter ornamented in chestnut brown and purple. Its leaves are very fragrant in drying and are used by the Indians who weave them in all kinds of baskets and plaques. In olden times it was the custom to scatter sweet-scented grasses before the churches of northern Europe on Saints' days; the vanilla grass, together with the less strongly scented vernal grass, being used for the purpose. Hence, it became known as holy grass that, even today, it is sometimes called.

There are some one hundred and fifty species of the genus *Paspalum,* an old Greek name for millet, perennial grasses with flowering spikelets arranged in one-sided racemes. Some species are tall and stout, others are low and spreading. Both the slender paspalum and the field paspalum are low-growing, with plump flowers that, even before blossoming, appear beaded with ripened seed.

218. Slender Paspalum

Among the grasses of late summer there are the aristidas, genus *Aristida,* from arista, a beard or awn, whose flowering scales each bear triple awns, a peculiarity that has given them the name of the triple-awned grasses. They are common in dry soil throughout the country. A familiar species is known as poverty grass, a tufted species with narrow, wiry leaves, and found in dry, upland meadows, pastures, and waste places. It is the smallest of the aristidas.

In many states the timothy is one of the most common of cultivated grasses, indeed, it is our chief hay grass, and perhaps of all our grasses is the one most generally known and easily recognized. Like many other grasses, it is a Eurasian species, but naturalized here, and by waysides and in fields its bright green bayonets, rising stiff and rigid, and tipped with cylindrical flowering heads, are a familiar sight from June to August. It is a perennial grass of the genus *Phleum,* Greek for a kind of reed, with stems often five feet tall and flower heads six to ten inches in length. Its name of timothy is that of its earliest cultivator, Timothy Hanson, a Maryland planter of the early eighteenth century.

Probably the first of the grasses to be cultivated as a forage plant was the ray grass of the genus *Lolium,* Latin for one of the species, a tufted, perennial grass long grown in Europe, but less cultivated here as other grasses are more suited to our soil and climate. Its slender, elongated spikes, appearing in midsummer, are beaded with edgewise placed spikelets, while the flowers are light green with pale pendent anthers. A closely related species is the "*infelix lolium*" of Virgil, that is supposed to have been the "tares among the wheat" mentioned in the Gospel of St. Matthew. Today it is known as the darnel, an annual occasionally found as a weed in grainfields and remarkable for its poisonous seeds.

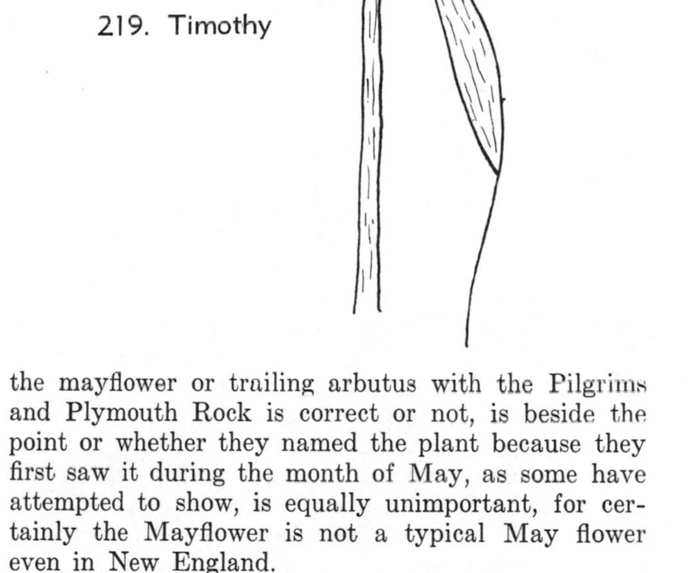

219. Timothy

76. Heath (Ericaceae)

The legend of the Mayflower, or trailing arbutus or ground laurel as it is variously known, beautifully given us by Whittier, is that after the Pilgrims' first dreadful winter, this flower was the first to greet them, and they took renewed hope and courage when

220. Flowers and Leaves of Trailing Arbutus

they saw such a beautiful plant blossoming so bravely amid the winter snows. Whether the association of the mayflower or trailing arbutus with the Pilgrims and Plymouth Rock is correct or not, is beside the point or whether they named the plant because they first saw it during the month of May, as some have attempted to show, is equally unimportant, for certainly the Mayflower is not a typical May flower even in New England.

Suffice, that it is one of our earliest spring blossoms, a most charming plant, and perhaps the most fragrant of our wild flowers. In his poem, "The Twenty-seventh of March," Bryant associates it with the hepatica, that is no doubt the earliest of spring flowers, when he says:

Within the woods
Tufts of ground laurel, creeping underneath
The leaves of last summer, send their sweets
Upon the chilly air, and by the oak
The squirrel cups, a graceful company
Hide in their bells a soft aerial blue.

The famed arbutus prefers dry ground, hillsides, and borders of rocky woods. It also grows in the vicinity of evergreens, and it is in such places that it is usually found with its open chalices scenting the air of the spring woods with a delicious, spicy fragrance that blends with the smell of pine and of damp soil being warmed into life. It seems a sacrilege, and in some states it is illegal, to pick the dainty blossoms that look as if they have been placed in our early spring woods as messengers of hope and gladness to those who have passed through a trying winter.

It is something of a shrubby plant, staying close to the ground beneath a cover of decayed leaves, with tough and light brown stems, the old leaves dull, light olive green, rusty-spotted, the new leaves that appear in June, lighter in color, the surface rough and netted with fine veins, and evergreen. The flowers are white or delicately pink-tinted, five-lobed, tubular, and with a frosty sheen, and are few or many in clusters at the ends of the branches. They are visited by the early queen bumblebees.

The trailing arbutus is not to be confused with the genus *Arbutus,* a small group of evergreen trees that includes the madrona, a shrubby tree with leathery evergreen leaves, satiny reddish brown bark, white spherical flowers in dense terminal cluster, and orange red fruits commonly found on hillsides and in canyons along the Pacific Coast; and the strawberry tree, a native of southern Europe and planted in the South as an ornamental. It is an evergreen shrub with sticky-haired branches, elliptical or oblong leaves, clustering white flowers, and beautiful rough, red berries. The name *Arbutus,* incidentally, is the Latin name of the strawberry tree.

The trailing arbutus, madrona, and strawberry tree are what are called ericaceous or heathlike plants, of the heath family, a large group of trees, shrubs, and herbs, of about seventy genera and fifteen hundred species that are largely evergreen and of very wide geographical distribution, though confined mostly to the north Temperate Zone. Generally the members of the family have simple, more or less leathery, opposite or whorled leaves, with perfect, often showy flowers, regular or highly irregular, solitary or in various kinds of clusters, quite often bell-shaped or urn-shaped, but sometimes of separate petals. The calyx is four to five-cleft or parted, the corolla four to five-lobed, the stamens are usually twice as many as the corolla lobes, and the ovary is two to five-celled with one style. The fruit may be either a berry, capsule, or drupe. The family includes many wild species, some valued for their fruits, and many ornamentals such as the azaleas and rhododendrons.

There are many who consider the creeping snowberry the daintiest member of the heath family, and there seems to be no argument with this view. It is a trailing and creeping evergreen whose roughish stems, often terra-cotta colored, stay close to the ground of the cold, wet woods that are its home. The dark, olive green leaves are stiff, tiny, oval or ovate and the small, shy flowers, hiding in early spring in leafy corners, are white and nodding and solitary on short stems that extend from the leaf axils. Later in summer, the flowers have given way to china white berries that are conspicuous against the green of the leaves. The berries have the refreshing flavor of sweet birch and so, too, have the leaves. The latter may be used in making a tea and the berries are a source of food to many a hungry bird. And so, also, are the berries of the bearberry.

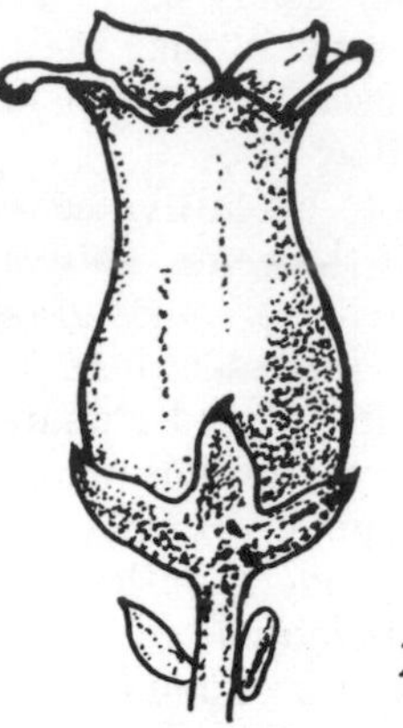

221. Single Flower of Bearberry

A trailing plant of a somewhat shrubby character, the bearberry is a lover of bare rocks, sandy soil, and hills, and in such places, it spreads its branches, sometimes to form luxuriant mats, in its manner of growth resembling the trailing arbutus and often mistaken for it. But its small, rounded, urn-shaped flowers are not pink or salver-form, but white and appear in May, the small dark, evergreen leaves are spatulate and not oval or oblong, and the fruit is not a capsule, but berrylike and red. The latter is dry and insipid and not edible, but game birds, especially grouse, and other birds, too, for that matter, as well as various mammals, such as the chipmunk and raccoon, are not adverse to adding the clusters of red bearberries to their scanty winter menu. Why the name of bearberry is not known unless the bear is supposed to feed on the berries. It doubtlessly does eat them, but it prefers blueberries, blackberries, and shadberries whereof it consumes great quantities.

Close relatives of the bearberry are the twenty-five species of shrubs and trees known as the manzanitas. They are common to the mountains of the Pacific Coast, have smooth, red-barked branches and trunks, elliptical or oblong, gray green evergreen leaves, white or pink urn-shaped flowers, and brownish red berries. The berries are used in making manzanita jelly and the Indians use them for making cider. The name manzanita is Spanish for "little apple."

"Where cornels arch their cool/ Dark boughs o'er beds of wintergreen," so wrote Bryant, though he did not know his plants too well, for the wintergreen more commonly occurs in the vicinity of evergreens not dogwoods. All of those who know the woodlands, are familiar with the shrubby plant that creeps upon or beneath the surface of the ground, its erect branches, three to six inches high, terminated by the clusters of evergreen leaves, the older ones dark green, the younger yellow green, glossy and leathery, oval to oblong, finely saw-edged, and very aromatic. The

222. Leaves and Flowers of Wintergreen

flowers are white, waxy, vase-shaped, and nodding; and they grow from the axils of the leaves. The fruit is actually a capsule, but enclosed by the calyx that thickens and turns fleshy, so that it appears as a globular red berry that has a pleasing spicy aromatic flavor. Certain birds, such as the ruffed grouse, are fond of the berries, and in some sections, they are used for making pies.

The young, tender leaves of the wintergreen make pleasant eating. The older leaves may be eaten too, but they are tougher and less agreeable. In many parts of the country, a pleasing tea is made by steeping the leaves in boiling water, hence, the wintergreen is also known as the mountain tea and teaberry. Twenty-five common names of the wintergreen are evidence that it is one of the best known of our wild plants. The leaves also furnish the wintergreen oil used in flavoring.

The wintergreen belongs to the genus *Gaultheria,* named for a Dr. Gaultier, a physician in Quebec, that also includes a slender, western, evergreen shrub with white or pink flowers known as salal. The spotted wintergreen, a distant relative, belongs to the genus *Chimaphila,* Greek for winter-loving in reference to the evergreen leaves. The spotted wintergreen is an inhabitant of dry woods with a more or less prostrate stem that is partly underground and that sends up occasional flowering and leafy shoots. The lance-shaped leaves, pointed, with distant teeth, are beautifully mottled with white along the veins, hence its name. The flowers are white and nodding and when short-lipped bees and flies are seen flying about them, it can be assumed that cross-fertilization will be effective.

A beautiful evergreen plant, "embodying," as someone once remarked, "the very essence of the woods," the pipsissewa or prince's pine is as familiar to those who know the wintergreen, for it, too, is found among the evergreens—the pines, the spruces, and the hemlocks. It is an evergreen herb, as is the spotted wintergreen, with buff brown stems, six to twelve inches high, dark green leaves, thick and shining, spatulate to lance-shaped, sharply toothed, in circles about the stems, and dainty, pale pinkish or waxy cream flowers, delicately scented, in terminal clusters. The pipsissewa is one of the latest of the fragile, early flowers that lend charm to the spring woodlands that have long since begun to surrender to the sturdier and more showy blossoms that are so typical of the late summer and early autumn landscape. The name pipsissewa is undoubtedly of Indian origin, but what it means has long been forgotten.

Some botanists consider the pyrolas and their relatives a separate family, while others consider it as part of the heath family as the author has done since their structural characteristics are essentially like those of the heath family. The pyrolas, the name a diminutive of *Pyrus,* the pear, because of the supposed resemblance of the leaves, are a group of hardy, low-growing

223. One-flowered Pyrola

perennials with spreading rootstalks, evergreen, roundish, basal leaves in clusters, and whitish, green, or purplish nodding flowers that may either be solitary or in terminal clusters (racemes), on a stalk having scale-like bracts. There are some fifteen species, of which, the following are the most familiar: the shinleaf, the round-leaf pyrola, and the small pyrola. The shinleaf, so-called because its leaves were once applied to every bruise on the human body as a sort of "shin plaster," is perhaps the most common of the pyrolas. It is a plant of the deep woods with rosettes of dark, olive green, elliptical leaves with margined petioles at the root surrounding a tall scape finished with a raceme of greenish white, waxy, fragrant flowers. A similar, but much taller species, the round-leaved pyrola has nearly round or very broad, oval leaves, more shiny and thicker, with leafstalks longer than the leaves, and flowers like those of the shinleaf. It grows in dry or damp, sandy woodlands. As its name suggests, the small pyrola is a somewhat smaller species of the northern woodlands, with ovate, pointed, deep green leaves in a circle near the base of the plant stem, and a one-sided row of small greenish white flowers, the flowerstalk often bent sideways as if weighed down by the flowers. Not exactly a true pyrola because it belongs to a genus all its own, the one-flowered pyrola is a very small plant, so small that it is apt to be overlooked until its surprising large blossom, somewhat like that of the shinleaf, appears above the ground. Its leaves are thin, deep green, shining, round or nearly so, and its solitary flower, at the end of a boldly curved flower stalk, reminiscent of a miniature shepherd's staff, is cream-colored or ivory white, half an inch or more across, with ten white stamens tipped with yellow anthers and a green pistil. When the flower has begun to fruit, the crooked stalk, that permitted the flower to droop from the end to protect its valuable contents from the rain, straightens out so that the little seed capsule may be held erect. The one-flowered pyrola may be found in pine woods and especially near a brook.

It is always thrilling to come upon a company of Indian pipes rising with ghostly grace from among the brown debris of the woodland floor, for they are such unusual plants, being waxy, cold, and clammy, and colorless in every part, that they almost seem to be visitors from another world. Parasites and saprophytes they are, obtaining their sustenance equally from the juices of living plants as well as from the decaying matter of dead ones. They are weirdly beautiful and decorative among the shadows of the forest.

The stem of the Indian pipe is thick, translucent white, and leafless, the leaves in the dim remote past having given way to scaly bracts. The flower is smooth, waxy, white (rarely pink) oblong, bell-shaped and nodding with eight to ten pale tan stamens. Merely a glance at this odd plant would hardly suggest that it is a relative on the one hand of some of the showiest and loveliest flowers in our flora, the azaleas, laurels, and rhododendrons, and on the other hand, to the modest, but no less, charming pipsissewa and wintergreen. Pick the plant and it quickly turns black as if in protest at such unseemly conduct.

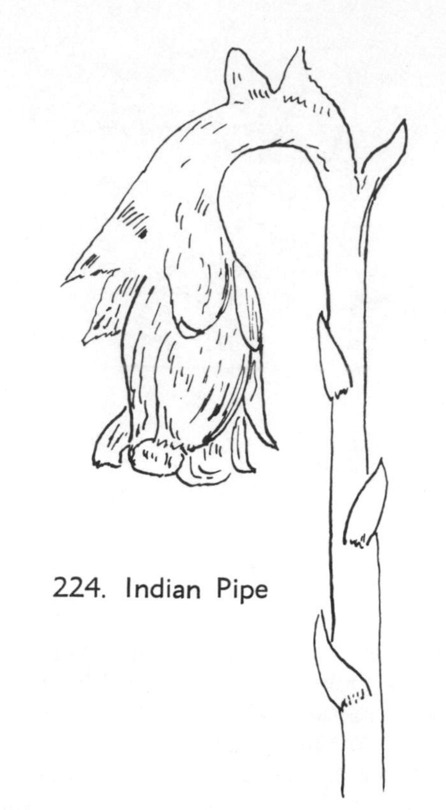

224. Indian Pipe

A somewhat similar parasitic plant that grows most generally over the roots of oaks and pines, is the false beechdrops or pinesap, that from a distance, appears in the cool, deep woods as a bunch of tall fungi. The stems, that grow in clusters, are rather whitish, a pale tan, or reddish and leafless, though with small bracts that are thin, papery, and yellowish red. The flowers are small, bell-shaped, brownish or a bright red, more or less touched with yellow, in a one-sided, terminal, slightly drooping raceme. Both the Indian pipe and the beechdrops would appear to be low in the scale of the plant-kingdom, instead of being members of a highly developed family, were it not for the flowers that they bear.

Of the shrubby members of the heath family, probably the best known are the azaleas and rhododendrons, that are perhaps not technically different from each other. The azaleas, the word azalea is Greek for dry in allusion to the old belief that the plants needed dry sites wherein to grow, have alternate, stalked leaves, more or less irregular flowers usually in terminal, umbellike clusters, and a fruit that is a capsule. The rhododendrons, the word rhododendron is from the Greek for rose and tree and the old Greek name for the oleander, but given by Linnaeus to the plants known as the rhododendrons, also have alternate leaves that are mostly stalked, flowers in terminal umbel-like clusters, and a fruit that is also a capsule. So it is seen that there is little difference be-

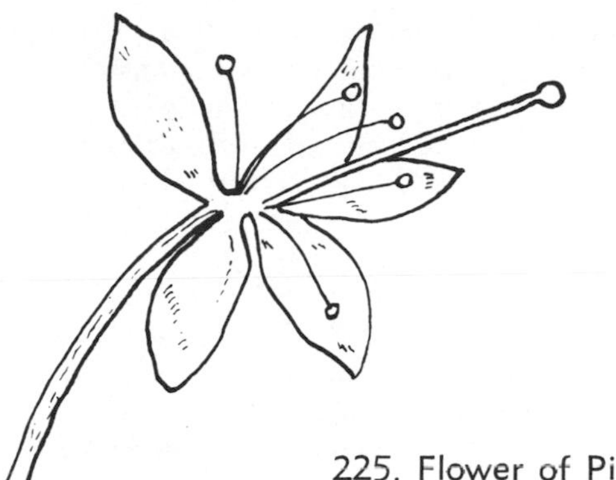

225. Flower of Pinxter Flower

226. Leaf of Flame Azalea

227. Flower of Flame Azalea

tween the azaleas and rhododendrons. Should one want to make a distinction it might be said that the azaleas are essentially deciduous shrubs with a long tubed corolla, while the rhododendrons essentially are evergreen with a short tubed corolla.

Someone once remarked that the pinxter flower is the most beautiful of all our azaleas. That, of course, may be open to dispute, though it cannot be denied that it is a beautiful plant. It is a native shrub of woodlands, thickets, and hillsides, two to five feet high, with golden, yellow green oblong leaves pointed at both ends, and crimson pink, purplish, or rose pink flowers in clusters and appearing before or with the leaves. Sometimes it is found in blossom in April, but it is usually in May that a sunny glen is aglow with its fragrant rosy masses. Merely a glance at its long protruding style tells us that it forms a convenient landing place for a visiting bee.

In books on wildflowers, one will find the pinxter flower, also called pink azalea and wild honeysuckle, though a honeysuckle it is not. The name pinxter is, however, the name whereby the plant is best known, and it was given to it by the early Dutch colonists who called it pingster-bloem because it flowered before or on Whitsunday (pingster).

Also called a honeysuckle, the swamp honeysuckle, though like the pinxter flower not a honeysuckle, or the white swamp azalea, as it should preferably be known, is a deciduous shrub, six to twelve feet high, of bogs and swamps. The entire plant is sticky and clammy, with many spreading branches, grayish bark and hairy twigs, obovate or blunt lance-shaped, yellow green leaves, and pure white or pink-tinged, fragrant flowers, the outside covered with ruddy, sticky hairs, the anthers yellow and the pistil pinkish and longer than the stamens. The flowers appear after the leaves and are visited by bees, butterflies, and moths, but the sticky-haired outer surface of the corolla tube discourages the unwanted, creeping insects.

One of our most magnificent shrubs and known in many horticultural forms, the flame azalea is a southern species commonly found in dry woodlands. It is a deciduous shrub, six to twelve feet high, with broadly elliptical, and generally hairy leaves, and yellow orange, or scarlet flowers that appear with the leaves and that are sticky-haired on the outside. Also a southern species and one of the most beautiful and fragrant of the azaleas, the tree or smooth azalea is a plant of the mountain woods with lustrous, light green leaves and funnel-shaped, pale or rose pink flowers with five prominent, exserted pink red stamens. A mountain form, native to the Catskills and Shawangunk Mountains and southward along the Alleghenies, the mountain azalea is a branching shrub with smooth twigs, oblong or oval leaves, and pink or white, funnel-shaped, nearly odorless flowers, expanding before or with the leaves.

There are several other native azaleas and, of course, many cultivated forms that glorify our gardens in early spring, and that are much the product of the gardener's art.

When the great laurel, or rosebay, covers a mountainside with its blossoms, one cannot help but be impressed with its magnificent beauty, for nowhere else does this gorgeous shrub attain such size or splendor though it will be found in woodlands, on hillsides, and by streamsides. It seems to be at its best on a mountainside where it towers among other trees and spreads its branches that interlock with others to form impenetrable thickets, and no less attractive are its showy blossoms against a background of dark, glossy leaves.

A near relative of the azalea, the great laurel is an

evergreen shrub or small tree, ten to twenty-five feet high, or even more, with lustrous, evergreen leaves, seven to ten inches long, thick and leathery, often rusty beneath, and bell-shaped flowers, that are rose pink, and spotted with yellow or orange, in large terminal clusters. A similar species, but not more than five feet high and with shorter leaves, the mountain rosebay, with lilac purple flowers, and of the high mountains from Virginia to Georgia is the parent of many of the finest horticultural forms of rhododendron. A beautiful little alpine evergreen, the Lapland rosebay is a dwarf species that hugs the rocky slopes of mountains. It is something of a prostrate, branching plant, with small oval or elliptical olive green leaves, rusty and scaly beneath and rose purple, bell-shaped flowers.

Emerson knew the rhodora well, indeed, he became so enamored of this beautiful shrub that the people in Massachusetts at one time called it "Emerson's flower."

> Rhodora! if the sages ask you why
> This charm is wasted on the marsh and sky,
> Dear, tell them, that if eyes were made for seeing,
> Then beauty is its own excuse for being.

The rhodora is a low, thin shrub, one to three feet high, of cool bogs, damp woods, and wet hillsides. It has dark green deciduous leaves, oval or oblong, and slightly hairy, and rose purple flowers in terminal clusters. They open with the leaves or just before them, are very showy, and resemble those of the honeysuckle with the upper lip slightly three-lobed, and the lower in two nearly separate sections. The name rhodora, incidentally, is from the Greek for a rose in allusion to the color of the flowers.

Peter Kalm, the Swedish botanist and pupil of Linnaeus, visited and travelled in North America early in the eighteenth century, and was more impressed by the beauty of the mountain laurel than any other flower. There is no gainsaying that it is a strong contender for being the most beautiful American shrub, for it is an exceptionally handsome plant when in bloom. It is a round-topped shrub, usually growing four to ten feet high, though sometimes becoming a small tree, especially in deep mountain ravines, and in some parts of its range forms impenetrable thickets. The stem and branches are irregular and angular in growth, the leaves are shiny dark green, elliptical, and evergreen, and the beautiful flowers, pink in the bud, waxy white and pinkish tinged when expanded, bowl-shaped with five lobes, are borne in large, dome-shaped clusters, that are very ornamental. Examine a single flower with a magnifying glass and one will find there are tiny pockets in the corolla wherein the anthers are tucked, the filaments being bent like a spring. When they are touched by foraging insects, the anthers are released with a snap, flinging out the pollen all over the body of the visitor. Touching them with a needle will have the same effect.

In his *Anabasis,* Xenophon tells us that his soldiers became sick from eating honey from a plant that was very likely a species of azalea or rhododendron. As that may be, fatal cases occurred in America in 1790 as a result of eating wild honey from the mountain laurel. Cattle, horses, sheep, and goats have been poisoned from eating the leaves, usually in early spring or in winter when the green leaves are the only conspicuous foliage available. Wildlife, however, seem to be immune to the mountain laurel's toxic properties, for ruffed grouse and deer feed extensively on the foliage, buds, and twigs without suffering any ill effects.

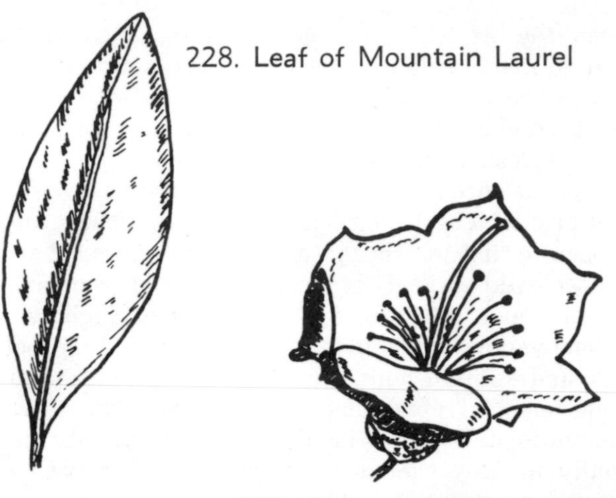

228. Leaf of Mountain Laurel

229. Flower of Mountain Laurel

The mountain laurel belongs to the genus *Kalmia,* named for Peter Kalm, a group of mostly evergreen shrubs with entire leaves that may be alternate, opposite, or whorled, and purple, pink, or white flowers that are, as a rule, showy and borne in terminal or lateral clusters. The flowers are flat or cup-shaped and five-lobed, with ten stamens that are caught in the corolla as in the mountain laurel. The fruit is a round, five-celled capsule.

Two other familiar species of the genus are the sheep laurel or lambkill, and the swamp or pale laurel. In writing of the lambkill, Thoreau says:

> The lamb-kill is out. I remember with what delight I used to discover this flower in dewy mornings. All things in the world must be seen with the morning dew upon them, must be seen with youthful, early, opened, hopeful eyes.

> Writing of the flower at evening he says:

> How beautiful the solid cylinders of the lamb-kill now, just before sunset—small, ten-sided, rosy, crimson basins about two inches above the recurved, drooping, dry capsules of last year.

He said that it is "handsomer than the mountain laurel," but in this few will agree with him.

The sheep laurel is a low shrub, only a foot or so

high, with narrow, evergreen leaves that are either opposite or in circles of three's, and lavender rose flowers in lateral clusters. Like the mountain laurel, it is poisonous to domestic animals, and children have been known to have been poisoned by it mistaking the first little pinkish leaves for those of the wintergreen. It is a food plant for the caterpillars of the brown elfin.

The sheep laurel is a plant of moist fields and swampy ground. The swamp or pale laurel, a similar, but smaller species, is at home in cold peat bogs. It is a straggling, little bush with mostly opposite leaves, long and narrow, with margins turned back, and few terminal, deep rose or lilac-colored flowers.

When most of our flowering shrubs have faded and are beginning to set their fruit, the sweet pepperbush opens its fragrant flowers to perfume the thickets that line slow-moving streams. It is a beautiful shrub, two to eight feet high, multi-branched, with dark brown bark, deep green, ovate or oblong leaves, pointed at the tip, wedge-shaped at the base, and small, white or pale pink flowers with a spicy odor in slender terminal clusters. A similar, but a southern species, that is often a small tree, the mountain sweet pepperbush is a plant of mountain woods. It has oval, pointed, finely toothed leaves and sweet-scented flowers in drooping racemes. An interesting feature of the sweet pepperbush is the character of the hairs found upon

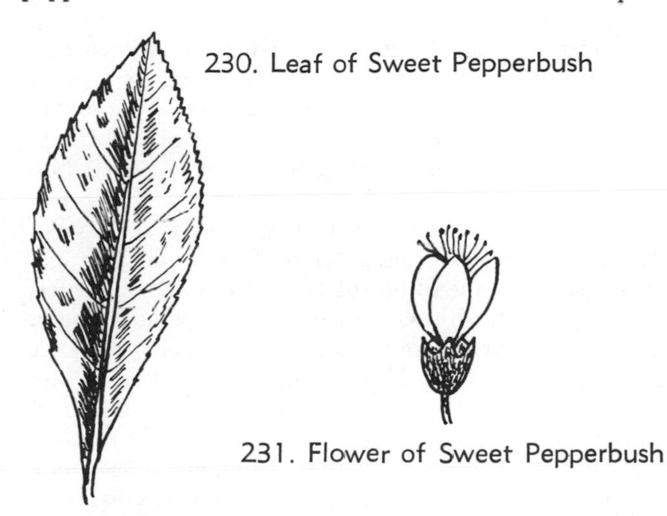

230. Leaf of Sweet Pepperbush

231. Flower of Sweet Pepperbush

the stem and flowers. Viewed with a magnifying glass they are seen to be arranged in very perfect stars. It is doubtful if anyone would claim that the leaves of the Labrador tea would make a good substitute for tea, though it is said that they were used for this purpose during the Revolutionary War, in spite of the fact that they are resinous, astringent, and bitter. Even though it may not serve any useful purpose, however, the Labrador tea is a beautiful evergreen shrub, from one to three feet high, found in swamps, bogs, and damp woods. It has slender, light brown ascending stems, extremely velvety-haired twigs, small, evergreen, leathery, oblong, entire leaves that when crushed, give off a pleasant tealike fragrance.

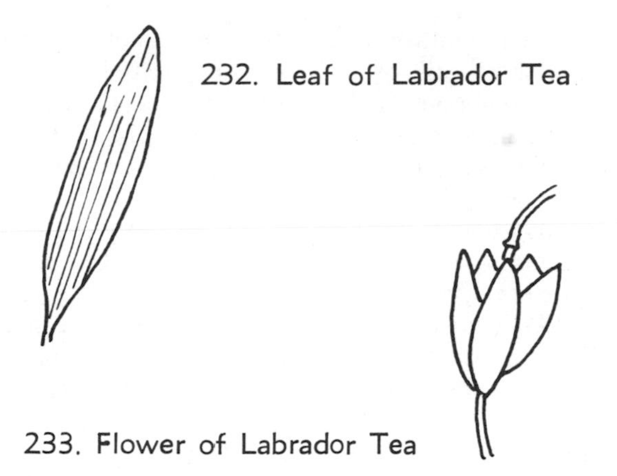

232. Leaf of Labrador Tea

233. Flower of Labrador Tea

The flowers are small, white, in terminal clusters, each with a five-toothed calyx, a corolla of five spreading petals, and five stamens that develop into five-celled capsules.

In his *Tour of Lapland,* Linnaeus writes of the wild rosemary that he placed in a genus he called *Andromeda*:

> This plant is always fixed in some turfy hillock in the midst of swamps, as Andromeda herself was chained to a rock in the sea which bathed her feet as the fresh water does the roots of this plant. . . . As the distressed virgin cast down her blushing face through excessive affliction, so does this rosy-colored flower hang its head, growing paler and paler till it withers away.

The wild rosemary, moorwort, water andromeda, to give it its various names, is a beautiful low shrub, two to twelve inches high of cold bogs, with small linear leaves, olive green above, very pale beneath, and small, dainty crimson pink and white nodding flowers. The fruit is a pear-shaped capsule. The bog rosemary is a similar species, but with larger leaves, and is also found in bogs as well as on lakeshores and riverbanks.

Of a different genus, but also called an andromeda, is the swamp andromeda or privet andromeda, a spreading shrub, seven to ten feet high, with elliptic or oblong leaves, toothless or indistinctly toothed, and small white flowers in terminal clusters on commonly leafless twigs. The latter continue to remain leafless so that when the capsules develop they appear as if on dry branches. A related species is the staggerbush, a beautiful, low-growing shrub, up to four feet high, with small, cylindric, five-parted, white or pink-tinted flowers, clustered at intervals along one side of nearly leafless twigs. When the glossy, oval leaves appear in early summer, calves and sheep seem to find them irresistibly attractive, and after feeding on them, stagger about from their poisonous effect. Another related, but evergreen species, is the fetterbush, a southern species of low wet woods and barrens. It is a plant, two to five feet high, with slender, erect, leafy branches, the leaves broadly elliptic, and white or pinkish flowers in terminal racemes. A similarly named shrub, the mountain fetterbush, is a southern erect shrub, with elliptic or oval leaves, pointed, minutely haired on the margin and white flowers in crowded, slender terminal clusters. It grows on moist, mountain slopes from Virginia to Georgia, and is a charming shrub for the garden.

The early botanists apparently were quite enamored with the Greek figures of mythology since they named many genera after them; Cassiope, Leucothoe, Cassandra, Andromeda, and Pieris, (the latter a name sometimes applied to the Muses from their supposed abode at Pieria, Thessaly) the genus whereof the mountain fetterbush belongs.

There are two native species of the genus *Cassiop* (Cassiope being the mother of Andromeda), the Cassiope or the moss plant, and the four-angled Cassiope. The Cassiope or moss plant is an arctic-alpine, tufted, evergreen shrub, one to four inches high, and moss-like. The leaves, a deep olive green, are linear, needle-shaped, curved and densely crowded and appressed on the branches. The flowers, bell-shaped with five lobes, are white or rose pink and nodding. The Cassiope grows on the summits of the high mountains of

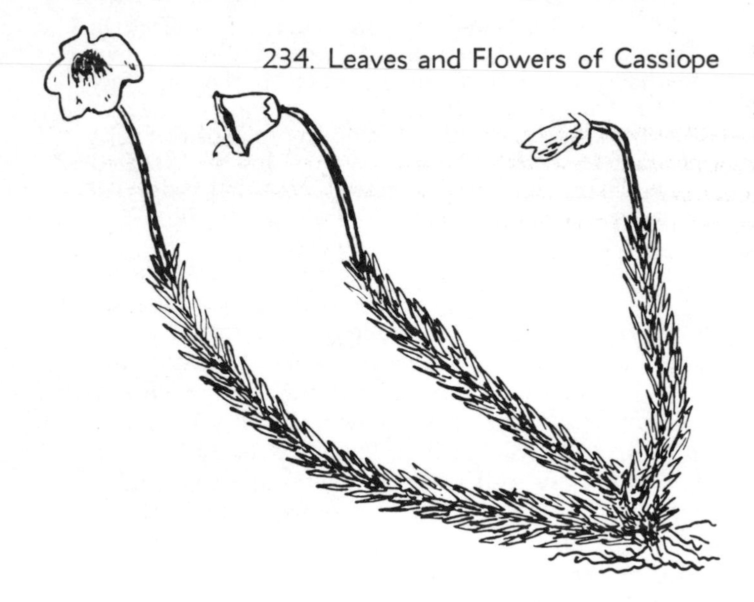
234. Leaves and Flowers of Cassiope

New England and New York, and throughout arctic America, Europe, and Asia. The four-angled Cassiope is a similar species except that the leaves are ovate or ovate-oblong and imbricated in four rows so that the branches appear four-sided or four-angled. It is a plant of arctic Asia and arctic America, touching the states only at Oregon.

Of our native shrubs the handsome leucothoes are much prized as ornamentals, especially the evergreen species. They belong to the genus *Leucothoe,* named for the daughter of Orchamus, mythological king of Babylonia, and have alternate leaves that may either be deciduous or evergreen and white, occasionally pink-tinged, urn-shaped flowers in clusters along, or at the tip of, the branches, the fruit being a dry, round, five-celled capsule.

A familiar species to those who know the outdoors, is the swamp leucothoe, an erect shrub of bushy habit, growing five to twelve feet high, and found in swamps and moist thickets. Its slender branches are covered with dark green, elliptical or broadly lance-shaped deciduous leaves that, late in autumn, assume a brilliant and beautiful scarlet color that lights up the swampy thickets wherein it grows. They are equally brightened in spring by the handsome, waxy, white, cup-shaped flowers that appear in long, erect, or slightly curved, terminal racemes.

Unlike the swamp leucothoe, Catesby's leucothoe is an evergreen shrub, three to six feet high, with slender, arching branches of southern range, but hardy in the North, and thus, suitable for northern gardens. Its leaves are large, broad, lance-shaped, very pointed, and spiny-toothed. Its flowers are white, often pinkish, waxy and bell-shaped, in drooping clusters along the branches. The plant blooms fairly early, the flower buds being developed the preceding autumn in the axils of the leaves. The buds, with close imbricate scales, look like cylindric cones about half an inch long, and as the weather grows colder, the scales turn a deep red, the same color passing into the ends of the zigzag stems. Sometimes the upper leaves acquire as rich a color as the flower buds while the lower leaves retain their lustrous deep green, thus, providing a beautiful contrast between the upper and lower parts of the plant.

A third species of leucothoe, also an evergreen and also of southern range, the downy leucothoe is a plant of low ground and moist woods. It has light brown, ascending branches, finely haired twigs, leathery, ovoid, lance-shaped, deep green leaves, and white flowers in dense spikes.

According to Greek mythology, Cassandra was a Trojan princess, the daughter of Priam and Hecuba. It is also the name of one of our more common shrubs, but why the shrub was named after the princess is not known, certainly there is nothing relative between the two. The name leatherleaf, whereby the plant is also known, is more appropriate, though somewhat more prosaic than Cassandra. It is a low, profusely branching shrub of bogs, two to four feet high, with small, evergreen, leathery, shining, elliptical leaves with scurfy scales on the lower surfaces, and small, bell-shaped, white flowers in a terminal leafy raceme. The Cassandra often blooms before the snow is gone, and as the little bells develop in form and texture, they droop upon the slender stems in a charming wandlike spray.

The Scottish heather has become naturalized on the island of Nantucket, but a similar species, also naturalized, has a larger range, though rather local, and may be found from Newfoundland south to New Jersey. It is a small, straggling shrub, five to fifteen inches high, usually growing in dense masses, with ascending branches and evergreen foliage. The gray green leaves are small, scarcely one-twelfth of an inch long, crowded and overlapping on the branches. The flowers are a delicate magenta pink or sometimes white, and bell-shaped in dense, terminal, one-sided spikes. The heather is widely cultivated for its evergreen foliage and profusion of flowers, with many horticultural varieties.

One of the many genera of the heath family is *Menziesa,* named for Archibald Menzies, an English surgeon and naturalist. It is a small genus of only seven species of small, deciduous shrubs with alternate leaves and bell or urn-shaped flowers in terminal clusters. A southern species, known as the minniebush or Allegheny menziesia, sometimes grows six feet high and occurs in the woodlands of the Allegheny mountains from Pennsylvania, south to Georgia and Alabama. The leaves are elliptical, twice as long as wide, sharp-pointed, with a glandular bristle at the apex, and a few drooping yellowish white or pinkish flowers. A similar, but western species, the smooth menziesia, is also a shrub, two to six feet high, with irregular, straggling branches, gray brown twigs, small, obovate, light green leaves, and tiny, bell-shaped, greenish, magenta-stained flowers in a terminal spreading cluster.

An odd little evergreen shrub that covers itself profusely with corymbs of tiny white flowers at blossoming time is the sand myrtle, a low spreading plant of southern, sandy, pine barrens. It has a scragged stem and branches, sepia brown bark, dark green leaves, and white flowers that are conspicuous because of their purplish anthers.

One can easily identify the sorrel tree or sourwood as a member of the heath family by its tiny, bell-shaped flowers and the dry capsules that follow them. It is a small, slender tree that reaches a maximum height of sixty feet and is essentially a southern species found in rich woods. The leaves are a deep green, alternate, stalked, broadly lance-shaped, and brilliantly scarlet in the fall. The flowers are white in one-sided, drooping clusters and are very handsome in midsummer. Honey made by bees from the nectar of the flowers is clear and delicious. The tree received its name of sourwood because the leaves and twigs are acrid or sour tasting. Somewhat surprisingly the deer eat the leaves.

In some books one will find the huckleberries, blueberries, and cranberries classified as a separate family, but the author has included them in the heath family because they differ only slightly from other members of the Ericaceae, some botanists viewing them only as a subfamily.

The huckleberries belong to the genus *Gaylussacia,* named for Gay-Lussac the French chemist, a group of North American shrubs with alternate, deciduous (except in one species) leaves that are usually without teeth and often resinous. The flowers are in small clusters (racemes) with a bell-shaped or urn-shaped corolla, having five, shallow lobes that are usually bent backwards and with ten stamens. The fruit is a berrylike drupe containing ten one-seeded nutlets, that distinguishes the huckleberries from the blueberries whose fruit is a true, many-seeded berry, not nutlets.

One of the more common species of huckleberry is the dwarf huckleberry, a plant, one to two feet high, mostly of sandy swamps, with variable, lustrous green leaves, resinously dotted and bristle-tipped, with white or magenta pink flowers in long, rather loose clusters, and purple black, berrylike drupes. A low, nearly prostrate, evergreen shrub of limited range, being found in dry woods from Pennsylvania and Delaware to Virginia and Tennessee, the box huckleberry is not over eighteen inches high with creeping stems that turn up at the tips. It has many elliptic leaves, that are a lustrous green above, paler beneath, without resin dots, white or pink flowers in small clusters, and blue berrylike drupes. Called the dangleberry or tangleberry, this species of huckleberry may easily be known by its large pale leaves that are covered with a bloom on the lower surface, and by its loose, drooping racemes of flowers or fruit. When not in flower or fruit, it may be recognized by the yellow reddish wood of the new growth and the peeling ashen gray bark. It is a spreading shrub, three to six feet high, found in moist woods and copses. The berrylike drupe is cadet blue with a grapelike bloom, and is sweet and edible. Also an erect shrub, but not over three feet high, the black huckleberry, also called highbush huckleberry, has many stiff gray brown branches, with the young growth being sticky and resinous. The leaves are somewhat oblong or elliptical, pale green, covered with many resin dots on the lower surface, and turn purplish, crimson, and orange in the fall. The flowers in one-sided racemes are pinkish or pale red, covered with many resin dots. They are followed by black or purplish black berrylike drupes that are sweet and pleasantly flavored and edible. The black huckleberry is often found in company with blueberries in rocky woods and thickets. Various game birds and songbirds feed on huckleberries, but not to the extent that they make use of blueberries.

There are about fifteen to twenty species of blueberries in the United States east of the central plains, the exact species often being difficult to identify. All belong to the genus *Vaccinium,* Latin for the whortleberry or blueberry, a large group of erect or prostrate shrubs, some one hundred and thirty species that range from the Arctic Circle to the summits of tropical mountains. The group includes, not only the blueberries, but also the cranberries and some other species such as the farkleberry and deerberry. The members have alternate, short-stalked leaves that often have fine hairs on the margins, small flowers, urn-shaped in the blueberries, but deeply four-parted with recurved corolla lobes in the cranberries, with eight to ten stamens, and true multi-seeded berries.

The farkleberry, essentially a southern species, is a tall shrub, seven to fifteen feet high, and sometimes a small tree, found on moist bottomlands, on the margins of streams and ponds, and often in the woods.

It has oval or oblong, entire, pointed, glossy leaves, evergreen in the far South, white, bell-shaped, pendent flowers, and a spherical black berry that is dry, sweet, and edible, but rather poor and insipid. The deerberry has a wide range and is found in dry woods and thickets from Maine to Florida and westward. It is a shrub, one to six feet high, with spreading, light brown, slightly finely haired branches, oval or ovate leaves that are pointed at the apex, round or heart-shaped at the base, whitish beneath, and many yellowish green flowers in graceful, leafy bracted clusters. The berry is pear-shaped, dull green or yellow, sour and inedible, though it is said that if the berries are stewed and properly sweetened, they make excellent pies.

The lowest and the earliest of the blueberries to mature its fruit, is the dwarf blueberry or low sweet blueberry of sandy or rocky soil. It can be distinguished from other blueberries by its close bunches of light blue, very sweet berries growing near the ends of the branches and attended by many leaves. It is a low-growing species, six inches to two feet high, climbing to rocky heights or carpeting dry, sandy places, with oval or elliptical small leaves, pointed at both ends, and finely toothed. The small, white or pink, bell-shaped flowers appear a little before the leaves and are followed by the large, pale blue, delicious berries that ripen by the last of June or early July. The berries are among the finest of our wild fruits and were much relished by the Indians. Henry Schoolcraft, the American ethnologist and the chief pioneer in the study of the American Indians, writes that while travelling down the Namokagum River in northwestern Wisconsin he noted that "both banks of the river are literally covered with the ripe whortleberry—it is large and delicious. The Indians feast on it. Thousands on thousands of bushels of this fruit could be gathered with little labor. It is seen in the dried state at every lodge. All the careful Indian housewives dry it. It is used as a seasoning in soups."

There often appears to be much confusion as to the names "whortleberry," "blueberry," and "huckleberry." The name whortleberry is an old name of an English species and once applied to nearly all of our native species. The name blueberry, is now strictly used for the species of the genus *Vaccinium,* the name huckleberry for the species of the genus *Gaylussacia.*

The berries of the dwarf blueberry, and other blueberries as well, are important to American wildlife, both to game birds, such as the grouse, and songbirds, such as the scarlet tanager, bluebird, robin, thrushes, and many others. Mammals, such as the bear and chipmunk, are also fond of them, while the deer feed on the foliage and branches.

In some respects, the low blueberry is similar to the swamp blueberry though it is somewhat larger, growing up to three feet high, with yellowish green warty branches and twigs. It prefers dry, sandy soil. The leaves are oval or broadly elliptical, bright green above with a slight bloom beneath and generally toothless. The flowers are greenish yellow or white tinged with pale magenta, and the berries are a dark blue with a bloom.

Of all the blueberries, the highbush blueberry is probably the finest of them all. It is a tall species,

235. Leaves and Flowers of Highbush Blueberry

236. Leaves and Flowers of Bog Bilberry

three to thirteen feet high, of swamps, wet meadows, and thickets, with spreading greenish brown branches and oval or oblong leaves, pointed at both ends, paler below than above. The flowers are white or suffused with magenta pink, and the berries are bluish black with a bloom. In fall, the highbush blueberry assumes a gorgeous coloring, becoming a most brilliant scarlet, a veritable burning bush of the swamps and meadows. This species is our late-market blueberry, becoming ripe in July and August, though in late seasons, may be gathered in September. The berries were used extensively in colonial times and the Indians often cooked them with meat, sometimes drying them before using them for this purpose.

Several species of the genus *Vaccinium* are known as bilberries. Thus, the bog bilberry is a stocky little shrub, four to twenty inches high with rough and gnarled stems, tiny olive green, obovate leaves, very small white flowers, and blue black berries, found on alpine mountain summits and lakeshores, and the dwarf bilberry is a somewhat similar species, not over eleven inches high, growing in rocky woods or on gravelly shores. The leaves are like those of the bog bilberry, but larger and wedge-shaped at the base and finely toothed, the flowers are also similar, white or pink, and the berries are a light blue with a bloom.

For every one hundred thousand persons who have seen the cranberry in the store and marketplace, probably not one has seen it growing in its native home in the bogs and marshes of northeastern United States. There are essentially two species of wild cranberries, the large and the small. The American, or large cranberry that is the commercial source of the cranberry, is a slender, creeping vine growing from six inches to two feet long with ascending branches and alternate, oval or oblong entire evergreen leaves, that are green above and rather whitish beneath. The flowers that are divided almost to the base into four petallike segments, are pale rose-colored, and the oblong or nearly round berries, at first green, but later red when they ripen in September and October, are from two-fifths to three-fourths of an inch long, but in cultivation, often become much larger. They are not eaten as fresh fruit, but made into a sauce and jelly.

The small cranberry is a smaller species with very slender erect stems, tiny leaves, olive green above, whitish beneath, similar flowers, the corolla lobes a pure red or rose pink, and similar berries, but smaller.

The name cranberry is believed to have been originally craneberry because the curve of the branches appeared like the crooked neck of the crane. It is believed that the Indians showed the Pilgrims how to prepare the berries for the table. Because the berries kept so well without deteriorating, ten barrels of them were sent to King Charles II as a gift, but whether they arrived safely after what was a long voyage in those days of sailing vessels is not known. Cranberries have a limited use to wildlife though the hudsonian godwit is said to be rather partial to the small cranberry.

237. Leaves and Flowers of American Cranberry

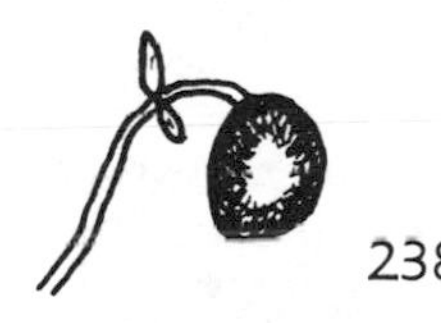

238. Fruit of American Cranberry

Two other plants with the name of cranberry, though they are not exactly cranberries, are the mountain cranberry, also called the cowberry, and the southern mountain cranberry. The former is a beautiful dwarf, evergreen, mat-forming shrub found wild from the mountain summits of New England to Alaska. Its leaves are small, elliptical or obovate, and leathery. The flowers are bell-shaped, white suffused with rose or deep pure pink, and the berries are spherical, crimson, slightly bitter, but very palatable and delicately flavored when cooked.

The southern mountain cranberry is a shrub, one to five feet high, found in the woodlands among the mountains from Virginia and West Virginia to Georgia. and Tennessee. It has slender, slightly angulate branches, lance-shaped, sharp-pointed leaves, bristly, finely toothed, solitary, deep crimson pink flowers, and, acid and insipid, dark, dull, magenta berries.

77. Holly (Aquifoliaceae)

The Romans used the holly in their festival of the Saturnalia. The ancient Teutonic tribes decorated their dwellings with the holly as a refuge for sylvan spirits from inclement weather. The Druids gave the holly a prominent place in their pagan rites. That the holly would eventually become associated with Christmas was a foregone conclusion and in the poetry and stories of England the plant is an inseparable feature of the merrymaking and greetings that so distinguish the Christmas tide:

On Christmas eve the bells were rung;
On Christmas eve the mass was sung;
That only night in all the year,
Saw the stoled priest the chalice rear.
The damsel donned her kirtle sheen;
The hall was dressed with holly green.
Scott in *Marmion*

The custom is ours, too, and a few days before Christmas, the stores are filled with holly for decorating our homes.

The holly of the English was the European holly. Our own native holly is the American holly, a handsome tree of moist woodlands. There are other native hollies to be sure, some fifteen species, all members of the holly family, a group of mostly evergreen trees and shrubs grouped in three genera that comprise some three hundred species, widely distributed throughout the temperate and tropical forests of both hemispheres. They have alternate, often evergreen, simple leaves that are spiny-margined in some species,

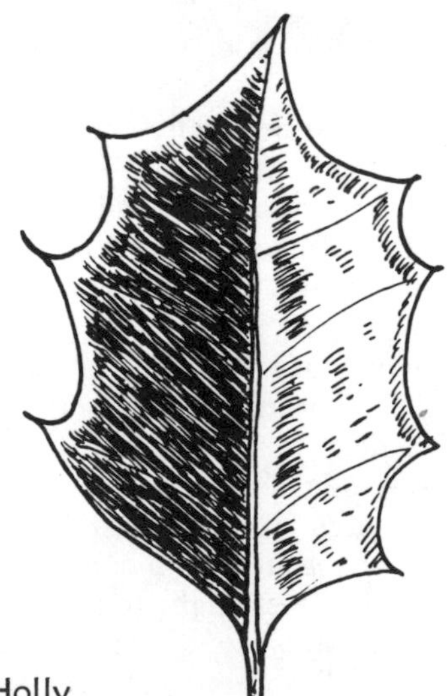

239. Leaf of American Holly

and small, inconspicuous, generally greenish yellow flowers, the individual flowers with four to five free or united sepals, four to five distinct and separate petals, sometimes united only at the base, and a three to many-celled ovary, each cell with one to two ovules. The fruit is a drupe.

The American holly of the genus *Ilex,* the ancient Latin name of the holly oak, rather than of the holly, sometimes also called white holly, is a spreading tree up to forty feet high, with a light gray or brownish gray, generally smooth bark, but becoming rough with age. The leaves are evergreen, leathery, elliptical, spiny, deep olive, or yellow olive green. The flowers are white, dioecious, in loose clusters, with the pistillate flowers developing into a scarlet or bright red fruit that persists on the branches through the winter. The European holly, that is somewhat similar, was introduced into America some years ago, growing best in Washington and Oregon.

From the leaves of the yaupon, an evergreen holly of the southern states and found in swamps and along streams, the Indians obtained their famous black drink, the leaves being dried and often toasted. The drink or tea appears to have been an article of trade among the Indians, being used by the tribes on both sides of the Mississippi River. If made very strong, the tea acts as an emetic. The shrub or small tree of spreading habit with very small oval, evergreen leaves and many red berries is a very close relative of the maté or Paraguay tea, from whose leaves a tea is made, to which the people of South America are as much addicted as those of North America are to tea and coffee. The tea has a remarkable stimulating effect upon the human system, and if indulged in to an excess has much the same effect as alcohol.

A similar shrub or small tree whose leaves are also used for making a tea is the dahoon, a plant of southern swamps with oblong evergreen leaves and red or yellowish "berries." The Creek Indians were said to have been very fond of the tea made from the toasted leaves. As both this and the preceding species are evergreen, the leaves may be gathered at any season.

A shrub similar to the yaupon and dahoon is the swamp holly with lustrous green, deciduous, obovate leaves that are round-toothed and hairy on the midrib beneath. It is also a southern species and grows in similar situations. A handsome shrub with a greater range, throughout the eastern half of the country, the black alder is common in swamps or on low grounds. It is a spreading shrub or a small tree, six to twenty feet high, with smooth, dull, gray ascending stems and branches, with deciduous, dark green, elliptical or obovate leaves, and small, inconspicuous white flowers, and a beautiful fruit, brilliant orange-scarlet in scattered clusters, persisting on the branches long after

240. Leaf of Black Alder

241. Fruit of Black Alder

the leaves have fallen. In Lowell's words: ". . . With coral beads, the prim black alders shine."

This native holly is a shrub that is the equal in brilliancy and beauty of its fruit of any imported plant, and yet, it is not too well known. Its charm of course is in its "berries," that add a cheerful note to the sombre winter scene, and twigs that have decorated many a country household and urban, too, for that matter for they are often sold in stores or hawked on the streets. "I see," writes Thoreau,

> where a mouse, which had a hole under a stump, has eaten out clean the inside of the little seeds of the *Prinos verticillata* berries. What pretty fruit for them, these bright berries! They run up the twigs in the night and gather this shining fruit, take out the small seeds and eat these kernels at the entrance to their burrows. The ground is strewn with them.

Like the black alder, the smooth winterberry should be better known, and though it grows in swamps, it does not disdain a garden home. It is also a shrub and closely resembles the black alder, but its twigs are of a brown color and its scarlet orange "berries" are somewhat larger and ripen a little earlier. Also, whereas the leaves of the black alder turn black in autumn before falling, those of the smooth winterberry turn yellow.

On the low, sandy flats near the coast, the inkberry is somewhat tall and straggling, but under cultivation it becomes more compact in form. It is a rather slender, delicate-looking shrub, two to five feet high, with ashen gray, more or less downy twigs, light gray, spreading branches, and small leathery, evergreen, narrowly elliptical leaves that are pointed at both ends and with only a few rounded teeth near the tip. The flowers are small and white, the staminate in few-flowered cymes, and the pistillate generally solitary. The latter eventually develop into lustrous black drupes. The leaves, as well as those of the black alder, have been used as a substitute for tea.

Of a different genus, and the only species of the genus *Nemopanthus,* Greek for thread and flower in allusion to the slender flower stalks, the mountain holly is a slender, erect, gray-stemmed shrub, one to ten feet high, of swamps and damp, rich woods, with mostly ascending branches, grayish bark, and deciduous, elliptical leaves. The flowers are very small, white, on long slender stems, the staminate solitary or two to four in a cluster, and the pistillate solitary. The fruit is a berrylike drupe, light crimson red, on a long stem. As both its leaves and fruit fall early it is undesirable for decoration.

The fruit of some species of hollies are eaten by various birds and some mammals. The birds do not seem to care for the "berries" of the black alder, for instance, but they will eat those of the American holly, particularly certain songbirds such as the thrushes, mockingbirds, robins, catbirds, bluebirds, and thrashers. The evergreen species, too, are of value because of the protective shelter they afford.

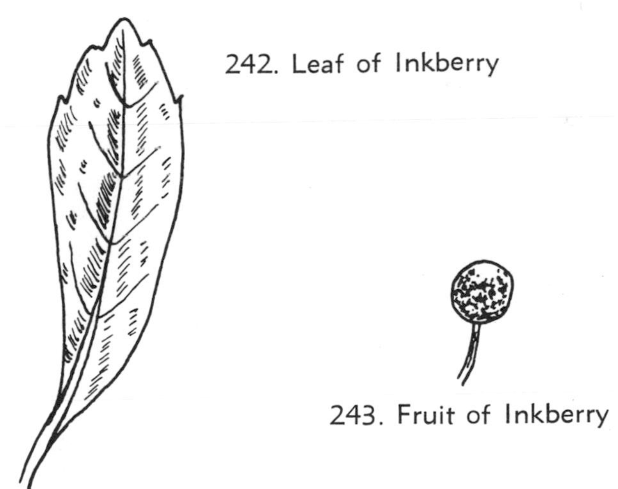

242. Leaf of Inkberry

243. Fruit of Inkberry

78. Honeybush (Melianthaceae)

A strong-scented, handsome, evergreen South African shrub, the honeybush is rather popular in southern California, though it is little grown elsewhere. It grows seven to ten feet high, has alternate, compound leaves that are nearly a foot long, the leaflets arranged feather-fashion, reddish brown flowers in a showy cluster almost a foot long, and a papery capsule as fruit. It is one of some six species of the genus *Melianthus,* Greek for honey and flower, in allusion to the sweet flowers. The genus is one of three genera of the honeybush family that contain some twenty species of South African shrubs. They have alternate, simple or compound leaves, regular or irregular flowers in terminal clusters (racemes), the flowers perfect with four or five sepals, four or five petals, four or ten stamens, and for fruit a capsule, as in the honeybush.

79. Honeysuckle (Caprifoliaceae)

There are one hundred and fifty or more species of honeysuckles found throughout the northern hemi-

244. Flower of Honeysuckle

spheres in both the Old World and the New. They are shrubs or woody climbers and belong to the genus *Lonicera,* named for Adam Lonicer, a sixteenth century German naturalist. Typically, they have opposite leaves that are usually entire and rarely evergreen, flowers that are tubular or bell-shaped, equally five-lobed or more often two-lipped, the upper lip composed of four lobes and the lower of one, and a fruit that is a fleshy berry.

Most of the honeysuckles do well in the garden, being of easy cultivation and thriving in almost any place, the tall forms suitable for shrub borders and general use, the lower ones of more value in the rock garden. The flowers that are often showy, are produced in abundance, and are frequently sweetly scented, and the fruits, white, orange, red, blue, or black, are not only ornamental, but a favorite food for birds, alone being reason enough to plant them.

The most common of the honeysuckles and a favorite, and deservedly so, is the Tartarian honeysuckle introduced into our country from Asia. It is an upright, vigorous shrub of pleasing habit, five to ten feet tall, with white to pink flowers and red fruit. An escape, it is often found on rocky shores and sheltered banks.

The fly honeysuckle or bush honeysuckle of our gardens is also an introduced species and also an escape. It is not unlike the preceding species, but with cream white or cream yellow flowers and scarlet berries.

Not all garden honeysuckles are foreign species, many of them are natives, indeed, there are many native honeysuckles that can be found on rambles throughout the countryside. In swamps and wet woodlands throughout the northern states, the swamp fly honeysuckle is quite common. It is a shrubby species with pale yellow flowers, two-lipped, the upper lip shallowly four-lobed, and the lower a single lobe. Reddish markings within serve as pathfinders for the bumblebee. Sometimes the flowers are in pairs and joined below like Siamese twins when the crimson berries become more or less united.

Generally throughout the same range, the American fly honeysuckle also prefers damp situations such as moist woods. It is a straggling, nearly upright shrub, three to five feet high, with smooth, light-brown branchlets, ovate leaves, bright green on both sides, the flowers greenish yellow, Naples yellow, or honey yellow, and the fruit, deep red berries. Another species also found growing in the same situations is the mountain fly honeysuckle, a small erect shrub, one to three feet high, with light, umber brown, upright stems, with shredded bark. The flowers are pale, honey yellow, in pairs at the base of the leaves. Eventually they become united and form a single cadet blue or gray black berry. It is said to be edible. So, too, are the berries of the twinberry that are eaten by the Indians, and at one time, by hunters and miners who probably still eat them. This honeysuckle is essentially a western species, a shrub of the Sierra Nevada and Cascade range, with pale yellow, sticky-haired, small flowers and dark purple berries.

Of a wider range than the preceding honeysuckles the coral or trumpet honeysuckle occurs from Massachusetts west to Nebraska and south to the Gulf States. It is a climbing vine, evergreen in mild climates, with oval to oblong leaves, shining green above, the upper leaves united around the stem by their bases to form a cup or disk, the flowers in terminal clusters, red or bright orange outside, yellow within, the fruit deep orange red berries. This is a species so often found growing in the garden, that when seen in the wild it is usually thought of as an escape. But it is most at home in thickets and copses where the flowers appear from May to October. As the longest tongued bumblebee is unable to plumb the depths of the slender-tubed trumpet honeysuckle, and as the night-flying moths cannot locate the flowers that have blended with the darkness that has descended when they begin to make their rounds, who then is responsible for most of the berries that follow the charming flowers. It is the ruby-throated hummingbird.

If the trumpet honeysuckle is such a familiar garden plant that when found in the wild it is thought of as an escape, the reverse is true of the sweet wild honeysuckle, or American woodbine as it is sometimes called, for this climbing vine is so common in thickets, waysides, and rocky woodlands that one forgets that it is a native of Europe, an immigrant, and an escape. The early settlers planted this honeysuckle in their gardens where the red berries attracted the birds who dropped the undigested seeds in woodlands. Here the seeds germinated quite as readily as in Europe, and finding their environment as favorable as in the old country, proceeded to grow and spread until everything else was "quite over-canopied with luscious woodbine."

245. Leaf of Bush Honeysuckle

246. Flower of Bush Honeysuckle

247. Fruit of Bush Honeysuckle

Not a true honeysuckle since it does not belong to the genus *Lonicera,* but called a honeysuckle anyway, the bush honeysuckle is an upright shrub, two to four feet high, with slender brown ocher stems and smooth or somewhat scaly bark. The leaves are opposite, oval and taper-pointed, and finely toothed. The flowers are yellow, small, fragrant, funnel-shaped, with five lobes and stamens, and usually three in a cluster. The fruit is a slender, beaked pod, dark brown, and three-quarters of an inch long. It is a common shrub of dry or rocky woodlands, or of hillsides, and is sometimes cultivated and useful under shade or in the wild garden, but it is not as important or as decorative as some related species as, for instance, the weigela.

The weigela, named for C. E. Weigel, is a shrub of distinction. It was discovered in China in 1844 by Robert Fortune who saw it growing in a mandarin's garden on the island of Chusan and described it as a bush covered with rose-colored flowers that hung in clusters from the axils of the leaves and the ends of the branches. "I immediately marked it," he wrote,

> as one of the finest plants in northern China and determined to send plants of it home in every ship until I should hear of its safe arrival.

Today the weigela is one of the most prized ornamental shrubs. There are ten species, all Asiatic in origin, and a number of horticultural forms.

All the honeysuckles and the weigela are members of the honeysuckle family, a fairly large aggregation of twelve genera and some three hundred and fifty species of herbs, shrubs, woody vines or trees, many of them evergreen, with a wide distribution, but mainly in the north Temperate Zone and in the mountains of the tropics. All of its genera are of garden interest.

In most of the genera the leaves are opposite and simple, but in one genus they are compound, with pinnately arranged leaflets. In some genera the flowers are regular and more or less tubular, while in others they are irregular, but always perfect. The calyx is of four to five teeth or lobes, the corolla is four to five-lobed, the stamens number four or five, and the ovary is two to five-celled. The fruit is either a fleshy berry or a dry capsule.

Many of the shrubs in the honeysuckle family are familiar plants of the roadside, thicket, and woodland, that is, familiar to those who are acquainted with plants and especially those growing in the wild, plants like the elder, snowberry, coralberry, and the many viburnums. Walking along almost any roadside in midsummer, one will find the elder "foamed over with blossoms white as spray" and perfuming the air with a honeylike fragrance. Unfortunately, the elder is such a common plant that it is largely ignored, except, perhaps, during the flowering season when it equals, if it does not surpass in beauty and effectiveness, the finest of our garden ornamentals. It is a most attractive shrub, some six to ten feet high, with pithy, dark maroon or green stems. Every whittling schoolboy of yore knew how easy it was to remove the pith to make a hollow reedlike stick for use as a popgun or fife. The leaves are compound, of mostly seven leaflets, oval, pointed, saw-edged, the odd leaflet at the end, and heavy-scented when crushed, though some will call them rank-odored. As in the Umbelliferae and Compositae, the small, cream white flowers are crowded together in flat-topped clusters, doubtlessly to attract numbers of bees, flies, and beetles who are also equally attracted by their honeylike odor. But no nectar awards the visitors, though there is pollen in ample supply, and so butterflies rarely stop.

Once the flowers have been fertilized, they begin to wither and the fruits begin to develop until the elder stands bowed with its burden of purple berries. They are round and very juicy, with three or four rough seeds, but not acid enough to be very palatable. But robins and catbirds, among others, like them and eat them before they are ripe and carry the nutlets to distant places. If the fruit is not quite to our taste when eaten raw, the berries have been used in pies and puddings and for making elderberry wine that is said to have medicinal properties.

I have often found a beautiful dark blue beetle with its wing covers one-third reddish-yellow or yellow on the flowers of the elder; its larvae bore into the pith of the stems.

There is another species of elder, the red-berried elder, found in rocky places and in dry woods. It is similar to the common or American elder, but may be distinguished from it by the shape of its flower cluster that is pyramidal and not broad and flat. Also, by its fruit that is a brilliant scarlet, and, if without flowers or fruit, by the brownish pith of the small twigs, the pith of the common elder being white.

Unlike the elder that is conspicuous both in flower and fruit, the snowberry is more in evidence during fruiting time when its clusters of white berries attract our attention. It is a rather small shrub, one to three feet high, with light brown upright stems, dull blue-green leaves, and small, white, bell-shaped flowers that are suffused with pink. It is found in rocky places and on riverbanks, but most people have seen it in the garden, for it is widely planted as an ornamental shrub.

What has been said of the snowberry can also be said of the coralberry or Indian currant, for this is also a garden ornamental, though a native of rocky slopes. It is an erect shrub, two to five feet high, with madder purple or brown stems, dull gray green, toothless leaves, and very small, white flowers suffused with pink. The smaller bees and honeybees are common visitors. At first, the small berries are a coral red, but finally turn a dull crimson magenta or a reddish purple.

The abundance of fruit is somewhat astonishing. The berries are in dense clusters that surround the stems and make a most effective display very attractive to the eye. Moreover, the berries have great staying power and persist into the winter without darkening or shrivelling. For some reason the birds do not like them and leave them alone.

Of a much stockier habit than the coralberry, the wolfberry is a western shrub, growing generally on rocky woodlands. The flowers are in small, nodding, dense, terminal clusters, the corolla is bell-shaped, white suffused with pink, and the berries are white, spherical, and in small clusters.

Included in the honeysuckle family is a group of shrubs and small trees known as the viburnums. Of the one hundred and twenty species, many are ornamental shrubs, evergreen and deciduous, grown in gardens and shrubberies for their showy flowers and decorative fruits as well as their handsome foliage that often takes on brilliant colors in the fall. Their distinguishing characters are their flat clusters of small white, rarely pink, flowers, and their showy clusters of one-seeded drupes that are often handsomely colored, persistent, and a favorite food of the birds. Sometimes the viburnums in bloom may be mistaken for the dogwoods, but the flowers of the viburnums are five-petaled, those of the dogwoods four-petaled.

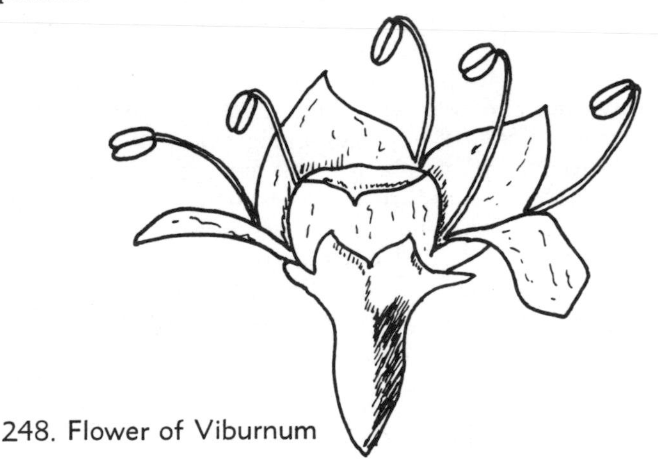

248. Flower of Viburnum

One of the best, if not the best, garden plants among the viburnums is the species known as the witherod or Appalachian tea. An inhabitant of swamps or moist thickets, it is a loose, straggling shrub, eight to twelve feet high, with brown gray stems and branches, in its wild home, but under cultivation, it is a compact, symmetrical plant with all the graces of civilization. The deep green leaves are rather leathery and thick. The flowers are in dome-shaped clusters (cymes) four or five inches across, are white, and are succeeded by many drupes that are at first a pale green, then turn into a bright rose, and finally darken into a blue black. The fruit of the three colors often appears at the same time. The fruit is sweet and edible and is eaten by grouse, pheasant, European partridge, and several species of songbirds. Deer and rabbits browse the twigs, and the dried leaves have been used for tea. The naked witherod is a similar shrub, but with glossy leaves and dark cadet blue fruits. It is also more southern in range.

Let the walker in the woods beware of the hobblebush, lest he suddenly find himself sprawling headlong on the ground. For the shrub has the habit of sending its straggling branches downward to the ground where, taking root, they form loops to trap the unwary. The hobblebush is a spreading shrub of the woodlands, commonly five, rarely nine feet high, with a dull madder purple or brown, smooth bark, its branches spreading horizontally, the young twigs scurfy-haired. The leaves are large, heart-shaped, finely toothed, light green, but turning a maroon in the fall. The flowers, in large flat clusters and white, are of two kinds and sizes, one large and showy and

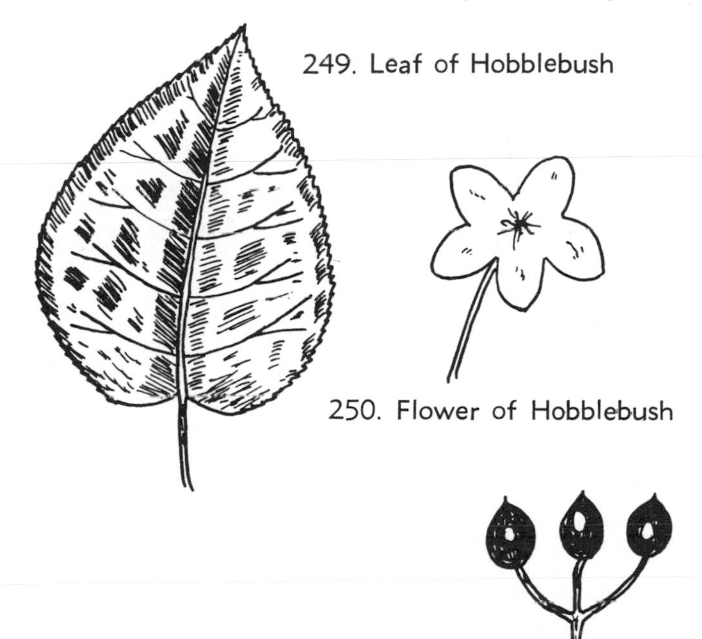

249. Leaf of Hobblebush

250. Flower of Hobblebush

251. Fruit of Hobblebush

sterile, lacking both stamens and pistil, the other within the larger inconspicuous, but performing the function of a flower. For what reason the sham? Doubtlessly to attract the insects that are, of course, responsible for the clusters of the bright scarlet or coral red fruits that later turn a dull dark purple. They are sweet and edible and the birds like them, often eating them before they have fully ripened.

Call it nannyberry, sheepberry, nannybush, sweet viburnum or wild raisin tree, and call it a shrub or small tree for it can be either, this viburnum of many names is greatly admired for "its compact habit, its lustrous foliage, its beautiful flowers, and its handsome fruit" reason enough for it to decorate parks and gardens. From Georgia northward, it grows in the rich, moist, soil of woodlands and the banks of streams, sometimes attaining the height of a tree, but more frequently merely a good-sized shrub. The large, showy, white flower clusters, appearing in May and seated among the deep green terminal leaves, indicate to migrating birds where a feast awaits them on their return flights, when the bluish black bloom-covered, sweet, edible "berries" ripen in October. It is said that the Indians liked the "berries" too. The prominent, wavy-winged margins of the leaf stems is a distinctive character that serves to distinguish this viburnum from the others.

Also at times an erect bushy shrub, at other times a small tree, the black haw has a short, crooked trunk and a gray-brown bark that was used at one time as a tonic and for other medicinal properties that it was said to have. Its leaves suggest those of a plum tree, its flowers are similar to those of the nannyberry, and its fruits are dark, cadet blue, on red stems. It has been said that the "berries" become edible after being touched by the frost, but this is not so, at least in the southern part of its range where they are good to eat long before the first frost appears. Foxes like them, so do the bobwhites and various songbirds. A similar species is the southern black haw that, as its name indicates, is more southern in range. The lower surface of the leaves are densely red-haired. Both species grow in much the same situations, as on hillsides, in open thickets, along fencerows and roadsides.

To know any one of the viburnums by their flowers is to know them all, for they all spread more or less flattened compound cymes of white flowers. It is their leaves that differ and serve to distinguish the species. Thus, for instance, the maplelike leaves of the mapleleaved arrowwood are unmistakable, indeed, this shrub, some three to five feet high that can be found growing on sandy or rocky hillsides and along the margins of woods is, when not in flower or fruit, so like a maple sapling that it might easily be mistaken for one. In a similar manner downy arrowwood may be recognized by the densely, velvety down on the lower surface of its leaves. It is a low, straggling shrub growing in rocky, wooded places, often on some high bank above a stream. Beetles and bees visit the flat-topped flower clusters in May, the short-tongued visitors licking up the abundant nectar and cross-fertilizing the small flowers as they make their way about the cyme. In autumn, the leaves of both species

252. Leaf of Arrowwood

253. Fruit of Arrowwood

take on deep, rich, tints of red and purple and are most attractive to the eye.

The species known simply as arrowwood is so-called because its wood was once used by the Indians for their arrows, but why its wood was better for such a purpose than that of the other arrowwoods is not known. It is a familiar shrub of low, moist ground with blackish sepia brown or ashen brown bark, the new twigs or shoots often straight and arrowlike, deep green, broadly ovate leaves, and flower clusters like the other viburnums, that later turn into shining blue "berries" that are eaten by birds, though it is difficult to understand why as they are dry, tasteless, and seedy.

In his book, *The Orchard and Fruit Garden,* E. P. Powell writes of the cranberry tree or highbush cranberry that "the fruit is very seldom used by human beings; but it makes excellent jelly, and a sauce fully as good as that from the real cranberry."

The cranberry tree or highbush cranberry, also known as the pimbina, is an exceedingly handsome shrub, four to ten feet high, with smooth, gray brown or buff branches, and grows in damp woods, along streams, and on the borders of swamps. The leaves are maplelike, commonly three-lobed, but sometimes five-lobed, light green, the flowers are white in broad, flat, dome clusters, and the fruit is a beautiful, bright, translucent, scarlet red drupe. Like the flowers of the hobblebush, those of the cranberry tree are also of two kinds: an outer circle of showy neutral blossoms to advertize for bees and flies, and small, perfect ones to reproduce the species.

The cranberry tree or highbush cranberry is somewhat northern in its range, and is also native to northern Europe and Asia. As a matter of fact, some botanists consider the American shrub only a variety of the European form. As that may be, it is a beautiful shrub and widely cultivated for ornament on the lawn and in the park. Its contribution to the beauty of the garden is, however, in its fruit that at first is a greenish yellow, touched with red, then later a bright red or scarlet. The "berries" are nearly round or slightly oval, about one-third of an inch in diameter, very sour, and persist into the winter. The grouse, pheasant, and some songbirds seem to like them despite their rather disagreeable taste. A variety of the cranberry tree is the common snowball of our gardens.

A much smaller, but very similar shrub and with a more northern range, the squashberry also produces sour red "berries" that, too, are used as a substitute for the cranberry.

An introduced species, the wayfaring tree is a native shrub of Europe and Asia and has long been in cultivation in our country. It is not unlike our native viburnums, with dark brown stems, dark green leaves, dome-shaped clusters of white flowers, and fruit that is at first red, but later turning black.

The American botanist, Asa Gray, tells us that the twinflower was a special favorite with "the immortal Linnaeus," and that "there is extant at least one contemporary portrait of Linnaeus on which he wears the tiny flowers in his buttonhole." Doubtless, the twinflower of the great taxonomist was the twinflower of Europe, but there is our own native species, a delicate and beautiful trailing vine common in the northern woodlands with a terra-cotta-colored, rather roughly wooded stem, and round, scalloped-toothed, evergreen leaves. The fragrant little bell-shaped, crimson pink flowers hang in pairs, each from its own tiny stalk, the two united below into a slender peduncle that grows upright from the trailing and creeping stems. It is a dainty floral beauty somehow seeming out of place in the cool woods.

The twinflower is essentially a herbaceous plant,

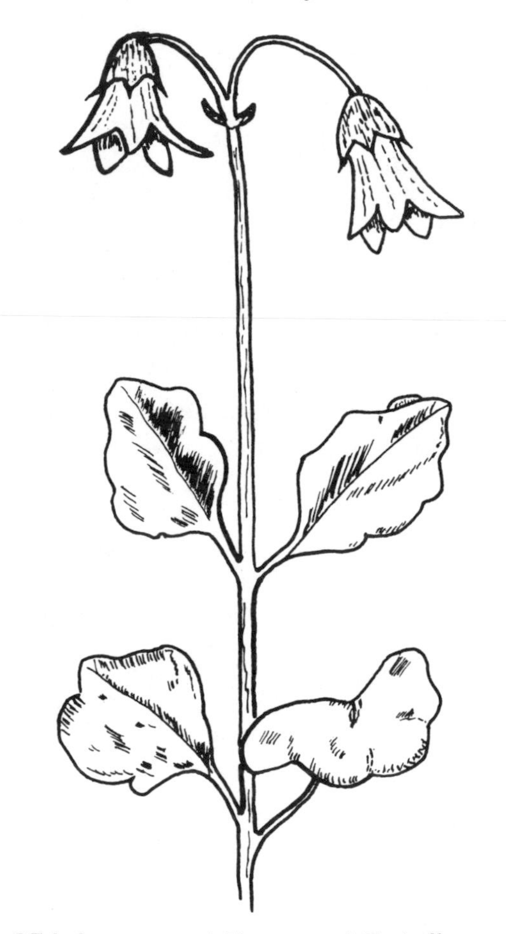

254. Leaves and Flowers of Twinflower

one of the few belonging to the honeysuckle family, and so too, is the horse gentian also called the feverwort, wild coffee, and tinker's weed, the last name in reference to its use by various medical practitioners who seem to have held it in high esteem for its medicinal qualities.

It is a coarse perennial fairly common in rich woodlands with a stout stem that is covered with fine hairs and that is also rather sticky and with opposite growing, light green, oval leaves. The flowers are inconspicuous purplish brown or madder brown and grow at the junction of the leaves with the plant stem. The fruit is a drupe, about half an inch long, orange scarlet, densely, finely haired, and contains three bony nutlets. The name horse gentian was undoubtedly given to the plant because the names of our larger domestic animals were often used as adjectives to indicate unusual size or coarseness in plants, and gentian because of its resemblance to the true gentians.

80. Hornwort (Ceratophyllaceae)

There are only one or two species of hornworts and they belong to a single genus *Ceratophyllum,* Greek for horn and leaf, in allusion to the hornlike leaves. They are submerged aquatic herbs, growing in ponds and slow streams, usually well out from the shore, with slender, widely branching stems a foot or two long, and narrow and whorled thrice-forked leaves divided into narrow rigid segments. The flowers are minute and monoecious, and are sessile in the axils. They lack sepals and petals, the staminate flowers with many stamens, the pistillate with a one-celled ovary containing one ovule. The fruit is an achene.

The flowers never rise above the water, but the stamens break off and float to the surface where the pollen is shed, and that "rains" down upon the pistillate flowers. In late summer the branches break, float for a while, and then sink to the bottom, partly because of their own weight and partly because of the weight of mollusks, insects and other animals that cling to them. The branches winter over on the bottom mud and in spring float up again. The hornworts form thick beds, so thick in fact that they leave no place for other plants wherein to grow. These hornwort forests are populated by a great assemblage of snails, worms, and other animals that feed on one another, or on the algae that grow on the hornworts.

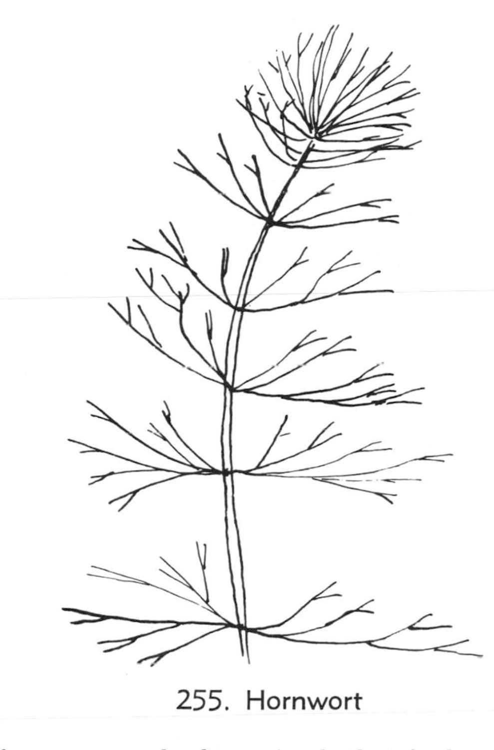

255. Hornwort

81. Horseradish Tree (Moringaceae)

The horseradish tree is so-called because of its pungent, edible root. It is a small, soft-wooded tree, not over twenty-five feet high, with a corky bark and thrice-compound, very feathery leaves. The whole leaf may be as long as two feet, but the leaflets are small and many. The flowers are fragrant and white, nearly an inch wide, in loose clusters (panicles) in the leaf axils, and the fruit is a nine-ribbed, cylindrical pod, sometimes fifteen inches long, with three-angled, winged seeds. The tree is a member of the genus *Moringa,* the Latinized version of the Malay name for the tree, the only genus of the horseradish family, a group of East Indian and African trees with pinnate leaves and irregular flowers with ten stamens, succeeded by a three-valved capsule.

The horseradish tree is cultivated in the South and has become established as an escape in southern Florida. Its seeds yield ben oil used for lubricating watches, the seeds being known in commerce as ben nuts.

82. Illicium (Illiciaceae)

The illicum family contains only a single genus of American and Asiatic evergreen shrubs or moderate-sized trees, with simple, entire, alternate leaves, perfect flowers with sepals and petals usually in several series, commonly of three, many stamens, and few carpels. The fruit is a follicle.

The family is represented in the United States by two species, the star anise and the purple anise, or stinkbush, that are found from Georgia to Florida and Louisiana. The star anise has small yellow flowers, the purple anise dark red that unfortunately smell like decaying fish. Both occur in low woods and swamps.

83. Indian Plum (Flacourtiaceae)

The Indian plum family is a group of about seventy genera and some five hundred species of tropical shrubs and trees having alternate leaves, regular, mostly perfect flowers with four or more sepals, usually no petals, many stamens, and a one-celled ovary. The fruit is a berry.

The Indian plum, wherefrom the family gets its name, is a native of Madagascar and southern Asia, and is grown in Florida and California where it is known as the ramontchi, governor's plum, or Batoko plum. It is not over twenty-five feet high and often shrubby, with alternate, oval or elliptic, toothed leaves, small, yellow flowers without petals, mostly in small clusters, and a pulpy, nearly globe-shaped, dark red fruit that somewhat resembles a plum. It belongs to the genus *Flacourtia,* named for Etienne de Flacourt, once a governor of Madagascar, a group of some fifteen species of Old World tropical, often spiny, shrubs or trees.

Two other members of the family, but of a different genus, *Dovyalis,* are also grown in Florida and California for their fruit. One of them is known as the kei apple or umkokolo, the other kitambilla or Ceylon gooseberry. Both have alternate, short-stalked leaves and inconspicuous small flowers, the staminate and pistillate on separate plants. The kei apple is a thorny shrub or small tree, ten to twenty feet high, the thorns long and stiff, with the leaves often clustered at the base of the thorn. The flowers are greenish and the fruit, being a berry, is nearly round, about an inch in diameter, yellow, and cranberry-flavored, but good only when cooked. It is a native of South Africa. The kitambilla, a native of India and Ceylon, is somewhat similar to the kei apple, but more branched and with a velvety, maroon purple berry that is less acid than that of the kei apple. Several Chilean species of the family are grown for ornament in California.

84. Iris (Iridaceae)

Look in the dictionary for the meaning of the word *iris,* and it will be evident why the ancients, always appreciative of the aesthetic, gave the name to the group of plants that is known so well, for *iris* means rainbow, and the irises are flowers of bright and varied colors.

There are over one hundred and fifty species of iris, all perennial herbs with long, narrow, generally sword-shaped leaves, and rhizomes or bulblike rootstocks. The flowers are in six segments and arise from spathelike bracts, the three outer segments being reflexed, the three inner ones usually smaller and erect. The fruit is a capsule.

The structure of an iris flower is something of a puzzle to an amateur botanist since the parts are so grown together. The perianth, ovary, and style all unite to transform the lower part of the blossom into a sort of stem, the upper part of the stem sometimes becoming tubular before it divides, at other times the division occurs directly. The perianth is of six segments, the outer ones of great beauty in both form and color, are bearded, many-veined, broad and reflexed. They are called the falls. The inner ones, called the standards, rise each on a claw, broaden, and then overarch.

If one looks directly into the flower, one will not at first see either the pistil or the stamens. What looks like three additional petals in the heart of the blossom is the style, its tip three-cleft, the three sections separating and curving outward. The stigmatic surface is the thin edge beneath the divided crest. It can be detected by its slightly shining surface, or by the pollen that might be adhering to it. The stamens, each with a short filament and large anther that is usually the color of the blossom, arise from the base of the outer segments of the perianth.

The floral structure of the iris is of more than passing interest because of the ingeniously contrived way it manages to secure cross-pollination by insects—specifically by bees, that seem especially attracted to it. Each of the three drooping sepals forms the floor of an arched passageway that leads to the nectar, while over the entrance to the passageway is the movable and outward-pointing stigma. With a reading glass one can watch a bee entering the passageway and how it brushes against the stigma that scrapes from its back the pollen previously collected from another flower. Guided by the dark veining and golden lines, the bee makes its way to the nectar. As it moves along the passageway its hairy back rubs against the pollen-laden anther above and becomes covered with a new supply, that it carries to another flower after it has sipped its fill of nectar and gone its way.

In addition to the many cultivated species and varieties of the iris there are a number that grow wild: the larger blue flag, a handsome and decorative plant found growing on the wet margins of ponds and in swamps; the slender blue flag, a slender-stemmed species with very narrow, grasslike leaves, occurring mainly near the coast in brackish swamps; the dwarf iris, a usually one-flowered small, slender-stemmed plant with grasslike leaves sometimes with white flowers growing on wooded hillsides; and the crested dwarf iris, with a lance-shaped leaf tapering at both ends that distinguishes it from all the others and a flower that is exceedingly delicate in color and dainty in form. It is common on hillsides and along streams.

The fleur-de-lis of France, the flower that Louis VII adopted for the emblem of his house and known at first as fleur-de-Louis, but later corrupted into its present form, is found all over Europe where it grows in marshy spots. It has splendid, clear yellow flowers on stems two to three feet high and a large, stout rhizome that is tough and fibrous and pinkish inside.

The irises, or "flags" as they are often called in America, are included in the iris family, sometimes also called the crocus or gladiolus family. It contains

256. Leaves and Flower of Iris

over sixty genera and a thousand or more species of perennial herbs, usually of brilliant coloring, that are scattered over most of the world. Over twenty genera are found in the gardens and greenhouses of our country.

The leaves are mostly basal, linear or sword-shaped, equitant—that is their bases overlap and bestride the leaves within or above them—and arise from bulbs, corms, or rootstocks. The flowers are usually showy, perfect, regular or irregular, emerging from a spathe of two or more bracts, with three sepals that are often petallike and not easily distinguished from the three petals, the six segments usually inserted at the top of a tubular receptacle. They have three stamens and a style with three stigmas that are sometimes expanded and colored to resemble the petals. The fruit is a three-angled capsule with many seeds.

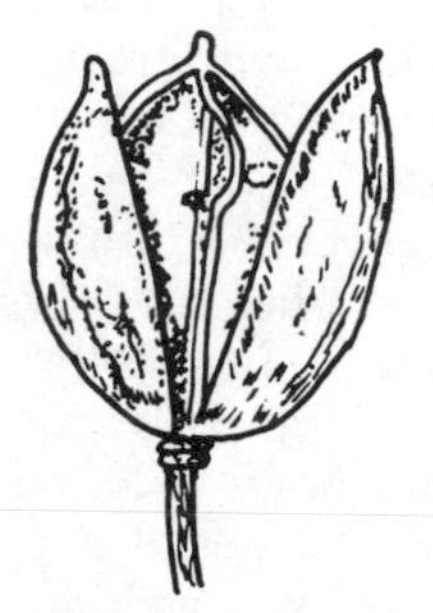

257. Fruit (Capsule) of Iris

The iris family includes many of the most popular and cultivated ornamental plants known: the iris, the crocus, and the gladiolus, as well as the freesias and many other horticultural species of wide cultivation. One species of iris produces the orrisroot of commerce, while a species of crocus produces the saffron, used to color and flavor food.

The crocus that gives us the saffron, known as the saffron crocus, is a fall-blooming species of southern Europe with white or lilac flowers, and it is said that more than four thousand flowers are required to produce one ounce of the coloring matter. Incidentally, the name crocus is from the Latin *saffron.*

North America does not have a native crocus, all the garden crocuses came originally from southern Europe and Asia. They are seemingly stemless plants arising from a corm (the crocus "bulb" of the stores), with grasslike leaves appearing before, or with, or after the flowers that bloom very early in the spring. Indeed to those who live in the North, the appearance of the first crocus in the garden is a sure sign that spring is on its way. The crocus, however, is not the only flower that foretells spring, for the blossoming hepatica among the decaying leaves of the woodland floor, the flowering spring beauty in the moist thicket, and the blooming shadbush by the roadside, also serve notice that winter is quickly disappearing.

There are a number of species of crocus grown in the garden: the Dutch crocus, with bright yellow flowers; the Scotch crocus, but a native of southeastern Europe and Asia Minor, with purple-tinged flowers, the outer segments purple-striped, and the throat yellow; the cloth of gold crocus, with white or lilac flowers; the common crocus, with flowers in a variety of colors; and the saffron crocus, with flowers white or lilac.

The gladiolus is a very popular, summer-blooming plant in home gardens. The name *gladiolus* is Latin for dagger, and the plant is so-called because of the shape of its leaves. Its ancestors are the native gladioli, brought into Europe from southern Africa about the middle of the eighteenth century. Two hundred or more species are known and several thousand varieties are today the result of the breeder's art.

The stem of the gladiolus is usually erect, unbranched and arises from a corm. The leaves are commonly sword-shaped, long, narrow, and handsome. The flowers are showy in a long, terminal spikelike cluster composed of leafy bracts. Between every two of these bracts there is a single, stalkless flower. The flowers are more or less funnel-shaped, but with a dilated tube, and usually curved upward. Of many colors, they bloom from the bottom upward.

The freesias have been mentioned as belonging to the iris family, and they are very fragrant and beautiful South African bulbous herbs. They are of greenhouse culture and extremely popular florists' flowers. Typically white or yellow, though some hybrid forms are pink, they have the structural characteristics of the family. Of more general interest are the blackberry lily and the blue-eyed grass, both misnomers, for the first is not a lily and the second is not a grass.

258. Leaves and Flower of Blue-eyed Grass

The blackberry lily will be found by the roadside. It is a Chinese plant grown in the garden for ornament, but has long been an escape. Its leaves are like those of the iris, and its flowers, borne in a loose terminal corymb, are a golden orange spotted with crimson and purple. The fruit is a fig-shaped capsule, about one inch long, that splits open in the fall, and exposes fleshy-coated, black seeds. They resemble the blackberry, hence, the name of the plant.

"Only for a day, and that must be a bright one, will this 'little sister of the stately blue flag' open its eyes, to close them in indignation on being picked." So writes Neltje Blanchan in her *Nature's Garden,* of the blue-eyed grass, a grasslike little plant, with stiff, pale blue green, linear leaves, and dainty violet blue flowers with a yellow center. As the perianth segments are regularly divided and spreading, the plant has also been called the blue star. It is common in fields and moist meadows, and is cross-fertilized mostly by bees and flowerflies. The fruit is a capsule, dull brown or purple-tinged, and sphere-shaped. There are a number of species that are all somewhat alike, but differing in minor details.

85. Jewelweed (Balsaminaceae)

With its orange yellow flowers spotted with reddish brown and hanging like jewels from a lady's ear, the jewelweed is a familiar plant of wet places, growing beside streams, ponds, and in moist dells. Its flowers

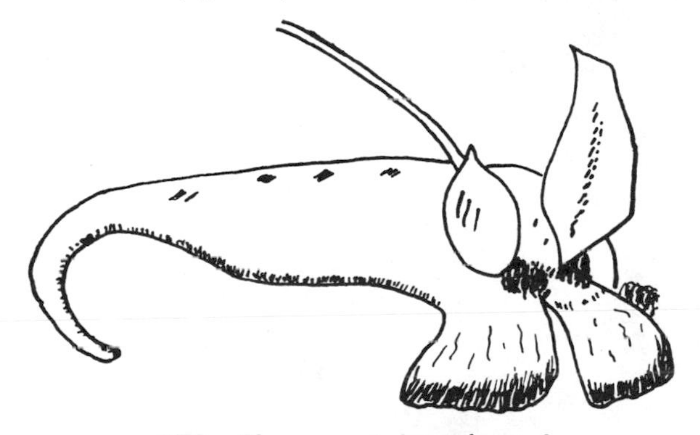

259. Flower of Jewelweed

are irregular and perfect, with three sepals and petals, but these are not easily distinguishable. One of the sepals is large, and sac-shaped and, contracted into a slender incurved spur, is admirably adapted to the long bill of the hummingbird. Of course, insects visit the jewelweed too, particularly the long-tongued bumblebees.

On a dewy morning or after a shower, the jewelweed can be found along a woodland brook or in the moist ground by the edge of a pond, with the notched edges of the drooping leaves hanging with dewdrops that sparkle like jewels in the sunshine. Could another name for the plant be more fitting? Still, the jewelweed does have another name—the touch-me-not. Touching the fruiting capsules when ripe, opens them with startling suddenness and they expel their tiny seeds to a distance of nearly four feet. One may well be frightened at the unexpected volley from the miniature machine gun.

The jewelweed of the preceding paragraphs is the spotted jewelweed with orange yellow flowers spotted with reddish brown. Another species, the pale jewelweed has paler yellow flowers, only sparingly dotted, and has its broader, sac-shaped sepal abruptly contracted into a short, notched, but not incurved spur. It is also a larger, stouter species, and occurs in the same situations.

The jewelweed family, also known as the balsam family, contains only two genera. One is *Impatiens* that comprises five hundred species including the common jewelweeds and a few garden balsams. They are largely succulent herbs of wide distribution, mostly found in tropical Asia and Africa, with stalked, simple, opposite or alternate leaves. The flowers are perfect, irregular, brightly colored, solitary, or in somewhat umbellike clusters, usually with three sepals, the two lateral ones small and greenish, the posterior one prolonged backward into a nectariferous tubular spur. They also have five petals, or three by the union of two pairs, five stamens, their filaments short and broad, and a five-celled ovary, with many ovules, the style short and produced into five stigmas. The fruit is a succulent capsule or berry, with several seeds, and in *Impatiens,* elastically explosive that discharges its seeds if touched when ripe.

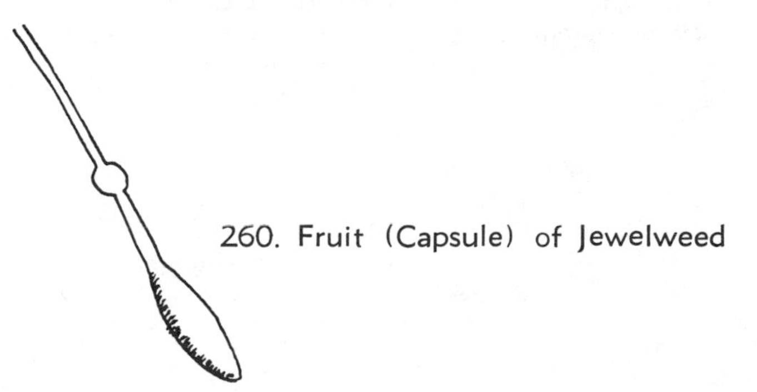

260. Fruit (Capsule) of Jewelweed

Aside from the two species of jewelweeds, the only other members of the family of interest are the species of garden balsams, such as the common balsam and the kind called sultana. The common balsam, a native of India, is a tender annual, tall, stiff, erect with a brittle stem, succulent, with smooth, lance-shaped leaves, and flowers clustered close to the stem. The flowers come in brilliant colors, some very double or camellia-flowered, in old rose, salmon pink, scarlet, yellow, purple, and white. The sultana, a native of eastern Africa and also a tender annual, is brittle and succulent, with smooth, alternate leaves, and solitary flowers on a short, slender stalk. In the original form the flowers are bright scarlet, but in the hybrids, they come in shades of pink, salmon, purple, and white.

86. Jipijapa (Cyclanthaceae)

The jipijapa family is a small group of tropical plants of no particular interest except for the jipijapa, the Panama hat plant, wherefrom Panama hats are made. It is a native of Ecuador, not a native of Panama, and is essentially a stemless, palmlike plant, with the leafstalks channeled, arising from the ground, and four to six feet long. The leaves are fanlike, the blade split into four main divisions, each being split again, and the segments drooping at the tip.

The Panama hat plant belongs to the genus *Carludovica,* named for Carlos IV and Ludovia, king and queen of Spain, and is a small group of perhaps thirty-five species of palmlike, but apparently stemless, tropical American, erect or climbing shrubs. They have palmlike leaves, that are long-stalked and fanlike, and flowers with floral structures in sets of four. The fruit is a syncarp.

The genus *Carludovica* is one of six genera that make up the jipijapa family, a small group of tropical erect palmlike plants with subterranean rhizomes, or epiphytic climbing lianes, and an inflorescence that is borne in a spadix.

87. Joint-Fir (Gnetaceae)

The joint-fir family is a group of trees and shrubs with mostly opposite leaves and small dioecious flowers, and consists of three genera, *Welwitschia, Gnetum,* and *Ephedra* that are, by some authors, considered to be separate families, namely the welwitschia family, the gnetum family, and the ephedra family. The genus *Welwitschia,* formerly *Tumboa,* contains only one species, the tumboa, a peculiar and rather weird plant of the South African desert, with a trunk less than a foot high, but often six feet in circumference, bearing two persistent leaves that grow at the base and die at the apex, and with conelike flower clusters.

The genus *Gnetum* contains a number of tropical trees and shrubs having climbing jointed stems, terminal spikes of flowers, and usually a drupaceous fruit. The name of the genus is said to be from *gnemon,* the native name of a species in the island of Ternate. There are thirty known species, commonly called joint-fir, in the genus *Ephedra,* an old Greek name probably for horsetail. They are chiefly low, or climbing, essentially leafless, desert shrubs with green twigs resembling the horsetail. The flowers are minute and very primitive, the staminate and pistillate on separate plants and consisting of only two to four very small, scarcely petallike organs. The staminate have two to eight stamens, while the pistillate have a single, upright ovule. The fruit is red and berrylike. Many species of the group have been cultivated for thousands of years in China where one of them yields mahuang, a source of ephedrine. One species is cultivated in the South, another is suited only to the deserts of the Southwest. This latter species is known as the Mexican or Mormon tea, and may be found growing with greasewood and cacti, forming dense, green bushes four or five feet high.

88. Katsura Tree (Trochodendraceae)

The katsura tree of the genus *Cercidiphyllum,* from *Kerkis,* Greek for leaf, is a Japanese tree, thirty to fifty feet high, frequently cultivated for ornament. Quite often it is divided into several trunks or stems. It has opposite, nearly round leaves, heart-shaped at the base, and mostly borne on short spurs. The staminate and pistillate flowers are on different plants. The flowers are small and without petals, and are borne on the spurs as the leaves unfold. The fruit is a splitting pod with many seeds. The katsura tree is a handsome foliage tree, the leaves purplish when they unfold, changing in the autumn from green to yellow to scarlet.

The genus *Cercidiphyllum* is one of three genera* of the katsura tree family, a small group of Japanese trees or shrubs with alternate or opposite leaves, perfect or dioecious, small flowers appearing before the leaves, without sepals and petals, the ovary being of separate carpels, and the fruit a pod. It is a group of unimportant trees or shrubs though several species are grown for ornament.

* Some books have made separate families of the three genera: The Cercidiphyllum Family (Cercidiphyllaceae; The Trochodendron Family (Trochodendraceae) and The Euptelea Family (Eupteleaceae)

89. Laurel (Lauraceae)

The sassafras is a most unusual and interesting tree. First there are the leaves. They may be ovate or obovate and entire, or they may be two-lobed, with one lobe at the side, making what are called "mittens," or

261. Leaf of Sassafras

they may be three-lobed. All three forms may occur on the same branch, a distinctive feature shared only by the mulberry. Then there are the dainty green buds of winter that one might nibble at, and of course, the aromatic bark of the roots that is distilled in large quantities for the oil, used in flavoring medicines, for scenting perfumery, for making candy, and for other purposes. It also has a medicinal value, and is often sold in drugstores. It also makes an excellent tea when served with sugar and cream. During the Civil War it was extensively used for this purpose, especially in the South, when tea from the Orient was unavailable. In the South the leaves and twigs, that yield a mucilaginous substance, are also used for thickening and flavoring soups, especially in creole cooking.

It is believed that the sassafras was first discovered by Bartholomew Gosnold, the English explorer, in 1602 on Cuttyhunk Island off the coast of Massachusetts, who sent several trees back to England where they were sold for three shillings a pound. Thus, in all likelihood, the sassafras was the first plant product to be exported from New England. During the colonial days, the sassafras was in great demand in Europe, where it was used for tea and for flavoring, and thus, became an important commercial article.

The sassafras of the genus *Sassafras,* from the Spanish *salsafras* in allusion to the medicinal value of its root bark, is usually a small, slender tree, fifteen to fifty feet high, with a flat or round, loose, open, irregular top, and common in rich woodlands. Both the yellow green twigs and leaves are finely haired when young, and also aromatic and mucilaginous. On young trees the bark is cracked into short blocks, while on older trees it is extremely rough, dark brown, reddish, scaly, and broken by shallow fissures into broad, flat ridges. The leaves, are alternate, without teeth, entire or lobed as has already been mentioned, and the insignificant flowers are greenish yellow and are borne in small clusters, (racemes.) The staminate and pistillate are on separate trees and appear with the developing leaves. The individual flowers have a six-lobed calyx, but no petals. The staminate have nine stamens, the pistillate have a one-celled ovary that develops into a deep, slaty blue drupe on a fleshy bright red stalk. The fruit is eaten by various songbirds and in this connection, it is interesting to note that the kingbird, crested flycatcher, and phoebe, all members of the flycatcher family and subsisting primarily on insects, are among those that eat it. In the fall the leaves turn to shades of yellow, tinged with red and

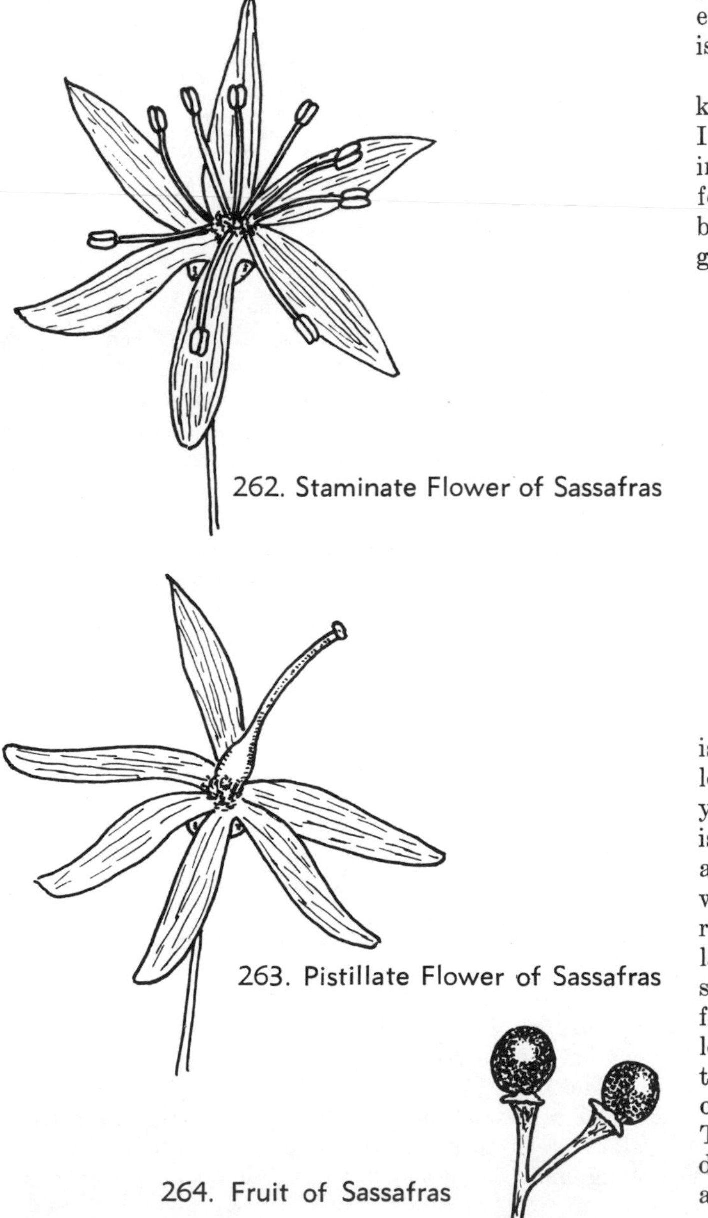

262. Staminate Flower of Sassafras

263. Pistillate Flower of Sassafras

264. Fruit of Sassafras

purple. In the summer they are eaten by the caterpillars of the spicebush swallowtail. The caterpillar, smooth, green with two pairs of orange, black-pupiled eyespots, will be found living in a one leaf nest, made by folding the edges of the leaf over it.

There are but three species of sassafras, one found in central China, the second in Formosa, and the third our own native sassafras of the eastern states. All are members of the laurel family, a large group of forty-five genera and one thousand species of mostly trees and shrubs, but also a few parasitic herbaceous vines of the tropics and warm regions, with a handful extending into the Temperate Zone. They generally have alternate, simple, leathery, and evergreen leaves in the tropical species, thin and deciduous in the temperate climate forms, and all very aromatic. The flowers are inconspicuous, yellow or green, perfect and imperfect, regular, without petals, but with a six-parted calyx, stamens in three or four whorls of three each, and a one-celled, one-ovuled ovary. The fruit is a berry or a drupe.

A familiar member of the family, one perhaps better known by its fruit, is the avocado or alligator pear. It is a tree native of tropical America, but cultivated in Florida and California. It is usually forty to sixty feet high, but less in the cultivated varieties, multi-branched, with oval or elliptic leaves and very small greenish flowers in dense, terminal clusters. The fruit

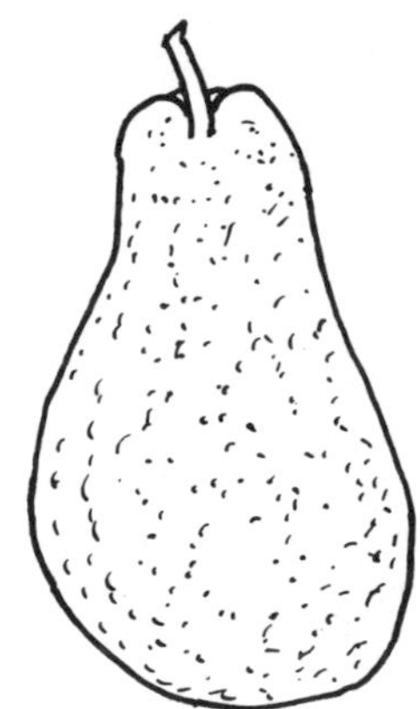

265. Fruit of Alligator Pear

is pear-shaped, or nearly round, three to six inches long, at first green, but later yellow when ripe, the yellow oily flesh being of delicious flavor. The avocado is one of fifty known species of the genus *Persea,* an ancient name for some Persian or Egyptian tree, whereof there are two native American species, the red bay and the swamp bay, both confined to the Atlantic and Gulf Coastal Plains. The red bay is a handsome, shapely, narrow-headed tree, fifty to seventy feet high, with a thick, red, furrowed bark, evergreen leaves, small, pale yellow or cream white flowers in terminal spikes, and a lustrous dark violet black ovoid berry. It is a tree of stream and swamp borders. The swamp bay is a similar, but smaller tree, with a dull brown bark and velvety-haired branchlets, characteristics that distinguish it from the preceding species, as do also, the long stalks whereupon the flowers and berries are borne.

A visitor to California cannot help but be impressed with the beauty and stateliness of the California laurel, also called spice tree, bay tree, and balm-of-heaven. It is a lover of wet soils and grows in rich bottomlands where it may reach a height of one hundred to one hundred and seventy-five feet, but elsewhere it is only a medium-sized tree rarely over twenty-five feet tall. A tree with a narrow crown and slim ascending branches, it is a handsome tree in a land of handsome trees. The willowlike, evergreen leaves are lustrous and rich in an aromatic oil, that is of a strong camphoric odor, pungent enough to produce sneezing when crushed, and the flowers, though small, are fragrant and bloom in December and January. The fruit resembles an olive, and is a fleshy, egg-shaped, yellowish green drupe. Burls are often formed on the trunks of older trees, and if devoid of defects, command a high price. The small ones are turned into many kinds of woodenware, the large ones are cut into thin sheets of veneer having considerable value and beauty. Incidentally, the California laurel is the only western species of the family.

Challenging the pussy willow, the alder and the shadbush, as the first to flower in the returning spring, the spicebush gilds the swampy March or April woods with knots of golden yellow flowers. If one wants to know why it is called the spicebush one need only break a twig and a spicy fragrance will be detected.

By looking carefully at several bushes one will find that some bushes are slightly more yellow than others. Ordinarily, staminate and pistillate flowers are found on separate plants. When both are examined with a magnifying glass or hand lens, the staminate flowers will be found to have, not only a yellow calyx, but yellow anthers as well. This makes them a brighter yellow than the pistillate flowers, since these have only the yellow sepals. The six-parted calyx takes the place of the corolla that is absent.

The spicebush, being aromatic in bark, leaves, and fruit, is a shrub seven to fifteen feet high, with smooth, dark brown bark, slender, brittle twigs, oval or oblong dark green leaves, and bright scarlet, oval drupes, that are relished by the thrushes, particularly by the wood thrush and veery, but are eaten sparingly by other birds. The leaves have been used in making a tea, as have also the twigs and bark. Michaux, the French botanist, while traveling through the wilderness wrote in his journal under the date of 9 February, 1796, upon stopping at a settler's cabin,

> I had supped the previous evening on tea made from the shrub called Spicewood. A handful of young twigs or branches is set to boil and after it has boiled at least a quarter of an hour, sugar is added and it is drunk like tea. I was told that milk makes it much more agreeable to the taste. This beverage restores strength, and it had that effect, for I was very tired when I arrived.

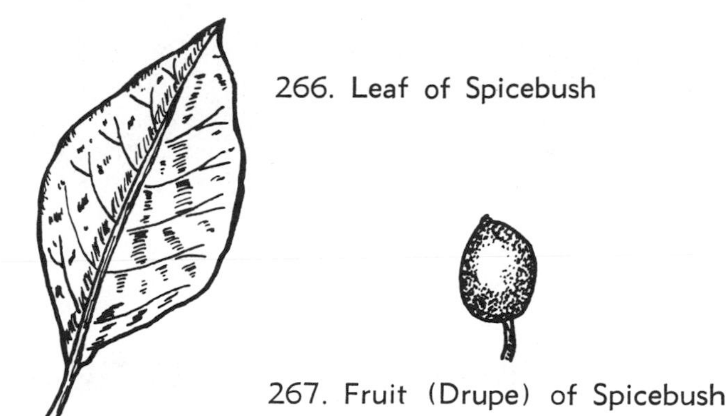

266. Leaf of Spicebush

267. Fruit (Drupe) of Spicebush

It is said that during the American Revolution the "berries" were dried, powdered, and used as a substitute for allspice. The shrub is used as a food plant for the caterpillars of the spicebush swallowtail who, not only eat the leaves, but make a nest of them as they do with the leaves of the sassafras.

Somewhat similar to the spicebush is the pond spice, a spreading shrub of southern swamps, six to nine feet high, with widely divergent branches and smooth, slender, zigzagged branchlets, willowlike leaves, pale golden yellow flowers, and spherical, crimson red, drupes. Both the crushed leaves and twigs are aromatic.

An important genus of the laurel family is *Cinnamomum,* the Old Greek name for cinnamon, for it contains some commercially valuable, aromatic evergreen trees, two of them widely grown as the source of camphor and cinnamon. The camphor tree, a native of China and Japan, grows forty to fifty feet high, has alternate, elliptic leaves, and yellow flowers. The fruit is a berry. The bruised foliage is camphor-scented and the commercial camphor is obtained by processing the bark. It is occasionally planted as a street tree in southern California and in Florida.

The cassia bark tree, a native of China, grows thirty to forty-five feet high, has opposite, oblong leaves, and

268. Leaves and Fruit of Camphor Tree

yellowish white flowers. It is largely grown in the Far East as a fraudulent adulterant of true cinnamon, and is occasionally planted in Florida and California. The true cinnamon, a native of India and Ceylon, is a small tree with stiff glossy leaves and small yellow flowers in silky clusters. Cinnamon is obtained from the bark of the younger branches. The cultivation of the true cinnamon tree is an important industry in Ceylon, but the tree is occasionally planted in Florida.

90. Leadwort (Plumbaginaceae)

Though the blossoms of the sea lavender may be tiny in themselves, when massed together and seen from a distance they appear as a mist blown in over the marshes from the sea. In some respects the sea lavender or marsh rosemary, as it is also known, is a curious plant with a slender multi-branched stem arising from a thick, woody root, and the branches somewhat erect. The leaves, narrowly oblong, tapering into long petioles and tipped with a bristly point, also come from the root. The very small flowers are light purple or lavender in color, and grow on one side of the naked branches. They have a curious tooth between each of the five tiny lobes of the corolla. When dried, the sprays make desirable winter bouquets.

There are about one hundred and eighty species of sea lavenders, though there are only about three native species. They are mostly annual or perennial herbs with generally basal, often tufted, leaves, and numerous small flowers in open loose panicles or in branching spikes. They are prevailingly lavender, rose pink, or bluish, but sometimes yellow or white with a calyx that is often membranous, colored and tubular, and a corolla of five nearly separate and often clawed petals. The fruit is a utricle. The sea lavenders are mostly salt marsh plants in the wild, but many are widely cultivated and are easily grown in the flower garden, preferably in a somewhat sandy soil.

The sea lavenders belong to the genus *Limonium,* Greek for meadow in allusion to the salt marsh habitat of many of the species. The genus is one of ten genera of the leadwort family, also known as the plumbago family, a group of three hundred widely distributed species of herbs, shrubs, and vines generally found growing on maritime shores and other saline situations, especially in the Mediterranean basin. They have alternate leaves or leaves grouped in a basal rosette, and regular, perfect flowers in dry bracted spikes, heads, or panicles. The flowers have a tubular or funnel-formed, calyx that is five-lobed or five-toothed, a five-lobed corolla, five stamens, and a one-celled ovary with one ovule, and five styles separated or united. The fruit is a utricle and is usually enclosed by the persistent calyx. Many species have long been favorites with gardeners, and are widely grown in gardens and greenhouses for ornamental purposes.

In addition to the genus of sea lavenders there are also the following genera: *Plumbago, Armeria,* and *Acantholimon.* The genus *Plumbago,* Latin for lead though the allusion in this instance seems obscure, consists of twelve species of subshrubs or herbs, generally known as the leadworts, sometimes climbing or trailing, mostly perennial, and natives of southern Europe, Africa, Asia and tropical America. A few, however, have been established in southern Florida. They have slender stems, alternate leaves, broadly lance-shaped, and blue, red, or white flowers in terminal spikes or clusters. They are grown mostly in a cool greenhouse, but may be placed in the garden during the summer months. They are highly desirable as pot plants. A common species of the greenhouse, the shrubby plumbago, a tender shrub of spreading habit is often turned out into the garden in late spring to bloom there until frost comes. The flowers are an azure blue.

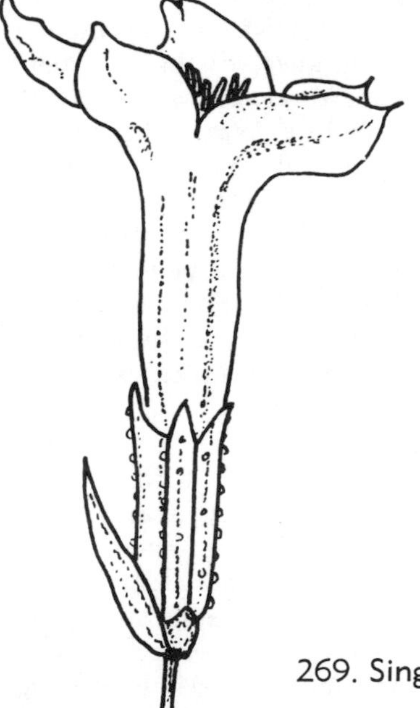

269. Single Flower of Leadwort

270. Leaves and Flowers of Sea Pink

There are only a few species in the genus *Armeria,* an old Latin name of the thrift, a name whereby the species are generally known, although they are also called sea pinks. They are low, perennial herbs, native of the sea cliffs and mountains of Europe, although there is one North American species found along the sea coast and on mountains from Labrador to Alaska, and south on the Pacific Coast to California. They have small evergreen leaves in basal rosettes and a dense globe-shaped head of chaffy flowers on a rigid, stiff flowering stalk. A familiar species for the general border and the rock garden is the common thrift with pink, purple, or white flowers in the horticultural varieties.

Some eighty species of low, evergreen, perennial herbs belong to the genus *Acantholimon,* Greek for spine, and limon, the sea lavender. They are all from the eastern Mediterranean region and only a few are in cultivation. They are generally known as the prickly thrift, as they have prickly or sharp-pointed, densely crowded, stiff leaves. The flowers are small, pink or purple, and are crowded in one-sided racemes or in a spike, and are usually hidden by the numerous bracts. They are chiefly rock garden plants, and require open sun and light, sandy soil.

91. Lily (Liliaceae)

Everyone knows the lily, the tulip, the daffodil, all three being members of the lily family, a huge aggregation, so huge that many botanists have divided it into several families or subfamilies. It consists of twenty-five hundred species of herbs, shrubs, and a few trees and climbers of worldwide distribution, most found in temperate and subtropical regions. It is a family extremely rich in plants of value, and of many uses as ornamentals and as food. The vast majority are perennial herbs, only a few become woody and treelike.

The leaves are various, basal, in rosettes or alternate on a well-developed stem, that in nearly all the herbs, arises from a bulb, usually simple, fleshy or dry and leathery, sometimes evergreen. The flowers are often showy, or small and inconspicuous, single or in spikes, racemes or panicles, typically of six segments, indistinguishable as to calyx or corolla, occasionally tubular, stamens usually six, ovary with three carpels and many ovules, remarkable for their simplicity and beauty. The fruit is a capsule or berry, usually with many seeds.

Most people when they think of the lily invariably have in mind the Easter lily but there are others,

271. Leaves and Flower of Easter Lily

over a hundred different species of the northern temperate zone, belonging the genus *Lilium,* an ancient Latin name of unknown meaning. The Easter lily is a native of China and Japan. It came to us by way of Bermuda and is often called the Bermuda lily. Actually it is a variety of a lily known as the white trumpet lily, the original type, and is also a species from the Far East. Until the early eighties, the Easter lily of America was the Madonna lily, a white flowered species from southern Europe and Asia, and believed to have been the lily so often referred to in the Bible. It did not take too well, however, to forcing, and florists were uncertain when it would bloom. To have the lilies ready for market a week or two after Easter meant ruin. So when our present Easter lily was introduced to the trade, it was hailed with enthusiasm, for not only did it take kindly to forcing, but it also took less time to bring it into bloom. Moreover, it had larger flowers, and more of them, than did the Madonna lily.

Our Easter lily, with its waxy white flowers, is such a familiar plant that it hardly needs description. All the lilies are erect perennial leafy-stemmed herbs with deep, scaly bulbs and of considerable garden importance. The leaves are scattered or in whorls, without marginal teeth. The flowers are extremely showy, erect, horizontal, or nodding, either solitary or in profuse clusters. The petals and sepals, often colored alike so that they are indistinguishable as such, are sometimes separate, or sometimes united when the flower is funnel-shaped. There are six stamens. The fruit is a multi-seeded capsule.

If the Madonna lily did not take well to forcing, it did take kindly to the garden, thriving in any good soil and in an abundance of light and air. Without a doubt, this is probably the oldest lily in cultivation, and for more than a thousand years, its exquisite form, snowy blossom and delicate fragrance have made it a symbol of beauty, purity, and love.

A very common lily in gardens is the upright lily. As a rule it is not over two feet high, with erect, orange red, somewhat dark-spotted flowers. It is a Japanese lily and much cultivated under a variety of forms, some with larger, apricot-colored flowers, some with salmon and unspotted flowers, and some in other color patterns. Also old garden favorites are the Japanese lily, one of the most beautiful lilies of Japan, with drooping fragrant white or bluish, spotted with rose red, flowers, now in many varieties; the golden-banded lily, also a native of Japan, with fragrant, drooping flowers, white, crimson-spotted and with a central, yellow band; and the tiger lily, probably the first lily brought to America, today in many varieties, with orange red or salmon red flowers that are spotted with black. Throughout the years the tiger lily has escaped from gardens and commonly may be found growing beside old farmhouses.

If one should drive anywhere throughout the eastern half of our country and as far south as Georgia, one will find the meadows of June alive with the nodding flowers of the meadow lily or Canada lily. The stem rises three feet high, has several whorls of bright green, lance-shaped leaves, and at the summit, divides into several flower stems, each with a pendulous blossom of buff yellow on the outside, orange or pale yellow within, and freckled with purple brown spots. The flowers are visited by the honeybee and the leaf-cutter bee who seek not only the nectar, but the brown pollen as well.

Though the meadow lily does not have the beauty of color found in the wood lily, or the subtle delicacy possessed by the Turk's Cap, it is unsurpassed in the graceful curves of its blossoms, and for this matchless quality alone, it is to be greatly admired. Undoubtedly, it is our most popular native lily.

On any midsummer's day, the wood lily stands like a torch of orange and scarlet among the grass and other herbs of the upland meadow, or gleams like a lantern in the shadows of a roadside thicket or woodland border. Without a doubt, this lily is the most beautifully colored of all our wild lilies, its flower cups opening upward, as if to catch the sunbeams that speed earthward from the distant sun, above the bright green leafy stems. It is our one lily whose petals do not recurve.

So unlike the wood lily in the manner of its petals, the Turk's cap lily, a beautiful and picturesque species, is remarkable for its completely recurved petals (or actually sepals), that completely expose the stamens with their brown anthers. Less common than our other native lilies, the Turk's cap grows in low meadows and marshes. It rises from five to nine feet high, its stem erect with light green leaves that alternate on the upper part, but are more or less in circles on the lower part. The flowers are nodding, commonly three to seven in a terminal group, and are a bright red orange, spotted with dark purple inside. They are largely fertilized by bees, but the monarch butterfly is a frequent visitor.

Aside from the lily probably no other member of the lily family holds such an exalted place in the garden as the tulip, a word derived from the Persian *toliban* for turban, that the inverted flower resembles. It is not known where the cultivated tulip originated. Its origin is lost in the mists of antiquity, but the Turk was responsible for its introduction to Europe. The Austrian ambassador brought the first seeds from the Sultan of Turkey to Vienna in 1554.

From Vienna the tulip generally made its way throughout Europe and appeared in England as early as 1599 where it became the flower of fashion. Parkinson, in his *Paradisus Terrestris,* about 1660, wrote that "tulipase do carry so stately and delightful a form, and do abide so long in their bravery, that there is no Lady or Gentleman of any worth that is not caught with this delight."

The tulips, genus *Tulipa,* a Latinized version of an Arabic word for turban, in allusion to the shape of the

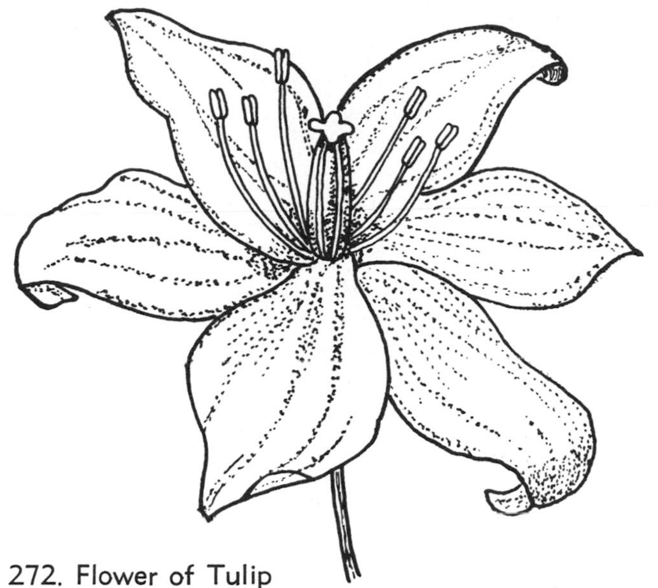

272. Flower of Tulip

flower, are bulbous herbs, with the bulb generally pointed. The stem is single. The leaves are mostly basal, though a few on the stem in some species, rather thick, bluish green with an entire margin. The flowers, usually solitary and erect, are bell-shaped or saucer-shaped, and are of many different colors. The fruit is a multi-seeded capsule. There are over sixty species of tulips and several thousand horticultural forms. The masses of dazzling color that the tulip can give to our gardens, lawns, and parks when other plants are as a whole still in the bud is perhaps its most endearing quality.

> And the Hyacinth purple and white and blue,
> Which flung from its bells a sweet peal anew
> Of music so delicate, soft and intense,
> It was felt like an odor within the sense.

So wrote Shelley of this perennial garden favorite, cultivated since ancient times and a native of the Mediterranean region. There are some thirty species of hyacinths, genus *Hyacinthus,* from the mythological character, but the chief species is the common garden hyacinth. It is a bulbous herb with a deep, large bulb, narrow, basal sometimes almost grasslike leaves without teeth, and fragrant flowers in a showy, terminal cluster (raceme) of many colors. The corolla is more or less bell-shaped, with its six lobes or segments spreading or turned backwards.

"Like a bunch of grapes" is the flower cluster of the grape hyacinth, an early flowering bulbous perennial that belongs in every garden as a welcome note to spring. There are forty-five species native to central Europe, with leaves four to six inches long, narrow and green, and flowers on a leafless stalk, in a terminal raceme, blue, blue and white, or pink, the individual flower urn-shaped and drooping, a great garden favorite being the species with blue flowers. The grape hyacinths are also known as muscari, from the Latin *muscus* for musk, in allusion to the musky odor of some of them.

Also an early-flowering bulbous plant and also a garden favorite, is the scilla or squill. As with the grape hyacinth, there are many species, genus *Scilla,* an old Latin name for the plants eighty to ninety known species in fact, with grasslike leaves and small blue, white, or purple bell-shaped flowers, in a terminal cluster at the end of a naked stalk. One of the

273. Flower of Squill

most planted is the Siberian squill whose blue flowers come out in March, but there are others also widely cultivated. They include the star hyacinth, the Spanish bluebell and the sea onion all with blue flowers, and all scillas, of course, despite their names.

Old-fashioned garden flowers that persist for years and that are of simple culture, are the bulbous herbs known as the Old World fritillaries, a name derived from the Latin *fritillus* for dicebox, in reference to the checkered markings of the petals. They are mostly unbranched plants with alternate or whorled leaves, sometimes in a terminal cluster above the flowers that are lilylike, pendent and generally early-blooming. Probably the best known is the crown imperial, a strong-smelling herb with a purple-spotted stem, and flowers nearly two inches long, and colored purplish, yellow red, or terra-cotta. Again, to quote Parkinson who writes

> the crown imperial, for its stately beautifulness deserveth the first place in this our garden of delight, to be here entreated of before all other lilies. The whole plant and every part thereof doe smell somewhat strong, as it were the savour of a foxe, yet is not unwholesome.

Popular in his day perhaps, but not quite so popular today, largely because the flowers have a smell that reminds us of the skunk cabbage.

Discovered by the Swiss botanist, Boissier, upon the heights of the Taurus range in Asia Minor, the chiondoxa, or glory-of-the-snow, has been cultivated since 1877 and has found wide acceptance in our gardens. It is a hardy, little, bulbous plant with narrow, pointed leaves and small blue flowers in tiny racemes at the end of the stalk.

Widely cultivated for its fine, persistent, but not evergreen leaves, and for its delicately scented, white, nodding flowers, the lily-of-the-valley is a cherished

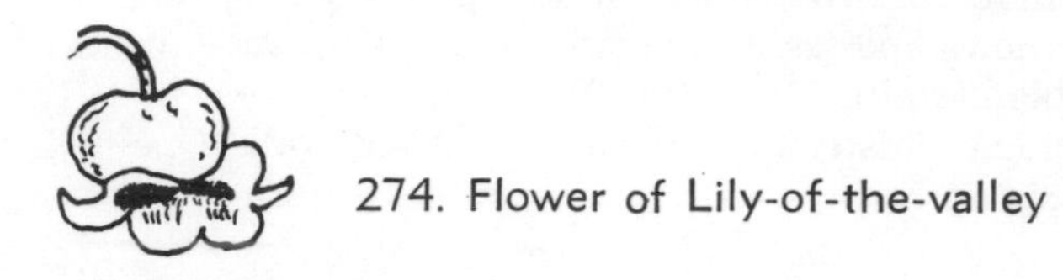
274. Flower of Lily-of-the-valley

plant of many gardeners, for few plants do as well in partial shade. It is a native of Europe, Asia, and America growing wild in the higher mountains from Virginia to South Carolina.

When one finds the star-of-Bethlehem growing in a field or meadow, one thinks of it as a wild flower. It is, though, essentially a garden plant. It is a slender ornamental with dark green, narrow and linear leaves,

275. Leaves and Flower of Star-of-Bethlehem

and twelve to twenty white, starlike flowers in a cluster on a stem six to eight inches long. The flowers are striped with green on the outside, and they open only in the sunshine. The generic name of the star-of-Bethlehem, *Ornithogalum,* is from the Greek meaning bird's milk, supposedly in allusion to the egg white color of the flower, and was given to the plant by Linnaeus because of the flower's likeness to the star in the East that guided the three Wise Men.

Few garden plants are of such striking appearance as the foxtail lilies, with their tall flower clusters often of several hundred flowers. They are magnificent Asiatic perennial herbs with cordlike or fibrous roots, basal somewhat narrow leaves, and spirelike clusters of yellow or orange, or white, or pink flowers that are rather bell-shaped. The generic name of the foxtail lilies is *Eremurus,* Greek for lonely tail, probably in reference to the solitary flower cluster. Hence, these plants are sometimes known as the eremuri.

Anyone who has eaten asparagus knows this vegetable by the young shoots that appear in early spring, for it is these that are eaten. Few people would likely recognize it in its more aesthetic dress, especially when found growing in the wild.

The asparagus has been a table vegetable for more than two thousand years. A native of southern Europe and western Asia, it was known to both the Greeks and Romans, but originally very different from our modern vegetable and probably no longer in cultivation. The asparagus, genus *Asparagus,* the Greek name for the vegetable, is a perennial herb with a stem rising from thick and matted rootstalks, two to three feet high when mature and multi-branched, but succulent and simple with fleshy scales when young. The leaves are small scales, the so-called leaves being narrow, threadlike branchlets that act as leaves. The flowers are small, greenish yellow, and set in the axils of the leafy branchlets. The fruit is a spherical red berry. Many people who grow the asparagus often use it as a decoration in their homes. It is hardly less decorative than the florist's asparagus fern (that is not a fern), used for trimming bouquets, and a feathery, fernlike climbing relative from South Africa widely grown for the trade.

One of the most unique plants often found in the garden is the plant variously called red-hot poker, flame flower, poker plant, or torch lily, but generally known to the trade as tritoma, all stressing the spectacular effect of spikes of blossoms that seem on fire, a flaming mixture of red, orange, and yellow. It is an African herbaceous perennial with thick, fleshy roots, long, linear basal leaves, and flowers in a spike or raceme above the leaves. There are several species and a number of fine horticultural forms.

The lily family is well represented in our flora with many native plants, the best known probably being the trilliums. They are handsome woodland plants with thick, short rootstalks wherefrom arise each spring, the flowering-stalks with scalelike sheathing leaves at the base. The three true leaves are arranged in a whorl at the top, the flowers are of various colors, solitary, on a short stalk growing from the center of the whorl of leaves, and the fruit is a berry.

A very common eastern species is the wake-robin, or birthroot or purple trilliam as it is variously known. It grows in rich woods with flowers from white to pink or brownish purple and that are ill-scented. Thus, the bees and butterflies leave them alone, but not the carrion loving greenfly, better known as the fleshfly, who finds the raw meat color of the flowers as acceptable as their odor. The berry is red and ovate.

In the stemless trillium, neither the flowers nor the leaves have stems. The somewhat fragrant flowers,

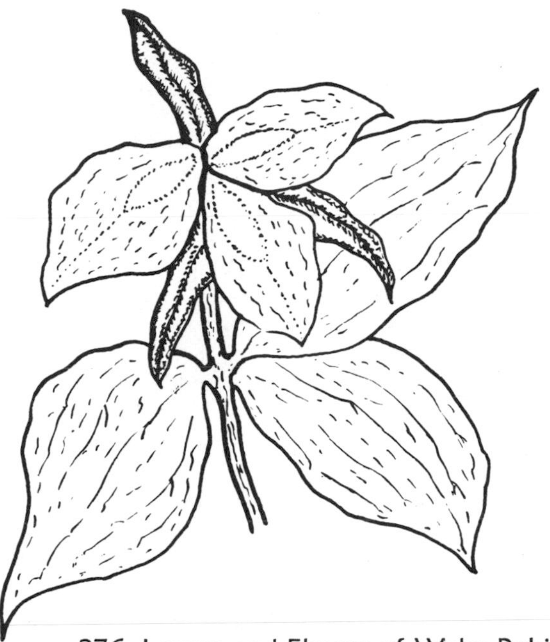

276. Leaves and Flower of Wake Robin

dark purple, purplish red, or greenish, are seated in a whorl of three egg-shaped, sometimes blotched leaves. The berry, as in the preceding species, is also red, but spherical rather than ovate. The word *trillium,* also the name of the genus, incidentally, is from the Latin for triple, in allusion to the leaves and flower parts being in threes, all the trillium flowers having three petals and three sepals.

The stemless trillium is found in moist woods, and along with it can be found the large flowering trillium, a handsome species with waxy white petals one and one half to two inches long, and sometimes even longer, and that curve gracefully backward. It seems that all the trilliums, prefer moist or rich woods, and the nodding trillium is no exception. Its white or pinkish flower droops from its stem until it is all but hidden beneath the leaves.

Most of the trilliums blossom in April, except the dwarf white trillium or early wake-robin, whose delicate little flowers, as white as the snow wherethrough they often have to grow, can be said to wake the returning robins of March into song. Doubtless, the most strikingly beautiful of all the trilliums is the painted trillium that comes into blossom a month later than its relatives. Its leaves are ovate, and taper to a sharp point. Its flowers are waxy white, veined and striped, with deep pink or wine color, and its fruit is a dark scarlet berry that falls at a touch.

Wherever one finds the purple trillium, in moist woodlands and along the brookside, one invariably will find the yellow adder's tongue or dogtooth violet. The name of adder's tongue is rather apt, for when the sharp, purplish point of a young plant pushes its way above the ground, it does not take much imagination to note the fancied resemblance to a serpent's tongue. As for the name of dogtooth violet, one can only guess. It neither looks like a dog's tooth nor is even a violet. Indeed, merely a glance at the solitary russet yellow, bell-shaped flower reveals its kinship to the tulip, lily, and trillium. Examine the flower more closely and it will be found that the perianth consists of six similar petallike segments, and that each of the six stamens stands before each of the segments. These are structural characteristics of a family of plants whose members are remarkable for the simplicity and beauty of their flowers.

The flower of the yellow adder's tongue is admirably adapted to long-tongued insects, and in all probability,

277. Leaves and Flower of Yellow Adder's Tongue

it is cross-fertilized by the early bees, especially by the queen bumblebees that have, but recently, emerged from their winter quarters. The common sulphur butterfly and the cabbage butterfly are occasional visitors, and possibly some flies too, are attracted by the mottled appearance of the leaves that are streaked and spotted with brown. The leaves are sometimes used for greens, and the bulbous root is edible when cooked, as are also the bulbs of the Canada lily. Thoreau relates that Indians ate the fleshy bulbs of the Turk's cap and often used them in thickening soups.

"Gray should not have named the flower from the Governor of New York," wrote Thoreau. "What is he to the lovers of flowers in Massachusetts? If named after a man, it must be a man of flowers." But DeWitt Clinton was a "man of flowers," a naturalist who often took to the fields and woods to escape the drudgery of his office, and to pursue in the open air, the study that filled him with delight and served as an antidote to the arduous cares of his governorship.

The Clintonia is a handsome plant of the woods and thickets with shiny, light green, large, oval to oblong leaves and with three to six cream or greenish yellow flowers on a slender flowerstalk about seven inches high. The berries, ripening about the middle of August, are a beautiful pure blue, a color that is rare and rather remarkable in nature.

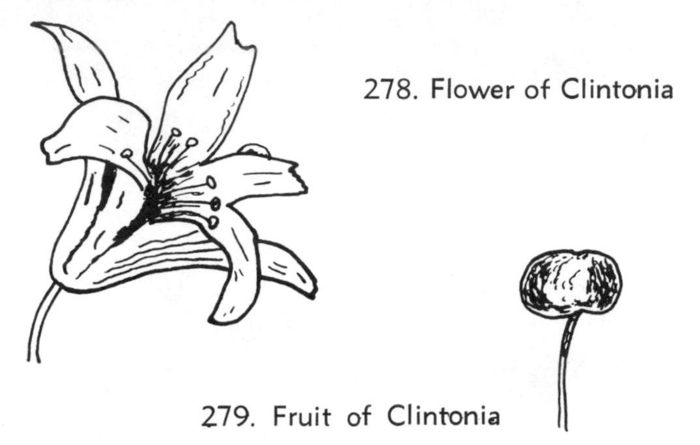

278. Flower of Clintonia

279. Fruit of Clintonia

An extremely pretty and graceful plant when grown in the garden, but hardly less so when found in the woods, thickets, and shady banks, the Solomon's seal grows to a height of one to three feet. Its light green leaves are alternately arranged on a slender stem and its tubular, bell-shaped whitish or yellowish green flowers droop in pairs beneath them. The fruit, at first a green berry with a whitish bloom, eventually turns a blue black, and is reminiscent of a small Con-

280. Flowers of Solomon's Seal

cord grape. After fruiting, the stem, rising from a multi-jointed, thick rootstock, withers and leaves a round scar. Hence, the age of the plant may be known by the number of scars on its root, as the age of a tree may be determined by the number of rings in its trunk.

There are other Solomon's seals, such as the great Solomon's seal, a capricious plant, sometimes only a foot high, at other times higher than the tallest man, and then there is a false Solomon's seal, two of them in fact, one with seven to twelve leaves the other with only three, both with pretty, starry clusters of white flowers and red berries. The three-leaved species grows in bogs or wet woods, while the multi-leaved species grows on moist banks and in meadows. Francis Parkman, the American historian, writes that the American Indians fed on the starchy root of the Solomon's seal and the roots of the same plant were used as food by the half-starved French colonists in America. It is said that the tender plant in spring is an excellent vegetable when boiled and served like asparagus.

There is a spikenard that is not a member of the lily family, and also a false spikenard that is. The latter is a beautiful woodland plant, somewhat resembling Solomon's seal, but with a spirallike cluster of fine white flowers at the tip of the stem. The flowers appear in May and develop into berries that are smaller than peas and that at first are greenish, then yellowish white speckled with madder brown. Finally, in late September, they turn a dull ruby red, are translucent and aromatic, and sometimes eaten. Some songbirds seem to like them.

A zigzag stem and twisted flower stalks, hence, its name, serve to identify the twisted stalk, a plant of moist woods, with light green leaves having a whitish bloom beneath clasping the stem. The flowers are curly-sepaled and greenish, and each hangs by a long, crooked threadlike stem from beneath the leaves. They are followed by a usually solitary berry, red, round, and nearly one half inch in diameter. A similar species differs from the preceding by having purple pink flowers and leaves that are not whitened by a bloom beneath. Both are visited by the bumblebees, the beelike flies, and the mining bees of the genus *Andrenidae,* and both belong to the genus *Streptopus,* Greek for twisted stalk, in allusion to their twisted stalks.

If one follows almost any woodland trail in May or June, one will be sure to find the Canada mayflow-

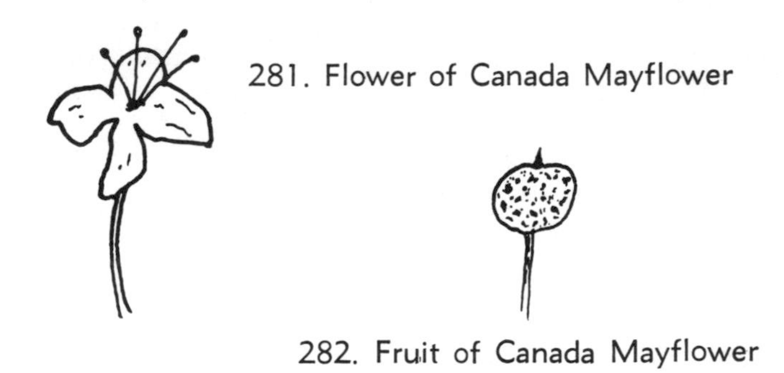

281. Flower of Canada Mayflower

282. Fruit of Canada Mayflower

er. It is a tiny plant with two or three light green, shiny leaves with a somewhat heart-shaped base, and small white flowers with a four-divided perianth and in a terminal spike or cluster. They are followed by yellow white berries, spotted with madder brown, that in early fall turn a translucent ruby red. Some birds, like the ruffed grouse, eat them.

When Linnaeus saw the little drooping flower of the bellwort, it reminded him of the uvula, the small piece of flesh hanging down from the soft palate in the back of the mouth, and so he named the genus of bellworts *Uvularia.* There are only four species, all hardy perennial herbs with thick, creeping rootstocks, light green alternate leaves without leaf stems, and yellow, bell-

283. Leaves and Flower of Bellwort

shaped, or tubular drooping flowers that develop into three-parted or three-celled capsules.

Occurring throughout the greater part of the East, the perfoliate bellwort, so-called because its leaves seem to be perforated by the stem, is fairly common in rich woods and thickets. Its flower consists of six petallike segments that are rough within to provide a foothold for the bumblebee as it seeks the nectar at the base of each of its six divisions. With a more westerly range, the similar large-flowered bellwort grows in the same situations as the perfoliate bellwort. Its flowers are somewhat larger, and open earlier when the queen bumblebees are flying about seeking the nectar of early spring flowers.

Similar to the perfoliate bellwort, the sessile-leaved bellwort differs, however, in having an angled stem, leaves that clasp the stem, and buffish cream-colored flowers. The fourth species of bellwort is a southern plant of the mountain woods and pine barren swamps and is known as the mountain bellwort. It is similar to the others, but with some slight differences. The roots of the perfoliate bellwort are edible when cooked, and presumably those of the others are as well.

The Indians relished the white, thick, tuberous root of the Indian cucumber or cucumber root, that is said

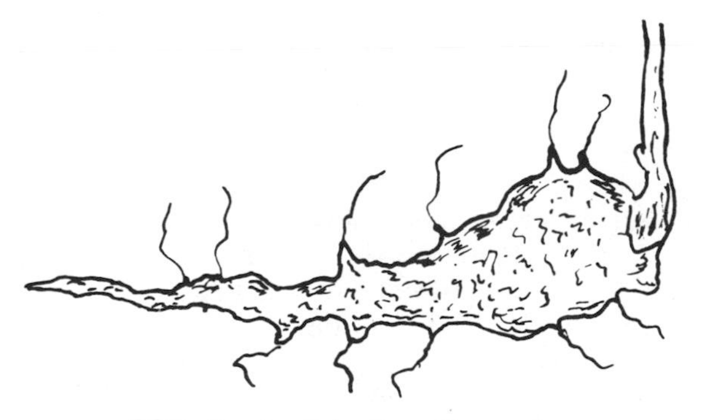

284. Root of Indian Cucumber

to greatly resemble the cucumber in taste and smell. It is a characteristic woodland plant, well adapted to subdued sunlight, and interesting in both flower and fruit. The stem is one to two and a half feet high, covered with fine wool when it first appears in the spring, but that later drops off, with two whorls of leaves, a lower whorl, about midway, of five to nine large, thin, oblong, taper-pointed leaves and an upper whorl of three to five small, oval, pointed leaves. The flowers are rather inconspicuous and greenish yellow,

285. Flower of Indian Cucumber

but accented by the terra-cotta color of the six stamens. They are set on fine curving stems in a loose cluster, and in September, are replaced by two or three purple black berries. At this time, the leaves take on rich tints and the Indian cucumber becomes a truly beautiful plant. It belongs, incidentally, to the genus *Medeola,* named for the sorceress, Medea, because of its supposed medicinal virtues.

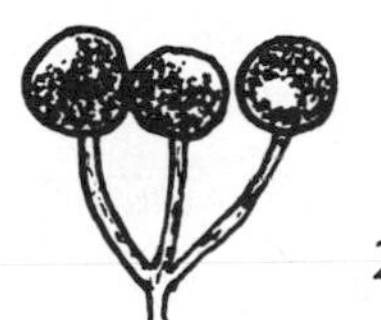

286. Fruit of Indian Cucumber

As the story goes, the devil's bit was once so highly regarded for its medicinal virtues that the devil, out of spite, bit off a piece of the rootstock, a degree of verisimilitude being given the story because the root ends as if it had been bitten off. Of such substance, are plants named. Anyway, the devil's bit is a plant of low grounds and swamps with mostly basal leaves, the stem with a few narrower ones farther up and terminated by a feathery spike, four to ten inches long, of small, fragrant, white flowers, the staminate and pistillate flowers on separate plants. The entire stem is wandlike from one to four feet high. Incidentally, the plant is of no medicinal value.

A plant of wet woods and meadows, the bunchflower is better named than the preceding, because its flowers are crowded or bunched at the end of a tall, unbranched stem four to five feet high. It is a stout plant with a thickish rootstalk and mostly basal leaves in a dense rosettelike cluster. The leaves are narrow and about twelve inches long. The flowers, small, greenish white, form a large panicle and are staminate, pistillate, and perfect on the same plant. After blossoming they turn brownish.

Borage and hellebore fill two scenes—
Sovereign plants to purge the veins
Of melancholy, and cheer the heart
Of those black fumes which make it smart.

Thus, Burton prescribed for madness in his *Anatomie of Melancholy*.

If the hellebore, however, had medicinal virtues, it also has its poisonous qualities. Horses and cattle have died from eating the young leaves in the spring; the seeds are fatal to poultry; and even people have lost their lives by mistaking its root for that of some harmless plant. This particular plant is the American white hellebore, genus *Veratrum,* an old name for the hellebore, a large, coarse, leafy perennial, but dangerously poisonous herb, that grows in wet meadows, low ground, and swamps everywhere. The leaves appear first in the spring, often as early as March, pushing up through the wet and sometimes frozen soil like spearheads. They are parallel-ribbed, broad, pointed, plaited, the lower ones usually more than a foot long and half as wide, broadly elliptic in outline, pointed at both ends, and clasping the stem with sheathing bases, the upper ones successively smaller as they near the top. The uninteresting large flower spike, six inches to nearly two feet long, is composed of pale yellowish green flowers, turning brown as they wither, with six spreading, oblong sepals, minutely toothed and fringed at the edge. They are visited by flies that must cross-fertilize them, as the anthers mature before the stigmas are ready to receive pollen, and develop into capsules nearly an inch long, ovoid, three-lobed, and containing many large, flat, broadly winged seeds that are carried by the wind and streams, since they float readily on the water.

In the popular mind, the onion and the lily appear to be two completely dissimilar plants and have little in common, but both have the structural characteristics of the lily family. The onion has been cultivated since remote antiquity, but where it originated is not known. Some believe its native land to be India, others Egypt. Most people know it as the bulb that is eaten. It is fairly large and nutritious. The leaves of the plant are hollow and tapering, and usually bluish gray; the flower stalk is also hollow. The flowers are white or lilac, and borne in a large, globose, terminal umbel.

The onion is a member of the genus *Allium,* Latin for garlic. It is a large genus of mostly onion-scented herbs and includes, besides the onion, the leek, garlic, chives, and shallot, as well as another group of perennial herbs grown for their ornamental flowers. All have bulbs, mostly basal leaves and typically hollow, and flowers from a few to many in a cluster (umbel). The fruit is a small capsule.

A native onion, the wild onion is fairly common on banks and hillsides where it grows from ten to twenty inches tall. The flower scape is bent or nodding at the top, the bell-shaped blossoms, that appear in July or August, are arranged in an umbel, the petallike segments being rose-colored or sometimes white. The bulb is very strongly flavored, but if parboiled, is very good to eat and is excellent for pickling. The leaves have been used to flavor soups.

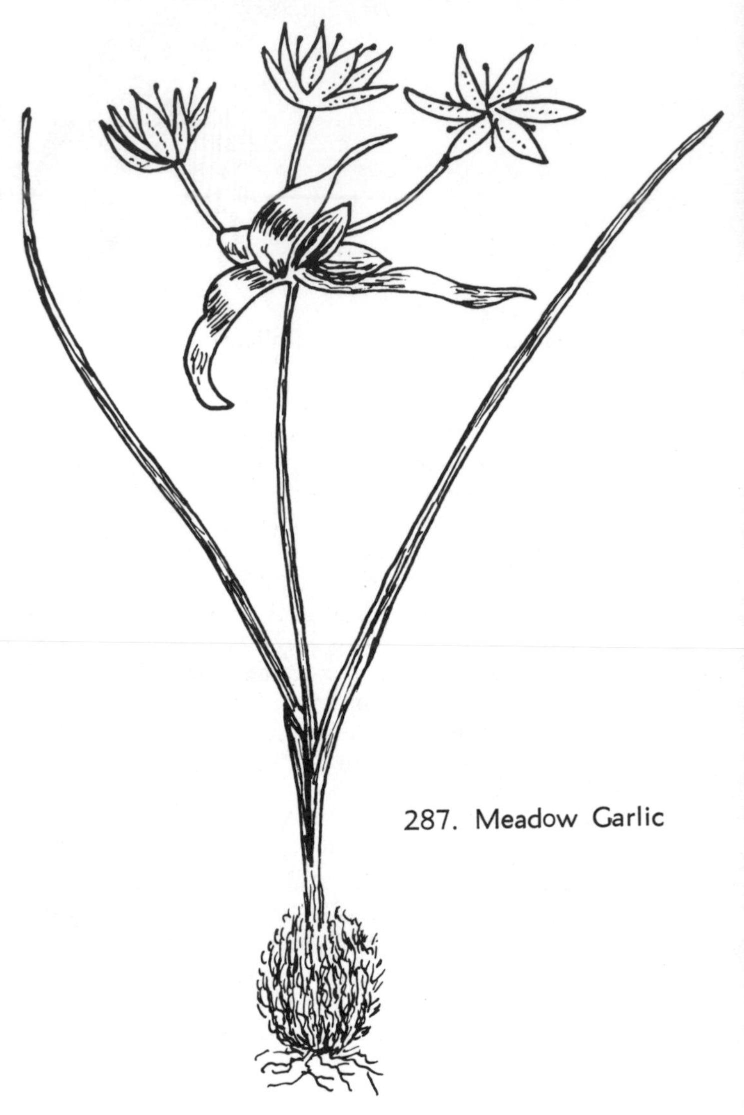

287. Meadow Garlic

A related species, the meadow garlic, growing in moist meadows, pastures, and thickets throughout the eastern half of the country, has a flower stalk eight inches to two feet high with a large umbel of many pink flowers, though they are sometimes white. The flowers are replaced by a cluster of bulblets that are slightly larger than grains of wheat or rye. Another species, the field garlic, is an immigrant from Europe, but an escape, and now found in pastures, waste places, and grainfields. Milk obtained from cows that have grazed in places where the wild onions (garlic) are common, may be contaminated with a garlic flavor. The cows do not have to eat the plants. All they need to do is to inhale the volatile substances from the flower clusters.

In rich woods, the leaves of the wild leek, two or three in number, appear in early spring from a coated bulb and grow to a height of less than a foot. Later the flowers appear, white or greenish white in a spokelike cluster from a spathe or leaflets at the top of a naked stem. The flowers are perfect, with stamens and pistils, and are six-parted with six green white sepals. The flowers are rich in nectar and are visited by the bees. In fact, all the alliums are assisted in cross-fertilization by bees, flies, butterflies and moths.

Thoreau compared the odor of the carrion flower with that of a dead rat in a wall. Not very complimentary of a plant that is otherwise beautiful and decorative. But the odor has its purpose. It attracts the little green fleshflies that share with many beetles the task of removing putrid flesh from the earth, playing the role of nature's scavengers, not a very pleasant task perhaps, but a needful one. Once the flowers have begun to wither and fruit has begun to set, one can approach the carrion flower without being offended by its odor and appreciate the beauty of the vine, growing to a length of fifteen feet in thickets and on riverbanks. The leaves are a beautiful, glossy green, egg-shaped, heart-shaped at the base, pointed at the tip, and take on resplendent tints in autumn. The long flower stem, proceeding from between the tendrils, that extend from the base of the leaf stalks, ends in a hemispherical flower cluster with spokelike stemlets whereupon are set the insignificant and ill-scented, greenish yellow flowers. They are followed by blue black berries that, it is said, were eaten by the Omaha Indians for their pleasant flavor.

Somewhat similar to the carrion flower, but differing from it in its angled stem that is covered with scattered prickles, its more rounded leaves, and its scentless flowers, the greenbrier or catbrier is a familiar plant to those who ramble outdoors. It is often found by the roadside or woodland thicket where it effectively bars further progress. Both the carrion flower and greenbrier are useful to wildlife as they provide good protective cover for rabbits and other small species (the yellow-breasted chat for one nesting in the tangles), and are a source of winter food, the berries being eaten extensively by such birds as the fish crow, ruffed grouse, mockingbird, and catbird. Both the carrion flower and greenbrier belong to the genus *Smilax,* an ancient Greek name for the plants.

Although the day lily is called a lily because it looks like one, it is not a true lily and belongs to an entirely different genus. It is a tall, robust plant, a native of Europe and Asia, but now an escape, usually found in meadows and upon the borders of streams. Its root is somewhat fleshy. Its leaves are long and narrow, almost two feet in length, tapering to a point, and light green. Its flowers, in loose corymbs of six to twelve at the summit of leafless scapes, three to five feet high, are tawny orange in color and open one or two at a time. They open for but a day, and so, the genus of the day lily is called *Hemerocallis,* Greek for beautiful for a day. All the books describe the flowers as tawny orange, but if one looks down into the cup one will see a heart of dull yellow that deepens at the point where the segments curve, and lightens again as the color runs to the tips. The result is a dull orange on a base of yellow.

There is about a dozen wild species of Hemerocallis and several hybrid forms. Less common than the day lily, the yellow day lily is also an escape. It is a more delicate species, with narrower leaves, and clear, pale yellow, fragrant flowers.

One of the most distinctive of the many genera that compose the lily family is the genus of *Yucca.* It consists of shrubs and trees with long, narrow, sharply pointed leaves, and huge flower clusters on towering stalks sometimes as much as ten feet high. Most of them are stemless with a basal rosette of sword-shaped, tough, leathery leaves, but two species, the Joshua tree and bear grass have distinct trunks. The flowers are white, waxy, cup-shaped, nodding, usually fragrant at night, some blossoming only at night, and borne in showy erect, terminal panicles. The fruit is usually a capsule.

The yuccas are striking plants that arrest attention, and are common in the southwestern United States where they culminate in the extraordinary and grotesque Joshua tree, frequently reaching a height of thirty or forty feet. The woody trunk branches to form an irregular head, and the terminal branches are bent and twisted in all directions. The entire plant is covered with an armor of short, stiff, narrow leaves. The Mormons are said to have named the tree, for when crossing the California deserts on their way to Utah, the bizarre branches appeared to them as the outstretched arms of a Joshua leading them out of the wilderness.

Other southwestern species are the desert candle and the Spanish dagger. The desert candle forms porcupinelike rosettes of basal, narrow leaves and the stout flowering stalk extends out from among them and ends in an immense cluster of pendant, creamy white blossoms. The Spanish dagger is a shrub or tree with a trunk twelve feet high, with stiff and sharp-pointed leaves, one to three feet long, and white, bell-shaped flowers. The fruit or seed pod is large and pulpy, three to six inches long, when ripe a dark purple, but also yellow. It has been compared to a short or stubby banana. It is said to have a peculiar, sweet taste when fresh, and is quite palatable. The Indians of New Mexico slice the ripe fruit and dry it in the sun for winter use.

Native from North Carolina south to Florida and west to Mississippi, Adam's needle, or bear grass, or Spanish bayonet, is the species most cultivated in the East. It is practically stemless, but the stalk of the flower cluster may be eight to twelve feet high. The leaves are two to two and a half feet long, about an inch wide, and thready on the margin. The flowers are white, or cream white, and about two inches long. Indians are said to have eaten the seed pods.

There is a remarkable interdependence between

the yucca plants and certain moths known as the yucca moths. The yuccas are entirely dependent on these moths for pollination, and without them, cannot set seed. The female moths are uniquely adapted for this purpose. The proboscis is not long and curled, as is usually the case in butterflies and moths, but instead, is short, spiny, and modified to collect pollen. After having gathered pollen from several flowers, the female moth thrusts her long lancelike ovipositor (or egg-laying apparatus) into the ovary of the flower and lays several eggs. Then she climbs the pistil and inserts the pollen mass into the stigma. The ovules develop into seeds, some being consumed by the larvae, though enough are left to perpetuate the species.

92. Linden (Tiliaceae)

No matter how isolated a linden may be, the bees are sure to find it, for the flowers are exceptionally fragrant enough to advertise the rich supply of nectar they provide for their insect visitors. The small, yellowish white flowers are borne in cymose clusters, pendulous at the end of a flowerstalk, attached for half its length to the vein of an oblong leaflike bract as long as the stalk itself.

The ancient town of Hybla was famed for its honey that the bees made from the nectar of the linden, or lime trees, that covered its sides and crowned its summit, and one may read that in obedience to Amphion's music:

> The linden broke her ranks and rent
> The woodbine wreaths that bound her,
> And down the middle, buzz! she went
> With all her bees around her.

The lindens have an honored place in classical literature, and Homer, Virgil, Horace, and Pliny all mention the lime tree and extol its virtues. Herodotus writes that "The Scythian diviners take also the leaf of the lime tree, which, dividing into three parts, they twine round their fingers; they then unbind it and exercise the art to which they pretend." It seems as if every part of the linden has been put to some use: its faggots for charcoal; its leaves as fodder for cattle; its flowers as nectar for the bees and for the perfumer; its seed balls for their oil in cooking and for table use; its inner bark for ropes, fishnets, and mats; and its wood, uniform in color and texture and free from knots and other imperfections, for the wood carver. Dryden describes it as, "Smooth grained and proper for the turner's trade,/ Which curious hands may carve and steel with ease invade."

It is said that Linnaeus, the Swedish botanist and taxonomist, had his name from a favorite linden tree that stood by his peasant father's house.

There are some thirty species of lindens of the genus *Tilia*, the old Latin name for the linden, native to the north Temperate Zone. They are handsome trees and make excellent shade trees, and are widely planted along streets and avenues. There was once a famous Berlin thoroughfare, "Unter den Linden." They have a mucilaginous sap, a tough inner bark, and a broad, dense head. Their leaves are deciduous, alternate, toothed, usually heart-shaped at the base, and with one side longer than the other. Their flowers are small, yellowish white, fragrant, and borne in long-stalked drooping clusters. Attached to the flowerstalk for about half its length is a thin, oblong bract that is one of the prominent characteristics of the group. The fruit is an almost spherical, woody, nutlike structure about the size of a small pea.

The author is not sure how many native lindens there are, for the authorities seem to differ, claiming from four to eighteen. Anyway, the American linden is one of them, a tall, symmetrical tree, fifty to seventy feet tall and sometimes taller, with an erect, pillarlike trunk, the branches spreading, often pendulous, forming a broad rounded head. It is common in rich woods, has a deep brownish gray bark with elongated fissures, and smooth, green or ruddy brown, often zigzag twigs. The leaves are somewhat heart-shaped, sharp-pointed, coarsely double-toothed, light green above, lighter beneath, and the flowers, appearing when the leaves are nearly full grown, are small, perfect, white or yellowish white, very fragrant, clustered on the end of the flower stalk borne on a narrow leaflike blade.

Perhaps no American tree has more abundant foliage than the linden, but unfortunately, the soft leaves attract numbers of insect enemies and the sticky surfaces catch dust and smoke. In fall, they turn a faded yellow, not as pleasing to the eye as the gold of the beech and hickory, and eventually, they are riddled with holes and, torn by the wind, they fall from the branches. In the winter, the linden may be recognized by its deep red buds. In some localities the American linden, incidentally also known as the basswood, is highly prized for its honey that has a slightly more acid tang than that made from the clovers. It is said that the Iroquois Indians made rope from the bark.

The European linden is similar to the American linden except that it is smaller and tapers to a lesser

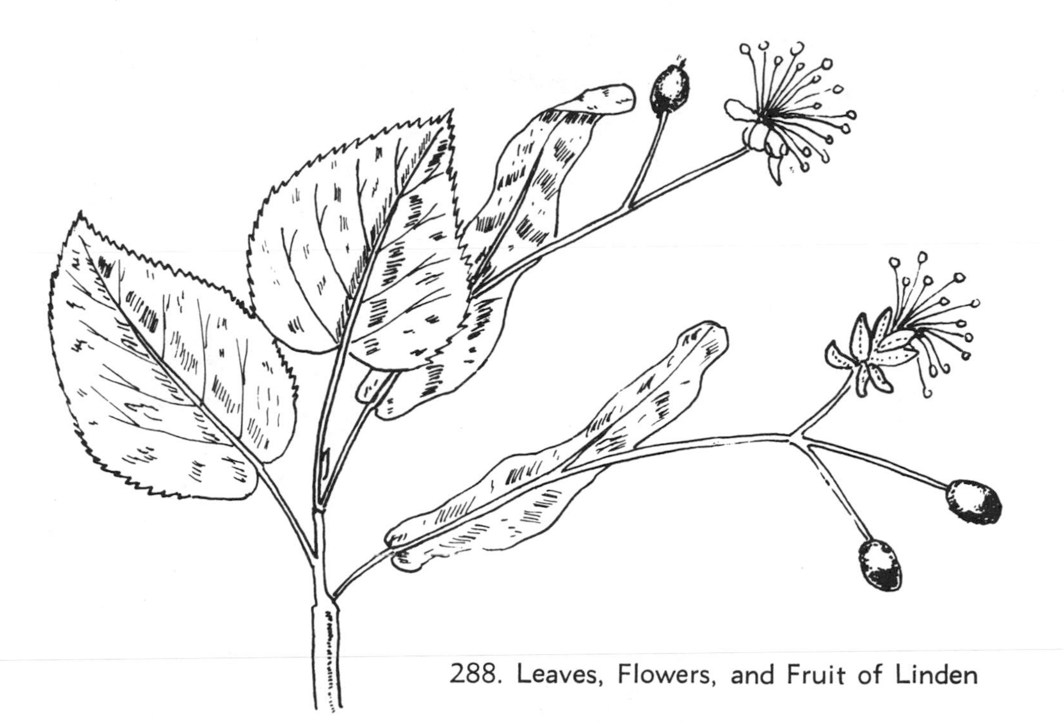

288. Leaves, Flowers, and Fruit of Linden

or narrower rounded top. There are also several other minor differences.

Another native species of linden is the downy basswood, a smaller and more southern tree, not more than sixty feet high, of swamp margins and streams, with smaller leaves that are thickly, rusty, finely haired and dull sage green beneath. A third native species is the white basswood, also a southern species, forty to sixty feet high, that shows a preference for the sides of mountain streams. Its leaves are similar to those of the American linden except for the lower surface that is covered with white, fine, downy hairs. As they flutter in the breeze, they make a dazzling display of white and green against a dark background of hemlocks and mountain laurel.

Several European lindens have been introduced into our country: the silver linden, a tree ninety feet or more high, rather pyramidal in habit, with upright branches and leaves whose lower surface is covered with silvery-white down; the broad-leaved or large-leaved linden, a tall, shapely tree growing up to one hundred and twenty feet high and often clipped to form a hedge—the alleys of the Tuilleries gardens were made of it; and the small-leaved linden, a shapely tree, ninety to one hundred feet high, with rounded leaves having tufts of brown hairs in the vein axils.

The genus *Tilia* is one of forty genera of the linden family, a fairly large group of four hundred species of trees, shrubs, and herbs widely scattered throughout the world, but most abundant in the tropics. They have a very fibrous inner bark, alternate, simple, deciduous leaves that are entire, toothed, or lobed, and regular, perfect flowers borne in clusters (cymes or corymbs). The individual flowers usually have five sepals that are united or separate, five petals, many stamens that are either free or united in bundles, and a two to ten-celled ovary. The fruit may be a capsule, berry, drupe or nut.

Aside from the lindens, the only other members of the linden family of any interest are the corktree or corkwood, the sparmania, and the jute plant. The corktree is a native of New Zealand and somewhat grown in southern California for ornament. It is a small tree, barely twenty feet high, with oval or heart-shaped leaves and white flowers. The sparmania is an African shrub or small tree, growing to twenty feet high, with five to seven-lobed, toothed leaves, and white flowers with prominent yellow stamens. It is an attractive and showy shrub, but can be grown outdoors only in frost free localities.

The jute is usually a single-stemmed annual growing up to fifteen feet high, with oblong, toothed leaves and small, yellow flowers. The commercial jute is a long fiber, six to ten feet in length, secured from the bark, but as a fiber, is inferior to that of flax and hemp. The jute is a native of India, but widely escaped in tropical countries. In 1870 the United States Department of Agriculture introduced the jute plant to several of the southern states, but met with indifferent success. A similar and related plant, called Jew's mallow, has become naturalized in Florida and adjacent states.

93. Lizard's Tail (Saururaceae)

An example of a perfect flower, that is, a flower with stamens and pistils, but without a calyx or corolla, is the lizard's tail, a perennial herb of marshes and swamps, with a slender rootstock, jointed stems, and heart-shaped leaves. The small, white flowers that are slightly fragrant, are on short pedicels, each with a little bract under it, and are crowded in a terminal spike, that gently nods and waves in every passing breeze. One can easily see a fanciful resemblance to the tail of a lizard, and so, its name.

The lizard's tail of the genus *Saururus,* from the Greek for the tail of a lizard, in allusion to the long, slender spike, is one of four species of a small family of the same name, a group containing only three genera of perennial herbs with alternate, entire, petioled leaves, small, perfect, incomplete flowers, without a calyx or corolla but with six to eight stamens, or sometimes fewer, and an ovary of three to four carpels. The fruit is either capsular or berrylike, that of the lizard's tail being slightly fleshy and strongly wrinkled when dry, of three to four indehiscent carpels united at the base.

Of the four species, two are Asiatic and two are native of North America, these being the lizard's tail and the yerba mansa, a perennial with basal leaves and a cone-shaped mass of densely packed brownish flowers found from California to western Texas.

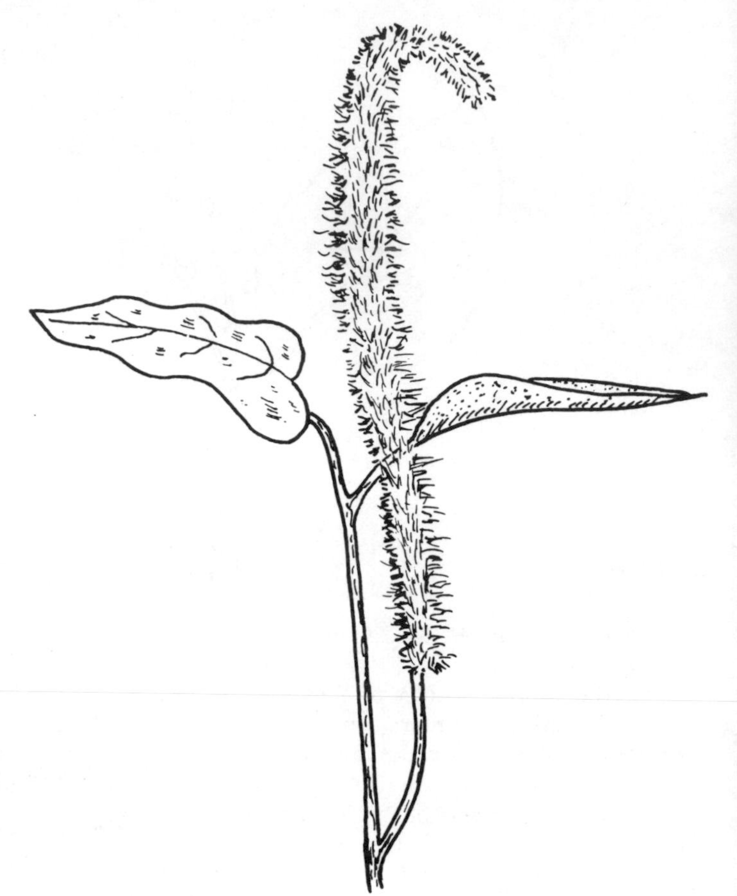

289. Leaves and Flowers of Lizard's Tail

94. Loasa (Loasaceae)

The loasa family is a group of thirteen genera and about two hundred species of herbs all native to the New World. They are erect or climbing branching plants, often with hooked stinging or sticky hairs, with alternate or opposite leaves, perfect and regular white, yellow, or reddish flowers in various kinds of clusters, the flowers with usually five sepals, usually five petals, many stamens, and a one-celled ovary, and a capsule as fruit.

The only genus of interest is *Mentzelia,* named for Christian Mentzel, a German botanist, and it contains about fifty species of western herbs. Three of them—the prairie lily, the blazing star, and few-seeded mentzelia—deserve more than passing notice. The prairie lily, also known as gumbo lily, is a biennial herb, a foot or more high, with much cut oval, lanceolate, or oblong leaves and white or yellow flowers that open towards night.

The prairie lily is a plant of the plains, as are also the blazing star, a stout perennial herb with long, narrow leaves, the edges wavy and with marginal teeth, and yellow flowers, and the few-seeded mentzelia, an annual, one to four feet high, single-stalked, or branched and straggling, with oval or ovate coarsely or finely toothed leaves, and bright yellow flowers that open in the evening and close the following morning. The last named species is often cultivated in the garden. By day, it is a rather unprepossessing plant, rough and thistlelike, but when the sun goes down and the buds open, the bright yellow flowers gleam in the dusk and fill the air with a delicate fragrance.

95. Lobelia (Lobeliaceae)

The deep red blossoms of the cardinal flower, a deeper red than the bird of the same name, are the same color as the hat worn by a prince of the Roman church. They almost seem to kindle into flame the moist thickets wherein the flower grows. Not many insects visit the blossoms, for their tongues are not long enough,

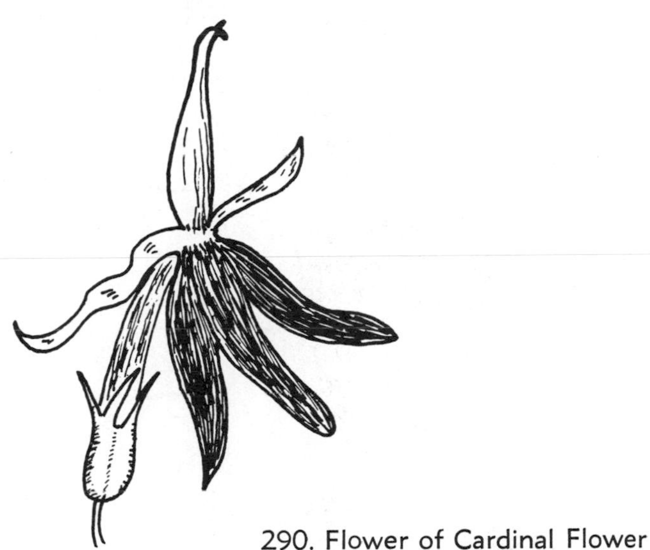

290. Flower of Cardinal Flower

but the hummingbird can reach the nectar. Why does the hummingbird not visit the cardinal's twin sister, the blue lobelia? The flowers are built on much the same plan, though the lobelia is slightly adapted for the bumblebee, it being a frequent visitor. Is it the color of the blossoms? Study the habits of the ruby-throat and one will find that its visits are confined chiefly to such flowers as the painted cup, Oswego tea, coral honeysuckle, columbine, and the garden salvia, fuschia, and phlox.

Who has not seen the cardinal flower in blossom along a brook, a meadow runnel, or even a roadside ditch? If it wanted to hide among the surrounding vegetation, it could do so no less than the scarlet tanager could conceal itself among the leafy branches of a tree. It is a stately flower of considerable beauty and one of our own plants, not an importation from some distant place, a plant strictly indigenous to America. More prosaically it is a tall, stiff plant, with toothed leaves, dark green, oblong to lance-shaped, with flowers in terminal clusters, the flowers tube-shaped and two-lipped, the upper two-lobed, the lower three-lobed of a rich, velvety color.

The cardinal flower is a lobelia, that is, it belongs to the genus *Lobelia*. According to Webster's dictionary the genus was named after a botanist and physician to James I by the name of Lobel, but according to various botany books it was named after Matthias de l'Obel, an early Flemish herbalist. All are correct, for it is the same gentleman who Anglicized his name when he became physician to the English king.

The genus *Lobelia* comprises over two hundred and fifty species of annual or perennial herbs, though some of the tropical ones are trees. There are several native species including the cardinal flower and the blue lobelia already referred to. The blue lobelia is also a tall, stiff plant, though slightly hairy with light green leaves pointed at both ends, and light blue violet flowers. It, too, like the cardinal flower seeks wet ground. Unlike its relatives, the pale spiked lobelia prefers dry ground, usually sandy soil. It is a smaller-flowered species with long slim spikes of pale blue violet flowers. Since the lobelias generally seem to delight in wet ground, it is not surprising to find one that not only grows beside water, but actually in it, and is often totally immersed. The slender, hollow, smooth stem of the water lobelia rises from a submerged tuft of round, hollow, fleshy leaves and bears at the top a scattered array of pale blue flowers.

The genus *Lobelia* is one of twenty genera of the lobelia family that includes some six hundred widely distributed herbaceous plants, and in the tropics, shrubs and trees. The herbaceous plants usually have an acrid or milky juice, alternate or basal leaves, and flowers in various sorts of clusters, chiefly in spikes and racemes. The corolla is irregular, tube-shaped, and mostly two-lipped, but the tube is sometimes parted to the base. The stamens are five in number with the anthers often united around the style, and the five-lobed or five-parted calyx is also united, the calyx tube adhering to the many-seeded capsule. The ovary is two to three-celled, and the fruit is a berry or a dry capsule.

How the Indians could have smoked and chewed the bitter leaves of one of the lobelias called Indian tobacco is a mystery. To chew even one of the green bladder pods is to experience a feeling of nausea. This plant is very poisonous. Grazing animals seem aware

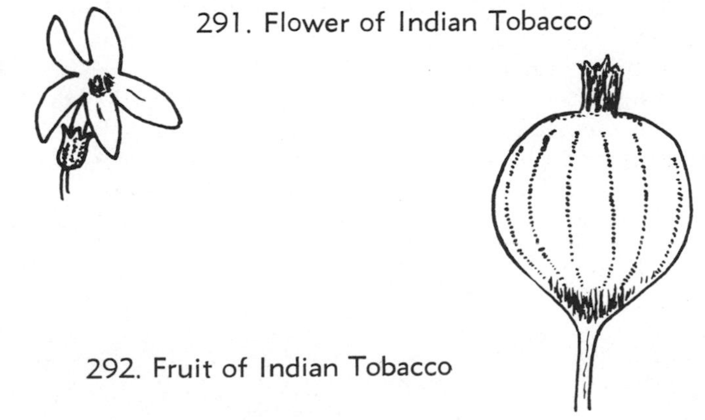

291. Flower of Indian Tobacco

292. Fruit of Indian Tobacco

of its toxic properties and leave it alone, but sometimes its young shoots are eaten and cause a sickness known as "slobbers." Like many other poisonous plants, it is, however, medicinally valuable, being used in the treatment of laryngitis and spasmodic asthma.

One of the most common of the lobelias, growing everywhere, in meadows, pastures, grainfields, and thickets, the Indian tobacco is also the least attractive of the tribe, for its flowers are small and few. It is an annual with a simple or branching, slightly hairy stem, oval, pointed light green leaves, and flowers that vary from light blue violet to pale lilac and even white. The fruit is a much inflated, rounded, ribbed, multi-seeded capsule.

96. Logania (Loganiaceae)

Famed throughout the South and dear to every southern heart, the Carolina or yellow jessamine heralds the approach of spring by opening its yellow blossoms in late February or early March. A beautiful vine, its yellow flowers a delightful contrast with its richly colored evergreen leaves, it grows to a length of ten to twenty feet, making its way from the ground up to the tallest trees of the pinelands, swamps, and sandhills of the Coastal Plain. The leaves are usually opposite, ovate or lance-shaped, small and shining, and the flowers are a bright yellow, very fragrant, in dense clusters (cymes), the flowers with a five-parted calyx, a tubular or funnel-shaped corolla, five stamens, and a two-celled ovary. The fruit is a capsule.

The Carolina jessamine of the genus *Gelsemium*, the Latinized version of gelsomina, the Italian for jasmine, is one of four hundred species of the logania family that contains about thirty genera. They are mostly tropical herbs, shrubs, and vines with opposite or whorled simple leaves, flowers usually showy and always in clusters with the characteristics of those of the Carolina jessamine, and the same kind of fruit, that is, a capsule.

There are only a few native species of the family. Besides the Carolina jessamine there is the Indian pink, a hardy perennial of rich woods, with spikelike clusters of red flowers, marked with yellow on the inside of the corolla, and whose roots are supposed to be of some medicinal value; and the mitrewort, an annual herb of damp soil, with lance-shaped or ovate leaves and small, white flowers in a terminal cluster.

An introduced ornamental is the butterfly bush of the genus *Buddleia*, named for Adam Buddle, a British botanist, a native of China and a beautiful plant with arching or pendulous branches, lance-shaped leaves, and lilac purple flowers in short, dense clusters.

Occasionally grown in extreme southern Florida is the strychnine, a medium-sized tree of India with

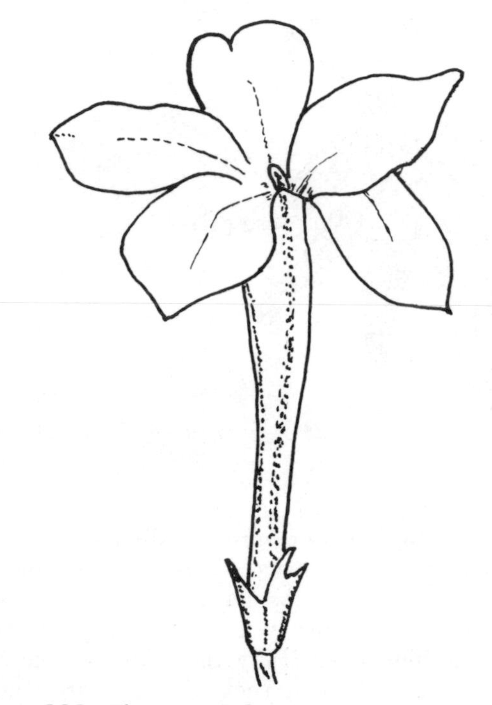

293. Flower of Carolina Jessamine

opposite, oval leaves, small yellowish white flowers in terminal, multi-branched clusters, and a fleshy and berrylike fruit. This is the plant whose seeds produce the drug strychnine. Another species of the same genus *Strychnos*, an old Greek name for a kind of nightshade, found in South America, is the source of the drug curare and of the arrow poisons used by native tribes. Also, sometimes grown in extreme southern Florida, is the natal orange, a spiny shrub, seven to ten feet high, with roundish leaves, yellowish white flowers in a compound cyme, and a yellow, berrylike fruit that is sweet and edible. It is a native of Central and South America.

97. Loosestrife (Lythraceae)

"With fantastic garlands did she come,/ Of crow flowers, nettles, daisies, and long purples," wrote Shakespeare of Ophelia, but the poet's knowledge of flowers was not what it should have been, for the long purples is not a spring flower, even in England. Perhaps it was the purple orchis that Ophelia wore.

At any rate, the long purples, also called the purple loosestrife, spiked willowherb, and red Sally, is a beautiful species naturalized from Europe and found growing in wet meadows and on the borders of swamps. It is an erect, smooth or slightly hairy, wandlike perennial with willowlike leaves and bright magenta or pinkish purple flowers in dense, showy, leafy, terminal spikes. The long-petaled flowers are trimorphic, that is, the stamens and styles are of three different lengths, and since six different kinds of yellow and green pollen are produced on the two sets of three stamens, one would expect that there would be eighteen ways wherein the insects could transfer the pollen. But only pollen brought from the shortest stamens to the shortest style, from the middle-length stamens to the middle-length style, and from the longest stamens to the longest style, can effectually fertilize the flower. The honeybee, bumblebee and many butterflies are common visitors, especially the yellow sulphur.

The long purples belongs to the genus *Lythrum*, Greek for blood in allusion to the color of some species, a group of annual or perennial herbs of the loosestrife family, comprising some two dozen widely scattered species. They have four sided stems, mostly opposite leaves, and flowers solitary in the leaf axils, in terminal spikes, or racemes. The flowers have a tubular, cylindrical calyx, usually six petals, six or twelve stamens, and a one to four-celled ovary with many ovules. The fruit is a capsule.

Two other fairly common species of the genus are the milk willow herb, an erect perennial of low moist ground, with the lowest leaves opposite, the others alternate, the leaves lance-shaped and pointed at the tip, and with solitary, purple flowers in the leaf axils; and the hyssop loosestrife, a smooth, branching annual of salt marshes, with a pale green stem and leaves, and pale purplish magenta flowers.

The loosestrife family comprises twenty genera and about four hundred and fifty species of herbs, shrubs, and trees of wide distribution, but most abundant in tropical America. The annual and perennial species are more common in temperate regions, the woody forms in warm climates. They generally have opposite, simple, entire leaves that are whorled in some species, perfect flowers that are mostly in the leaf axils, solitary or in clusters (panicles or cymes). The flowers are regular or irregular, with a more or less bell-shaped calyx that is mostly four to six-toothed or lobed, and sometimes with secondary teeth between, four to six petals inserted at the throat of the calyx, four to eight stamens, and a two to six-celled ovary with many ovules. The fruit is a capsule containing numerous seeds.

294. Leaves and Flowers of Loosestrife

Several other native herbaceous plants of the family that might be mentioned are the swamp loosestrife or water willow, the water purslane, the clammy cuphea, and the tooth cups. The swamp loosestrife is a somewhat shrubby plant of swampy places with reclining or recurved stems, three to eight feet high and rooting at the tips. The leaves are opposite or in threes and willowlike. The flowers are small, bell-shaped, magenta, in a dense, nearly stalkless cluster, with half of the ten stamens protruding, and the others hidden. The fruit is a three to five-chambered capsule. The clammy cuphea is a clammy, hairy, branching, rather homely annual of dry, sandy fields, with ovate lance-shaped, dull green leaves and small, magenta pink flowers. A related species is the cigar flower of Mexico, a popular pot plant and often used for summer bedding. It is a shrub, but often grown as a tender annual, eight to fifteen inches high, with lance-shaped leaves and solitary flowers. The calyx tube is red, but with a darker ring near the tip and a grayish white mouth, that suggested the name. The cigar flower, though tender to frost, makes an excellent border plant because it blooms continuously. The water purslane is an aquatic plant, rooting in the mud of ponds, with solitary greenish flowers and the tooth cups is another swamp dwelling member of the family with two to four purple flowers in the axils of the opposite, narrow leaves.

Several tropical, woody species of the family are of more than passing interest, such as the mignonette tree of Africa, Asia, and Australia whose leaves are the source of the red dye, henna; and the crape myrtle, an Asiatic shrub and tree widely cultivated in the South for its showy bloom and where it is often found as an escape. It is shrub or tree, up to twenty feet high, the twigs four-angled, the leaves elliptic to oblongish, the flowers pink in clusters four to nine inches long, and the fruit a capsule. There are several varieties, one with white flowers, one with purple flowers, and one with red flowers.

98. Lopseed (Phrymaceae)

The lopseed family has only one genus, *Phryma*, of unknown origin and only one species, the lopseed, a perennial herb of moist and open woods. It has slender branching stems and coarsely toothed ovate leaves, small, opposite purplish or rose-colored flowers in slender, terminal spikes, the flowers with a two-lipped cylindrical calyx, the upper lip of three bristle awl-shaped teeth, the lower short and two-toothed, a two-lipped corolla, the upper lip notched, the lower much larger and three-lobed, four stamens, and a one-celled, one-ovuled ovary, and a one-seeded achene.

295. Flower of Lopseed

99. Madder (Rubiaceae)

One of the most delightful of the early spring flowers is the bluet. By itself, it is not a showy flower, but when massed together in countless numbers, bluets trace a milky path beautiful to see in fields and meadows. A bluet may be light blue, pale lilac, or nearly white. It has a yellowish center and is funnel-shaped, with four pointed, petallike, oval lobes. Now there are two kinds of flowers that, strangely enough, do not occur in the same patch. Should one select a flower from one patch and examine it closely, as with a magnifying glass, one may find that the stamens are situated in the lower part of the corolla tube, with the stigmas exserted. But a flower from another patch may have the stamens elevated to the mouth of the corolla, with stigmas below the opening—in other words, just the reverse of the first flower.

The reason for the two kinds of flowers is to secure cross-pollination. An insect visiting form *a*, for instance, becomes dusted with pollen from the anthers

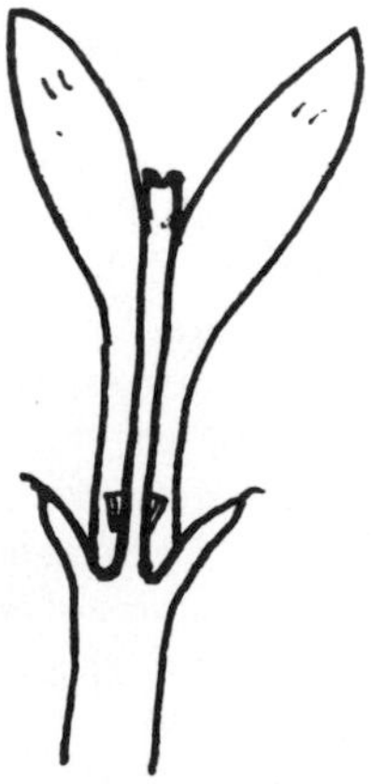

296. Flower of Bluet, Form **A**

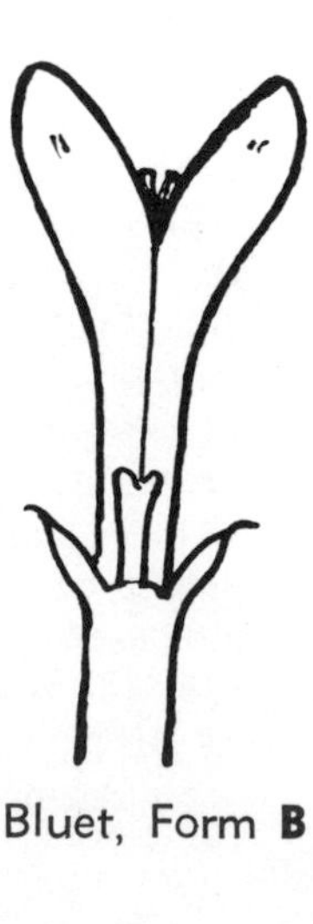

297. Flower of Bluet, Form **B**

in the middle of the tube, and this pollen is brushed off by the stigmas of form *b*. Conversely, an insect visiting form *b* gets dusted with pollen that is removed by the protruding stigmas of form *a*. Mining bees are the chief agents of cross-pollination, but the smaller butterflies—the clouded sulphur, the meadow fritillary, and the painted lady—are also instrumental in doing so.

The bluets, or Houstonia, or quaker ladies, or innocence, are a familiar flower of waysides and meadows. The plant is a perennial, and forms dense tufts of oblong, lance-shaped, tiny, light green root leaves and slender, threadlike stems set sparingly with small, opposite leaflets. The flowers appear from April to July and sparsely throughout the summer. A pot of roots gathered in the fall and placed in a sunny window will likely produce a little colony of flowers throughout the winter. Some books call the bluets Houstonia, the generic name of the plant given to it by Linnaeus for Dr. Houston, an English botanist who collected it in South America.

298. Leaves and Flowers of Partridge Berry

Whether it is in June when its little evergreen leaves are sprinkled with pairs of waxy, cream white, pink-tipped, lilac-scented flowers, or in autumn or winter when its coral red "berries" gleam in the shade of the woodland floor, the partridge berry is always one of the loveliest sights in the woods. It is a little trailing vine with dark green leaves, occasionally with white veins, heart-shaped, and on short stems, and with two kinds of flowers as in the bluets. In some plants, the stamens extend beyond the corolla and the pistil is shorter than the tube, while in others, the stamens are short and the pistil extends beyond the flower, a device of nature to insure cross-pollination, accomplished by the same insects that visit the bluets. The flowers grow in pairs, hence, the name of twinflower as the plant is sometimes known, and so united at the base that it takes the two blossoms to form one berry, or as the poet Choate expressed it:

Made glad with springtime fancies pearly white,
Two tender blossoms on a single stem
In their sweet coral fruitage close unite
As round head cut from a garnet red.

The bright red "berries," actually drupes, persist on the vines all winter and are edible, but are rather tasteless. They are eaten by some game birds such as the bobwhite quail and ruffed grouse, the latter sometimes called partridge, hence, the name of the plant, but they are not of too much importance to wildlife even when birds and various mammals often have trouble finding enough to eat.

The bluets and partridgeberry are members of the madder family, one of the largest groups of all the flowering plants with some five thousand species of herbs, shrubs, and trees most being tropical. It is an economically important family, for it includes the coffee and quinine plants as well as many others that produce substances useful as dyes and medicines.

The leaves are opposite or whorled, that is in circles, simple, and usually without teeth. The flowers are mostly perfect, solitary or in clusters. The calyx is two to six-cleft, united or parted, the corolla is funnel-formed, with four, sometimes five, lobes and as many stamens and the ovary is mostly two-celled. The fruit is a capsule, berry, or drupe.

Scratch grass, grip grass, cling rascal, catchweed, hedgeburs, sweethearts, beggar-lice, clover grass, stick-a-back, gripgrass are some of the English folk names given to a plant commonly known as the cleavers or goose grass, and that serve to indicate its chief char-

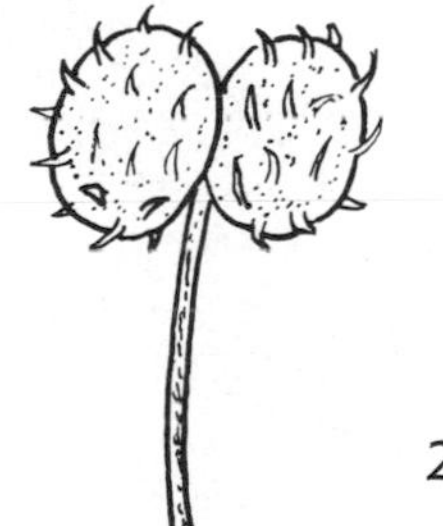

299. Fruit of Cleavers

acteristic—twin burs that steal a ride on every passing animal, including man. The cleavers is something of a worthless weed and occasionally a serious pest to the sheepherder because the tiny burs that become fastened in the wool of sheep mar the quality of the fleece. It is an annual with a weak, reclining stem, two to five feet long, the stem too weak to support itself so it climbs over other plants whereto it clings by means of backward-turning prickles. The light green leaves, one to two inches long, bristle-pointed, the margins and ribs rough with short, stiff hairs, are set in whorls of sixes or eights. About two tiny white flowers are borne on a stalk. They have a four-lobed corolla with four stamens inserted on the tube and two styles and are followed by small, twinned, globular burs beset with many short, hooked bristles. The cleavers grows in shady thickets and roadsides.

The cleavers belongs to a genus of plants commonly

known as the bedstraws, and itself is sometimes called a bedstraw. There are some twenty or more of these plants in the United States, most of them native to the South and East. They are low plants with four-angled stems, leaves in whorls, and small white flowers in open clusters. The sweet-scented bedstraw is a common species found in rich woodlands and whose foliage is fragrant after drying. The yellow bedstraw, naturalized from Europe, has yellow flowers instead of white, and grows in dry waste places along the borders of fields. The rough bedstraw, so-called because the leaves are prickly rough on the edges and ribs, prefers damp soil and its flowers are peculiarly odorous, perhaps unpleasantly so. And the northern bedstraw, more northern in its range than the others, is a smooth species and occurs in bogs, mossy woods, and wet shores.

An interesting sidelight on the bedstraws is that the name of the genus is *Galium,* Greek for a plant mentioned by Dioscorides, a Greek physician of the first century and the author of a standard work on *materia medica.*

With a fragrance that suggests the jasmine, the buttonbush, genus *Cephalanthus,* Greek for flower

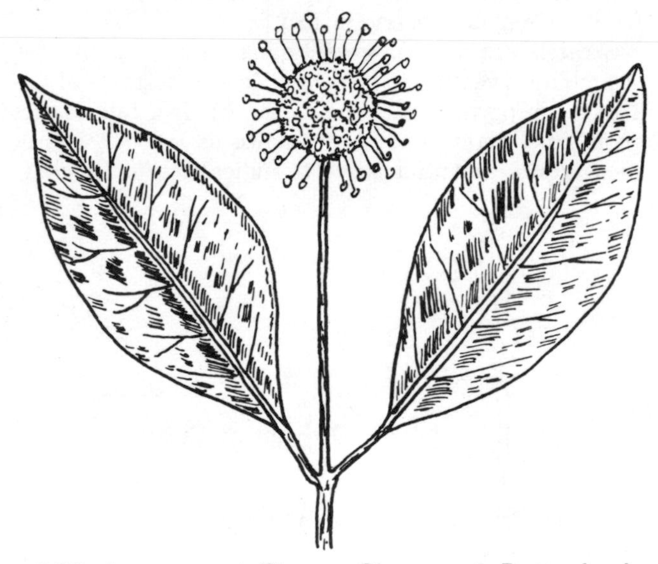

300. Leaves and Flower Cluster of Buttonbush

head, is a most attractive shrub, though its distinctive claim for attention is its dense, creamy white globes of blossoms that have aptly been called little "cushions of pins." To appreciate the perfection of detail in one of these "cushions," one has to examine it with a magnifying glass. But should one first, out of curiosity, start to count the number of florets, one may count a hundred or so and then become confused and tire of the exercise. By looking carefully at one of the florets, one will discover that its long tube is primarily adapted for the long-tongued insects, hence, butterflies are the most abundant visitors. But there are many others, such as the honeybees, bumblebees, mining and carpenter bees, flies, wasps, and beetles.

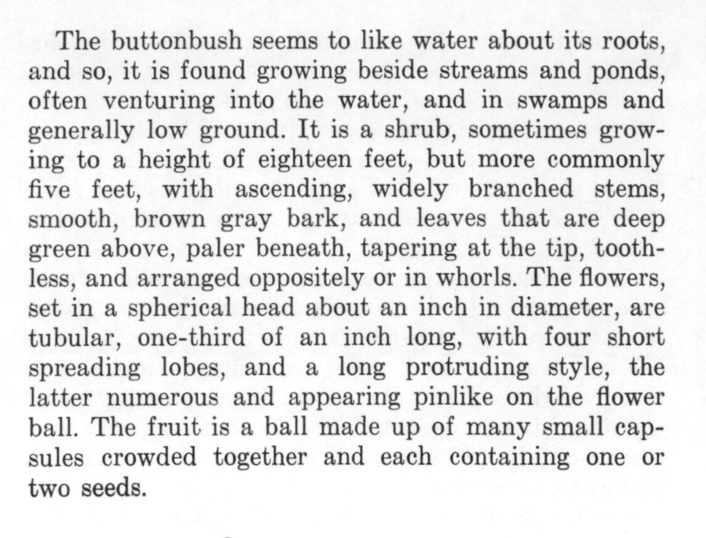

The buttonbush seems to like water about its roots, and so, it is found growing beside streams and ponds, often venturing into the water, and in swamps and generally low ground. It is a shrub, sometimes growing to a height of eighteen feet, but more commonly five feet, with ascending, widely branched stems, smooth, brown gray bark, and leaves that are deep green above, paler beneath, tapering at the tip, toothless, and arranged oppositely or in whorls. The flowers, set in a spherical head about an inch in diameter, are tubular, one-third of an inch long, with four short spreading lobes, and a long protruding style, the latter numerous and appearing pinlike on the flower ball. The fruit is a ball made up of many small capsules crowded together and each containing one or two seeds.

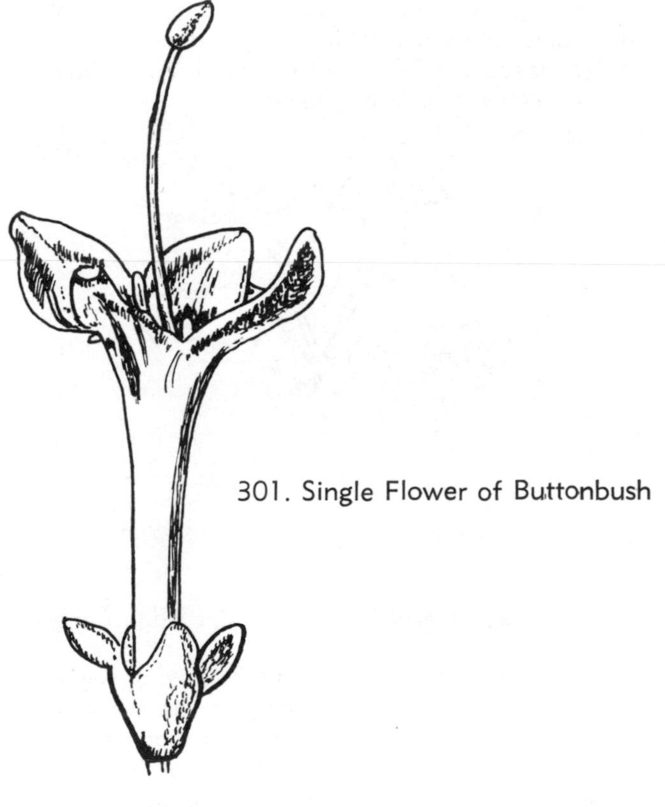

301. Single Flower of Buttonbush

Several European species of the madder family are old favorites in the garden. These are the sweet woodruff, a low-spreading perennial herb with white flowers and whose dried leaves are fragrant; the Dyer's woodruff, a red-rooted perennial with red or pinkish white flowers; and a species of bedstraw variously called galium, white bedstraw, baby's breath, and wild madder. It is a smooth-stemmed herb, erect or arching, with leaves in whorls of six or eight, and small, white flowers in practically leafless clusters.

As has already been said, the madder family is essentially a group of tropical plants many being of economic importance as the coffee, cinchona, and madder. The coffee plant, genus *Coffee,* the source of our beverage, is a shrub with bright green, oval leaves, clus-

302. Leaves and Fruits (Berries) of Coffee

303. Flower of Coffee

ters of fragrant white flowers, and dark red berries, the size of a cherry, and containing two seeds buried in a fleshy pulp, the seeds being processed and roasted to produce the beverage coffee. It is a native of Abyssinia, the name coffee likely to have been derived from the province of Kaffa, as in that country the coffee plant is known to grow wild. Coffee was in use in Abyssinia in the fifteenth century, but beyond that time, its history becomes somewhat obscure. However, it appears that about the beginning of the seventeenth century it was used as a beverage in Egypt, Arabia, and Turkey, but later prohibited by the Koran as an intoxicating drink. Despite the ban it became the national drink of Arabia. Somewhere about this time the use of coffee spread to Europe, and as a beverage, appeared in the coffee houses that began to spring up in London about 1652. Many regarded it as a public moral evil, but in spite of the efforts of prohibitionists, the coffee houses continued to survive and played an important role in the social and literary life of England during the following two centuries.

For a long time most of the coffee came from Arabia, but towards the end of the seventeenth century, the Dutch began to cultivate the coffee plant in the East and West Indies. From the West Indies it eventually spread to Brazil and Mexico, Brazil today producing most of the world's supply.

The quinine tree that for a long time was considered to be the best source of quinine, but now replaced by Javanese hybrids, is known scientifically as *Cinchona officinalis.* It is one of a number of species belonging to the genus *Cinchona,* the name given to a group of South American trees because the wife of the Count of Chinchon, a Spanish official in Peru in 1630, was cured of malaria by quinine obtained from the bark. The Cinchonas are shrubs and trees native to the Andes with leathery, evergreen leaves and fragrant, white or pink flowers. Today relatively little quinine is obtained from Peruvian trees, most of it being produced on plantations in India and Java.

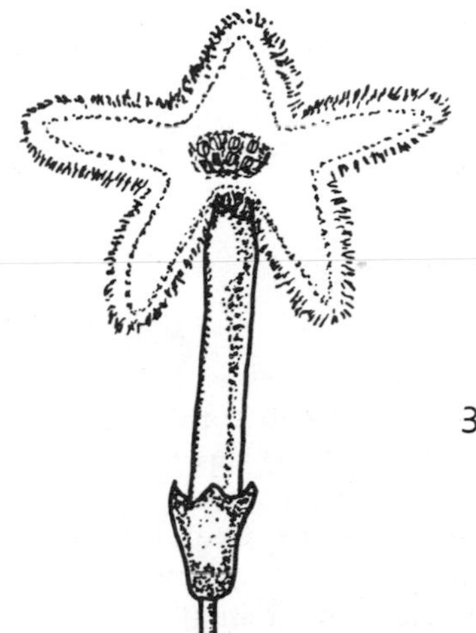

304. Flower of Cinchona

Until 1869 when the dye alizarin was synthetically made, the dye was obtained from the madder plant, a plant known from antiquity and widely cultivated. It is a European herb, with leaves in whorls of four to six, lance-shaped with prickly hairs on midribs and margins, and small greenish yellow flowers in branching clusters. The fruit is a small berry. For centuries it was the source of a brilliant red dye and of considerable commercial importance, and thus, the name of the family whereto it belongs, became known as the madder family for the scientific name of the family is *Rubiaceae* from the Latin *rubia,* madder, akin to *rubeus,* red. The madder, wild madder, or common madder, has been planted in gardens for the dye that is obtained by grinding the roots, and from our gardens it has escaped and may now be found growing in damp shady places. Lastly, mention should be made of the cape jasmine. It is the florist's gardenia, widely planted in the gardens of the South, and a native of China, but long supposed to have come from the Cape of Good Hope. Hence, its common name, though it is probably better known simply as the gardenia, named for Dr. Alexander Garden, a Charleston physician and friend of Linnaeus. When most people think of the gardenia, they have in mind the fragrant, roselike, white flower with waxy petals, not the shrub itself, which is a plant two to five feet high with glossy, thick, leathery green leaves.

There are several arborescent species of the family in the United States, such as the fevertree, a small tree, seldom more than twenty-five to thirty feet high, with a narrow, rounded crown of numerous, slender, horizontal branches, of stream margins, wet bottomlands, and swamps. The leaves are opposite, simple, deciduous, and elliptical to ovate, dark green above, and pale green below. The flowers are complete, trumpet-shaped, with a rose-colored calyx and a greenish yellow corolla, the calyx being the showier, and a light brown, papery, nearly globular, capsule. In early colonial days a decoction made from the bark of the fevertree was used in treating malaria. Actually it is a close relative of the Peruvian cinchona tree, wherefrom quinine formerly was obtained.

Several tropical species reach their northern limit in southern Florida. These are the Caribbean princewood, a shrub or small tree whose bark was once used in the preparation of a tonic to reduce fever, the roughleaf velvetseed, a small evergreen tree of dry and sandy soils near water; and the Everglades velvetseed, an even smaller tree found in hammocks on the Everglade Keys and neighboring islands.

100. Madeira Vine (Basellaceae)

The Madeira-vine family is a small group of tropical, herbaceous vines contained in six genera of wide distribution in the tropics. They have tuberous rootstalks and rather fleshy, alternate, entire leaves, regular flowers in racemes, the racemes sometimes branched, the individual flowers with two sepals, five petals that are separate or partly united, five stamens, a one-celled ovary with a single ovule, and a small fruit that is usually enclosed within the withered flower.

The Madeira vine, wherefrom the family derives its name, is a tall creeper with oval or heart-shaped, taper-pointed leaves, small white flowers in slender, drooping racemes, that are often twelve inches long and showy, and a fleshy fruit. Tubercles are produced in the axils of the leaves and, if planted in moist sand, will produce new plants. The Madeira vine, also known as mignonette vine from the fragrance of its flowers, of the genus *Boussingaultia,* named for J. B. Boussingault, a French chemist, is a native of tropical America, but is grown for ornament in warm regions and is an escape in Florida and Texas.

Another member of the family whose leaves are used for greens in the tropics, but cultivated in the north for ornament, is the Malabar nightshade. Belonging to the genus *Basella,* it is a moderately tall, but rampant climbing vine, with greenish purple stems, succulent leaves that are nearly round and somewhat heart-shaped at the base, and reddish flowers in short spikes. It is a native of tropical Asia and is mostly grown here in greenhouses, but may be started as a tender annual and planted outdoors after the danger of frost is past.

101. Magnolia (Magnoliaceae)

The magnolias are an ancient tribe, tracing their lineage to the Cretaceous period, some sixty million years ago. They are not unique in this respect, of course, for others, such as the sycamore, sweet gum, the birches, the oaks, as well as a number of others, have equally as long a history.

The magnolias are of more than passing interest and may easily be recognized because of several features. They have the largest flowers of any trees in cultivation. They may not be the showiest, for other trees, such as the apple or dogwood when in blossom, may be completely covered with blossoms. But few trees can match the size of the magnolia flower that often is six inches or even a foot in diameter. Then, too, the flowers of most magnolias are deliciously fragrant, the petals notable for their texture, thick, waxy, and lustrous and exquisite in their coloring. The leaves are also somewhat extraordinary, for they may be a yard or more in length and the mass of foliage is luxuriant and tropical in appearance. Finally their fruits are a curious conelike aggregation of follicles or samaras, and in summer and autumn, make the trees unusually attractive. They assume rosy tints as they ripen and when they open later, the

305. Leaves and Flower of Magnolia

scarlet seeds are suspended on slender, elastic threads.

The magnolia family comprises some ten genera and about one hundred species of trees and shrubs distributed in the temperate and subtropical parts of America and eastern Asia. Their leaves are commonly large, alternate, simple, often leathery, deciduous or persistent, and stipulate in the arborescent forms. Their flowers are large, often very showy, solitary, terminal or axillary, perfect, with many sepals and petals that are often similar, many stamens, and many carpels that are arranged in close spirals over the elongated conelike or spirelike axis. Their fruit is a group or aggregate of follicles or samaras, the seeds suspended on slender filaments.

306. Fruit of Magnolia

The magnolias belong to the genus *Magnolia,* named for Pierre Magnol, a French botanist of the seventeenth century. The genus includes some seventy to eighty species, eight being native to the United States. Only one of these eight species is an evergreen, the great-flowered magnolia, also known as the southern magnolia or evergreen magnolia. It is a tree of noble proportions, sixty to eighty feet high, sometimes higher, with a straight bole and somewhat pyramidal head, of southern swamps and bottomlands, but usually found on drier sites. It is often cultivated in the streets and gardens of the South.

The bark is brown gray, rough, with short thin scales, and the branchlets and buds are rusty-woollen when young. The leaves are large, elliptical or oblong, a lustrous deep green above, rusty-downed beneath and the flowers are lilylike or cup-shaped, white, very fragrant with usually six petals. The fruit is about four inches long, ovoid, and rusty-woollen.

Called sweet bay because its foliage is somewhat like that of the bay tree of the Old World, the sweet bay is a shrub in the North, but a slender tree up to seventy feet tall in the South and somewhat remarkable for its range, extending from Massachusetts along the coast to Florida and west to the Gulf States. It is a species of wet and swampy places, and hence, also called the swamp bay. It also has other names, such as white bay and beaver tree, the latter name because the beavers eat the fleshy roots and use the soft wood to build their lodges.

The bark of the sweet bay is light brown with small scales, the branches a gray brown, and the leaves are smooth, lustrous bright green with a bloom beneath and at first silky, soft-haired, oblong or elliptical, falling in autumn in the North, but mostly evergreen in the South. The flowers, appearing with the leaves are about two inches long, creamy white and very fragrant. The ovoid or oval fruit is dark red or deep magenta pink. One needs only to glance at the leaves, the flowers, or the fruit to know that the sweet bay is a magnolia. Which calls to mind these words of Whittier:

> Long they sat and talked together, . . .
> Of the marvellous valley hidden in the depths of Gloucester woods,
> Full of plants that love the summer, blooms of warmer latitudes,

Where the Arctic birch is braided by the tropic's
flowery vines,
And the white magnolia blossoms star the twilight
of the pines.

Magnolias are essentially Southern trees, but the cucumber tree, the hardiest of them all, is found as far north as the Canadian border. Somehow it seems out of place in our northern forests, for its large green leaves betray the tree's tropical character and are a sharp contrast to the smaller leaves of the oaks, elms, and maples. Although the yellowish green flowers become fairly large, measuring about two inches long, they are virtually lost among the leaves. Unlike the flowers of most magnolias, they are neither beautiful nor fragrant. The fruit at first is green and shaped like a cucumber, but it soon flushes pink and later turns red as autumn approaches. Then, as the fruit matures, the red berries within break through the skin and hang for a time on long white threads, eventually eaten by the birds or torn off by the wind.

The cucumber tree is a large, handsome, pyramidal tree of spreading habit, fifty to sixty feet high, occasionally higher, of rocky uplands, moist rich woodlands, often on mountainslopes, with a grayish brown bark broken into small, thin scales, slender ascending branches, oval leaves, and tulip-shaped flowers. The cucumber tree makes less of a show than other magnolias at blossoming time, for the flowers have little to commend them and the fruits are often distorted in form from the failure of the many carpels to set seed, and yet, the cucumber tree has many virtues not the least being that it grows vigorously from seed, and may readily be cultivated.

A relatively rare tree found in scattered or detached groups throughout the South, where it usually grows in rich valleys protected from the wind, the great-leaved or big leaf magnolia is notable for the size of its leaves and flowers. The leaves are almost a yard long and no simple leaf approaches it outside of the tropics. The flowers, too, are unusually large, creamy white, with a dash of purple at the base. Neither the leaves or the flowers, however, have a beauty to match their size, for the leaves are quickly torn to ribbons by the wind and whatever leaf or twig touches the petals mars them with a brown bruise.

Also rather rare and local throughout the South, the umbrella tree is a small tree growing to forty feet high with spreading branches and open head of wooded, hillsides and swamp borders, its closely thatched dome of glossy leaves suggesting an umbrella. Its bark is thin, gray, smooth with small blisterlike warts, and its leaves are very large, eighteen to twenty inches long, thin, pale green and softly haired beneath. Its flowers, appearing with the leaves, are cup-shaped, white, ten inches across, and have an unpleasant odor. The flowers are surrounded by an umbrellalike whorl of leaves and have the distinctive feature of three recurved sepals. This magnolia is perhaps the trimmest and neatest of them all.

The ear-leaved umbrella tree, a slender tree, twenty-five to thirty feet high, of mountain woods and streamsides of mountain valleys, is similar to the preceding species, except that its flowers are fragrant and its leaves are somewhat smaller and ear-shaped at the base.

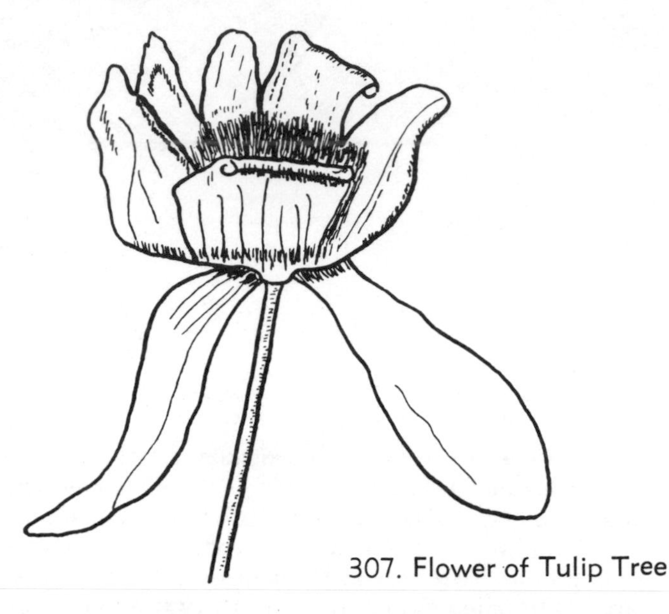

307. Flower of Tulip Tree

Place a garden tulip on a branch of the tulip tree and it could almost pass for the blossom of the tree itself. Large and showy, with dashes of red and orange on its greenish yellow, tulip-shaped corolla, the tulip blossom in no way fails to compare with the cultivated *Tulipa*. But one has to look closely at the flower to appreciate its distinctive charm. In the forest the tree reaches a height of one hundred and ninety feet, a magnificent size, but when it stands alone, it attains its finest growth, for then its trunk rises like a Corinthian column into a long, narrow, pyramidal crown of pleasing symmetry. Outlined against the sky, it towers over neighboring, oaks, elms, and maples.

The tulip tree, one of the most distinctive trees of eastern North America and one of the most valuable hardwoods from the forester's viewpoint, is not of the genus *Magnolia*, but of the genus *Liriodendron*, Greek for lily and tree, in allusion to the shape of its flowers. There are only two species, our native tulip tree and a species found in central China. The tulip tree is also known as the yellow poplar, because its peculiarly notched leaves on long, slender petioles tend to tremble in the slightest breeze like those of the poplar, and whitewood from the color of its wood. At one time it was known as the canoewood since the Indians habitually made their dugout canoes of its trunk.

The bark of the tulip tree is brownish gray, often with dark or sepia brown confluent ridges, the leaves are smooth, with four lobes, cut nearly square across the top, rounded toward the base, toothless and long-stemmed, dark green and lustrous, and turning a rich russet in the fall, and the flowers are terminal, solitary, showy, tulip or lilylike, with six greenish white

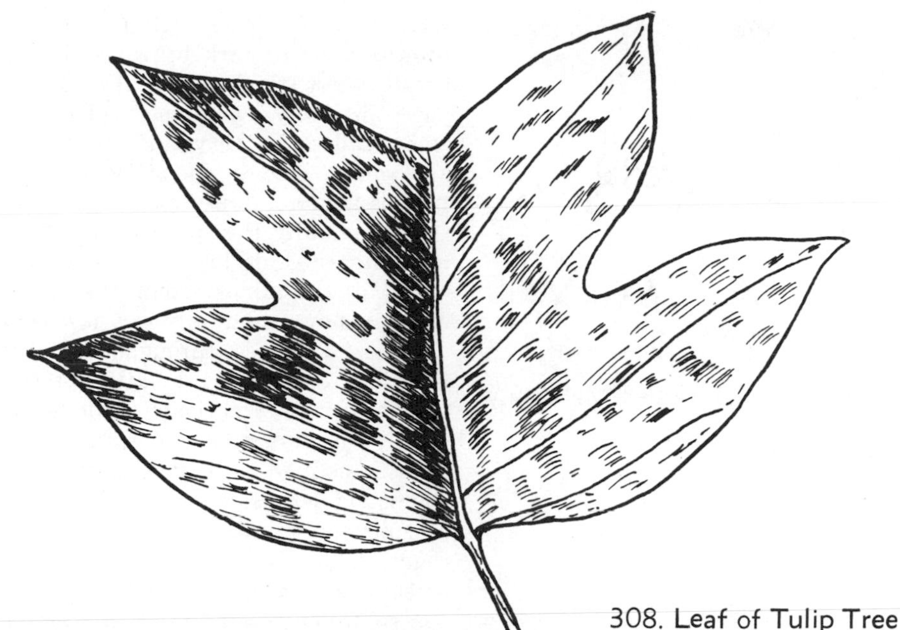

308. Leaf of Tulip Tree

petals, stained orange at the base, three greenish sepals, many stamens, and a one-celled ovary. The fruit is a narrow, light brown cone of many samara-like carpels, each enclosing one or two orange seeds that remain for awhile on slender filaments before falling.

The tulip tree, easily recognized when in leaf by the "chopped-off" ends of the leaves, is of interest and beauty whatever the season, and makes a magnificent lawn and shade tree.

The seeds of the tulip tree are eaten by several species of birds and squirrels and so, too, are the berrylike seeds of the magnolias. Generally speaking however, the wildlife value of the tulip tree and the magnolias is not very considerable. The caterpillars of the spicebush swallowtail number the tulip tree among their food plants, and the caterpillars of the palamedes swallowtail include the sweet bay in their dietary.

Several exotic magnolias are widely planted as ornamentals, three of them being the yulan magnolia, the purple magnolia, and the starry magnolia. The yulan magnolia, a native of China, that grows up to fifty feet high, has pure white, fragrant flowers that are bell-shaped and fully six inches across. The purple magnolia, also a native of China, is not much more than a shrub with relatively small, almost scentless flowers, the outside being purple. The starry magnolia, a native of Japan, is a multi-branched, spreading shrub and the earliest of the magnolias to blossom, the flowers appearing in March. Unlike most magnolias, the flowers are star-shaped, opening out flat instead of forming cups or bells, and when they open, measure three inches across. They have twelve to eighteen sepals and petals, twice as many as most magnolias have. It is a very fine species for the lawn, though the flowers appear so early they are sometimes destroyed by a late frost.

102. Mahogany (Maliaceae)

The mahogany family consists of some fifty genera and eight hundred species of trees, shrubs, and woody annuals most being tropical, and includes some of the finest known cabinet timbers. The leaves are mostly alternate and pinnately compound, deciduous or persistent, and often very large. The flowers are not showy except in certain species, and usually occur in many-branched clusters. The flowers are usually perfect and have a four to five-lobed calyx, the edges of the lobes overlapping, four to five petals, free or united to the stamens, eight to ten stamens, usually united into a tube, and a mostly two to five-celled ovary. The fruit is a capsule, drupe or berry.

Undoubtedly the best known representative of the family is the mahogany tree whose heavy, brownish red wood is highly valued for elegant furniture. There are some five or six species of mahoganies, of the genus *Swietenia,* named for Gerard van Swieten, an

309. Leaflet of West Indies Mahogany

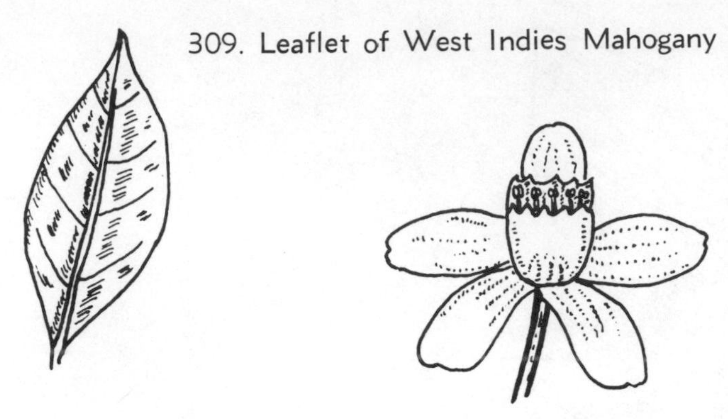

310. Flower of West Indies Mahogany

311. Fruit of West Indies Mahogany

Austrian botanist and physician, all native to tropical America, though there are many African and Philippine Island substitutes. The traditional mahogany is the West Indies mahogany whose northern range extends into southern Florida. It is a tall tree, although in Florida it rarely attains a height of more than forty to sixty feet, with a moderately thick, dark brownish red to dark brown bark, that is broken into broad, short, thickened scales. The leaves are alternate, pinnately compound and persistent, the four to eight leaflets leathery and toothless, and the flowers are small, inconspicuous, whitish in panicles. The fruit is a large, five-valved, dark reddish brown capsule. Originally the species occurred quite abundantly in southern Florida, but excessive cutting has resulted in the almost complete extinction of the tree and it is now found only in a few very localized areas.

Almost anywhere throughout the South, one may see the Chinaberry tree, a native of the Orient and introduced into American gardens long ago. It is a mostly deciduous tree of spreading habit, sometimes growing as high as fifty feet, with alternate, pinnately compound leaves, purple or lilac, fragrant flowers in conspicuous, terminal, compound panicles, and nearly round yellow berries, three quarters of an inch in diameter, and hanging after the leaves fall.

Also frequently planted in Florida and California as an ornamental is the cigar-box cedar, so-called because its wood is used for cigar boxes, or Spanish cedar. It is a tropical American species, growing one hundred feet high, with alternate, pinnately compound leaves, the leaflets ten to twenty-two, four to seven inches long, and remotely toothed, and yellowish green flowers, the clusters shorter than the leaves. A related species and a native of China, though not quite so tall and with greenish white flowers, is hardier and may be planted further north.

103. Makomako (Elaeocarpaceae)

The makomako family comprises seven genera and a hundred and twenty species of trees and shrubs that are mostly tropical. Some species yield excellent timber, and the drupaceous fruit of others is used as a pickle or in curries. Two species of the genus *Aristotelia,* named for Aristotle, are cultivated in southern California for ornament, one a Chilean shrub, and the other also a shrub or small tree known as the makomako, but also called the New Zealand wineberry. Both have opposite, evergreen leaves and polygamous flowers, with four to five sepals, four to five petals, many stamens, and a pulpy, edible fruit. The makomako is a shrub or small tree, that in New Zealand forms hazellike copses, eight to twenty-five feet high, with oval or nearly oval leaves, small rose-colored flowers in profuse clusters, and a fruit that is at first red, but later changes to a purplish black

104. Mallow (Malvaceae)

Because its fruit is round in form like that of a cheese, and when green, is mucilaginous and sweet and eaten by children of another day, the plant is known by the name of cheeses. Like other wild flowers it has other names, too, such as the common mallow, the

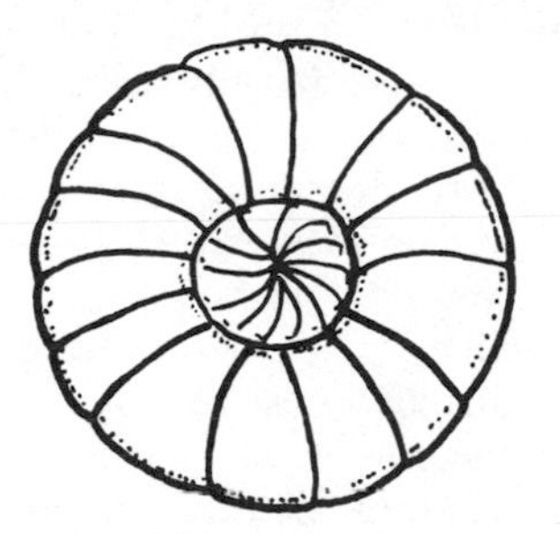

312. Fruit or Seed Vessel of Common Mallow

running mallow, the dwarf mallow, and the round-leaved mallow. A native of Europe, it is an escape like so many other plant immigrants, and today, it is common in waste places and in dooryards and gardens where it is considered a weed. It has deep, branching roots that seem to spread nearly as far

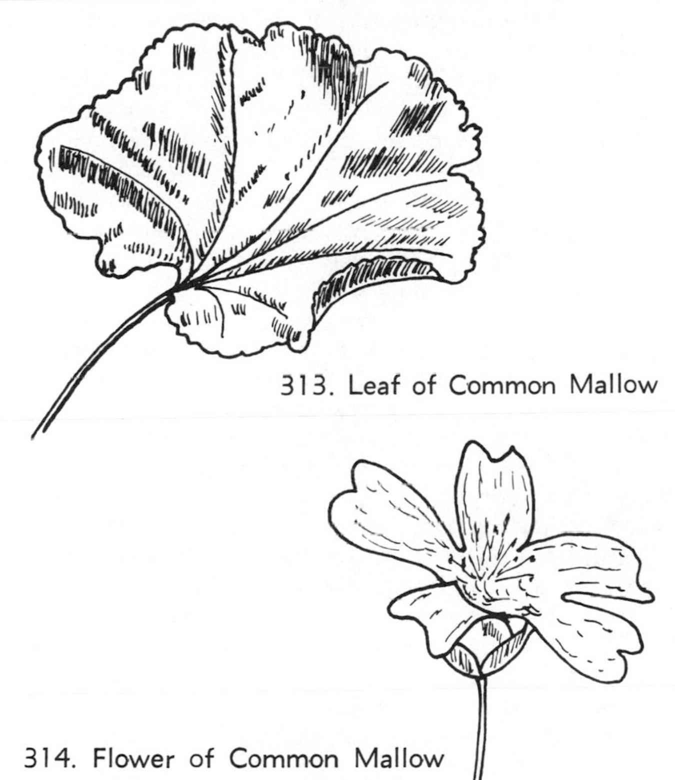

313. Leaf of Common Mallow

314. Flower of Common Mallow

beneath the ground as do its creeping stems above it. The stems are six inches to nearly two feet long, with rounded or kidney-shaped, dark green leaves that usually have five, shallow, scalloped-shaped lobes, irregularly toothed, and on long stalks. The flowers are pale pink, veined with deeper pink, and are clustered in the leaf axils. If one looks at them with a magnifying glass one will find that they bear a close resemblance to hollyhocks. Pythagoras valued it as a spinach, as did other Greeks and Romans as well. It is cultivated as a potherb in Egypt, and in France and Italy the tender shoots are eaten as a salad.

There are thirty species of mallows of the genus *Malva,* Greek for emollient, in allusion to the mucilaginous juice of some species. They have alternate, usually angled, lobed, or dissected leaves, solitary or clustered flowers, usually with three or two involucre bracts beneath them, a five-cleft but united calyx and five petals. The fruit is a collection of united carpels forming a cheese-shaped cluster.

A species of mallow whose seeds are eaten by country people and that is also a desirable vegetable when boiled is the high mallow, also an adventive from Europe and found growing on roadsides and in waste places. It has an erect, branching stem, slightly finely haired, though sometimes smooth, with leaves a fairly light green, long-stalked, toothed, and five to seven-lobed, and rose purple flowers, wherefrom the French have derived their word mauve, first applied to this plant. A very similar, but a more handsome plant, and an escape from gardens, the musk mallow is so-called because its medium green leaves that are deeply slashed or cut have the delicate odor of musk when crushed. Its white or very pale magenta pink flowers, nearly two inches broad and flat, are borne in terminal clusters. The mallows are fairly rich in nectar and are visited by various insects, cross-fertilization being effected mostly by bees and butterflies. The species of *Malva,* the true species of mallows, serve as food plants for the gray hairstreak (butterfly).

Growing in the salt marshes of the Atlantic Coast from Massachusetts to New Jersey where it relieves the drab monotone of the marsh with its pinkish flowers, the marsh mallow is of more than passing interest because its thick, mucilaginous root is used in Europe, in the making of confectionery. It is a perennial herb, three to four feet high, more or less downy, with usually three-lobed, generally ovalish or heart-shaped leaves, and flowers that are solitary, or at best a few together, in the leaf axils.

The marsh mallow is a very close relative of the hollyhock of old fashioned gardens, both belonging to the same genus *Althaea,* the Greek name of the marsh mallow. The hollyhock is a native of China and long in cultivation in that country when discovered by Europeans. It is the Holy Mallow brought into western Europe, it is believed, by the Crusaders.

The hollyhock is essentially a perennial herb, but grown mostly as a biennial and even as an annual. It is a plant of vigorous growth and ornamental character with an erect, leafy stem, spirelike and hairy. The leaves are generally round, rough, wavy-angled on the margin, or shallowly five to seven-lobed, and the flowers, essentially stalkless, are in long, stiff, wandlike terminal clusters. They are large, typically single, and red or white, but in the horticultural forms, they are of many colors and often double. The life of a hollyhock blossom is from three to five days and usually there are a number of blossoms in different stages of bloom on a well-grown stalk. The caterpillars of both the painted lady and the checkered skipper include the hollyhock in their dietary.

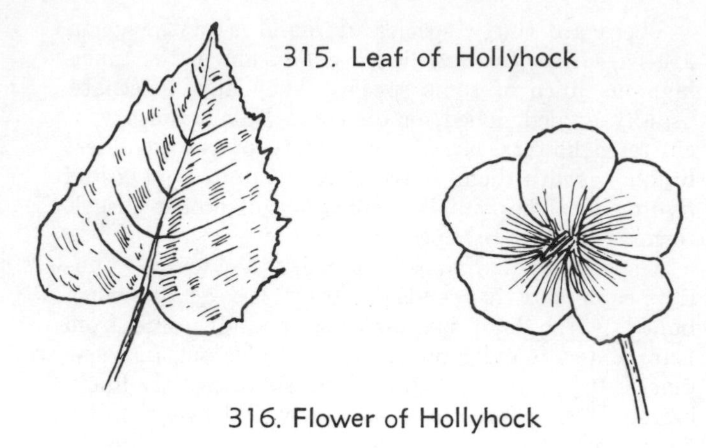

315. Leaf of Hollyhock

316. Flower of Hollyhock

The rose mallow, or swamp mallow, or sea hollyhock, to give the plant its various names, can be transplanted from the wild into the garden with every hope of success, though it does best in the salt marshes, where it grows among the tall sedges and cattails. Strangely enough this plant, used to living in a salty, wet soil, does uncommonly well in ordinary garden soil and with no more moisture than it can receive from showers and the garden hose.

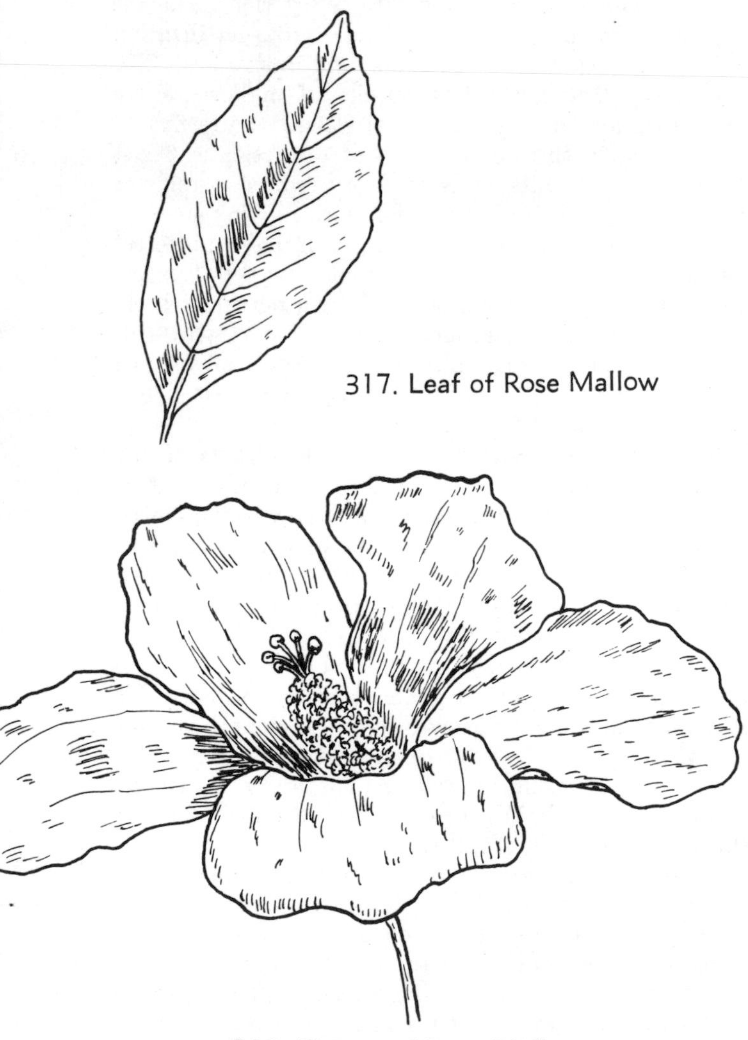

317. Leaf of Rose Mallow

318. Flower of Rose Mallow

The rose mallow is a perennial herb, three to seven feet high, with stout, shrublike stems, olive green leaves, bright above and densely white beneath, ovate pointed on long stalks, and flowers four to six inches across, with five broad petals, pale crimson pink or white, borne singly, or in scant clusters.

The rose mallow is one of about two hundred species of the genus *Hibiscus,* a group of herbs, shrubs, and trees. It is a genus of considerable horticultural interest because it includes many garden annuals, plants that yield perfume, a few food, many showy perennials, some shrubs, and brilliantly colored tropical trees. They have alternate leaves, with pinnate veins, and that are sometimes lobed or parted, usually large flowers, mostly bell-shaped, of five petals and sepals, the sepals sometimes united to form a five-toothed calyx, the stamens united to form a tubular structure that surrounds the style, and a dry, five-valved capsule.

Some of the many species of the genus, in addition to the rose or swamp mallow, include the crimson-eye rose mallow or white rose mallow, a plant of the coastal marshes, with white flowers marked with a crimson or purple center, and perhaps the source of the eye in the improved mallows; the halberd-leaved rose mallow, a smooth species found on the banks of rivers and small streams, with flesh pink flowers, and sometimes with a dark magenta center; the rose-of-China or China rose, a gorgeous, tropical Asiatic shrub, with flowers typically a rose red, flaring and spectacularly showy because of its petals and the long column of stamens, widely planted in the frostfree parts of the United States; the rose-of-Sharon, the only really hardy shrub of the genus and highly valued in the garden for its late bloom, its flowers broadly bell-shaped, red, purple, violet, or white, and exceptionally showy on dark or cloudy days; the flower-of-an-hour, a garden annual, with pale yellow or yellowish white flowers with a dark eye; the cotton rose, sometimes called Confederate rose, a native of China, a shrub or sometimes treelike, with opening white or pink flowers, but soon changing to a deep red, a common bush in the South; and okra or gumbo, a garden vegetable grown for its immature, mucilaginous pods, that are ribbed and beaked, four to twelve inches long.

The mallow family, whereof all the foregoing plants belong, is a large family of some forty to fifty genera, and perhaps one thousand species of herbs, trees, and shrubs that are scattered throughout the world excluding the cold regions. They have alternate leaves that may be entire or deeply lobed or cut, sometimes dissected, and often have showy flowers that are usually perfect and regular, commonly with five sepals and five petals, both free, or the sepals sometimes united, many stamens that are united to form a tube that surrounds the styles, and five or many carpels, one to several ovules in each carpel. The fruit is typically a dry capsule, breaking into segments as it ripens. The family includes many garden ornamentals, a few foods, and others of considerable economic value, such as cotton.

One of the many genera of the mallow family is *Malvastrum.* It is a large genus of American and South African summer-blooming herbs. One species, the red false mallow or prairie mallow, is a woody-based perennial herb, eight to twelve inches high, silvery haired as a rule, with leaves oval in outline, but deeply three to five-lobed, the segments again incised, and brick red flowers, about an inch wide, in close, terminal clusters. It is a plant of dry prairies, hillsides, and wild pastures in the central states, but sometimes grown in the flower garden. The yellow false mallow is another species with narrowly elliptical leaves and yellow flowers growing on dry rocky hillsides from Alabama northward into Kansas. On the Pacific Coast there are also several species, such as the attractive desert five spot, with hairy rounded leaves and rose pink petals each marked with a darker spot and the bush mallow, a tall shrubby species with pink flowers found on the hillsides of southern California.

Rather well-liked for their showy flowers are the poppy mallows, natives of the prairie regions of the Central States. A representative species, often planted in the garden, is the purple poppy mallow, a perennial with a deep rootstock, leaves three to five-parted, finger-fashion, the segments mostly wedge-shaped, and large, showy purple crimson or magenta flowers.

Essentially tropical American hairy shrubs or shrubby herbs, only a few species of the wax mallows, as they are known, are grown for ornament outdoors in the deep South. They have alternate, more or less heart-shaped, somewhat lobed leaves, and showy red flowers, somewhat resembling a fuchsia. The fruit is at first berrylike and sticky, but later becomes dry and separates into segments.

Originally from India, the Indian mallow has become well-established in cultivated ground, waste places, farmyards and similar situations. It is a coarse herb with a stout, erect, round, softly haired stem, three to five feet high, with hairy, roundish or heart-shaped, tapering leaves, and yellow solitary flowers. The inner bark of the plant yields, a fine, strong fiber that may be made into twine, rope, or paper, hence, it is also called the American jute that, of course, being a misnomer.

The Indian mallow belongs to the genus *Abutilon,* called the flowering maples because of the similarity of their leaves to those of the maples. There are some hundred species of these plants, also known as Chinese bellflowers. They are tropical shrubs, only a few being grown in greenhouses. Those that are herbs are used as bedding plants in frostfree areas.

Another immigrant from the tropics, but now an inhabitant of fields, pastures, and waste places, the prickly sida gets its name from a spinelike protuberance found at the base of some of its leaves. It is a low, branched plant, with stems eight to twenty inches high, erect, downy-haired, with scalloped-toothed, ovate to lance-shaped leaves, and small yellow flowers on short axillary peduncles. The prickly sida occurs throughout the central and eastern states, and a few related species are found in the southeastern states, one of them being known as the paroquet bur because of its hooked carpels that catch in the fur of animals.

The globe mallows, another group of the mallow family, are mostly natives of tropical and subtropical America, but there is a native western species of stream margins called the maple-leaved mallow. It is an erect perennial with circular leaves, palmately five to seven-lobed or cleft, and pink or white flowers, on short stems, clustered in the upper axils or in terminal spike-like racemes. It has a very limited range occurring from the Dakotas west to Nevada and British Columbia.

Most people are inclined to think of the cotton plant as a single species, but actually there are some thirty species, some of them woody herbs, but others treelike, such as the tree cotton, an erect plant eight to ten feet high, whereof Pliny writes:

> The upper part of Egypt toward Arabia produces a shrub which some call gossopion and others xylon, whence the name xylina given to the threads obtained from it. It is low growing and bears a fruit like the beaded nut, and from the interior of this is taken a wool for weaving. None is comparable to

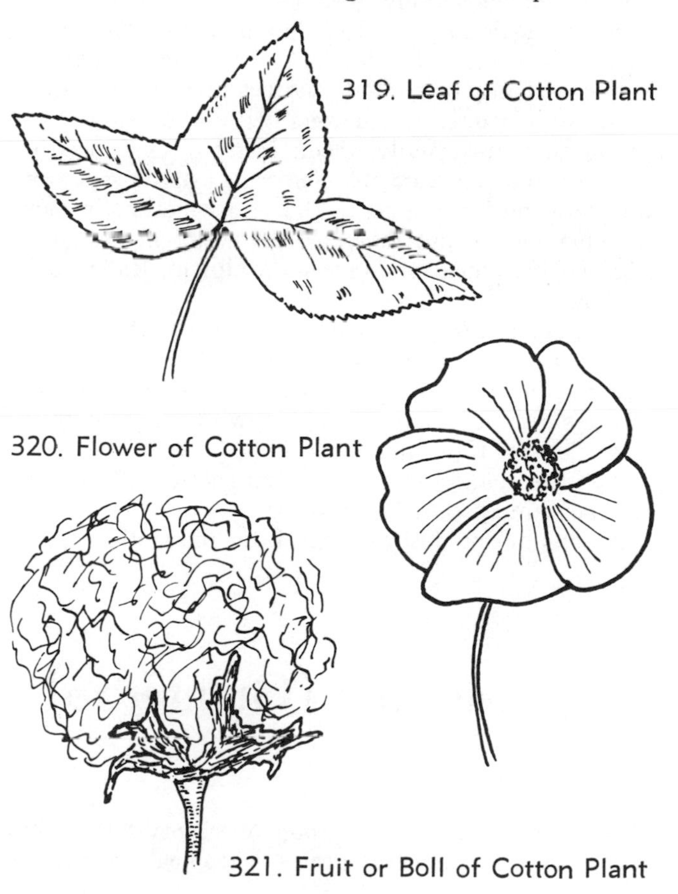

319. Leaf of Cotton Plant

320. Flower of Cotton Plant

321. Fruit or Boll of Cotton Plant

this in softness and whiteness. The cloth made from it is used by preference for the dress of the Egyptian priests.

The various species of cotton plants are of the genus *Gossypium,* a very old name for the cotton plant. They have alternate, more or less lobed or ribbed finger-fashion, often dotted, leaves, and large, showy, generally white, yellow, or pinkish purple flowers that often change color after opening, and usually are solitary in the leaf axils. The fruit is a capsule (the cotton boll) containing the seeds that are attached to the fibers (cotton).

It is not known when the cotton plant was first cultivated. Cloth made from it has been found in five thousand year old ruins in India, but the plant was not cultivated in the Mediterranean region until the beginning of the Christian era. The true value of cotton as a textile fiber did not become known until the Middle Ages. Columbus found the American aborigines using cotton cloth, and the Spaniards found cotton growing in Mexico, Central and South America. Coronado found it under cultivation in what is now Arizona and New Mexico.

In the early days of this country, cotton was not widely used as it was a laborious process to remove the fibers from the seeds by hand. With the invention of the cotton gin, however, cotton growing became a staple and a paying crop. About two-thirds of the world's supply today comes from the United States, nineteen of the states considering it a leading crop. In addition to the fibers, the cottonseed oil extracted from the seeds, is used in making margarine butter substitutes, table oils, and soaps. The "cake" left after the oil has been removed is used as cattle food. Another important use of the cotton plant is as a source of cellulose acetate products, such as plastics used to make films, molded articles, and foamed insulation.

105. Malpighia (Malpighiaceae)

The malpighia family is a tropical group of about fifty-five genera and over six hundred and fifty species of trees, shrubs, and vines. They have mostly opposite leaves and nearly regular, perfect flowers in various kinds of clusters, the flowers with five sepals, five petals that are usually unequal and often fringed or toothed, and ten stamens. The fruit usually separates into three nutlike segments, that are sometimes winged in some genera, but more or less fleshy in the genus *Malpighia,* named for Marcello Malpighi, Italian naturalist.

The genus *Malpighi* comprises over forty species, one being the Barbados cherry, grown in southern Florida for its acid fruits that are used in preserves. It is a shrub, five to ten feet high, with oval or narrow, spiny leaves, rose pink flowers in short-stalked, axillary clusters, and red drupes about the size of a cherry and containing three stones.

A popular and desirable shrub for foundation planting is the thryallis, one of fifteen species of the genus of the same name, that are scattered from Texas and California to Brazil. It is a neat, low shrub, three to five feet high, with opposite, entire, more or less oblong, bluish green leaves, yellow flowers in slender-stalked clusters, and a three-valved capsule. Though the flowers are not conspicuous, they do add a bit of color to any planting while the fine, wiry, glossy, red brown stems contrast nicely with the foliage.

Another member of the family, known as the butterfly vine, is also cultivated in the Deep South. It is a slender-stemmed, woody, twining vine of the genus *Stigmaphyllon,* Greek for stigma and leaf, in allusion to the leaflike stigmas, with smooth, heart-shaped, hairy-margined, bright green leaves, large bright yellow flowers in umbellike clusters growing in the axils of the leaves, and a one to three-celled winged fruit. It is a native of the West Indies and South America.

106. Mamey (Guttiferae)

The mamey family is a group of twenty-five genera and about five hundred species of tropical, resinous or aromatic trees and shrubs, with opposite, entire leaves and rather showy, sometimes waxy flowers that are solitary or in clusters. The flowers are mostly dioecious usually with four sepals and four petals, but sometimes two or six, many stamens, and a two or more-celled ovary. The fruit is fleshy, and usually edible. Many species yield useful timber, the resins of others afford pigments, and the fruits of the mangosteen and the mamey, or mammee apple, are important.

The mangosteen, highly prized in the Indo-Malayan region where it is native, and grown occasionally in extreme southern Florida, is a handsome tree up to thirty feet tall, with opposite, thick, leathery, shining green leaves, rose pink flowers nearly two inches wide, and a reddish purple fruit, its outer rind thick, but the five to seven orangelike, white segments are juicy and of superb flavor. It belongs to the genus *Garcinia,* named for Laurence Garcin, a French botanist.

Like the mangosteen, the mamey or mammee apple of the genus *Mammea,* from mamey, the West Indian name for the cultivated species, is also prized for its fruit in the tropics where it is cultivated for this purpose, and to a limited extent in the warmest parts of southern Florida. It is a tree, forty to sixty feet high, with opposite, thick, glossy leaves, white, fragrant flowers about an inch wide, and a nearly round drupe, the rind russet, the pulp yellow, sweet in the best varieties, but somewhat acid in the poorer ones. The fruit is nearly six inches in diameter and contains one to four large seeds.

Another useful member of the family is the calaba, also called Maria and Santa Maria tree, a very beautiful shade tree in the tropics, but of value only in warm frostfree regions and somewhat grown in the most sheltered parts of Florida. It is a tall branching tree of the genus *Calophyllum,* Greek for beautiful leaf, in allusion to the handsome foliage, up to one hundred feet high, providing a very dense shade, with oval-oblong, bright shiny green, leathery leaves, white, fragrant flowers in small clusters (racemes), and a fleshy fruit about an inch in diameter.

107. Mangrove (Rhizophoraceae)

The Florida Everglades have trails that lead into mangrove swamps, and it is here that one will see close at hand a most remarkable tree. Saltwater is generally lethal to land plants, but the mangrove is not only able to live in it, but is nourished by it as well.

Another odd facet of this most peculiar tree is that a mangrove seedling does not develop in the ground—the seeds germinate within the ovary on the tree where the seedling grows like a twig until it is about a foot or more long, then dropping like a plumb bob into the water. This is an illustration of *vivipary,* a phenomenon uncommon among plants.

A close look at the twiglike embryo plant reveals it to be a complete plant, tinged with photosynthetic cells. The twig is actually a woody tube and the future trunk, the pointed tip the radicle or the beginning of the root system, and the opposite end the potential leafy shoot. The pointed tip is admirably adapted to pierce the mucky soil and to anchor the seedling. Once the seedling is firmly secured it begins to send forth arching roots that soon develop into a maze of prop roots and an aerial leafy stem above the high water level. The development of the mangrove seedling is a most unusual and interesting example of a plant's adaptation to its environment.

The prop roots, aside from their normal function of anchoring and supporting the plant, also act as a sort of strainer and collect all manner of debris from the sea—silt, mud, driftwood, seaweed, shell fragments, and pieces of coral. Such materials, gathered by the roots of many trees, in the course of time build up a solid terrain. Then the mangroves migrate farther into the sea and the island forests advance and take over the ground formerly occupied by them, as lakes and ponds are similarly transformed into dry land.

The mangrove swamp forests that rim the southwestern coast of Florida, are living breakwaters that protect the shore from erosion and by collecting mud, peat, and other substances, extend the continental land mass into the sea. Most of the mangroves are small trees or spreading shrubs, but there are some, especially those that grow in the estuary of the Shark River, that are more than two feet thick and sixty feet high. These swamp forests provide a sanctuary and nesting sites for terns, pelicans, wood storks, egrets, spoonbills and herons. Various species of epiphytes or air plants grow on the tree branches, and certain oysters that have become adapted to living out of water at low tide and called "coon oysters" because raccoons are especially fond of them, cover the partially exposed roots with their white shells. Crabs, too, may be seen scurrying among the branches as well as ant nests that mark the level of high tide.

Sometimes when a seedling falls into the water, it fails to take root and then drifts with the water currents. It can remain viable for a long time, and if eventually it reaches a sandbar or mud flat soft enough for the pointed end to become embedded, it will begin to send out arching roots that take hold in the ridge of land. Soon a maze of roots will begin to collect the debris of the sea, and before long, a small island, perhaps not more than a few square feet in area, will be built.

If swirling currents and hurricanes do not uproot the mangrove before it matures, it will eventually become festooned with seedlings that will drop into the water and either be carried away by the currents to colonize other sandbars or mud flats, or become embedded in the debris collected by the parent tree where they will sprout and grow and send out arching roots that will collect more debris and add to the little island.

There are three species of mangroves, all small trees, distributed along the tidal shores of the tropics. They belong to the mangrove family, a group of about twenty genera and sixty species of tropical, evergreen, maritime trees and shrubs. They have opposite, evergreen, commonly leathery leaves, perfect flowers, in axillary clusters, with four or five or more sepals, four to five petals, the number of stamens equal to or more numerous than the petals, and a two to six-celled ovary, and a fruit that may be either a berry, or a dry or somewhat fleshy capsule.

The mangrove that is found on the coast of Florida is the red mangrove. It is a round-topped tree, fifteen to twenty-five feet high, but in the tropics, much higher, with a columnar bole and drooping aerial roots. The bark, that yields tannin, is reddish brown or gray, irregularly broken by shallow fissures. The leaves are simple, ovate to elliptical, toothless, dark green and shining above, and paler beneath. The flowers are perfect, axillary, two to three on a short stalk, yellow, and hairy inside, while the fruit is a leathery, rusty brown, conical berry.

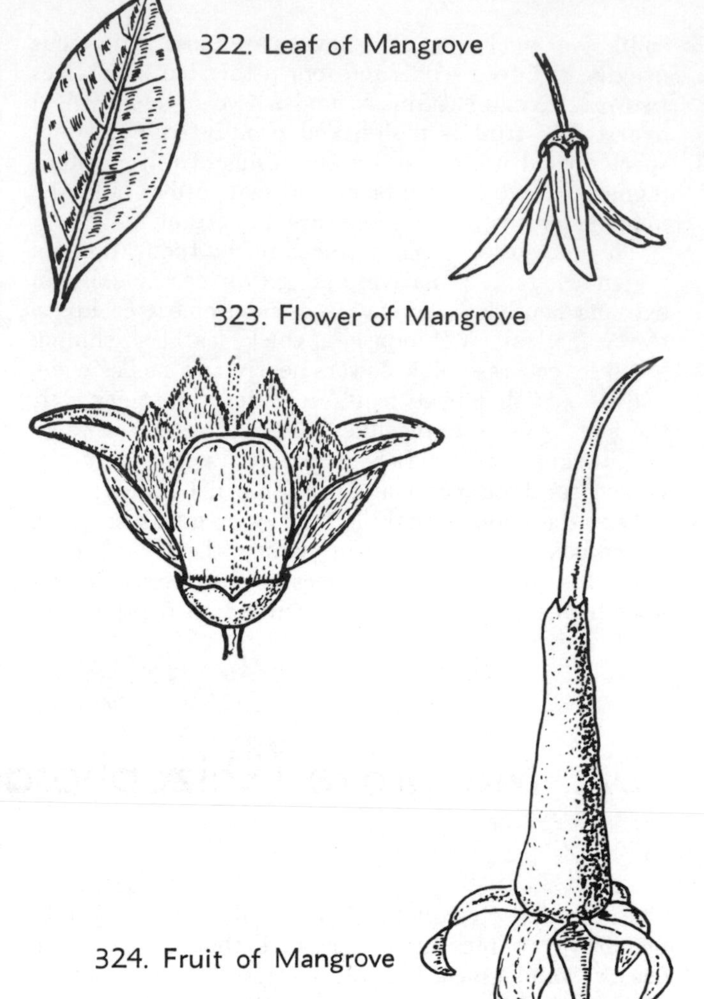
322. Leaf of Mangrove

323. Flower of Mangrove

324. Fruit of Mangrove

108. Maple (Aceraceae)

The signs of a returning spring are many, but for the author, spring does not really arrive until the red maple is in blossom. It is because he has lived in New England for most of his life where the tree is an exquisite barometer of the seasons. In the winter its red twigs against the whitened landscape form a pleasing contrast; in the spring its crimson or scarlet blossoms flash against the blue sky; while in summer its leaves swing on crimson and scarlet stems. Even the new growth is aflame in the same colors. Then in September, its leaves begin to show touches of crimson heralding the approach of autumn—a season wherein it stands preeminent amidst the brilliancy and splendor of Nature's annual pageant.

If one is observant one may notice that some red maples appear very red when in blossom, while others appear yellowish. This distinction in color is due to the flowers. The red maple is essentially dioecious, that is, it bears staminate and pistillate flowers on separate trees, but this is not a strict rule, for a branch with staminate flowers can be found on a tree with pistillate flowers, and pistillate clusters may occur on a tree with staminate flowers. Even perfect flowers may sometimes be found. If one were to examine the flowers closely one would find that the reddish flowers are the fertile ones, whereas those fringed with yellow stamens and orange colored are the sterile ones.

The red maple (few trees are more aptly named) is a spreading symmetrical tree, forty to fifty feet high, of swamps, low, wet woods, and stream borders, but widely used as an ornamental and shade tree. Its bark is smooth, light gray brown on young trees, very dark gray brown and furrowed into long ridges and sometimes shaggy on old trees. Its leaves, remarkably red in the bud, are three to five-lobed, deep green or light green above, pale or whitish green beneath, double-toothed or triple-toothed, on long stems usually bright red above, turning a rich red in autumn. The flowers appear before the leaves and the fruit is a

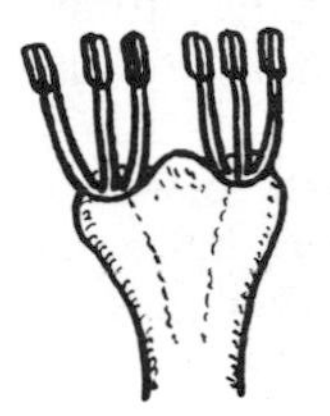

325. Staminate Flowers of Red Maple

326. Pistillate Flowers of Red Maple

two-winged samara that is bright red when young, in long-stemmed drooping clusters and ripening in spring.

Emerson knew the red maple:

> The scarlet maple-keys betray,
> What potent blood hath modest May.

And so too, did Lowell:

> The maple puts her corals on in May,
> While loitering frosts about the lowlands cling,
> To be in tune with what the robins sing,
> Plastering new log huts 'mid her branches gray;
> But when the autumn southward turns away,
> Then in her veins burns most the blood of spring,
> And every leaf, intensely blossoming,
> Makes the year's sunset pale the set of day.

The maple family has only two genera, the genus *Dipteronia,* comprising two small trees of central China, and the genus *Acer* the ancient Latin name of the maple, consisting of one hundred and forty-eight species of trees and shrubs widely scattered throughout the northern hemisphere, the greatest number of species occurring in the eastern Himalaya Mountain Range and in central China.

Their leaves are deciduous, opposite, simple or palmately lobed or pinnately compound, and their flowers are in axillary or terminal corymbs, cymes, racemes, or panicles. They are small, commonly unisexual, with four to five sepals and petals, four to ten stamens, and a two-celled ovary with one or two ovules. The fruit is a two-winged samara.

As early as the flowers of the red maple may appear, those of the silver maple precede them. This is a large and ornamental tree, sixty to eighty feet high, sometimes as much as one hundred and twenty feet, with beautifully incised leaves and with a trunk that soon divides into three or four stout, upright, secondary stems, forming a wide spreading head with drooping branches. In its poise and outline it somewhat suggests the elm. It is a tree of rich, moist soil, found on riverbanks or in lowland woods, with a smooth, gray bark with a brown tinge on young trees, ruddy brown, rough and scaly on old trees, and with five-lobed leaves, the incisions often very deep and sharp at the base, deep green above, and silvery bluish, green white beneath. The leaves turn upward with every passing breeze and give the foliage a broken whitish color, hence, the name silver or white maple, as it is also called, and that serves to distinguish it from other maples. They turn yellow in the fall. The flowers lack petals, but are otherwise similar to those of the red maple, and the samara has widely spreading wings and is downy when young. It ripens in early spring.

In the autumn one never knows what the sugar maple will do. It may have leaves of yellow or crimson, scarlet or orange, green with a spot of crimson, or crimson with a spot of pink. The leaves may even have a patchwork of yellow, purple, and scarlet.

The sugar maple is a commercially useful tree, in the northern part of its range producing the sugary sap wherefrom maple syrup and maple sugar are made. At times it produces the peculiar grain of "curly maple" and "bird's-eye maple."

The sugar maple is a large handsome tree, fifty to seventy feet high, and in the forest sometimes one hundred and ten feet high, branching profusely with many upright branches to form an oval or oblong head. It is a tree of rich woods and rocky hillsides, with a light brown bark, frequently ocher-tinged, deeply furrowed into long coarse, shaggy flakes or plates, and slender, smooth, yellowish brown, dotted twigs. The leaves are broad and glossy green, of three to five lobes with toothed margins, and the flowers are greenish yellow appearing with the leaves. The fruit, a samara, or maple key as it is often called, ripens in the fall. It is somewhat horseshoe-shaped, with nearly parallel or slightly divergent wings, and is borne on slender stems in short terminal clusters.

There is a southern sugar maple and also a black sugar maple. The southern sugar maple is a somewhat smaller tree with a more spreading habit, and with

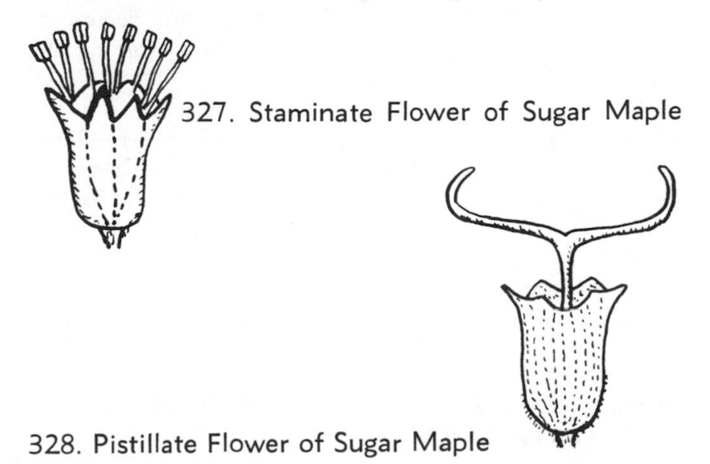

327. Staminate Flower of Sugar Maple

328. Pistillate Flower of Sugar Maple

smaller leaves that have a somewhat undulating margin. It grows along swamps and streams throughout the Gulf States. The black sugar maple, also known simply as the black maple, is also similar to the sugar maple. Its bark, however, is more corrugated, and its leaves are usually three-lobed, somewhat pubescent, and with a drooping habit, as if the stout petioles were too weak to support their burden. Perhaps the most distinguishing all year-round character is the orange color of the stout branchlets that, incidentally, are stouter than those of the sugar maple, with conspicuous warty lenticels. It is a tree of rich soil and a sugar tree.

The Indians knew of the properties of the sugar maples. Even before the settlers came to America the Indians, especially the Ottawa Tribe, collected the sap and made sugar-making a festive and ceremonious occasion. One of the earliest accounts, written about 1700, tells how the Indians gathered the sap in bark or skin vessels, and boiled it down in a rather crude way. Schoolcraft visited the Indians along the south shore of Lake Superior in 1823 while they were making sugar from the sap of the sugar trees. They kept the sap in large oxhide vats, but did the boiling in kettles. They appeared to be so fond of the sugar that they kept very little of it beyond the sugar-making season. Eventually the primitive methods of the Indians were displaced by the scientific processes of the white man.

The striped maple can easily be recognized by its striped bark, by the striking white streaks that run lengthwise on the somewhat smooth, ruddy brown or dull greenish trunk. They are more evident on young trees, but persist on comparatively old trees. The striped maple is a small tree, fifteen to thirty feet high, sometimes not much more than a shrub, and common in cool, rocky or rich woods, often on mountainslopes. It has a short trunk, slender, upright branches, smooth, olive green twigs, and extremely large leaves, four to seven inches long. They are about as broad as they are long with three triangular lobes, the lobes with sharp points, dark green above, lighter beneath, conspicuously three-nerved, and finely toothed. They turn a bright yellow in the fall. The flowers are yellow, bell-like, in long, graceful drooping racemes, and are succeeded by large, showy keys with pale green, widely divergent wings.

The striped maple is also known as the moosewood because its twigs and leaves are a favorite food of the moose as well as the white-tailed deer. Emerson says that in their "winter beats," the tree is found generally completely stripped. Doubtless both appreciate the young, tender twigs for the saccharine juice that they contain.

The striped maple is a mountain tree and so, too, is the mountain maple, being found on hillsides and mountainslopes. Both form much of the undergrowth of our northern forests and of lower Canada.

The mountain maple is a small tree, twelve to twenty-five high, but more commonly a tall shrub with a slender, short trunk, slender ascending branches, a ruddy brown or dull, gray bark, and lustreless leaves with three, or indistinctly five lobes, that are coarsely toothed. It is our one maple that bears its flowers in upright racemes, but when the keys appear, the racemes droop. The flowers are greenish yellow, the keys are red in summer, but turn a dark brown in autumn when the leaves take on brilliant shades of yellow and scarlet.

There are other native maples: the chalk maple of our southeastern states, a small tree with a light gray bark, broad three to five-lobed leaves, yellow flowers, and leaves that turn a bright scarlet in the fall; the dwarf or Sierra maple, a shrubby tree rarely more than twenty feet high growing on the mountainslopes of the far western states, with three to five-lobed leaves and small flowers in drooping clusters; the vine maple, a shrub or small tree of the Pacific Coast, with more often a vinelike habit; and the big-leaved maple or Oregon maple, a handsome tree of riverbanks and bottomlands and widely cultivated. It has three to five-lobed leaves, one foot or more across, dark green almost leathery, pale green beneath, turning brilliant yellow or orange in fall, yellow, fragrant flowers in hanging compound panicles, and broad-winged samaras, with widely divergent wings.

Unlike the preceding maples that have simple leaves, the box elder or ash-leaved maple has compound leaves. It is a fast-growing tree, thirty to fifty feet high, of riverbanks and the margins of ponds and streams, and though planted as a shade tree, it is not particularly decorative and, moreover, is short-lived. The bark is light brown gray, with shallow, narrow furrows, and short, flattopped ridges, the twigs are olive green, and the leaves are pinnately compound with three to five leaflets and bright green. The flowers are yellow green, the staminate in flat-topped clusters (corymbs), the pistillate in hanging terminal clusters (racemes), and the keys have wings set at an acute angle and are usually incurved.

In addition to our own native maples there are a number of introduced species that are planted for shade and ornament. Two of them, the sycamore maple and the Norway maple, are from Europe. The sycamore maple is a large vigorously-growing tree, fifty to sixty feet high, with five-lobed leaves, dark green and smooth above with a slight bloom beneath that resemble those of the sycamore, yellowish green flowers in drooping, compound terminal clusters, and keys with wings spreading at an acute or right angle.

The Norway maple is a round-headed tree of dense foliage that turns yellow in the fall. It is a medium-sized tree, forty-five to sixty feet high, but often attaining a height of one hundred feet in its native habitat in northern Europe. The bark is dark brown. The leaves are five-lobed with marginal teeth and bright green. The flowers are greenish yellow, in erect, multi-flowered, stalked, flat-topped clusters (corymbs). The samaras are drooping with horizontally spreading wings, and very large, the wings almost two inches long. A distinguishng character of the Norway maple is that a broken petiole exudes a milky juice that quickly coagulates.

During ancient Rome's most luxurious period, workers in maplewood ranked with jewellers and goldsmiths, and made tables of curly maple with the most beautiful colors and patterns as revealed by their polished tops. It is said that Cicero paid ten thousand *sesterces* for a table that showed "spots and maculations" that imitated the colors and shapes of panthers and tigers. Also, one of the Ptolemies had a circular table three inches thick and four and a half feet in diameter, for which, he paid fifteen hundred thousand *sesterces.*

Besides the two importations from Europe, there are a number of maples from Japan whose gardeners have developed a great number of beautiful garden varieties. They are, for the most part, shrubs and usually spreading in habit as if to display to best advantage their wonderful forms and the exquisite coloring of their leaves and fruit.

The seeds of the maples, as well as the buds and flowers, provide food for many kinds of birds and mammals, the squirrels and chipmunks often storing the seeds in caches after removing the hull and wings. Many birds use the leaves and seedstalks in nest building.

109. Mayaca (Mayacaceae)

The members of the mayaca family are slender; branching mosslike, aquatic plants with narrowly linear, sessile, entire leaves, notched at the tip, solitary, white, perfect, regular flowers with three sepals, three petals, three stamens, and a one-celled ovary with several or many ovules. The fruit is a capsule.

There is only one genus in the family, *Mayaca,* the aboriginal name of the plants in Guiana, that contains about seven species native of warm and tropical America, only the following being found in the United States and known as mayaca. It occurs in freshwater pools and streams of the South and has little branched, tufted stems, linear-lanceolate, translucent leaves that densely clothe the stems, axillary flowers, but borne near the ends of the branches, and rarely more than one on each branch, and an oblong-oval capsule.

110. Meadow Beauty (Melastomaceae)

The meadow-beauty family is a large family of trees, shrubs, or herbs with three to nine nerved, opposite leaves, sometimes handsomely colored, flowers often very showy and in magnificent clusters, sometimes with brightly colored bracts, the anthers with thickened or appendaged connectives, the petals inserted on the throat of the calyx, and the fruit a berry or a capsule. It contains over one hundred and seventy genera and nearly three thousand species practically all being tropical, especially plentiful in the Amazon Valley. Only one genus, *Rhexia,* of perhaps twelve species, occurs in the United States.

They are pretty little perennial herbs of opposite leaves with three to seven veins and perfect, regular flowers with four petals and as many calyx lobes. There are four to eight stamens, the stigma is far in advance of the anthers, and the flowers are cross-fertilized, chiefly by butterflies and bees. The fruit is a four-celled capsule.

Our common species is the meadow beauty or deer grass. It is a nearly smooth-stemmed plant, the stem rather square, with light green leaves, oval, acute at the tip, the margins saw-edged, and seated on the stem, and with magenta or purple pink flowers in a few flowered cluster (cyme). They are followed by seed vessels that, according to Thoreau, are "perfect little cream pitchers of graceful form." The seeds within are coiled like snail shells. The meadow beauty is a plant of sandy swamps and marshes. A slender, hairy-stemmed species with narrower leaves and paler petals occurs in the pine barren bogs of the South.

The family furnishes some of our most beautiful greenhouse plants, but generally speaking, they are not much cultivated, though some species have found more favor than others.

111. Mezereum (Thymelaeaceae)

An unusual feature, almost a peculiarity, of the leatherwood, is the character of its thick, porous bark. It is soft and pliant, and yet, tough and tenacious. Break a stem from the shrub and though the wood breaks easily the bark will neither give nor yield. It moves easily, but the fibers will simply not part. The Indians knew of the peculiar quality of the bark and used it for bow strings, cordage, and baskets. It also contains poisonous properties and taken internally will cause vomiting and purging. Externally it may irritate the skin and produce a blistering in susceptible people.

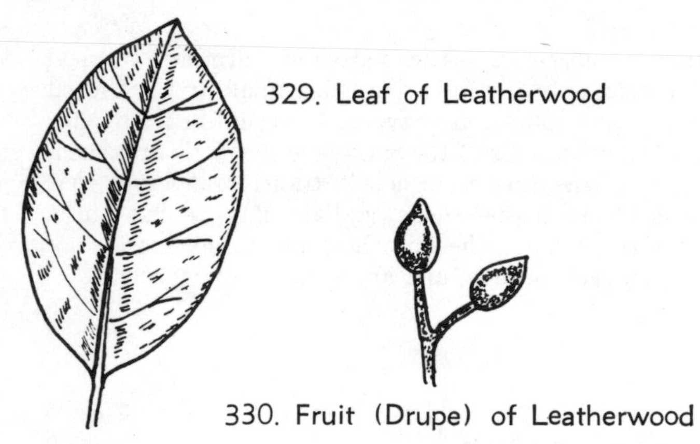

329. Leaf of Leatherwood

330. Fruit (Drupe) of Leatherwood

The leatherwood is a rather stocky shrub, two to five feet high, with a thick, somewhat gnarled, often solitary, trunklike, brown stem, and ascending smooth, yellow green twigs that are clustered at the ends of the branches. The leaves are a dull light green, alternate, oval to elliptical, toothless, and the flowers are small, pale yellow, appearing with or before the leaves, with a six-parted calyx, no petals, eight stamens and a one-celled, one-ovuled ovary, three to four in a cluster. The fruit is an ovoid, red, berrylike drupe. The leatherwood is commonly found in damp rich woods and thickets, and is sometimes cultivated.

The leatherwood is our only native representative of the mezereum family, a group of some forty genera and over four hundred and fifty species of trees and shrubs of warm and temperate regions, most of them being native to South Africa and Australia. Their leaves are usually alternate, but sometimes opposite and entire, their perfect, regular flowers are small, but attractively clustered in heads, panicles, or spikes, the flowers with four to five sepals, no petals, or what petals there are, are reduced to scales, mostly four to five stamens, and a one-celled ovary. The fruit may be fleshy, nutlike, or dry.

Widely distributed and common over nearly all of Europe and northern Asia and for centuries a favorite garden plant in the Old World, the mezereum or daphne is often found as an escape in this country. It is an upright shrub, one to three feet high, with light brown stems, broadly lance-shaped or obovate, bright green leaves, and lilac purple or rose purple, fragrant flowers in clusters of two to five, the flowers later developing into an oblong-oval red drupe.

The mezereum is one of fifty species of the genus *Daphne,* the Greek name for the true laurel and of no application to these plants, native to Europe and Asia, many of them common in American gardens, as, for instance, the spurge laurel, an erect, bushy shrub,

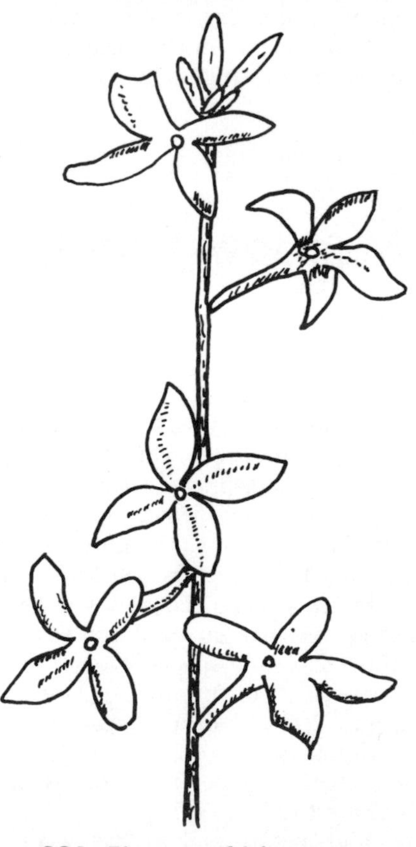

331. Flowers of Mezerum

eighteen to thirty inches high with evergreen foliage. Its leaves are somewhat oblong, gradually narrowed towards the base, its flowers are yellowish green in nearly stalkless racemes, and its fruit is bluish black.

Two other cultivated members of the family are the paper tree or paper bush and the rice flower. The paper tree, a native of China and Japan where its bark has been used in paper making, is sometimes found in southern gardens. The yellow, fragrant flowers in dense headlike clusters appear before the leaves. The rice flower is usually grown as a cool greenhouse plant, but may also be cultivated in the garden. There are some eighty species, native to Australia and New Zealand. The one most commonly planted in southern gardens is a multi-branched shrub, two to three feet high, with oval leaves, and rose pink flowers in round heads. The fruit is like a very small plum.

112. Mignonette (Resedaceae)

Without its fragrance, the mignonette would just be another flower of no importance, but as it is, its fragrance has made it a popular garden flower of long-standing.

In its native home, which is northern Africa, the mignonette is a weed. Several hundred years ago it appeared in Italy where it became a great favorite and received its name of mignonette—"little darling." It appeared in England in 1742 and eventually came to our shores.

A blooming stem of mignonette consists essentially of green sepals, a white fringe, and reddish anthers. The white fringe is the petals, the reddish anthers provide the color and that is about all that can be said for the flowers, which are neither showy or colorful, except for their fragrance. It is for this quality that the plant is extensively grown as a garden flower.

The mignonette is one of some sixty-five species of the family of the same name that are grouped in about six genera. They are annual or perennial herbs, with a few woody plants, with watery juice, and are largely indigenous to the Mediterranean basin. Their leaves are alternate, simple, entire, often pinnately divided and their flowers are usually irregular, perfect and small in terminal racemes or spikes. The flowers have a four to seven-lobed calyx, four to seven petals that are entire or cleft, three to forty stamens, and a one-celled ovary. Their fruit is a capsule, usually three to six-horned or angled, and opening at the top when ripe.

The mignonette of the genus *Reseda,* the Latin name of a plant, from the word meaning to heal, is a little plant growing from ten to eighteen inches high with red, white, and yellow finely cut flowers in a dense spike. With us the plant is an annual. The gardener's ideal of mignonette has been a plant with as large a spike as possible, and so, the mignonette has a number of varieties, one variety being a large garden form. As the color and form of the flower has thus been secondary in importance to size and abundance of spikes, the flower itself has varied little from its wild progenitor. It is only the anthers, in size and color, that have shown any change.

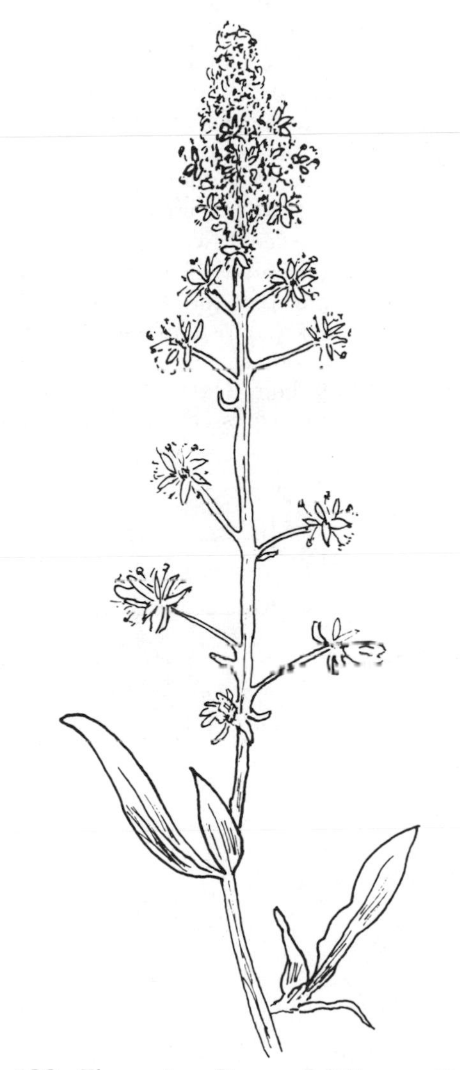

332. Flowering Stem of Mignonette

113. Milkweed (Asclepiadaceae)

About eighteen species of milkweeds are found in the United States. These are a small part of the hundred and fifty species of milkweeds and a still smaller part, indeed but a fraction, of the two thousand or so species that comprise the milkweed family with its two hundred and twenty genera. The milkweeds are milky-juiced herbs, vines, or shrubs of wide geographical distribution, most numerous in the warmer regions and comparatively rare in cold climates.

The leaves are opposite or whorled, rarely alternate, simple, and usually without teeth. The flowers are perfect, regular, in terminal or axillary umbels or cymes, and have five sepals that are separate or nearly so, a five-lobed or five-parted corolla, the lobes usually bent backward, five stamens, the filaments usually united into a column, a crownlike organ (corona), usually between the stamens and petals, and two pistils. The fruit is composed of two follicles, often widely divergent, each with many seeds crowned with a tuft of silky hairs. The anthers and stigma are remarkably connected and the pollen coheres in waxy or granular pear-shaped masses (pollinia) that often become attached to the feet of insects and sometimes ensuring their death.

The milkweeds are surpassed only by the orchids in

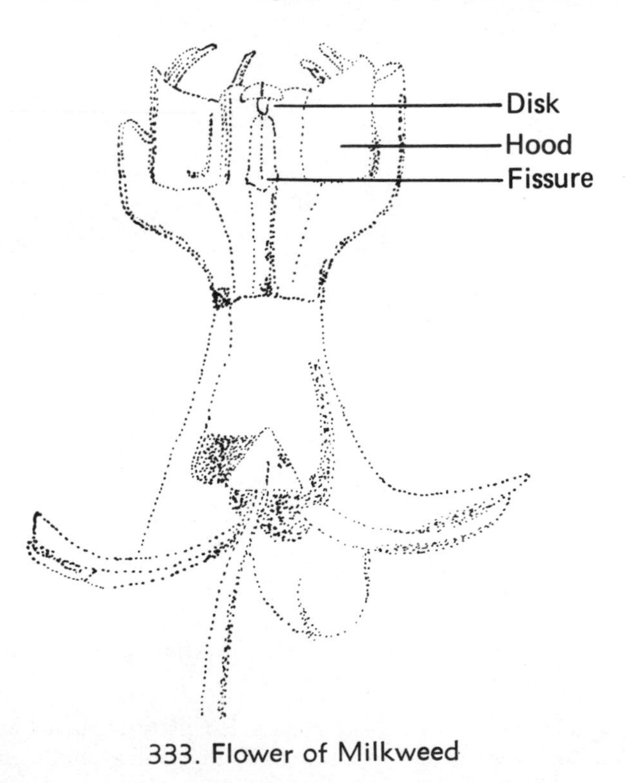

333. Flower of Milkweed

the ingeniousness of their flower structure. The pollinia are so designed that when an insect steps upon the edge of the flower to sip nectar, its legs slip between the peculiarly shaped nectariferous hoods situated in front of each anther. As it then draws its legs upward, a claw, hair, or spine catches in a *v* shaped fissure and is guided along a slit to a notched disk that becomes attached to the leg. Since the pollinia are each connected to the disk by a stalk, they are carried off when the insect leaves. Upon the insect's arrival at another milkweed flower, the pollinia are easily introduced into the stigmatic chamber. The struggle of the insect at this time breaks the stalk of the pollinia and the insect is relieved of its load. Sometimes an insect loses a leg or is permanently entrapped.

Nearly everyone is familiar with the common milkweed. With its cloyingly sweet, somewhat pendulous flower clusters, it is one of the more familiar of our summer wildflowers, and from June to September it is found blossoming along roadsides and in fields and waste places, its richly nectar laden flowers a veritable banquet table for all kinds of flying insects.

A native perennial, it grows from three to five feet tall. It has a stout, sturdy stem, light yellow green leaves in opposite pairs and that are oblong, tapering at both ends, and very finely haired, and numerous flowers, dull, pale greenish purple pink, brownish pink, pale lavender brown, crimson pink, or even yellowish in globular umbels. They later develop into two thick, warty pods, that when mature, split open on one side, releasing many yellow brown seeds, each with a tuft or parachute of silky, white, fluffy hairs that catch in the wind. The seeds sailing through the air is a familiar sight in early autumn. They also float on water so, lacking transportation by the wind, they can still obtain wide distribution through this agent.

The milkweeds are distasteful to animals, and grazing animals leave them alone as a rule, but sometimes the leaves are eaten and poisoning results, often proving fatal, sheep seeming to be the most susceptible. What may seem rather strange in view of the nature of the leaves and stems, the plant is good as a potherb. If collected when young and tender, that is, when only a few inches high, the crisp, succulent shoots make an excellent "dish of greens" if cooked like asparagus, or if the leaves are used, they have after being cooked, a taste somewhat like that of spinach. It is said that a good brown sugar can be made from the flowers and Fremont, the American explorer and soldier, relates that he found the Indians of the Platte River country cooked the pods with buffalo meat.

The milkweed is the chief, indeed, practically the only, food plant of the monarch butterfly, and during the summer the caterpillars, with their alternate bands of yellow, black, and white may usually be found feeding on the leaves.

The purple milkweed is a misnamed species for its flowers are not purple, but a crimson or crimson magenta. Nevertheless, it is a handsome plant common in dry fields and thickets. Unlike the purple milkweed, the swamp milkweed is at least well named, for it grows in swamps and wet places. It is a smooth species, with narrow or lance-shaped leaves, and small, dull crimson pink flowers. This plant yields a tough fiber, finer than that of hemp, and of considerable tensile strength, but for some reason has never been commercially exploited.

If the soil wherein it grows is to its liking, the poke milkweed may grow as high as six feet. It is common on the borders of thickets and woods, has rather large leaves sometimes as much as nine inches long, and drooping or spreading umbels of cream white flowers. In contrast, the four-leaved milkweed is a much lower-growing species, one to two feet high. It has a slender stem, with about two circles of four leaves in the middle and two pairs of opposite smaller leaves on the upper part whereby it may readily be recognized. It is a somewhat delicate plant with few clusters of magenta pink flowers, and may be found in woods and copses. An exquisite and ethereal member of the milkweed clan, the whorled milkweed is an extremely small narrow-leaved plant with a slender stem, the leaves in circles of four to seven, and small umbels of delicate greenish white flowers. It is common on dry hills.

The most handsome of all the milkweeds is undoubtedly the orange milkweed, perhaps better known as the butterfly weed, with its brilliant clusters of bright reddish orange flowers that literally set dry fields ablaze with color, and above which, butterflies hover, sip, and sail away. These include the spicebush, tiger, the pipe-vine and black swallowtails, the common white or cabbage, the regal fritillary, the pearl crescent, the coral hairstreak, the bronze copper, the delicate, tailed blue, and the splendid monarch. It is said that the Indians ate its tuberous root and also used it for various maladies, though it is not likely they knew of the alleged healing virtues of the genus, to which Linnaeus gave the name *Asclepias,* a Latinized corruption of Asklepios, the legendary Greek physician and god of medicine.

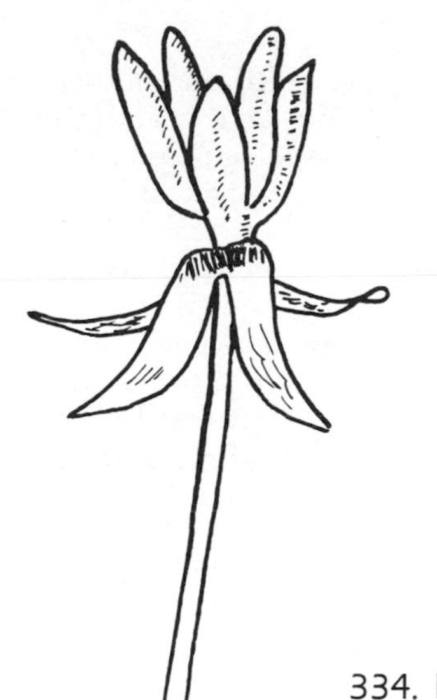

334. Flower of Butterfly Weed

114. Milkwort (Polygalaceae)

I can still recall the time when I first saw the fringed polygala. I was walking along a country road and one moment chanced to look down at a stone wall when my eye caught a bit of crimson and on looking closer saw a most dainty little flower, almost hidden from view by its surrounding leaves.

The fringed polygala is common in damp, rich woods, but it is such a low-growing plant that one really has to search for it. The flowering branches spring from prostrate stems and roots that are sometimes a foot long. They have a few broad, ovate, bright green leaves crowded at the top, (a few lower ones reduced to mere scales), that persist through the winter when they turn a bronze red. The flowers are crimson or magenta, rarely are they white, of a rather peculiar shape, suggesting an orchis. The calyx consists of five sepals that are unequal, two of them being broad, winglike, and colored like the three

335. Leaves and Flower of Fringed Polygala

petals. The petals are united in a tube, the lowest ending in a pouch containing the pistil and anthers, somewhat larger than the others and beautifully fringed. This last petal serves as a landing platform for bees that depress the pouch by their weight, the rigid pistil and stamens being forced out through a slit at the top of the pouch and thereby coming in contact with the plant's visitors. Thus, cross-fertilization is effected by the pollen carrying bee brushing against the exposed stigma of the next flower it visits. Honeybees and mining bees are the chief visitors. Besides the visible flowers, the plant also bears cleistogamous flowers that are hidden beneath the surface of the ground.

The fringed polygala belongs to a genus of plants called the milkworts, indeed, itself is often known as the fringed milkwort, and takes its more common name from the genus known as *Polygala,* Greek for much milk, in reference to the superstition that some species, if eaten by cows, would increase their supply of milk. The genus is a large one comprising about five hundred species. They are essentially herbs with alternate, lance-shaped leaves, and flowers in terminal clusters or spikes, in some species showy, and of various colors. The calyx consists of five sepals, three being small and the other two being large, petallike, and colored. There are only three petals that are connected with each other in a tubelike form, the lower one often crested at the tip. The stamens number six or eight and are united into one or two sets and, in part, coherent with the lower petal, but free above. The ovary is two-celled, the style prolonged and curved. The fruit is a two-seeded pod.

The genus *Polygala* is one of ten genera of the milkwort family whose structural characteristics are essentially those of the milkworts. It is a large family of seven hundred species most occurring in the tropics. Only a few are found in the United States and all of them are of the same genus, *Polygala.*

In moist and sandy fields one might look for the common milkwort, also known as the field or purple milkwort. It is a branching and leafy species, six to fifteen inches tall, with narrowly, oblong leaves, and numerous very small flowers, bright magenta, pink, or almost red, clustered in a globular clover head that gradually lengthens to a cylindrical spike. The cross-leaved milkwort is an attractive species, with dull magenta pink flowers in cloverlike heads and with leaves generally arranged in clusters of four, hence, its name. It grows in sandy swamps. A similar species, but with a more slender stem and shorter leaves, the short-leaved milkwort is common on the borders of brackish swamps. The whorled milkwort may easily be recognized by the circles of four to five leaves on the stem and common everywhere in fields and on the roadsides. Its greenish white or magenta-tinged flowers are crowded together in conic spikes. Less common, and then only locally, the racemed milkwort has tiny, aesthetic, dull crimson flowers in delicate long clusters at the tips of leafy stems, and like the fringed polygala, also has cleistogamous flowers. And though seneca snakeroot is not called a milkwort, it is of the same species. It is the only white milkwort, one of the tallest, and grows in rocky woodlands. The plant has some medicinal value and was probably used by the Seneca Indians after whom it was named.

336. Leaves and Flower Cluster of Common Milkwort

337. Leaves and Flower Cluster of Seneca Snakeroot

115. Mint (Labiatae)

Someone once remarked that music charms the "savage beast." Or is it "breast?" It really does not matter for whatever it may be it calls to mind the Hindu snake charmer and his performing cobras. Those who have seen the performance will recall that the snake charmer, sitting cross-legged, moves his body from side to side in time to the strange and crooning music that he elicits from his reedlike instrument and that the reptiles, with hoods spread widely and their eyes fastened with glassy monotony upon the charmer, also sway to the refrain.

The snake charmer's act is always fascinating to watch. To the uninitiated, it would appear that the snakes are "charmed" or hypnotized by the strange music, but as a matter of fact, neither hypnotism nor music figure in the business. The swaying of the snakes is due wholly to the movements of the Hindu snake charmer that the snakes follow as they alter their position in aiming to strike. The shrill notes of the reed are merely stage effects designed to appeal to the imagination of the spectators for snakes show no interest in music of any kind, and were the weird intonations to cease, the dance, if so it can be called, would continue without interruption.

The performance of the snake charmer calls to mind how many wild animals are attracted to lures and scents obtained from various wild plants. Bears, raccoons, and other animals throw all caution to the winds when they get a whiff of musk or rhodium, and many of the larger cats, such as lions, tigers, leopards, and jaguars are irresistably attracted by lavender. Most people know what a little catnip or catmint will do to our pet cat. Schuyler Mathews in his *Fieldbook of American Wildflowers* tells of how his pet Manx cat would walk a mile every other day or so, to a spot where catnip grew in plenty, notwithstanding the way was through the woods and over a hill of no small difficulty. As a matter of fact, it is rather curious how cats will seek out the hoary-haired plant in the waste places where it grows and become half-crazed with delight over its aromatic odor. It is, incidentally, an excellent tonic for both cats and kittens. Most cats prefer it dried. It seems that they cannot go on a real catnip jag on the fresh article—it must be aged like wine and whiskey.

Dried catnip or catmint can be bought in the drug store, but it can be found growing wild in the woods, and be taken home and dried. It is an exceedingly common wildflower with a stem and deeply round-toothed leaves that are densely downy and sage green in color. The pale lilac, or lilac white and spotted flowers, that appear from June to October, are also downy and are gathered in small terminal clusters. If there is any doubt as to the identity of the plant all one has to do is crush some of the leaves, when it will be recognized by its strongly aromatic odor.

The word catmint suggests a number of other mints, such as peppermint, spearmint, water mint, wild mint, and mountain mint. Except for the latter, all belong to the genus *Mentha,* from the Greek *Minthe,* meaning nymph. Most people have eaten peppermint candy and are familiar with its taste, but few would likely recognize the plant, wherefrom the characteristic taste of the confection is in part derived, as it grows along

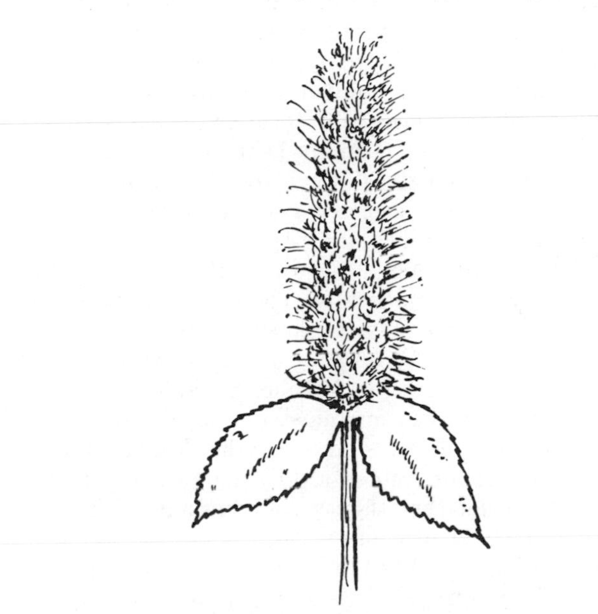

338. Leaves and Flower Cluster of Peppermint

brooksides and ditches and in wet places. It is a perennial, spreading by creeping roots and with a purplish stem eighteen to thirty-six inches tall. The leaves are two to three inches long, wrinkled, veiny, and toothed and the purplish pink blossoms occur in narrow, loose, disconnected terminal spikes and often on a long stem proceeding from between the plant stem and leaf stem. One should have no difficulty in identifying the peppermint—all one needs to do is chew the leaves—their intensely pungent, aromatic taste, resembling that of pepper, is unmistakable. One will also discover that the taste is accompanied by a peculiar sensation of coldness.

The peppermint is commercially cultivated for its essential oil, used in flavoring extracts, in confectionery, in medicines, and in the manufacture of the cordial *creme de menthe*. Some people use the leaves for seasoning, and grow the plant in their gardens for this purpose.

It has been said that in the Middle Ages spearmint was used as a charm against the bite of serpents, scorpions, and mad dogs. Of a more practical use, it was employed in the making of cheese, incorporated with pennyroyal in puddings, and boiled with green peas, the latter a custom that still prevails in England. Today, spearmint, like peppermint, is cultivated commercially for its essential oil that is distilled from the leaves, and used as a flavoring for chewing gum and candies. It is also frequently grown in gardens for household use being served with vinegar as a sauce for roast beef. And sprigs of the plant are often used in making the "seductive and intoxicating drink known as mint julep," and in mint jelly.

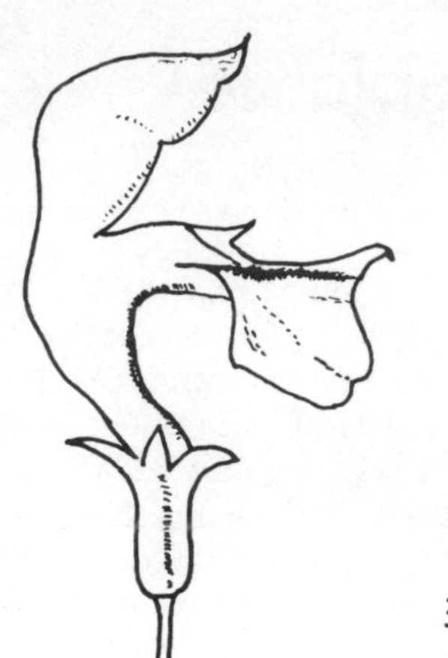

339. Flower of a Mint

Like the peppermint, the spearmint is also an escape, growing in wet places and along roadsides everywhere. It, too, is a perennial with a stem one to two feet high from creeping rootstocks, leaves two to three inches long, wrinkled, veiny, and coarsely toothed, and small, pale purple, bluish, or whitish flowers that are clustered in whorls and forming terminal, interrupted spikes. The spearmint is in bloom from July to September when the flowers are visited by various flies (some nineteen species), bees, and wasps.

Both the spearmint and peppermint are imports from Europe and Asia, but there is a native mint, and the only one incidentally. The wild mint or American mint is found in wet places. It is more or less hairy throughout with leaves that taper conspicuously from the center towards both ends, and lilac white or white flowers that are clustered in circles around the plant stem. It is said that the Maine Indians roasted the leaves before a fire and ate them with salt in the belief that they were nourishing. The wild mint has the odor of pennyroyal, that is, the true pennyroyal of Europe that is not to be confused with our species of pennyroyal. The latter is a small annual, usually found in dry pastures, with an erect, finely haired stem, small, light olive green leaves, and tiny, pale violet, or lavender flowers that are fertilized mostly by bumblebees, honeybees, and the smaller bees. The taste and odor is much like that of the true pennyroyal. There is also another pennyroyal, the false pennyroyal, a slender branching annual with lance-shaped, toothless or nearly toothless leaves and small, pale blue flowers. An unusual characteristic of the flowers is that the five sharp lobes of the bell-shaped calyx, and the five rounded spreading lobes of the corolla, are of equal length, contrary to the usual floral structure of the mint family whereto it belongs, as do all the mints. Another feature is that the pistil greatly exceeds the stamens in length, evidence that the flower is cross-pollinated, the chief visitors being the bumblebees, the honeybees, and the smaller butterflies such as the white cabbage and the common sulphur.

There are a number of other mints, that is, plants with the word as part of their names: the mountain mint, a stout and stiff-stemmed species, its lilac white purple spotted flowers in somewhat flat-topped clusters and found in dry fields, or pastures, or on the borders of thickets; the horse mint of roadsides and field borders; and the two less common species, the corn mint and water mint both liking wet places, such as swamps and roadside ditches or moist fields.

The mint family is a very large group of plants of about three thousand species, and one hundred and sixty genera. They are mostly aromatic herbs and shrubs, a few are trees, worldwide in distribution, and found under a great variety of climatic and soil conditions. Superficially, they all resemble each other, and are somewhat difficult to identify as to species, but all have the family characters of a square stem and opposite, simple leaves that are covered with tiny glands containing a strong scented, volatile oil of a peppery nature. The flowers are perfect, usually small, tubular, with the corolla entire or with a two-lobed upper lip and a three-lobed lower lip, hence, the name of the family, *Labiatae,* Latin meaning lips. The calyx is regular or two-lipped, usually five-toothed or five-parted, the stamens are four in number, and the ovary is deeply four-lobed, the style arising from the center, with a two-lobed stigma. The fruit consists of four one-seeded bony nutlets.

Many people are acquainted with the mint family through such plants as thyme, sage, and marjoram that have been grown for generations in American herb gardens. Thyme, an erect woody perennial with a strong, mintlike odor, was highly prized by the Romans as a seasoning for various kinds of dishes, and Ovid, Pliny, and Virgil all speak of it in relation to bees, the honey made from the nectar obtained from its small lilac or purplish flowers apparently having a pungent and aromatic flavor. The Greeks, too, found the honey much to their favor and regarded the plant as denoting graceful elegance and as an emblem of activity. "To smell of thyme," was high praise indeed, and given only to those with a faultless life style.

Although sage is probably not used as extensively as it once was, this woody, perennial, undershrub growing to a height of two feet has for centuries, been highly ranked because of its pleasant odor and certain medicinal qualities. A tea made from the leaves was once a cure for nervous headache and a mixture of sage and honey was a standard treatment for cankers of the mouth. Writing in 1596, Gerarde says that

> sage is good for the head and brain; it quickens the memory and the senses. The juice of the leaves mixed with honey is good for those who spit blood. No man needs to doubt of the wholesomeness of Sage.

This was some four hundred years ago. Today there are a plethora of remedies for all sorts of ailments.

Virgil and Pliny both speak of the marjorum or sweet marjorum to distinguish it from the pot or wild marjorum probably better known as origanum, for among the Romans, and the Greeks too, it was used as a decoration at marriage feasts, when it was woven into wreaths to crown the young married couple. Sweet marjorum is a fragrant perennial herb or undershrub with a pleasing odor and an aromatic, bitterish taste that makes it highly valued as a seasoning for soups, stews, and dressings. Origanum is also a perennial herb with aromatic leaves and much used as a seasoning.

Lest some reader should take offense for omitting them, the author should also mention sweet basil, an annual of extremely fragrant foliage, long in cultivation and used to flavor soups and occasionally salads; summer savory, also an annual and once highly esteemed as a remedy for many ailments, but today, simply a slender-stemmed plant, extremely aromatic in stem, leaf, and flowers and used as a seasoning; rosemary, a hardy perennial evergreen shrub and a garden favorite because of its strongly scented leaves and flowers; and the sweet or common lavender, also a shrub, and also a garden favorite, and prized for ages because of the pleasant aromatic fragrance of its leaves.

Mention of the garden calls to mind that many people may not realize that the garden coleus with its brilliantly colored leaves is a member of the mint family. But its stem is square, its leaves opposite, and its flowers two-lipped.

The salvia or scarlet sage, one of the most brilliant red-flowered plants in cultivation, and found in almost every garden, is also a member of the mint family. An unusual feature of this flower is that the calyx and bracts, that are normally green in color, in most flowers are red, thus, making the corolla a brilliant mass of color that pleases the eye and attracts pollen-carrying insects. Indeed, what is more eye-catching than a well-grown bed of salvias. It defies description and in the intensity of color is almost barbaric.

Should one examine the flowers from the outside, one would find that the nectar wells lie so deep that it would seem impossible that they could be reached by any creature except a moth, a butterfly, or a hummingbird. There is no platform for the bees to alight upon, and the tube is too long to be fathomed by their tongues. The bees are clever, however, and can adapt themselves to various floral structures. To learn how they secure the nectar, watch a bee enter a salvia blossom. It is a most interesting performance and will repay the time spent in doing so.

The blue salvia bears flowers that are as vividly blue as those of the scarlet sage are scarlet. Unlike the latter, which is an annual, it is a perennial growing to a height of two feet or more, with short, sticky hairs, arrow-shaped, toothed leaves and gentian blue flowers, two inches long, in pairs, and in widely spaced racemes. Both salvias, as well as the garden sage mentioned above, belong to the genus of the same name, from the Latin to be healthy, in reference to the medicinal properties of some of the species.

340. Leaves and Flowers of Ground Ivy

The mint family is well represented among the wildings, though many of them are escapes from colonial gardens, as the spearmint and peppermint, that were brought over from Europe by the early settlers. One of the first to blossom, in April, is gill-over-the-ground, or ground ivy, a small creeping plant forming a dense mat with deep green, kidney-shaped, or heart-shaped, scalloped leaves and small, pale purple flowers. At one time in Europe, the aromatic leaves of this little creeper were used in fermenting beer and was then known as gill ale. Gill, it is said, was derived from the old French word, *guiller*, to ferment or make merry. Hence, its popular name. It is also said that the poor in England once made a tea from its leaves and perhaps still do. It is not known if the leaves have ever been used in this country for such a purpose, but a tea is sometimes brewed from the leaves of the Oswego tea or bee balm.

341. Flower of Oswego Tea

The Oswego tea is a brilliant and showy wildflower whose scarlet red color, of a deeper red than most summer flowers, is strongly relieved by the background of shady woodland where it grows in the wildwood tangles of mountain streams. It is a strong growing perennial with sombre, dark green leaves and flowers clustered in the solitary, terminal rounded heads of dark red calyxes. Although commonly visited by bees, the flowers are peculiarly adapted to the visits of butterflies, of which, the common sulphur and the monarch are frequent visitors According to those who have tried it, a decoction made from the leaves is only a little inferior to the true tea. At any rate, it was used as a substitute by the colonists who were taught by the Indians how to brew it.

The wild bergamot is a similar species, but of a lower stature, and with flowers that are variable in color though generally crimson pink or rose red, but often lilac or lavender. It grows in open woods or thickets where one may often see butterflies fluttering about the flower heads. Occasionally one may see a ruby-throated hummingbird flashing about the bright patches for an instant, and an instant only, and then will be gone.

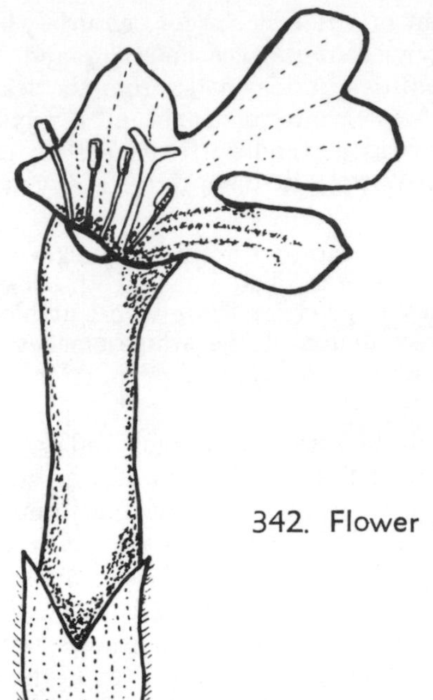

342. Flower of Bugleweed

In the rich woods of midsummer, the horse balm opens its yellow, lemon-scented flowers and many bees, large and small, arrive to sip the nectar. They then leave, pollen-dusted, for some other flower, and thus, cross-pollination is effected. At about the same time, the bugleweed opens its tiny white blossoms clustered at the bases of the leaves and flowerflies and small bees feast at their banquet tables. It grows in wet places, and in similar situations, will one find the water horehound, which is not the horehound of commercial use.

Of special interest are the various species of skullcaps because unlike most members of the mint family they are bitter instead of aromatic. These plants belong to the genus *Scutellaria,* from *scutella,* a dish, in allusion to the peculiar hump on the upper section of the calyx, that, however, does not even remotely resemble a dish. But Linnaeus, who named the genus, apparently thought otherwise. It is a widely distributed genus of plants with blue and violet two-lipped flowers, some small to the point of insignificance, others showy enough for the garden, but all rich in nectar and eagerly sought by bees, and with opposite,

343. Flower of Larger Skullcap

simple leaves though sometimes cut into lobes. One of the most popular of the skullcaps is mad-dog skullcap, common in damp and shady places throughout the country. The flowers are borne on one side of the stem that later is decorated with odd little hoodlike green calyxes containing four white nutlets. The old herb doctors, who professed to cure hydrophobia with it, are responsible for its English misnomer.

Those who are familiar with our wild flowers, know the blue curls, found in dry, sandy fields, and blooming throughout the summer. It is an annual species with a somewhat woolly sticky, stiff stem and lance-shaped leaves that have an aromatic balsamlike odor. The pale violet or magenta flowers are remarkable for

344. Flower of Blue Curls

the extraordinary length of the violet stamens that extend in a curving line far beyond the five-lobed corolla or flower cup, hence, its name. There is another member of the mint family also called blue curls, but perhaps better known under the name of self-heal or heal-all. This plant is found everywhere, sometimes dusty and stunted by the roadside, at other times truly beautiful in its fresh purple, violet, and white when perfectly developed under more favorable conditions. It is a low perennial generally with a single stem, oblong leaves, and tubular, two-lipped hooded flowers in a spike, that somewhat resembles a clover

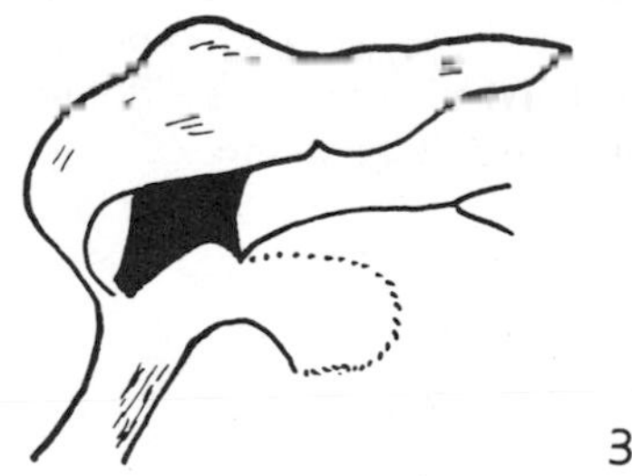

345. Flower of Self-heal

head, and that is frequently visited by several species of bumblebees; the common sulphur butterfly is also an attendant.

The old name of the plant, *Brunella,* is said to have been derived from the German word, *braune,* the quinsy, that it was supposed to cure. With the plant, in fact, old herbalists pretended to cure every ill that the flesh was prone to. Doubtless of more value is the horehound, a white-woolled, bitter, but aromatic perennial with scalloped-toothed leaves, and small, tubular, dull white flowers, found as an escape in waste places. The familiar horehound candy, used as a remedy for colds, is made from the juice obtained by boiling the plants. In England a beer is made from it or it was at one time.

Weeds are plants that grow where they are not wanted, and yet, many of them become so domestic and familiar and crowd about our dooryards, that some of them have come to be regarded with real affection. Such is the motherwort, a perpendicular growing decorative herb with deeply cut leaves and tiny pale lilac flowers, whereto bees are invariably seen clinging. The motherwort is an immigrant from Europe, now an escape and common everywhere in waste places. So, too, are the henbit and hemp nettle, but often returning to the garden as weeds. The henbit has rounded, deeply scallop-toothed leaves on slender stems six to eighteen inches high, and flowers in small axillary and terminal clusters. The calyx is hairy, with five erect, awllike teeth, the corolla tube slender, with the upper lip erect, entire, and bearded, dark red, the lower one three-lobed, white spotted with purple. The flowers contain much nectar and honeybees are frequent visitors.

The hemp nettle, genus *Galeopsis,* Latin for nettle, sometimes reaches a height of three feet, has a stem with spreading branches, ovate, hairy and pointed leaves, and several encircling clusters of small, pale magenta flowers gathered at the stems of the floral leaves. It is common in waste places everywhere, and is cross-fertilized by the bumblebees and small bees.

There are other nettles, but not of the same genus, such as the hedge nettle, genus *Stachys,* a coarse weed with a stem one to four feet tall and with downward pointing hairs, oblong to lance-shaped leaves, and with tubular bell-shaped flowers, magenta purple, clustered in circles, and forming a terminal spike.

A rather curious flower is the false dragonhead or obedient plant, that prefers wet places wherein to live, such as ditches, moist meadows, and brooksides. It has a simple or branched leafy stem, lance-shaped leaves that are mostly toothed, and showy flowers crowded in terminal leafless spikes. They are pinkish pale lilac, often variegated with white, and funnel-shaped, and have the odd habit of remaining in any position wherein they are placed. For instance, with a slight puff one can blow the blossoms to the opposite side of the spike where they will remain unless further disturbed. The reason why they remain in position is because of the friction of the pedicels against the subtending bracts. If these are removed the blossoms become limp. Doubtless this peculiarity is an adaptation to protect the flowers in stormy weather in turning the mouth of the flowers against a driving rain.

The wood sage or American germander is probably as pretty a plant as any that can be found. It is covered with soft down, has light green leaves, and purple, though sometimes magenta or a pinkish white, fairly large flowers in a terminal spike. The corolla is unique in that it does not seem to have an upper lip. This is because the four upper lobes are nearly equal, oblong, and turned forward. The lower lobe is much larger and prominent, and forms a convenient landing for visiting bees. The plant grows in moist, thicket borders or marshes.

And lastly, the hyssop should be mentioned, a coarse stiff, aromatic perennial from Europe widely grown for ornament, and used as an herb for flavoring, but an escape to roadsides and waste places. It is slender stemmed, lance-leaved, with tubular, pale violet flowers with projecting stamens.

116. Mistletoe (Loranthaceae)

The mistletoe was a ceremonial plant among early European peoples and was held in special veneration by the Druids, especially when they found it growing on an oak tree that it rarely does. It was considered as something of a magic plant, and to some extent, it is still regarded as a charm or fetish, as witness to our custom of hanging a sprig over a doorway at Christmas time, a custom that originated centuries ago.

The mistletoe of Europe is not our mistletoe. Ours is of a different species and genus, one of some five hundred species as a matter of fact. It is a multi-branched, shrubby evergreen, a yellowish green plant parasitic on various broad-leaved trees. It has opposite, oblong or narrowly oblong, toothless leaves that are broadest above the middle, thick and leathery in texture, and small and inconspicuous yellowish green flowers, blooming in October and November. The fruits, that are berrylike, round, waxy white, one-seeded, with a sticky-gummed pulp, do not mature until the following fall. Driving almost anywhere in the South during the winter, one will find them most conspicuous on the leafless trees. They are considered poisonous and should not be eaten.

The mistletoe has chlorophyll and makes its own food, and in this respect is not a parasite, but instead of having aerial roots that serve only to anchor the plant like true epiphytes, the mistletoe has specialized rootlets, called haustoria, that extend into the conducting tubes of the host, and thus, obtains its supply of water.

The mistletoe family is a group of mostly parasitic, evergreen shrubs growing upon trees, or rarely are they terrestrial trees and shrubs. The five hundred species, contained in twenty-one genera, are widely dispersed, mainly in the tropics.

The mistletoes have opposite or whorled, simple, entire, often leathery leaves that are sometimes reduced to mere scales, perfect or imperfect flowers with two to three sepals, two to three petals, two to three stamens, the sepals and petals free or united, and a one-celled ovary, and a fruit that is berrylike or drupelike.

There are some twenty-two species of mistletoes native to the United States: the American mistletoes of the genus *Phoradendron,* from the Greek for thief and tree, in allusion to their parasitic habit, and the dwarf mistletoes of the genus *Arceuthobium,* from the Greek for juniper and life. The former are parasitic on broad-leaved trees, the latter on conifers. The dwarf mistletoes are essentially western species. Various birds, such as cedar waxwings, the phainopepla, and bluebirds feed on the berries, and by so doing, doubtless aid in the dispersal of these plants.

Leaves, Flowers, and Fruit of Mistletoe

117. Moonseed (Menispermaceae)

The moonseed family is a group of fifty-five genera and one hundred and fifty species of climbing, or twining, woody or herbaceous vines, mainly of tropical distribution, but a few extending into the Temperate Zones. They have alternate, entire or lobed leaves and small dioecious flowers in racemose or cymose clusters. The flowers have four to twelve sepals, six petals, about the same number of stamens as petals, and generally six carpels. The fruit is a one-seeded drupe.

There are only a few native species, which are the Carolina moonseed, the Canada moonseed, and the cupseed. The Carolina moonseed is a slender vine, trailing or climbing to the height of several feet, with a smooth or pubescent stem, broadly ovate leaves that may be entire or lobed, greenish flowers in axillary and terminal panicles, and a red drupe. It belongs to the genus *Cocculus,* from the Greek *coccus* for berry, and is found on riverbanks. The Canada moonseed is also a climbing vine, making its way over bushes and walls in woods and along streams. The leaves are very broadly ovate, entire or with three to seven lobes, the flowers are white or greenish white in loose panicles, and the drupes are round, black, and covered with a bloom. The drupes, when ripe in September, look like frost grapes and add color to the tangled growth along the streams. The Canada moonseed is a member of the genus *Menispermum,* Greek for moon and seed. The cupseed, like the other species, is also a vine climbing to the tops of trees in rich woods. Of the genus *Calycocarpum* Greek for cup and fruit, it has a smooth or slightly hairy stem, deeply three to five-lobed large leaves, greenish white flowers in long racemose panicles, and a black drupe.

An East Indian species is very poisonous and is used in India for stupefying fish to facilitate their capture, and in the form of an ointment is used for destroying vermin.

118. Morning Glory (Convolvulaceae)

The morning glory can become a rampant weed. It will trail itself over the rarest rosebush, smother the finest perennial, in short become a nuisance, and yet, who does not want the morning glory in the garden, to have the pleasure of feasting one's eyes upon its opening blossoms, fresh with the dew of night upon them, and to remain for but a transitory moment under the spell of their delicate and ethereal beauty. Is it not worth a few moments time to keep it under control?

The morning glory's home is South America, but it has long since been cultivated in our gardens wherefrom it has escaped and become one of our wild flowers, and so, it is found along the roadside, in fields, and thickets trailing its vine about the surrounding herbage and opening its blossoms in the morning for every passerby to see.

The morning glory has a number of near relatives and some not quite so near, found in the garden and in the wild: the moonflower, the manroot, the cypress vine and the sweet potato of the market. The moonflower is a milky-juiced vine with somewhat prickly stems, large heart-shaped leaves, and salver-shaped white flowers sometimes with greenish lines. The flowers begin to open about sundown and do not close until the next morning is well on toward noon. Often regarded by farmers as a dangrous weed, but producing very beautiful white morning glory flowers, the manroot, also called man-of-the-earth, and wild sweet potato vine, is a native perennial vine of sandy fields and riverbanks. It can achieve a massive root, often weighing one hundred pounds and being sweet and edible. The Indians called the plant *Mecha-meck* and appears to have been one of their favorite foods. The herbaceous vine dies each autumn, but the root remains in the ground well below the frost line and each spring sends up a new stem with leaves nearly as broad as they are long, and large, handsome blossoms sometimes three inches in diameter.

An annual twining vine with feathery foliage and small, bright red flowers and a native of tropical America, the cypress vine grows to a height of some twenty feet and is useful in the garden as a covering for screens and trellises. It is mostly planted in the southern states where it has become naturalized.

Originally a tropical American morning glory, the modern sweet potato is distinctly a warm season crop and is grown commercially only in the South, but good crops can be obtained as far north as southern New

Jersey. It is a sprawling perennial vine, rooting at the joints, with lobed or unlobed leaves, or sometimes divided finger-fashion, and with funnel-shaped, rose pink, or violet rose flowers that rarely appear in the United States. The edible tubers grow close together under the crown and, unlike the common potato, they do not have eyes since they are botanically roots, whereas the white potato is a stem, its eyes being buds.

The morning glory family consists of some forty-five genera and a thousand or more species of mostly annual or perennial herbs widely distributed throughout temperate regions with a few shrubs and trees in the tropics, where most of the members are abundant. The majority of them are trailing or climbing vines. They have generally alternate, simple leaves, perfect, funnel-shaped or long, tubular flowers, a calyx of five lobes, a corolla also of five lobes with a flaring rim, five stamens inserted deep in the corolla, mostly a two-celled ovary, each with two ovules, and the fruit is a capsule.

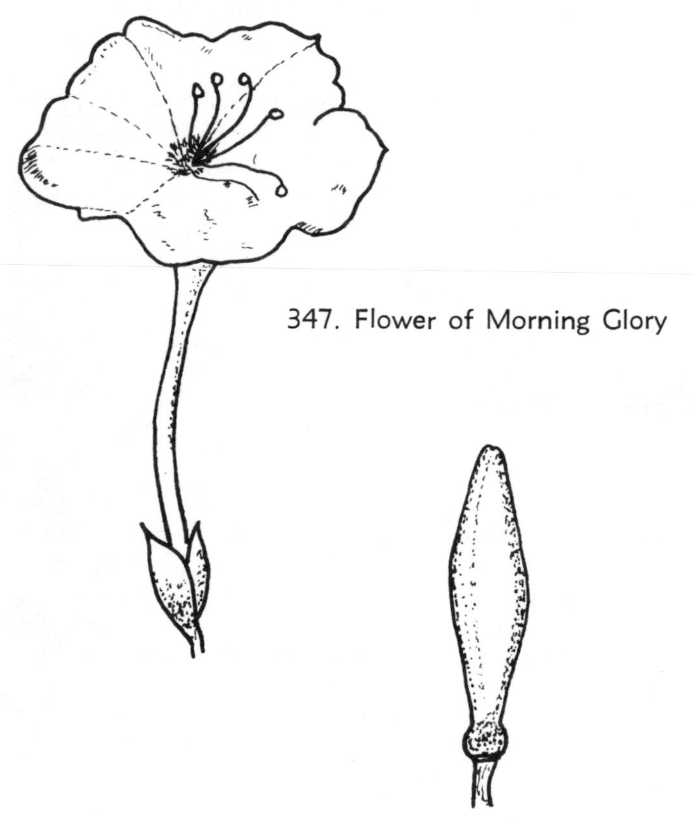

347. Flower of Morning Glory

348. Fruit of Morning Glory

To see a vigorous vine trailing its way over a stone wall, or winding about the shrubbery of a wayside thicket with pretty bell-shaped pink and white flowers, and to think of it as a morning glory is not to err too badly, for it is kin to the climbing vine of the garden trellis. It is a bindweed, however, of which there are several species, one hundred and seventy-five to be exact, and all with the same family characteristics as the morning glory, cypress vine and the others that have been mentioned. They all belong to the genus *Convolvulus,* from the Latin *convolvo,* meaning to roll around or twine. The name bindweed suggests a somewhat sinister meaning, a kind of evil plant that grips and seizes other plants, and if weaker than itself, perhaps with disastrous results. To be sure, the bindweeds are sometimes pernicious and obnoxious weeds.

Common along fencerows, wayside hedges, and in thickets and fields, the hedge bindweed has a smooth stem three to ten feet long, triangular or arrow-shaped leaves, with light pink, white-striped, or all white, bell-shaped flowers. A small edition of this bindweed, the field bindweed is an introduced species that has escaped to fields, meadows, and waste places. It may be distinguished from its more robust relative by the fact that the calyx is without leafy bracts, that of the hedge bindweed being enclosed in two ovate bracts. More insects seem to visit it than the other bindweed, perhaps because they are attracted by its peculiar fragrance. A third species, the upright bindweed, is a small, erect, or slightly twining plant, hardly a foot long, with oval, light green leaves, heart-shaped at the base, and funnel-formed, white flowers about two inches long.

One may usually find on any of the bindweeds during the summer months the goldbug, clinging like a drop of molten gold to the leaves. Few insects have the matchless beauty of this tortoise beetle. The American entomologist, Harris, in writing about it says that:

> when living it has the power of changing its hues, at one time appearing only of a dull yellow color, and at other times shining with the splendor of polished brass or gold, tinged sometimes also with variable tints of pearl. The wing-covers, the parts which exhibit a change of color, are lined beneath with an orange colored paint, which seems to be filled with little vessels; and these are probably the source of the changeable brilliancy of the insects.

There are several "goldbugs," but the species that Harris writes about is known as the golden tortoise beetle.

Like a living skein enlacing/ Coiling, climbing, turning, chasing, describes the common dodder, a parasitic member of the family, often troublesome in gardens, but more often found in meadows, ditches, streamsides and other low, damp, shady places where it weaves its way among and about other plants, however low or however high, for it is a climber, twisting itself about their stalks and extracting their juices through a thousand tiny suckers. Its threadlike stem, varying in color from dull yellow to dull orange, is crowded with bunches of tiny, dull white, bell-shaped flowers. There are other dodders all much alike and all starting at first from the ground, but after securing a convenient plant whereupon to climb, their roots die and they henceforth, live a parasitic existence.

119. Moschatel (Adoxaceae)

Only a single species makes up the moschatel family, the muskroot or moschatel, a small perennial herb found in shaded rocky places from Arctic America, south to Iowa and Wisconsin, and in the Rocky Mountains to Colorado, and, also, in northern Europe and Asia. It has a tuberous root stalk, basal and opposite ternately compound leaves, and small, green flowers in terminal heads. The flowers have two to three sepals, four to six petals, a rotate corolla, four to six stamens, and a three to five-celled ovary, an ovule in each cell. The fruit is a small drupe with three to five nutlets. The name moschatel is a diminutive meaning musk, in reference to the odor of the plant.

120. Mulberry (Moraceae)

Many people associate the mulberry tree with the silkworm that feeds on its leaves. The silkworm industry is not what it used to be, now that there are all kinds of synthetic fibers, but the mulberry and the silkworm still remain inseparable.

There are twelve to fifteen species of mulberry trees of the genus *Morus,* the old Latin name of the mulberry. They have alternate, often lobed leaves, and small, greenish flowers, without petals, but with usually four sepals, in stalked, hanging catkins, sometimes on different trees. The fruit is edible, berrylike, but it is actually a multiple fruit consisting of many nutlets, each covered with the fleshy sepals from several flowers. It resembles the blackberry.

There are two native species of mulberries in the United States, the red mulberry and the Texas mulberry, and two naturalized species, the white mulberry and the black mulberry. The red mulberry, discovered in Virginia, held out hope to the early colonists of a profitable silkworm industry, "apt to feede Silkeworms to make silke, a commoditie not meanely profitable." But the red mulberry proved a poor substitute for the white mulberry, whose leaves are the chosen food of the silkworm.

The red mulberry is a tree up to sixty feet high, of rich open woods with a scaly, brown bark and light green leaves, somewhat variable in shape, but usually oval or ovate, often with two or three round-notched, abruptly pointed lobes, and coarsely toothed. The staminate and pistillate flowers are on the same

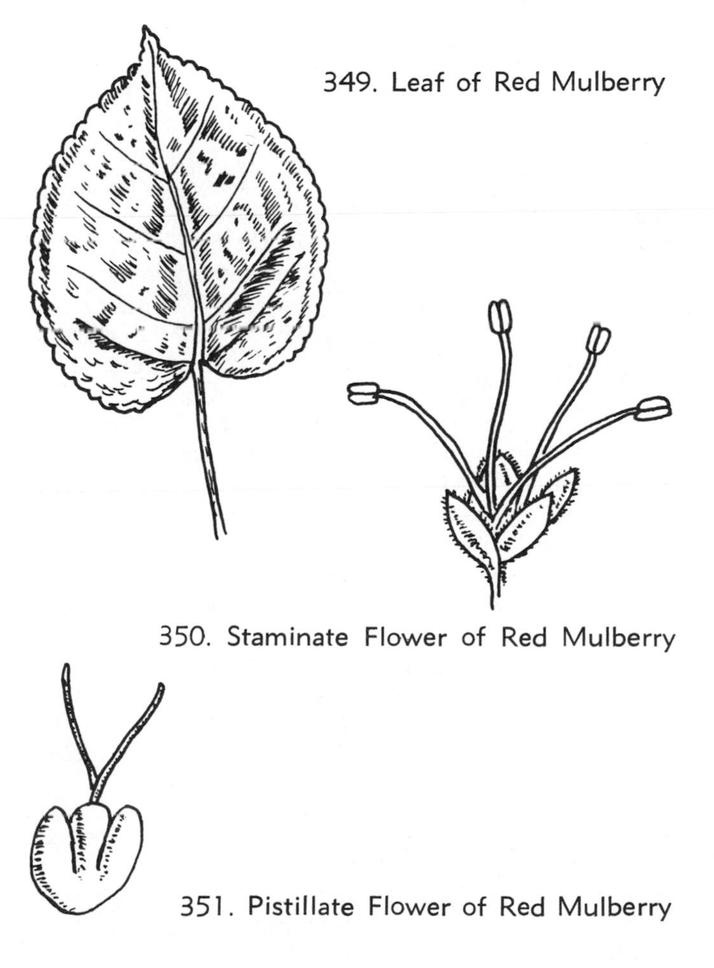

349. Leaf of Red Mulberry

350. Staminate Flower of Red Mulberry

351. Pistillate Flower of Red Mulberry

tree, not infrequently on separate trees, the staminate in drooping, greenish catkins, the pistillate in narrow spikes, with the latter developing into a juicy, cloyingly sweet, succulent, dark purple fruit, of which, the songbirds are very fond as, indeed, they are of all mulberries. It is said that the Indians made ropes and a coarse cloth from the bast fibers of the bark.

The Texas mulberry grows as a small tree or shrub in the arid Southwest. Its leaves are similar to those of the red mulberry, but much smaller, and its fruit is somewhat spherical and black.

When the red mulberry proved a poor substitute for the white mulberry, the latter, a native of China, where it had been cultivated from ancient times for the purpose of rearing silkworms, was introduced into this country, and though several attempts were made to raise silkworms they proved fruitless chiefly because of high labor cost. But the tree has remained with us and it is now found as an escape along roadsides, fencerows, and in waste land generally, being most abundant east of the Appalachian Mountains.

The white mulberry is a tree not over fifty feet high with a light gray brown bark, broadly oval leaves, usually lobed and coarsely toothed, and bright green above, and an insipid, sweetish, usually white fruit, though sometimes it is pinkish violet. A variety, known as the Russian mulberry, is smaller, more hardy, and usually has red fruit. It has been widely distributed in the West, where it has been recommended by the Forest Service for planting.

The earliest records of antiquity tell of the black mulberry. Ovid mentions the tree as the one introduced in the story of Pyramis and Thisbe, and Pliny speaks of it, saying that "of all cultivated trees the mulberry is the last that buds, which it never does until the cold weather is past and it is therefore called the wisest of trees." The Greeks dedicated it to Minerva and about 1605 the tree was introduced into England for the purpose of raising silkworms, but the attempt was doomed to failure for the silkworm refused to eat the leaves. The black mulberry, probably a native of Persia and cultivated in Europe for its fruit, was introduced early in the Pacific and southern states for its large, dark-colored juicy fruit, wherefrom the tree gets its name in part, and also from its sombre foliage.

Children of another generation played a game, with the refrain that some readers may remember:

As we go round the mulberry bush,
The mulberry bush, the mulberry bush
As we go round the mulberry bush,
So early in the morning.

Called a mulberry, but not a true mulberry, the paper mulberry is one of three known species of an Asiatic genus that is cultivated in this country, and an escape to roadsides, as an ornamental of considerable popularity because of its luxuriant foliage. It is a small tree, twenty-five to forty feet high, with a light gray brown bark, large leaves that are similar to those of the red mulberry, and with the staminate and pistillate flowers on different trees, the staminate in hanging catkins, the pistillate in small, compact heads. The fruit is a syncarp, in other words, an aggregate of many carpels, the product of the ovaries of several flowers. The name of paper mulberry is due to the fact that in Asia its bark is manufactured into paper. It was formerly planted as a street tree in our country, but recently, less so because of its troublesome habit of throwing up suckers.

The mulberry family, sometimes also known as the fig family, and of interest because of its edible fruits, is a large group of perhaps fifty-five genera and some one thousand species of trees, shrubs, herbs, and vines with milky juice that are largely tropical in distribution, though there are numerous species in the temperate regions of the world. They have alternate, usually simple, often deeply lobed leaves, regular flowers, without much distinction between sepals and petals that together are usually four in number, and with a one to two-celled ovary, small and inconspicuous, the staminate and pistillate separate, sometimes on different trees. The fruit is an achene, drupe, or nut, sometimes the true fruits being enveloped by or imbedded in a thickened, fleshy perianth as in the mulberry, or in a greatly enlarged axis as in the fig. The family includes, in addition to the mulberries, such plants as the osage orange, noted for its extremely hard wood and its large, orangelike fruits, hemp, hops, and various species of figs, as well as the common "rubber plant" of our conservatories, and the banyan tree of India.

The osage orange was named in part because its fruit looks like an orange, and in part, because it appears to have been introduced to the early settlers of St. Louis by the Osage Indians. Later, when it came to be used as a hedge (its spiny branches made an effective barrier) in regions where fencing material was difficult to obtain it became known as hedge apple, and by this name it is known to many people of the prairie states.

The osage orange, originally restricted to an area from Kansas and Missouri, to Oklahoma, Arkansas and Texas, but now planted as an ornamental and a hedge throughout the East, is a handsome, roundheaded tree, thirty to fifty feet high, with a very light brown bark, twigs set with spines or thorns, and ovate to oblong-lanceolate leaves, abruptly pointed, and lustrous, deep green above, paler beneath. The staminate flowers are in loose, short, rounded clusters, the pistillate in a dense spherical head, both staminate and pistillate flowers greenish and on separate trees. The orangelike fruit is a peculiar, spherical, green multiple of drupes, four to five inches in diameter. When crushed, it exudes a bitter milky juice. The wood, which is a bright orange color, and that yields a yellow dye when extracted with hot water, is used for making bows, hence, the name bowwood whereby the tree is sometimes known.

The largest of the many genera that comprise the mulberry family is *Ficus,* the classical Latin for the fig, a huge group of chiefly tropical trees, shrubs, and vines, some being epiphytes at first, others in the tropics being strangling trees, and still others producing aerial roots. They have the characters of the family.

There are two native species, the Florida strangler fig, and the shortleaf fig, both found in southern Florida. Both are epiphytes at first, but then become trees, the Florida strangler fig, also known as the golden fig, a broad, round-topped tree, fifty to sixty feet high, with oblong leaves, narrowed at both ends, and nearly stalkless yellow fruit, sending down strong aerial roots and becoming banyanlike, the short-leaf fig, forty to fifty feet high, with broadly ovate leaves and spreading branches, occasionally developing aerial roots and forming an open irregular head. An introduced species, the common fig, is cultivated for its fruit in the southern states, sometimes being found as an escape. It is an irregularly branching tree, usually not over twenty-five feet high, with leaves deeply three to five-lobed, rough above, hairy beneath. The fruit is the edible fig. The tree is a native of the Mediterranean region.

The flowers of the fig tree are produced on the inside of a hollow pear-shaped structure so that the "fig wasps" that fertilize them have to crawl in through a small opening. The dependence of the fig upon the wasps is an interesting and baffling chapter in the romance of science.

The fig has been in cultivation in the Old World since before 3000 B.C. and has figured prominently in the history of the peoples of Asia Minor for centuries. It was cultivated in California over a hundred years ago, many fig trees having been grown in the various mission gardens, there having been such trees in the Santa Clara Mission in 1792.

Also members of the same genus are the common household rubber plant, naturally a large tree, but usually grown as a pot plant, with oblong-elliptic, green, glossy leaves, and a yellowish fruit, the sacred peepul tree of the Hindoos, also called the bo tree, sometimes planted in Florida and California, with roundish or ovalish leaves and a purplish fruit; and the banyan tree of India, a widely spreading tree with many aerial roots that ultimately become additional trunks, broadly oval leaves, and a red fruit. There is a banyan tree in India that covers nearly an acre—in time a banyan tree can become virtually a forest.

When Captain Bligh sailed from England in the ship *Bounty* for the South Sea Islands his mission was to obtain some specimens of the breadfruit tree for transplanting to the West Indies. His mission was doomed to failure, but a second expedition was successful and he returned with over a thousand young trees. But the fruit did not prove as useful to the English as had been hoped.

352. Leaves and Flower Cluster of Common Fig

353. Leaf of Breadfruit

355. Leaves and Flowers of Hemp

354. Fruit of Breadfruit

The breadfruit tree of Polynesia is a tree up to fifty feet high with a striking and dense foliage. The leaves are eighteen to twenty-four inches long, thick and leathery and somewhat oval, but deeply lobed with sharp segments. The staminate flowers are in spikes, six inches long, yellowish, drooping or curving downward; the pistillate flowers in nearly globe-shaped clusters. The fruit is four to eight inches in diameter, yellow, the outside slightly spiny, the flesh delicious. In the true breadfruit the fruit is seedless, but in a variety, known as the breadnut, the fruit does contain seeds. The breadnut is cultivated for the seeds that are cooked.

Of considerable economic importance, the hemp plant is a valuable species of the family. It belongs to the genus *Cannabis,* the classical name of the hemp, and incidentally, is the only species. A native of central Asia it is widely grown for its fibers, but it is also found as an escape, on wasteland in many parts of the world. It is known to have been under cultivation for several thousand years before Christ. It was first planted in Kentucky in 1775, which still remains the chief hemp producing state, though the plant is also cultivated in other states.

To obtain the fibers, the stems of the plant are cut and allowed to stand until dry. They are then spread out on the ground and exposed to cool, moist weather, or to alternating freezing and thawing. The process, called retting, separates the bark from the inner tissues and promotes the decay of the softer tissues that surround the fibers, These fibers are removed and woven into cordage, twine, rope, sacking or sail cloth.

The hemp is a strong-smelling herb, with a rough, almost woody stem, pinnately compound leaves, and small, inconspicuous, green flowers, the staminate and pistillate flowers on separate plants. The fruit is a compressed, ovoid achene. The seeds frequently form a common ingredient of "bird seed." Both the young leaves and the pistillate flowers contain a powerful narcotic substance or drug called hashish though in the United States it is known as marijuana.

Another important member of the mulberry family is the common hop. It is a native of Eurasia, but widely cultivated as an agricultural crop and found as an escape on alluvial banks, rubbish heaps, and similar places. It is a tall-growing perennial vine with leaves generally three-lobed, the middle lobe larger than the others, rough above, but not so rough below. The staminate and pistillate green flowers occur on separate plants, the staminate in catkinlike racemes with a five-parted calyx and five stamens, but without petals; the pistillate in pairs, each pair beneath a large bract.

The bracts at maturity form a conelike body (the hop), and are dotted with resinous glands that contain a substance known as lapulin. It is this lapulin that is used in brewing beer, in making a malt extract, and before the days of yeast cakes, it was also used in raising bread. The fruit is a small achene, of no economic value. Our commercial hop fields are found chiefly in New York and Oregon. Another species of the hop genus *Humulus* is a quick growing, stem-climbing, annual vine useful for covering fences or unsightly buildings.

The hemp and the common hop have been placed in the nettle family (Urticaceae), the hemp family (Cannabinaceae), and the mulberry family (Moraceae). Some authors consider these three families together with the elm family (Urticaceae) to be subfamilies of a single family (Urticaceae).

121. Mustard (Cruciferae)

When one has a dinner of corned beef and cabbage, or when one has a dish of cole slaw served with an entree, one may well pause for a moment and reflect that thousands of miles away, the ancestor of our cultivated cabbage still may be found growing wild on the sea cliffs of England, Wales, and Europe. Indeed, it is the ancestor also of Brussels sprouts, the delicate cauliflower, broccoli, and turnip that are grown with so much care in our gardens. Not only is the wild cabbage the ancestor of these varieties, but of hundreds more found in almost every country.

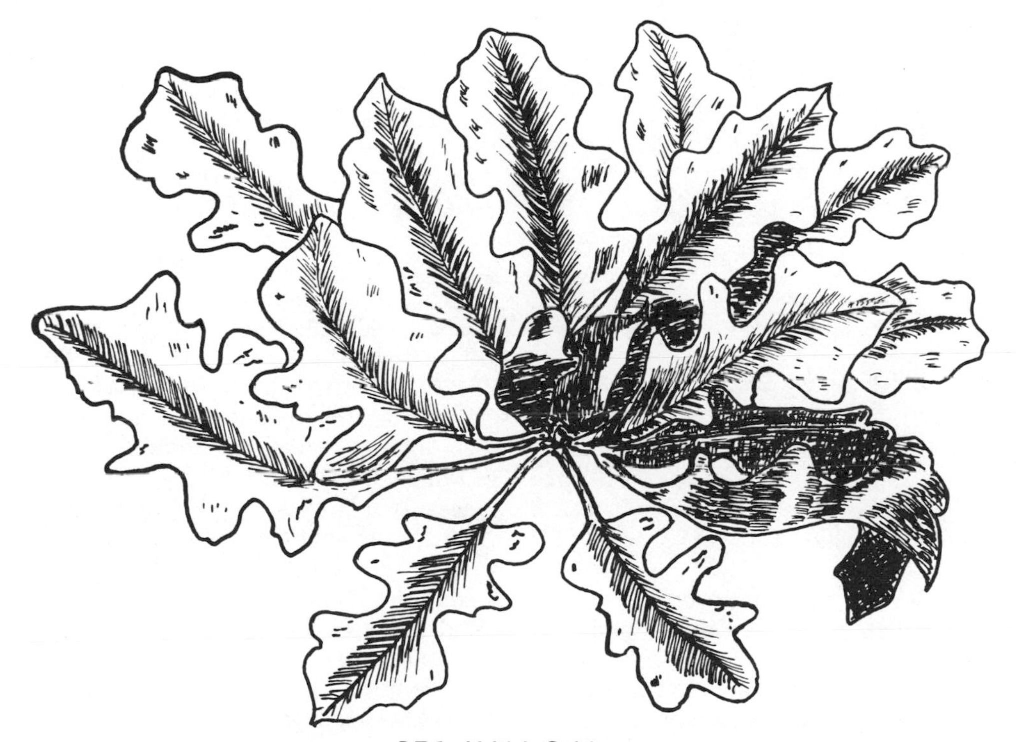

356. Wild Cabbage

Of this primitive cabbage, known as the sea cabbage or wild cabbage, Anne Pratt wrote in *The Flowering Plants of Great Britain* that

> few plants are more conspicuous on the sea cliffs of England than this cabbage; from May till the end of summer it is one of the loveliest ornaments of the cliffs. Much of it is verdant when all around is fading; and dark purplish, red-tinted leaves mingle with those which are green, and with others which are of the deepest yellow, and when the hoar-frost spangles them they seem enriched with glittering diamonds.

It is not often that a primitive plant figuratively is found growing side by side with our cultivated forms. The wild cabbage is an exception. Merely a glance at the wild cabbage, with its white or yellow flowers, and few scattered leaves of green or reddish purple, and growing in its bleak and barren environment, and one may well wonder how culture could have brought about such remarkable changes. Because as a rule, our domestic plants vary so much in form from their wild forbears, that it is difficult to recognize the original stock, which, not infrequently, has already disappeared.

357. Flowering and Fruiting Stem of Cabbage

The cabbage has no great value as food yet it has a place on our table. And, incidentally, the next time cabbage is served at home, taste the core. It has a somewhat pungent flavor like that of the mustard spread on "hot dogs." The radish, that was grown at the time of Christ, also has a sharp flavor and so, too, has the watercress, highly prized for its pungent-tasting young leaves.

All these plants are related to one another and all belong to the group of plants commonly called the mustard family. It is a large aggregation of about two thousand species, and two hundred genera of what may seem widely diverse plants that are widely distributed, but more abundant in the temperate and cold parts of the world, rather than in the tropics. Many are of considerable value to man either as food or as ornamentals. They are annual or perennial herbs, rarely shrubby, with a watery juice that is often acrid. The leaves are prevailingly alternate, simple, but often deeply lobed or pinnatifid, and frequently bitter, but never poisonous. To anyone unfamiliar with botanical classification, the placing in the same group of such seemingly different plants as the cabbage and its numerous varieties, mustard, water cress, pennycress, and such garden flowers as the sweet alyssum, candytuft, and stock would appear to be an arbitrary and confusing procedure. But these plants have been grouped together because they all have the same floral structure (all plants as well as animals are classified according to structure), and the same type of fruit.

Examine the pale purple flowers of the radish or the small white flowers of the sweet alyssum or candytuft, and compare them with the small, pure yellow flowers of the black mustard, that blooms in the fields from June to September, and one will find that they are perfect, regular and alike in having four sepals and four petals, that are arranged in the form of a cross, and six stamens. Later on, when the flowers have gone to seed, examine the fruit and one will find that they, too, are much alike, being a pod that, when the seeds are ripe, opens by two valves, and thus, leaving the seeds free to fall. Such a seed pod, that may be globular or oval, flat or greatly elongated, is known as a silique when longer than broad, and as a silicle when about as broad as long.

The scientific name of the mustard family is *Cruciferae,* from the word *crucifer* meaning crossbearer, and was adopted because of the crosslike arrangement of the petals and sepals. The crucifers are, by no means martyrs, despite their name, but are a group of vigorous plants adapted to more than hold their own in the never-ending struggle for existence. Looking at any one of them one will observe that they make one stem serve for many flowers, usually arranged in a raceme, and that they do not permit a flower to develop at the tip of the stem, for to do so would be to put a stop to its upward growth.

Consider, for instance, the shepherd's purse, a common and abundant wildflower that may be found growing along the roadsides and in fields and waste places. It is a slender little plant, and an importation from Europe, with small white flowers in a long, loose raceme that are followed by triangular and notched seed pods. Look at the plant closely and, bearing in mind what has just been said about the stem, observe that one stem does service for all the flowers, and that not one of them is found at the tip. Note also that the flowers, though small individually, are not entirely inconspicuous. This is because they are clustered together so that collectively, they attract those insects, such as the flowerflies and houseflies, that aid in pollination, something that they might not succeed in doing were they arranged separately by themselves on the stem. And should the flies fail, then the long stamens standing on a level with the stigma are well calculated to self-pollinate the flowers.

This is the reason why the shepherd's purse has been able to march around the world and compete successfully with other plants. Notice, too, that as summer passes into autumn, the plant continues to blossom until frost covers the ground, and even during the winter, when snow blankets the landscape, it will flower in sheltered and secluded nooks. By thus extending its flowering season far beyond that of any native flower, it avoids the fierce competition for insect trade that it would otherwise have to contend with were it to flower during a shorter period. Observing the places where it is found growing, one will discover that it is not a proud plant, but will take root wherever it may. It is not choosy, but is easily satisfied with unoccupied waste land and other localities where other plants refuse to grow. Is it any wonder, then, that the shepherd's purse has been eminently successful?

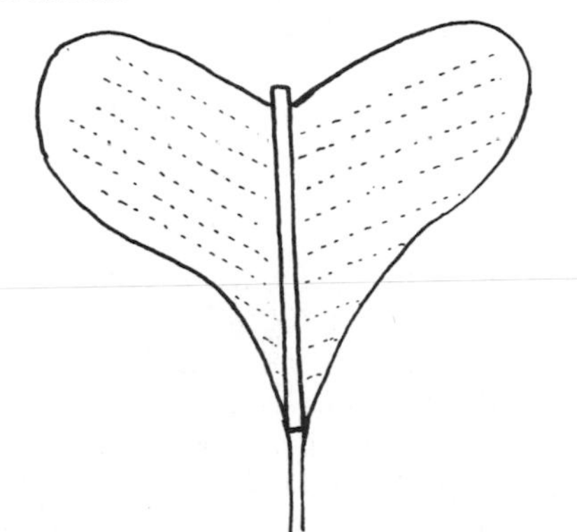

358. Fruit (Silicle) of Shepherd's Purse

Those who follow the byways of nature know the shepherd's purse well with its peculiar triangular pods, silicles, that are supposed to resemble the purse once carried by shepherds. The peppergrass, another wild crucifer that may be found along roadsides and in waste places, growing literally side by side with the shepherd's purse, is equally familiar.

When a boy I used to place the seed pods, which cluster thickly about the flowering stem in a cylindrical curving column, between my teeth and break them for their peppery flavor. I still do on occasion when I wander through the fields and woods.

As a rule, the early spring flowers are small, shy, and seemingly delicate little plants, and may be found more by accident or by painstaking search. There are exceptions, of course, such as the yellow rocket that sends upwards from a single root crown, a dozen or more sturdy stems a foot or more high, each bearing a showy, panicled spike of brilliant, yellow flowers that are frequently visited by the early bees and handsome flowerflies. This is the plant that brightens meadows and the banks of neglected runlets in early April when it gives color to a landscape not yet painted by nature's paintbrush.

Although the yellow rocket is one of the earliest of the mustard group to appear, it is not the earliest, this honor going to the common or vernal whitlow grass that blossoms in March, though some books have it in February. It is a small, rather insignificant plant, not more than five inches high, though some-

359. Flower of Whitlow Grass

times not more than an inch high, with four white, deeply notched petals and small, hairy, lance-shaped, toothless leaves that are clustered at the base of the flowering stems that are themselves leafless. It is also an immigrant from Europe, and common throughout our range in barren fields and along the roadside.

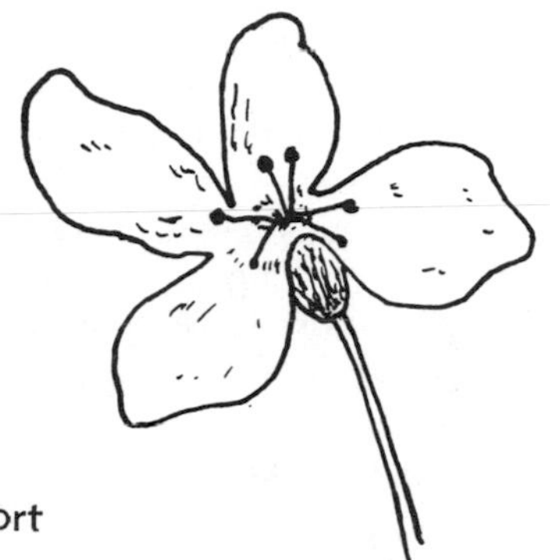

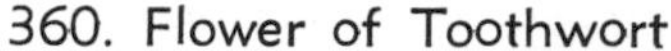

360. Flower of Toothwort

Among the anemones, bloodroot, claytonia and other spring flowers one may find the toothwort or crinkleroot, if one searches diligently. It is a low woodland plant having two opposite leaves, each of three ovate and toothless leaflets on the stem, and inconspicuous white flowers with many yellow stamens. The long root is wrinkled, toothed (whence the plant's name), and is edible with a pleasant pungent flavor like watercress.

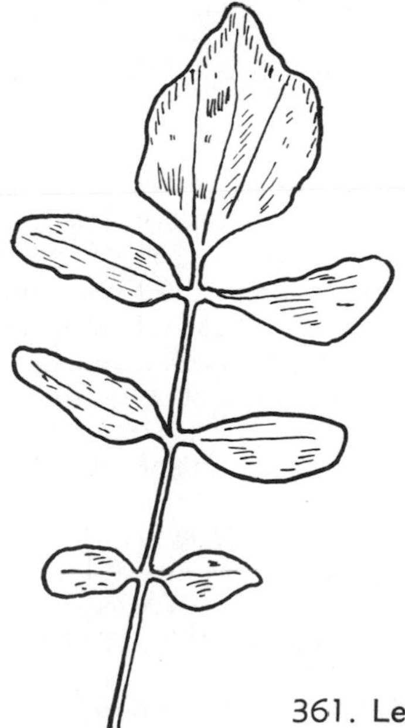

361. Leaves of Watercress

And speaking of watercress, here is one of the aquatically inclined members of the mustard family, for it delights in brooks and small streams. This extremely hardy, European perennial has for centuries, been highly prized for its pungent leaves, and the tender shoots, both having been used for greens and salads since ancient times. They are also much used as a garnish. It is said that Xenophon recommended the plant to the Persians, and the Romans considered it an excellent food for mentally ill people. It is cultivated in many parts of Europe, and in the United States, great quantities are produced for the market where it is sold in such closely packed bunches that it is hardly recognized as the plant found floating on the water in the wild. As Tennyson put it in "The Brook": "I linger round my shingly bars/ I loiter round my cresses."

There are other cresses, such as the spring cress, the bitter cress, and the rock cress to name a few. The spring cress, like the watercress, delights to grow in water, so it is found in wet meadows, low ground, and springs. The flowers are visited and cross-pollinated by the mining bees, and the leaves serve as food for the fulcate orange tip, an exquisite, little, white butterfly with a dark yellow, triangular spot across its wings. The bitter cress, too, likes watery, low-lying ground. Known also as the ladies' smock or cuckoo flower, its white blossoms open in April and May when "Lady-smocks all silver white" appear in our meadows, as they did in Shakespeare's England, though it must be admitted that our flowers seem to have more of a rose color. Its showy flowers, that are larger than those of the spring cress, attract bees, flies, butterflies and other insects. Indeed, it is said that because of its showy flowers and the large amount of nectar that they contain, this plant is visited by more insects than any other crucifer. This may explain, in part, why it has triumphantly marched around the world.

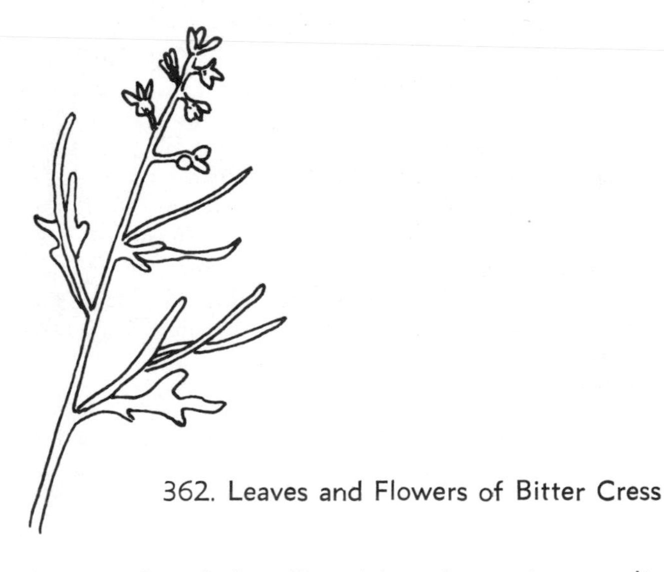
362. Leaves and Flowers of Bitter Cress

Unlike the spring and bitter cresses, the rock cress, as might be suspected from its name, prefers rocky banks and stony pastures. It is a generally hairy little plant with white or greenish flowers that are visited by small bees and flies, and that are followed by very narrow, flattened pods two inches long or less. At a glance it might be mistaken for the early saxifrage, but its floral structure should quickly serve to identify it. The cresses do not belong to a single genus, but are found among several genera. Thus, the watercress belongs to the genus *Radicula,* the spring cress and bitter cress to the genus *Cardamine,* and the rock cress to the genus *Arabis.*

The mustard cultivated for its brown seeds, is the black mustard. It also grows wild, in such places as fields and along the roadside. It is a tall annual, with stiff hairy, mostly green foliage, and yellow flowers in many short clusters that appear throughout the summer, from June to September. A similar, but rarer species, is the white mustard. It is more or less hairy, with yellow flowers that are a little larger than those of the black mustard, and with bristly pods that are constricted like a necklace, its brown seeds somewhat lighter than those of its relative. The seeds, too, are used as a condiment, but being considerably milder, are more often mixed with those of the black mustard. Incidentally, both of these mustards are naturalized from Europe and both are members of the genus *Brassica.*

There are other wild mustards, notably the hedge mustard, and the field mustard or charlock. The former is a homely, straggling plant with tiny, light yellow flowers and light green, smooth leaves. The pods are awl-shaped and are closely appressed to the stem. This mustard abounds in waste places and blooms from May to September though it has been seen blossoming in November.

The field mustard, found in fields and waste places and widely distributed throughout the northern states, grows two to three feet high and has oval, coarsely saw-toothed leaves that stand boldly out from the stem. The lower leaves, however, are more irregular and lobed at their bases. All are rough to the touch, however, and conspicuously veined. The light yellow flowers, measuring more than half an inch across, have six stamens, like all members of the mustard family, but two of them are shorter than the other four. Honeybees, the brilliant flowerflies, and other small insects visit the flowers regularly for their nectar. The insects insert their tongues between the stamens, and thus, cross-fertilize the flowers. On stormy days when insects are not abroad, the stamens curve upward and bring the pollen-covered anthers in contact with the stigma, for it is better for the flowers to set self-fertilized seed than none at all. The seeds of the mustards are on the pungent side, but those of the field mustard are milder than the others. It is doubtless for this reason that they are eaten by various birds, both game birds and songbirds. Even some mammals like them.

"What is a weed?" This is a question often asked, and what is surprising is that it admits of many answers, when in truth there is only one, and a simple one at that. A weed is a plant growing in a place where it is not wanted. Thus, to give an extreme illustration, the most exquisite orchid would be a weed if found growing in a tomato patch. The mustards that have been mentioned, often invade our gardens, and thus, become weeds. So, too, do many other plants. But this invasion is not a one way affair for many cultivated flowers have escaped from gardens and taken to growing in the wild. The radish is one such plant, and it is quite frequently found growing in old fields. This plant, whose root is a familiar food item on the table, has been held in high esteem since even before the Christian era, when a volume was written on it alone. It is said that the ancient Greeks, in offering their oblations to Apollo, presented turnips in lead, beets in silver, and radishes in vessels of beaten gold. And Pliny tells of a radish eaten in Rome that had such a transparent root that one could see through it.

The horseradish is another crucifer that has escaped from the garden and now may be found growing in waste places, especially along brooks. For centuries the horseradish has been cultivated to satisfy the jaded appetites of the rich and near rich. The root is parsniplike and white, and always branched below. When bruised, it gives off a volatile oil of strong pungent odor, and has a hot, biting taste, that gives the plant its antiscorbutic value. The flowers are white, borne on an ascending, terminal raceme, and the silique is globular, but for some strange reason the plant invariably fails to set viable seed in this country. Hence, the horseradish is propagated by root cuttings. As a condiment, the root is grated and mixed with vinegar.

The edible members of the mustard family—cabbage and its varieties, mustard, turnip, radish—have been under cultivation since time immemorial. The turnip and cabbage are known to have been grown for as long as two thousand years, the radish was known at the time of Christ, and the horseradish for about as long. But the ornamentals that are grown in our garden, appear to have a shorter history.

Of these, the sweet alyssum is undoubtedly one of the most favored. One has only to glance at it to see that it has all the characteristics of the family; alternate leaves, the raceme of flowers, and the cross-like arrangement of the corolla, as well as the tiny pointed spherical pod, whose juice is biting to the taste. Look down at a flower cluster, and it becomes quite apparent why the sweet alyssum blossoms all summer. In the cluster there is a great number of buds in all stages of development and produced from the side of the stem. The growing point at the center does not bring forth a flower, and so, as the lower and outer flowers mature and ripen seeds, this growing point lengthens and produces more buds. This continues until frost whitens the ground. It may be found in blossom as late as Thanksgiving, often much later.

One of the most pleasing of our garden annuals is the stock or gilliflower. It belongs to the genus *Mathiola,* named for Peter Mathioli, an Italian writer on plants. The genus contains many species native to the Old World, all herbaceous plants with stiff, hairy stems, and flowers collected in a dense terminal mass and ranging in color from white through rose, crimson, purple, and parti-colored. Only two species, however, are in common cultivation.

"What's in a name?" is a remark that is often heard and that can be taken several ways. Consider, for instance, the candytuft, a favorite garden flower with many people. One might be excused for thinking that the name was given to the plant because of some resemblance to a confection or that it was at one time used in the making of a sweet. This, however, is not the case. The name is derived from the island of Crete, the old English name for this piece of land in the Mediterranean being Candy or Candia, wherefrom seeds were brought into England some three hundred years ago that produced the plant known as the purple candytuft. Gerarde, an early English botanist, relates that in 1587, he received some seeds that produced in his garden, flowers that were "sometimes blue, often purple, sometimes flesh-colored and seldom white."

There are some twenty-five species of candytuft, but only three seem to have found much favor. They are the rocket candytuft, a more or less erect annual about twelve inches high, with fragrant white flowers; the purple or globe candytuft that is the leading annual garden species, a more or less erect plant some eight to fifteen inches high with flowers in close clusters and that are pink, red, lilac, or violet, but not fragrant; and the evergreen candytuft, a most delightful plant that spreads its shrubby stems in a thick mat over the ground and opens its white flowers in April. It is a robust perennial, and for this reason this candytuft is much preferred.

A garden ornamental that is often planted, not for its flowers, that are ordinary, but for its curious flat seed pods, is the honesty or lunaria. It is an erect, bushy plant and rather coarse, and the pink purple flowers have little to commend them, but it is redeemed by its seed pods that have been called the "Pope's Money." They are extensively used in winter bouquets of dried plants. The seed pods vary between oval and orbicular in outline, and when the valves fall off, the seeds remain attached to a thin, pearly membrane wherefrom they soon slip away. This thin, translucent, partition disk has given the plant the name of moneywort. But it is also called lunaria, from luna, the moon, that the color and shape of the fruit suggest.

Honesty holds the record of being a favorite in English gardens for more than three hundred years, but the dame's violet or sweet rocket, as it is also known, can match this record. Early English garden-

363. Flowering and Fruiting Stem of Honesty

ers have written that the ladies of the time were fond of having it in their apartments because of its fragrance, hence, its name. But its generic name is *Hesperis,* of Greek derivation meaning evening. It very likely received this name because the flowers of some species are more fragrant in the evening. There are two species of honesty, one being an annual, though cultivated as a biennial, the other a perennial. There are, however, about twenty-five species of *Hesperis,* all Old World biennial or perennial herbs. Dame's violet is a vigorous, hardy perennial forming clumps two to three feet high, branching from the base, and covered with showy terminal racemes of white, lilac, pink, or purple flowers. Both honesty and dame's violet are escapes from our gardens, and may be found growing in the wild.

The mustard family has relatively few garden ornamentals. Besides those that have already been mentioned, there is also the wallflower, a garden favorite from remote antiquity. It is a hardy plant, though the idea prevails that it is not hardy, and grows to a height of twelve or thirteen inches. It is covered with minute, forked hairs, and has narrow leaves that are often clustered beneath the flowers that are yellow or orange brown, and fragrant, and arranged in terminal racemes. Originally a native of southern Europe, it has been much improved so that it now comes also in reddish or reddish black shades. The wallflower does not like wet or slushy winters, probably accounting for the belief that it is not hardy. As its name implies, it is useful in the rock wall, but it is also used for the border and as a bedding plant. It is essentially a garden plant with us, but in Europe it "sings among the ruined walls/ and covers with light the grieving stones."

122. Myoporum (Myoporaceae)

The myoporum family is a small group of five genera and ninety species, chiefly Australian and Pacific Island shrubs and trees. They have opposite, alternate, or scattered, usually entire leaves, flowers with a more or less five-parted calyx, a tubular or funnel-shaped corolla, four stamens, and a two-celled ovary, though sometimes two to four-celled, and a berrylike fruit.

The genus typifying the family is *Myoporum,* Greek meaning that the leaves have resinous dots, a group of thirty species of shrubs or trees, having small, axillary, white flowers, with a five-parted, bell-shaped calyx. Several species resemble heaths, and only about three are cultivated for ornament, mostly in the greenhouse, and are otherwise little known. Probably the best known member of the group is the bastard sandalwood, also called naio, a tall tree in its native region of Hawaii, sometimes as much as sixty feet high, with oblong leaves mostly crowded at the ends of the twigs, white or pink flowers in clusters of five to eight, and a white fruit.

123. Myrsine (Myrsinaceae)

The myrsine family, with thirty genera and about six hundred species of trees and shrubs, is distinctly a tropical group, with a few species in southern Florida, having alternate, glandular leaves, white or pink flowers, and a one-celled indehiscent fruit.

One of the species found in Florida is a plant known as the marlberry. It is one of two hundred species of trees and shrubs of the genus *Ardisia,* Greek for a point, all being tropical and often known as spear-flower. They have alternate, leathery, often shining, green leaves, and small white or reddish flowers in cymes or panicles. The flowers have a five-lobed calyx, a five-lobed corolla, five stamens, and a one-celled ovary. The fruit is a drupe. The marlberry, occurring in Florida, is a shrub or rarely a small tree, up to about twenty-five feet in height, and found in moist hammocks and sandy soils near the sea. It has ovate, oblong, or obovate leaves, perfect, minute, fragrant, rusty red flowers in terminal panicles, and a lustrous black, globular fruit. Another member of the same genus, but a native of Asia, is often cultivated as a greenhouse pot plant. It is a bushy shrub, with oblong leaves, red flowers and scarlet berries. When in fruit it makes a fine shrub for Christmas decoration.

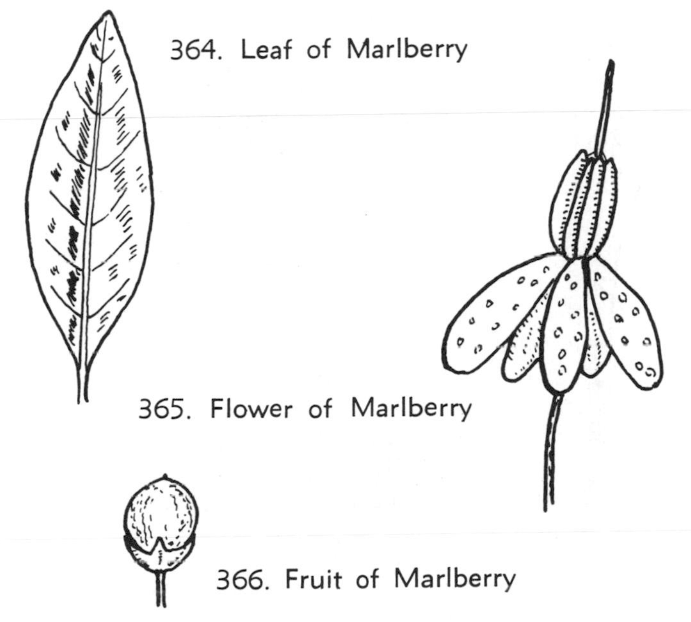

364. Leaf of Marlberry

365. Flower of Marlberry

366. Fruit of Marlberry

A second species of the family that extends into southern Florida is a shrub known as guiana rapanea. It is a member of the genus *Rapanea,* a tropical group of some one hundred and fifty species of evergreen trees and shrubs that are widely distributed throughout the tropical regions of the world. Some species found in South America are used extensively for pulpwood, furniture, and construction, and their barks yield tannin extracts.

The guiana rapanea is not much more than a tall, slender shrub, though sometimes it might become a small tree, with simple, persistent, leathery, obovate to oblong leaves that are often crowded toward the tips of the twigs, minute, perfect or unisexual, white, striped by purple, flowers in axillary clusters, and a globular, dark, blue black drupe. It grows on hammocks, flatwoods, and riverbanks.

124. Myrtle (Myrtaceae)

Guava, allspice, bay rum, and eucalyptus are familiar, almost household words, and all are trees of the myrtle family, a large group comprising about seventy-two genera and three thousand species of chiefly tropical, aromatic shrubs and trees, many of high economic value because of their edible fruits, aromatic oils, and their use as ornamentals.

Their leaves are mostly opposite, usually evergreen, thick and glandular dotted, and generally toothless. Their flowers, often very showy, are solitary or in clusters, perfect and regular, with four or five sepals, the same number of petals, many stamens, a one to many-celled ovary with one to many ovules in each cell. Their fruit is a berry, capsule, pod, nut or drupe.

The highly esteemed guava jelly is made from the fruit of the guava tree. There are one hundred and fifty species of guava trees, genus *Psidium,* the Greek name for the pomegranate that the fruit rather resembles, but only three are widely cultivated in warm countries for their fruit. These are the strawberry

367. Fruit of Guava Tree

guava, a shrub or small tree not over twenty-five feet high, with a smooth, grayish brown bark, elliptical, leathery leaves, white flowers about an inch wide, and a roundish, purplish red berry with white flesh; the common guava, a shrub or tree not over thirty feet high, with four-angled twigs, brownish green, scaly bark, oblong leaves, white flowers, and an egg-shaped or pear-shaped, yellow berry, the flesh whitish yellow or pinkish; and the Brazilian guava, a shrub, usually not over eight feet high, with oblong leaves, flowers in clusters of two to three, and a greenish yellow berry with white flesh. All three species are tropical South American plants, but the common guava is cultivated in a limited area in California, and in peninsular Florida where it has become naturalized in many places. The strawberry guava is also grown in protected areas of the Gulf Coast and southern California.

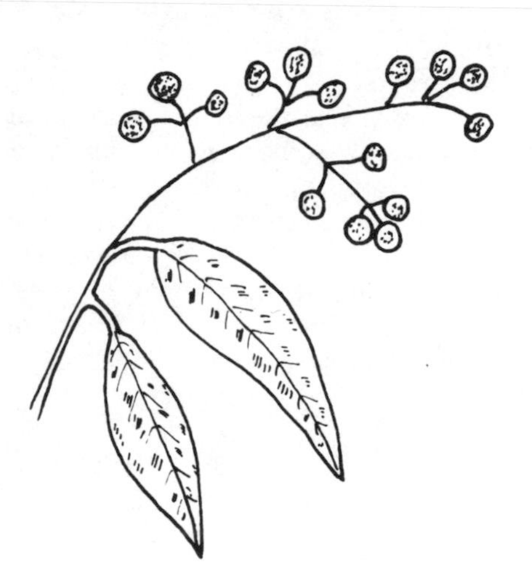

368. Leaves and Fruit of the Allspice Tree

The condiment allspice is made from the dried unripe "berries" of the allspice tree, one of five species of the genus *Pimenta,* the Latinized version of the Spanish pimento and is now the English name for the fruit of the allspice tree, but should not be confused with pimiento that is the fruit of certain peppers. The allspice is an aromatic tree twenty to forty feet high, of the West Indies and Central America, cultivated in Florida as an ornamental, with oblong leaves, small white flowers, and a dark brown, berrylike drupe. Another species of the same genus is the true bayberry, also called the bay rum tree, from whose leaves the aromatic product is obtained. It is a very aromatic tree, thirty to forty-five feet high, with elliptical leaves, small, white flowers, and a dark brown, berrylike drupe.

One of the most noteworthy genera of the family is the genus *Eucalyptus,* Greek for well and to cover, in allusion to the lidlike arrangement of the flowers, an enormous genus of over three hundred and sixty-five species of chiefly Australian, often gigantic, very aromatic, evergreen trees, many widely planted in the South for their striking flowers and foliage. They are especially popular in California where many species have been introduced from Australia, and where they are prized because of their freedom from insect pests, and for their use as bee trees.

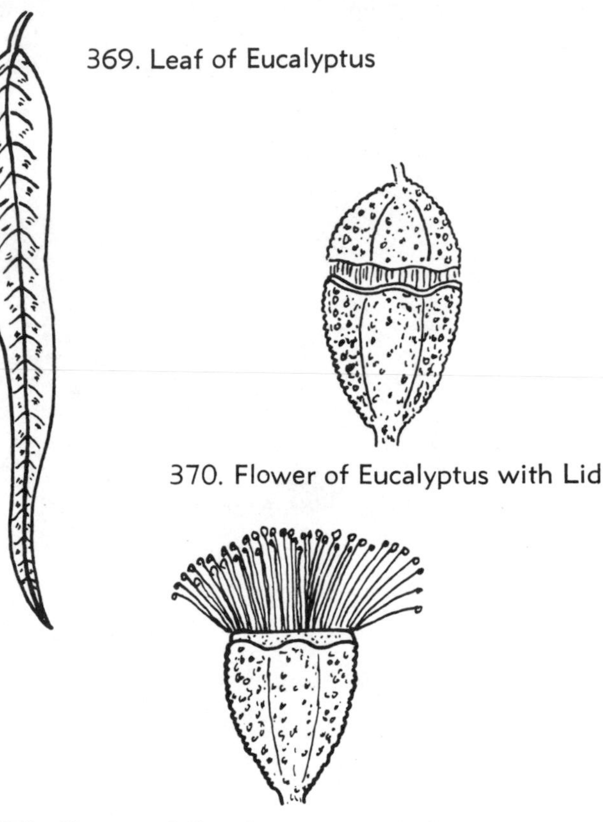

369. Leaf of Eucalyptus

370. Flower of Eucalyptus with Lid

371. Flower of Eucalyptus with Lid Removed

They have prevailingly alternate, toothless leaves, flowers usually in the leaf axils, often in small umbels, but at other times, in branched clusters, and white, yellow, or red, the calyx bell-shaped or turban-shaped, the calyx lobes and petals forming a lid at flowering time, and separating from the calyx tube, and a capsule as fruit, that opens at the top by three to six valves. The capsules are often colored and many are sent East for winter decorations.

The eucalyptus trees have much to commend them. They develop readily from seeds, grow with astonishing rapidity and vigor, and sprout indefinitely from a stump. They have a graceful habit, a handsome bark, beautiful evergreen, narrow and tapering, or sickle-shaped leaves that hang in graceful pendant clusters, and curious looking fruits, all features designed to enhance their popularity. Many of them have a very hard wood, and, durable under water and in soil, they have a number of uses. The flowers, leaves, and bark are aromatic, and yield such products as gums and resins useful in medicine and in the arts, such as the oil of eucalyptus. Finally, growing dense and tall, they serve as windbreaks, particularly about citrus groves, and are widely planted as street trees, arching over as the elms of New England.

Over seventy species of eucalyptus trees, also known as gum trees, are cultivated in this country, mostly in California, and of these the following should be mentioned: the peppermint gum, its foliage peppermint-scented; the blue gum, one of the most widely cultivated, with narrow, pointed leaves six to twelve inches long and creamy white flowers; the red gum, with a beaked or conical pointed fruit; and the scarlet flowered gum, with showy masses of white, pink, or scarlet flowers.

The "cloves" of commerce, used as a very aromatic, pungent spice, are the dried, unopened flower buds of the clove tree, a native of the Moluccas or Spice Islands of the Malay Archipelago. It is an evergreen, growing to the size of an apple tree with crimson flowers, that are not permitted to mature under cultivation. Each clove is a dried bud, the long, cylindrical part being the ovary of the flower, surmounted by the four, toothlike lobes of the calyx, the infolded petals being within. The clove tree belongs to the genus *Eugenia,* named for Prince Eugene of Savoy, a patron of botany, a group of perhaps a thousand species of chiefly tropical, generally aromatic, shrubs and trees, of considerable economic and garden value for their often edible fruits, and as ornamentals in tropical and subtropical regions, some of them grown in Florida and California.

The classic or true myrtle is an evergreen, aromatic shrub, three to nine feet high, a native to the Mediterranean Basin and western Asia, widely planted outdoors in the South and as a pot plant in the northern states. It has oval to lance-shaped, shining green leaves, white or pinkish flowers, and a bluish black berry that is white in the horticultural forms.

There are four species of the family that are native to southern Florida; two species of nakedwoods and two species of lidflowers. The two nakedwoods are medium-sized or small trees; the Simpson nakedwood with oblong to elliptical leaves, white flowers, and a one-seeded red berry; the twinberry nakedwood with mostly oval to obovate leaves and a small, black, aromatic berry. The lidflowers are both small trees; the pale lidflower with ovate to oblong sessile leaves and an oblong to nearly globular, reddish brown berry; the myrtle-of-the-river lidflower with elliptical, nearly sessile leaves, and a small, globular berry depressed at the apex.

Several species of the same genus as the clove tree, occurring in tropical America, have pushed their way northward, and may now be found in lower Florida and the Keys. They were formerly known as the stoppers, but are now called the eugenias. The white stopper eugenia is a small tree with mostly oval, ovate, or elliptical leaves, small white flowers, and a sweet

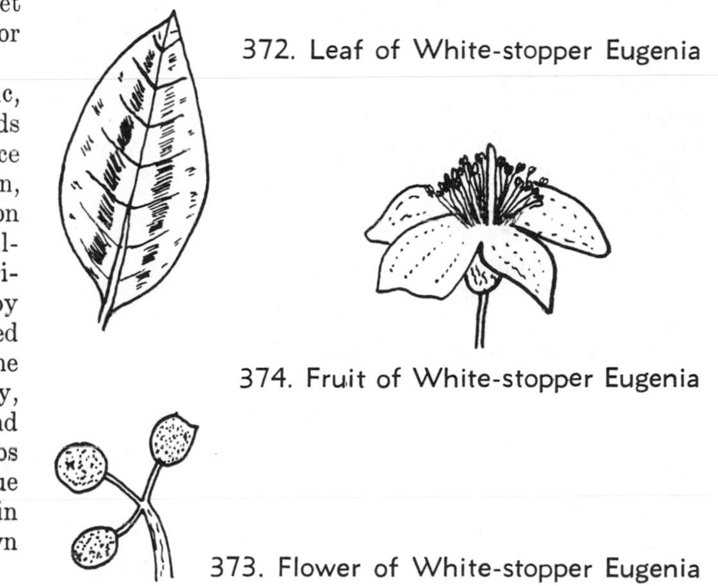

372. Leaf of White-stopper Eugenia

374. Fruit of White-stopper Eugenia

373. Flower of White-stopper Eugenia

and juicy, berrylike fruit. The spiceberry eugenia is also a small tree with an orange, berrylike fruit that is tinged with red or black, and the boxleaf eugenia is hardly more than a shrub with nearly sessile, obovate leaves and a small black, elliptical fruit. The largest of the eugenias is the redberry eugenia, often growing as high as sixty feet. It has a scarlet fruit and extends southward from Biscayne Bay over the Keys, to many of the West Indies.

125. Naiad (Naiadaceae)

The naiad family is a group of thirty or forty species of slender, branching, annual aquatics all belonging to the genus *Naias,* from the Greek for water nymph. They have opposite, alternate or whorled leaves, that are more or less linear, dioecious or monoecious flowers that are solitary or in axillary clusters. They lack a perianth, the staminate flowers consisting of a single stamen enclosed in a tiny membranous bract, the pistillate of a single pistil containing one ovule. The fruit is an achene.

About a half dozen species occur in North America. The large naias may be found in lakes from central New York south to Florida, and west to California, the slender naias in ponds and streams throughout nearly all of North America, and the threadlike naias in pools and ponds from eastern Massachusetts to Delaware, Pennsylvania, and Missouri.

126. Nasturtium (Tropaeolaceae)

About three hundred years or so ago, a plant known today as the dwarf nasturtium, appeared in England, having come to the island kingdom from Peru by way of Spain and France. For a hundred years thereafter, it was cultivated for its seeds, that were pickled, and for its leaves and flowers, that were used in making salads. Just when, is not known, but after a while it came to be called Indian cress.

After a century or so since the appearance of the dwarf nasturtium in England, the plant known today as the common nasturtium, arrived to join its larger relative. It, too, came to stay and few plants have become so popular in the garden.

375. Leaf of Nasturtium

376. Flower of Nasturtium

The common nasturtium is one of about forty-five species of the only genus, *Tropaeolum,* of the nasturtium family. They are annual or perennial, soft-stemmed herbs, most are climbing, and natives of the cooler parts of South America. The perennial species have tuberous underground stems, otherwise they are like the annuals. All of them have a watery juice of pungent odor and aromatic taste, and alternate leaves, more or less round and light green, with strongly marked veins radiating from the center whereto the leafstalk is attached. This latter is fleshy and sensitive, and curls around any object that it may come in contact with. The flowers are showy, solitary, and irregular, grow from the axils of the leaves, and may be pale yellow, orange, scarlet, crimson, or dark red. The calyx consists of five sepals, united at the base, and on the upper side of the flower, are extended into a long, descending spur. The corolla also consists of five petals, the two upper more or less different from the others, and inserted at the mouth of the spur, the three lower clawed and fringed at the base. There are eight stamens, the filaments usually turned downward and curving. The ovary has three lobes that surround the base of the style, and the fruit is composed of three distinct rough carpels with a seed in each. The name of the genus, *Tropaeolum,* is from the Greek meaning a trophy, and was so named because the nasturtium was considered as representing ancient trophies, with its shield-shaped leaves and its flowers suggesting gilded helmets spattered with blood and punctured with lances.

Looking down into the heart of a nasturtium flower, one will see various markings all pointing down into the spur so that no visiting bee can mistake the direction to the nectar. The petal fringe is enough to discourage any wandering ant from going farther. Once fertilization has been effected, both the calyx and corolla wither and the flower stem begins to curve, to draw the fruit down into the cool, shady retreat beneath the leaves where it can mature unharmed by the sunshine.

127. Nepenthes (Nepenthaceae)

The nepenthes family occurs from South China to Australia, being most abundant in Borneo, whose only genus *Nepenthes,* from the Greek for removing all sorrow, probably in allusion to the assumed narcotic properties of the plants, contains a number of species called pitcher plants, that are among the showiest of all insectivorous species, and that are cultivated as novelties in the greenhouses and homes of the United States. They are climbing, often tree-perching, or bog herbs, sometimes a little woody at the base, with alternate, usually long leaves that are prolonged at the tip into a long tendril, that ends in a hollow, pitcher-like, winged structure with a thickened rim, a lid, and on the inside of the pitcher, several honey glands. Usually the pitchers, as brilliantly colored as tropical flowers in various shades of yellow, green, purple or blue, the margin often blood red and the inside a light blue, are suspended upright at the end of the long tendrils. At a distance they look like flowers and are mistaken as such by the insects that are attracted to them, not only by the brilliant colors, but also by the drops of honey deposited on the rim of the pitchers by the honey glands. Unsuspecting insects venturing beyond the rim, find within, a waxy surface, and down they slide into a pool of liquid, secreted by the glandular cells of the leaves. Upon trying to climb back up the slippery walls, they find it impossible to do so, and fall back into the pool of liquid that also contains water, and are drowned. Whereupon they are digested by juices secreted by the plant and their nitrogenous food is absorbed by the cells lining the cavity.

The flowers of these pitcher plants are inconspicuous, the staminate and pistillate occurring on different plants. They lack petals, but have three to four sepals, the staminate with four to sixteen united stamens, the pistillate with three to four carpels, with numerous ovules in each carpel. The fruit is a leathery capsule.

128. Nettle (Urticaceae)

The nettle family is a group of five hundred species of herbs or undershrubs and sometimes vines grouped in about forty genera. They are very widely distributed and are most abundant in the tropics. Many of them have stinging hairs on the leaves and twigs.

The leaves are alternate or opposite and simple. The flowers are small and inconspicuous, unisexual, and are borne in various kinds of clusters. The staminate flowers have a four to five-parted calyx, no petals, and four to five stamens. The pistillate flowers have a tubular or three to five-parted calyx, also with no petals, and a one-celled ovary. The fruit is a one-seeded achene or somewhat drupaceous because of the large and succulent calyx that surrounds the ripened ovary in some species. Most of the nettles are noxious weeds, a few have a slight value as ornamentals.

There are over thirty species of nettles of the genus *Urtica,* the Latin name of the nettles and is from *urere,* to burn. The tall, or slender, nettle is a native perennial of barnyards, roadsides, and waste places, with an erect stem two to seven feet high, opposite, lance-shaped, toothed, dark green leaves, both the stem and leaves sparingly set with stinging hairs, and small, greenish flowers that are sometimes dioecious, but more often on the same plant, the staminate ones near the top, the pistillate in the axils below, hanging in long, compound clusters.

Perhaps it is just as well that the great, or stinging nettle is not more common, because its stings are so venomous that anyone coming in contact with it is apt to suffer considerable discomfort for some time. It is a bristly herb, two to four feet tall, of roadsides and waste places with oval or heart-shaped, deeply toothed leaves, densely covered with stinging hairs, and small, greenish flowers in large, compound clusters from the axils of the upper leaves. Neither of the two nettles is poisonous, but both can produce an intense itching. Surprisingly, perhaps, the stinging nettle, a Eurasian species and naturalized in the United States, has long been used for food in Europe. The tender tops are frequently eaten in Belgium, Germany, and other countries, and in Scotland the young, tender tops, in the spring, are often boiled and eaten as greens. Sir Walter Scott relates that it was once cultivated in Scottish gardens as a potherb. A larger and stouter plant than either of the above species occurs in the west. Called the western nettle it grows in waste places and along the borders of streams.

Not a true nettle, as it is of a different species, the false nettle, genus *Boehmeria,* named for G. R. Boehmer, a German professor, resembles the stinging nettle, but its foliage is stingless. It is a taller and coarser plant with a stem from six inches to more than two feet in height, of moist shaded places. Its leaves are long ovate to lance-shaped, and its greenish flowers are borne in densely, crowded, elongated, axillary spikes. A related species is widely grown in China, and to some extent in southern United States, for the fine fiber that is obtained from the inner bark. It is known as the ramie or Chinese silk plant. As cultivated in Florida and Louisiana, it is a bushy plant, four to six feet high, with hairy stems, broadly oval, coarsely toothed leaves that are rough above, felty white beneath, and green flowers crowded in small heads that are grouped in a spikelike cluster.

Several other members of the family are the richweed or clearweed, a smooth, low nettle, three to eighteen inches high, of cool and moist shaded places, with ovate, coarsely toothed leaves; the wood nettle, a plant with stinging hairs, two to three feet high, of rich woods, with ovate leaves and greenish flowers; and the pellitory, a low, annual, simple or sparingly branched, minutely-downed plant of shaded, rocky banks, with oblong-lanceolate leaves, and staminate, pistillate, and perfect flowers intermixed in the same cymose axillary clusters.

A prostrate, mosslike plant, native to Sardinia and Corsica, is grown in the cool greenhouse for its minute foliage. Called baby tears or Corsican nettle, it is an extremely delicate plant with numerous, somewhat inequilateral, mosslike leaves of very unequal size, none over a quarter of an inch long, and generally roundish, and extremely minute, greenish, solitary flowers. In warm regions it is used to cover rock walls.

129. Nightshade (Solanaceae)

In his *Survival of the Unlike,* Liberty Hyde Bailey, the American botanist and horticulturist, wrote that

> the modern petunia is a strange compound of the two original species which were introduced to cultivation less than three-quarters of a century ago. The first petunia to be discovered was found by Commerson on the shores of the La Plata in South America, and from the dried specimens which he sent home the French botanist, Jussieu, constructed the genus *Petunia* and named the plant *Petunia nyctaginiflora* in allusion to the four-o'clocklike flowers.
>
> This plant appears to have been introduced into cultivation in 1823. It is a plant of upright habit, thick, sticky leaves and sticky stems, and very long-tubed white flowers which exhale a strong perfume at nightfall. This plant nearly or even wholly pure is not infrequent in old gardens, and fair strains of it can be had in the market.
>
> This old-fashioned petunia is a coarse plant and is now but little known.
>
> The second species of petunia first flowered in the Glasgow Botanical Garden in July, 1831, from seeds sent the fall before from Buenos Ayres by Mr. Tweedie, and in 1831 an excellent colored plate of it appeared in the Botanical Magazine, under the name *Salpiglossis integrifolia.* This is a more compact plant than the other, with a decumbent base, narrower leaves, and small red purple flowers, which have a very broad or ventricose tube scarcely twice longer than the slender calyx lobes. This little plant has been known under a variety of names. Lindley was the first to refer it to the genus *Petunia* and called it *Petunia violacea,* the name which it still bears.
>
> *Petunia violacea* early hybridized with the older white petunia, *Petunia nyctaginiflora,* and as early as 1837 a number of these hybrids—indistinguishable from the common garden forms of the present day—were illustrated in colors in the Botanical Magazine. Sir W. J. Hooker, who described these hybrids, declared that it must be confessed that here, as in many other vegetable productions, the art and skill of the horticulturist had improved nature. Here then our common petunias started as hybrids, but the most singular part of the history is that the true old *Petunia violacea* is now lost to cultivation.

Such is a brief history of the petunia, one of our most common and one of our most cherished garden flowers.

There are about a dozen species of petunias. They are annual or perhaps perennial, weak, straggling, clammy or sticky herbs with soft, flabby leaves and variously colored flowers, the corolla funnel-shaped, and are widely used for bedding, window boxes, or for the border. Nothing more needs to be said about these popular plants.

The petunia belongs to the nightshade family, a family of herbs, climbing vines, and a few small trees, consisting of two thousand species of wide geographical distribution in temperate regions, and very abundant in the tropics. It is basically a family of narcotic-poisonous plants and is easily one of the leading horticultural groups, for it includes vegetables of worldwide cultivation, narcotics, drugs, tobacco, and a large number of garden flowers.

The leaves of the nightshade family are alternate, entire or variously lobed or pinnately compound. The flowers are regular, or rarely irregular, perfect, solitary in a few, but mostly in various sorts of clusters, built on a plan of five with five lobes to the calyx and corolla and five stamens, the ovary two to five-celled, the style and stigma one. The fruit is a capsule or berry.

The nightshade family contains so many well known and useful plants that it is difficult to decide what to consider next, but this author has selected the potato for some rather obvious reasons, it being such a common food and so highly regarded for its high starch content and nutritional value. It is not without good reason that it has become the staple vegetable crop of the temperate world.

The economic usefulness of the plant lies in its peculiar habit of developing slender, white underground stems that gradually enlarge at the free end and produce the tubers so familiar to us. Most people think of these tubers as roots, but they are actually stems as can be seen by the so-called "eyes" that develop, and that are, in reality, leaf buds.

According to an early Spanish chronicler, the potato originally grew far up in the Andes and the Incas valued it highly. Eventually, the Spaniards took it to Spain and from there it went to England. The rest is history. The story goes that at first it was rejected in France, the French viewing the plant with scorn, until Louis XV took to wearing a bouquet of its flowers that, incidentally, are beautiful, golden-centered, white stars. Most people do not realize that the potato is a poisonous plant. The tubers are the only edible part, as the leaves, buds, and fruit contain solanine, a poisonous substance. Animals have been poisoned by feeding on the potato sprouts, and tubers growing at the soil surface where they usually turn green by being exposed to sunlight are unfit for food. There are a number of cases on record where people who have eaten the green tubers have done so with fatal results.

For a long time the tomato, or "love apple" as it was once called, was regarded as poisonous, and people refused to eat it. Until some daring soul, throwing caution to the winds as it were, decided to eat it with no dire results. This was in 1850, or thereabouts. Today the tomato is one of the most important of our agricultural products and is cultivated in every part of the world where it will grow.

As with so many others of our food plants, the tomato had its origin in Peru where it was cultivated and developed by the preIncan races centuries before America was discovered. Since the fruits of the members of the nightshade family are poisonous, that the Indians could have developed some of them and transformed them into fruits fit for human consumption,

377. Flowers of the Potato Plant

378. Flowers of the Tomato Plant

seems nothing short of the miraculous. Indeed, not only did the Peruvian races develop all of our present-day varieties of the tomato, as well of the peppers, but they also had a number of varieties of the tomato that are unknown to us, but still cultivated in Peru. Our tomatoes are tender, delicate plants and quite sensitive to frost, but in Peru they grow tomatoes that are unaffected by the heaviest frosts, and that can even withstand freezing temperatures. At an altitude of eight to ten thousand feet, the most common tomato is a small fruit about the size of a plum, being borne on vines that wind their way over trees and buildings. And the yellow tree tomato thrives best in the severe weather of its Andean home almost three miles above sea level.

The different varieties of the garden tomato show a variation in the leaves, but the flowers have varied little from the primitive: the five or six pointed, yellow stars, the corolla lobes alternating with the green lobes of the calyx. The large anthers grown together, form the conical tube in the center, and from the opening at the summit, the tiny, green style protrudes with its green stigmatic head, surrounded by the spreading tips of the anthers. The fruit may be red or yellow, and is a multi-seeded berry.

If the tomato were a long time in becoming accepted as a food, this was certainly not the case with the peppers. Almost from the time that the Spaniards set foot on the new world both the sweet or "bull" peppers, and the hot or "chili" peppers, have not only been popular, but in great demand. Oddly enough, the name "chili" is the old Aztec name of the pepper plants and is used today to designate the very hottest of peppers, whereas in Peru, the name is applied to the big sweet or mild peppers because "chile" means cold or cool in the Quechua Indian language.

Our American or garden peppers are likely varieties of a single species. They have simple, alternate leaves, white or greenish flowers that are usually stalked and solitary, or in two to three flowered clusters, and a podlike berry with a thickish rind, but much diversified as to shape and color. Thus, the fruit may be roundish or heart-shaped, red, yellow, or purplish and very pungent, erect or recurved, as in the cherry pepper, or narrow, slender, three inches long, pencil-thick, red, very pungent and erect as in the red cluster pepper. It can also be long and tapered to a point, four to ten inches long and red or yellow as in the long pepper, or it may be cone-shaped or cylindrical, one to two inches long, erect, and red, yellow, or purplish as in the cone pepper.

The red, or cayenne pepper, and paprika are both obtained from our American peppers, but the white and black peppers are obtained from an entirely different plant, the true pepper, belonging to the pepper family. What may seem surprising is that both the fiery cayenne pepper and the mild paprika are obtained from the same chili pepper pods. However, if the entire pods and the seeds are dried and ground, the product is cayenne or red pepper, but if only the rind or skin is dried and pulverized then the resulting powder is known as paprika.

Though somewhat less popular than the potato or tomato, the eggplant is largely cultivated in tropical countries since it needs heat for its proper development. It is native to India and is a rather woody, herbaceous plant with large, lobed leaves and a fruit that is among the largest in the family, a pear-shaped berry often six inches in diameter. It may be boiled entire or sliced, and fried, though there are several other ways of preparing it for the table.

Plants are used for many different purposes, a list of the many ways would seem endless. Some of the purposes are strange indeed, but probably none is so strange as the use of tobacco. It is interesting to note that the word "tobacco" actually means the pipe wherein the leaves of the plant are smoked, and not the plant itself. For when the Spaniards first saw the Indians smoking tobacco in the West Indies and asked them the name of the plant they misunderstood and thought that the white men wanted to know their name for the pipe that they said was "toba-go." And so, the Spaniards called the plant tobago, which, in the corrupted form *tobacco* it became known to the English speaking people.

It is something of a mystery how or why the Indians discovered the use of tobacco or acquired the habit of smoking its leaves, for the tobacco in its natural state, is not particularly suitable for smoking. Indeed, the dried leaves of almost any other plant could be used as well. For to bring out the natural flavor and aroma of tobacco, the leaves must be subjected to a long, complicated and delicate process of drying, fermentation and curing. The finest of tobaccos are quite worthless unless properly prepared. Moreover, the method whereby they are prepared, varies greatly depending on their eventual use. In other words, whether the tobacco is to be used for cigars, cigarettes, for pipe smoking, for snuff, or for chewing. And strange as it seems, the Indians somehow had learned how to prepare tobacco for all these uses long before the white men put in an appearance in America, for the Indians of North, Central, and South America smoked cigars, cigarettes, and pipes, and many of them used snuff and chewed tobacco ages before Columbus set out on his memorable voyage of discovery.

However, it must be pointed out the northern tribes did not use tobacco as universally as is generally supposed. Most of the Indians who lived in what is now the United States looked upon tobacco as somewhat semi-sacred, and used it mainly for ceremonial purposes. Moreover, in places such as New England, the districts about the Great Lakes, and on the plains, tobacco was scarce and hence, valuable. However, a certain amount was grown in the rich valleys of New England and a great deal was acquired by trade with southern sections of the country.

The tobacco plant is essentially a tender, tropical

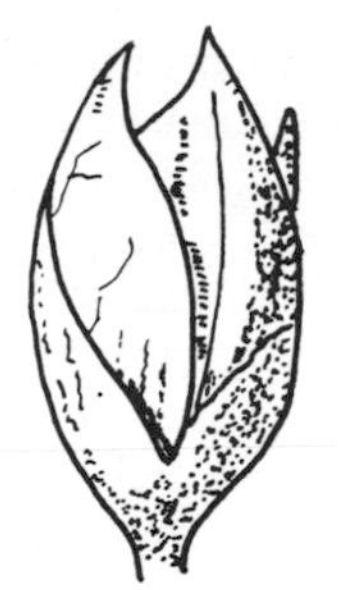

379. Fruit (Capsule) of the Tobacco Plant

annual, reaching a height of five feet and with large, funnel-shaped and five-lobed flowers that are usually greenish or purplish white in color. The fruit is a capsule.

There are many varieties under cultivation today on all the continents, in subtropical and temperate climates. Some species of tobacco are cultivated by the American Indians and are sometimes found naturalized in various states. A native species may be found growing in the canyons of the Southwest, where also may be seen the tree tobacco, a species introduced from Argentina. It is a small tree with long, tubular, yellow flowers.

As is well known, all tobacco leaves contain a poisonous substance called nicotine, widely used as an insecticide. The word nicotine comes from the name Nicot. Nicot was an ambassador to Portugal in the sixteenth century, and is said to have introduced tobacco to the courts of Portugal and France. To commemorate his name, the genus whereto the tobacco plant belongs is called *Nicotiana,* a name familiar to those who have a garden, for it is the name of a very popular perennial of easy culture. The nicotiana is an erect, slender plant, more or less covered throughout with fine, soft hairs, its leaves up to four inches long and either blunt or pointed at the tip, and with tubular, fragrant flowers that open toward the evening and perfume the air,

> from buds that keep
> Their odor to themselves all day,
> But when the sunlight dies away
> Let the delicious secret out
> To every breeze that roams about.

380. Flower of Nicotiana

Undoubtedly, the petunia and nicotiana are the two most popular flowers of the nightshade family to be found in the garden, but there are others favored by many gardeners, such as the salpiglossis, datura or trumpet flower, browallia, matrimony vine, and butterfly flower. The salpiglossis is a striking annual, somewhat resembling the petunia in appearance, but taller with large, funnel-shaped, showy flowers, varying in color through purple, blue, brown, yellow, red to cream white, all having a velvety look and generally veined with gold. It is a beautiful plant for the garden and easy to cultivate.

Known in American gardens since 1895 when it was found in South America by an orchid collector, the datura or trumpet flower is a rather attractive plant with large, showy, trumpet-shaped flowers, six to seven inches long, often two or three well-defined trumpets, one within another. The flowers are followed by globular, prickly pods containing large, flat seeds.

Named in honor of Dr. John Browall, a Swedish botanist and a friend of Linnaeus, browallia is a plant of midsummer flowering, one to two feet high, with numerous flowers, in racemes above the mass of foliage, that give to the flower bed an exquisite blue through the long, hot August days. With neither tendrils nor a twining habit, the matrimony vine seems to be ill named, for a vine it is not. Rather it is a somewhat spiny shrub with long, slender arching branches. Somewhat ornamental in appearance, it can be trained upon a wall, or upon the side of a house, or over arbors and fences. The name matrimony was suggested by the manner wherein the pale purple flowers grow, side by side in the axils of the leaves. The fruit is an orange red oval berry and is very decorative in autumn. Sometimes called the poor man's orchid, the butterfly flower is a strong-growing, albeit, dainty plant, two feet high or more, with leaves cut into many fernlike green segments, and showy flowers in terminal multi-flowered clusters of many colors. It is a splendid garden annual.

In almost any thicket or along the stone wall or fencerow during the summer, the graceful violet pur-

381. Flower of Nightshade

ple flowers of the nightshade will likely be seen. They are a joy to behold, but no less so than the drooping cymes of bright red berries that appear in autumn. It is a most decorative plant, and useful too, for migrating birds find the berries attractive. The hard, indigestible seeds pass through them unaltered, and are voided many miles distant, a nice refinement to ensure wide distribution. Examine a flower closely and note the deeply, five-cleft corolla and the yellow conic center formed by the five anthers that are fused together. The wilted leaves of the nightshade are poisonous, and cattle, horses, and sheep have been poisoned by eating them. So, too, are the berries that are attractive to children who should be warned against them.

The nightshade, also known as the bittersweet, a name also applied to another plant, is an immigrant from Europe that has made its escape into the wild, but there is our own native nightshade called the black nightshade from the color of its fruit. It has an erect, smooth, branching stem with ovate, wavy-toothed leaves, white flowers in small side clusters, and jet black berries on thin, drooping stems. This nightshade is a plant of fields and waste places and open woods, on loamy or gravelly soils. The wilted leaves and unripe berries are poisonous, but the ripe berries are often cooked for preserves or jams or used in pies. Boiling the berries apparently destroys their toxic properties. The fruit in some improved forms of this nightshade, known as garden huckleberry, sunberry, and donderberry, is used for making pies.

As a point of interest, the generic name of the nightshades (there are a number of species), is *Solanum,* derived from *solamen,* meaning solace or consolation, in allusion to the relief afforded by the narcotic properties of some of them.

So variable that the early botanists thought it must have several distinct species, the Virginia ground cherry is a common plant of open ground, rich, dry pastures, and hillsides. It is a branching and erect-stemmed species with ovate, lance-shaped leaves that taper towards both ends, and a dull, pale yellow flower with five brown purple spots. Honeybees and mining bees are frequent visitors. The fruit is an edible, reddish berry.

A related species, called the strawberry tomato, is probably an escape. Anyway, it is a downy plant with angular leaves, and a light green yellow flower, brown-spotted at the throat, with violet anthers. The fruit is a golden yellow berry. This plant has probably been in cultivation for over a hundred and fifty years, for it bears many berries that are excellent for preserves and sauce, and for making pies. They are often sold in the market.

When the settlers returned to the deserted foundations of Jamestown, in 1609, Captain John Smith reported the presence of a weed that is now known variously as the Jamestown weed, Jimson weed, or thorn apple. Asiatic in origin, but of worldwide distribution, carried everywhere by the gypsies, a favorite medicine of theirs for ages, it is a plant of waste places and vacant lots. It grows three to five feet high and has a smooth, green, stout stem. The leaves are large, rather thin, egg-shaped in outline, the edges irregularly wavy-toothed or angled, and of a rank odor. The white trumpet-shaped flowers, about four inches long, with a light green calyx less than half the length of the corolla, having five sharp-pointed lobes, open late in the afternoon to welcome the sphinx moths whose tongues are long enough to seek the nectar. But mischievous bees, and flies and beetles too, often manage to squeeze into the flowers as they begin to unfold, not for the nectar, but for

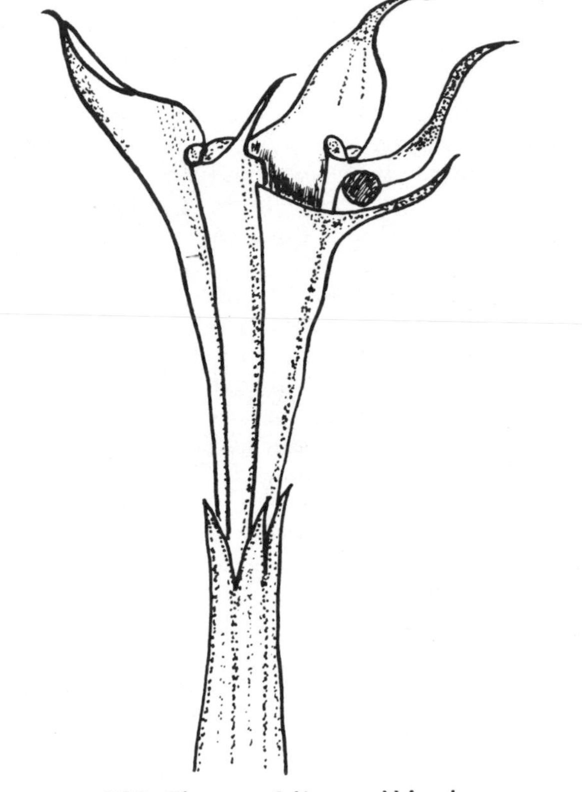

382. Flower of Jimson Weed

the pollen that they sometimes entirely remove before the moths arrive. The fruit is a densely, prickly, egg-shaped capsule with seeds that contain a powerful narcotic poison. Children are known to have eaten the unripe seed pods with disastrous results, and hence, they should be warned against them. Cattle, horses, and sheep have also been poisoned by eating the tops.

There is a similar species with a more slender stem, and darker green leaves, and pale lavender or violet flowers, whence, its name of purple thorn apple. This plant has become so at home in fields and waste lands throughout the eastern part of the country, that it is sometimes forgotten that it belongs to tropical America.

Lastly there is the Jerusalem cherry, a popular greenhouse pot plant or house plant, much grown by florists for its persistent, scarlet or yellow fruits, that are globe-shaped and about one half of an inch in diameter. These ornamental, cherrylike berries are toxic, lest anyone be tempted to eat them.

130. Nolana (Nolanaceae)

Often grown in the garden for their colorful, tubular, blue flowers, the nolanas are erect or prostrate perennial herbs, grown as annuals. There are some twenty species comprising the genus *Nolana,* from the Latin for a little bell, in allusion to the shape of the flower, one of three genera of the nolana family, a small group of South American herbs or undershrubs.

Natives of Chile and Peru they have an angular stem, that is sometimes spotted and streaked, smooth or sticky, spoon-shaped leaves that are usually in pairs, and bell-shaped, solitary stalked, blue or purple flowers borne in the axils of the leaves. The nolanas are suitable for rock gardens or barren hillsides, and can also be utilized for hanging baskets.

131. Nutmeg (Myristicaceae)

The nutmeg family is a small group of about a hundred species and eight genera of mostly evergreen, broad-leaved tropical trees, native to India, Australia and various Pacific Islands.

They have alternate, simple, entire leaves, small dioecious flowers with a three-lobed calyx, but no petals, two to thirty stamens, a one-celled ovary with one ovule, and a fleshy one-seeded fruit.

The common nutmeg of commerce is a tall tree, growing in hot, moist valleys of the tropics, with a striking aroma. The outer covering of the seed is known as mace, used as a spice, and the seeds as nutmeg that, when ground, is the grated nutmeg used in flavoring food. Both mace and nutmeg were unknown to the Greeks and Romans, and apparently were first used in Constantinople about A.D. 500. Apart from this one species, the other members of the family are of little interest.

383. Leaves and Fruit of the Nutmeg Tree

132. Ochna (Ochnaceae)

The ochna family is a group of tropical trees or shrubs with seventeen genera and about two hundred species, with thick, shining, parallel-veined leaves and with flowers in panicles.

The only cultivated plant of the family is a shrub of the genus *Ochna,* from the Greek for a pear tree, in allusion to the pearlike leaves, of tropical Africa, grown for ornament in southern California. It grows three to five feet high with alternate, leathery, oblong, toothed leaves, regular, yellow flowers, with a calyx of five separate, petallike sepals and five petals, and a drupe as fruit.

133. Ocotillo (Fouquieriaceae)

One of the most bizarre of our desert plants is the ocotillo, that may be found inhabiting the gravelly and rocky slopes of the southwestern deserts and into Mexico. It belongs to the genus *Fouquieria,* named for Pierre E. Fouquier, a French professor of medicine, a small group of trees and shrubs, and the only genus of the family, natives of Mexico and the southwestern United States. They have brittle wood and spiny stems, with the flowers borne on naked branches.

The best known species of the ocotillo family is the plant of the same name, though also called coachwhip, vine cactus, candlewood, or Jacob's staff. It is cactuslike and consists of several very prickly, rigid stems, eight to twenty feet high that seem leafless, though there are clusters of small leaves in the axils of the spines in early spring. The flowers are scarlet, tubular, about an inch long, with five sepals, a five-lobed corolla, and ten to seventeen stamens that extend beyond the lobes of the corolla. The flower clusters are very showy, six to ten inches long. When the ocotillo is in flower, with each waving stem surmounted by a scarlet plume, it is one of the memorable sights of the desert. The fruit is a capsule. The ocotillo makes a showy hedge plant, fine for the desert garden, or in a greenhouse. It is not hardy in the North.

134. Oleaster (Elaeagnaceae)

The oleaster family is a small group of three genera and thirty to forty species of erect shrubs or low trees found mostly in the north temperate and subtropical regions of both the Eastern and Western hemispheres. Their leaves and young twigs are often thickly covered with golden brown or silvery, shield-shaped, or star-shaped scales.

The leaves are alternate or rarely opposite and entire. The flowers are perfect or unisexual, often dioecious, not very showy, solitary or in small axillary clusters, the flowers with a tubular, two to four-lobed calyx, no petals, two to four stamens, or twice as many, and a one-ovuled ovary. The fruit is often incorrectly called a berry, because of the fleshy receptacle that, however, encloses the true fruit, the latter a dry nut or achene.

Of the three genera, the genus *Elaeagnus,* Greek for olive and the chaste tree, is by far the most important. It is a handsome group of perhaps forty species of trees and shrubs, hardy and wind resistant and useful as windbreaks in the prairie states, and several cultivated for their ornamental foliage and for their decorative or edible fruits. Only one species, the silverberry, is native to North America.

The silverberry, also known as the wolfberry, is a tall, erect shrub of the Northwest, six to twelve feet high, with ascending branches and scurfy, brown twigs that eventually become silvery gray. The green gray leaves are elliptical or lance-shaped, silvery scurvy on both sides. The flowers are tubular, fragrant, silvery without, pale yellow within, and the fruit is drupelike, ovoid, silvery, dry and mealy, and edible. It is eaten by various songbirds.

Of the several imported cultivated species the oleaster, or Russian olive or garden elaeagnus, as it is variously called, and a native of southeastern Europe and western Asia, is one of the best. Believed to

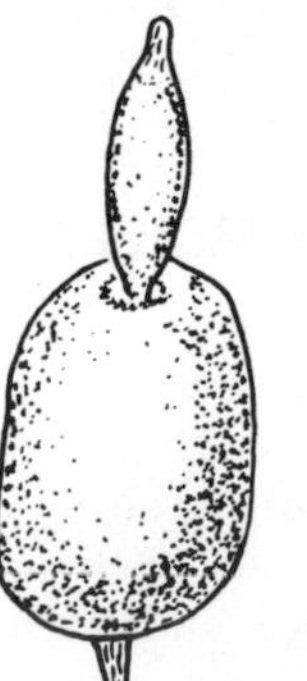

384. Drupelike Fruit of Russian Olive

be the wild olive of classical authors, it is a sometimes spiny tree ten to twenty feet high with willowlike leaves, hence, it has often been called the Jerusalem willow. The leaves are silvery beneath, the silvery whiteness of the foliage making the tree unusually conspicuous. The flowers are fragrant, silvery without, yellow within and the silvery yellowish fruit is egg-shaped with a mealy, but sweet flesh. The Russian olive is now widely established in the West, particularly the Northwest, where its fruit, persisting on the branches throughout most of the winter is an important food for such birds as the cedar waxwing, robin, and evening grosbeak.

Another introduced species, is the long-stemmed elaeagnus or gumi. It is a spreading shrub, four to nine feet high, with elliptic leaves, yellowish white flowers, and a red, scaly fruit of a pleasant acid flavor.

A somewhat interesting plant of the family and a native of the Far West is the buffalo berry, also known as the wild oleaster and silverleaf. It is a spiny or thorny shrub, or small tree, ten to eighteen feet high, with scraggly, angularly set, light gray twigs and narrowly oblong or elliptical, toothless leaves that are silvery on both sides. The flowers are small, yellow, in spikes, opening with the leaves, the staminate and pistillate on separate plants, the latter developing into an egg-shaped, yellowish or red fruit. It is sour but edible. The "berries" are about the size of currants, and are especially valuable for jelly. The Indians dried the "berries" for winter use, and often cooked them with buffalo meat, hence, their common name.

George Catlin, the American artist and traveller, describes the buffalo berry in his travels among the Indians about 1833, at the time growing abundantly about the mouth of the Yellowstone River, as follows:

> The buffalo bushes which are peculiar to these northern regions, lined the banks of the river and defiles in the bluffs, sometimes for miles together; forming almost impassible hedges so loaded with the weight of their fruit, that their boughs were everywhere gracefully bending down and resting on the ground.
>
> This shrub which may be said to be the most beautiful ornament that decks out the wild prairies, forms a striking contrast to the rest of the foliage, from the blue appearance of its leaves, by which it can be distinguished for miles in distance. The fruit which it produces in such incredible profusion, hanging in clusters to every twig, is about the size of ordinary currants, and not unlike them in color and even in flavor; being exceedingly acid and almost unpalatable, until they are bitten by the frosts of autumn, when they are sweetened, and their flavor delicious; having to the taste much the character of grapes, and I am inclined to think, would produce excellent wine.

Once when he was feasting with an Indian chief at another place he writes that a bowl

> was filled with a kind of paste or pudding made of the flour of the *pomme blanche,* as the French call it, a delicious turnip of the prairie, finely flavored with the buffalo berries, which are collected in great quantities in this country and used with divers dishes in cooking as we in civilized countries use dried currants, which they very much resemble.

A similar, but thornless species, the Canadian buffalo berry is a shrub not over eight feet high, with elliptic or oval leaves, green above, silvery beneath, green yellow inconspicuous flowers in small spikes, and egg-shaped, yellowish red "berries" that are almost tasteless.

135. Olive (Oleaceae)

Gerarde, the quaint herbalist, recounts a traditional belief concerning the ash tree. Citing the authority of Pliny he says:

> Serpents dare not so much as touch the morning and evening shadows of the tree, but shun them far off . . . Being penned with boughes laid round about (they) will sooner go into the fire than come near the boughes of the ash. It is a wonderful courtesie in nature that the ash should floure before the Serpents appeare, and not cast his leaves before they be gon again.

According to another tradition the ash is particularly susceptible to being struck by lightning. As an ancient folklore rhyme has it:

> Beware the oak it draws the stroke,
> Avoid the ash it courts the flash,
> Creep under the thorn, it will save you from harm.

Prehistoric man made his club out of the ash tree, finding that the wood is light, strong, and elastic. Achilles, it is said, fought with a spear made of ash wood, and the arrows that Cupid used were also made of it. The North American Indians preferred the wood of the ash tree for their bows and paddles.

There are sixteen species of ash trees in the United

States. They belong to the genus *Fraxinus*, the classical Latin name for the ash, and have deciduous, opposite, compound leaves, the leaflets entire or serrate. The flowers are small, greenish or whitish, perfect or uni-

385. Fruit (Samaras) of Red Ash

sexual, generally without petals, and with or without a four-lobed calyx, the stamens usually are two in number, and usually with a two-celled ovary. They appear in early spring, before or with, the unfolding of the leaves. The fruit is a samara with an elongated terminal wing.

Of all our ashes, the white ash is doubtless the most beautiful and most important, its common name being derived from the pale, sometimes silvery, under surface of the leaves. It is a tall, stately tree, sixty to seventy-five feet high, with a straight, columnar trunk and a high, pyramidal or round head of erect stout branches. The twigs are smooth and brittle, the bark is a dark gray or gray brown, deeply and somewhat furrowed into short, perpendicular channels and confluent narrow ridges, and the leaves are a lustreless green above, paler or silvery green beneath with usually seven leaflets. In autumn they turn a brownish purple fading into yellow. The flowers are green, with a four-lobed calyx, in panicles, the staminate and pistillate on separate trees, the samaras, one to two inches long, the wing bluntly lance-shaped. They

386. Staminate Flower of White Ash

387. Pistillate Flower of White Ash

are borne in drooping panicles and remain on the bare branches until midwinter. In the winter, the white ash can be recognized from almost any distance—depending, of course, on how good one's eyesight is—by the color of its buds that are rusty to dark brown, a few are even black, and hence, conspicuous on the winter landscape.

The white ash is a tree of rich, moist, cool woods, though it also grows in fields and on riverbanks. The black ash, however, is most commonly found on the borders of swamps and on the banks of streams. It is a slender tree, fifty to sixty feet high, with a narrow head of slender upright branches, and may generally be distinguished from other ashes at any time of the year by its slender bole that generally measures less than eighteen inches in diameter. When young, it looks like a dark gray granite column, its bark is so even and close-textured. Later, however, it becomes scaly. Its buds, too, are a recognizable character, being dark brown to nearly black, though at times they may be confused with those of the white ash, but the scaly bark of the black ash will serve to distinguish it from the white species if any other identifying character is needed. The black ash may also be known among other ashes by the fact that its leaflets are sessile with the exception of the terminal one. In other words, the leaflets of our other ashes have petioles or stems, those of the black ash do not. The leaves are so dark green that, from a distance, they look almost black. From early times the Indians have made baskets from the wood of the black ash. They split the wood into sticks an inch or so wide and two or three inches thick, and then bend them over a block. The strain, breaks the loose tissue that forms the spring wood, and separates the bands of dense, tough, summer wood into thin strips. These they use in weaving their baskets.

The red ash appears to have derived its name from the red, inner layer of the outer bark of its branches. It is a medium-sized tree, thirty to fifty feet high, with an irregular, compact head of twiggy branches, and found in low, moist, rich soil on riverbanks or on the margins of swamps and ponds. It closely resembles the white ash, but may be distinguished from it by the velvety downiness of its branchlets and twigs, and by the red, under surface of the bark on the branches. Its bark, too, is not quite so deeply furrowed as that of the white ash.

A variety of the red ash is known as the green ash, and differs essentially from the typical species in its entire smoothness of leaf and twig. The leaflets are also somewhat narrower than those of the red ash, and more sharply toothed. What has been said about the red ash and green ash, is what will be found in most books on trees. The green ash has been considered by some authorities as a variety of the red ash, by others as a distinct species. Now it seems as if they are both considered to be the same species, with the name green ash. So much for the vagaries of taxonomy.

If one wants to know how the blue ash got its name, crush some of the inner bark in some water. The water will turn blue from the dye that the bark contains. This is one way of identifying the tree, but there is still an easier way. Look at the twigs toward the tips. Are they four-sided? If they are, the twigs are those of the blue ash. The perfect flowers also help. These two characteristics set the blue ash apart from the other ashes.

The blue ash is a tall, slender tree, sixty to seventy feet high, with a narrow head of slender upright branches, light brownish gray bark, and yellow green leaves that turn pure yellow in the fall. It grows in moist woods on rich limestone hills of the Ohio and upper Mississippi Valleys. When travelling through the Allegheny Mountains the French botanist and explorer, Michaux, became enamored with the tree, and sent seeds back to Europe to be planted in European gardens.

There are other ashes, as for instance, the Biltmore ash, a species with a limited range through the Appalachian mountain region and similar to the white ash except that its leaves and young twigs are densely coated with fine hairs; the pumpkin ash, a tall slender tree of river swamps and the wet margins of ponds with the largest leaves and keys of any of our ashes; and the water ash, a tree up to forty feet high, common in the river swamps and wet coastal regions of the South where it is often the companion of the bald cypress.

In the West there are the dwarf ash, which is unusual in having simple leaves; the Arizona ash, found in the canyons from California to Utah; and the larger Oregon ash, a broad-crowned shapely tree, seventy-five to eighty feet high, with a stout trunk, erect branches, reddish gray or grayish brown bark, and pale green leaves, of rich, moist soil.

Then there are the importations; the flowering ash and the European ash, both of Eurasia. The flowering ash, the most common species in cultivation and the showiest, is a round-headed tree, not over fifty feet high, usually much less, with seven oblong or oval leaflets, and whitish, fragrant flowers in dense terminal clusters. The European ash is a large tree, taller than our white ash, with nine oblong, dark green leaflets, and a very narrow samara, the wing often notched at the tip.

In Europe the ash has been a tree of considerable value. From the early days of agriculture and mechanics, its wood was used in making all kinds of implements and tools, as well as providing the soldier of old with spears, lances, shields, pikes, and bows. Before paper came into use, its inner bark was used to write upon. Evelyn considered the wood of the ash next to that of the oak in importance for general use, and William Cobbett recommends the fruit be used in fattening hogs. "The seeds of the ash," he says, "are very full of oil, and a pig that is put to his shifts will pick the seeds very nicely out from the husks."

In Evelyn's time, the kernels of ash seeds were believed to have medicinal value and apothecaries carried them on their shelves, calling them *Lingua avis* because they were "like almost to (sic) divers birds' tongues." Gerarde gives a quaint recipe for reducing; "Three or four leaves of the ash taken in wine each morning doe make those lean that are fat." Parkinson endorsed it as "a singular good medicine—with fasting a small quantity—for those already fat or tending thereunto, to abate their greatnesse, and cause them to be lancke and gaunt." Though today, both would be viewed as quacks, they were estimable gentlemen, Gerarde being regarded as an authority and Parkinson being the King's Apothecarye. And no doubt "fasting a small quantity" was the effective ingredient of the prescribed treatment for reducing.

The ashes have only a moderate value to wildlife, the seeds being eaten by a number of birds and mammals and the twigs and foliage by such browsers as the mule and white tailed deer. The caterpillars of three species of butterflies include the ashes among their food plants; the Baltimore, the harvester, and the tiger swallowtail.

The genus *Fraxinus* is one of twenty-two genera of the olive family that includes about five hundred species of trees and shrubs that are widely dispersed in the temperate and tropical regions of the Old and New World. In addition to the ashes, it includes the olive tree, that furnishes the olive and olive oil of commerce, and such widely planted shrubs as the lilac, forsythia, and privet.

The leaves are prevailingly opposite, simple or commonly pinnately compound, and sometimes evergreen. The flowers are small and inconspicuous in some genera, handsome and showy in others, regular, perfect or imperfect, with or without a four-lobed calyx, with or without a corolla that, if present, may be tubular, and four-lobed or of four separate petals, usually two stamens, the ovary mostly two-celled with two ovules in each cell, and usually borne in profuse clusters. The fruit is either a true drupe, a dry and winged samara, or a dryish berry.

One of the most beautiful of our ornamental shrubs is the fringe tree, widely cultivated for its showy white flowers with four linear petals nearly an inch long, purple dotted at the base within, in loose, drooping, graceful clusters about seven inches long with a fringelike snowy appearance; the genus *Chionanthus*, whereto it belongs, means snow flower in Greek.

The fringe tree is a small, slender, southern tree, twenty to thirty feet high, of moist copses and riverbanks, when isolated, forming a round-topped figure with rather stout irregular branches. Its bark is brown, its leaves are oval or oblong and dark green, and its fruit is a one-seeded drupe, dark blue with a slight bloom, and quite ornamental. The fringe tree is normally a tree in the South, but in the North, it is more often a shrub. The charm of the tree is in its

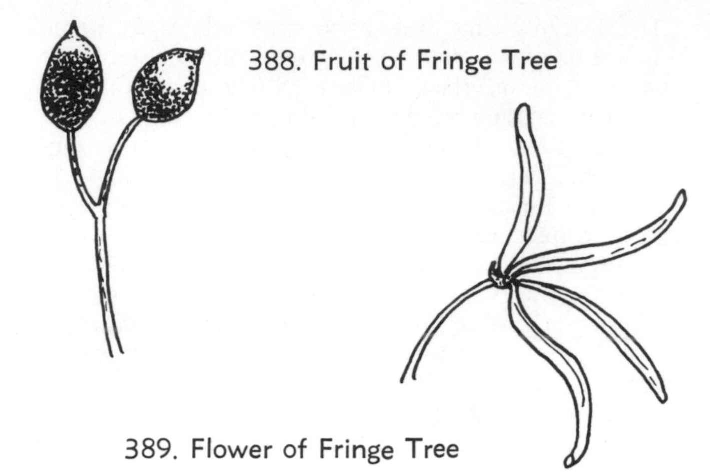
388. Fruit of Fringe Tree

389. Flower of Fringe Tree

singular flowers. The light and graceful clusters are like feathery masses of snowflakes and when they appear as the leaves are half grown, the contrast of soft green and pure white, or the combination of both, gives it a sort of ethereal beauty.

The common lilac, a handsome widely planted shrub, growing to twenty feet high, and known to almost everyone, is said to be a native of the mountainous regions of central Europe and appears to have come to the attention of European botanists toward the end of the sixteenth century. From that time on, it spread rapidly throughout European gardens as it was a showy and hardy shrub and easy to cultivate.

It is uncertain just when the lilac appeared in America, but it is known that about 1650, it seems to have graced many a cottage yard. And when the log houses began to spring up along the western frontier many of them had the lilac growing alongside.

The lilac is such a well-known shrub that a description seems almost unnecessary. The leaves are oval to heart-shaped, smooth, dull green, pointed and without teeth. The flowers are fragrant, tubular, with four spreading lobes and two stamens in a cluster known as a thyrse, and the fruit is a capsule. The flowers are usually lilac, but in the many horticultural forms, they may be white, pink, or purple.

Almost as well known perhaps as the lilac, is the forsythia, a handsome, spring blooming shrub, widely planted for its profuse yellow flowers that bloom before or with the unfolding leaves. The golden bells gleam in the April sunshine and transform the brown bush into a glowing mass of color.

The forsythia, a native of China, appears to have been introduced in England about the end of the eighteenth century, and first bloomed in the royal gardens at Kensington whose director was William Forsyth, after whom it was named. Just when the forsythia appeared in America seems rather uncertain, but once it did appear, it has ever since been included in the list of our ornamental shrubs.

There are four species of forsythia, all inclined to be arching or spreading, some varieties rooting at the ends of the pendulous branches. The leaves are stalked, but the flowers are practically stalkless. They have a four-lobed calyx, a corolla that is bell-shaped below, but split into four strap-shaped lobes that appear like four separate petals, and are borne in clusters from one to six in the axils of the leaves. The fruit is a two-celled capsule.

Few shrubs make such a desirable hedge plant as the privet and it is almost exclusively used for this purpose, though it is also planted as an ornamental shrub. It has many virtues. It does well in a smoke polluted atmosphere, as in an industrial city, it is remarkably free from both insect pests and plant diseases, and its foliage is a dark, handsome green just missing being evergreen, though in the South,

390. Leaf of Privet

391. Fruit of Privet

sometimes so. The flower cluster is a thyrsus, in fact a miniature copy of the lilac cluster, and the individual flower in shape is much like that of a small lilac blossom. The fruit is a small, black, berrylike drupe. The privet or prim, as it is also known, may sometimes be found as an escape along thickets and roadsides, having originally come from Europe. The privet belongs to the genus *Ligustrum,* the classical Latin name for the privet, comprising fifty species all from the Old World. They are mostly shrubs with generally ovalish leaves, that are evergreen in some species, and small sometimes fragrant flowers in clusters that often are not produced on clipped hedge specimens. The fruit, usually black or bluish, is a one to four-seeded berrylike drupe. The so-called California privet, really a native of Japan, is the most widely planted of the hedge privets. The wax privet, an evergreen shrub with oblong-oval leathery leaves, is widely planted in the South and is not hardy northward.

The swamp privet, of another genus, *Forestiera,* named for Le-Forestier, a French naturalist and physician, is a native shrub, rarely a small tree, found growing in the swamps or along riverbanks of the Middle West. It has a dull, brown bark, leaves commonly elliptical and conspicuously narrowed and pointed at both ends, green, yellow flowers in inconspicuous small clusters, and a berrylike wrinkled, dull purplish blue, drupe.

The olive tree, whose stone fruit is the olive of commerce, has been cultivated since prehistoric times. It is a native of the Mediterranean region and was brought to North America from Spain, and to California from Mexico by the mission fathers. It is a hardy subtropical evergreen tree, twenty-five to seventy feet high, of high heat requirement, and thus, the commercial culture of the tree is limited in our country to the Pacific Southwest and to a lesser extent in Florida. Its leaves are elliptic or oblong, green above, silvery and somewhat scurfy below, and the fragrant

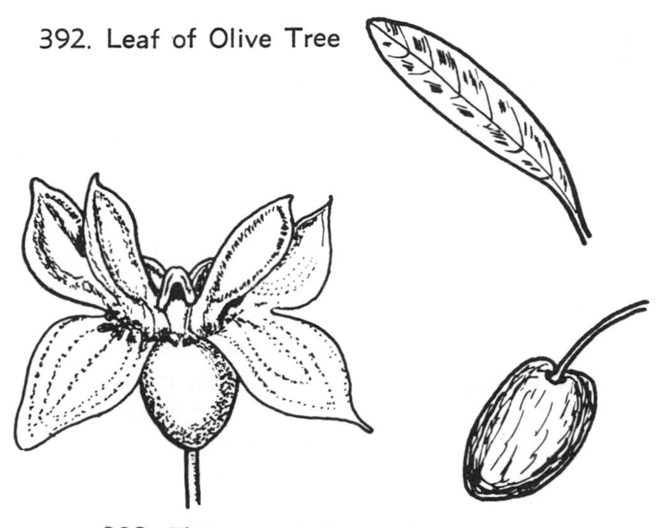

392. Leaf of Olive Tree

393. Flower and Fruit of Olive Tree

flowers are borne in clusters. Unlike most fruits, the olive is rarely eaten fresh because of its bitter flavor. Instead, most of the fruit, which is a drupe, is picked before ripe, treated with lye or soda, and then pickled in a salt solution. If left to ripen before pickling, the resulting fruits are black instead of yellowish green.

The tea olive, also called sweet olive, is not a true olive, but a member of a different genus *Osmanthus,* Greek for fragrance and flower in allusion to the fragrance of the common tea olive, the true olive belonging to the genus *Olea* the classical Latin name for the olive. There are ten known species, all, but one, Asiatic and Polynesian in origin, the lone exception being the native American devilwood. The tea olives are evergreen shrubs or small trees with flowers in cymes or panicles, and the fruit a drupe. The common tea olive, the species most widely planted, has oval or oblong leaves, and very fragrant white flowers. It is a very popular shrub in the warmer sections of the country, though it is occasionally grown under glass in the North. The devilwood is a small southern tree, twenty to forty feet high, with elliptic leaves that are shining green above.

Widely cultivated and esteemed for their attractive, fragrant flowers, the jasmines, also called jessamines, comprise a large genus, *Jasminum,* the Arabic name for jasmine, of shrubs and vines, chiefly tropical and subtropical found in Eurasia and Africa. They are climbing or spreading shrubs with compound, opposite or alternate leaves, sometimes with only one leaflet, and frequently green, angled stems. The flowers occur in multi-flowered clusters, have a bell-shaped calyx and a tubular corolla, and the fruit is a small berry. The Spanish jasmine is a white-flowered species widely cultivated as an ornamental and for perfumery. The common white jasmine also has white, fragrant flowers and is an attractive shrub that has been cultivated for centuries. The Italian jasmine has yellow flowers and so, too, have the primrose jasmine and the winter jasmine that blooms all winter in mild places. The Arabian jasmine with white flowers that turn purple with age should also be mentioned. They are very fragrant, and for this reason, the shrub is widely cultivated in the East. The Chinese use jasmine in flavoring tea.

136. Orchid (Orchidaceae)

The orchids, often called the elite of the plant kingdom, probably show the most advanced development of floral structure for insect pollination, the flowers having been modified to function as spring traps, adhesive structures, and hair triggers attached to explosive shells of pollen. Indeed, in some instances, the flowers have become so modified that they look like insects themselves.

An orchid flower consists of three similar sepals, two lateral petals, and a third differentiated into a conspicuous lip that is often spurred and always conspicuously colored and that secretes nectar. In most orchids, there is only one stamen, united with the style into a single, fleshy body known as the column, placed at the axil of the flower and facing the lip. The style, instead of ending in the stigma as in most flowers, terminates in a beak, the rostellum, at the base of the anther or between the two anther cells. The stigma, that is gummy or viscid, is located directly below the rostellum. Two stemmed, pear-shaped pollen clusters called pollinia are produced in the anther cells. Each of the pollinia is composed of several packets of pollen that are tied together by elastic threads that run together and form a stem ending in a glandular or sticky disc contained within the rostellum. It is the disc that becomes attached to an insect and ensures the transfer of the pollen to the stigma of another flower.

If one can find the showy orchis at the moment a bumblebee is visiting it, one can see just how an orchid manages to secure cross-pollination by an insect. Or perhaps it would be better to find the plant first and then wait for the bumblebee. It requires a little patience, of course; unfortunately, the insects cannot be commanded to perform when it is desired that they do so. Incidentally, a reading glass would be most useful in this case.

394. Flower of Showy Orchis

The showy orchis grows in moist, rich woods, and frequently in hemlock groves. The flowers, few in number on the stem, have the sepals and petals lightly united in a magenta or dark rich crimson hood with a conspicuous, almost white, lip prolonged into a spur wherein the nectar is secreted. When the bumblebee, generally a queen, alights on the projecting lip, she thrusts her head into the spur, and as she sips the nectar, brushes past the rostellum at the top of the column, thus rupturing the thin membrane covering it and exposing the sticky round discs attached to the pollinia. The discs immediately adhere to the bee's face or forehead and are carried away by her as she flies to another orchis blossom, where the pollen grains are scraped off on the viscid stigma.

The showy orchis, with another more northern species known as shinplasters, or the small, round-leaved orchis are our only true orchis, though there are seventy or more species widely distributed throughout the world. The showy orchis has two basal, shining, somewhat oblong leaves and a flowering stalk nearly twelve inches high, the flower cluster bracted. The small, round-leaved orchis, found mostly in the mountains, has a single leaf nearly circular and magenta flowers with a white lip, the stalk of the cluster naked. Both have tuberous roots.

The showy orchis and the small, round-leaved orchis belong to the genus *Orchis,* the old Greek name for the seventy odd species. Other species, also called orchis, are members of the genus *Habenaria,* from the Greek for rein, in allusion to the long, narrow spur of some species. It is a large genus of four hundred widely distributed species that are commonly called fringed orchis from the beautifully fringed flowers of some of them. They have fleshy or tuberous roots, leafy stems, in our species, the leaves usually with a sheathing base, and greenish, white, purple or yellow flowers, mostly in spikes or racemes. The sepals are equal or nearly so and spreading, either separate or united at the base. The petals are mostly smaller than the sepals. The lip is entire, toothed or fringed laterally, spreading or drooping, and three to five-cleft. The spur is shorter or longer than the lip.

One of the more handsome and more striking of the habenarias is the large, purple-fringed orchis that may be found in meadows and rich woods. It grows twelve to thirty-six inches high, has oval or lance shaped leaves, with their bases clasping the stem, and fragrant purplish violet flowers in a spike sometimes twelve inches long, the lip deeply three-parted and fringed, prolonged into a long, threadlike spur, the upper sepal and petals close together or united. The smaller, purple-fringed orchis grows in wetter grounds, being quite common in swamps and wet woods. It resembles the preceding species, but the flowers are pinkish or lilac, smaller, and less fringed. The spur is shorter too. In both species, cross-fertilization is effected by moths and butterflies whose heads and eyes are frequently decorated by the pear-shaped pollen masses. A more southern species and with flowers truly purple, the purple orchis, of wet meadows, has a fan-shaped lip that is toothed, but not fringed. Its leaves are somewhat narrower than those of the large, purple-fringed, and the smaller, purple-fringed orchis.

Away from the questing eyes of man, in places where the water snake and bittern are at home, the beautiful, milk white orchid may be found growing in the muddy waters of bog and swamp. It is a dainty little orchid with a stem twelve to twenty inches high, its lower leaves lance-shaped and quite long, its upper leaves bractlike, and with a lip that is variously cut and fringed giving the entire spike of milk white flowers a soft, lacelike appearance. It is white perhaps so that the night-flying moths might see it in the dusk of an evening. Certainly its long spur is adapted to the sphinx moths.

Another exceedingly handsome and stately species is the yellow-fringed orchis of meadows and wet sandy barrens. It is a slender plant with leaves similar to those of the white-fringed orchis, and a large multi-flowered spike of showy golden or orange yellow flowers with narrow fringed petals and a deeply fringed lip. Its yellow flowers, too, are beckoning signals to the sphinx moths after darkness has settled over the landscape, but the butterflies by day find them also. A similar, but southern species, the yellow-crested orchis occurs in bogs from New Jersey, southward.

Common in all wet places from Maine, south and

west to Minnesota, is the small, pale-green orchis. It has a stout stem, with several lance-shaped leaves and small flowers with yellow green sepals and petals, the blunt lip toothed on either side. Often found growing with it, the ragged fringed orchis is a species remarkable for its lacerated, three-parted flower lip that is finely cut into threadlike divisions that attract the passing butterfly. It is ten to twenty inches high with lance-shaped leaves and a long flower spike crowded with inconspicuous deeply-spurred, greenish yellow flowers that are so ingeniously designed that an insect visitor cannot obtain the nectar without the pollen disc becoming attached to it.

Not as common as some of the other orchises, the green, round-leaved orchis can easily be recognized by its two large, nearly round leaves, sometimes as much as seven inches across or the size of dinner plates, that lie flat on the ground. They are light green and shining above and silvery white beneath. The orchis, a plant of rich woods in the mountains, has a bracted stem, and greenish white or whitish, yellow green flowers in a loose cluster, the upper sepal short and rounded, and the lateral ones ovate. The lip is narrow, drooping and long, almost three times the length of the small, lance-shaped petals, and the spur is slender, curved, enlarged at the base, nearly two inches long, and is peculiarly adapted to the tongue of the lesser sphinx moths.

The orchid family is an enormous group of probably more than five hundred genera and somewhere between seventy-five hundred and fifteen thousand species of low, erect, sprawling, or climbing herbs scattered all over the world, but most abundant and most showy in the tropics. They are perennials with bulbous, tuberous, or thickened, fleshy stems and roots with the stems occurring in many varied forms. Most are terrestrial, but there are many epiphytic, and there are even some saprophytic species. Without question, the family has the most spectacularly beautiful flowers of the entire plant kingdom.

The leaves are usually alternate, simple, entire, thin, or often thick and fleshy, linear, oblong or round. In the tropical species, the leaves often have a swollen base (a pseudbulb), that stores water over the dry season when the blade often falls away. In many tropical genera, the air roots of the epiphytes are covered with a whitish film that absorbs atmospheric moisture.

The flowers are small, greenish, and inconspicuous in many species, often showy because of size and color in others, sometimes grotesquely irregular and oddly splotched in the cultivated species, and present extreme specialization to insect fertilization. Did not Darwin devote two volumes to the subject? The flowers also present great difficulty in separating the various genera. Typically the orchid flower is usually perfect and highly irregular, with three sepals and three petals, one of the latter forming a lip or spur that is extremely varied as to form and color. The ovary is usually long and twisted and may be one-celled or three-celled with many ovules. The fruit is a capsule containing many small seeds.

Most people have heard of the lady slippers and many have seen them growing in the wild. They are of the genus *Cypripedium,* a name given to them by Linnaeus in 1753, and literally meaning "slipper of Venus" that well describes the shape of the saclike lip of the flowers. They have fleshy, usually brittle root systems, sometimes fibrous-rooted, leaves that are usually plaited like a fan, multi-nerved, and generally with a sheathing base, and few or solitary flowers, with three sepals, two of them often united, erect, and showy, the petals mostly spreading, oblong or narrower, the lip very striking, much enlarged, and sac or pouchlike.

395. Lady's Slipper

Of our native lady slippers, the moccasin flower, or stemless lady slipper, is supposed to be the most common of them all. The author is not sure that this is so any longer, since many thoughtless people have picked it indiscriminately, and it is rapidly vanishing from the scene. But if one should ever find it amidst the brown carpet of the woodland floor, swinging balloonlike in the murmuring wind, observe how ingeniously it compels insects to carry its pollen. For once an insect has feasted in the large banquet chamber, it must first rub along the sticky overhanging stigma with its rigid, sharply pointed papillae that comb out the pollen it has brought from another flower, and then it must pass an anther that, drawn downward on a hinge, covers it with more pollen before it can emerge out into the open. The moccasin flower can easily be recognized, for it has two basal, ovalish leaves and a solitary, somewhat fragrant, flower at the end of a long, scurfy-haired stem, the pouch or lip crimson pink, veined with a deeper pink, the petals yellow green.

A lady's slipper of bogs and wet meadows, the white lady's slipper is a handsome plant with three to four, light green, narrow, elliptical leaves, and a flower with two wavy and twisted, narrow green petals, three broader, green, purple-blotched sepals, and a lip or pouch open at the top by a fissure, white outside, and purple-streaked inside. With its flaunted beauty and decorative form, the large yellow lady's slipper of woods and thickets attracts our eye no less than that of the little bees, though doubtless they are also attracted to it by its heavy, oily fragrance. Its leaves are large, broad, and pointed, three to five inches long, on a stem two feet high. At the end of the stem is the large, showy, solitary flower, with two of the sepals united, greenish or yellowish, striped with purple or dull red, two of the petals, long and narrow, brown, twisted, the third a much inflated sac, nearly two inches long, a pale yellow striped with purple. The similar small, yellow lady's slipper, a delicately fragrant orchid about half its size, has a brighter yellow pouch, with its sepals and petals sometimes purplish.

With the shape of the spurred lip and the sepals around it suggesting a ram's head, the ram's head lady's slipper is the smallest and rarest of our native lady's slippers. It is a species of cold swamps with three or four elliptical leaves on a stem six to twelve inches high, and a single drooping flower with purplish brown sepals and petals and a red and white lip.

As beautiful as all these lady's slipper may be, the most beautiful of all is doubtless the showy lady's slipper. It hides away in inaccessible swamps and peat bogs where it grows to a height of two or three feet. Its stem is stout and downy, its leaves are ovate and strongly-ribbed, and its flower is fragrant with white sepals, petals, and pouch, though the latter is stained with light crimson magenta. It is not easy to find, and only the most zealous will venture to penetrate the morass wherein it grows for a glimpse of its beauty.

A marsh orchid and the last of the orchids to flower, the ladies' tresses has a peculiarly twisted or spiral flower spike with small, translucent, yellowish white, or variably cream white flowers that are cleverly contrived for insect pollination. It is a plant seven to eighteen inches high, with clustered, fleshy roots, narrow leaves, almost grasslike, from or near the base, and flowers in rows of threes, quite close together, and slightly fragrant, the two lateral sepals free, the upper one united with the petals, and all with the oblong spreading lip, crinkle-edge, forming a bugle horn-shaped tiny blossom. There are other species of ladies's tresses—the grass leaved and slender—all being known by the twisting of the flower spike and all being fertilized by the smaller bees, moths, and butterflies. Unlike the other members of the tribe that prefer moist places, the slender ladies' tresses curiously selects dry fields, hillsides, open woods, and sandy places as a habitat.

Being among the smallest of our orchids, the twayblades are often overlooked in their native haunts of moist woods and swamps. There are several American species. The heart-leaved twayblade, for instance, is a delicate plant not more than four to ten inches high, with a very slender stem, two opposite light green stemless leaves with a shape that resembles the *ace of spades,* and a loose cluster of greenish or madder purple tiny flowers, the flowers without a spur, but with a very long, two-cleft lip with nectar in a furrow. The delicate rostellum above the lip explodes at the slightest touch of a visiting insect and ejects a viscid fluid that strikes the pollen masses situated just over the crest of the rostellum. Thus, the insect comes in contact with both the fluid and the pollen masses. A similar species, the broad-lipped twayblade, has roundish leaves and flowers with a wedge-oblong lip that is much longer than the narrow sepals and petals. Both species are visited by the smaller bees and the tiny beelike flies.

Of a diffrent genus, the large twayblade is a small, but showy species, three to ten inches high, of rich, moist woods, with rather large, shiny leaves at the base of the flower scape that expand into large, loose sheathing bases, and brownish or madder purple flowers in a raceme that is sometimes five inches long, the sepals and petals reflexed, the petals exceedingly narrow, the lip large and standing out from the flower.

Arethusa was a nymph who was changed into a fountain by the goddess Diana, so that the infatuated river god, who had fallen in love with Arethusa on seeing her at her bath, could no longer pursue her. Evidently knowing his mythology, Linnaeus, fancying a large, single-flowered, delicately scented orchid of bogs as a maiden living in a wet place where presumably none could follow her, named it Arethusa. One of the prettiest of our orchids it has a bulblike root wherefrom arises a scape, six to ten inches high, that is at first leafless, except for two or three sheathing bracts at the base, but from the upper bract a solitary, linear leaf later develops. The flower is one to two inches long, the sepals and petals, that point upward like the fingers of a half-open hand seen in profile, are purple or rose pink, the lip recurved and spreading, rounded and toothed at apex, fringed, spotted with purple, with three white ridges running down its surface. It forms a convenient and colored landing platform for visiting insects.

Another orchid of cool peaty bogs, and also named after a nymph, the Calypso is equally as elusive as the Arethusa. It is a small plant hardly six inches in height, with a single, rounded basal leaf and a showy solitary flower, variegated with purple, pink, and yellow, the ascending wavy sepals and petals suggesting the outstretched fingers of a hand, the lip broad, swollen, sac-shaped, two-parted at the apex, whitish tinged with rose, and crested with yellow, woolly hairs that are like spun glass.

Among our loveliest bog orchids is the handsome calopogon or grass pink. It has a bulblike root, a scape twelve to eighteen inches high, with a single, linear, bright green leaf issuing from a sheathing base, and

396. Flower of Grass Pink

a few purplish pink, sweet-scented flowers, the lower in bloom while the upper are still in bud, the flowers with a long, spreading lip crested with yellow, orange, and magenta hairs. The name calopogon is from the Greek for beautiful and beard, in reference to the handsome bearded lip.

The calopogon is not a rare plant and it can be found rather easily enough if one wants to venture into the bogs where it grows. In such places, in company with it, one will also find the rose pogonia or snakemouth, a most delicate little orchid and far less pretentious than its relative. It is a low plant, six to nine inches high, with a fibrous root and a small, lance-shaped leaf halfway up the stem, a tiny one just below the blossom, though sometimes a long-

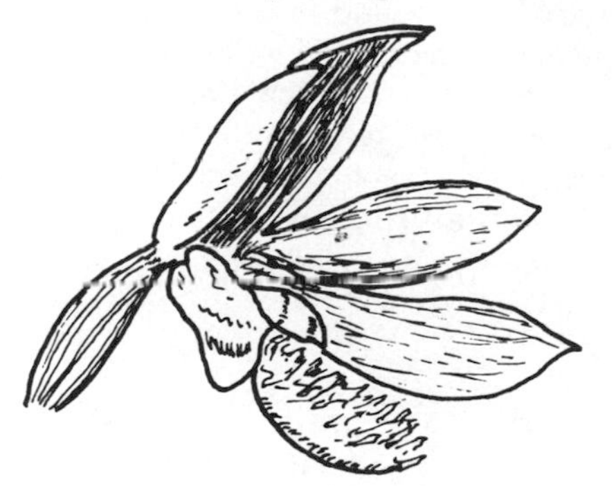

397. Flower of Snakemouth

stemmed leaf arises from the root. The flower is terminal, rather large, pale rose pink, or crimson pink, with a fragrance like fresh red raspberries, and an alluring landing place, fringed and crested, all designed to woo a passing bee with the prospect of refreshment in the nectary beyond. A local species and less showy, the nodding pogonia is remarkable for its dainty, pendulous, pink or pale purple flowers that are considerably smaller. Still rarer than the nodding pogonia, at least in the East, is the whorled pogonia, distinguished from the others by its circle of five light green leaves at the summit of the stem. Unlike the rose pogonia, the last two species are woodland plants.

The beautifully marked little leaves of the lesser rattlesnake plantain that carpet the ground beneath the spruces and hemlocks of our northern woods will attract the eye quicker than will the spires of the insignificant flowers. They extend from the base of the stem, tufted, or ascend the stem for few inches, and are conspicuously veined or blotched with silvery white. The stem is scaly, slender, slightly woolly and set on one side only with translucent, greenish or creamy white small flowers; the saclike lip of the flowers with a recurved wavy margin. The downy rattlesnake plantain, a more common species in the East and found in woodlands generally, is usually a taller plant, with larger cream white, globular-lipped flowers and on both sides of the stem. It is glandular-haired throughout, and its dark grayish green leaves are more heavily netted with white.

To most people the word orchid suggests some exotic tropical plant with gorgeous blossoms, marvelous in form and exquisitely painted with tints from nature's palette; hence it might seem difficult to believe that the insignificant little coral root with its leafless stem and small dull flowers of indefinite shade is an orchid too. And yet, a glance at its flower and it becomes evident that it is a member of the orchid family. But without leaves, one may ask, how does it manage to survive, for all orchids have leaves to serve as food factories. But the coral root has departed from the general norm of the orchids and has taken to living on dead or decaying vegetable matter, in other words has become a saprophyte.

The coral root was so named from its irregular branching root that resembles coral. There are several species of coral roots, one being the early coral root

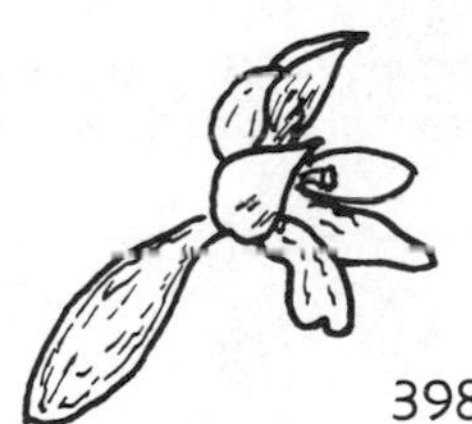

398. Flower of Early Coral Root

that blossoms in May. It is commonly found in evergreen woods, has a straight, yellowish brown leafless, but scaly stem, and small uninteresting madder purple flowers, three to twelve in number, in a raceme, the sepals and petals tiny, the lip whitish and red-dotted. Then there is the small-flowered coral root, also of evergreen woods, that blossoms in July. It is a generally taller species, with a purplish brown flower stem, six to twelve inches high, leafless, but with one or two sheathing scales, and very small, dull purple brown blossoms that droop on a stiff stem, the sepals and petals marked with purple lines, the lip white and spotted. The large coral root, common in spruce woods, is an even taller species, with larger and many more flowers. They are a dull brownish purple, as is the stem, the lip white with purple lines and spots.

A somewhat curious orchid of the woods is the species known as Adam and Eve, or puttyroot. The name Adam and Eve was given to it by some imaginative person because of its two singular corms, barely hidden below the ground, that are united by a slender stalk. But usually, several old corms are attached, as one little bulb is produced each year that sends up in late summer or early fall a broad leaf, that generally lasts over winter. The next season the flowers appear. They are yellowish brown and purple, each on a short pedicel, in a few-flowered, loose, bracted raceme. In the meantime, the old corms retain their life, perhaps to help nourish the young one taxed with sending up the flowering scape. A strong, glutinous substance obtained from the corms has been used as a cement, hence, its name of puttyroot.

A small orchid of cold woods or bogs, the green adder's mouth has tiny, white green flowers in a small cluster (raceme). The sepals are oblong and spreading, the petals are threadlike or linear, also spreading, and the lip is three-pointed. An oval or roundish leaf clasps the slender stem about halfway up. In the white adder's mouth, the single elliptical leaf is at the base of and sheaths the stem. The flowers are small and white. This species grows in damp, shady woods.

The crane-fly orchid, so named because of a fancied resemblance of its flower to a crane fly, is a southern species, but found as far north as New Jersey and Ohio. It occurs in the woods where its long-petioled, broad, acute, plaited leaf, springs from the bulb in the fall long after the flower has perished. It often lives through the winter. The greenish purple flowers appear in July on a scape with several small scales at the base, nodding on pedicels without bracts, in a terminal loose raceme. The sepals and petals are long and narrow, and the lip three-lobed, prolonged backwards into a threadlike spur twice as long as the flower.

The culture of orchids began in the British Isles about 1732 with species collected in the Bahamas by Peter Collinson, a collecting botanist. By 1800, some fifty species were under cultivation and many hybrids raised from seed. Many collectors about the same time began exploring tropical America and other regions for these exotic plants, as rare specimens brought high prices. Eventually, interest in orchids spread to other countries, and today, orchid culturists can be found almost anywhere. To discuss or even to name the many species of orchids in cultivation would serve no purpose, and be beyond the scope of this book, but it might be said that the species of *Cattleya* are to most people, *the* orchids, being the most widely grown of the florists' orchids. They are tropical American plants, most growing in the trees (true epiphytes), with club-shaped, thickened stems wherefrom arise one to three thick, fleshy leaves. The flowers sometimes solitary, but often two or three or even more together, are large and very showy, the petals and sepals are alike, or the petals are broader than the sepals, the lip is three-lobed, the middle lobe spreading and larger than the lateral ones. In color, these orchids vary from a rich purple and rose to lilac and white.

The only orchid of any economic importance is the vanilla plant, the source of vanilla, but now much reduced in value because of the production of synthetic vanilline. The plant is native to tropical America and is a tall-climbing, fleshy-stemmed vine with oblong, thick, fleshy leaves. The flowers are greenish yellow, grouped in a raceme, the sepals and petals are narrow, the lip is trumpet-shaped, shorter than the petals, its lobes rather scalloped. The fruit is a slender beanlike pod, and is the vanilla bean of commerce. The commercial production of the vanilla plant is now chiefly centered in the East Indies. A mature vine will bear forty to fifty bean pods a year for many years. The unripe pods are picked and dried, the vanillin crystallizing on the outside of the pods.

137. Orpine (Crassulaceae)

By exerting a little lateral pressure with the fingertips and then blowing between the two loosened surfaces the leaf of the live forever can be inflated like a paper bag. Country children of another generation took great delight in blowing up the leaves.

The live-forever is an unusual plant in some respects. A somewhat niggardly bloomer, seeds being seldom produced, it spreads almost entirely by its tuberous rootstalks and by rooting at the nodes, forming thrifty tufts even in places where the soil seems to offer little in the way of nourishment. Broken stalks become slips, that develop roots and form new plants.

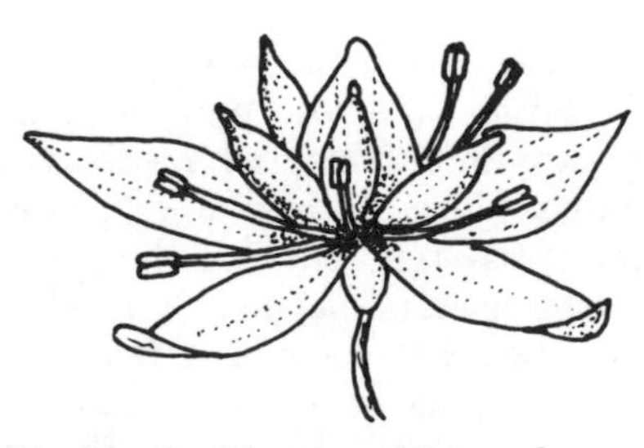

399. Single Flower of Live-forever

The live-forever is very tenacious of life, and even though the mowing knife will lay it low, it will continue to thrive on the juices stored up in fleshy leaves and stem as proof that it lives up to its name.

The live-forever, an immigrant from Europe and once a garden favorite and still found in gardens, has long been an escape and it is now found in fields and waysides and similar places. It has a stout, smooth, erect stem generally greenish, but sometimes purplish, six inches to two feet high, very smooth, fleshy, light green, bluntly toothed leaves, and purple flowers in a dense, compound cyme that appear so sparingly, that some people suspect that it never blooms at all.

The live-forever is a sedum, in other words, it is one of a group of plants known as the sedums and also as stonecrops. They are low-growing, annual or perennial, fleshy herbs, and probably number about three hundred species of the genus *Sedum,* from the Latin to sit, in allusion to the way they grow on rocks and walls, found throughout the temperate and colder regions of the northern hemisphere. The sedums are of diverse habits, thus, there are some that creep, with the stems rooting at the joints or trailing; there are some that are tufted; there are others in rosettes; and still others that are upright or erect. Their leaves may be either alternate, opposite, or in whorls, and their flowers, in terminal clusters, have a calyx of four to five sepals, a corolla of four to eight petals with the number of stamens double that of the petals, the flowers developing into four to five, one-celled follicles, each with several seeds.

Found in stony woodlands and on rocky ledges, the wild stonecrop is a small species with small, toothless, fleshy and rather wedge-shaped leaves, and little five-petaled white flowers on horizontally spreading branches. It is a typical example of the sedums. A garden immigrant, but found in fields and roadsides, the mossy stonecrop, also known as the wall pepper and golden moss, is a low, creeping perennial with tufted stems, yellow green, smooth, fleshy, triangular leaves and bright yellow flowers in terminal clusters. Another immigrant, also found in the garden, is the yellow stonecrop. It has fleshy, narrow, cylindrical leaves and golden yellow flowers, and forms a carpet. The widow's cross or flowering moss has similar leaves, but purplish flowers. It is a native species found on rocks, but also cultivated in the garden. The caterpillars of the variegated fritillary and the buckeye include sedums among their food plants, and both the mountain beaver and the pika are said to feed on them.

The sedums constitute one of the twenty genera of the orpine or stonecrop family, a group of nine hundred species of mostly succulent, or fleshy herbs, or undershrubs of wide distribution in the warm, dry regions of both hemispheres. They have opposite or alternate, simple or pinnate, often extremely succulent, leaves, perfect flowers, nearly always in clusters, often in cymes, the individual flowers small, but the clusters often showy with a floral structure like that of the sedums, and the fruit is a dry, one-celled follicle, also like that of the sedums. The family includes many species and varieties that are extensively used for borders in landscape decoration and in rock gardens, as well as some grown as potted plants.

There are several other members of the orpine family that should be included in our account of the family such as the jade plant, a native of Africa and Asia and cultivated as a greenhouse or pot plant because of its foliage, that is often made up of thick leaves of an unusual shape; the river leek, a queer, little, mud-loving plant with opposite, elliptical leaves and greenish white flowers; the life plant or live leaf, a strange plant of tropical America, Asia, and Madagascar whose fleshy, lobed leaves on falling to the ground will produce young plants from the notches along their margins, the several species of live forevers, of the genus *Echeveria,* some being found in the rocky canyons and on the stony slopes of southern California with yellow or red flowers in terminal clusters on tall flowering stalks, and gray green rosettes that form a striking contrast to the reddish brown vertical cliffs whereon they seem to cling so precar-

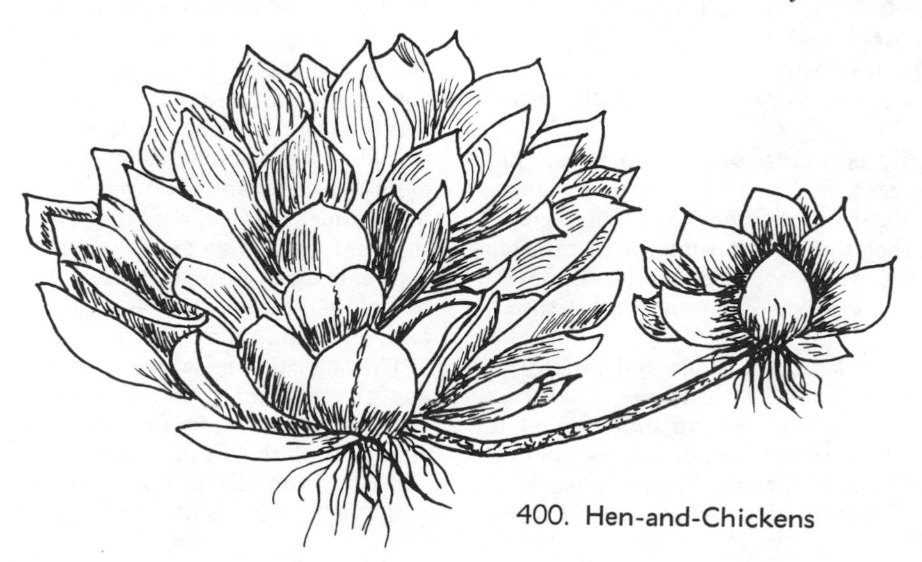

400. Hen-and-Chickens

iously; the ditch stonecrop, a familiar plant of ditches and swamps with insignificant, greenish yellow or yellow green flowers in slender, bending clusters; and the house leek or hen-and-chickens.

The house leek, so-called because it has been found growing on the thatched roofs of houses in Europe, or hen-and-chickens because the plant increases by little rosettes that are sent out from the parent plant, is more widely cultivated than other species of the group whereto it belongs, and that are generally called houseleeks and hen-and-chickens. Largely used for carpet bedding, the plant appears as a rosette of thickened leaves, that are wedge-shaped and tipped with a soft prickle, and has a flowering stalk about a foot high that divides into one-sided clusters of pinkish-red flowers.

There are some sixty species of house leeks, of the genus *Sempervivum,* Latin for live forever, in allusion to the lasting quality of some species. They are mountain plants of Europe and Asia that were designed by Nature to cling to rocks, to cover sandy, rocky hillsides, and to withstand drought. The plants form compact rosettes of stemless fleshy leaves, that are often spotted with red towards the tip, and from the center whereof an erect flowering stalk produces flowers usually in some shade of yellow or rose purple, each flower having six sepals and six petals, thus differing from the four and five parted flowers of the related orpines. The many species vary in the size of rosettes, color of leaves, and character of flowers.

138. Palm (Palmaceae)

To those people who live in the Temperate Zone the palms may seem to be curious and exotic plants, used as ornamentals in the streets of southern cities or as decorative plants in hotel lobbies, but in the tropics, they are as familiar to the inhabitants as the oaks, maples, and hickories are to New Englanders and what is more, there is not a human need that they do not supply.

They provide a grateful shade in the deserts where few plants can survive; they furnish material for building houses and their leaves are used to thatch the roofs and walls; they supply fibers for weaving cloth and for making rope, fish nets and lines, mats, fans, and hats; and they produce edible fruits such as cocoanuts, dates, and others that yield chocolate and valuable oils. Their sap gives wine, sugar, and wax; their stems fresh salads and sago for food, and wands for basketwork and furniture. They also bear the important tropical betel nut wherefrom a mastic is made that is chewed by one tenth of the world's peoples.

The palm family is a large one, of about one hundred and forty genera and more than twelve hundred species of trees, shrubs, and vines. Most of them are real denizens of the tropics, but some of them extend somewhat into the warmer parts of the Temperate Zone. A few of the cultivated species can withstand considerable frost.

Their stems or trunks are slender and usually unbranched, are often covered by old leaves or the bases of the leafstalks, and have a crown of leaves at the top. Unlike the stems of our common trees, they are soft with very little woody tissue. They are much like huge cornstalks. The stem of a palm is largely a mass of pithy tissue with scattered vascular bundles, and with fibrous or corky tissues on the outside and, lacking a growing layer of cells (the cambium), is unable to grow indefinitely in width.

Their leaves are evergreen, stiff, often very large, some of them reaching a length of fifteen feet, usually palmate or pinnate, and are clustered in a dense, terminal crown. Their flowers are small, usually greenish, and inconspicuous, unisexual or perfect, and usually occur in a huge, conspicuous, multi-branched spadix among the leaves. The spathe covers the bud and may be persistent, woody, and boat-shaped, but it does not enclose the mature inflorescence. The flowers have three sepals and three petals, six or more stamens, and a three-celled ovary. Their fruit may be a one-seeded berry, drupe, or nut of various size and nature. The most familiar are the large drupes of the coconut palm and the berry of the date palm.

The species of the palm family are generally, but not technically, divided into two groups by their leaves—that is, whether their leaves are pinnately or palmately compound—in other words, the feather palms' leaflets are arranged feather-fashion, and the fan palms' leaflets are arranged like the fingers of the hand. The first group includes such plants as the hog cabbage, coconut palm, date palm and royal palm, the second the cabbage palm, palmettos, saw cabbage palm, silver palm and needle palm.

The hog cabbage, also known as the buccaneer palm and the Sargent palm, is a rather small palm, not growing over twenty-five feet high, and is a native of a few of the Florida Keys. Of the genus *Pseudophoenix,* from the Greek, meaning false date, it has a diameter of about twelve inches and is usually bulged near the middle. The leaves, in a terminal crown, are about four to six feet long, the larger leaflets or segments sixteen to eighteen inches in length, and the cherrylike fruits are orange red. They hang ripe among the leaves in May and June. The hog palm is often used as a substitute for the royal palm, though it is not as decorative.

The royal palm, more specifically the Cuban royal palm as there are several species of royal palms, of the genus *Rorstonea,* named for General Roy Stone, an American engineer in Puerto Rico, magnificent feather palms from tropical America, is also a native of some of the Florida Keys as well as the Everglades. It is one of the noblest of tropical trees, usually not over seventy feet high, with the trunk swelling near the middle or just above it and tapering above and below the swelling. The leaves are ten to fifteen feet long in a terminal crown, each leaf bending outward and downward with a singular grace. From the base of the cluster, hang the golden yellow flowers in branched spikes, about two feet long, and are succeeded by violet, oblong berries. The royal palm is widely planted in Florida and may be seen along the avenues of many a Florida city.

Probably the most widely distributed of all tropical trees and certainly the most important of all cultivated palms, the coconut palm, a native of the tropical Old World and introduced in tropical America by the Spanish and Portuguese, is a tall slender tree, growing to a height of eighty feet or more, with a trunk diameter of only one or two feet. It is highly wind resistant and along the Florida seacoast, for instance, the tree may be seen in many picturesque slanting attitudes, the result of being buffeted by tropical winds. The open crown of leaves, also, presents a ragged and windswept appearance. The coconuts are borne in clusters among the leaves. The fruit of the coconut palm is technically a drupe, what is generally called the nut being merely the seed of the drupe. The seed that is the coconut of commerce is surrounded by light, impervious, fibrous tissues that enable it to keep afloat for a long time, hence, the fruit is capable of being dispersed by water.

The meat of the coconut is the plant embryo and is the solid part of the stored food. The liquid part is the "milk" that is a rich solution of sugar with oil held in suspension. The oily meat of the nut—termed copra when dried—is the source of the dessicated coconut and of the coconut oil used extensively in cooking oils, blended fats, and soaps. The coir fiber of the husks is used for cordage, coarse matting, and brushes, and the shells are made into household utensils and a high grade charcoal. Even the leaves of the plant are used for thatching and for mats. The coconut palm is probably unexcelled in importance among the world's tree fruits and is planted to the extent of millions of acres, most of the cultivated acreage being in Asia and the East Indies. The three "eyes" at the end of the coconut suggest a monkey's face that give it the name "cocos." This is the name of the genus whereto it belongs and is the Portuguese word for monkey.

Theophrastus gave the name *Phoenix,* the Greek name for date, to the genus whereto the date palm belongs, a group of some dozen Asiatic and African species. They are majestic trees that grow to a height of a hundred feet, with a stout trunk, long, gracefully arching leaves that form a dense, vase-shaped crown, the leaves of several hundred leaflets that are folded lengthwise, yellow flowers in clusters of many thousands in leathery spathes, and fleshy drupes in large pendant clusters that are at first red or yellow, but that turn to deep purple or black as they ripen.

The date palm is a tall plant, usually producing suckers at the base, with erect and stiff leaves when young, but drooping with age, the segments or leaflets,

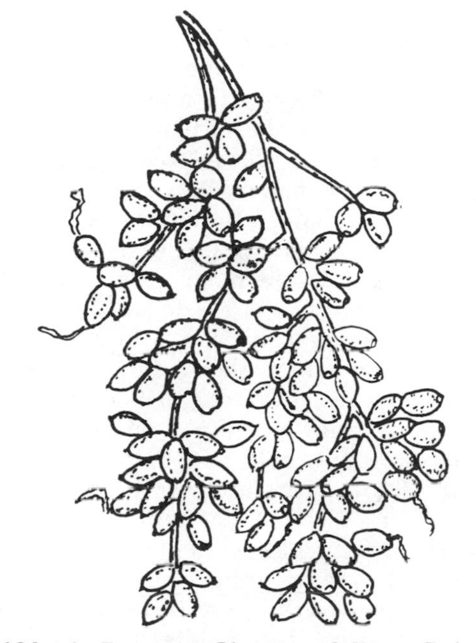

101. A Fruiting Cluster of Date Palm

twelve to eighteen inches long, bluish green, the lower ones spiny. The staminate and pistillate flowers are borne on separate trees and the fruit is rather oblong with a sweet pulp. The date palm has been in cultivation for some four thousand years. The center of date culture is Mesopotamia, Arabia and northern Africa as the tree will produce fruit only in the hot, dry deserts. Spanish priests introduced the date palms into Florida and California during the sixteenth and seventeenth centuries, but growing the date on a commercial scale has been successful only in the Coachella Valley of Southern California. Like the coconut palm, the date palm has many uses beside its edible fruit, furnishing material for the construction of buildings, baskets, ropes, and alcoholic drinks.

Another species of the genus is the Canary Island palm, a handsome, ornamental tree, fifty to sixty feet high, but much less in cultivation, widely planted in the South and used for avenue planting in California.

It is a more slender, graceful tree than the date palm, but very much similar to it in appearance, with darker green foliage and more arched and drooping foliage. The fruit is egg-shaped or roundish, and yellowish red. As its name implies, it is a native of the Canary Islands.

One of the characteristic features of the southeastern coast is the cabbage palmetto, a tree often reaching a height of seventy to ninety feet, with a crown of spreading fanlike leaves at the top of the trunk. The trunk is clothed for a considerable distance from the top with the persistent leaf bases that eventually split and give the trunk the appearance of being encased in a kind of basketwork. The small, yellow flowers are borne in great multi-branched clusters, that are usually longer than the leaves, and thus, hang below them, and are followed by black drupes that ripen in autumn, but persist into the following summer.

The cabbage palmetto is one of twenty species of the genus *Sabal,* a group of palms with moderately tall trunks, though the dwarf palmetto is apparently stemless, the leafstalks arising from the ground with small, greenish white flowers in a cluster from among the fanlike leaves, and a roundish or pear-shaped, dark-colored drupe. Several species of the genus, one of the hardiest genera of palms, are the dwarf palmetto, a low plant that grows in swamps and along streams; the delta palmetto of the Louisiana bayous; and the scrub palmetto of the Florida pinelands.

Common from the Carolinas to Louisiana, the saw palmetto is a shrubby plant of sand dunes and pinelands and sometimes cultivated for ornament. It is the only species of its genus, *Serenoa,* named for Sereno Watson, an American botanist, and usually has creeping stems, though some are erect and several feet high, sprawling over the ground like big black snakes. The leafstalks are prickly, the leaves fan-shaped and divided to or below the middle into about twenty rather stiff segments that have two teeth at the tip. The perfect flowers are borne in clusters that are usually longer than the leafstalks, and the fruit is egg-shaped, or roundish, black and one-seeded.

A fan palm of southern Florida, the Bahamas, and possibly from Cuba, the saw cabbage palm, also known as the Everglades palm, is a handsome palm often forty feet high when full grown, with long-stalked green leaves, the stalks spiny and the leaves cut nearly halfway down into many divisions, which, in turn, are split, small yellowish flowers, two or three together, or often only one, and a resinous, black drupe. This palm belongs to the genus *Acoeloraphe* though it formerly belonged to the same genus as the saw palmetto.

A small genus of unarmed palms, *Thrinax,* Greek for fan, in allusion to the fanlike leaves, of southern Florida, the West Indies and Central America includes the silvertop palmetto, widely planted for ornament. It is twenty to thirty feet high with fanlike leaves that are nearly two feet wide and densely silvery on the lower surface, but pale green above and divided into many, deeply cleft, pointed and stiff segments. The perfect flowers are borne in clusters from among the leaves. The fruit is white. The silvertop palmetto grows in the pinelands and on hammocks. A related and similar, but somewhat more slender species, the Florida thatch palm, also grows on hammocks and along sandy shores.

Another native palm of southern Florida is the silver palm, of the genus *Coccothrinax,* a palm with a short trunk or sometimes stemless, with fanlike, nearly round leaves that are two to three feet wide, pale green above, silvery beneath, perfect flowers in a cluster just at the crown of the leaves, and a berry-like, round, black fruit. This palm is often cultivated for ornament.

The most sharply spined of our native palms is the well-named needle or porcupine palm, found in low, moist places from the coastal plain of South Carolina to Florida and Mississippi. It is a nearly stemless palm armed at the base with many spines or needles seven to fifteen inches long. The leaves are cut into many stiff, narrow, prominently ribbed segments that are toothed or cut at the tip, and both the flower and fruit clusters are smothered in the leaf sheaths and needles near the base of the plant. The needle palm, of the genus *Rhapidophyllum,* Greek for leaf and *Rhapis,* the latter a genus with multi-parted leaves, grows in low, dense masses.

A striking feature of the Colorado desert and of canyonsides in the neighboring mountains is the California fan palm, a palm with a massive trunk growing sixty to eighty feet high, the upper part of the trunk being covered with the long-persistent, withered or dead hanging leaves. The living leaves are fanlike, long stalked, grayish green, and are cut nearly to the middle into many narrow, drooping, thready segments. The perfect flowers are in drooping clusters and the fruit is an elliptical black drupe. The California fan palm is one of two species of the genus *Washingtonia,* named after George Washington, the other species being somewhat taller with a more slender trunk. Both species are widely planted throughout central and southern California where they are the characteristic street and lawn tree.

The fruit of the California fan palm occurs in clusters that may weigh about ten pounds. The Indians eat the fleshy pulp and then grind the seeds into a meal that some think is as good as coconut.

Among the introduced ornamental palms may be mentioned the dwarf fan palm, a native of southern Europe, and a dwarf and bushy species with spiny leafstalks and rigid, deeply lobed leaves; the Mexican blue palm, a native of Mexico and similar to the California fan palm in its habit of having the old dead leaves form a drooping thatch around the trunk, with bluish green leaves, waxy, and deeply cut into nearly fifty segments; and the windmill palm a low-growing Asiatic fan palm, a very popular cultivated species because of its hardiness.

In addition to the above palms, many others are of no little economic importance. Thus, for instance, the raphia palm, of Madagascar and West Africa, yields the raphia of commerce, a strong wrapping cord or tape obtained from the leaf epidermis; the betel palm, of tropical Asia, produces the betelnut, whose seeds are mixed with pepper leaves to form the "chewing gum" of tropical peoples; the sugar palm, of India, supplies a sap that is converted into sugar; the oil palm, of Africa, furnishes palm oil; and the Brazilian ivory palm bears fruit whose seeds contain the vegetable ivory, a hard substance wherefrom buttons are made.

139. Papaya (Caricaceae)

The papaya family is a small group of twenty to thirty species and two genera of mostly succulent-stemmed trees with soft wood and milky juice, that are essentially confined to the subtropics and tropics of the western hemisphere. Commonly known as papaws, they have a straight, stout, unbranched stem that grows fifteen to twenty feet high, and that is crowned at the top by a mass of large palmately lobed leaves. The flowers are commonly dioecious, sometimes perfect, in few-flowered racemes, with a five-lobed, inconspicuous calyx, five petals, ten stamens, and a one-celled ovary with many ovules. The yellow or reddish staminate flowers are in long stalked clusters, with the petals united to form a slender tube, the pistillate flowers, also yellow or reddish, are in nearly stalkless clusters, with the petals nearly separate. The fruit is a large, pulpy berry with many seeds and suggests a yellow or orange melon. It may weigh as much as fifteen pounds. The fruit, that may be either oblong or spherical, has a thick wall of pulpy tissue enclosing a cavity with many dark seeds, and is often eaten as a vegetable. The milky juice from the fruit when dried and powdered becomes the papain of commerce, used as a digestant.

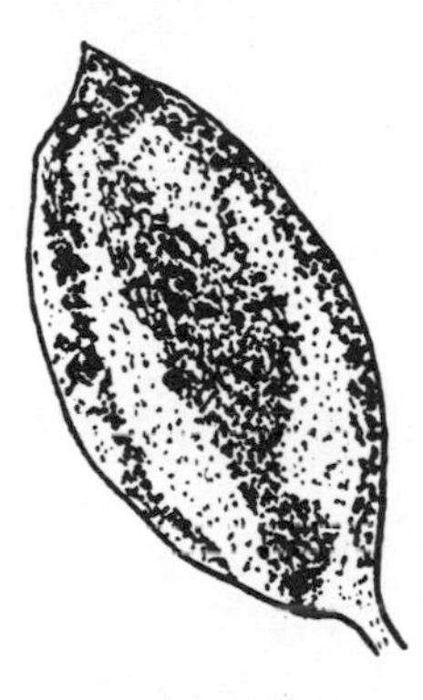

402. Fruit of the Papaya

Only one member of the family is found in the United States, the papaya of the genus *Carica,* Latin for a kind of dried fig, actually from Caricus pertaining to Caria that is a division of southwestern Asia Minor, but neither the fig nor papaya is native there. In the tropics the papaya is a tree often twenty-five to thirty feet high, but in the hammocks, pinelands, and cutover lands of lower Florida, is rarely more than twelve to fifteen feet high. It has alternate, simple, palmately-lobed leaves, bright green above, paler below, fragrant staminate flowers in a multi-flowered axillary, cymose panicle, the pistillate in one to three flowered cymes, and a yellow green to orange, ovoid to elliptical berry about four inches long and three inches wide, with a thick skin and a sweetish pulp. The papaya is widely cultivated throughout the tropics for its fruit, highly regarded for its food value. Its juice makes a spicy, thirst-quenching beverage.

140. Passion Flower (Passifloraceae)

It is said that missionaries in South America named the passionflower because they fancied they saw in the flower the symbols of our Saviour—"the crown of thorns in the fringes of the blossom, nails in the styles with their capitate stigmas, hammers to drive them in the stamens, cords in the tendrils." This is the western world's sole contribution to the symbolic flowers of Christendom. Who would gainsay that the flower does not deserve a place beside the mystical roses and trefoils of ecclesiastical decoration.

A weed in southern cotton fields and common along the roadsides and in waste places of the South, the passionflower belongs to the family of the same name, a group of some three hundred and fifty species and about eighteen genera of usually tendril-climbing herbs or woody plants of the tropics and subtropics, most abundant in South America. They have alternate, simple, entire or lobed leaves, somewhat bizarre, often showy, perfect, regular flowers, with three to five sepals that are often petallike, and sometimes tubular, three to five petals or sometimes none, mostly five stamens that are free and united in bundles, and a one-celled ovary with many ovules. Within the flower there is usually a fringed, often differently colored corona or crown composed of many free filaments. The fruit is a capsule or berry.

The passionflower, native to the southern states, is a strong perennial vine that grows to a length of fifteen to thirty feet, generally low-climbing or trailing, and often covering bushes, and is supported by long tendrils that grow from the axils of the leaves. It has three-lobed leaves, the lobes finely toothed, and blossoms that are about two inches in diameter. The sepals are green without, tinged lavender within. The petals are white, and the filaments that compose the crown or corona are purple or pink. The fruit is a yellow edible berry, about the size and shape of a hen's egg, and is called maypop by the people of the South. Captain John Smith writes that the plant was cultivated by the Indians of Virginia and says that the fruit is pleasant and wholesome. Many people make a jelly of the fruit.

The yellow passionflower, somewhat more northern in range than the passionflower, being found as far north as Pennsylvania, is a slender, rather delicate vine that climbs among the bushes of thickets and hillsides, sometimes ten or twelve feet high. It has long, threadlike, sensitive tendrils, three-lobed leaves with entire margins, greenish yellow flowers, and a dark purple fruit that has a rich purple juice with a rather peculiar flavor.

Both passionflowers belong to the genus *Passiflora*, Latin for passionflower, in allusion to the flowers suggesting the Crucifixion, a genus that contains a number of other tropical species whose fruits are used in making soft drinks, sherbets, jam, and marmalade. Some of these are the purple granadilla, with three-lobed leaves, the lobes toothed, white flowers, the crown white and purple, and a dark purple fruit; the yellow granadilla, also called water lemon and Jamaica honeysuckle, with unlobed leaves, white, red-spotted flowers, the crown white and violet, and a yellow fruit; and the giant granadilla, the leading passionflower cultivated for its fruit, that is greenish yellow and nearly ten inches long. It is a strong-growing vine with winged stems and unlobed leaves, and fragrant white flowers, about three inches wide, the crown purple and white. This species is also grown merely for ornament.

141. Pea (Leguminosae)

Long a garden favorite and still a garden favorite, the sweet pea is probably the most highly developed of all our annuals. As a commercial cut flower it is exceedingly popular, and growing it for its seeds has become an important business.

The white variety of our garden sweet pea is said to have come from Sicily, the pink-and-white, known as the painted lady, from Ceylon. It is not actually known who was the first to cultivate the sweet pea, though it is generally believed to have been Father Franciscus Cupani, a devout Italian monk and an enthusiastic botanist. This was in 1699. Anyway, he was known to have sent seeds to England and various other places. Some thirty years later sweet pea seeds were an article of commerce, and in 1793, a London seed catalogue listed five varieties. Other varieties were added from time to time, but it was not until 1876 that any marked improvement was made in the culture of the sweet pea. At that time Henry Eckford, of Shropshire, England began work upon the plant and working patiently he succeeded, by cross-fertilization and selection, in obtaining new colors and larger flowers, in fact, more than two hundred varieties. Throughout the following years to the present, more than three thousand varieties have been recorded, though hardly more than three hundred are now in cultivation.

The flower of the sweet pea is rather peculiar in appearance and has been imaginatively likened to a butterfly. Hence, it is called papilionaceous, that is,

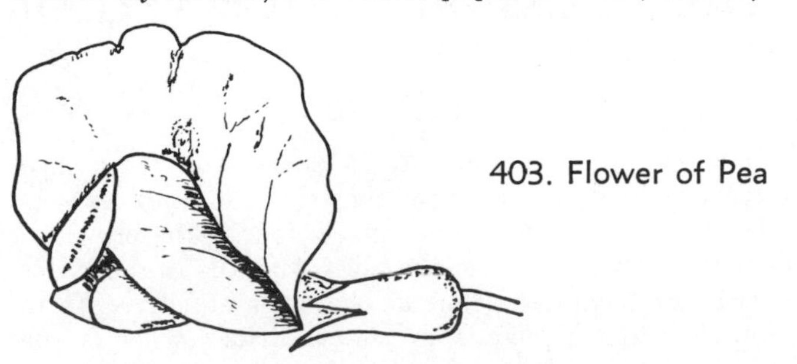

403. Flower of Pea

"butterflylike." It consists of five petals. The upper or odd one is called the standard or bearer. It is larger than the others, encloses them in the bud, and usually turns backward or spreads. The two side or lateral petals are known as the wings and the two lower ones are united to form what is called the keel and that encloses the stamens and pistil. There are usually ten stamens that commonly are in two groups, one free, and the other nine united into a tube.

404. Leaves and Flowers of Kidney Bean

The sweet pea is a member of the pea family, an enormous cosmopolitan group including more than four hundred and thirty genera, and between five and ten thousand species of herbs, shrubs, trees, and many climbing plants. They occur under extremely varied soil and climatic conditions, but are more abundant in temperate and warm regions.

The leaves are alternate and typically compound, rarely simple, the leaflets in some species pinnate, featherlike, in others digitate, fingerlike; the flowers, like those of the sweet pea, are usually very showy, in spikes, heads, racemes, or panicles; and the fruit is a legume or pod, the pea and bean being good examples.

The pea family includes a great many species and varieties of considerable economic importance. They contribute a variety of products such as foods (peas, beans, peanuts, and lentils), forage crops (alfalfa and clover), drugs (licorice), dyes (haematoxylon, so useful in microscopic technique), and ornamentals (sweet pea, wisteria, and locust).

It goes without saying that the members of the pea family most familiar to everyone are the garden sweet pea, the garden pea, and the garden bean, all so well known, that it seems hardly necessary to say anything about them. The common sweet pea is an annual vinelike plant, four to six feet high, with oblong or oval leaflets, and fragrant flowers of many colors, the fruit a pod two inches long containing globe-shaped seeds that are hard and grayish brown. There is also a perennial pea, often called the everlasting pea, one of the hardiest and most easily cultivated species, doing well in almost any location. It climbs by tendrils up to nine feet or more, has ovate leaflets, and produces several rose pink and white blossoms in a long-stalked cluster, followed by a pod up to five inches long. This perennial pea is so rampant, that uncontrolled, it can easily become a nuisance.

The garden pea is a tall-growing pea with smooth foliage and oval or oblong leaflets, the flowers are usually white, and the pod is two to four inches long containing smooth or wrinkled seeds. There are several varieties of the plant. Two are the field pea with pinkish purple flowers and grown only for forage, and the early dwarf pea that grows only from eight to twenty-four inches high and has shorter pods than the tall kind.

Our garden bean is believed to be of American origin, as it was unknown before the discovery of America, although other species of beans have been cultivated since ancient times. To those who grow the bean in the garden, the word bean generally means two types—the string bean, often called the snap, kidney, or stringless bean, and the lima bean. The string bean is the one generally grown because of its delicious, edible, green pods and its very nutritious seeds. It is a tall twining annual, with a hairy stem, broadly oval, pointed leaflets, and yellowish white flowers. The pod is slender, slightly curved, four to eight inches long, with seeds of many colors in the numerous horticultural forms. The lima bean is somewhat similar, though it has a more robust growth, large pods, and large, flat seeds. The wax or butter bean, a variety of the string bean with yellow pods, and the scarlet runner bean, grown for ornament, should also be mentioned. The latter is a tall-growing vine with broadly oval leaflets, scarlet showy flowers, and nearly foot long pods, the seeds almost an inch in width.

There are many other members of the pea family whose fruits serve as food for human consumption, but the two best known are the lentil and peanut. The lentil is not much seen in our country, but it is rather extensively grown in Europe. It is a branching annual, ten to eighteen inches high, with compound leaves, the leaflets arranged feather-fashion, and pealike, whitish blossoms. The pod is fairly long with two seeds that are lenslike and dark-colored.

Everyone knows the peanut, but few know that its native home is Brazil. There are ten known species, but only one, the common peanut sometimes known as the goober or groundnut, is cultivated. It is of considerable economic importance, and is grown all over the world where the climate permits.

It is an interesting annual herb, twelve to eighteen inches high, with alternate, compound leaves of four oval leaflets, and two kinds of flowers: one kind showy, yellow, pealike, and sterile; the other also yellow, but fertile and on a recurved stalk that touches the ground, penetrates it, and carries the fertilized ovary beneath the surface where it ripens into a pod (peanut) whose oily nutritious seeds when roasted are eaten by children and adults alike.

Although the soybean has been grown as a food crop in China and Japan for thousands of years, it is grown in our country chiefly for forage, for green manuring and for its seeds. It is an erect, bushy annual, three to six feet high, with alternate, compound leaves of three oval leaflets, and pealike, inconspicuous white or purplish flowers. The pod is short-stalked, drooping, two to three inches long, hairy and brownish with globe-shaped seeds that are variously colored in the many varieties that are now being cultured. The seeds are extremely nutritious, and besides their high food value, their oil is used in the manufacture of paint, linoleum and soap, and the vinegar from one variety is the basis of Worcestershire sauce.

Equally as well known as the preceding species, the clovers are also useful members of the pea family, especially as forage. There are three hundred species of clovers, nearly all from temperate regions. In all of our species the leaves are compound with three, usually toothed leaflets, most of them stalkless, whence the name of the genus *Trifolium,* Latin for three leaves. The flowers are small and crowded in dense heads. Examined with a magnifying glass each is seen to be perched at the tip of a little stalk, and each has a tubular calyx with five delicate points and a corolla that resembles the sweet pea. All the clovers, indeed as well as all the members of the pea family, have tubercles on their roots, containing bacteria that are able to convert the free nitrogen of the air into nitrates so valuable as plant nutrients. It is this ability that makes clover, alfalfa, and vetch so desirable as cover crops. The Romans plowed these crops into the ground to enrich the soil, a practice continued throughout the centuries, though it was not until quite recently that the reason why these plants are of such value was established.

Doubtless the most common of our clovers is the red clover whose tufts may be found wherever grass grows, such as in fields, meadows, roadsides, everywhere in fact. In this species, the individual blossoms or florets are closely packed together in a somewhat pyramidal, globular cluster that ranges through crimson or magenta, to paler tints of the same colors. Each floret contains an abundance of nectar, but it is of little use to the honeybees for their tongues are too short to reach down into the nectar wells.* The butterflies with their long tongues can do so easily and often drain the nectaries without giving any service in return, for they are not sufficiently heavy to depress the keel, and thus, expose the anthers. Only the burly bumblebees can do so, and hence, are the plant's chief benefactors. Years ago, farmers in Australia planted red clover and had a bounteous crop, but failed to get any seed to plant for the following year simply because they had neglected to import the bumblebees. Once they had taken care of the oversight, all went well. If one should look at the rather soft, dull bluish green leaflets, one will find that they are conspicuously marked by a whitish or yellow green triangle. Towards evening, the leaflets fold downward.

For ages the clover has been regarded as a sort of mystic plant, and the superstitious believed that good luck would attend anyone finding a variation in the number of leaflets that had more than the usual number. How often, as a child, did one look for a four-leaved clover.

If not quite as common as the red clover, the white clover is not much less so. This ubiquitous plant, whose creeping branches send up solitary, globular heads of white or pinkish flowers on erect, leafless stems from May until frost covers the ground or even later until snow begins to fall, occurs in fields, open wasteland, and on our lawns since it is a con-

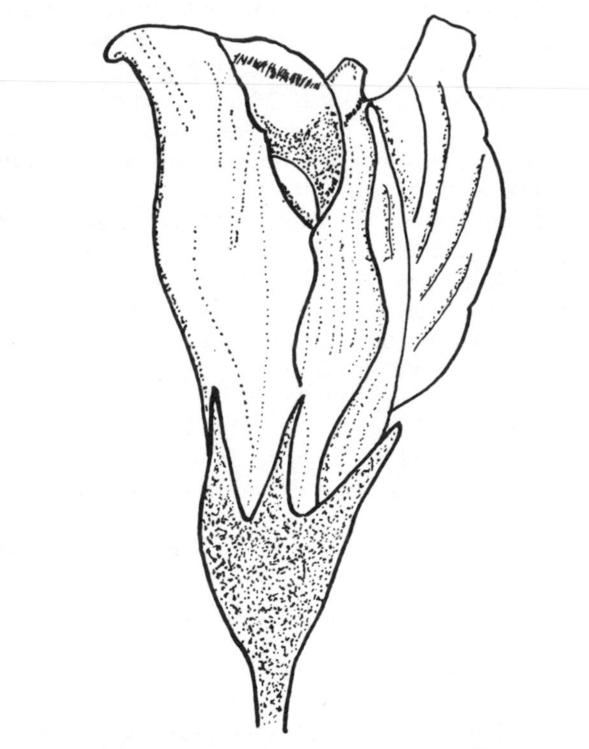

405. Single Flower of White Clover

* It has been shown, however, that honeybees can be induced to pollinate red clover.

406. Flower Cluster (Head) of White Clover

stituent of grass mixtures. Bees attend the flowers and the caterpillars of the orange sulphur butterfly feed on the leaves. Every year on the 17th of March, Saint Patrick's Day, it is sold by street vendors as the "shamrock."

Somewhat similar to the white clover, the alsike clover differs in having a branching, stout, and rather juicy stem and in having flowers that vary from pinkish cream to crimson pink. It may be found in meadows, on roadsides, and in waste places. Also in waste places and in old fields, may be found the rabbit-foot clover, remarkable for its oblong, fuzzy, flower heads, the corolla greenish white and the calyx green with pink tips, the overall effect being a sort of grayish pink.

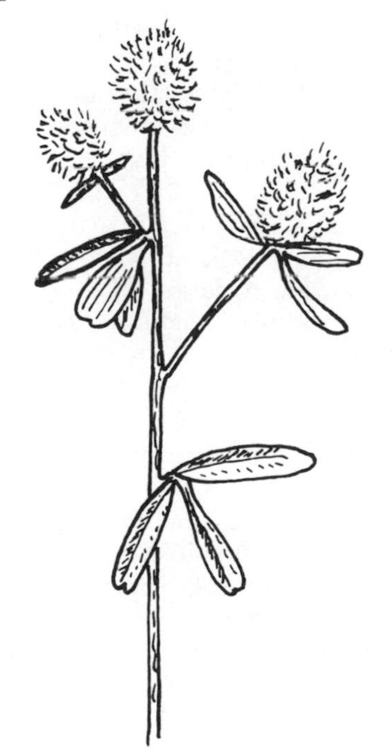

407. Leaves and Flower Cluster of Rabbit-foot Clover

What may be somewhat surprising there is also a clover with yellow blossoms. It is the yellow or hop clover, a small annual species with a smooth stem and light green, narrow and long leaflets that do not look much like the typical clover leaf. When the florets wither they turn brown so that the entire head resembles a small dried hop.

Once occupying a place among sweet herbs because its leaves are fragrant when dried, the sweet clover or white melilot, today is a weed and an escape to the roadside and waste places. The author does not know why it is called a clover; a better name is the melilot, its generic name being *Melilotus,* Greek for honey and lotus in allusion to the fragrance of its foliage, and its similarity to the genus *Lotus.*

408. Leaf and Flowers of Yellow Melilot

There is also a yellow melilot. It is similar to the white species, except for the color of its blossom, and found in the same situations, indeed, it often grows with it. Both species have an erect stem, three to six feet or more tall, compound leaves of three leaflets, very small pealike blossoms in a narrow, spirelike cluster, and the fruit is a small, egg-shaped pod with one or two seeds, not at all pealike. The blossoms, present from June through August, are rich in nectar and attract innumerable bees, wasps, flies, moths, and beetles.

A rather interesting little plant that may be seen growing in dry soil and in open, sandy places, is the trailing bush clover. It has perpendicular branches that arise from a stout horizontal stem, small, clover-like leaves, and tiny pealike blossoms that are magenta pink or a light purple magenta.

The trailing bush clover is not the only bush clover, there are others. Some of them are grown for their showy flower clusters while one species is grown in the South for forage. They are also known as *Lespedezas,* from the name of the genus whereto they belong, a name given to the genus by Andre Michaux, a French botanist, for his friend a Governor Lespedez, once Spanish governor of Florida.

The bush clovers or lespedezas are all annual or perennial herbs with alternate compound leaves, with mostly three leaflets, small, but showy flowers in profuse clusters and in many species, of two kinds. One kind is pealike, showy, and usually sterile, the other is fertile, but without petals. The fruit is a very short pod with a single seed. Some species may be seen in the garden, while others may be found growing in the wild. They include the white bush clover, also called the rounded-bush clover, with dense clusters of yellowish white flowers, visited by the leaf-cutter bee among others, and whose leaves serve as food for the caterpillar of the eastern-tailed blue butterfly; the hairy bush clover, common on dry hillsides as is the preceding species, with yellowish white flowers, a purplish rose spot on the standard petal that serves as a pathfinder to the nectary, in oblong spikes an inch and a half long, the entire plant often downy to the point of silkiness; and the violet bush clover, perhaps the most common of them all, in dry soil, and on the borders of copses everywhere, with an upright and tall stem, sometimes only a foot high, at other times three feet, and flowers that are small, purple or violet purple.

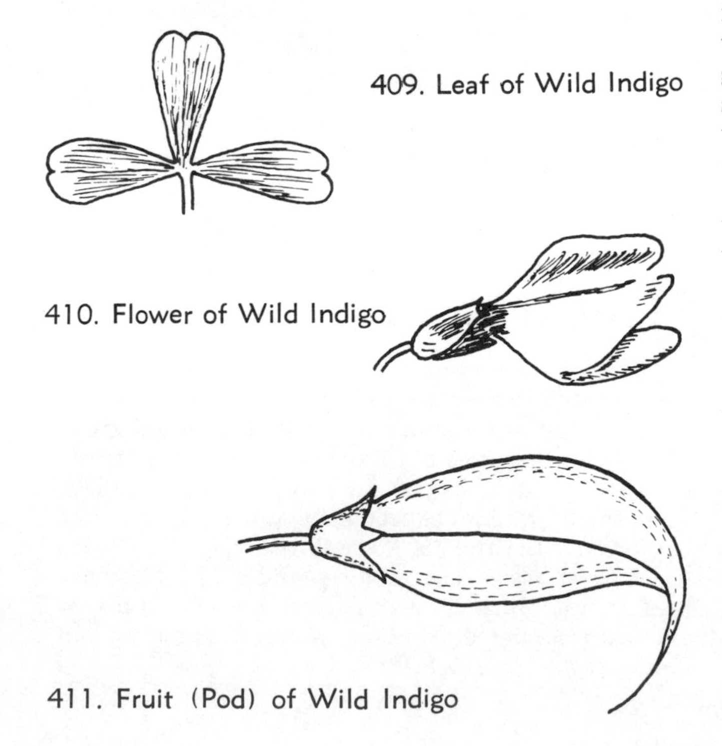
409. Leaf of Wild Indigo

410. Flower of Wild Indigo

411. Fruit (Pod) of Wild Indigo

One of the most curious of plants is the wild indigo that turns black upon withering when it looks as if it had been charred by fire. It grows in dry, sandy soil and is a smooth and slender plant with deep, gray green, compound leaves of three leaflets, wedge-shaped and covered with a slight bloom, and small pealike, pure yellow flowers that are visited by bumblebees, flowerflies, the leaf-cutter bee, and bees of the genus *Halictus,* the latter probably the most efficient agents of cross-fertilization. The caterpillars of many butterflies such as the frosted elfin, the eastern-tailed blue, and the wild indigo dusky wing feed on the leaves. Why is it called the wild indigo, or false indigo as it is sometimes known? Simply because some related species yield a poor sort of indigo dye.

There is also another false indigo called the blue false indigo because of its blue or violet-colored blossoms. It grows in rich, alluvial soil from Pennsylvania, west and south, and because it is ornamental both in flower and foliage, it is often found in gardens.

In the past, children took great delight in shaking the pods of a plant that was found growing in the wild, just to hear the seeds rattle. Called the rattlebox, it is not a particularly common plant, though it is widely distributed, with somewhat insignificant yellow flowers and sepia black pods that are boxlike, inflated, and seedy.

In his book on American wild flowers, Schuyler Mathews in writing of the blue or wild lupine says that, "this is one of our most charming so-called blue flowers; but it rings all the changes on violet and purple and scarcely touches blue." As that may be, it is a charming plant that is content with sterile wasteland, and so it is found in sandy fields, steep, gravelly banks, and exposed sunny hills where it spreads far and wide in large colonies and mirroring the blue of June skies so that, as Thoreau said, "the earth is blued with it."

The word lupine comes from the Latin word *lupus,* the wolf. But why should the plant be named after the wolf? Because farmers at one time mistakenly believed that the plant preyed on the fertility of the soil and made it unfit for planting their crops.

There are over one hundred species of lupines, of the genus *Lupinus,* seventy being native to North America. All are of similar appearance, with palmately or finger-shaped compound leaves, showy flowers in terminal spikes or racemes, and blackish, flattened pods that contain five or six kidney-shaped brown seeds. Some are treelike, others are herbaceous perennials, and still others are annuals. Many are garden favorites. The species known as bluebonnet, a showy silky haired annual with blue flowers, and a native of Texas, is the state flower of the Lone Star state.

When one travels about the country and sees the same plants growing here and there, often in rather remote areas, one may often wonder how they ever got there in the first place. It is no mystery for plants have devised many ways to effect a distribution of their seeds. Some have designed silken parachutes, as the thistles and dandelions, others wings and darts, as the maples and ashes, that are carried by the wind near and far; others creep along the ground, in a tortoise pace perhaps, but as in the fable a winning one; and still others have enclosed their seeds in bright and fleshy berries that attract the birds that deposit the seeds in distant places. Was it not Darwin who raised over sixty wild plants from seeds carried in a pellet of mud taken from the leg of a partridge? And then there are those whose seeds are furnished with hooks and barbs that catch in the fur of animals, and thus, receive free transportation to places far from their place of origin; plants such as the tick trefoils for instance.

There are many species of tick trefoils of the genus *Desmodium,* from the Greek for a chain, in reference to the jointed pods. They are somewhat weedy plants of indifferent garden interest with the characteristics of the pea family, namely compound leaves of usually three leaflets, small and pealike flowers, and a prominently jointed pod, the segments becoming detached separately and often catching in clothing.

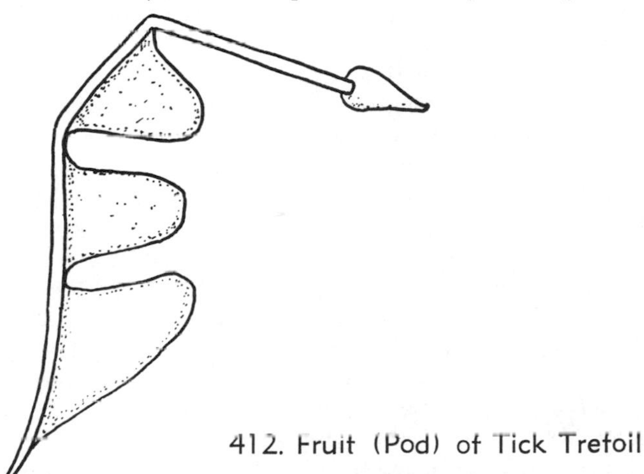

412. Fruit (Pod) of Tick Trefoil

One of the more common of the tick trefoils is the species known as the naked-flowered, tick trefoil. It grows in woods where it lifts its narrow, few-flowered panicles of rose purple blossoms about two feet above the ground, the flowers being fertilized by honeybees and mining bees. The caterpillars of such butterflies as the eastern-tailed blue and the hoary edge, find the leaves much to their taste. To distinguish this species from other trefoils, note that the flowerstalk is usually leafless and that the shorter leaf bearing stem rises from the base of the plant and has its leaves all crowded at the top.

Another tick trefoil, the pointed-leaved tick trefoil, has for its distinguishing feature, a cluster of leaves high up on the same stem, wherefrom arises a stalk bearing a number of purple flowers. These flowers are larger than those of the preceding species. They appear in June and may be seen until September in dry or rocky woods.

Unlike either of the two species that have just been mentioned, the prostrate tick trefoil lies outstretched for two to six feet on the dry ground of open woods and copses, the others having erect stems two to three feet high. Aside from its prostrate habit, it can easily be recognized by its nearly perfectly round leaflets, by its rather loose racemes of deep purple flowers, and by its three to five jointed pod that is deeply scalloped on its lower edge.

Of all the various tick trefoils one is likely to find on outdoor rambles, the Canadian tick trefoil is undoubtedly the most showy of them all, its flowers larger than those of the others, varying in color from magenta to magenta pink, and crowded in clusters terminating a tall, stout, and hairy stem. It is common on the borders of copses and on riverbanks.

In waste places everywhere there may be found a plant with downy, procumbent stems, with three, wedge-shaped leaflets having a bristle tip, with yellow flowers in small, short spikes, and with almost black pods, kidney-shaped, containing, but one seed, known by the rather odd name of nonesuch, a name bestowed on it because of its reputed superiority as a fodder. Of somewhat more interest perhaps, it is said that for many years the nonesuch was recognized in Ireland as the true shamrock. It still passes for a shamrock. The nonesuch is of no garden value, but is sometimes used as a forage. The Indians of southern California at one time greatly relished the seeds, and probably still do.

Of more importance as a forage plant, a related species known as alfalfa is widely grown in the West for this purpose, but in the East it is an escape, and may be found in dry fields and sandy wastes. It is

413. Leaves and Flowers of Alfalfa

a smooth perennial, from one to three feet high, with a long taproot, purplish flowers in short terminal clusters, and pods that are slightly downy and twisted.

When the June sun shines brightly and soft breezes blow over the landscape, decorated with countless blossoms of many colors, the vetches, in fields and thickets, mirror the sky with patches of blue. Or perhaps, not exactly blue, but rather a purplish or violet, though whatever the color, the bees reap a harvest. There are about one hundred and fifty species of vetches—some are annual, some are perennial, some are climbing plants others are erect, but all have compound leaves, the leaflets from one to twelve pairs, those in the climbing species having the end leaflet modified into a tendril whereby they climb. The flowers are blue, violet, purplish, yellowish or white, pealike of course, usually small, in short racemes, and the fruit is a flat pod. The caterpillars of the common sulphur butterfly feed on the leaves, and the Indians once ate the starchy seeds, that resemble those of the common pea, though smaller. Europeans, also, often eat them.

The vetches of the genus *Vicia,* the classical Latin name, are found distributed throughout the northern hemisphere and South America, and the most widely distributed of them all is the species known as the cow vetch, but it has also been called bird vetch, cat fitch, cat peas, titters, and tine grass. Anyway, it is a graceful climbing plant, characterized by a fine, downy hairiness. The flowers are at first blue, but later a violet, and the color of the foliage is rather gray, olive green. Altogether it is a showy plant and grows on the borders of thickets and in cultivated fields.

The cow vetch is an immigrant from Europe and so, too, is the common vetch usually grown for fodder, but also an escape, a survival of former cultivation by means of self-sown, dormant seeds, and so it may be found growing in meadows, waste places, and along the roadside. It is a climbing plant with purple or purplish flowers that grow in pairs or singly at the junction of the stem with the leafstalk. There is a third species that grows in the same situations and called the hairy vetch because it is covered all over with long, soft, persistent hairs. It resembles the cow vetch in form and habit.

In company with the vetches one may find some of their relatives, such as the ground nut, the wild bean, and the wild or hog peanut. Whittier, as a barefoot boy, knew that: "Where the freshest berries grow,/ Where the ground nut trails its vine,/ Where the wood-grape's clusters shine," there was something good to eat beneath the surface soil. For in the root system of the ground nut, a slender vine that climbs over bushes, ferns, and sedges in moist thickets, there are from one to one dozen tubers connected by narrow fibrous strands that are sweet and edible. Edward Winslow, one of the founders of the Plymouth Colony, relates that during the first hard winter the Pilgrims "were forced to live on ground nuts." They were important in the dietary of the American Indian, and during colonial days, the Swedes on the Delaware ate them for want of bread. Were it not for the popularity of the potato, the ground nut would likely be cultivated for the American table. As the American botanist Asa Gray once said, had civilization begun in America instead of in the old world, the ground nut would have been the first edible tuber to be developed and cultivated. The ground nut is undoubtedly one of our very best wild foods. The tubers may be eaten raw, but they are better boiled or roasted.

One needs only to glance at the ground nut to recognize it as a member of the pea family. The compound leaves are composed of three to seven toothless, pointed leaflets, and the flowers are pealike, brownish red, chocolate brown, or reddish purple, paler without than within, fragrant, indeed, somewhat cloyingly sweet, numerous, in racemelike clusters from the leaf axils. The fruit is a leathery, slightly curved pod, two to four inches long. The ground nut is a beautiful plant, worthy of cultivation, though words can only begin to convey the beauty of its blossoms, "of the dull, deep, lurid hue of the inner part of the banner and wings, or the curious red, dusted with gray, of the exterior."

The older naturalists whose botany was "learned of schools" often confused the wild bean with the ground nut, but without excuse, as the wild bean has loose racemes of smaller, purple flowers and leaflets in threes. Indeed it best resembles our garden bean. It is a perennial vine that climbs over bushes or trails upon the ground in thickets where it sometimes reaches a length of twelve to fourteen feet. The six to four brown seeds, contained in a short-stalked, drooping pod about two inches long, flat and slightly curved, may be prepared and cooked like the garden variety. The seeds were highly regarded by the North American Indians who could keep them in the dried state for use when other vegetable foods were scarce.

The Indians also considered the fruit of the hog peanut an important article of food. In autumn and early winter, the women would rob the nests of white-footed mice and other rodents obtaining "big piles of them," and the Dakota Nations, when taking the pods from the nests of animals, always left corn or other foods in exchange.

The hog peanut is a dainty, slender vine, three to eight feet long, usually found in rich woodlands, but also growing along roadsides and fences. It trails on the ground, or twines and climbs over low shrubbery. The leaves are formed of three smooth, angularly, ovate-pointed leaflets, and the perfect, lilac or magenta lilac, narrow blossoms occur in small drooping clusters. They are succeeded by many small pods about an inch long generally containing three mottled beans.

In addition to these flowers there are others near the roots on threadlike creeping branches. They are rudimentary in form without petals and with only a few free stamens. These flowers are self-fertilized, and produce pear-shaped pods containing one big, light brown seed the size and shape of a peanut. These seeds are more agreeable in taste than a raw peanut and, as a matter of fact, are very good eating. At times, the large seeds are quite abundant and may be found beneath the dead leaves, usually under the surface of the ground. They may be gathered without injury to the plant. Hogs, running in the woods, often root about in the moist soil where the plant grows in search of them, hence, its name of hog peanut.

At any point along the Atlantic Coast from Maine to New Jersey, the sturdy clumps of the beach pea will be seen growing beyond the reach of the tides in dunes and sandy wastelands. It is a sprawling perennial plant with compound leaves of six to twelve oval leaflets that are bristle-tipped, and with ruddy purple, or violet purple flower clusters of five to twelve pea-shaped florets. An unusual feature of the plant is the pair of conspicuous halberd-shaped stipules that are as large as the leaflets.

Although the licorice that is known so well in the form of a candy and as a cough remedy is obtained from a cultivated species, there is a native wild species, known simply as the wild licorice or American licorice. The cultivated species is a European perennial herb, known as the common or Spanish licorice, two to three feet high, with compound leaves of four to eight pairs of oval leaflets, blue flowers that are pealike, in short clusters, and a flattish pod three to four inches long. Our native species, occurring from Hudson Bay to British Columbia, south to Missouri, northern Mexico, and California, has compound leaves of fifteen to nineteen, oblong, lance-shaped leaflets that have a pointed tip, whitish flowers in short spikes, and oblong pods beset with hooked prickles. The long, fleshy roots, that closely resemble the roots of the cultivated species, are sweet and are eaten by the Indians.

Under the general term of senna are grouped a large number of herbs, shrubs, and trees that some botanists have placed in a separate family known as the senna family, but that are more generally included in the pea family. They are mostly tropical plants, but a few herbs, such as the wild senna, partridge pea, and the wild sensitive plant, occur in the Temperate Zone. All are members of the genus *Cassia*, the old Greek name for the plants.

The wild senna is a showy and decorative plant found growing in alluvial, moist, or rich soil, in swamps and along the roadside. It is a stout perennial with compound leaves of twelve to eighteen broad, lance-shaped leaflets of a somewhat yellow green tone. The flowers are yellow in racemes some being in the leaf axils, others terminal, and are followed by narrow, flat, curving pods three to four inches long. The caterpillars of the cloudless sulphur, the orange barred sulphur, the sleepy orange, and the little sul-

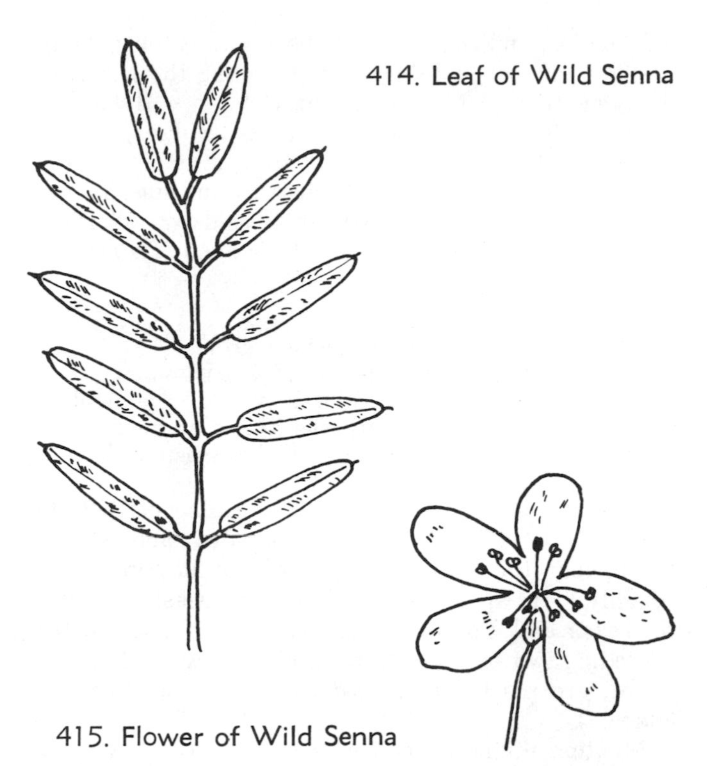
414. Leaf of Wild Senna

415. Flower of Wild Senna

phur butterflies all use it as a food plant. Grazing animals leave the plant undisturbed, or if forced to browse its leaves by a lack of forage, they suffer "scours" as it has a strong cathartic action. The leaflets, stripped from their stalks at flowering time and carefully dried, have been sold in the drug market.

Like the wild senna, the partridge pea is also strongly cathartic, and when eaten by cattle and sheep the animals "scour" very badly. Grazing horses, too, get sick in the same manner. The plant is low and spreading, sometimes branching wider than its height of one or two feet. The leaflets, twenty to thirty in number, are blunt, lance-shaped and each is tipped with a sharp bristle. They are sensitive and "go to sleep" at night, that is, they fold together and droop against the stalk. The flowers are large, showy, and yellow, an inch or more in diameter. Sometimes the petals are spotted with purple at the base, and usually six of the anthers are purple, the remaining four yellow. The pods are slender, about two inches long, somewhat hairy, and when mature, they split open with a slight twisting action so that the seeds are expelled quite some distance. The net result, of course, is that where one plant grew this year there will be a number of them the following year, in what one might call, a patch of them. The caterpillars of the little sulphur butterfly feed on the leaves.

The partridge pea grows in dry, sandy soil, in meadows, fields, pastures, roadsides, and waste places. In similar situations, one will also find the wild sensitive plant or sensitive pea, a somewhat similar species, but with small and inconspicuous, yellow flowers. Shelley's famous poem, "The Sensitive Plant," refers to a tropical species whose leaves are more sensitive than our species, and that close at the slightest touch. Only a slight movement follows our touching the leaves of our sensitive plant, but a sharp blow will have a quicker effect. Like those of the partridge pea, the leaves of the sensitive plant "go to sleep at night," and also partly close in the bright sunshine, in the first case to protect them against cold by radiation, and in the second case to protect them against the fierce heat of the day.

> A sensitive plant in the garden grew,
> And the young winds fed it with silver dew;
> And it opened its fan-like leaves to the light,
> And closed them beneath the kisses of night.
> —Shelley, "The Sensitive Plant"

A native of tropical America and also of the Old World, the coffee senna has invaded and spread throughout the United States, so that it now occurs roughly in an area from Virginia to Kansas, and south to Texas and Florida. It is somewhat similar to the preceding two, except that the seeds, known as Magdad coffee, are roasted and used as a substitute for real coffee.

Mention of the coffee senna leads to the Kentucky coffee tree whose seeds were ground and roasted by the early settlers, particularly those who lived in the interior part of the country where commercial coffee was difficult to obtain. This was especially true in Kentucky, a remote region before the Revolution. It is said that the seeds were also roasted and eaten by the Indians.

The Kentucky coffee tree is one of the last trees to come into leaf in the spring and one of the first to lose them in the fall. It is so long without leaves that the French Canadians were led to call it *chicot,* meaning "stump" and its generic name, *Gymnocladus,* from the Greek, means "naked branch." The tree prefers a rich, moist soil, and usually grows in lowlands along streams, but as a forest tree, it is rather rare and is more commonly seen in cultivation.

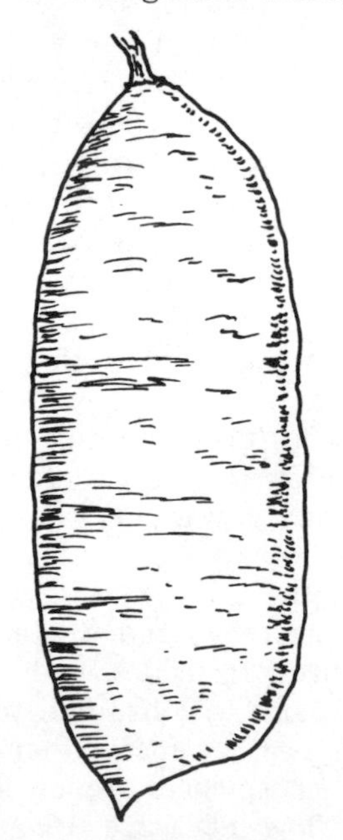

416. Fruit (Pod) of Kentucky Coffee Tree

It is a remarkably rough, coarsely branched tree with crooked, coarse twigs, some seventy-five to a hundred feet tall. The bark is gray, deeply furrowed, the leaves are dull deep green, doubly compound, one to three feet long, the flowers are greenish white, the staminate in short racemelike corymbs three to four inches long, the pistillate in racemes ten to twelve inches long on separate trees. The fruit is a pod or legume (as the pod of the pea family is sometimes called), thick-walled, leathery, purple or dark reddish brown, six to ten inches long, with a sweetish pulp wherein are contained six to nine bony, globular seeds, and persisting on the tree all winter. It is no small wonder how the pioneers ground the seeds for coffee, for they are extremely hard. How they could have brewed a "coffee" with such an unbelievably bitter taste, is also a matter of wonder, but then, they were a hardy folk.

Like the Kentucky coffee tree, the honey locust is also more often seen in cultivation than in its native wilds. It is a strikingly handsome tree with a brown bark that appears "alive-looking" and with slender and wiry branches that shine as if they had been polished. Tolerant of many soils, its preferred habitat is rich woods or bottomlands where it may often attain the astonishing proportions of one hundred and forty feet in height and a trunk diameter of six feet. Its foliage is graceful and plumy, of a clear intense emerald, and the pods in midsummer show shades of red and green and are beautiful, not only in form, but in color and texture as well. Unlike other members of the pea family, the flowers are not pealike, but are regular. They are greenish, in inconspicuous, nearly pendant clusters, the staminate and pistillate separate and often on different trees, or sometimes combined in one cluster.

Undoubtedly the most striking peculiarity of the honey locust is its thorns that are modified branches arising from the pith, and not skin deep, as are the prickles of the locust or yellow locust or false acacia, as it is also sometimes known. Many a honey locust is literally covered, both trunk and branches, with the thorns, from two to six inches long, sometimes in clusters, often three-pronged, and ruddy brown in color, that provide a formidable defense against both man and beast.

The honey locust derives its name, not from the honey of the flowers, that are, incidentally, fragrant and nectar-laden, but from the half ripe fruit that contains a sweet edible pulp, pleasing to the taste, but that later dries and turns bitter. The long, flat, pendulous pods hang in clusters from the branches, and contract in drying when they so twist and curl that they are easily rolled by the wind over the frozen ground of winter until they eventually find a place to lodge. Here the seeds eventually soften and germinate, and saplings spring up far from the parent tree.

The honey locust is a beautiful tree when young, and growing rapidly, reaches a height of seventy feet and a trunk diameter of three or four feet only in a few years. Unfortunately, it is not a handsome tree in old age, because its twigs and branches are brittle and easily broken by the wind, so that in winter, it has an unkempt, ragged appearance. And, since the pods hang on all winter, "chattering in the wind," they only serve to call attention to its unsightly aspect as a feature of the landscape.

Why some trees grow faster than others is a question that might well be asked, and, indeed, the answer is in the character of the roots. A tree with shallow roots, that is, roots that lie and extend just beneath the surface, grows faster because the soil there is richer than at a great depth, but unless the roots have ample space wherein to spread, they can quickly exhaust the soil and the tree will be slowed in its growth. A tree whose roots penetrate deeply into the ground and whose roots also spread widely, will grow more slowly, but more steadily, and other factors being equal, will attain a greater height.

As ugly as the locust may appear in winter, it redeems itself in the spring when the delicate leaf sprays, appearing at first in yellow green and silvery with down, change to a dark green as the handsome clusters of cream white flowers unfold and scent the air with a delightful perfume. Then they become alive with bees.

The locust is a native tree, originally confined to a relatively small area (if Pennsylvania to Georgia and west to Iowa and Oklahoma can be called a small area), but since has been planted over a much wider range. Originally a forest tree it can now be found growing in groves, along fencerows, and on woodland borders. It has the interesting habit of folding its leaflets and drooping its leaves on rainy days and as night approaches, doubtless for the purpose of avoiding excessive loss of moisture and heat.

417. Fruit (Pod) of Locust

Unlike the thorns of the honey locust, the prickles of the locust or false acacia are skin deep, that is, they are part of the bark and will come off with it when the latter is peeled. They are actually stipules of the leaves, but they persist and can become quite formidable.

The seed pods are thin, brown, smooth, two to four inches long, and contain from four to eight seeds. The seeds were gathered and cooked by the Indians. They are slightly acid and oily, but when cooked, as peas and beans are cooked, they lose their acidity and become a pleasant, nutritious food. The pods can be dried and preserved for winter use should other vegetable food become scarce.

The water locust, a small, flat-topped, irregular tree, found in the swamps of the Mississippi basin and the South, should also be mentioned. The leaves and flowers are similar to those of the honey locust. The tree can probably best be recognized by its brown, polished thorns that are three to five inches long, pointed and stiff, and sometimes flattened, like the blade of a sword.

According to legend, the redbud tree, glowing red in the spring time, was said to have blushed when Judas hanged himself from it. It was not our redbud, but an Asiatic species, and yet, our own native tree bears the blistering name of Judas-tree.

Our redbud awakens with the shadbush and the

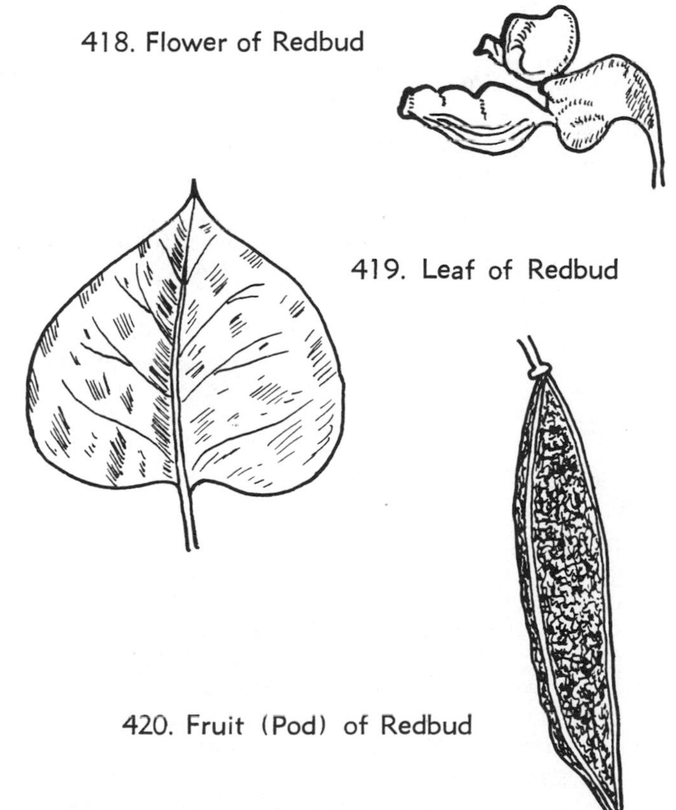

418. Flower of Redbud

419. Leaf of Redbud

420. Fruit (Pod) of Redbud

wild plum, and covers its still naked twigs with a profusion of rosy magenta, pealike blossoms in clusters that hug the branches closely. It is a small, slender, southern tree of rich bottomlands, but also skirting the woodlands in the North, and in common cultivation as an ornamental throughout the East and South. After the flowers, come the leaves, glossy, round or heart-shaped, with veins arranged finger-fashion, simple and not compound like most members of the pea family, and beautiful in their dark green, but no less so than the pods that follow, dainty pale green, but later of a purple hue.

Rarest of the trees in eastern North America is the distinction of the yellow wood or virgilia. It is native to the bottomlands and mountainslopes of Tennessee, Kentucky, and North Carolina, where it is distinctively local and rare, but it has been rather extensively cultivated and may be seen in private grounds throughout the East. It is a beautiful, medium-sized tree, thirty-five to forty feet high, with a wide, graceful head of slender pendulous branches, gray bark and smooth like that of the beech, but often wrinkled and knobby in places when it resembles the hide of an elephant. The leaves are compound, a foot long, of seven to eleven oval leaflets, bright yellow green, but turning a clear yellow in the fall. The flowers are large, white, pealike, and fragrant; and the pods are thin, smooth, with a few small, flattened seeds. Such are the botanical characters of the yellow wood, but a prosaic description fails to do it justice. The tree must be seen to be appreciated. And why is it called the yellow wood? Because its wood is yellow and its sap gives a yellow dye. As for the name virgilia, this has been its garden name.

According to tradition, the woadwaxen was introduced into the United States by Governor John Endicott of Salem who was something of a pioneer of American horticulture. It was at one time collected by the poor people in England who sold it to the dyers for the yellow dye that it yielded. Hence, it is also known as dyer's greenwood.

The woadwaxen is a low-branching shrub, one to two feet high, with creeping rootstalks and upright branches. It is a native of Europe and has escaped from gardens and is found on dry hillsides within a rather small area, from Maine to New York. The leaf is called a compound leaf, but has only a single leaflet that is a contradiction in terms, and thus, would seem to pose a botanical puzzle. It is really not a puzzle, however, for if one would look at the petiole or leaf stalk, it would be seen that there is a joint that indicates that the leaf is compound, but that all the leaflets except one have become aborted. The flowers are bright yellow, pealike, in terminal clusters, and the pods are slender, flat with several seeds. There are several horticultural forms of the woadwaxen that vary mostly in habit.

A second shrub, also naturalized from Europe, is the Scotch broom. It grows two to nine feet high and has wiry, rigid, olive green, ascending stems that are smooth and angular. The leaves are small, compound, of three leaflets that are a light green, though sometimes there is only a single leaflet. The flowers appear in May or June, are a bright yellow, and pealike. A Scotch broom in blossom is like a fountain of gold, and is extremely attractive. The pod is flat, one to one and a half inches long, curved, smooth on the surface, long-haired on the edges, and brown. As an escape, the Scotch broom grows on sandy soil and in waste places from Nova Scotia south to Virginia and somewhat beyond.

Another immigrant from Europe, the furze or gorse is a multi-branched, twiggy and spiny shrub of two to three feet high, and established along the coast, usually in waste places. The leaves are needlelike or thornlike, very spiny in appearance, a third to two-thirds of an inch long, and dark green. The flowers are small, pealike, a light golden yellow, appearing from May to July, and the pod is only about one half of an inch long, brown and hairy. The plant is of some use as a sand binder.

Although the preceding species have come from the old world, there are some native species, too, as the false indigo, for instance, a bushy shrub that grows from five to twenty feet high on the borders of streams from Connecticut south to Florida, and west to the central states. The leaves are compound with eleven to twenty-five leaflets that come out of the bud a pale green, but later change to a bright yellow, green when full-grown and a pale yellow in autumn. The flowers, appearing from May to July in terminal clusters, show an interesting case of arrested development. The corolla begins as a pealike blossom, but when the bud opens, all that is visible of the blossom is the broad banner or standard that is closely wrapped about the stamens and pistil. The banner, filaments, and style are a deep rich purple, the anthers a brilliant orange, with the net result that the entire flower cluster glares vividly in the sunlight. The shrub is ornamentally effective, and is cultivated rather extensively; as a matter of fact, there are pale blue, white and crisped-foliaged horticultural varieties.

That the lead plant was supposed to indicate the presence of lead ore wherever it grew, was a belief once held by many and based essentially on the leaden color of the plant. This color is due to presence of silvery hairs or down that densely cover the entire plant. It is a spreading bush, one to four feet high, with light green leaves and blue flowers in terminal spikes, and occurring on the western prairies. An odd and attractive shrub, its leaves, in silken gray, contrast delightfully with the blue of its spikes, and it is a desirable addition to the garden in the shrub border or as a specimen plant.

Easily cultivated and highly prized for its beauty, the rose acacia is a native of the woodlands from Virginia and Tennessee to Georgia and Alabama, but also found as an escape northward to Connecticut. It is a fairly large shrub with the twigs, branches, and leaf stems so conspicuously covered with bristly hairs, that the entire plant looks mossy. Hence, it is also called moss locust, bristly locust, and pink locust, the name locust given to it because it belongs to the same genus as the locust that has already been described, and pink locust because of the color of its flowers that are exquisitely rose colored and pealike in loose clusters. As a point of interest, the name of the genus *Robinia* commemorates the botanical labors of Jean Robin, who was a herbalist to Henry VIII, and director of the gardens of the Louvre.

The genus *Acacia* is an enormous group of quick-growing shrubs and trees found all over the tropical world, and a few in subtropical regions. They normally have twice compound leaves, the leaflets very numerous and small, very small flowers crowded into dense, finger-shaped or globular clusters that may be solitary, but more often arranged in variously branched sprays, and consequently very handsome, and a fruit like a pea pod, but often woody when ripe and sometimes twisted.

Planted in the gardens of the warmer parts of the world and growing wild in the Rio Grande Valley, the huisache is a small, spiny tree with graceful spreading branches, and pendulous twigs that are covered with feathery, twice pinnate leaves. Its numerous, bright yellow flowers are very fragrant, and occur in rather dense clusters. Another species of the same genus is the green wattle, a tree often fifty to sixty feet high, and widely planted in California. It has twice compound leaves and globular, cream-yellow flower heads that are arranged in a raceme. A third species is the cat's claw of western Texas to California, a thorny shrub of rocky canyons and mesas, with twice pinnate leaves and yellow flowers that are borne in fingerlike, close racemes. Some members of the genus lack leaves and have, instead, the leafstalks broadened and flattened to form leaflike phyllodes, such as the kangaroo thorn, a spiny shrub with globular, solitary flower heads, about the size of a pea, but very numerous and yellow; and the golden wattle, a shrub or small tree, without prickles, and with lemon yellow flowers in finger-shaped clusters. Prized for their delicate, fernlike foliage and sunny yellow flowers, many species of acacias are widely grown in the South and planted as ornamental street trees.

One of the wonder plants of arid and semi-arid regions, the mesquite or screw bean is a tree sixty feet high along the rivers of southern Arizona, but elsewhere, as in the gray gravelly stretches of the Southwest, it is a low, spiny shrub. In either case, it has compound leaves that are subdivided into delicate and frail-looking leaflets, small, greenish white flowers, in axillary spikes, and straight or twisted pods that are often used as food by animals and the native Indians, the latter either drying and grinding the pods into a meal or cooking the green pods that, when cooked, are considered a great delicacy.

Some other members of the pea family found in the arid wastes of the Southwest are the smoke tree, a stout and spiny shrub or small tree, conspicuously gray because of its hoary twigs and spines, with inconspicuous leaves and small, violet blue flowers; the ironwood, a small tree, with hoary, spiny twigs, and locustlike purple or violet flowers, leaves, and seed pods; and the paloverde or green-barked acacia, a desert shrub that is green-barked, spiny, and leafless for much of the year, the leaves, when present, twice compound, and with showy, yellow flowers.

Several other trees of the pea family might also be mentioned in passing, such as the horse bean, or ratama, a native in the valleys of the Rio Grande and the Colorado. It is a small, graceful tree with drooping branches that are covered with spines, long leaf stems with many pairs of tiny leaflets, and bright, yellow, fragrant flowers in clusters four to six inches long. Another tree is the wild tamarind, a small tree with stout, spreading branches, compound leaves whose leaflets are again subdivided into more delicate leaflets, and small, white or greenish flowers, and found in Florida. And a third species is the mimosa, or silk tree, one of the South's outstanding ornamental flowering trees for roadside planting. It is a medium-sized tree, not over thirty feet high, with a broad, spreading crown, foliage of graceful, feathery, double compound leaves, light pink flowers in slender-stalked, compact heads, and a flat pod. It is incorrectly called mimosa, the name of another group of mostly tropical American herbs, shrubs, and trees of the pea family. One of these, called the sensitive plant, has become naturalized in Florida and along the Gulf. It is a low, somewhat woody perennial, more or less hairy and slightly spiny, with twice compound leaves, and rose purple or lavender flowers. Upon the slightest irritation, all the leaflets immediately fold up and the whole leaf collapses.

In addition to the herbs, shrubs, and trees, the pea family also includes a number of woody vines. Perhaps foremost of these woody vines is the wisteria, one of the best and one of the most common of hardy climbers, a strong grower, long-lived, and healthy, with an airy and graceful foliage and long, drooping racemes of purplish pealike blossoms, the flower clusters often a foot long.

There are two introduced species, the Chinese wisteria and the Japanese wisteria, and two native ones. The Chinese species is a high-climbing vine, in old specimens reaching the tops of trees and houses. It has compound leaves of seven to thirteen leaflets, bluish violet flowers in clusters about a foot long and fragrant, and pods five to seven inches long and densely velvety. The Japanese wisteria is similar, but with thirteen to nineteen leaflets and flower clusters eighteen to twenty inches long. These two are the usual ones in cultivation.

Our two native wisterias are the American and Kentucky. The American wisteria, found wild from Virginia to Florida and Texas, climbs twenty to thirty feet high and has nine to fifteen, elliptic leaflets. The lilac purple flowers occur in clusters that are hardly over four inches long, while the pods are smooth, somewhat knobby, and beanlike. The Kentucky wisteria is somewhat like the American, but with minor differences. It occurs in the floodplains of the Mississippi River drainage. The wisterias serve as food plants for the caterpillars of the silver-spotted skipper, the long-tailed skipper, and Horace's dusky wing. The two skipper butterflies also find other members of the pea family equally as desirable as food plants, including both the locust and honey locust.

Of a rather notorious reputation because of its rampant growth, the kudzu vine is a native of Japan and China. It is a somewhat woody plant, hairy-stemmed, and often climbing to a great height. The compound leaves have three, oval or nearly round leaflets that are hairy and have a small point, the flowers are pealike, purple, fragrant, in dense, upright clusters nearly ten inches in length, and the pods are hairy, flat, somewhat oblong, and up to four inches long. The kudzu vine grows quickly and is used for providing a dense shade over arbors, but unless controlled, it is likely to take possession of our gardens. In the South it is widely grown as a forage crop. It is not hardy in the North, and even in the South, it is often killed to the ground, but it quickly grows again from its thick, starch-yielding roots that have been used in Japan as food.

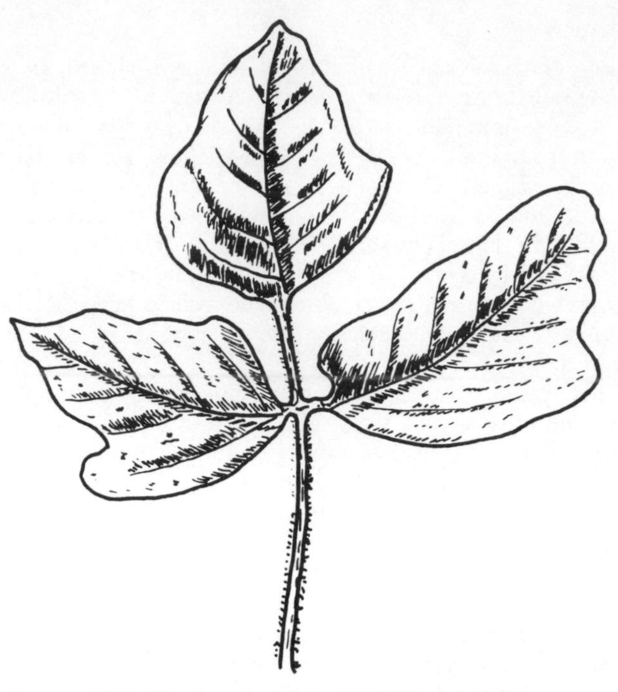

421. Compound Leaf of Kudzu Vine

142. Pepper (Piperaceae)

The pepper family, best known as the source of black and white pepper, is a group of ten to twelve genera and about twelve hundred species of tropical and subtropical regions. They are mostly herbs or shrubs, but some are trees and others are woody climbers. Their leaves are large, alternate, opposite, or whorled, and entire, often thick or fleshy and usually minutely, but distinctly, spotted with small dots. Their flowers are very small and crowded in dense spikes or racemes, they are perfect or unisexual, and without either a calyx or corolla, but have two to six or more stamens, and a one-celled, one-ovuled ovary. Their fruit is berrylike with a fleshy, thin or dry pericarp, which is the ripened wall of the ovary.

The familiar black pepper is made from the dry fruits or ground dry fruits of a woody vine of the East Indies. The white pepper is the same fruits with the outer skin removed. The pepper was once one of the most expensive products brought to Europe by sailing ships or caravan routes. The cubeb is another pepper and the "betel" leaves, a favorite chewing material of the natives of India, are the leaves of the betel, a climbing woody vine.

One genus *Peperomia,* Greek for pepperlike, in allusion to its close relationship to the true pepper,

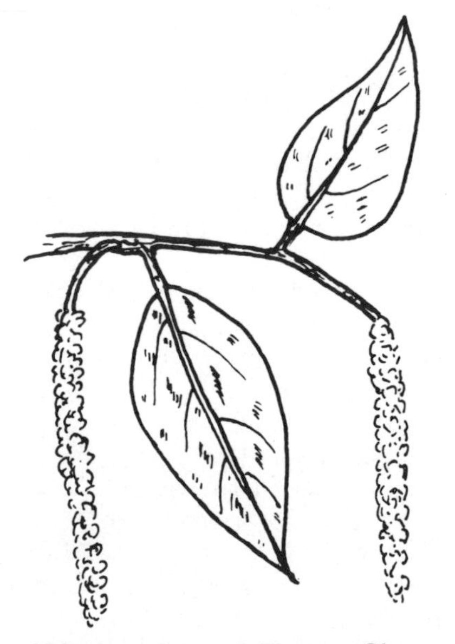

422. Leaves and Flower Cluster of Pepper

423. Fruit of Pepper

genus *Piper,* the classical Latin name for pepper, consists of more than five hundred species containing a number that are grown for ornament. One of them is a Brazilian stemless herb very widely grown as a foliage plant by florists. It has thick, fleshy, oval, entire leaves, and minute flowers crowded on a dense, slender, usually curving spike. It is a very popular little pot plant, but requires a tropical greenhouse and plenty of moisture. It makes a good plant for the living room.

Of the seven hundred members of the genus *Piper,* only three are likely to be found in cultivation. They are the cubeb, a woody vine, ultimately becoming treelike, yielding the cubebs of commerce; the true pepper, a woody vine with aerial roots that produces the black and white pepper as mentioned above; and the third, an ornamental climber grown for its handsome foliage. All three species, incidentally, are greenhouse plants, it being difficult, if not impossible, to grow them outdoors in the United States for lack of heat and moisture, though possibly it may be done in lower Florida.

143. Phlox (Polemoniaceae)

The phlox family, or rather some members of the family, need no introduction to most gardeners, nor do some of them need any introduction to those who are fairly well acquainted with our wild flowers. The family is essentially a North American family, being rare in Europe and Asia. It consists of twelve genera and over two hundred species of mostly annual and perennial herbs, with few woody plants of the nature of low trees or climbing shrubs. The leaves are opposite or alternate, simple or compound, and usually entire. The flowers are perfect, regular, or nearly so, often crowded in heads or corymbs, and have a five-lobed calyx, a corolla of five united petals, that are tubular or funnel-shaped with spreading lobes, five stamens, and an ovary that is usually three-celled, with generally a three-branched style. The fruit is a capsule.

Among the wild species of the phlox family may be included the wild blue phlox, the downy phlox, the wild sweet william, and the ground or moss pink, all members of the genus *Phlox,* from the Greek for flame in allusion to the flowers. The wild blue phlox, common in moist, thin woodlands, has a somewhat finely haired, sticky stem with deep green, ovate, lance-shaped, pointed leaves, and pale lilac blue, slightly fragrant flowers in loose, spreading clusters. The downy phlox has a soft, downed stem and leaves, the leaves deep green, linear or lance-shaped, without teeth and stemless, and flowers that vary from pale, crimson pink to purple and white. Both are perennials as is the wild sweet william, a plant of low woods, but also found along dry and dusty roadsides. It is an erect species growing to a height of three feet, with purple-spotted stems, lance-shaped leaves, and pink or purple flowers in loose racemes. It is not the sweet william of the garden, nor is it even related to it.

Useful in the rock garden and often planted there, the ground or moss pink is a charming little plant that is found growing in dense, evergreen mats on dry, sandy, and rocky hillsides. It has tufted stems, small, yellow green leaves, awl-shaped and barely one-half inch long, and clusters of flowers that range in color from white through crimson pink to light magenta. Another low ground species, the creeping phlox has larger pink, purple, or white flowers.

Our garden phlox, at home in woods and thickets, is a tall species with a stout, smooth stem, growing to a height of four feet, broadly lance-shaped, thin, dark green leaves, and flowers, varying in color from pink and lilac to white, in large spreading clusters. It, too, is a perennial with many varieties that are widely cultivated. The perennial phloxes (there are some annual species), usually bloom in early summer and the butterflies appear to be the principal visitors, although the long-tongued bees and even flies can sip their nectar, and are often seen hovering above the flower clusters.

424. Flower of Greek Valerian

A charming plant growing in thin woods, the Greek valerian is often planted in gardens. It is a smooth perennial with weak stems, alternate pinnate leaves, and light blue flowers in loose clusters. A related and rare species, known as Jacob's ladder and also planted in gardens, is found only by the mountain streams and in the swamps of the north. It has a stout, horizontal root, many rootlets, and erect smooth stems that grow up to three feet. The leaves are compound, as in the preceding species, and the flowers, bright violet or blue, are numerous in a somewhat long cluster.

Widely grown in the South for its quick growth and showy bloom when treated as an annual, the cup-and-saucer vine is a tropical American species. Unlike the trumpet vine and Virginia creeper, that climb by means of aerial rootlets, it climbs by means of tendrils at the apex of its pinnate leaves. They are forked and end in delicate branchlets, each bearing a minute, double hook at its tip, and being flexible, the slightest breeze will sway them about so that the little hooks can catch hold of any slight irregularity, whereupon, the tendrils then curl around and make the attachment permanent. The flowers are large, about two inches long, violet or greenish purple, and quite showy.

Essentially western species, the gilias are showy plants. There are nearly a hundred of them, and they are especially abundant in the Rocky Mountain region. Indeed, one can hardly climb the mountain paths without seeing one at every turn. One of them, known as prickly phlox, is an erect, widely branched shrub with prickly, palmately lobed leaves and dense clusters of pink flowers, and grows in the chaparral of southern California. Another, the scarlet gilia, found throughout the western states, has a basal rosette of pinnate leaves, and terminal clusters of scarlet flowers, and a third species, the bird's-eyes, or tricolored gilia, has finely dissected leaves, and lilac purple flowers. It is a California annual and an extremely popular garden flower.

144. Pickerelweed (Pontederiaceae)

It is said that someone once saw the pickerel lay its eggs among the leaves of the pickerelweed, and so, the plant received its name. But the pickerel also lays its eggs among the sedges, arums, wild rice, and other water plants, as do other fish as well. Of more importance, perhaps, the pickerelweed, at home in the shallows of pond and stream, suggests to many, long summer days, when trout and pickerel and other fish swim slowly among the weeds that line the water's edge, when turtles and frogs view us with wondering eye, when dragonflies dart swiftly through the air, when whirligig beetles play on the water's surface, and bees, and flies, and other insects congregate about the flowers that give color to the shoreline.

When in blossom, the showy flower spike of the pickerelweed, crowded with ephemeral, violet blue flowers above the rich, glossy spathe, invests this water-loving plant with a unique charm. Each flower, marked with a distinct yellow green spot, lasts only a day, but it is followed by a succession of blooms on the gradually lengthening spike, so that the plant blooms well into fall. The flower cup is funnel-formed and six-divided, the upper three divisions united and the three lower ones spread apart. Three of the six stamens are long and protruding, the other three are short and often abortive, and the blue anthers are placed in such a way that it is impossible for an insect to enter the flower without brushing against them.

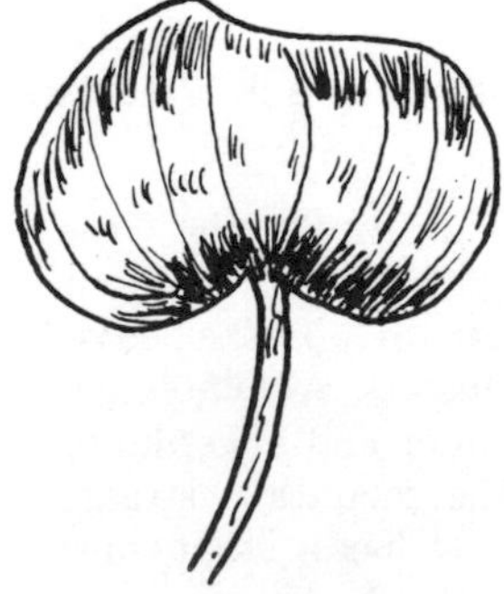

425. Leaf of Mud Plantain

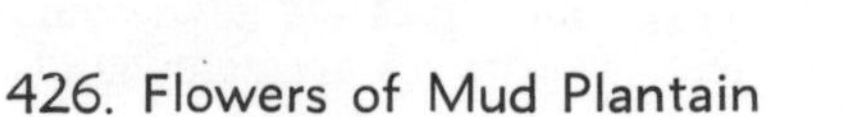

426. Flowers of Mud Plantain

Unlike the pickerelweed, a tall plant growing to a height of four feet, the mud plaintain is a small plant, not more than twelve inches high, and grows in mud or shallow water. It has deep green, floating, round, kidney-shaped leaves on long stems, and two to five white, or pale blue, perfectly developed flowers that are, like those of the pickerelweed, short-lived. The flower tube is six-divided and there are three stamens, two with yellow anthers, one with a greenish anther.

The pickerelweed family is a small family of erect or floating aquatic herbs that occur in marshes and the quiet waters of ponds and streams. The leaves are alternate on long stems, with the stems sometimes dilated to form a bladder. The flowers are perfect, irregular, with a six-parted perianth, three or six stamens, and a three-celled ovary, and borne upon a spadix that issues from the spathe that is soon outgrown. The fruit is a multi-seeded capsule, or an achene.

The pickerelweed family includes only about twenty species, and in addition to the two already discussed, only the water hyacinth, a South American plant that has invaded Florida, needs mention. It is a floating plant with roots that hang downward into the water and with rounded leaves whose stems are inflated and pitcherlike, thus, acting as buoys and helping to keep the plant afloat. The flowers are funnel-shaped, six-lobed and pale violet, lilac, or white, and are borne in small clusters. They are quite beautiful, actually so beautiful that they have been called water orchids. In Florida and in several tropical countries, the water hyacinth has become a troublesome pest, completely choking otherwise navigable streams, as the St. John's River for instance. It has been such a costly matter to keep this river open to navigation, that in Florida, the water hyacinth has become known as the "million dollar weed." In the southern states the water hyacinth is often grown for ornament in shallow pools or even in tubs of water.

145. Pine (Pinaceae)

The pines and their relatives—the spruces, firs, hemlocks, and cedars—are a familiar feature of the landscape and are beautiful trees at any season, but perhaps no more so, than when they become laden with snow. As the snow cover deepens, and food becomes increasingly inaccessible to the wildlife, the pines become important because the food they provide in bark, leaves, and seeds, is above even the deepest blanket of snow, and thus, available to a variety of birds and mammals.

The pines have an ancient lineage, survivors from a prehistoric age when their contemporaries were the lycopods, the sigillarids, and the cycads whose remains form the coal deposits. The visible signs that distinguish the pines and their relatives from other trees give little hint of the structural difference that places them remotely from their companions in the present-day forest. The architecture of the pine flower is simplicity itself and typical of the plants that formed the vegetation of the earth long ago. Looking at a pine flower today one looks back in time millions of years.

The fruit of the pine is a cone, and thus, the tree is called a conifer or cone-bearer. The fruit of the spruce, fir, hemlock, and larch is also a cone, and hence, these trees are also known as conifers. As are the cedar, the yew, and the juniper, though their fruit is called a berry. All the conifers are commonly known as "evergreens" because, unlike the oaks, maples, and hickories, their leaves remain on the branches throughout the winter, that is, all except the larch and the bald cypress.

The conifers are a hardy tribe. They are at their best on mountainslopes and highlands where other trees find the rigors of winter too severe for survival, but they are also able to grow on sandy, rocky, or otherwise poor soil not suitable for other tree growth. There are five hundred million acres of forest in the United States today, and of the trees that make up this vast acreage of plant life, the conifers are by far the most abundant. That they are one of our major natural resources is beyond debate. One can hardly dispute their value in furnishing building materials, pulp wood, naval stores, and a number of other products.

427. Cluster of Staminate Flowers of White Pine

428. Cluster of Pistillate Flowers of White Pine

429. Seed of White Pine

430. Leaves of White Pine

The pine family contains thirty-three genera and over two hundred and fifty species of evergreen trees and shrubs (except the larch and bald cypress), many having a resinous and aromatic sap. Their leaves are more or less needlelike or awl-shaped in the pines, spruces, firs, hemlocks and larch, but scalelike and pressed tightly into twiglike or fan-shaped clusters in the junipers, arbor vitae, and a few other genera. Their flowers are not the conventional type of flower, but instead, there are naked male and female reproductive organs borne above or below, but always between, small, often woody, scales. The conical aggregation of these scales of the female flowers results in the cone, best typified by the common pine cone. The male and female flowers are generally on different twigs and in some instances on different trees, but the female cone, that eventually produces the seeds, is the only permanent one. The male flowers produce an abundance of pollen that is always windblown. Between the mature scales of the female cone, the naked, often winged, seeds develop. Upon sprouting, the seeds send up more than two seed leaves or cotyledons, and thus, the plants are said to be polycotyledons. As all other plants have only one or two cotyledons, the members of the pine family are somewhat unique in this respect.

To the armchair naturalist every conifer is a pine. Travelling about the countryside, one may find many a farmhouse, roadside inn, or tea room displaying a sign that might read *The Pines, The Pines Inn,* or some such name, when there is no pine to be seen, though such places, to be sure, may stand in the midst of sombre spruces or nestle in the shade of stately cedars.

The pines of the genus *Pinus,* the old Latin name for the pines, may be distinguished from other conifers by the fact that their needlelike leaves are borne in clusters of from two to five, and rarely solitary. In most pines the trunk is continuous with whorls or tiers of branches. The staminate flowers consist of naked, catkinlike or conelike masses of anthers that produce a considerable amount of pollen. The pistillate flowers consist of naked ovules between the bases of woody scales, the scales collectively forming the familiar pine cone. In some species the seeds are edible and, perhaps more importantly, they serve as food for many game birds and songbirds. The leaves, too, are consumed by some species of grouse and by several browsers. Porcupines and small rodents also feed on the bark and wood.

There are some ninety species of pines, a third being native to North America. Most aristocratic of all is doubtless the white pine, a stately and beautiful tree with its tall, straight trunk and lustrous, blue green foliage. Its leaves are in clusters of five. So, too, are those of the sugar pine, of the Cascade and Sierra Mountains, the largest of our western pines, whose cones are the largest known, sometimes measuring twenty inches or more in length. Other pines with five needles in a cluster are the whitebark pine, a western low-growing timber tree; the limber pine, also a low-growing pine of the desert mountainslopes; the foxtail pine; the hickory pine; and the western white pine, a slender pine of the Far West, found on the higher mountains of the Pacific Coast states.

Perhaps the most virile and the hardiest of the pines is the pitch pine, a tree that will grow under the most adverse conditions, "clinging like a limpet to a rock," as someone once remarked, to any substratum wherein it can gain a foothold. Witness the trees that grow in the sand dunes of Cape Cod, Massachusetts, where they are battered into all kinds of grotesque shapes by the storms that come up the coast. It is a rugged tree, its contorted branches with their coarse, yellow green needles giving it an exotic appearance, and yet, not without a certain picturesqueness. It is useless no doubt for timber or firewood, or for any other purpose, except that as a windbreaker for exposed, windswept dunes of the Atlantic Coast.

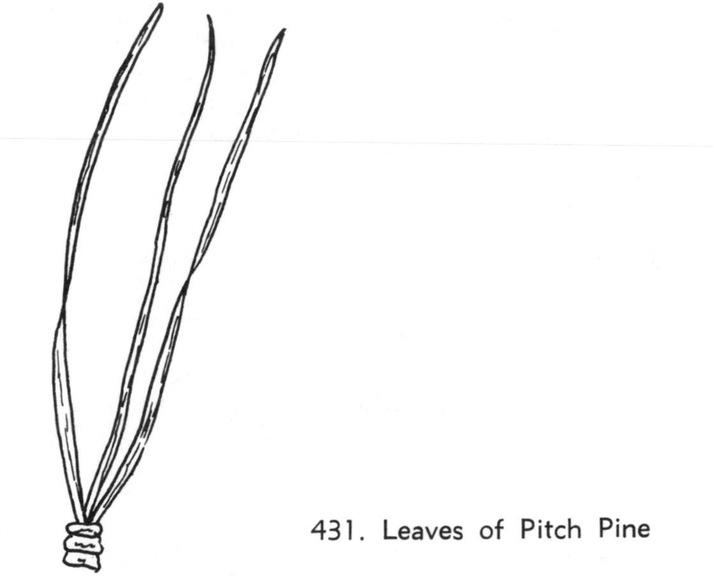

431. Leaves of Pitch Pine

The pitch pine has three needles in a cluster, as does the western yellow or ponderosa pine, a valuable forest tree that forms some of the most extensive pine forests of the continent. It is a majestic tree, its trunk rising like a spire and lifting its head into the sky, and clothed with short, leafy branches as far down as there is room. With the longest needles of any conifer, the leaves twelve to fifteen inches long, the long-leaved pine of the southeastern states is well named. It is a tall, straight tree, sparingly branched, with bright olive green needles, and of great commercial value as its resinous secretions produce turpentine and various pitch products. The slash pine, that forms miles of open forest on the southern coastal plain, is also a three-needled pine. Other western pines with three leaves in a cluster are the digger pine, growing on the sun-baked foothills of western California, and of more than passing interest because of its peculiar sparse foliage; the bigcone pine, with the heaviest cones of any pine, weighing from five to eight pounds, and hanging like old-fashioned "sugar loaves" on the stout branches; and the knobcone pine, whose cones persist on the branches for an indefinite period, from ten to fifty years.

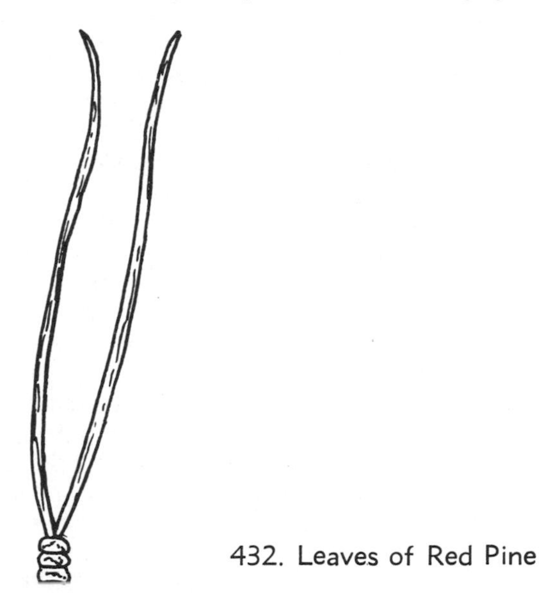

432. Leaves of Red Pine

Of incomparable beauty, with two leaves in a cluster, the red pine is a pyramidal tree, with glossy green leaves, egg-shaped or conical cones, and with conspicuous broad, flat ridges of the bark that are reddish brown. It occurs scattered through the woods from Nova Scotia to Pennsylvania, and being quick-growing, is useful in ornamental plantings. Also with two leaves in a cluster are the Jack pine or scrub pine, a scrubby little tree with the shortest leaves of any pine, so short that they bear a remote resemblance to those of the spruce and fir; the eastern yellow pine or shortleaf pine, a distinctively southern pine of imposing proportions, whose leaves are short only in proportion to those of the long-leaved pine; the shore or scrub pine, a stunted, gnarled, round-shouldered tree of bogs and sand dunes from northern California to Alaska; the lodgepole pine, a tall and slender tree of the western mountains wherefrom the Indians cut poles for their lodges and tepees; and the Scotch pine and the Austrian pine, both introduced species, the former a long-lived rugged tree with twisted, blue green leaves, the latter a handsome tree and darker green than any other evergreen except the red cedar.

A group of pines known as the nut pines have a variable number of needles. The nut pine or pinon pine has only one leaf in a cluster. It is a small tree, with the form of an old apple tree, growing on the sun-baked mesas and the arid mountainslopes of the west, its single, cylindrical leaf, pale grayish green like the rest of the desert foliage, setting it apart from other pines. The Arizona pine, a dwarfed and bushy tree of the canyonsides of Arizona, has dark green leaves in clusters of two or three; the Rocky Mountain nut pine, a small tree of the dryer mountain regions, generally has two leaves in a cluster; the four-leaved pine, as its name suggests, has four leaves in a cluster. No other pine has this number of leaves in a cluster, and thus, it is easily recognized. It inhabits the mountains of southern California.

The seeds of the nut pines are edible and are an important source of food for the Indians. They are sweet and pleasant and very palatable, especially after having been roasted, and for the Indians of the southwestern mountains they serve as a substitute for wheat. They are pounded or ground into meal, then baked in cakes, or cooked as gruel. In seasons when the trees are fruitful, an industrious Indian family can gather as many as fifty bushels a month. The seeds of the various nut pines are also sold in the markets. They are also a source of food for the wildlife, the pack rat in particular gathering them and placing them on a pile at the bottom of its conical nest for its supply of winter food.

The spruces are not as well known as they should be. Most people, perhaps, do not know them at all except maybe for a few cultivated species. The spruces of the genus *Picea,* the classical Latin name for the spruce, are usually majestic, sometimes gigantic evergreen trees. There are only about thirty-nine species, all from the northern hemisphere, and as a rule, from the cooler and moister areas. They are tall, pyramidal in form, with a single, unbranched trunk, and with tiers or whorls of branches. The leaves, usually about an inch in length, are generally four-sided in cross-section, small, numerous, needlelike, each on a tiny footlike cushion or stalk, and grow out from the twigs on all sides. The staminate flowers consist of naked anthers only, the pistillate of naked ovules between the scales of what is eventually the cone. The fruit is a pendant, woody cone, with two winged seeds beneath each scale. A resinous sap that exudes from the bark is chewed as spruce gum and has some medicinal properties.

There are three eastern species of spruce, the white, red, and black. The white spruce, a handsome evergreen, has a pale bark and silvery, blue green foliage that accounts for its name and serves to identify it in the woods, though it may also be distinguished from the red and black by the rather unpleasant odor of its leaves. The red spruce, so named doubtless by the lumberman because of the color of its wood, though the twigs during their first winter are a bright red and there is a distinct tinge of red in the bark, is common on the uplands and may be recognized by its hairy twigs. It is, perhaps, the most cheerful of our eastern spruces primarily because its foliage is a shining, yellowish green, whereas that of the other two is a blue green. The black spruce closely resembles the red spruce, though its foliage is not of the same color. It is rather ragged and uneven in appearance with the cones persisting for several years so that the tree is often burdened with the empty husks of several crops, a peculiar habit that gives the tree a rather unkempt appearance.

Among our western species, the Engelmann spruce is a tall tree that crowns the lower and higher peaks of the Rocky Mountains and the Cascade range in Washington and Oregon, often climbing to altitudes between one and two miles above sea level. It is a splendid species and often planted in the Eastern States where it thrives, though the disagreeable odor of its leaves is a point against it. The Colorado blue spruce, a beautiful, silvery, pale gray blue evergreen, the lightest colored of all the spruces, is a native of the mountains of Colorado and Wyoming, and is familiar throughout the United States as a park and garden ornamental. A very large massive spruce of the far northwest, the Sitka spruce is a lover of swamps and moist places. In such places, the graceful sweep of its long, drooping branches and the lustrous sheen of its leaves, provides a constant and delightful interplay of light and shadow. An introduced species, the Norway spruce is probably the most widely cultivated evergreen tree in America. It is remarkable for its long, sweeping, pendulous branches and dark foliage that give it a rather sombre appearance. Many horticultural forms of the tree have been developed for special purposes. As with the pines, the spruces provide food for the wildlife. Various species of grouse feed on the leaves, and certain songbirds, as the crossbills, eat the seeds, that are also consumed by the squirrels and chipmunks. The young twigs and leaves of the black spruce and red spruce are used in making the famous beverage, "spruce beer."

Unlike the leaves of the spruces that are four-angled, those of the firs are flattened. The firs of the genus *Abies*, the classical Latin name for the fir, are magnificent evergreen trees, generally with a symmetrical, pyramidal habit, and whorls of wide-spreading, horizontal, graceful branches. The leaves are usually flat, narrow, stalkless, and persistent when dry, unlike the spruces that shed their leaves when dry, and are arranged in two rows on the twigs, thus, the spray of foliage is flattened. The flowers are in axillary, scaly catkins, the pistillate erect on the upper branches, the staminate on the undersides of the branches lower down on the tree. The cones are erect (they hang in the spruces), the scales falling off at maturity.

There are thirty-five to forty species of firs, all from the north Temperate Zone. Many of them are among our finest cultivated evergreens. The fir of our northeastern states is the balsam fir, rather easily recognized by its blistery bark. It is a slender tree of damp woods and swamps, but it is also found on the mountains, climbing to the summits where it becomes dwarfed, and is no larger than a mere shrub. Its bark is smooth, but warty with "balsam blisters," small excrescences or vesicles that are laden with a resinous substance called Canada balsam. Its leaves are dark green above, paler beneath, and its cones are two to four inches long. The Canada balsam, having a resinous, but not an unpleasant flavor, was much used as a chewing gum before the days of the pleasantly flavored, commercial chewing gums. Today, Canada balsam is refined and used in the arts, especially in the mounting of microscope slides.

Other firs are the Fraser fir, a southern representative of the balsam fir and similar to it, differing only in technical characters; the alpine fir or Rocky Mountain fir, a tall tree of the western mountains with a narow spire of thickly crowded branches and purplish cones; the giant fir, a magnificent tree up to three hundred feet high with greenish cones and also a western species; the white fir, of the Pacific Coast mountainslopes, and so-named because of its pale foliage and the gray bark of its branches; the silver fir, of the cascades, with silvery white bark; the noble fir, whose home is in the mountains of Washington, Oregon and northern California, with blue green foliage, and stout, brown or purple cones; and the red fir, another tall western fir, with reddish brown bark, red twigs, and reddish purple staminate flowers.

The well-known Douglas fir, or Douglas spruce, one of the most valuable timber trees of the northwest, is neither a spruce nor a fir, distinctly differing from both of them, and yet, superficially resembling the latter, but of a different genus. It is a large tree, probably the tallest in America next to the redwood, with bright green, drooping foliage and handsome pendant cones.

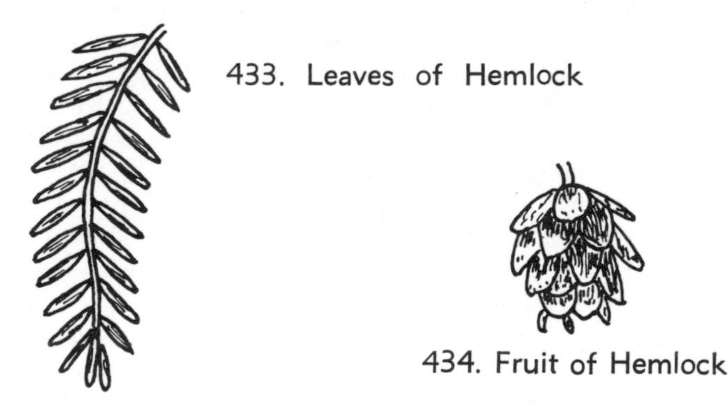

433. Leaves of Hemlock

434. Fruit of Hemlock

A spruce or fir is frequently mistaken for a hemlock, but the hemlock may easily be distinguished from either by its distinctly stalked leaves. There are only about ten species of hemlocks that occur in North America and eastern Asia. They are generally magnificent trees with horizontal or pendulous branches, deeply furrowed bark, and are pyramidal in habit. The leaves are short, flattened needles, usually an inch or less in length, minutely grooved on the upper surface and with two white lines on the lower surface, numerous, and grow in two opposite rows on the stem. The flowers are monoecious and solitary, and the cones are usually small and oval (except one) with thin, somewhat woody scales. Unlike most other plants, the name of the genus *Tsuga*, whereto the hemlocks belong, is not of Greek or Latin origin, but is the Japanese name for one of the Asiatic hemlocks.

The common hemlock of the eastern states, the eastern hemlock, is a tree that does as well in cold swamps as on mountainslopes, ravines, and rocky woods. With its feathery foliage and graceful pendulous twigs, the eastern hemlock has a certain degree of airiness not found in any other evergreen, and thus, is often seen in parks and on private grounds.

The Carolina hemlock, found in the mountains and on dry, rocky hillsides from Virginia to Georgia, is similar to the eastern hemlock, but a much smaller tree. The greatest of the hemlocks and a tree of noble stature, the western hemlock is a Pacific Coast evergreen where it dominates the forests in size as well as in numbers. Often attaining a height of two hundred feet with a trunk diameter of as much as ten feet, with drooping, horizontal branches and a dark green foliage, it is the most superb of the hemlocks, a delight to the artist and to the lumberman too. Also a western species, the mountain hemlock, common on the mountainslopes from Alaska southward through the Sierras of California, resembles the western hemlock in general aspect, and is often confused with it. The bark of the hemlock contains from seven to twelve percent tannin, and in the United States, that of the eastern hemlock was one of the principal commercial sources of this substance for many years. At one time, the Indians made a tea from hemlock leaves and apparently relished it.

The small winged seeds of the hemlocks are an important food for such birds as the pine siskin, the crossbills, and chickadees and several rodents including the red squirrel. The porcupines in New England are fond of the hemlock and often kill young trees by stripping the bark and wood. Hemlock groves are favorite nesting sites for several species of northern birds, such as the veery, the black-throated blue warbler, the black-throated green warbler, and the junco.

Unlike the pines, spruces, firs, and hemlocks, the larches lose their leaves in the winter. They are tall, pyramidal trees, with a cone-shaped head, and with few horizontal branches. Their leaves are awllike, in clusters or fascicles on lateral spurlike growths, except on new growth where they appear singly and are spirally arranged. The staminate and pistillate flowers are borne separately on the same tree. The cones are woody, solitary, erect, and sessile on the twigs.

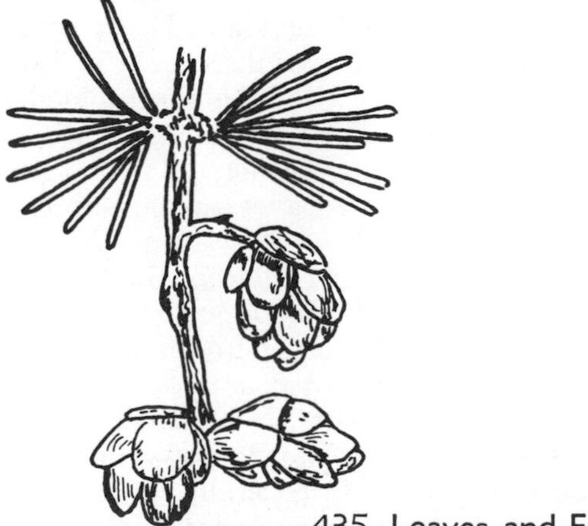
435. Leaves and Fruit of Larch

There are three North American species of the larch of the genus *Larix,* the ancient name of the larch: the eastern larch, also called the tamarack and hackmatack, the western larch, and the alpine larch. The eastern larch is a tree of the swamps and lowlands whose fibrous roots were the Indian's thread. It was tough and fine as a shoemaker's "waxed end" and was used to bind together the canoe of birch. Hiawatha sang:

> Give me of your roots, O Tamarack!
> Of your fibrous roots, O Larch Tree!
> My canoe to bind together
> So to bind the ends together
> That the water may not enter
> That the water may not wet me.

The eastern larch is a hardy tree, living farther to the north than any other conifer. It is, moreover a picturesque tree, with its tall, thin trunk, and pyramidal head that, together with its bright blue green feathery foliage, makes it highly desirable as an ornamental for parks and gardens.

The western larch is the tallest of the larches and is abundant on the slopes of the Cascades and the Blue Mountains of Washington and Oregon. The mature tree is rather remarkable for its immensely tall trunk, short branches, and sparse foliage. The alpine larch is a much smaller tree, also found on the western mountains. The European larch, an introduced species, is in common cultivation in this country. Its twigs are stouter than those of the eastern larch, its needles are longer, and its cones are larger.

Anyone who has visited the southern swamps knows the bald cypress, a tree of rather stately proportions that grows to a height of fifty to eighty feet, sometimes as much as one hundred and fifty feet. Often completely surrounded with water, its trunk rises perpendicularly for forty feet or more without a single branch extending outwards from it. At its base long, thick roots stretch out with branches that go down to anchor the tree in its soggy, muddy bed, while above the roots smooth, conical "knees," rise several feet above the water to function as aerating organs, bringing air to the root tissues that might otherwise be suffocated by the water over the root system.

The bald cypress of the genus *Taxodium,* from the Greek meaning taxuslike, far more widespread in the United States in former geologic periods than it is now, is pyramidal in habit with a thick, ruddy brown bark and horizontal or pendulous branches. The leaves are light green, about three-quarters of an inch long, very numerous, the foliage graceful and feathery, orange in the autumn just before the leaves fall. The flowers are monoecious and small, the staminate in loose, drooping panicles, and the pistillate globose, scaly, near the ends of the twigs. The cones are woody, almost spherical, tan brown, with thick, angular, somewhat shield-shaped scales.

Common in the swamps and marshland of the Atlantic coastal plain, the white cedar of the genus *Chamaecyparis,* from the Greek for ground cypress, is a slender, fast-growing evergreen, in the East twenty to forty-five feet high, but in the southern part of its range, sometimes reaching a height of ninety feet, with short, slender, ascending or nearly horizontal branches and a delicate and feathery foliage. Its summit is spirelike or narrowly conical, its bark is distinctly shredded, on young trees detached in broad, thin, red brown strips, on old trees less shredded and grayish brown, its leaves are small, scalelike, opposite in four rows, blue green or light olive green, and aromatic, strong-scented when crushed. The flowers are monoecious, small, terminal of four to six scales, the staminate are red or yellow and abundant, the pistillate are greenish and few. The cones are small, spherical, one-quarter of an inch in diameter, at first blue green, but later becoming brown.

Two related western species are the yellow cypress and the Lawson cypress. The yellow cypress, found from Alaska to Oregon, is similar in some respects to the white cedar, but with branchlets and twigs in a fanlike growth. The Lawson cypress, growing along the coast from Oregon to northern California, is a beautiful, deep-toned evergreen and in common cultivation.

Extremely valuable as ornamentals and as timber trees, the arbor vitaes that grow in the wild are magnificent forest evergreens. They are pyramidal in habit, with twigs densely covered with tiny, scalelike leaves in flat, fanlike sprays or fronds. The flowers are monoecious, the staminate are a spherical cluster of stamens, the pistillate are a cone of scales, the latter green and keeled.

There are only six species of arbor vitae of the genus *Thuja,* found in North America and eastern Asia, though there are innumerable horticultural varieties. The American arbor vitae, growing in the swamps and on cool, rocky river and pond shores from Nova Scotia south and along the mountains to North Carolina and eastern Tennessee, is a medium-sized tree, the trunk buttressed at the base, with a shredded, reddish brown bark that separates into long narrow strips, and smooth, red, and shining branches. The leaves are a dark green or a golden-green in overlapping scales, the frondlike, fan-shaped, leafy twigs being extremely handsome. The cones are small, about one-half inch long, pale green at first, later reddish brown with six to twelve pointless, thin, oblong scales. They persist on the twigs throughout the winter.

The American arbor vitae is a pygmy compared to a western species, the giant arbor vitae, a massive tree found singly or in small groves on the moist lowlands of the Pacific Coast, though it climbs the mountains to a level more than a mile high. The Indians use the biggest specimens for their totem poles and for their hollowed-out war canoes. They also split planks from the trunks of the tree for building their lodges and use the inner bark to make blankets and ropes.

Everyone has heard and read about the big trees of California and many have seen them. It is difficult to realize that they are merely remnants of an ancient group that was widely spread over the northern hemisphere. Fossil remains have been found in Europe and Asia, and in the United States, fossil cones have been found as far east as New Jersey. Some botanists believe that the trees that make up the Petrified Forest National Park of Arizona were members of what was once a mighty race. They are known as the sequoias, named after Sequoiah, a Georgian Indian, and inventor of the Cherokee alphabet.

Today there are only two survivors of an almost extinct group of trees, and both gigantic evergreens: the big tree or giant sequoia, and the redwood. The big tree, the largest cone-bearing evergreen in the world, commonly two hundred and fifty to two hundred and eighty feet high, but sometimes over three hundred feet tall, with a trunk diameter of seventeen to twenty-four feet eight feet above the ground, is pyramidal in habit when young, but later round-topped. The bark is a light cinnamon brown, the outer plates often a dull lavender gray, extremely coarse and thick. The leaves are blue green, awllike, sharp-pointed, and overlapping. The staminate and pistillate flowers are on the same tree, terminal, conical, scaly, the staminate with broad scales and abundant pollen, the pistillate with twenty-five to forty, needle-tipped scales, with three to seven ovules under each. The cones are dark brown, woody, the scales, before opening, resembling the wrinkled surface of a pineapple. The big tree is found only in a few groves of the Sierras, many of these groves having been secured and protected as national parks and forests such as in Yosemite and Sequoia National Parks and the Sierra National Forest. The largest known specimen of the big tree is the famous General Sherman tree in Sequoia National Park. It measures nearly three hundred and twenty feet in height and has a trunk diameter of thirty-five feet. It was discovered in 1879 by a trapper who named it after the general under whom he had served in the Civil War.

In many ways the redwood is not different from the big tree, but its spreading leaves on the terminal twigs give it a more graceful, feathery spray than do the awllike blades of its relative. The pistillate flowers, too, have fewer scales, the buds are more scaly, and the cones are smaller. It does not quite attain the size of the big tree, but with its lustrous leaves, its ruddy bark, and gracefully curving branches it is a more attractive tree. The redwood grows in dense stands at moderate altitudes, generally in the fog belt, along the coast of Oregon and California, a favorite location is in the sheltered mountain basins. Several of the best known groves are preserved in the California State Redwood Park in Santa Cruz County and the Muir Grove in Marin County. The burls of the redwood are often sold by florists, as they make interesting growths when placed in water.

Possibly no evergreen of our flora has such a wide distribution as the red cedar, found from the Atlantic coastline west to the Rocky Mountains, and from the Canadian border almost to the gulf of Mexico. Tolerant of all types of soil, it grows as easily on the gravelly slopes and rocky ridges of the North, as it does in the swamps and rich alluvial bottomlands of the South. Examining a branchlet of the tree, one will most likely find that there are two kinds of leaves —awl-shaped and scale-shaped. There does not seem to be any set rule that determines their occurrence, except that the awl-shaped leaves always appear on young trees, but both may occur on the same branchlet of a mature plant. The awl-shaped leaves are rigid, long, sharply pointed with a white bloom on the upper surface. The scale-shaped leaves are minute, closely appressed, acute or obtuse, and usually bear a glandular disk on the back.

The red cedar is columnar or pyramidal in habit when young, but becomes broader and more spreading with age. It is a rather small tree, from twenty-five to forty feet high, but sometimes as much as ninety feet or more, with slender branches and a reddish brown bark that peels off in long strips. The inconspicuous flowers appear on separate plants, the staminate of four to six scales, each with several pollen sacs, the pistillate of minute, paired, bluish, fleshy scales with two ovules. The latter develop into blue berries with a slight bloom, about the size of a pea, and containing one to four seeds. Sometimes in spring there appear rather peculiar, brown, pulpy, jellylike masses on the branches of the tree and wherefrom extend numerous, long, thin, bright orange tendrils or horns that twist about like petals of a flower. They are called cedar apples and represent one stage in the life history of the apple rust, a disease of the apple.

436. Leaves and Fruit of Red Cedar

The red cedar is not a cedar, nor is the white cedar previously mentioned, but is a juniper, of the genus *Juniperus,* the old Latin name for the junipers. The genus is a large one, of evergreen shrubs and trees widely scattered throughout the northern hemisphere. They vary in habit from low, prostrate shrubs to tall, slender trees, and are extensively cultivated. Their leaves are of two kinds, as described above in our description of the red cedar. The flowers are insignificant, the staminate are in small oval clusters (catkins), the pistillate are composed of small scales whereon the ovules are borne and develop into a berrylike fruit.

One of our more abundant junipers, that has been developed into a number of horticultural varieties, is the common juniper. It is mostly a shrub, though sometimes a low tree. In the northeastern states, it forms prickly thickets in open pastures. A variety, the prostrate juniper, is a wide-spreading, low-growing form that rarely exceeds three or four feet in height. It is widely distributed and occurs in the poor, rocky soil of pastures. The creeping juniper is a prostrate, trailing shrub with long spreading branches. It grows on the barren, rocky, or sandy margins of swamps. The Rocky Mountain red cedar, or simply western red cedar, is a round-topped tree somewhat similar to the red cedar. Another similar species is the western juniper found on arid hills, dry plains, and at high elevations among the mountains. This species, as well as other western junipers, have berries that are often eaten by the Indians. At times they are ground and made into cakes. An aromatic oil is obtained from the berries of various species and is used to flavor gin.

The junipers and cedars are a hardy group of plants that thrive on sterile soil, are indifferent to unfavorable climates, and seem to love the sun and wind. They are important to wildlife throughout the country except in the Prairies. Hoofed browsers feed extensively on their twigs and foliage, and their berries form an important part of the diet of many mammals and birds, especially the cedar waxwing that, living up to its name, is one of the principal users of the berries. Besides their value as food, the junipers and cedars also provide important protective and nesting cover. The cedars are a favorite nesting site for chipping sparrows, song sparrows, robins, and mockingbirds, and such species as the juncos, myrtle warblers, and various sparrows make use of their dense foliage as roosting cover.

The true cedars or traditional cedars, the most famous being the cedar-of-Lebanon and the deodar, belong to the genus *Cedrus,* an old Greek name for a resinous tree. There are only four species, handsome evergreen trees with stiff, needlelike, four-angled leaves that are scattered or that occur in small, dense clusters. The staminate and pistillate flowers occur on the same tree and are wholly without petals or sepals. The cones have closely appressed scales, with broadly winged, triangular seeds between them. The cedar-of-Lebanon is a massive, profusely branched tree, up to one hundred feet high, with dark green leaves and brown cones, two to four inches long. The deodar, a taller and more graceful tree, has dark bluish green leaves and reddish brown cones, three to five inches long. Both are widely cultivated in the South as ornamentals.

Like the red cedar and the white cedar, that are not true cedars, the bald cypress that has already been mentioned, is not a true cypress. The true cypresses are members of the genus *Cupressus,* the classical name of the Italian cypress, and are magnificent evergreen trees, mostly from the warmer parts of the Old World though a few occur in western North America. There are twelve known species, half being commonly cultivated in the warmer parts of the world. The leaves are very small, opposite, scalelike, pressed against the twigs and aromatic. The cones are nearly globe-shaped and woody, with six to twelve scales, each with many narrowly winged flattish seeds.

Visitors to California know of the famed Monterey cypress and native Californians are well acquainted with it. This rugged, picturesque evergreen, fighting a last-ditch battle for survival, is found only in the Carmel Bay region; no where else in the world will it be found, except in cultivation. Fortunately, it is a hardy species and does well under cultivation, and thus, it is known, with its several horticultural forms, in many parts of the world. Also of restricted distribution is the Arizona cypress, a medium-sized tree of pyramidal habit with pale bluish green leaves and cones of the same color. The hardiest of all the cypresses, the Macnab cypress, a bushy tree with resinous, dark green leaves, is limited to the mountainous parts of California and Oregon. The Italian cypress, the cypress of history and that is mentioned more often in classical literature than any other conifer, and whose sombre foliage was the badge of grief, is a tree up to seventy-five feet tall, with erect or horizontal branches, flattened twigs, dark green foliage, and cones nearly an inch and a half in diameter.

> Nor, when you die, shall any of the trees you have planted, save only the mournful cypresses, follow their master.
>
> Horace

Along the streets and in the parks of California, Florida, and some other southern states, one will find some exotic members of the pine family. They are stately, symmetrical trees of the genus *Araucaria,* from Araucana, an old name for the southern part of Chile, with branches in regular tiers, leaves that are evergreen, prickly and scalelike or expanded into flat, thick, leathery blades, always stiff, and flowers and seeds that are borne between the scales of an egg-shaped or globelike cone that falls apart with age. There are twelve species, and they are found in Australia, New Guinea, and Chile.

One of the more curious of these trees is the monkey puzzle, so-called from its habit of inextricable branching. The straight, central trunk has spreading, stout branches in tiers, but they twist and turn to form an impenetrable and very prickly growth. The leaves, a bright green, and prickle-pointed, ovalish, flat, leathery, very stiff, persist for years. A native of Chile, it is often planted for its grotesque growth.

Another species that has been introduced from Australia to California and Florida is the bunya bunya. It is a tall tree up to one hundred and fifty feet high, but less in cultivation, with stiff, oblong-oval, sharp-pointed, glossy green leaves in distinct rows. Another introduced Australian evergreen tree of about the same height is the Moreton Bay pine, also called the hoop pine, with stiff, sharp-pointed, needlelike leaves. More commonly cultivated than any of the preceding, is the Norfolk Island pine, from the island of that name. It is an extremely popular florist's plant and is usually grown by the million as a pot plant two to ten feet high, but in its native home, it often reaches a height of two hundred feet. Its horizontal or slightly upturned branches, set in tiers of five to seven, are thickly beset with sharp-pointed leaves that are curved at the tip.

146. Pineapple (Bromeliaceae)

The pineapple family is of obvious economic importance because of the pineapple plant, and is of interest because of some ornamental species. The family is a fairly large one of forty genera and perhaps a thousand species. All are scurfy, somewhat shrubby herbs, often epiphytic (i.e., air plants), mostly native to tropical America with elongated, swordlike, entire or sometimes spiny-margined leaves that are commonly in rosettes. In some genera the rosette and leaf bases hold so much water, that they are mosquito breeders in the tropics.

The flowers are regular and perfect with three sepals

and three petals, the latter free, or united to form a tubular perianth with erect or spreading lobes, six stamens, and a three-celled ovary with many ovules. They are often very showy and occur in spikes, heads, or panicles. The fruit is a berry or capsule, surrounded or crowned by the persistent calyx, with small, hairy or winged seeds.

The pineapple plant comprises a single species of the genus *Ananas,* the Latinized form of the South American name for the plant, and appears to have come originally from the Amazon Valley. It has leaves typical of the family, and from the basal rosette, rises the flowering stalk, leafy and bracted, two to four feet high, and bearing at the top a tuft of leaves that later becomes the crown of the fruit, and when detached will grow into a new plant.

The flowers, without stalks, are violet, or reddish, and sterile in the cultivated varieties, but functional in the wild plant where the ripened ovaries develop into berries that are the true fruit of the pineapple. What is called the fruit, the pineapple that is eaten, is not a true fruit at all, but the result of the fusion of the fleshy stalk of the inflorescence (the juicy part of the mature fruit), with all the sepals, petals and the sterile ovaries of the flower cluster, the sterile ovaries forming the berries, that is, the angular segments embedded in the fruit. The pineapple is what the botanists know as a multiple fruit, or syncarp, that is the product of the ovaries of several flowers. The pineapple is cultivated by means of cuttings.

The natives of Mexico and Peru cultivated the pineapple long before the arrival of the Spaniards who introduced it to the Indies, Asia, and Africa. Today, the pineapple is common in most tropical countries, and either fresh or canned is known and highly prized as a dessert fruit almost everywhere. At one time, the pineapple was cultivated in Florida, but several factors combined to make the growing of the pineapple rather unprofitable. It is, however, still cultivated there, but on a much smaller scale. Most of our supply of pineapples today is almost wholly from the West Indian and Hawaiian Islands.

A familiar sight throughout the southeastern United States is the lichenlike plant, known as Spanish moss, hanging in streamers and festoons from the trees, especially the live oaks and cypresses. Its stems and leaves are gray and threadlike, the dense festoons often hanging over twenty feet from the branches, giving the trees a weird and somewhat unearthly look. From a distance they appear like a dense fog.

The Spanish moss is one of a large group of the genus *Tillandsia,* named for Elias Tillands, a Swedish physician and botanist, consisting of chiefly epiphytic plants, nearly all from tropical America. There are three hundred and fifty species of various habit. Some are stemless and with a basal rosette of narrow leaves, others, as in the Spanish moss, have long, trailing, threadlike stems, and very narrow, threadlike leaves, almost mosslike, that have led to the popular misconception that these plants are mosses. In color, the leaves vary from gray green or grayish to a violet or reddish tinted green, and in some species the violet, rose, white, or yellow flowers are showy, while in others such as the Spanish moss they are small and inconspicuous.

Frequently cultivated in extreme southern Florida for its unique habit and showy flowers is the pinguin, a native of the West Indies. It has a rosette of spiny, sword-shaped leaves, four to six feet long, red or purple flowers with long petals and protruding stamens in a dense panicle on a mealy, or whitish, stout stalk. Many other species of the family are cultvated in greenhouses either for their showy flowers, or as extremely handsome, often brightly colored, foliage plants.

147. Pink (Caryophyllaceae)

From June to September, the bouncing Bet, or soapwort as it is also known, may be seen blossoming along roadsides, banks, and waste places. A native of Europe, but an escape from colonial gardens, its pink or white blossoms, scallop-tipped, with an old-fashioned, spicy odor, remotely resemble those of the pinks. Butterflies, delighting in bright colors, have little interest in the flowers, but the night-flying moths can see them in the darkness and are attracted to them by their fragrance, that becomes more pronounced after sunset. Colonial housewives bruised the leaves in water and made a cleansing soaplike lather that served them well. Hence, its name of soapwort.

The soapwort is a stout, strong-growing perennial, reaching a height of three feet, not greatly branched, with ovate or broadly lance-shaped leaves and flowers in dense clusters. Long ago a garden favorite, it is still a garden plant and useful for the border. A close relative, also a garden plant, and also a wilding, growing in grain and alfalfa fields and waste places, the cow cockle is an annual, three feet high, with smooth, broadly lance-shaped leaves and deep pink flowers in loose clusters. Both belong to the genus *Saponaria,* Latin for soap, including twenty species

437. Leaves and Flowers of Bouncing Bet

with opposite, simple leaves, and flowers having a five-lobed calyx, a corolla of five petals that alternate with the sepals, and five stamens, the flowers developing into a capsule.

One might do well to give the bladder campion more than a passing glance, for its greatly inflated, pale green, and beautifully veined flower cup is not unlike a miniature citrus melon. The white flowers gleam in the darkness and guide the moths that are seduced by the strong perfume that scents the night air.

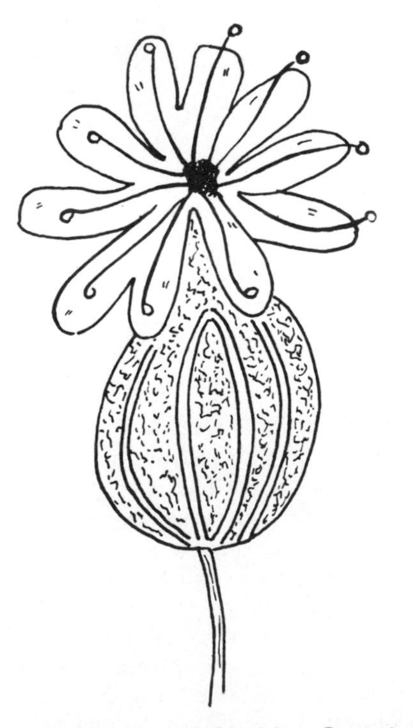

438. Flower of Bladder Campion

The bladder campion is a delicately beautiful plant, a native of Europe, but naturalized in America where it may be found growing in fields, roadsides, and on alluvial banks. Its stems, thickly tufted, pale green and smooth with a white bloom, grow six to two feet high. Its leaves are a deep green, oblong, and pointed. Its flowers are white, drooping, in loose, open panicles, and have a pale green and inflated calyx, a corolla of five petals that are deeply cut, and ten long stamens that are tipped with brown anthers.

The bladder campion has many close relatives because it belongs to a genus that contains three hundred and fifty species of both wild and cultivated flowers. The members of the genus *Silene,* the Greek name of one of Bacchus' companions, are of erect, tufted, or spreading habit, with opposite, simple, entire leaves, and solitary or clustered flowers with a tubular calyx of five teethlike lobes, a corolla of five separate petals, and ten stamens. Included in the genus are such wild flowers as the starry campion, a beautiful and delicate wilding visited by the common yellow butterfly and many moths, with lance-shaped leaves and white flowers in a loose, terminal spike, the flowers, star-shaped with fringed edges; the sleepy catchfly, a curious plant of woodlands and waste places, with swollen, sticky joints on the stem that prevent creeping insects, that are useless as pollen carriers, from stealing the nectar, and whose small, pink flowers open only for a short time in the sunshine; the night-flowering catchfly of fields, meadows, and waste places, with a beautifully marked calyx that resembles spun glass and delicately fragrant white blossoms that open only at dusk and close the following morning; and the wild pink, a very low species growing only from four to ten inches high in dry, sandy, gravelly places, with a sticky-haired stem just below the flowers to discourage would-be pilferers.

The flowers are crimson pink with somewhat wedge-shaped petals.

Of the garden species of the genus there are the sweet William catchfly (the catchfly of old gardens), an erect, bright green annual of two feet high, with broadly lance-shaped leaves and light or deep pink flowers in terminal clusters, quite often an escape; the cushion pink, a tufted perennial not more than two inches high, with lance-shaped leaves, and purplish red, solitary flowers; the sea campion, a grayish green perennial a foot tall, with broadly lance-shaped leaves, and white flowers in one to four-flowered clusters; the moss campion, a perennial of spreading habit, growing up to six inches high and covered with short, soft hairs, with small, lance-shaped leaves in rosettes, and rose or purple flowers, one or two on each stalk; and the fire pink, a showy perennial, six to ten inches high, with thin leaves somewhat oblong or finger-shaped, and few flowers in a loose cluster, the petals being a deep crimson with two lobes.

Like the night-flowering catchfly, the evening lychnis opens its flowers toward evening and closes them during the following morning. Its white, sweet-scented flowers, also, closely resemble those of the catchfly, but they have five styles, whereas the latter have three. The evening lychnis is a charming plant, naturalized from Europe, and growing in grainfields, meadows, and waste places. It has a thick, fleshy root, wherefrom is sent up several slender, branching stems, one to two feet high, and somewhat hairy and viscid. The leaves are long ovate to lance-shaped and dark green, and the flowers are white, the petals deeply cleft and the calyx inflated, tinged crimson along the hairy ribs.

The evening lychnis belongs to the genus of the same name, *Lychnis,* Greek for lamp in allusion to the flame-colored flowers of certain species, a large genus of annual or biennial, but mostly perennial herbs having a calyx of five teeth, five petals, and ten stamens. It includes some old garden favorites that have been cultivated for centuries, such as the scarlet lychnis or Maltese cross, a hairy perennial herb, eighteen to thirty inches high, a native of Russia, with oval or lance-shaped leaves, and scarlet flowers, about one inch wide, in dense, terminal clusters; the cuckoo-flower or ragged robin, also a hairy perennial herb, twelve to twenty inches high, sticky towards the top, with narrow, lance-shaped leaves, and red or pink flowers in loose panicles, the petals with four narrow segments, an escape to meadows and waste places and often a pernicious weed; the mullein pink, also known as dusty miller and rose campion, a white, woolly biennial or perennial herb, eighteen to thirty inches high with oval or oblong leaves and solitary, terminal, crimson flowers about one inch wide; and the rose-of-heaven a widely planted garden annual, twelve to fifteen inches high, with narrow leaves, and rose pink, solitary, terminal flowers about the same width as those of the mullein pink.

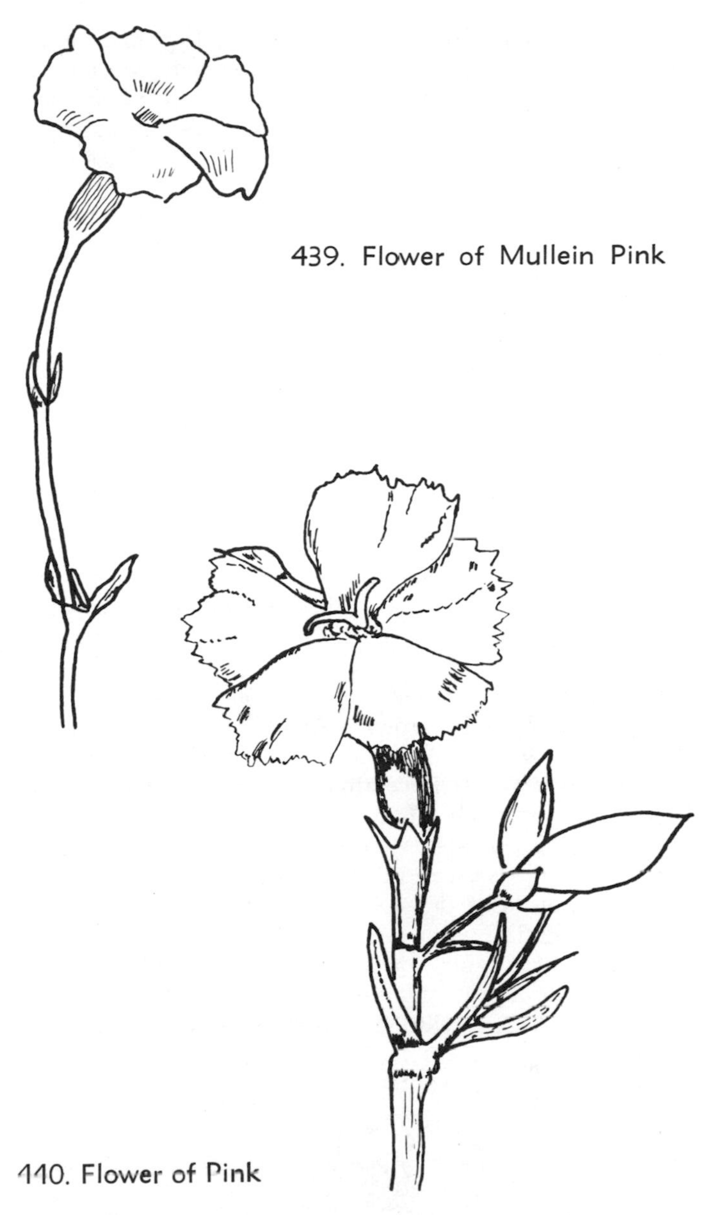

439. Flower of Mullein Pink

440. Flower of Pink

All the preceding genera and species belong to the pink family that contains seventy genera and about twelve hundred species of annual and perennial herbs that are largely confined to the north Temperate Zone and the colder regions of the earth. The leaves are simple, entire, opposite and are attached to swollen nodes. The flowers are regular, perfect, solitary, or in clusters, with four or five sepals and petals, eight to ten stamens, and a three- to five-carpelled or one-celled ovary. The fruit is a capsule that contains many seeds. The family includes many of our most beautiful garden flowers such as the pinks and carnations, in addition to those already mentioned.

As Shakespeare has Biron in *Love's Labor Lost* exclaim "Allons! allons! sow'd cockle, reap'd no corn," the farmers of his time had trouble similar to the farmers of today, with the corn cockle, a pernicious weed of the grainfields whose seeds are poisonous. When ground with wheat, the flour is unwholesome and even dangerous as a food. Poultry and other animals have been killed when fed with screenings composed largely of cockle seeds. In some of the principal wheat-growing states, the loss to farmers may amount annually to several million dollars.

The corn cockle is a densely haired, straight-branched plant with lance-shaped to linear leaves, and terminal flowers, often an inch and a half broad, on long, hairy peduncles, with five spreading, reddish purple petals that are slightly notched at the outer edge and darkly spotted near the claw. A native of Europe, this vigorous invader may well outstrip many of our native plants in the struggle for existence. Bees and the long-tongued flies are often seen about the blossoms, but only gather pollen, for it is the butterflies and moths that serve the plants as agents of fertilization.

Appearances are often deceiving, to use an old cliche, but it certainly holds true in the case of the common chickweed, probably the most persistent, and hardiest plant on earth, in spite of its frail appearance.

It is widely distributed throughout North America, and one may find it growing almost anywhere, in fields, meadows, waste places, lawns and gardens. One may also find it blossoming almost any time of the year, except in severely cold weather. During a January thaw, its green stems with buds, flowers, and seeds, may usually be seen in some sheltered place. This is one reason why it may be found almost anywhere throughout the world. Another, is that it can readily adjust to changing conditions, and will quickly take possession of any unoccupied place. A third reason is that it can fertilize itself when it is too cold for insect pollinators to be abroad, producing abundant seed even in the winter.

441. Leaves and Flowers of Chickweed

Every gardener knows the chickweed, the familiar, weak-stemmed, low-lying weed, with its small, ovate-pointed, light green leaves, slightly woolly stems, minute white flowers with five petals almost cut in two, and five, larger, green sepals, much longer than the petals, and that if left alone, would eventually usurp every inch of garden soil. It belongs to the genus *Stellaria,* Latin for star in allusion to the shape of the flower, that contains several species known as the stitchworts. One of these, the long-leaved stitchwort, is a tall, very slender plant with many branches, a stem with rough edges, light green, lance-shaped leaves, and tiny, starlike white flowers, the five petals cut so deeply that they appear as ten. A similar species, the lesser stitchwort, has smaller leaves. Both grow in fields, meadows, and grassy waysides.

Small low herbs, some found growing in the wild, others cultivated in the garden, are the sandworts. They are well named since a few of them grow in sandy places, such as the thyme-leaved sandwort, a tiny, multi-branched, and spreading little plant with small ovate leaves and miniature white flowers. It is not an aggressive weed, but it still does its best to cover sterile soil, thus having its place in the scheme of things.

Another similar, tiny plant with somewhat larger flowers, the petals white, translucent and notched at the tip, is the mountain sandwort. It likes cooler temperatures than its relative, and thus, is found mostly on the mountains. It is a tufted plant and makes patches a foot wide, and is quite suited to the rock garden in cool regions, often being cultivated for this purpose. It has numerous small leaves and flowers in groups of one to five. A third species that lives up to its name is the pine-barren sandwort, a species of the seacoast growing in sandy places where nothing else seems to grow. It is a tufted plant, with short and grooved stems, leafless branches, the lower leaves tiny, awl-shaped, and densely overlapping, and with a few, dainty, white flowers about half an inch wide.

Another plant of sandy places, but not a close relative of the sandworts, the sand spurry is a low, smooth species with frail stems, linear leaves, and tiny, crimson pink flowers, with glandular and hairy sepals. It grows in dense company along roadsides and in waste places.

In grain and cornfields, and in cultivated ground generally, the corn spurry can often become an undesirable weed, for its growth is so rapid, that a field of young turnips or carrots may swiftly be smothered with it. Its stems, six to eighteen inches tall, are slender, erect, and bright green. Its leaves, one to two inches long, are linear or awl-shaped, and apparently whorled at the joints of the stem, but actually in two opposite clusters of six to eight. Its white flowers are in terminal cymes and open only in the sunshine.

One of the many genera of the pink family is the genus *Cerastium,* Greek for horn, in reference to the shape of the follicle. Some of the species are wildings, some are cultivated, and some are wildings grown in the garden for their beauty, such as the field chickweed, sometimes called the starry grasswort, a densely tufted perennial with very narrow leaves found in dry, rocky places, usually on hills in the southern part of its range, in meadows and pastures in the North, but often grown in the garden for its white, starry flowers, more than half an inch wide, borne in graceful, terminal clusters.

Also popular in the garden is snow-in-summer, a prostrate plant not over six inches high and forming large patches with many conspicuously, white-woolled leaves, and white flowers. In flower, the blossoms fairly overwhelm the leaves, and later when the flowers have faded, the white and velvety leaves take their place. A close look at the flower will reveal its chickweed character of divided petals, though it may not be of the same genus as the true chickweed, the two genera being separated on minor details.

As for the common mouse-ear chickweed, this species is a plant of fields, meadows, roadsides and waste places, and a weed should it invade the garden. It has a hairy and clammy stem, oblong leaves, and clustered white flowers with no pretensions to beauty.

A beautiful plant in itself, but grown chiefly in the garden because it is an exquisite accompaniment to other flowers, and thus, often used by florists in making up bouquets, baby's breath is a familiar and popular garden flower. The perennial species grows from one to three feet high with continually dividing branches, smooth, sharp-pointed, light green leaves, and many small, white flowers on graceful feathery stalks. The annual species is a lower growing plant with somewhat similar leaves and flowers that may, however, be pink as well as white.

The true pinks, such as the sweet William and the deliciously spicy clove pink, and natives of Europe, were once regarded as sacred to Zeus, and hence, were named *Dianthus* by Theophrastus about 300 B.C., literally meaning, in Greek, flowers of Zeus, or of Jupiter, the Roman god equivalent to Zeus. They are members of a genus of the same name, a genus that contains two hundred species with opposite, usually narrow leaves, swollen joints, and terminal flowers that are solitary in the carnation, and a few others.

Several species, such as the Deptford pink and maiden pink, have become naturalized in our country, and are found growing in fields, waysides, and waste places. The Deptford pink, a rather insignificant plant, but not without a certain charm when seen growing in patches and attended by various butterflies, grows six to eighteen inches high, with a stiff, erect, finely haired stem, light green narrow leaves, and small clusters of crimson pink, white-dotted flowers, the petals toothed or jagged-edged. The maiden pink is a turf-forming plant, the leaves scarcely one inch long and very narrow, with little, crimson pink or white pink flowers that bloom singly. They resemble the sweet William.

The sweet William is a popular garden plant and when in flower, a bed makes a gorgeous display, the color scheme varying through an infinite variety of tints and combinations of tints. It is a plant with many named forms, but the typical form is a smooth herb, twelve to twenty-four inches high with green, flat leaves that are somewhat broader than in most pinks, and red, rose purple, white or sometimes varicolored flowers in dense, flat cymes. They are not fragrant.

A native of China and Japan, and introduced into Europe by a French missionary early in the eighteenth century, the China pink is a beautiful flower, though lacking somewhat in fragrance, with an extended blooming season that adds to its value, from early summer until snow falls. It is a tufted plant, twelve to eighteen inches high, its stems erect and somewhat stiff, with linear leaves and red, white, or lilac flowers that may be either solitary or in sparse clusters.

To many people the carnation is a florist's flower, but where the winters are mild it can be grown successfully outdoors. The Romans knew of the carnation or clove pink. It was cultivated by them and it has been under cultivation from that time, giving rise to forms that number in the thousands. It is a smooth-tufted herb, one to three feet high, with somewhat stiffish stems, grayish foliage, and fragrant, usually solitary flowers.

An old garden favorite, known in many forms and varieties, the garden or grass pink is a mat-forming herb with erect stems, nine to eighteen inches high, with bluish gray leaves and fragrant flowers, rose pink to purplish, or white, or with variegated colors. A row of the grass pink is a worthwhile acquisition to any garden.

148. Pipewort (Eriocaulaceae)

The pipewort family is a group of nine genera and about five hundred species of bog or aquatic perennial herbs widely distributed in the warmer and tropical regions of the earth, though a few extend into temperate parts, but mostly found in South America. They have fibrous, mostly knotted, or spongy roots, alternate, tufted, densely grasslike, basal leaves, minute, unisexual flowers in small, densely crowded, terminal heads, on long, slender scapes. The flower head is subtended by a scaly, bracted involucre, the individual flowers are in the axil of a thin, dry membranaceous bract and have two or three sepals, two or three petals, four stamens, and a two to three-celled ovary with two to three ovules. The fruit is a two-to three-celled capsule.

Only a few species are native to the United States, most being the pipeworts of the genus *Eriocaulin,* Greek in allusion to the wool at the base of the scape in some species. They occur in pine barren ponds, swamps, and tidal flats along the Atlantic coast and Great Lakes regions. They are stemless, or short-stemmed, submerged plants, with awl-shaped or grass-like leaves that rise from a rosette, or bunch from a very short stem, and with long stiff flowering stalks that project above the surface of the water. The minute, whitish flowers are borne in small, terminal heads, the latter looking like little buttons set on the tops of slender stalks, hence, the pipeworts are also known as buttonrods.

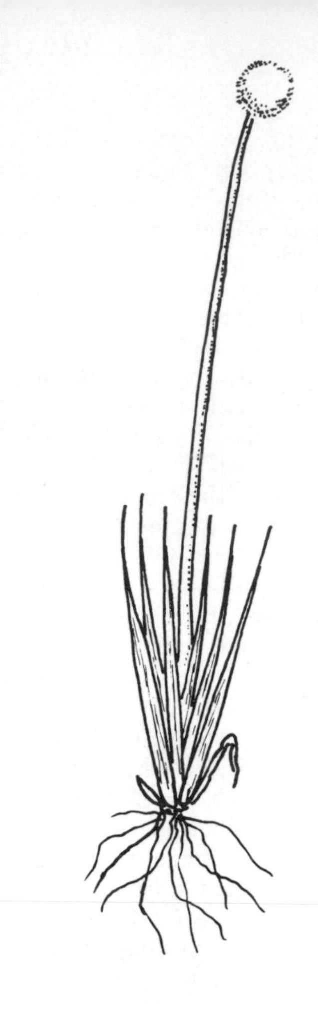

442. Leaves and Flower Cluster of Pipewort

149. Pitcher Plant (Sarraceniaceae)

If one has never seen the pitcher plant, one may well pay a visit to a bog or swamp, in May or June where it plays its villainous role of luring insects to their death. Few plants are so curiously fascinating. The general form of the flower and its rather peculiar coloring is designed to attract the carrion flies that are especially fitted to play their part as agents of cross-pollination, though it is not the flower, but the leaves that draw our attention. For they are hollow and pitcher-shaped, exquisitely designed to trap the insects whereon the plant partly depends for its sustenance. Examine one of them and it will be found that it holds water and, also, the remains of the insects that came to feast, but instead, found a watery grave. Once inside, the hapless victims find no means of escape, their only avenue to the outside world bristling with downward-pointing, stiff hairs that form an effective barrier. Try as they may to fly or crawl out of their prison, every attempt is hopeless, and soon exhausted, they fall into the water to give their lives, so that plant may live. For the pitcher plant lives in soil that is usually deficient in nitrogen compounds, and without the insects to make up the deficiency, it could not form the proteins that are essential to its survival. The pitcher plant calls to mind:

> What's this I hear
> About the new carnivora?
> Can little plants
> Eat bugs and ants
> And gnats and flies?

There are seven North American species of pitcher plants of the genus *Sarracenia,* one of three genera of the pitcher plant family and named for Dr. Sarrasin, a physician and naturalist of Quebec. They are all herbaceous perennials, found only in bogs and wet places where the soil consists of sandy peat and sphagnum moss, with creeping underground stems, basal, pitcherlike or tubular leaves, and solitary flowers, on leafless stalks. The flowers have a calyx of five sepals, many stamens, and a pistil with an umbrellalike stigma. The fruit is a capsule. These characteristics typify the family.

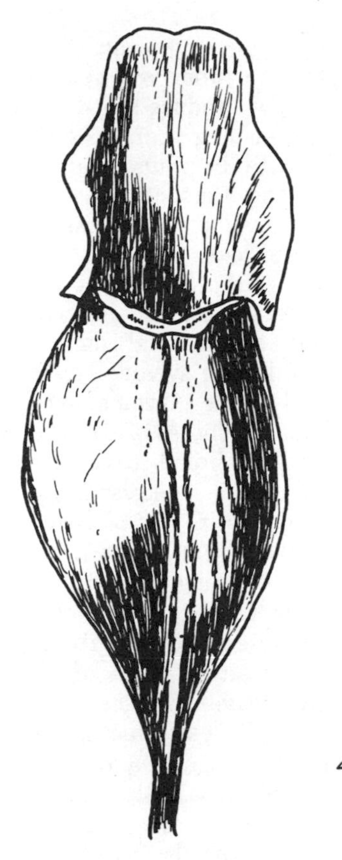

443. Leaf of Pitcher Plant

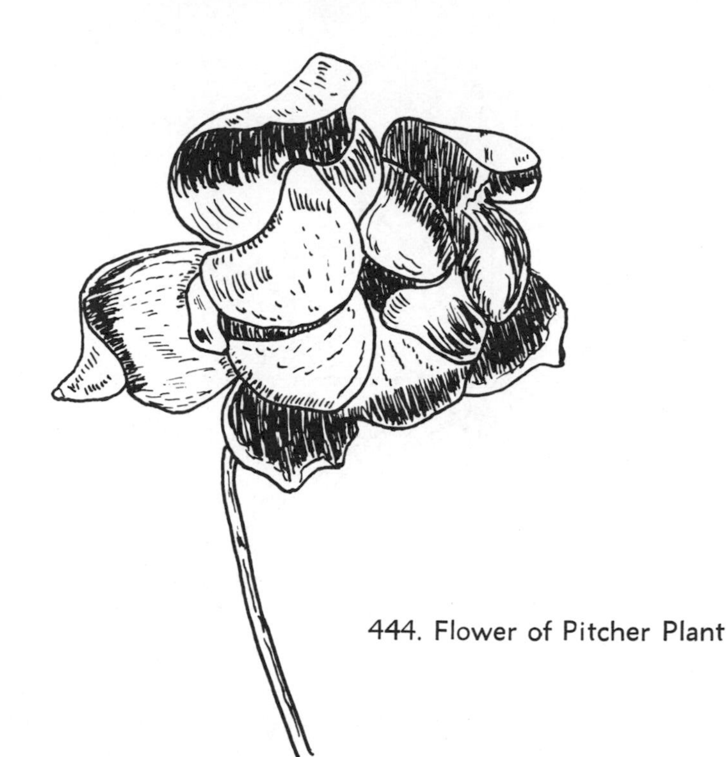

444. Flower of Pitcher Plant

The common pitcher plant of the East, also called sidesaddle flower, huntsman's cup, and Indian dipper, has leaves a foot long and a flower stem of about the same length. The general coloring of the plant is green, with red purple veining. The sepals are madder purple, greenish on the inside, the petals are a dull pink, and the style is green.

In the South there is a somewhat larger pitcher plant called the trumpets. It has elongated, trumpet-shaped, nearly erect leaves that are sometimes nearly three feet long, and light ochre or dull yellow flowers. A corresponding western species of our eastern pitcher plant is similar in appearance, but somewhat more slender, and with a hood over the top of the "pitcher." The flowers are yellowish purple.

150. Plane Tree (Plantanaceae)

Clear are the depths where its eddies play,
And dimples deepen and whirl away;
And the plane tree's speckled arms o'ershoot
The swifter current than mines its root.

So wrote Bryant of the plane tree, better known as the sycamore.

All trees shed their bark. They have to, so that the trunk may increase in width. But most trees do so unobtrusively, the bark falling off in plates or scales of varying size. The shagbark has no compunction in advertising the process, the silver maple hardly less so, and the sycamore not at all, for the bark of this tree flakes off in great irregular masses, leaving the surface a crazy patchwork of greenish white and gray and brown. The smaller branches sometimes appear as if they had been given a coat of whitewash. The reason for the tree flaking lies in the rigid texture of the bark tissue that entirely lacks the expansive ability common to the bark of other trees, so that it is unable to stretch to accomodate an increase in the growth of the wood, and thus, the tree has to slough it off. The appearance of the bark is characteristic of the tree and is a means of distinguishing it even at a distance and at any time of the year.

The sycamore is a tall, stately tree, fifty to eighty feet high, but often higher, with a tall trunk, a loose, broad head, and mottled green and white branches. The bark is a dark reddish brown, furrowed at the base of the trunk, but smooth and scaly above, peeling

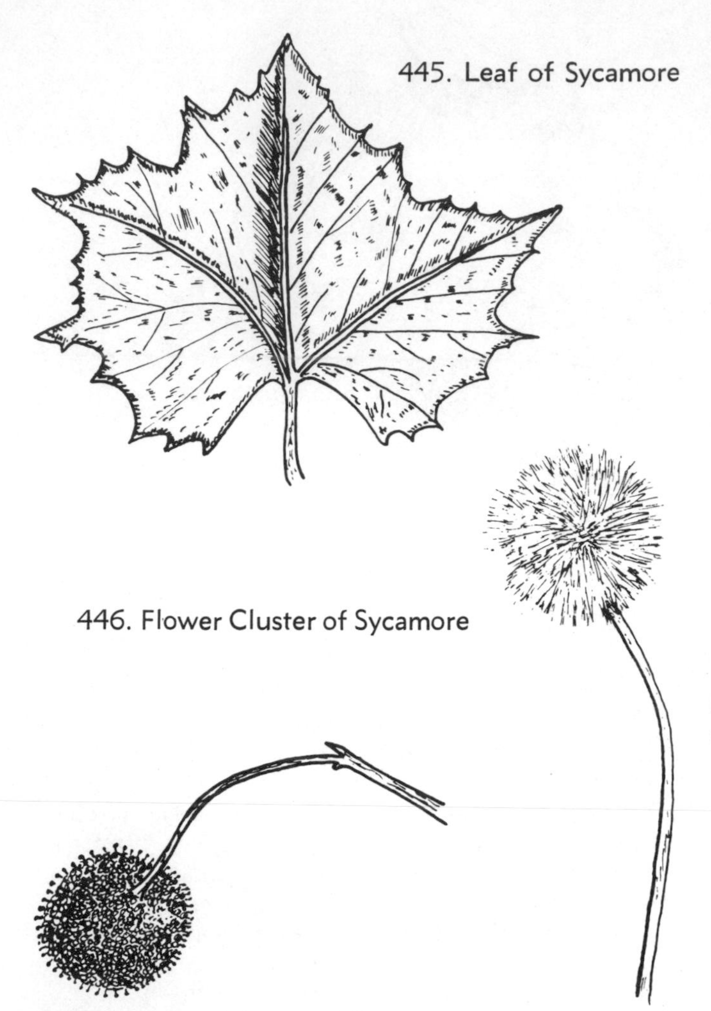

445. Leaf of Sycamore

446. Flower Cluster of Sycamore

447. Fruiting Cluster of Sycamore

off in thin, brittle layers, leaving patches of buffish white, brown gray, or pale tan-colored inner bark exposed. The leaves are large, four to nine inches wide, with three to five lobes, coarsely toothed, alternate, yellow green and turn a russet brown in autumn. As they unfold, they cast off a fuzzy covering of branched hairs that can become quite irritating to the mucous membranes of the eyes and throat, and an annoyance to some people. This moulting period, however, is very brief and quickly passes. The flowers are small and inconspicuous, in dense, ball-like, stalked clusters, the staminate and pistillate in separate clusters on the same tree, the staminate a deep red, the pistillate a pale green, tinged with red, both flowers with three to eight sepals and petals, the staminate with an equal number of stamens, the pistillate with three to eight pistils, each one-celled. The latter develop into a globose head of elongated, obovoid achenes, each surrounded at the base by a circle of brown hairs and usually bearing at the apex a small, curved spur. They persist on the branches throughout the winter and serve to identify the tree at this season.

Looking at the branch of almost any tree in early August, one will find the developing buds that will produce the leaves of the following spring. Then looking at a branch of the sycamore, one will not find such buds. Will the tree, then, not produce any leaves the next year or in any of the ensuing years? Detach a leaf from the twig and it will be found that its stem ends in a hollow cone, and that on the twig, there is a little bud that was developing all summer under the protecting base of the leaf.

The sycamore, also known as the buttonwood, is common throughout the eastern states, and is found growing along the banks of streams and on rich bottomlands. It is one of three native sycamores, the other two being western species, the Arizona sycamore and the California sycamore. The Arizona sycamore, probably the most abundant, broad-leaved tree in the Southwest, grows along stream banks and canyon walls, and can be recognized by its five to seven-lobed leaves and its racemose clusters of achenes. The California sycamore bears its fruiting clusters, or buttonballs, in a series of four to six, strung on a tough fibrous stem. It is a medium-sized tree of stream banks and the foothills of the western slope of the Sierra Nevada Mountains with leaves that have the same general outline as those of its eastern relative, but the lobes are more slenderly triangular and somewhat more deeply cleft.

There are only six or seven species of sycamores. They belong to the genus *Platanus,* the classical Greek name for the plane tree, and the only genus of the plane tree family. So the characteristics of the sycamore are those of the family. Two of the non-native species are the London plane tree, generally known as the London plane, and the Oriental plane tree, also known as the Oriental plane. The London plane, presumably a hybrid between the American sycamore and the Oriental plane, is a tall, widely spreading tree up to one hundred and forty feet high, with three to five-lobed leaves and fruiting clusters in groups of two. The Oriental plane is not as tall as either the American sycamore or the London plane, and has five to seven-lobed leaves and ball-like, fruiting heads in clusters of two to six. The latter is not cultivated in the United States, but the London plane is widely planted on the streets and in the parks in most American cities in the temperate region, for it withstands abuse, smoke, dust, and wind probably better than any other tree.

The Oriental plane is the plane tree of the Greek writers, a tree held in high esteem by the ancients for its stateliness and beauty. Plato, it is said, held discourse in groves of the tree, and Xerxes is presumed to have held up his army for days so that he might contemplate the beauty of the tree. That he might never forget it, he had its form wrought upon a medal of gold. The "Apothecarye of London," John Parkinson, quaintly and wisely writes that the plane trees "are planted by the waysides and in market places for the shadowes sake onely."

151. Plantain (Plantaginaceae)

Everyone knows the plantain, a persistent intruder in yards and lawns, and like many other weeds, an immigrant from Eurasia. It is such a familiar plant that it hardly needs description. Its leaves are all basal. They are long, ovate, entire, obtuse, rounded at the base, with five to seven prominent, lengthwise veins that gather together into a thick, channeled petiole. If one looks at the plantain closely, one will observe that the outer row of the spreading tuft almost hugs the ground to conserve moisture for the clustered fibrous roots, as well as to crowd out other growth so that it may have adequate living space. The flowers are small, dull, greenish white, perfect, with a four-lobed corolla and are densely crowded on a slender, cylindrical spike, three inches to a foot or more in length. The threadlike style matures before the corolla opens, the stamens maturing later, thus, cross-fertilization is ensured. The seed capsule is a small urn or pyxis, that opens near the middle and that contains five to sixteen seeds. The seeds, when wet, develop a coat of mucilage that helps in their dispersal.

There are about two hundred and thirty, widely dispersed herbaceous, annual or perennial species, and three genera in the plantain family that should not be confused with the banana and pineapple families, also known as plantains in certain parts of the world. The leaves are alternate, opposite or basal and simple. The flowers are small, perfect, or polygamous in dense terminal spikes or heads on elongated scapes. The calyx is four-parted, the corolla is usually four-lobed, the stamens are two or four on the tube, or throat of the corolla, and are alternate with the lobes, and the ovary is one to four-celled, commonly two-celled, with one or more ovules in each cell. The fruit is a capsule (pyxis) or a nutlet.

In addition to the common plantain found on our lawns and in our dooryards, there is the red-stemmed plantain, and the narrow-leaved plantain, the latter also known as the English plantain or ribgrass. The red-stemmed plantain is similar to the common plantain, but the leaves are thinner with rather long, slender petioles that are crimson at the base, and the flower spikes are less dense. These two plantains

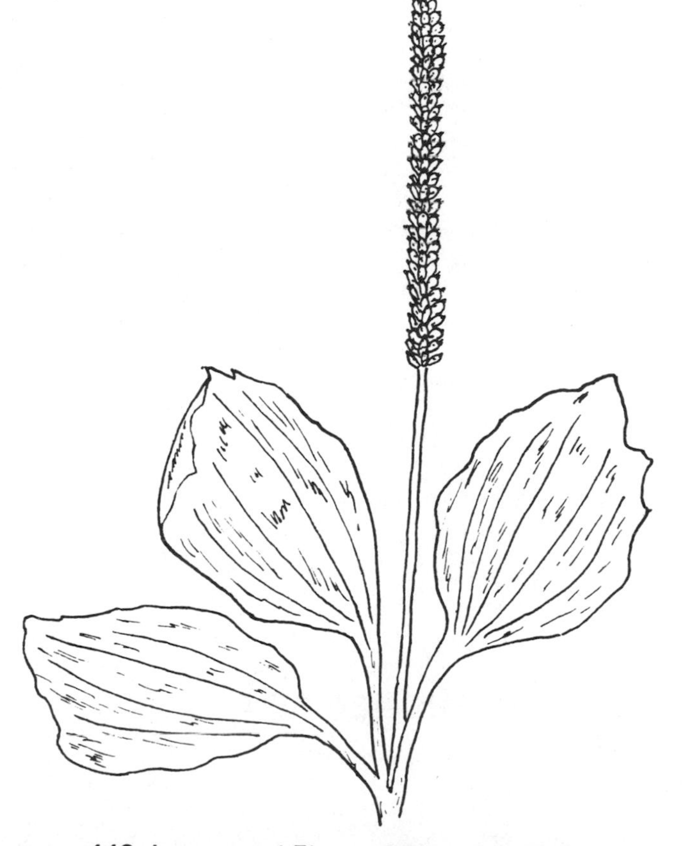

448. Leaves and Flower Spike of Plantain

are often found growing together. The narrow-leaved plantain or English plantain, is a more pernicious weed than its two broad-leaved relatives, and whereas, they seem to prefer the yard and roadside, it overruns meadows and pastures. It has a short, thick rootstock, with many branching rootlets, and thickly tufted leaves, oblong, lance-shaped, thick, hairy on both sides, with three to seven ribs tapering to the petioles. The scape is very slender, strong, wiry, and hairy, and the flower spike is dense and short with dull white flowers. An average plant will produce about a thousand seeds that are a very common impurity of grass and clover seeds. The mucilaginous seeds (ispaghul seeds) of an Old World plantain are used in making a beverage. All of our plantains belong to the genus *Plantago,* the Latin name for the plants.

152. Pokeweed (Phytolaccaceae)

"All on fire with ripeness," said Thoreau of the pokeweed when in fruit, and a rather accurate description, for when the dark purple berries are mature (in August), the leaves and even the footstalks take on bright tints of crimson lake. The berries ripen when the birds are beginning to move southwards and are very popular with such species as the robin, towhee, rose-breasted grosbeak, catbird, bluebird, and particularly the mourning dove. Birds have become intoxicated from gormandizing on the berries.

The pokeweed, also known as pokeberry, pigeonberry, and inkberry, is a handsome plant, albeit a dangerous one, for its root is very poisonous and children have suffered from eating the berries. Yet, both have been used extensively in the preparation of certain drugs, and as a household remedy for skin diseases and rheumatism. And if the young shoots are cooked thoroughly and the water changed several times, they make excellent greens or substitutes for asparagus.

The pokeweed is a plant, four to ten feet tall, of fencerows, thickets, and waste places with a stout, smooth, usually red or purplish stem and oblong lance-shaped, rather thick, deep green, entire leaves, that have an unpleasant odor when bruised. The flowers are white in terminal clusters (racemes), and the fruit, in drooping clusters, are dark purple and contain a purplish crimson juice that is sometimes used as ink, hence, its name of inkberry.

The pokeweed belongs to a family of twenty-two genera and perhaps one hundred species of mostly tropical herbs, trees, and shrubs. Only one genus *Phytolacca,* Greek for plant and French or Italian for *lac,* in allusion to the staining berries, is represented in our flora, and only with two species, the pokeweed and the umbra, the latter a handsome, evergreen tree much grown for ornament in southern California. The members of the family have alternate, simple, entire leaves, and flowers in racemes, either terminal or in the axils of the leaves. The flowers are regular, the staminate and pistillate on separate plants, and have a tubular calyx, four stamens, and a several-celled ovary that forms a berry in fruit. There are no petals.

A species of the genus *Rivina,* a group of tropical American perennial herbs, and known as the bloodberry or rouge-plant is common throughout Florida and the Gulf states and is cultivated there. It is also grown in the warm, temperate greenhouse northward. It is a spreading herb, one to three feet high, with alternate, ovalish leaves, small white or pinkish flowers, without petals in erect racemes that are far more attractive in fruit than in flower, the fruit being a pea-sized, red berry.

153. Pomegranate (Punicaceae)

The Carthaginians cultivated the pomegranate and introduced it into Europe where it became known as *"malum punicum"* or apple of Carthage, the ancient city being once called Punica. It is one of only two species of the pomegranate family, the other being a little-known plant from the island of Socotra. Both belong to the same genus, *Punica*.

The pomegranate is an Asiatic shrub or small tree, ten to twenty feet high, often with spiny-tipped branches, and opposite, short-stalked, simple oblong to oval, shining leaves. The perfect flowers occur in small clusters of one to five that are set on short shoots borne in the leaf axils. They have a leathery, partly tubular calyx of five to seven lobes that persists on the fruit, a corolla of five to seven separate, wrinkled, orange red petals, many stamens, and an ovary of four to fifteen united carpels. The fruit is a thick-skinned or leathery, several-chambered, brownish yellow or red berry, about the size of an orange, containing numerous seeds that are surrounded by a bright red, juicy pulp. This is the commercial pomegranate.

The pomegranate, one of the oldest of cultivated fruits and well known to Theophrastus, has been prized for centuries abroad, but does not seem to have found much favor in the United States, although it is cultivated in the South where it is not widely grown. Many cultivated varieties are recognized, one being a dwarf, double-flowered form often seen in greenhouses and florist shops, and grown for ornament. A few varieties are sweet, wonderful, and acid.

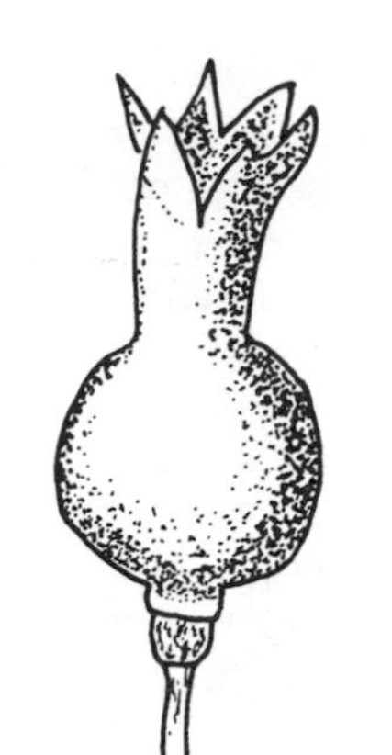

449. Fruit of Pomegranate

154. Pondweed (Potamogetonaceae)

Anyone that has ever stood by the edge of a pond or on the shore of a stream has undoubtedly seen the pondweeds, for they are common members of any freshwater community and, incidentally, are probably more important to the welfare of water animals than any other group of aquatic seed plants. There are about forty species native to North America, most growing in great abundance, providing both food and shelter for many aquatic animals, as well as food for many birds, being, as a matter of fact, one of the principal foods of wild ducks. Bullheads, perch, and pickerel use them to hide in, and many snails get their living from the leaves. Pondweeds grow rooted to the bottom and their long shifting leaves extend through acres of pond and lake water. Most of them are wholly submerged, except for their flowers, and in some species, the uppermost leaves float on the surface.

The pondweeds belong to the family of the same name, a group of five to nine genera and about seventy-five species of perennial, aquatic herbs of saline or fresh water, many being submerged. They are of very wide distribution, in freshwater rivers, ponds, lakes and in brackish water along the seacoast, of temperate parts of the world. Most of the American species are in the northern states, the Northeast being particularly rich in these plants in both number of species and in abundance. They are the dominant vegetation in the lakes and ponds of the Great Lakes region, and eastward into New England. Most of them are freshwater species, but a few thrive in moderate salinity or alkalinity, and can be found in ponds, lakes, streams, and bays at various depths, depending on the clarity and constancy of the water.

The pondweeds are plants with the main stem usually a branching rhizome that spreads through the bottom mud and rooting at the nodes. They have alternate leaves, though the uppermost are sometimes opposite, sessile or with a petiole, frequently with a sheathing base, the blade either expanded or capillary, or reduced to scales and often of two kinds, either submerged and reduced, or broadened and floating. The flowers are small, inconspicuous, perfect or unisexual, in terminal spikes projecting above the water or in axillary clusters. The flowers have a perianth of four herbaceous distinct segments, or membranous and tubular or cup-shaped, one to six stamens, and one to nine, one-celled ovaries, each with one ovule. The fruit is nutlike or drupelike.

Of the many species of pondweeds of the genus *Potamogeton*, Greek in allusion to the aquatic habit, the following three are among the most common: the common floating pondweed, the curled-leaved or ruffle-leaved pondweed, and the sago or fennel-leaved pondweed. The common floating pondweed, found in ponds and streams throughout North America, except in the extreme north, has floating leaves with oval blades and long leafstalks, the submerged ones incomplete and threadlike. The curled-leaved pondweed, found in ponds and lakes and often in brackish water, has long, very translucent, brown leaves with crispy, wavy edges. This species is such a favorite food of wild ducks that it is planted in game ponds. The sago pondweed is a very common species of slow streams and lake borders, and often of brackish or limy water. It is one of the earliest to appear in the spring, and grows entirely submerged at a depth of one to six feet. Some of the plants that grow in deep water are often eight feet long. Muskrats, beaver, moose, and deer feed consistently on pondweeds, as well as a number of species of waterfowl. The larvae of a moth, *Nymphula,* commonly feed on the leaves, the caterpillars of the moth, *Acentropis,* live in dense, silken webs that they make on the leaves, and the caterpillars of the moth, *Paraponyx,* live in cases that they make from the leaves. Wherever pondweeds are abundant, they support the larvae of midges, that in turn become important food for fishes.

Although the pondweeds form the dominant group of the family, there are other species such as the horned pondweed, a plant with needlelike leaves rising from stipuled nodes and found in brackish or fresh ponds, pools, and ditches, several species of naiads, slender, branching aquatics that are wholly submerged, and the two marine plants, ditch grass and eel grass, that most people mistake for grasses. The ditch grass of the genus *Ruppia,* named in honor of Heinrich Bernhard Rupp, a German botanist, has long threadlike forking stems and narrow, grasslike leaves that wave in masses beneath the surface of the water, and flowers in spikes that open above the water. It is common in brackish estuaries and along the Atlantic and Pacific Coasts. The eel grass is another submerged plant with a slender rootstalk, branching stems, and ribbonlike leaves, of the genus *Zostera,* Greek in allusion to the ribbonlike leaves. The naked flowers occur in rows on a leaflike spadix that is hidden in a leaf sheath. The eel grass is found in bays, streams, and ditches along both the Atlantic and Pacific Coasts.

155. Poppy (Papaveraceae)

In hidden copse and shaded thicket, in woodland border and on low hillside, the warm sunshine of April quickens into life the buds of the bloodroot and one

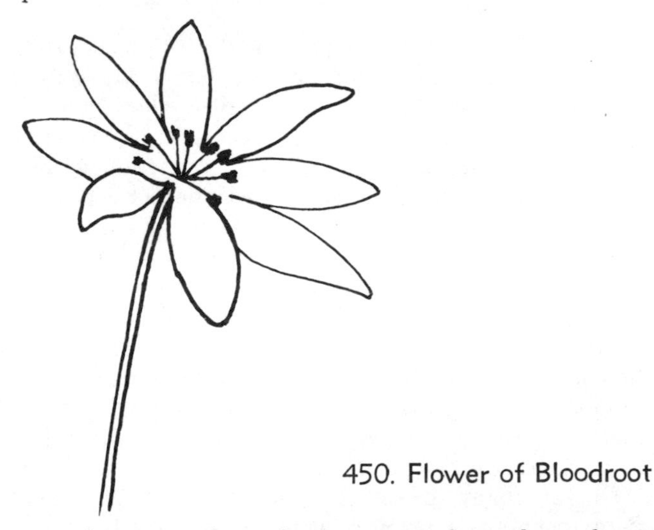

450. Flower of Bloodroot

can almost see them slowly emerge from the embrace of silvery green leaf cloaks, to expand into white-petaled, golden-centered flowers of evanescent beauty, for on the morrow, the delicate blossoms, unable to withstand the spring winds, fall, and are gone.

Observe that the golden orange anthers mature after the two-lobed stigma, that shrivels before the pollen is ripe, and that the outer stamens are somewhat shorter than the inner ones in an advanced flower. Also that the stigma is prominent in a newly opened flower so that cross-pollination may be effected by such insect visitors as the hive bees, bumblebees, mining bees, and the beelike flies.

The bloodroot is a beautiful, but fragile flower, one and a half inches broad, with generally eight, brilliant white petals, set on the end of a naked scape. A single, large round leaf, usually with seven irregular lobes, enfolds the bud and expands with it. Both the flower and leaf come from a thick rootstock. Eventually the flower becomes an oblong, dry pod with many light yellow brown seeds. Puncture not only the root, but any part of the plant, and there flows a red orange juice that stains whatever it touches, and that the Indians found serviceable as a war paint. The bloodroot has a bitter and acrid taste, as it contains an alkaloid that is toxic.

Gray relates that the generic Greek name of the celandine (*Chelidonium*) meaning a swallow was given it because its flowers appear when the returning swallows are seen skimming over ponds and streams and freshly ploughed fields, but quaint old Gerarde writes that it was not given its name

> because it first springeth at the coming in of the swallowes, or dieth when they go away, for as we have saide, it may be founde all the yeare, but because some holde opinion, that with this herbe the dams restore sight to their young ones, when their eies be put out.

451. Leaves and Flowers of Celandine

Whatever the origin of its name may be, the celandine, a native of Europe, has invaded our shores and found hospitable ground in waste lands, fields, roadsides, and in the gardens about our dwellings. It is a plant some twelve to thirty inches high, with deeply divided or cleft leaves, distinctly, pale bluish green beneath, the terminal one the largest, and small, yellow flowers with two sepals and four petals, many yellow stamens and a green style, in small, stalked clusters that are umbellike. The plant has a yellow juice that, like that of the bloodroot, stains whatever it comes in contact with.

There is a western woodland species also with a yellow juice called the celandine poppy, although it is not a celandine in that it does not belong to the same genus. It does, however, resemble the celandine in flowers and foliage, although the flowers are about double the size.

Although called poppies, the prickly poppies are not true poppies. They are tropical and American herbs of perhaps ten species, one grown widely as an annual, with a yellow juice, cut leaves, toothed and spiny-margined, and large flowers with two or three sepals, and four to six rather showy petals, that develop into prickly pods. One species, the Mexican prickly poppy, with yellow flowers, has become naturalized on hillsides and on cultivated ground along the Gulf States and throughout the Southwest. A second species, the Carolina poppy, with white flowers, is found in the southern coastal plain region, and a third species is the common prairie prickly poppy also with white flowers. There is still another species, the western prickly poppy. It, too, has white flowers, and occurs in the mountain valleys and canyons from California to Texas.

It might be well to mention at this point, a rather attractive California herb known as creamcups. It is an erect plant, six to twelve inches high, with linear leaves, weak, hairy white stems, and white or yellowish flowers whose six petals form a cup-shaped blossom. Mention of the creamcups calls to mind two western species of the poppy family that are tall, bushy plants. One is the Matilija poppy that grows in gravelly stream washes and is sometimes eight feet high. It has compound leaves and large, white, short-stemmed flowers four or five inches in diameter. The second is the tree poppy, often cultivated for its handsome yellow flowers. It is common on the dry slopes and ridges of the lower foothills of California where it grows from four to eight feet high.

What may almost seem to be a queer quirk of nature is that the California poppy, so commonly grown in Eastern gardens and perhaps the best known western wild flower, should be confined to a rather limited range. Eastern gardeners know it well, for it is one of our most popular garden annuals and makes a fine display, but hardly comparable to the colorful sight of a treeless expanse on fire with millions of its yellow flowers, so crowding each other that no earth or foliage is visible.

The early Spanish explorers, as they sailed along the California coast, observed the flame of fire upon the hillsides and called the coast the "land of fire" and "sacred to San Pascual since his altar cloth is spread upon all its hills." Later, in 1815, when the Russian expedition, under Kotzbue, explored the coast, the fields of yellow poppies impressed the Russians so much that the naturalist of the expedition, Chamisso, in reporting the plant, gave it the name of Eschscholz in honor of the surgeon of the expedition. And so, it became known as *Eschscholtzia californica* seemingly a somewhat unfortunate name as it is not too easy to pronounce.

All the foregoing species are members of the poppy family that comprises about twenty-five genera and one hundred and fifty species of mostly annual or perennial herbs of wide distribution, but being most common in the north Temperate Zone. Most of them have a milky or colored juice (latex) and alternate or basal leaves that are simple, entire, and often lobed or divided. The flowers are usually solitary, regular and perfect, often large and showy, with typically two sepals and four to six petals, the latter falling early, indeed, the entire flowers generally wither very quickly when picked, and an ovary of two or many united carpels. The fruit is a many-seeded capsule opening by pores or valves. The family includes many popular garden flowers, a number of wild species, and the opium poppy that has been cultivated since ancient times.

The true poppies, whereof there are probably a few hundred species, belong to the genus *Papaver*, the classical Latin name for the plant. They are found mostly in the temperate regions of Europe and Asia, a few in western North America, and vary in height from six inches to four feet. Their leaves are basal, hairy, generally many, and as a rule, deeply segmented. Their flowers are solitary, on a long scape, nodding when in bud, but straightening out when they unfold, with two sepals and five petals that are vividly colored, red, violet, yellow or white. Their fruit is a capsule that is covered with a shieldlike cap, beneath which, are small pores through which the seeds are dispersed.

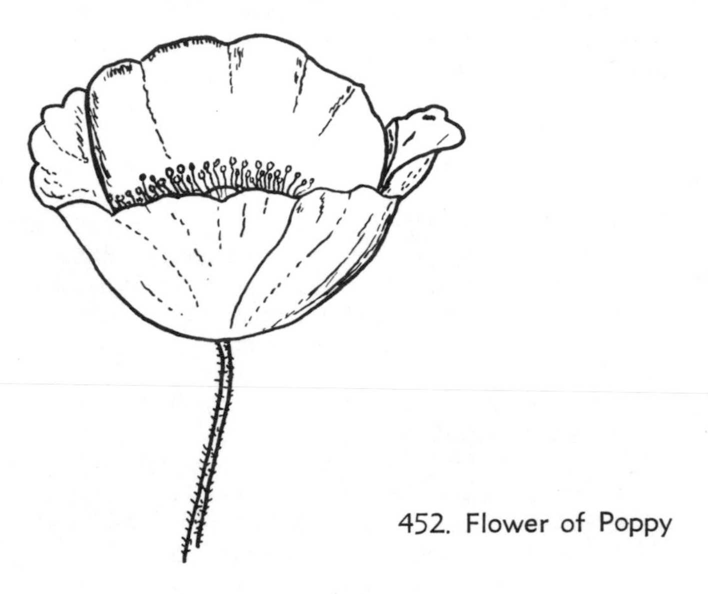

452. Flower of Poppy

Of the different species of poppies, the most favored for the garden are the two perennial species, the Iceland and Oriental, and the two annual species, the corn and opium poppies. The Iceland poppy is a hardy plant found throughout the arctic regions where it is one of the most common and colorful of the wild flowers. It is a robust plant with hairy or smooth, deeply cut leaves, and sweet-scented flowers on a flowering stalk fifteen inches high. The colors range through white, red, yellow, and orange. The Oriental poppy, perhaps the most superb and barbaric member of the poppy group, a native of Persia and the Mediterranean region and introduced into England in 1714, is a strong-growing species, with a deeply growing rootstock, covered all over with stiffish hairs, pinnately cut leaves, and flowers the largest of all the poppies, often six inches in diameter, scarlet, the petals purplish black marked at the base.

The corn poppy, or scarlet poppy as it is also known, is the poppy of English literature and immortalized in the famous poem "Flanders Field" of the first World War. A native of Europe it has branching, wiry stems, hairy and finely lobed, yellow green leaves, and small flowers of red, scarlet, deep purple, and occasionally white. It was from this species that the Shirley poppy was developed in England, in the town of the same name in 1886.

The narcotic character of the opium poppy was known to the Greeks as far back as 300 B.C. Gerarde writes that "being of many variable colours and of great beautie, although of an evill smell, our gentlewomen doe call them Jone Silver-pin." The wild opium poppy is native to the northern coast of the Mediterranean, and is a strong-growing plant three to four feet high, with grayish green, coarsely lobed and toothed leaves, and white, pink, red, or purple flowers three to four inches in diameter. It is widely naturalized in Europe, extensively cultivated in China, India and Turkey, and often planted as an ornamental in our gardens wherefrom it has escaped, and is sometimes found growing wild in waste places.

The dried, milky juice obtained by slitting the unripe capsules and allowing the juice to flow out is opium. It is the source of a large number of alkaloids, the most important being morphine, a colorless or white, odorless substance with a bitter taste, and codeine, a nearly transparent, odorless substance also with a bitter taste, but not quite so pronounced as that of morphine. As is well known, opium possesses sedative powers and has been smoked, eaten, and chewed. It is said that "it has been the source, by its judicious employment, of more happiness and, by its abuse, of more misery than any other drug employed by mankind." Strangely enough, the seeds of the poppy are harmless and have no injurious effect on the body. They are frequently sprinkled on bread, buns, and cookies.

156. Primrose (Primulaceae)

Quaint old Parkinson, in reference to the common loosestrife, writes that

> it is believed to take away strife, or debate between ye beasts, not onely those that are yoked together, but even those that are wild also, by making them tame and quiet . . . if it be either put about their yokes or their necks, which how true, I leave to them shall try and find it soe.

This common wild loosestrife of Europe, a rather

stout, downy species with terminal clusters of yellow flowers was once a garden favorite in our Eastern states, but long since having passed from favor, made its escape and is now found in fields and along the roadsides. There are a number of native loosestrifes. They have leaves variously arranged without marginal teeth and flowers solitary or in clusters, sometimes in the leaf axils or sometimes terminal, with a bell-shaped or wheel-shaped corolla, and a capsule for fruit. A delicate and pretty species, common in open woodlands, thickets, and roadsides is the four-leaved or whorled loosestrife. It is a slender, erect plant, one to three feet tall with light green, lance-shaped, pointed leaves that are arranged generally in a circle of four, but sometimes three and six. Slender long stems, each bearing a single star-shaped, light golden yellow flower, dotted with terra-cotta red, that often extends in faint streaks over the corolla lobes, project from the bases of the leaves. Self-fertilization rarely occurs as the stigma extends far beyond the anthers.

Another common, but lower growing species, and found in low ground, the bulb-bearing loosestrife has a distinctly characteristic flower spike. It is an aggregation of slender-pedicelled, yellow flowers, streaked or dotted with reddish. The leaves are opposite, lance-shaped and black-dotted, and in the axils, there often appear little elongated bulblets that are actually suppressed branches. Sometimes no flowers are produced, only these bulblets. All the loosestrifes are visited by various bees, including the bumblebees and honeybees, presumably for the purpose of collecting pollen. The loosestrifes belong to the genus *Lysimachia,* named after King Lysimachus of Thrace, a name derived from two Greek words meaning *release from strife.*

Long a favorite of hanging baskets and urns, the moneywort may often be found in moist soil near dwellings, an escape from gardens. It is an extremely beautiful trailing vine with a creeping, not a climbing habit, and, being a rampant grower, will run all over the place. The leaves are dark green, small, shining, almost round, and opposite, and from the axil of each leaf with the stem, there extends a short flower stem bearing one rather large, golden yellow flower.

In the moist shade of woods and thickets there may be seen a delicate little plant, with a long horizontally creeping root that sends upward a thin stem, three to nine inches high, and ends in a circle of sharp-pointed, lance-shaped, tapering, light green leaves. Above this whorl of leaves is airily poised on a wiry stalk a fragile, white starlike flower, hence its name of starflower. The beelike flies are its benefactors.

Not of the genus *Lysimachia,* but so much like a loosestrife as to be called a loosestrife, the fringed loosestrife is a rather handsome plant of moist, thin woods and the borders of thickets, with leaves in pairs and light golden yellow flowers that are almost of a pure yellow tone.

A long time ago the Greek philosopher Theophrastus called an entirely different plant by the scientific name of dodecatheon, from the Greek *dodeka,* twelve, and *theos,* gods, in other words the twelve gods. Theophrastus never saw the American cowslip, but apparently Linnaeus did, and fancied he saw in the flower cluster a congress of gods seated around a miniature Olympus. And so, he gave the name *Dodecatheon* to a genus of American wild flowers, of perhaps some thirty species, of which, the American cowslip, also known as the shooting star, is one. It is a handsome wild flower of the western prairies, hillsides, and open woods, and is often cultivated in the wild garden or rock garden.

453. Flower of Shooting Star

The American cowslip or shooting star appears as a rosette of oblong green leaves wherefrom arises a simple stem bearing a cluster of nodding, pointed flowers that suggest a small cyclamen. The flowers are light magenta, pink magenta, or white, and have a five-lobed corolla, the lobes long and strongly turned backwards, reddish-yellow stamens that are clustered together, and a conelike collection of purple anthers. The flowers appear to have a frightened look, or to the fanciful, appear like shooting stars.

In pools and ditches and in shallow stagnant water there grows a rather peculiar water plant called the featherfoil, its odd appearance due to the cluster of inflated flowerstalks that are about half an inch thick, constricted at the joints, and practically leafless. The leaves, divided into threadlike divisions, and distributed on the floating and rooting stems, are beneath the water. The flowers are white and insignificant, and are in whorls about the joints of the flower stems, forming an interrupted raceme.

A plant of many names—scarlet pimpernel, pimpernel, red chickweed, poison chickweed, week-a-peep, shepherd's clock, poisonweed, burnet rose, poor man's weather glass—the scarlet pimpernel is a low-spreading, annual herb adventive from Europe and found in waste places, dry fields, and roadsides. Called the scarlet pimpernel from the color of its flowers, the flowers are not scarlet, but of a reddish copper, or terra-cotta, indeed, there is no other flower of exactly the same peculiar shade. The color of the flower, however, is actually variable, and may also be blue, or purple or white. The leaves are stemless and toothless, are opposite and in pairs on the stem.

Not without reason is it called poisonweed or poison chickweed, for it has "pronounced diuretic and narcotic properties." Cattle usually leave it alone, but horses have died from eating it. And not without reason has it been called the poorman's weather glass, for before a storm or when the sun goes behind a cloud or on a dull day the blossoms close. They open only in sunshine and close at four o'clock. Birds like the seeds and doubtless many of them are dropped undigested, for the scarlet pimpernel is one of the most widely distributed species known.

All the aforementioned plants belong to the primrose family, a family of thirty genera and seven hundred species of annual or perennial herbs, widely distributed, but most abundant in the north Temperate Zone. Their leaves are alternate, opposite, or whorled and in some genera, they are in a basal rosette. Their flowers are regular, perfect, solitary or in various kinds of clusters, and have a tubular or bell-shaped calyx, a funnel-shaped or tubular corolla that is abruptly expanded above and generally of five lobes, five stamens that are inserted on the corolla tube opposite the petals, and one style and stigma. Their fruit is a capsule with many seeds.

The common primrose of our gardens is the English primrose, a wild flower of England, but early transferred to the garden and there transformed into many varieties. It reoccurs in English poetry. Thus, Shakespeare says "the primrose path that leads to the eternal bonfire," and Milton writes, "the rathe primrose that forsaken dies."

The common primrose belongs to the genus *Primula,* that contains three hundred species of low-growing, herbaceous perennials and a few biennials. They have a short stem or none at all. The leaves are crowded, stalked, long and narrow, or rounded and tufted. The flowers are on leafless stalks, solitary or in umbels, or whorled in tiers, or in rounded heads, in various shades of red, white, yellow, blue, pink, and purple. The floral structure and fruit are typical of the family. Of garden interest the following species are hardy perennials: *auricula, cortusoides, denticulata, elatior* (oxlip), *veris* (cowslip), and *vulgaris* (the common primrose).

157. Purslane (Portulacaceae)

As if mindful of economy, the spring beauty, like its cousin the portulaca, opens its starry blossoms only in the sunshine, when its benefactors the bees and flies are about, and closes them at other times to protect the store of pollen and nectar from unwanted pilferers. It is a charming little flower, pink with veinings of a deeper pink, though sometimes white with pink veinings. The blossoms, on long, slender, fragile stems, are turned mostly in one direction. Pick them, and the petals close and the whole plant droops as if in rebellion against such an indignity.

454. Flower of Spring Beauty

If one examines the flower with a magnifying glass, one will find that there are five petals, but only two sepals. One will also find that the golden stamens develop before the stigma matures, so that self-fertilization cannot take place. When the anthers have shed their pollen, only then do the three stigmatic arms open outward to receive pollen dust from a young flower. This is carried to them by female bumblebees and the little brown bombylius, as well as other friendly visitors that, in fair exchange, seek the nectar whereto the pink pathways guide them.

The spring beauty or claytonia, as it is sometimes called, the latter name after Dr. John Clayton, an early American botanist, is one of our early spring flowers, appearing in March in open, moist woods. It is a small plant, six to twelve inches high, with deep green, grasslike leaves, and at best an evanescent beauty. As Bryant says: "And the spring beauty boasts no tendered streak/ Than the soft red on many a youthful cheek."

The spring beauty is a member of the purslane family that comprises twenty genera and about two hundred species of mostly fleshy or succulent, often prostrate herbs, and some tropical shrubs, largely native to the western hemisphere, though a few are widely distributed in the Old World. Their leaves are alternate or opposite, simple, entire, and sometimes united at the base. Their flowers are regular, perfect, solitary or clustered, showy in some genera, but quickly fading in most, and opening only in the sunlight in a few, with usually two sepals, four or five petals, sometimes slightly united at the base, often notched at the tip, an indefinite number of stamens, the ovary one-celled, ovules one to many, and the style two to three-parted. Their fruit is a multi-seeded, three-valved capsule. The purslane family is a somewhat notorious group of plants because it contains the pestiferous "pussley," and yet, it also includes several beautiful garden plants such as the portulaca.

Every gardener and farmer knows the pestiferous "pussley" perhaps more generally known as the purslane, because it is a most troublesome weed. An immigrant from Europe, naturalized in the East where it is generally found in cultivated ground, in waste places, and about dooryards, with several indigenous relatives in the west and southwest, it is a trailing, fleshy leaved, prostrate annual with reddish, fleshy stems, the joints producing roots when in contact with the soil, dark green, thick, spoon-shaped leaves, and small, bright yellow, solitary flowers. No matter how obnoxious a plant may be, it is sure to have some redeeming qualities. The purslane is no exception. It was used as a food more than two thousand years ago in India and as potherb in Europe for centuries. In America it has not found much favor. The plant is cooked and served like spinach, and in England, the young stems that are sometimes pickled, together with the leaves, are frequently used as a summer salad. In southern Europe it is sometimes used in soups. In Mexico it is often sold with other vegetables in the market. The caterpillars of the variegated fritillary include it in their dietary. There are two horticultural varieties, one with thicker and mostly erect stems used by some people as "greens," the other with large, double flowers.

The portulaca, a garden favorite with many, is a native of the hot plains of southern Brazil and was described in the *Botanical Magazine* of 1829. Since that time, it has been in general cultivation and has varied into many garden forms, some being among the gayest-colored of all flowers. It is a trailing annual, multi-branched and fleshy, with spoon-shaped leaves, and terminal, showy flowers, an inch in diameter, and ranging in color from white, pink, yellow, red, or purple. The portulaca or rose moss, as it is also known, is suitable in the garden for dry, sunny rockeries, dry banks, and border edges. A bed will often perpetuate itself, and in some places, it persists about old gardens.

455. Leaves and Flower of Portulaca

In the West there are several members of the purslane family, such as pussy paws, a perennial plant with a basal rosette of fleshy leaves and small, white or pink flowers borne in a small spike; and bitterroot, the state flower of Montana, a stemless, fleshy plant with a basal rosette of succulent leaves, about an inch long, the flowering stalk about as long, and rose pink or white flowers. Its starchy root is edible in the spring, but becomes very bitter by midsummer. It was once used by the Indians as food.

A third western species, known as Indian lettuce or miners' lettuce, is a somewhat odd and dainty plant. It is an annual herb, coarse, green, often reddening with age, with a stem, six inches to a foot high, arising from a bunch of basal leaves that vary greatly in shape. The flowers are small, white on stalks one foot high with two disklike stem leaves beneath them. It is said that the Indians are very fond of it and eat it raw as a salad or cook it as a potherb. In the early California gold-mining days, when fresh fruits and vegetables were not always available, and the miners often falling victim to scurvy and other diseases, they ate the plant as a preventative, probably having learned of its use from the Indians or possibly from the Spaniards along the coast. The plant has been introduced into Europe, where it is cultivated under the name of winter purslane and used for salads and as a potherb.

158. Quassia (Simaroubaceae)

The ailanthus, or tree-of-heaven as it is also known, has something of a history. In 1751 Jesuit missionaries to China, where it was a favorite ornamental and shade tree, sent seeds to England believing that it could become acclimated there and that its leaves could be used as a food for a certain kind of silkworm. The experiment failed, if it can be called an experiment, but the English were, nevertheless, delighted with the stately and graceful tree, and planted it extensively in parks and similar places. In 1784 the tree was brought to this country from China, and again in 1820, it was brought over from England. The saplings were planted first near Philadelphia, then later in Rhode Island, and also abundantly in New York City and Brooklyn. At first it proved highly popular, but eventually came blossoming time and its popularity waned for the blossoms had a rank odor and the pollen was annoying to people suffering from catarrh or what is today called hay fever. Then, too, the appearance of the inchworm did not help matters, for the caterpillars kept falling on passersby. And so the trees were cut down. After awhile, however, it was discovered that only the staminate flowers were malodorous and that pistillate trees, whose flowers were without odor, could be supplied by nurserymen by taking cuttings of pistillate trees. And so, the ailanthus again came into favor and has increased in popularity ever since. As for the inchworm, the birds took care of that little matter.

The ailanthus has many virtues, particularly as a city tree for it seems to be immune to smoke and dust, and it retains its bright, green foliage throughout late summer when so many other trees become ragged and unsightly. It is, also, a rapid grower, as are all trees whose roots grow near the surface of the ground, and it is not unusual for the growing stems of young plants to reach a length of four to six feet during the summer. Moreover, the seeds will germinate in the most unlikely places and young saplings may frequently be seen growing on rubbish heaps and in the cracks and crannies of pavements.

The ailanthus is a medium-sized tree, forty to fifty feet high, with widely spreading, stout branches and coarse twigs, and a somewhat irregular and picturesque head. The bark is brown gray, the twigs ocher yellow or tan brown with a fine, velvety down. The leaves are alternate, pinnately compound, with thirteen to forty-one leaflets that are ovate-lanceolate and with one or two blunt teeth at the base. The flowers are small, yellowish green, in large terminal clusters, the staminate and pistillate flowers on separate trees, and the fruit is an oblong, twisted samara. The samaras are borne in clusters that are pale green or often magenta pink tinged, but later become brown when dry, and they remain in that condition on the branches well into the winter.

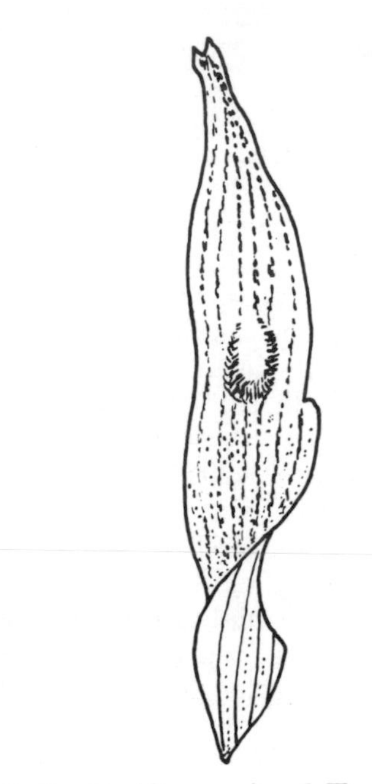

456. Fruit (Samara) of Tree-of-heaven

A young ailanthus tree may often be mistaken for a sumach because there is a superficial resemblance, but the differences between them are easily recognizable. Thus, the growing shoot and last year's wood of the sumach is velvety, whereas those of the ailanthus are smooth. Moreover, the margin of an ailanthus leaflet is entire except for the one or two blunt teeth at the base, while that of the sumach leaflet is toothed. Again, the lower surface of the ailanthus leaflet is pale green, that of the sumach leaflet is whitish. When autumn comes, the difference between the two is unmistakable, for the leaf of the ailanthus either turns a lemon color or falls unchanged, whereas that of the sumach becomes scarlet and orange.

The Quassia family, whereof the ailanthus is a member, is a group comprising thirty genera and about two hundred, chiefly tropical, and subtropical trees and shrubs, with alternate, pinnately compound leaves, mostly small, regular, dioecious or polygamous flowers, with three to five sepals and petals, as many or twice as many stamens, and one to five-celled ovaries, with a drupelike fruit.

The family is represented in the United States by only three genera, and each with a single species. The three species, the Mexican alvaradoa, the bitterbush, and the paradise tree, are all small trees that are found only in southern Florida. The paradise tree, as its name implies, is one of the most beautiful trees in tropical gardens. It has alternate, persistent, pinnately compound leaves, with ten to fourteen leathery leaflets, and staminate and pistillate flowers that are borne on separate trees. Its fruit, about the size of a wild plum, is deep purple or scarlet. Quassin, a bitter, tonic drug, is obtained from this tree, as well as other members of the genus, there being three others that are restricted to the tropical forests of the New World.

One of the best-known tropical species of the family is the quassia or bitterwood, of northern South America, from whose wood is obtained a water-soluble bitter principle used commercially in the manufacture of insecticides and in medicines.

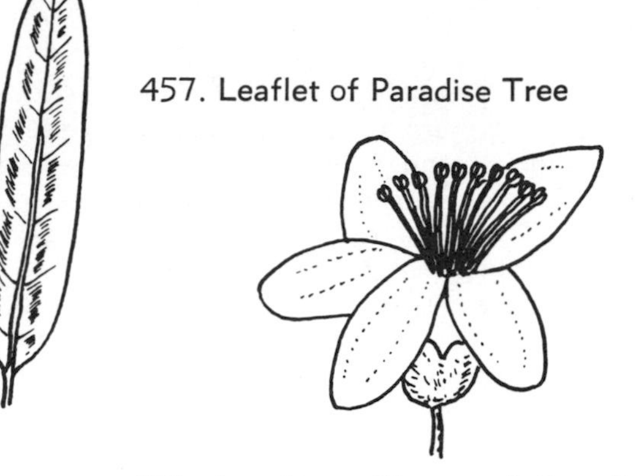

457. Leaflet of Paradise Tree

458. Staminate Flower of Paradise Tree

459. Pistillate Flower of Paradise Tree

460. Fruit of Paradise Tree

159. Rafflesia (Rafflesiaceae)

The rafflesia family is of interest because it contains the largest flowers in the world. It is a small group of seven genera and about twenty-five species of parasitic herbs found chiefly in the warm regions of the Old World. They grow on the roots or branches of other plants, and have leaves that are ordinarily scalelike. The flowers are unisexual with four to ten sepals, no petals, many stamens, and four to eight carpels containing numerous ovules. The fruit is a berry.

The flowers of the species *Rafflesia Arnoldi,* occurring in the Malay Archipelago, often measure three feet in diameter and weigh as much as ten to twenty pounds. They are the largest of all flowers.

The only representative of the family in the United States is a species that is a parasite on the twigs of certain shrubs of the pea family, and is found near the mouth of the Colorado River in California and Arizona. It measures only five to six millimeters, including the stem underneath the flower.

160. River Weed (Podostemaceae)

The riverweed family comprises twenty-one genera and about one hundred and seventy-five species of small, aquatic, freshwater, mostly annual, fleshy herbs found mostly in the tropics. Their leaves, as a rule, are poorly differentiated from the stem, and the whole plant commonly resembles the thallus of an alga or hepatic, in other words, looking much like a seaweed or moss. Their flowers are minute, perfect, without a perianth, and subtended by a spathelike involucre, with two to many stamens, and a two to three-celled ovary. Their fruit is a many-seeded, ribbed capsule.

The family is represented in North America by the riverweed of the genus *Podostemum,* Greek for stalked stamens, in allusion to the stamens being raised on a stalk by the side of the ovary. It is a dark green, rather stiff plant, firmly attached to stones in shallow streams. It is densely tufted, the leaves narrowly linear, commonly split above into almost threadlike segments or lobes, and the flowers are nearly sessile in the spathelike involucre. The capsule is oblong-oval and eight-ribbed. The plant resembles a seaweed and is tenaciously attached to loose stones by fleshy disks or processes, in place of roots.

161. Rockrose (Cistaceae)

In its own manner, the frostweed is as unique as any flower one will find. The plant thrives in dry fields and sandy places, and here the solitary flowers, with

461. Leaves and Flower of Frostweed

their showy yellow petals, open only for a day—and the day must be a bright and sunny one. The next day, the petals fall, having served their purpose. The stamens drop too, but the club-shaped pistil, having been dusted with sufficient pollen, remains to develop into a rounded, ovoid pod. Another flower succeeds the first one, and the second is succeeded by a third, and so the succession continues for weeks. Then, as summer begins to wane, smaller flowers appear that have no petals, and that are clustered at the bases of the leaves. The pods of these flowers are no larger than pinheads.

The plant ends its blooming before the first frosts of autumn. Why, then, is it called the frostweed? On some cold, November morning, examine the plant, and ice crystals that might easily be mistaken for bits of glistening quartz will be found about the base of the stem or in the cracked bark of the root, where the sap has oozed out and frozen solid. At times, the frozen sap assumes a feathery, whimsical form that stimulates the imagination.

The frostweed is an erect plant about a foot high, with lance-oblong, dull green leaves, hoary with fine hair on the lower surface. A related species is similar in habit, but more hoary-pubescent, while a third species has its flowers in corymblike clusters at the summit of the stem or branches. Several species, natives of southern Europe, are cultivated in the rock garden.

The frostweeds are members of a genus called *Helianthemum,* Greek for sun and flowers in allusion to their blooming only when the sun shines. They are usually prostrate or sprawling, woody plants or herbs, with usually opposite, entire leaves and flowers prevailing yellow, either solitary, or in few-flowered clusters, and sometimes are also called the sunroses. Only a few are cultivated in the garden, the most common species being known as the rockrose. It is a very pretty, low-growing, evergreen, forming broad clumps that during the flowering season are rather hidden by the mass of bloom. A Eurasian plant, it has long been in cultivation and comes in many varieties.

The rockrose family is a small one with only eight genera and one hundred and fifty species. They are low shrubs or woody herbs mostly from the north Temperate Zone, generally with ephemeral, but often showy, flowers. They usually have opposite leaves, but in the genus *Helianthemum,* the upper ones are alternate. They also have regular flowers with five sepals, the three inner often persistent, the two outer ones bractlike, smaller and often wanting, mostly five petals, many stamens, and a one-celled, or rarely incompletely five or ten-celled ovary with two to many ovules, and a style with a three-lobed or single stigma. Their fruit is a capsule.

A number of Mediterranean shrubs of the genus

Cistus, usually known as the rockroses, have been long known in Old World gardens, but are less known here possibly because they cannot endure northern winters. They are handsome garden plants with evergreen, or nearly so, leaves and large flowers that somewhat suggest a single rose. Unable to stand slushy, severe winters and requiring open sunlight, they are mostly grown in the South and in California.

On the sands of the seashore, in pine barrens, and on the beaches of rivers and lakes, there grows a bushy, little evergreen shrub, densely coated with soft, white, woolly hair, its leaves awl-shaped, and scalelike. Like the showy flowers of the frostweed, the small, bright yellow flowers of the beach heather open only in the sunshine, and then for a day only. But the little shrub, by growing in large colonies and continuing a succession of bloom, nevertheless, gives a charming color to the sand dunes that would otherwise appear a drab and monotonous feature of the seabeach. A somewhat similar species, the golden heather, may also be found on the seabeach as far south as Virginia.

Common to dry, sterile soil are the various species of pinweeds, slender, erect-branched, perennial herbs developing leafy shoots from the base, with the leaves of the basal shoots elliptical, oblong, or narrowly lanceolate to linear. The flowers are very small, greenish or purple, and occur in panicles. There are more than a dozen eastern species, most of them growing along the coastal plain, though there are a few that thrive as far west as the prairie states.

162. Rose (Rosaceae)

It was Shakespeare who said that "a rose by any other name would smell as sweet." But would it then be a rose? As Gertrude Stein would have it "a rose is a rose, a rose, a rose."

Everyone knows the rose, of course, undoubtedly the best loved of all our garden flowers whose recorded history dates back to pre-Babylonian times. No flower figures so prominently in literature, ancient and modern, sacred and profane, as the rose does. In Rome it was often placed over the door of a public or private banquet hall, and whoever passed beneath it bound himself not to reveal whatever was said or done within. Hence, the expression *sub rosa* that has prevailed to this day.

The rose family, whereto the rose itself has given its name, is an extensive group of trees, shrubs, and herbs of very diverse habit and highly esteemed for its most beautiful flowers and luscious fruit. Today, most botanists include in the rose family two groups that were formerly classified as separate families and known as the apple family, but sometimes called the pear or quince family, and the peach family, often called the plum, almond, or cherry family. The basis for once having separated the three groups, that is, the rose, apple, and peach, is found in certain differences that the ovary and fruits exhibit in the three groups. But now, all three are included in a heterogeneous family that embraces about one hundred genera and twenty-five hundred species of very wide distribution.

Typically the leaves in this large family are usually alternate, simple or compound, with stipules or small, leafy formations at the base of the leafstalk. The flowers are regular and perfect with characteristically five petals and five sepals that are inserted on the rim of the receptacle that may be of various shapes. The stamens are numerous, and are commonly borne in several cycles of five on the receptacle. There may be one or many pistils, with one, few, or many carpels and with two or more ovules in each carpel. The fruit may be a pome, drupe, follicle, achene, or an enlarged, hollow, fleshy hip often mistakenly called a berry.

There are two hundred species of the rose, though one authority has claimed that there are over four thousand, and thousands of varieties, all members of the genus *Rosa,* the old Latin name for the rose. The differences between the species are mainly technical. Roses are prickly shrubs or vines, the prickles being straight or hooked, with alternate compound leaves, the leaflets arranged feather-fashion, that is, pinnately, with the odd one always at the end. The flowers are solitary or in small clusters, typically with five petals in the wild, single species, but much doubled in most of the horticultural forms, and nearly always fragrant. Both the stamens and pistils are many in number, the latter enclosed at the base in a cup-shaped receptacle that enlarges in fruit, becoming fleshy and berrylike (the familiar rose hip), and enclosing the true fruits that are bony achenes. The hips of our native wild roses were often eaten by the Indians and those of the sweetbrier and other species have been used in making a jelly. Many birds favor the hips, and thus, aid in the dispersal of the seeds.

The prickles that the roses are provided with were likely designed by nature to prevent the foliage from being eaten by various animals, but they do not seem to be too effective as antelope, mule deer, and the white-tailed deer eat both the twigs and the foliage. Some species, to guard against crawling pilferers, have their calyxes covered with fine hairs or coated with a sticky secretion that appears to be more effective. The brightly colored petals of the flowers advertise, to various insects that aid in cross-fertilization, an abundance of pollen, but insects that come seeking nectar will be disappointed, as the roses do not secrete this sugary fluid. Some species of bees and a species of flower beetle seem to depend upon certain wild roses exclusively for pollen to feed themselves and their larvae. Occasionally, self-fertilization does occur, as during periods of cloudy or rainy weather when bees and other insects are not abroad.

It hardly seems necessary to say that all of our beautiful horticultural forms have been derived from the wild species of roses, mostly by rather complex methods of breeding and selection, although the parentage of some of our finest cultural varieties is somewhat in doubt. There are probably about thirty-five species of wild roses in the United States. Foremost of these is, perhaps, the prairie rose, a climbing or arching species and, incidentally, our only native climbing species, found in thickets and on the prairies. It is the parent stock of several valuable, double-flowering horticultural forms and is the only native rose that has been cultivated. Not without good reason either, for it is a beautiful species with its handsome foliage and full clusters of flowers, deep rose pink when they first expand, but later turning a paler color, and with autumn colors that paint the leaves a bewildering confusion of green and purple bronzes, heightened by pink and rose and dull red, with yellow and orange.

One of the most abundant of our wild roses is the swamp rose, sometimes only a foot high, at other times taller than a man. At flowering time it seems to take possession of the swamps and waste lowlands when its bright pink blossoms, two to two and a half inches in diameter, conspicuously advertise their wares to bees and other insects. Later, the scarlet hips glisten in the bright autumn sunlight.

Equally as abundant is the pasture rose, a low, slender-stemmed species that one will find blossoming in dry or rocky soil throughout the month of June. And those who know the wild flowers, know the meadow rose that one may recognize by its low habit, its unarmed stems, and its broad, dilated stipules. It grows in moist, somewhat rocky places, and blossoms in June, its flowers pink as are those of all the native roses. To be sure, white flowers do occur occasionally, but as an inconstant variation rather than as a specific character.

Several immigrants have escaped from gardens and become naturalized, such as the dog rose that is Shakespeare's cankerbloom, a lovely plant that spreads its long, straggling branches along the roadsides and banks, and covering waste places with its smooth, beautiful foliage, and in June and July with pink or white flowers; and the sweetbrier, the eglantine of Chaucer, Spenser, and Shakespeare.

> I know a bank whereon the wild thyme blows,
> Where oxlips and the nodding violet grows;
> Quite over-canopied with lush woodbine,
> With sweet musk roses and eglantine.
>
> Shakespeare

The sweetbrier blossomed in Pilgrim gardens long before the end of the seventeenth century. It is a multi-branched shrub with a pleasant aromatic foliage, but its long, prickly branches that sprawl, scratch and catch in clothing, makes it undesirable for lawn or garden, and for this reason has been banished to the roadsides, rocky pastures, and fields where grazing cattle give it a wide berth, apparently fearing the hooked prickles and not liking its fragrance. Sheep, too, seem to avoid it for "With brambles and bushes in pasture too full,/ Poore sheepe be in danger and loseth their wull."

Not a few members of the rose family have found wide appeal as ornamentals. The pearl bush, a spreading shrub five to ten feet high and an importation from China, is a very popular garden plant. Its white flowers in a showy, terminal cluster resemble cherry blossoms, but, whereas the cherry blossom shows a yellow center, that of the pearl bush is curiously green. It is at its best around the middle of May when the flowers come out, for there is then a delightful contrast between the tender green of the young leaves and the snowy white of the many blossoms that crowd upon the ends of every twig.

Equally as popular is the fire thorn or pyracantha, a low, thorny, evergreen shrub, three to eight feet high, very pretty in the spring with its many clusters of white flowers, but especially handsome in the fall when loaded with bright red fruit that persist on the branches all winter if not eaten by the birds who are very fond of them. The word pyracantha is from the Greek *pyr* for fire and *acanthus* for thorn, in allusion to the color of the fruit and to the thorns. It is a native of Europe and has been cultivated in the United States for a long time. It is used for hedges and for borders in parks and gardens, and for planting on rocky slopes, in sunny rockeries, as well as for wall coverings. It bears pruning well and is easily trained into any desired shape. Like many other immigrants, it has escaped, and has established itself in thickets and on roadsides throughout the South.

A third, very popular ornamental shrub, is the Japanese quince or Japanese flowering quince, one of the first to bloom. It is a rather low shrub with somewhat spiny branches, scarlet red flowers, and yellowish green fruit, sometimes used for preserving.

For sheer beauty, few of our ornamental shrubs can surpass the flowering almond. It is a dwarf bush, rarely more than four feet high, and in early spring the slender branches, before the leaves appear, burst into bloom with rosy red flowers that appear so profuse that they transform the branches into flowery sceptres. Unfortunately, its surpassing loveliness lasts for one short week and then fades quickly from memory, until another spring calls forth its fleeting beauty.

Every homeowner who has a few ornamental shrubs about his house is sure to have a spirea or two, for the spireas, whereof there are eighty known species and many hybrids, all of the genus *Spirea,* from the Greek for wreath or garland, are among the most widely used shrubs for the garden. One look at any of them and the reason becomes fairly obvious. For they are of unsurpassing beauty, with a grace and charm distinctively their own. What is equally important, they thrive in a variety of soils and under all exposures, requiring a minimum of care. Probably the best known is the bridal wreath, long a garden favorite in the United States. It is a slender shrub, four to six feet high, with arching, somewhat angled branches, leaves pointed at both ends, and pure white flowers in nearly stalkless clusters. In the early autumn, the leaves turn a scarlet and orange, with a tone of brown that gives a depth and richness to the coloring. Another popular garden spirea is Thunberg's spirea, a handsome twiggy shrub with pure white flowers and narrowly shaped leaves that turn scarlet in the fall. The most cultivated spirea, however, is Van Houtte's spirea, primarily because it can withstand smoke and city conditions better than most others. It, too, is a slender shrub with beautifully arching branches and pure white flowers in multi-flowered clusters.

There are two wild species of spirea that are familiar to anyone acquainted with our wildings; the meadowsweet, and the steeplebush or hardhack. The meadowsweet is a small, upright shrub, with light green leaves, and a beautiful, pyramidal flower spike crowded with flesh pink and white blossoms with prominent pink red stamens. They resemble apple blossoms. Small bees, flowerflies, and beetles are among the many visitors that come in great numbers seeking the accessible pollen and nectar, the latter secreted in a conspicuous, orange-colored disk. The fruit, like that of all the spireas, is a follicle or pod wherein are contained the seeds.

The meadowsweet elects to grow along the roadsides, in rocky pastures, and in waste places, but its relative the steeplebush prefers wet places, roadside ditches, swamps and the like. The steeplebush is much like the meadowsweet in general habit, but may be distinguished from it by its terra-cotta red, woolly stem, and leaves, the latter olive green of a dark tone above and very whitish and woolly beneath, and its slender steeplelike flower spike crowded with tiny, deep rosy pink flowers. The steeplebush would be a handsome plant were it not that the succession of bloom is slow and downward so that the top of the spike is usually in a half-withered condition, detracting from its appearance.

The spike of the steeplebush attracts our attention no less than the countless bees, flies, and beetles that seek the abundant pollen, either for themselves or their progeny, and thus, become agents of cross-fertilization. However, most spireas are able to fertilize themselves should the insects fail them. Anyone interested in insects and in collecting them will find both of these plants, that is, the meadowsweet and steeplebush, good hunting grounds.

Why are the lower surfaces of the leaves so woolly? Doubtless, as a protective absorbent to prevent the pores (stomata) in the leaves from becoming clogged with the vapors that must rise from the damp ground wherein the plant grows. Every plant, like every animal, must breathe, that is take in oxygen and give off carbon dioxide waste (a process known as respiration), and must also be able to get rid of any excess moisture taken in by the roots (a process called transpiration.) All this is effected by the pores in the leaves and stems, and were they to malfunction, as by becoming clogged with dust, dirt and the like, the plant would suffer, and in extreme cases, die.

The ninebark is a fairly common shrub found on rocky banks and riversides, with ascending, light brown stems, bright tan yellow, slender twigs, dark green leaves, and white flowers in terminal, nearly spherical clusters. The fruit is an inflated, bladderlike pod and herein lies the plant's charm, for when the fruit is plentifully set at the ends of the long branches and have taken on a rich purplish or reddish hue, the ninebark becomes undeniably decorative. Partly for this reason and partly because it is easy to grow, it is often found in the garden.

While the trees are still leafless and cool breezes still blow over the landscape, though the sun may be shining brightly, the silvery white chandeliers of the shadbush give notice that spring has come. Named shadbush by the early settlers of the eastern states because it blossoms at the time the shad begin to ascend the tidal rivers, it is also known as the juneberry because its fruit ripens during the month of June, rather early for fruits to mature. The shadbush may be a shrub or small tree, with gray bark varie-

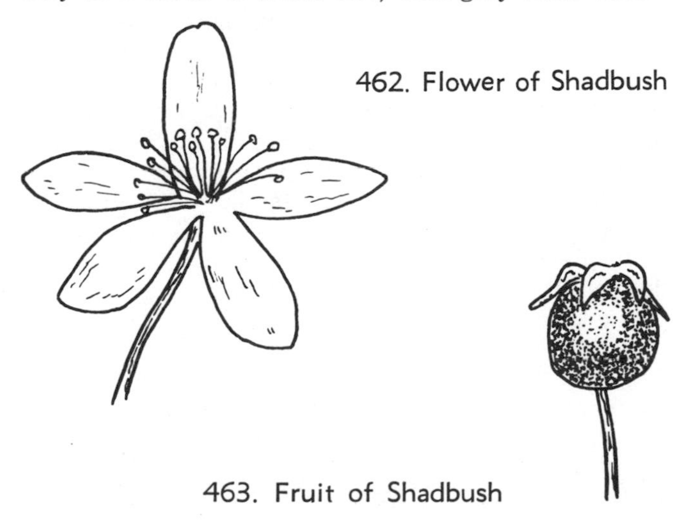

462. Flower of Shadbush

463. Fruit of Shadbush

gated with sepia brown striping, on old trees narrowly furrowed into flat, scaly ridges, on young trees greenish gray and quite smooth, and with deep green leaves that turn a rusty red in the fall. The flowers are white and hang in long, loose clusters, and on hillside and open woodland, mining bees gather and buzz their gratitude for early favors. Later, the flowers give way to a fruit that is like a miniature apple, but bony inside and that varies in color from crimson through magenta, to dark purple.

There are a number of species of shadbushes or juneberries—the low juneberry, the round-leaved shadbush, the long-rooted shadbush—all separated on technical differences. In the popular mind, however, all shadbushes are alike, and to various birds, also, the birds having preempted their fruit for as long as man can remember. Sometimes one can succeed in getting to the berries before the birds do. The berries are of a sweet and pleasant odor and may be eaten directly or made into jellies for winter use. They are also excellent for pies.

Aside from being bitten by a deerfly or getting a rash of poison ivy, there is nothing more annoying than getting caught in a thicket of brambles. There is no need for it, of course, but sometimes one inadvertently becomes ensnared in the tangles. The word bramble is applied to a prickly shrub, usually known as the raspberry or blackberry. It is a compromise between a perennial herb and a shrub. The stems are indeed woody, but instead of living on from year to year and bearing an indefinite number of fruits, as the currant or gooseberry, they live a year or so and then die after maturing their fruit. The roots, of course, live on indefinitely. The young stems quickly grow until they have reached their normal size and then stop growing. These stems, known as canes, are quite evident in any healthy black raspberry or blackberry bush, but not quite so apparent in the red raspberry.

The genus *Rubus,* the old Latin name of the brambles, whereof the raspberry and blackberry belong, is a large one, comprising five hundred or more species. Most of them are from the north Temperate Zone, but a few are found on tropical mountains and some even in the Arctic Circle. They are erect or trailing plants many of them with biennial canes, in other words, leafy the first year, but flowering and fruiting the second year, and then dying. All have the characteristics of the rose family, with alternate leaves that are simple, but more usually compound with pinnate leaflets, and with prevailingly white flowers, though occasionally purplish pink, of five petals and numerous stamens. The fruit is a collection of small, sometimes dry, drupelets, the "berry," that is, the part that is eaten, being of two kinds. In the blackberry and dewberry the mass of fleshy drupelets, usually called seeds, remain attached to the fleshy receptacle that is part of the "berry" and eaten with it. In the raspberry, however, the mass of drupelets or what is incorrectly called the berry, separates from the receptacle when picked, and the fruit is consequently hollow. To put it differently, it is the receptacle in the blackberries that is eaten, but in the raspberry it is left on the stem. It all sounds rather complicated. The loganberry, of hybrid origin, has fruits that are generally regarded as being of the blackberry type, although one of its parents is the raspberry.

There are both cultivated and wild raspberries, and blackberries. One of the more common of the wild species of raspberries is the purple-flowering raspberry whose name is a misnomer, as the rose family is not capable of producing a true purple flower. The flower instead, is at first, a deep crimson pink, eventually fading to a magenta pink.

The purple-flowering raspberry is a showy, shrubby plant that is found growing along the roadside, in copses, and in stony woodlands, with stems covered with short red or brown bristly hairs, large, maple-like leaves of three to five lobes, and roselike flowers nearly a foot across that are visited by bumblebees, among other insects. The fruit is insipid and looks like a flat, red raspberry.

One of the best of our wild fruits is undoubtedly that of the wild red raspberry, whose fruit is light red, juicy and most pleasing to the taste. It is a low, erect shrub with white, roselike flowers, and occurs on rocky hillsides and along fencerows. This is a native species and the progenitor of all our cultivated varieties. Growing in much the same situations as the red raspberry, the black raspberry is also the source of many cultivated varieties. It is an erect, prickly plant with very bluish or even bluish purple stems that ultimately arch over and root at the top, thus, making an impenetrable thicket. The flowers are whitish, small, in dense clusters and appear in May as do the flowers of the other raspberries. They are sparingly visited by insects, but self-fertilized as they usually produce an abundance of purplish black "berries" that are hollow, like a thimble. For this reason, the plant is also known as the thimbleberry.

Everyone knows the blackberry or common brier because of its tendency to take over any neglected place and appropriate it for its own. Of wild, luxuriant growth, beautiful in leaf and flower, glorious in its rich, autumnal coloring, and abundant in fruit, the common brier flourishes equally on seashore, mountainside, woodland border, roadside, and in grassy fields.

The tangled blackberry crossed and recrossed, weaves
A prickly network of ensanguined leaves.

Lowell

For those who do not know the common brier, it is a straggling, prickly bush with stems both erect and recurved, alternate, compound leaves of three to five leaflets that in autumn, turn a rich, vinous red, varying to bronze, purple, or fading to orange, and white, showy flowers in terminal, leafless panicles. The flowers of the blackberry are better adapted to their insect visitors than those of the raspberries, as they have much larger petals that are spread out flat to attract the insects, as well as to provide space for the stamens to spread away from the stigmas, an arrangement that gives freer access to the nectar secreted in a fleshy ring at the base. The fruit is an aggregate fruit of many small, black, shining drupes borne on a long, white axis that is the elongation of the receptacle. It is sweet and aromatic and when at its best, few wild fruits have greater appeal. As Whittier wrote: "For my taste the blackberry cone/ Purpled over hedge and stone." From this wild blackberry has come the cultivated blackberry.

There are other blackberries such as the leaf-bracted blackberry, the thornless blackberry, the mountain blackberry, and the low running blackberry, more commonly known as the dewberry. The latter is a strong-growing, prickly plant that trails its woody stem by the dusty roadside, in dry fields, and on sterile, rocky hillsides. Walt Whitman saw beauty in the vine when he wrote: "The running blackberry would adorn the parlors of heaven." He saw more than beauty in the plant, however, for he also comments on the fruit that are shining black, at first sour, but becoming sweet when fully ripened. Unlike the dewberry, the swamp blackberry is insignificant as to fruit that is sour and worthless, but it is not without its redeeming features for few trailing plants combine a better effect of flower and foliage. Its greatest charm, however, is in winter when its still persistent leaves, in rich autumnal reds, glow among the dry deadweeds and grasses of low woods and swampy meadows.

The annual pageant that nature puts on in the fall is a magnificent panorama of color, but an apple orchard at the time of flowering, especially a commercial orchard that extends as far as the eye can see, is nothing less than a breath-taking spectacle. In the apple-growing regions, such as the apple belt that extends from Massachusetts westward through Michigan, orchards attract thousands of visitors each year to view the sea of blossoms. It is one of nature's finest displays.

A streamlined, well-tended orchard, such as one where apples are grown for the market, is a scene of symmetry and form, and one that appeals to our sense of the practical, but an old-fashioned orchard of venerable trees with their great twisted and gnarled branches, lifting aloft their round heads and casting irregular shadows on the ground, is more picturesque, with an indefinable charm of its own. It has, too, a touch of the wild and untamed, and when the trees are in flower they sketch on nature's canvas a picture of unrivalled beauty. Even when naked and outlined against the winter sky, they delight the eye with their quaint ruggedness.

The ancestral parent of the apple tree in the orchard is the wild apple of southern Europe and Asia. How or when this species first came under cultivation is lost in antiquity, though it is known that the fruit was eaten by the prehistoric Swiss lake dwellers. For more than three thousand years the apple has been cultivated and improved and many varieties developed, more than a thousand in fact, including the Baldwin, the McIntosh, the Winesap, the Cortland, the Delicious to name a few. From myth, folklore, and the written record, one can learn how closely the apple has been interwoven with the fabric of civilization. From the classics, both the ancient and the modern, one can learn how extensive a role the apple has played in history, and the effect it has had upon human destiny. What course would history have taken had not Adam, and Helen of Troy bitten into the fruit? Perhaps we might still wonder what holds us on the planet earth had it not fallen on Newton. And what would have been the legend of William Tell. The apple is the fruit of fruits.

464. Leaf of Apple Tree

The apple, like many other introduced species, has escaped from cultivation and has established itself in many localities. It is especially common among watercourses. It is a low tree of twenty to forty feet high with wide-spreading, heavy limbs and gnarled branches beset with thornlike growths. The bark is gray brown, rough, and irregularly scaly, the leaves rigid, leathery, dark olive green above, paler beneath, the flowers white, pink-striped with five petals and many stamens, thus showing its kinship to the rose, and the fruit is a pome, commonly green, green yellow, or red. Both the apple and the crabapple belong to the genus *Malus*, the ancient Latin name for the apple.

Thoreau found the wild apple much to his liking for in his essay on "Wild Apples" he writes:

> To appreciate the wild and sharp flavors of these October fruits, it is necessary that you be breathing the sharp October or November air. The outdoor air and exercise which the walker gets, give a different tone to his palate, and he craves a fruit which the sedentary would call harsh and crabbed. They must be eaten in the fields, when your system

is all aglow with exercise, when the frosty weather nips your fingers, the wind rattles the bare boughs or rustles the few remaining leaves, and the jay is heard screaming around. What is sour in the house, a bracing walk makes sweet. Some of the apples might be labeled, "To be eaten in the wind."

That the European or Asiatic forebear wherefrom our cultivated apple was derived was any less harsh in taste or any larger in size than our own native crabapple seems most unlikely. That leads us to believe that, should all our cultivated apples by any quirk of nature be completely swept out of existence, it would be possible to regain the cultured varieties by the cultivation of the crabapple.

Peter Kalm, a Swedish botanist who visited our country two hundred years ago wrote, in his *Travels in North America,*

> that crab-trees are a species of wild apple-trees, which grow in the woods and glades, but especially on little hillocks, near rivers. In New Jersey the tree is rather scarce; but in Pennsylvania it is plentiful. Some people had planted a single tree of this kind near their houses on account of the fine smells which its flowers afford. It had begun to open some of its flowers about a day or two ago; however, most of them were not yet open. They are exactly like the blossoms of the common apple-trees except that the color is a little more reddish in the Crab-trees; though some kinds of the cultivated trees have flowers which are very near as red; but the smell distinguishes them plainly; for the wild trées have a very pleasant smell, somewhat like the raspberry.
>
> The apples or crabs, are small, sour and unfit for anything but to make vinegar of. They lie under the trees all winter and acquire a yellow color. They seldom begin to rot before spring comes on.

There is little that can be added to Kalm's description of the crabapple. Probably no plant satisfies our sense of the aesthetic more than the wild crabapple in blossom, for then, the tree is invested in a rose-colored bloom and fills the air with a spicy, stimulating fragrance. Linnaeus probably never saw more than a dried specimen, but he named the tree most worthily *coronaria* "fit for crowns and garlands."

Our native American crabapple is a small, round-topped tree somewhat similar to the common apple with light brown gray bark, dark green leaves, white and pink flowers, and fruit that is somewhat spherical, but flattened at the ends and yellow green. They are of more value than for merely making vinegar, as Kalm says, for when made into a jelly the jelly has a wild tang, an indescribable piquancy of flavor that a jelly made from the common apple cannot match.

The peach orchard at the height of blossoming is not as flamboyant as an apple orchard and is less spectacular, but it has a subtle quality that makes it no less appealing to our sense of the aesthetic. One will likely catch a first glimpse of it as a rosy mist on the distant landscape as one drives along the highway, but as one gets closer, it will deepen in brilliance and leave one with a mental image that one will always be able to recall, even though it may be years later.

Like the apple, the peach is a fruit of antiquity. It is mentioned in a poem written about 1800 B.C., leading us to suspect that it must have been cultivated in Asia much earlier. It first appears to have been introduced into Europe about 500 B.C. Later it was brought to Mexico by the Spaniards and eagerly adopted by the Indians. From then it was only a matter of time before it appeared in our country.

The peach is a small tree and one may well wonder how it can bear sufficient fruit to make it profitable to an orchardist. A peach tree bears an abundance of fruit, however, and is often so loaded that it seems as if it cannot stand up to the load it carries. Branches sometimes break under their weight, but not often, for crowded peaches lead to stunted growth and professional orchardists will see that this does not often happen, and will thin it judiciously.

465. Leaf of Peach Tree

Like the apple, there are many cultivated varieties, and like the apple, the peach is also an escape and may often be found growing on bottomlands and in thickets. It is a small tree, with rough, scaly, gray brown bark and madder-purple twigs, shiny, bright green leaves, the flowers light crimson pink. The fruit is a drupe, velvety-skinned with a juicy pulp, and hardly needs description.

With few exceptions, the cultivated varieties of pears grown in America are of European origin, having been derived from the species known as the European pear. As in the case of both the apple and peach, the pear occasionally runs wild in thickets and open

466. Leaf of Pear Tree

woods in the eastern states. It is a broad-headed tree, sometimes reaching a height of forty-five feet, with leaves that are rather roundish, olive green above, paler below, and with white flowers that appear with the leaves. The fruit is yellow green, and a fleshy pome, obovoid in shape, in other words, what is known as pear-shaped.

With many people the cherry is as popular a fruit as any other, and like the apple, peach, and pear, is of very ancient culture. There are many horticultural varieties and a few native species. The largest is the wild black cherry, reaching a height of forty to sixty feet, or as much as ninety feet in the southern parts of its range. It is a tree with a stout, sturdy trunk, spreading branches, and a round head, and is common in rich woods and along the roadsides. The bark on young trees is smooth, dark ruddy brown with horizontal markings, the outer layer thin, papery, and translucent, the inner bright green and aromatic, on old trees blackish brown and rough and scaly with reflexed edges. The newest twigs are a golden ocher, the older are light or shiny, red brown. The leaves are long, sharp-pointed, dark lustrous green, paler beneath, and with fine, rounded teeth. The flowers, in drooping clusters, are white and rather fragrant, though sometimes unpleasantly so, and appear after the leaves in May. The fruit is dark red when immature, lustrous black when ripe, about the size of a pea, in clusters, and much relished by the birds. At one time, it was often used for its tonic properties when added to rum or brandy, making what was known as "cherry bounce." The bark, too, has been used as a medicine and as a flavoring extract. It is in its wood, however, that the chief value of this cherry lies, for it is beautiful enough when polished

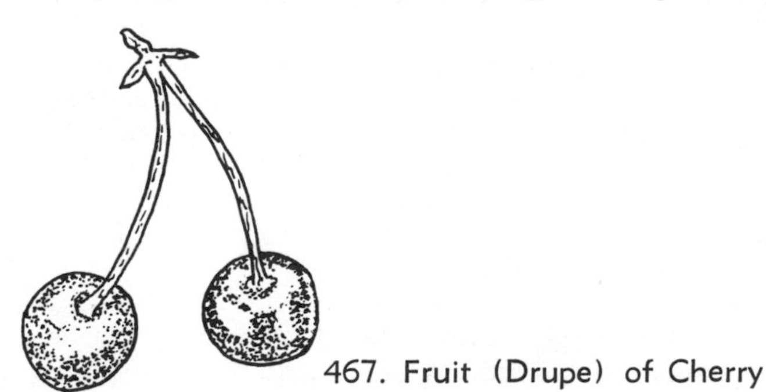

467. Fruit (Drupe) of Cherry

to compete with mahogany and rosewood. Its rich, lustrous brown deepens and softens with age, and for this reason, has been used in making fine furniture and the interior finish of houses. More prosaically, it is also used for tool handles and the like.

Unlike the wild black cherry whose range extends throughout the eastern half of the United States, the wild red cherry is more northern and goes south only along the mountaintops. It is a slender tree with regular, horizontal branches and a round head. The bark is smooth, shining reddish brown on the branches, broken into thin, curling horizontal plates on the trunk. The twigs are red. The leaves are slender, pointed, finely toothed with wavy edges, and are a bright lustrous green, turning yellow in the fall. The flowers are small and white, in long-stemmed, lateral clusters appearing before or with the leaves in May, and are laden with nectar. Then comes the harvest, when the tree, in rocky woodlands and on rocky hillsides, along fencerows and by stone walls, stands gemmed on all its branches with tiny, translucent ruby cherries. They are small and too sour for our taste, but the birds are especially fond of them and pick every one. Not without good reason is the tree also known as the bird cherry.

Unlike its relatives, the chokecherry is a miniature tree, rarely growing much higher than a lilac bush. It is very common and will be found growing besides roads, in rich woods and copses, and along riverbanks. Sometimes it is mistaken for a young, wild black cherry sapling, but, whereas the leaves and bark of the latter are aromatic, the chokecherry exhales an odor that is rank and disagreeable. Then, too, its leaves are twice as broad as those of its relative, and its cherries are harsh and bitter and very astringent so that anyone who is apt to taste them will not do so a second time. They pucker the mouth and affect the throat so that one is likely to choke on them. The birds like them, however, and will strip a tree in no time. That they are effective agents of seed dispersal, goes without saying.

All of our cherries, as well as the plums, are subject to a fungus disease known as black knot. The disease manifests itself by black excrescences or swellings on the branches. The chokecherry is especially subject to attack, and one will rarely find a tree without the knotlike growths.

Mention of the plums calls to mind that there are wild plums, as well as the cultivated varieties. One of these is the native wild plum, that is sometimes a shrub, but more often, a small, graceful, little tree with numerous branches that are somewhat thorny and with a rather thick, rough bark. The ovate or oblong leaves have long, tapering points, are rounded at the base, and are sharply, often doubly saw-toothed. In April or early May, the tree is covered with masses of white flowers, and bees hang over the nectar-laden blossoms as if intoxicated. The flowers are followed by round, red, sometimes yellow fruits that are pulpy, with a rather tough skin, and are pleasant to the taste. George Catlin, the artist, in writing about his travels among the Indians in the region of Oklahoma about 1837, says that

> the next hour we would be trailing through broad and verdant valleys of green prairies, into which we had descended; and oftentimes find our progress completely arrested by hundreds of acres of small plum trees of four to six feet in height; so closely interwoven and interlocked together, as entirely to dispute our progress, sending us several miles around; when every bush that was in sight was so loaded with the weight of its delicious wild fruit, that they were in many instances literally without leaves on their branches and bent quite close to the ground.

The wild plum is excellent for preserves and jellies and is frequently in great demand. Many cultivated varieties have been developed from it.

Some excellent cultivated varieties have also been developed from the Canada or red plum, a species very similar to the wild plum, found growing along the road and streams, in thickets, and fencerows. It is a small tree, in general form distorted like that of the apple tree, with branches armed with stiff spines. Its deep red fruits, an inch in diameter, are very sour and generally used in making preserves and jellies, though sometimes they are stewed and even eaten raw.

A third native plum whose fruit is unexcelled for jellies and preserves is the Chickasaw plum. It is a small tree, southern in its distribution. The small, round fruit is soft and sweet and more like a cherry than a plum, and is sold in the markets of the South.

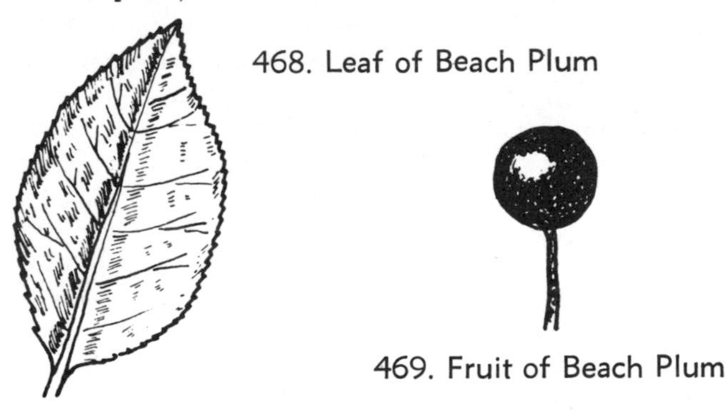

468. Leaf of Beach Plum

469. Fruit of Beach Plum

The fruit of the beach plum is also gathered and sold in the markets, being excellent for jellies. The beach plum is a low shrub with a straggling habit and is found on the sea beaches and sand dunes from New Brunswick to Virginia. The brown stems are crooked, the leaves olive-green above, paler and finely haired beneath, and the flowers are white in profuse clusters. They are followed by equally profuse clusters of handsome, globular, purple or scarlet fruits that become ripe in August and September. The skin is rather thick and tough but the pulp is sweet and juicy. They are good eating.

Another straggling shrub, rather limited in its range, occurring in copses and sandy barrens from Connecticut to the mountains of Pennsylvania, is the American sloe or Porter's plum. The ripe fruits are dark purple with a bloom and slightly acid. The people that live where the sloe is found use them for jams, jellies, and pies. Incidentally, all the cherries and plums belong to the genus *Prunus,* the classical Latin name for the plum tree.

At this point several other members of the rose family, namely, the sand cherry and the three species of chokeberry, should also be mentioned. The sand cherry is a prostrate spreading shrub, sometimes lifting its branches six inches above the ground, though sometimes three or four feet, with rough, brown gray, scraggy stems and bright red brown new twigs. The leaves are blunt, lance-shaped, bright olive green above, and paler beneath. The flowers, appearing in May with the leaves, are white, in lateral, few-flowered umbels, and the fruit is a dark red or dark purple drupe, nearly black, and somewhat insipid. One will find the sand cherry rather common on sandy gravelly shores along the sea coast from New Brunswick to New Jersey and along the shores of the Great Lakes.

The chokeberries are wild plants growing in various situations and are most attractive because of their color, hence, they are often grown in informal shrubberies. Three species are now recorded—the red, black, and purple—and their specific differences lie in the fruit rather than in the leaves or flowers. The red chokeberry is the species most abundant in the south, where it may be found growing in moist situations and in the woods. The other two are northern in range. It is an erect shrub, two to eight feet high, with slender branching stems and grayish brown bark, and with clustered, slender, brown stems and twigs. The leaves are elliptical, finely toothed, deep olive green above, smooth and densely woolly beneath. The flowers are white or crimson magenta tinged in mostly terminal clusters, and the fruit is a small berrylike pome, a bright garnet red and astringent.

The black chokeberry closely resembles the preceding in habit, but is somewhat smaller. The leaves and flowers are also somewhat similar, but the fruit is a very dark purple or quite black, and not quite so astringent. The purple chokeberry is the tallest of the three, sometimes reaching a height of twelve feet. The leaves and flowers are much like those of the other two species, but the fruit is a rather deep purple. Both the black and purple chokeberries are found in much the same places as the red species, that is, in wet situations, though the black chokeberry is often found growing on rocky uplands. Although the fruits of the chokeberries are widely available and persist throughout much of the winter, they do not seem to have much of an appeal to birds and other wild life that continually pass them by.

For sheer beauty, few trees can surpass the mountain ash with its handsome foliage and showy clusters of flowers and fruits. It is a small tree, fifteen to twenty feet in height, the branches slender and ascending, the bark a light or brown gray, smooth on young trees, rough on older ones, the leaves compound with bright green leaflets, the flowers white in broad, flat-topped clusters, and the fruit bright, shining coral red or deep scarlet, about the size of a pea on slender, spreading dark red stems.

In her tree book Julia Rogers writes that the way to see the American mountain ash at its best is

> to take a leisurely October drive through the wooded uplands of New England or lower Canada. Along the borders of swamps, or climbing the rocky bluffs, with the wild plums and the straggling beeches, this frail, scarlet-berried ash leaps up like a yellow flame, and the broad discs of its fruit gleam among the leaves like red embers in a grate. There is no handsomer leaf at any season than this one, on its red stem, its pointed leaflets dainty and slim as a willow's.

The mountain ash that is planted as an ornamental on many lawns is not the American species, but the European mountain ash, that seems to do better under cultivation than does its wild cousin of the hills. In Europe the mountain ash is known as the rowan tree. Strange legends and superstitions, centuries old, have become associated with it and at one time, any part of the tree, a leafy spray, a cluster of berries, a piece of its wood, was believed to be an effective charm against evil spirits. The tree was planted at the gates of churchyards and by cottage doors, and leafy twigs were hung over the thresholds. As an ancient song has it:

> Their spells were vain; the hags returned
> To the queen in sorrowful mood,
> Crying that witches have no power
> Where is roan-tree wood.

Before passing on to the herbaceous members of the rose family, it might be well to pause for a moment and to consider a group of shrubs and trees variously known as hawthorns, thorns, or thornapples. It is an extensive group of one thousand species found in the north Temperate Zone, but most common in eastern North America.

The hawthorns are mostly flat-topped trees, branching rather irregularly, the shrubby form with generally ascending, the tree form with spreading, branches. The leaves are alternate and always toothed or lobed. The flowers are white, though red or pink in some horticultural forms, nearly always in small clusters and have an unpleasant odor. They bloom in May and are followed by a fruit that resembles a miniature apple and that ripens in October. The fruit is commonly deep red or scarlet red, but never pure scarlet. Sometimes it is suffused with orange becoming then a dull orange scarlet. The hard seeds are surrounded by a dry or pulpy flesh that is edible in some species, but in most cases, not very palatable. Birds and other wild animals do not seem to think so either, for they eat them rather sparingly.

The hawthorns belong to a group or genus known as the *Crataegus,* an old Greek name for these plants and that means strength in reference to their hard wood. The following are some of the more common species: the cockspur thorn, a large shrub or small tree with numerous thorns whose leaves turn a bright orange and scarlet in the fall and whose red fruit hangs on the branches all winter, generally found along the margins of swamps or near streams; the white thorn, a low tree with crooked-spreading branches that grows in thickets, in upland woods and rocky pastures, or near the borders of streams; the scarlet haw, a small tree with a straight trunk, spreading and contorted branches, partial to the margins of swamps and the banks of streams, perhaps the handsomest of the hawthorns, and with a fruit of rather pleasant flavor and the most edible of all; the black thorn or pear thorn, a shrub or small tree that prefers alluvial soil and with a fruit that is pear-shaped; and the dotted haw, a tree of mountain regions with a flat-domed crown and low, wide, spreading, thorny gray branches, small flowers with pink or white anthers, and a large fruit that is orange yellow or red and dotted.

The showy flowers and fruits, the vivid color of the autumn foliage, and their striking character as revealed in winter by the rigid branches and menacing thorns, make the hawthorns attractive at all seasons and many of them are excellent for lawn planting.

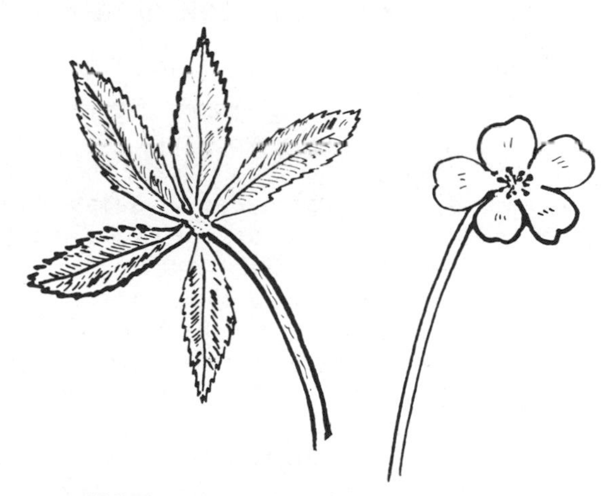

470. Leaf and Flower of Cinquefoil

> The fair maid who, the first of May,
> Goes to the fields at break of day
> And washes in dew from the hawthorn tree
> Will ever after handsome be.

If one should ramble about the countryside in April, one will see the common cinquefoil decorating meadow and pasture, or weaving its embroidery over the stony and barren roadside, its small yellow flowers peeping among the leaves. One might easily mistake it for a yellow-flowered strawberry as many have done and who have called it a wild strawberry.

One needs only to glance at the cinquefoil flower to observe its resemblance to a rose. Indeed, looking at it through a magnifying glass, it could easily be taken for a yellow rose. Compare it with any rose, wild or cultivated, and one will see they have much in common, five sepals, five petals, and many stamens in multiples of five.

There are three hundred species of cinquefoils found everywhere in the north temperate and subarctic regions. They are all perennial herbs or small shrubs with creeping or erect stems, compound leaves with three or many leaflets, hairy flowers in small loose clusters, yellow, red, or white, and with a fruit that is a dry, one-seeded achene. The name of the group or genus of cinquefoils is *Potentilla,* the diminutive of the Latin *potens* that means powerful. The cinquefoils were once believed to have strong medicinal powers. This was in the Middle Ages, when people were more credulous than they are now.

The common cinquefoil is also known as five-finger because its compound leaves have five leaflets. The leaves of other cinquefoils, however, also have five leaflets, as the marsh five-finger, known as the purple cinquefoil because its flowers are of this color. It is the only purple-flowered cinquefoil, incidentally, and so, is easily distinguished from the others. The stem is stout, somewhat reddish, and a little woody at the base, the leaflets are blunt-tipped and sharp-toothed, and the flowers are magenta purple within, and somewhat greenish without. It grows in swamps and cold bogs.

In dry sterile fields, one should find the silvery cinquefoil, a small species remarkable for its silvery aspect. The stem is covered with silky, white wool and so are the leaves, or more specifically the leaflets, on the lower surface, presumably an adaptation to prevent loss of precious moisture. As in the common cinquefoil, the flowers are yellow, rather small, and loosely clustered at the ends of the branches.

All three of the species that have been mentioned, are herbaceous plants or herbs, but the shrubby cinquefoil is a woody perennial or shrub. It has nearly erect stems, tan brown in color, with a bark that has a tendency to peel off in shreds. The leaves are olive green with five to seven leaflets that are unlike those of other species, in that they are toothless. The deep yellow flowers with rounded petals are generally an inch broad, either solitary or in cymes at the tips of the branches, and on a bright sunny day, they attract many visitors out after pollen and nectar. In their greed the beetles often devour the anthers. Swamps and wet places are the shrubby cinquefoil's habitats.

Anyone who has never tasted a wild strawberry, has missed a most delightful taste experience. It has a delicate flavor all its own that the cultivated varieties are unable to match. As someone once said: "I had rather have one pint of wild strawberries than a gallon of tame ones."

Our common wild and native strawberry, genus *Fragaria,* Latin for fragrance in allusion to the plea-

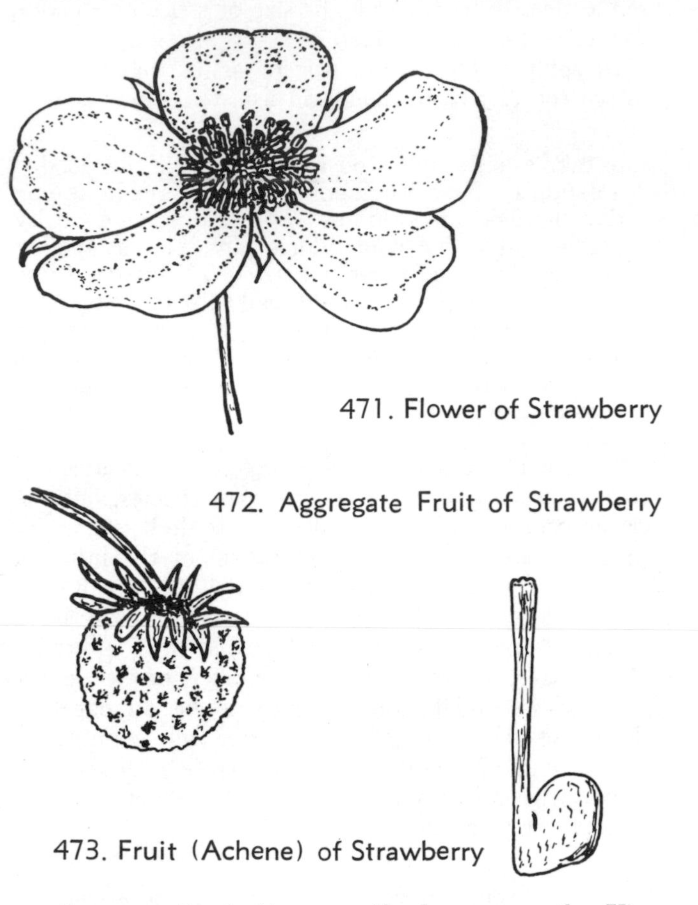

471. Flower of Strawberry

472. Aggregate Fruit of Strawberry

473. Fruit (Achene) of Strawberry

santly aromatic fruit, generally known as the Virginia strawberry, is at home in fields and pastures. The compound leaves of three leaflets, with saw-toothed edges, are on long stems that are covered with soft hairs and that come from the roots. The flowers, that appear in April, are white, five-petaled, with many orange yellow stamens, and are borne on stems shorter than those of the leaves. They are followed by the scarlet fruit that ripens in June or July. It is the much-enlarged, juicy, very fleshy receptacle wherein or whereon are embedded the true fruits that are commonly small achenes, but incorrectly called seeds. The "berry" is fragrant and undoubtedly the most delicious of all our wild fruits.

How did the strawberry get its name? Some say that in early times it was named from the straw that was laid between the rows to keep the fruit clean. Others claim that in earliest Anglo-Saxon it was called *streowberie,* and later *straberry,* from its peculiar straying suckers that lay as if strewn on the ground.

There is another native strawberry known as the American wood strawberry. It occurs in rocky woodlands and pastures and is a slender species with thin leaflets that are more ovate and less wedge-shaped than in our more common species. The scarlet fruit, too, is more conical and the seeds are borne, not in pits as in the Virginia strawberry, but upon the shining, smooth surface.

A delicate woodland plant of the rose family is the creeping dalibarda with a white blossom like that of the wild strawberry and with densely woolly, or finely haired stems and leaves that are dark green, heart-shaped, and wavy or scallop-toothed, in form closely resembling those of the common blue violet. The one to two white flowers, about one-half of an inch in diameter and that appear in June, are borne on long, fuzzy, sometimes ruddy, stems. It is said that they are fertilized in the bud before opening, but that appears to be rather questionable. Like the violets, the dalibarda has one or two cleistogamous flowers near the base that never open, and that need no insects to fertilize them.

One of the more common herbaceous members of the rose family is the agrimony, found in thickets and along woodland borders. It has a glandular, hairy stem and compound leaves with a hairy stalk that have a spicy odor when crushed. The five-petaled, yellow flowers with orange anthers, are not showy, and with good reason, for they contain no nectar and, thus, to advertise for insects would serve no purpose. To be sure, occasional visitors do alight on the blossoms and may distribute a certain amount of pollen, but the little blossoms chiefly fertilize themselves. They develop into pretty seed urns that, encircled with a rim of hooks, catch in clothing and in the fur of every passing mammal.

There is a white avens, a purple avens, a yellow avens, and a long-plumed avens, as a matter of fact, fifty species that belong to the genus *Geum* that many, who do not know better, call "gums." They are all perennial herbs with chiefly basal leaves, flowers solitary or in corymbs, yellow, white, or red, and the fruit a collection of silky-plumed achenes that are often as showy as the flowers. A few are grown in the border or rock garden for ornament, others grow wild, such as the white avens that may be found along the woodland border and in shaded roads. It is a rather tall, finely-haired plant with an angular-branching stem, three-divided leaves except the simple uppermost ones, the root leaves of three to five leaflets, and with small white flowers that are succeeded by a ball of achenes, each ending in an elongated, hooked style that catch in clothing like those of the agrimony.

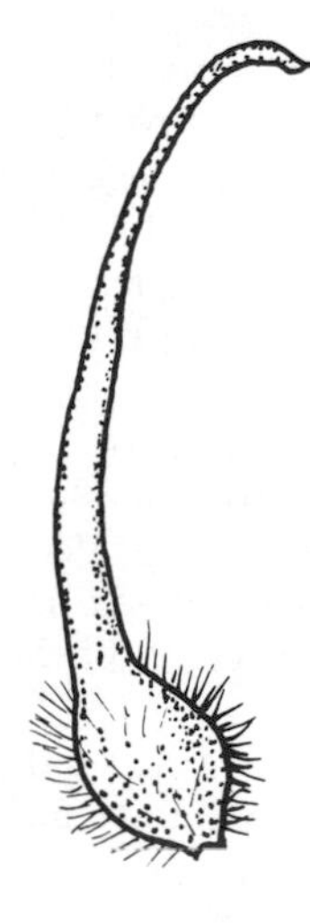

474. Fruit (Achene) of Avens

Unlike the white avens, the purple avens prefers wet places such as bogs and meadows, wherein to grow. The flowers, one inch broad, terminal, solitary and nodding, are purple with some orange-chrome, and are visited by bumblebees that, in return for their feast of nectar, carry pollen from old flowers to the maturing stigmas of younger ones.

Like the purple avens, the yellow avens is also partial to wet places, moist meadows, swamps and the like. Here, after the golden yellow blossoms of the marsh marigold have faded from the scene, those of the yellow avens twinkle in their place and, in the autumn, the seed clusters steal a ride on every passing animal in the manner of the agrimony, burdock, tick trefoil, and others that have devised such a way to scatter their seeds far and wide.

In contrast to the three preceding species, the clustered achenes, feathered with long, silvery hairs, of the long-plumed avens, are more noticeable than the flowers. This species is an exceedingly pretty and graceful plant, with a decorative, deeply cut leaf and a ruddy flower stalk that usually bears three, dull crimson red flowers. It grows in dry or rocky soil and is rather rare.

The goat's beard is another handsome species with a compound flower spike, formed of many little spikes, and that reaches a height up to as much as seven feet above the rich soil of its woodland home. Staminate flowers occur on one plant, pistillate on another, being an exception to the general rule of the family. Many more beetles effect a transfer of pollen than any other visitors.

Although the queen-of-the-prairie is seen at its best in the low, moist meadows of the Ohio valley, it may also be found throughout the East where it is mostly an escape from our gardens. It is a stately, beautiful plant with fragrant, deep pink or peach-blossomed colored flowers and cut-lobed, deep green, smooth, large leaves of sometimes seven divisions. Both butterflies and bees swarm about the flowers, though mostly butterflies, that appear to have a special fondness for pink, as the bees have for blue flowers.

163. Rue (Rutaceae)

From its name of prickly ash anyone would conclude that the plant is prickly and is an ash, or bears some resemblance to an ash. It is prickly, of course, but it is not an ash, though its leaves resemble those of the ashes. That is the extent of its likeness.

The prickly ash is a bitter, aromatic shrub or small tree, eight to twenty-five feet high, of woods, thickets, and riverbanks with ascending stems and twigs, a light gray bluish, and usually very prickly, compound leaves of five to eleven ovate pointed, toothless, deep

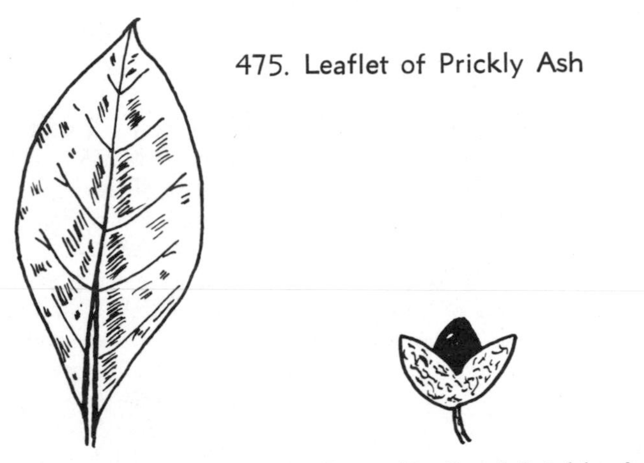

475. Leaflet of Prickly Ash

476. Fruit (Pod) of Prickly Ash

green leaflets, small, inconspicuous, greenish yellow flowers, and dry, reddish brown pods. The shrub is pungent and aromatic in all its parts, and when the leaves are crushed, they give off a strong lemonlike odor. It is also called the angelica tree and toothache tree, the latter because it was once used as a remedy for toothache, rheumatism, ulcers and a variety of other ailments.

Another similar species is more southern in range. Known also as Hercules'-club and toothache tree, it is more of a tree than a shrub with very sharp, prickly stems and found growing in sandy soil. Both species of prickly ash belong to the genus *Zanthoxylum*, the Greek for yellow and wood, in allusion to the yellow wood of some species, comprising one hundred and fifty species of aromatic, prickly shrubs and trees that are mostly tropical. They have alternate, pinnately compound leaves, and small, greenish flowers that develop into pods. A third member of the genus is the wild lime, a native to the West Indies, but naturalized on the Florida Keys. It is generally a shrub, but sometimes is a small tree.

The genus *Zanthoxylum* is one of about one hundred genera with about a thousand species of mostly tropical trees and shrubs that constitute the rue family. They have alternate or opposite, simple or compound, glandular, aromatic, often evergreen leaves, mostly regular, perfect flowers with four to five sepals and petals, and eight to ten stamens, a four to five-celled ovary, and a fruit that may be berrylike, drupaceous, or leathery and dry, sometimes winged.

There are only a few native species of the family. Besides the two prickly ashes there are the torchwood, a shrubby tree found only in southern Florida with compound leaves of only three leaflets, small, white flowers in open clusters, and a black fruit; the turpentine broom, a strongly scented desert shrub, with yellowish green, glandular stems, narrow leaves, and small dark purple flowers; and two species of trees known as the hop trees. The eastern species is a small tree or often merely a shrub, ten to twenty feet high, with a dark brown bark, compound leaves of only three leaflets, the leaflets pointed with a few rounded teeth and generally with a disagreeable odor, small greenish flowers, and a hoplike fruit, a multiveined, nearly round, very bitter samara, occasionally used as a substitute for hops. It is a tree of rocky woods. The western hop tree of the coast ranges and the California Sierras has similar leaves and fruit.

A number of species of the family are introduced garden plants, such as the rue, for instance. It is an aromatic, perennial herb or undershrub of southern Europe, with twice-compound, alternate, evergreen leaves, dull yellow flowers in a terminal cluster, and a four to five-lobed capsule. The plant has been cultivated for centuries and was once called herb of grace because it was associated with repentance.

Another perennial herb of Eurasia, long cultivated for ornament, is the plant variously called gas plant, dittany, fraxinella, and burning bush; so-called because the strong odor of its foliage and flowers will ignite, though faintly, if a lighted match is held near it. It is a somewhat woody herb, two to three feet high, with compound leaves, the leaflets oval and leathery, white flowers in a terminal cluster, and a hard, almost woody capsule.

A desirable lawn specimen is the cork tree, an ornamental and somewhat picturesque tree of Northern China, forty to fifty feet high, with gray, deeply fissured, corky bark, compound leaves of five to thirteen oval or ovate leaflets, small, yellow green flowers, and a black berrylike fruit.

Of all the genera of the rue family, sometimes called the orange family, the genus *Citrus,* a classical name of some other tree, but used by Linnaeus for the citrus fruits is, without question, the most important, as it contains the various kinds of "citrus fruits" such as the orange, lemon, lime, tangerine, and grapefruit. All of these trees came originally from tropical or subtropical Asia or Indo-Malaya, but are now spread throughout the world as cultivated fruits or as escapes, many being quite worthless. They are, or were before modified by cultivation, rather small, often spiny, highly aromatic trees with compound leaves of only one leaflet, thus, appearing as simple leaves, a distinctly winged leafstalk in most of them, flowers either solitary or in small clusters with five white, often waxy flowers, and a kind of fruit known botanically as a hesperidium. The fruit is spherical or lemon-shaped, usually with a number of compartments separated by thin partitions, and a white, stringy, central cord, pulpy juice, and a spongy outer covering or rind containing aromatic oil glands. Some of the more important economic species are the following: the sweet or common orange, a medium-sized tree, fifteen to twenty-five feet high, with a few blunt spines or none at all, oblong-oval leaves, white, fragrant flowers, and a sphere-shaped fruit with a sweet pulp; the sour orange, a small tree with long spines, oval or oblong leaves, fragrant flowers, and an acid, bitter, flattened, globe-shaped fruit; the King orange, an almost spineless tree with broadly oval leaves, and

477. Leaf of Sweet Orange

478. Flower of Sweet Orange

a fruit that is orange or reddish, flattened at each end, and with a loose, thin, skin, and sweet or slightly acid fruit; the tangerine, also a nearly spineless tree, with oval, lance-shaped leaves, and a smooth, orange reddish fruit; the citron, a shrub or small tree with stiff, short spines, toothed leaves, white flowers purplish on the outside, and an oval, lemon yellow fruit, with a rough, thick skin, highly aromatic, and that is candied to make commercial citron; the grapefruit, a tree usually without spines and a fruit having a finely grained, moderately acid pulp; the lemon, a small tree, with short and stiff spines, oblong or elliptic leaves, white flowers that are pinkish outside, and an egg-shaped fruit with a nipplelike projection at the end and very sour; and the bergamot, a spiny

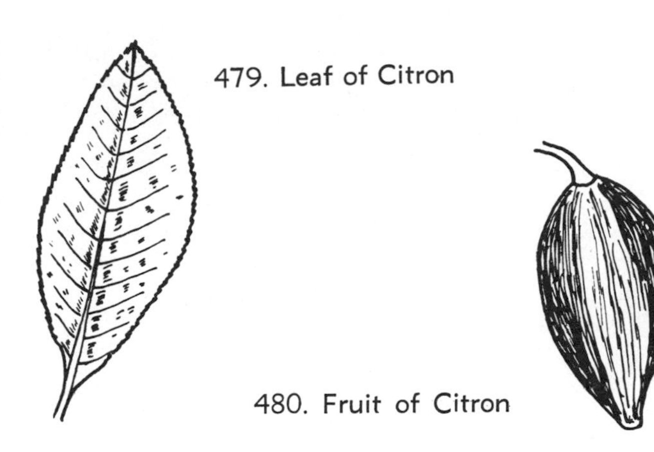

479. Leaf of Citron

480. Fruit of Citron

tree with oblong-oval leaves and a sour, pear-shaped fruit, the tree yielding an essential oil called bergamot oil. The navel or "seedless orange" is a mutant found a number of years ago in Brazil, but since improved and grown in California.

The familiar kumquat is a member of another genus, *Fortunalla* named for Robert Fortune, an English traveller who introduced the kumquat into Europe. It is a dwarf, evergreen tree with elliptical leaves, flowers like those of the orange blossom, and a small, orangelike fruit the shape and size of a very small plum. As the skin and the entire contents are edible, it is eaten whole.

The sour orange appears to have been the first orange introduced to the western world. It had been cultivated in China and India as long ago as 3000 B..C, but did not appear in Europe until the twelfth century of the Christian era. It was perhaps the first fruit tree to be brought to the United States, being planted in Florida by the Spaniards. In Spain, it is cultivated under the name of the Seville orange, and is the finest orange for marmalade. The peel is also candied and the flowers yield a perfume known in the trade as oil of Neroli.

Sometime during the fifteenth century, the sweet orange was brought to Europe from China by Portuguese sailors, and soon became a cultivated fruit. Just when the sweet orange arrived in Florida is not known, but groves were found in various places in the state as early as 1800. Since 1895, when a killing frost crippled the orange growing industry in the northern part of the state, all of the Florida oranges have been grown in the southern, frost free sections.

Orange growing in California began much later than in Florida. The Spanish Jesuits planted the first orange trees at San Gabriel Mission in 1804, and about forty years later, the first commercial grove was established in what is now the city of Los Angeles. Today, California is the leading orange growing state, Florida the second.

The lemon, a native of India, was introduced to Spain by the Arabs sometime during the twelfth century. The islands and coasts of the Mediterranean are a favorable region for the cultivation of the lemon, very sensitive to frost, and most of the European crop is grown there. Lemons were grown successfully in Florida until 1895 when a severe cold spell combined with a fungus disease, practically put an end to lemon-growing in that state. Today, California produces most of our lemons. The citron, though cultivated in Florida and California, is essentially a crop of the Mediterranean region.

The grapefruit, so-called because the fruits grow in clusters like large bunches of grapes, spread from its home in Polynesia to the West Indies, and from there, was taken by the Spaniards to Florida. Until recently, the grapefruit was cultivated only sparingly, but has increased in popularity, and today, is grown somewhat more extensively in such states as Florida, Texas, Arizona, and California. Most of our limes come from Mexico and the West Indies, though lime trees are grown to a limited extent on the Florida Keys.

164. Rush (Juncaceae)

In olden days, the people of England covered the floors of their dwellings with rushes and the king could require, as William the Conqueror did, his subjects to furnish "straw for his bed-chamber. . . . and in summer straw rushes." Erasmus did not think highly of the custom saying that such a floor-covering was a source of pestilence, as the lowest layer of rushes was often left untouched for years. It also has been said that in those days the route of a procession was made green with scattered rushes and that they were used in covering the stage in Shakespeare's time.

The rushes are grasslike herbs common by the roadside and in all marshy places where their green leaves, together with those of the grasses, carpet the earth with a green covering. They are often mistaken for grasses, but belong to a distinct family and, as a matter of fact, are closely related to the lilies. This becomes quite evident if their tiny green blossoms are examined carefully, when they are seen to be essentially lilylike in form.

481. Leaf and Flowers of the Rush

There are three hundred species of rushes, contained in eight genera, widely distributed throughout the temperate and cold regions of the earth. They are annual or mostly perennial herbs with creeping, underground stems, grasslike leaves that are often stiff and round or swordlike and jointed, the leaf sheath with free margins, though sometimes reduced to membranous sheaths, and small, perfect flowers, often borne in dense spikes or heads, though sometimes loosely aggregated or single, the flowers with three sepals, three chafflike, greenish or brownish petals, three stamens, and a three-celled or one-celled ovary. The fruit is a capsule with few to many, often tailed, seeds.

Our native rushes are chiefly of the genus *Juncus*, and the largest of our common species is the bog rush, found in clumps in moist places such as low meadows, though it may also be seen by the roadside. It is a tufted perennial, two to four feet high, with stiffly erect round stems, and with a small cluster of brownish green flowers near the tip that appears to be attached laterally, though actually the upright part above the blossoms is a leaf of the inflorescence. As this leaf seems to be a continuation of the stem, it looks as if the flower cluster burst right out from the stem. The bog rush is one of the few meadow plants that stay green until late autumn, and even in winter, low tufts of the green stems may often be seen along the brookside. Its fibrous stems are often woven into mats.

A few of our other common rushes are the grass-leaved rush, a species of moist, sandy places, with an erect and somewhat flattened stem, long, flat, and grasslike leaves, and an inflorescence that is composed of three to twenty, small, brownish green heads of

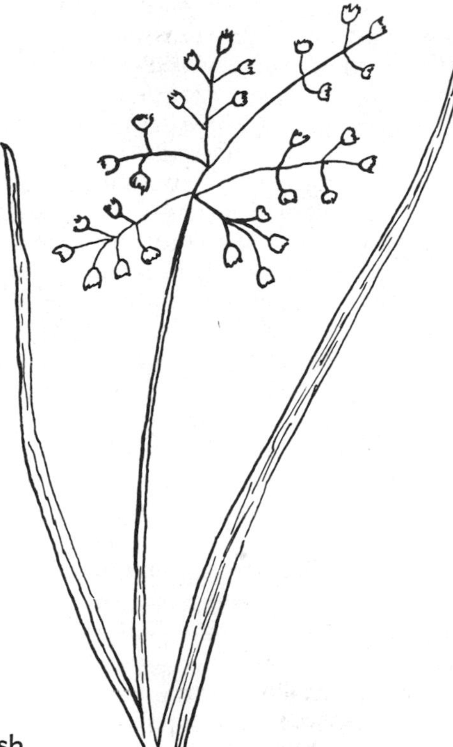

482. Bog Rush

flowers; the yard rush, common in country dooryards and by pathways and that grows in low-spreading clumps of wiry, gistening stems, its tiny flowers, like pale stars, appearing throughout the summer; the toad rush, an odd little plant, rarely eight inches tall, spreading in tangled mats over low ground by the roadside and on the borders of dried-up pools; and the black rush, common in the brackish marshes from New Jersey to Texas, its stems stout and cylindrical and its brownish flowers in loose terminal clusters.

A second genus of the rush family, *Luzula,* derived from the Italian word for glow-worm, in allusion to the shining seed capsules, contains a number of species known as the wood rushes. They are perennials that are often hairy and that are usually found in dry ground, with flat and soft, usually hairy leaves, and spiked, crowded, or umbeled flowers. Two common species are the common wood rush and the hairy

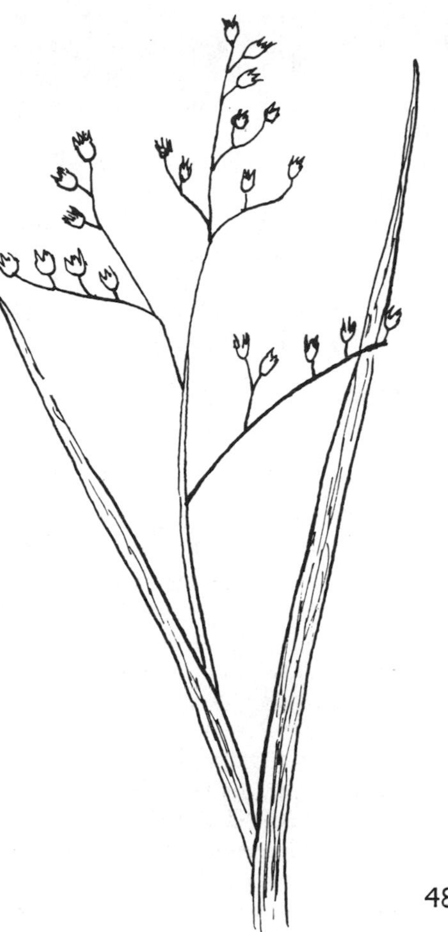

483. Yard Rush

wood rush. The common wood rush, one of the earliest flowering plants and often appearing while the ground is still brown from winter's frosts, grows in tiny tufts in various locations, from dry, open woodlands to low marshes. Its leaves are linear, flat, and hairy, and its blossoming umble is composed of short branches that bear small, densely flowered spikes. The hairy wood rush, found on dry wooded banks, may be recognized by the many long hairs on the leaves, and by the fact that its perianth is shorter than its capsule.

165. St.-John's-Wort (Hypericaceae)

At one time, the golden yellow flowers of the St.-John's-wort were hung in windows by superstitious people to avert the evil eye and the spells of the spirits of darkness. The plant was named from the belief that the dew that fell upon it the evening

484. Flower of St.-John's-Wort

before St. John's day, the 24th of June, was useful in protecting the eyes from disease, so on that day, the plant was collected, dipped in oil, and made into a balm. There is no evidence that the balm was good for the eyes or any other ailment.

Were it not for the brown petals of the withered flowers that remain on the flower stems, the St.-John's-wort might be an attractive plant, but it always seems to have an unkempt, untidy look. An immigrant from Europe, but an escape and now well established throughout most of the country, occurring in fields, upland pastures, waste places, and waysides, it is a pernicious weed with juices when young, that are so acrid and blistering that no grazing animal will eat it. It has a stem, ten to thirty inches tall, multi-branched, rather stiff, with opposite leaves, ovate, pointed or oblong, light green, thickly dotted with sepia brown, and shiny, deep golden yellow flowers in terminal clusters, the flowers with five petals, many stamens, and three styles. The fruit is an ovoid capsule with small, rounded, oblong seeds.

The St.-John's-wort family includes forty-five genera and seven hundred species of trees, shrubs, and herbs that are largely tropical in distribution. There are only about thirty species of the St.-John's-wort genus, *Hypericum*.

The leaves are opposite, simple, and toothless, often prominently spotted with resinous dots. The flowers are perfect, regular, sometimes solitary, but typically in branched clusters (cymes), with two to six sepals and petals, numerous stamens, often in three or five distinct clusters, two to five styles, the ovary two to many-celled. The fruit is fleshy and berrylike, or a dry, multi-seeded capsule.

Our species of St.-John's-wort are much alike, differing only in minor details. There is, for instance, the shrubby St.-John's-wort with yellow blossoms, about half an inch in diameter, that are provided with so many stamens that the many-flowered terminal clusters have a soft, feathery effect; the great or giant St.-John's-wort, a tall and showy plant whose deep lemon yellow blossoms brighten riverbanks and meadows in midsummer; the spotted St.-John's-wort, a plant remarkable for its spottiness and its golden yellow blossoms marked with thin, blackish lines; the Canada St.-John's-wort, that has linear leaves, tiny, deep golden yellow flowers that are succeeded by tiny pods that are conspicuously ruddy and exceed in length the five-lobed, green calyx; and the marsh St.-John's-wort, common in swamps, with pinkish, flesh-colored flowers, magenta seed pods, the entire plant frequently in late summer with a crimson or pinkish tone.

One can easily become confused with the many species of St.-John's-worts, varying as they do only in slight details, but to the maidens of another time any one of them had magical powers that would bring a husband within a year, if successfully cultivated in the garden.

Two other members of the family, but of another genus, *Ascyrum*, that are found growing wild, are the St.-Andrew's-cross and St.-Peter's-wort. Because the four, pale yellow petals of the St.-Andrew's-cross are arranged in pairs in the form of a cross, the plant was given its name by Linnaeus in 1753. It is a low, branching, smooth plant with small, deep green leaves and may be found in dry, sandy soil and in pine barrens. The St.-Peter's-wort is a similar plant growing in the same places and in bloom at the same time, but it has larger flowers in small clusters at the tips of its upright branches, the flowers of the St.-Andrew's-cross occurring at the leaf axils.

166. Sandalwood (Santalaceae)

The sandalwood family is perhaps of interest only because of the sandalwood, much used in ornamental carving and cabinetwork such as chests and boxes, as its fragrant odor drives away insects. The wood is that of the sandalwood, a member of the sandalwood family, a group of herbs, shrubs, or rarely trees, comprising about twenty-six genera and two hundred and fifty species, mostly tropical. They have alternate or opposite, toothless leaves, and staminate and pistillate, or polygamous flowers on separate plants. The flowers have a four to five-cleft calyx, no petals, the stamens equal in number to the lobes of the calyx, and a one-celled ovary. The fruit is either a nut or a drupe. Most of the species are root parasites.

The family is represented in the United States by only four genera and a very few species, three shrubs and the others herbs, the latter belonging to the genus *Comandra*, Greek for hair and man in allusion to the hairs on the calyx lobes that are attached to the anthers. There are four species, a common one being known as the bastard toadflax. It is an erect, wiry, usually branched herb with oblong, stalkless, bright green leaves, white flowers in terminal bunched clusters, and a small dry fruit that is crowned by the withered, but persistent flower. The bastard toadflax grows in dry, sandy places and thickets, and is partially or wholly parasitic on the roots of trees and shrubs.

485. Leaf of Buffalo Nut

486. Flowers of Buffalo Nut

487. Fruit of Buffalo Nut

The first of the three shrubs is the buffalo nut or oil nut. It is an upright, but often straggling shrub, three to twelve feet high, with many slender, light brown stems, elliptical or narrowly obovate leaves pointed at both ends, dark green above, paler beneath, and pale green flowers in short terminal spikes, the pistillate developing into a pear-shaped, leathery, yellowish green fruit containing a solitary large oily seed. The shrub is a root parasite of other shrubs and trees, and all parts of the plant, but particularly the seeds, contain a bitterly pungent and poisonous oil. It belongs to the genus *Pyrularia,* a diminutive of *Pyrus,* from the shape of the fruit, and is common in rich woods and on rocky, wooded slopes from Pennsylvania south to Georgia and Alabama.

The second of the shrubs is Nestronia of the genus *Nestronia,* said to have been derived from a Greek word for Daphna. It is a low shrub, one to three feet high, and a rather rare species, parasitic on the roots of various broad-leaved trees and shrubs, in dry woodlands from western Virginia, south to northern Georgia and Alabama. It has oval to egg-shaped, opposite leaves, small greenish flowers, and a berrylike, egg-shaped, one-seeded, yellowish green fruit. The third shrub, and also a rare species, growing on cliffs or bluffs along streams in the southern Appalachians where it is a parasite on the roots of hemlock and possibly other trees, is the buckleya. It has lance-shaped to narrowly egg-shaped, opposite leaves, small greenish flowers, and a berrylike, ellipsoid, one-seeded, yellowish green to dull orange fruit. The shrub is a member of the genus *Buckleya,* named after S. B. Buckley, an American botanist.

167. Sapodilla (Sapotaceae)

The sapodilla family is a group of thirty genera and over four hundred species of tropical shrubs and trees, mostly with a milky juice, only two genera being represented in the United States. They have alternate, entire leaves, that are often rather thick and leathery, the flowers never very showy, often solitary or a few clustered in the leaf axils, the perfect, regular flowers with four to six sepals, the corolla united, its lobes with small appendages or slightly fringed, the same number of stamens as the lobes of the corolla, the ovary four- to twelve-celled, an ovule in each cell, and the fruit a berry.

Aside from our own native species the following are of rather more than passing interest: the sapodilla, the gutta percha tree, the marmalade plum, the star apple, and the canistel. The sapodilla is a tree not over sixty feet high with evergreen leaves, white flowers, and an apple-shaped, russet berry that is widely grown throughout the tropical world for its sweet and delicious fruit, but of equal importance is the milky latex, harvested in Yucatan and neighboring regions for chicle used in chewing gum, hence, often called the chicle tree.

Sometimes cultivated in southern Florida for interest, the gutta percha tree is an Indo-Malayan, milky-juiced tree not over forty feet high, with leathery leaves, small, white flowers, and a small, egg-shaped berry. It produces gutta percha a substance resembling rubber, and that has a variety of uses. The marmalade plum is grown in the tropics for its fruit. It is a tree up to sixty feet high with the leaves usually clustered at the ends of the twigs, small, white flowers, and a russet, somewhat pear-shaped fruit with a thick skin and firm, spicy, reddish pulp.

Because the cross-section of its fruit suggests a star in the arrangement of its seeds, hence its name, the star apple is an ornamental fruit tree of tropical America, twenty-five to thirty feet high, that is grown chiefly for its fruit, although the lower surface of the leaves, that are deep green above, golden and felty beneath, is very beautiful. The flowers are small and inconspicuous and the fruit is apple-shaped, thick, smooth, greenish purple, the flesh white and sweet when ripe. A related species, both being members of the genus *Chrysophyllum*, Greek for golden leaf, in allusion to the golden hairs on the under surface of many of the leaves, and one of the two genera with members growing in the United States, is the satinleaf, a small, round-headed, tree, rather compactly branched, and found in southern Florida. The leaves are a deep green above with a coppery undersurface that is very showy, the flowers are small and inconspicuous, and the fruit is a deep purple and usually found at most seasons, due to the irregular flowering period.

The canistel, also known as the eggfruit, is a small tree up to twenty-five feet high grown for its edible fruit in the tropics and in southern Florida. It has leathery leaves, small, white flowers, and a roundish or egg-shaped, orange yellow berry with an orange-colored, pasty, sweetish pulp.

The second genus represented in the United States is *Bumelia*. It contains twenty-five, chiefly tropical American shrubs and trees, but four of them occur in the South. They have alternate, entire, persistent and green, but not evergreen, leaves, minute white flowers in small, many flowered clusters, and a berrylike fruit. The southern buckthorn, as one of the four species is known, is a shrub or small tree, ten to thirty-five feet high, with a brown gray, rough bark, elliptical, olive green leaves, small white flowers in dense, globular clusters, and a black, cherrylike ovoid fruit. It is found in moist thickets and on lowlands. A similar shrub or tree, ten to sixty feet high, also grows in moist thickets and on lowlands. It is variously known as the woolly bumelia, chittamwood or shittimwood, and false buckthorn. Its leaves are leathery, its flowers are in smaller clusters, but its fruit is like that of the preceding species. The small bumelia is much like the woolly bumelia, but the tough bumelia, that may be found in dry, sandy pinelands and dunes near the coast, has top-shaped leaves or broadest, well above the middle, lustrous green above, but coated beneath with copper-colored or golden brown, silky hairs, and a black, oblong fruit.

168. Saxifrage (Saxifragaceae)

If one were to explore the woodland or hillside in early spring, one would be sure to find rosettes of fresh, green leaves rooted in the clefts of rocks. And if one were to examine the rosettes carefully, one would observe small, finely haired balls in the center

488. Flower of Early Saxifrage

of the leafy tuffets. These little balls soon expand into branching, downy stems, bearing many little white, perfect, starlike flowers with ten yellow stamens. They are visited by the early bees and such butterflies as the mourning cloak and tortoise shell, and are succeeded by rather odd, madder purple, two-beaked seed vessels (capsules). Why the downy stems? The hairs are sticky and, thus, guard the flowers against unwanted pilferers, such as the crawling ants whose feet become ensnared in them.

The plant just described is the early saxifrage, the name saxifrage coming from the Latin meaning "rock-breaker," having been interpreted as meaning that the saxifrage grows in rocky crevices that it has broken open in order to gain a foothold, or that it was once considered to be a cure for gallstones. There are four hundred species of saxifrages found chiefly in the temperate regions of Europe and America. They are of very diverse habit, but are, for the most part, low-growing, spreading or creeping, the rootstalks spreading by lateral shoots or runners. Their leaves may either be thick and fleshy, or soft and mosslike, sometimes arranged in a rosette, and vary in shape from round or spoon-shaped to oval, with usually toothed margins. Their flowers are gathered in terminal clusters on long stalks, each with a calyx of five sepals and a corolla of five or more petals. They may either be pink, white, purple, or yellow, and are succeeded by two-celled, multi-seeded capsules.

A number of saxifrages make useful plants for the border and for the rock garden. There are also several wild native species such as the swamp saxifrage, and the lettuce saxifrage in addition to the early saxifrage already mentioned, and that appears in April. The swamp saxifrage is a much larger plant than the early saxifrage, growing one to two feet high. It has a somewhat stout, sticky-haired stem, large, blunt, lance-shaped leaves four to eight inches long, and greenish white flowers with very narrow petals. It is found in bogs and on wet banks. The yellow mountain saxifrage, occurring in the cold mountain regions of Vermont, New York, and Michigan, has its stems growing in tufts. The stems are three to five inches high, with fleshy leaves, mostly from the root, and long, narrow, and lance-shaped. The flowers, in few or several clusters, are yellow, and spotted with orange. It is a rather pretty species. The lettuce saxifrage, an inhabitant of the borders of cool mountain streams and swampy places, has its leaves in a thick mat at the base of the stem. Sometimes nearly a foot long, they are rounded at the top and taper downward in a margined petiole. The flowers are white in a loose panicle that terminates the flower scape, and are followed by sharp-pointed seed pods. In some places the plant is used for salads.

The saxifrages belong to the family of the same name. The family is a fairly large one that some botanists have split into several smaller ones, such as the hydrangea and gooseberry families, but here they are all included in the saxifrage family. It consists of eighty genera and about one thousand species of herbs and shrubs of wide distribution in temperate and cold regions. In the herbs, the leaves are often in a basal rosette, but in the shrubs they are alternate or opposite. The flowers, of four or five sepals and petals, four or five or eight or ten stamens, and with an ovary of two to four segments, are regular and showy, and profuse in many of the shrubs, but smaller in the garden and wild species. The fruit is either a berry or capsule.

Linnaeus fancied that he saw in the fruit of the mitrewort or bishop's-cap a resemblance to the mitre, a tall, pointed, folded cap worn by bishops during sacred ceremonies, and so he named the plant *Mitella,* the diminutive of *mitra,* a cap, that is the name of the genus whereto it belongs. The mitrewort is a hairy plant of rich woods with leaves of two kinds. One kind occurs on the rootstalks or runners. They are heart-shaped with three to five lobes and toothed, and are set on slender petioles. The second kind are two in number and occur opposite and sessile halfway up the flower stem. The flowers are white, the petals beautifully cut and resembling a snow crystal, and are borne in slender, graceful racemes. A much smaller and daintier species and occurring in cool woods and mossy bogs is the naked mitrewort or bishop's-cap that can be distinguished by its naked stem, that is, without the two leaves on the flower stem, and its fewer flowers.

Known also as the false mitrewort, the foamflower is a decorative little plant of the moist woodland floor, its fuzzy, bright white foam flowers appearing conspicuously above the leaves in late spring or early summer. The stem of the plant is a rootstalk wherefrom the long-petioled, rounded or broadly, heart-shaped leaves and flower scapes grow less than and up to a foot high.

A stout-stemmed plant with heart-shaped, scalloped leaves, and small, bell-shaped, whitish green flowers in terminal clusters and growing to a height of three feet, the alumroot is found in the rocky woodlands of most every state from coast to coast. Of an entirely different habitat, living in cold bogs or in wet shady places, the golden saxifrage, of no relation to the true saxifrage, except rather remotely and that only because it is a member of the same family, is a slender, low-growing species with its stem lying along the ground. Its leaves are roundish, finely scalloped and its flowers are yellowish or purplish green with orange anthers. The latter are small and scattered with short stalks, or none at all, in leafy clusters. The name of the plant is misleading, for it is not predominately yellow.

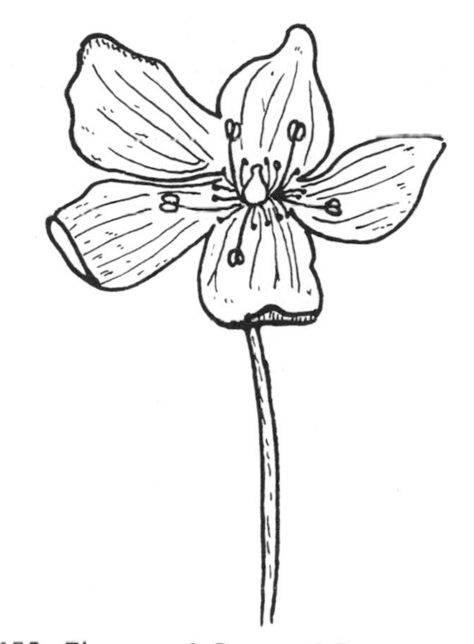

489. Flower of Grass of Parnassus

If the name of the golden saxifrage is misleading, so also is the name, grass of Parnassus, a native perennial herb, for it is neither a grass nor did it ever grow in the meadows about the home of the Muses. Doubtless, it was given the name whereby it is known because its European counterpart is fabled to have grown on Mount Parnassus. At any rate, it is an interesting plant with solitary, cream white flowers delicately veined with green, and about an inch broad. The flower has five petals and five fertile stamens with straw yellow anthers that alternate with the petals. There are also about fifteen imperfect stamens that encircle the pistil. The flowers are visited by bees, the skippers (butterflies), and the common yellow and white cabbage butterflies. A single ovate leaf clasps the flowering stem. The others spring from the root and form a basal cluster.

Several species of the genus *Astilbe,* Greek for not shining, in allusion to the leaflets, are widely grown as border perennials and much favored by florists who often, but incorrectly, sell them as spirea, since they are spirealike in appearance. A wild species, known as false goat's beard, is found in the southern woodlands. It is a coarse plant, four or five feet high, with twice or thrice-compound leaves, the leaflets thin, heart-shaped, lobed, and toothed, and white or yellowish flowers in a large, compound panicle. Mention of garden flowers calls to mind the crimson bells or coral bells, a very popular garden perennial. It grows from one to two feet high with basal leaves and a flowering stalk crowned at the top by a loose, often somewhat one-sided cluster of small, red, bell-shaped flowers. The red bell is all calyx, the corolla being represented by five tiny points that may be seen between and a little below the lobes of the bell. It is a close relative of the alumroot, both belonging to the same genus *Heuchera,* named for Johann Heinrich von Heucher, a German botanist whereof there are several species and all known by the general name of alumroots, because of their acrid and astringent roots.

490. Flower of Mock Orange

There is a mock orange that is a garden ornamental, and there is also a mock orange that grows in the southern mountains. The garden mock orange is a beautiful shrub that has been in cultivation a long time, so long as a matter of fact, that its origin is unknown, though doubtless it is a native of Europe. Gerarde had the shrub in his garden, "in the suburb of Holborne in verie great plentie." It is a shrub that grows to a height of ten feet, with dark brown bark, oval to oblong leaves that are pointed, and creamy white, fragrant flowers in terminal racemes. The fruit is a many-seeded capsule.

Our native mock orange is a shrub five to eight feet high with ascending, light buff brown, somewhat roughish stems, and arching branches with the newer twigs, light tan yellow. The thin bark of the shrub is more or less shredded or peeling. The leaves are light green, ovate, sharp-pointed, occasionally toothed, and the flowers are white, solitary, odorless and terminate the branches. A taller, but otherwise similar southern shrub, with somewhat larger flowers, the large-flowered syringa grows along streams. It might also be added that another species occurs in the rocky mountains of the southwestern states, and on the hillsides of the Northwest.

Like the mock orange there is a garden hydrangea and a wild hydrangea. The garden hydrangea is such a well-known and familiar shrub that it hardly needs describing. It is a treelike shrub, eight to twenty-five feet high, with oval or elliptic leaves that are rounded or wedge-shaped at the base, and with flower clusters eight to twelve inches long, the flowers at first white, but later changing to pink and purple. The magnificent bloom appears in August and does not complete its color scheme until October. It then remains on the branches until torn off by winter storms. Its very handsome flower clusters and its ease of growing and cheapness has made the garden hydrangea one of the most popular of shrubs, though it has found disfavor with some people because it has become so widely planted.

Our wild hydrangea, found growing mostly on the rocky banks of streams, is an upright, somewhat open, straggling shrub, three to five feet high, with slender, light brown twigs. The leaves are deep green, rather oval in outline, somewhat heart-shaped at the base, pointed, and coarsely sharp-toothed. The flowers are white in rather flat clusters. The name *Hydrangea* is from the Greek for water vessel, in allusion to the shape of the fruit, which, is a two-horned, ribbed capsule containing many seeds.

Another native shrub of the saxifrage family is the itea or Virginia willow. It is a shrub of the swamps, four to nine feet high, with alternate, rather narrow, pointed, sharp-toothed leaves that resemble the willow, hence, the plant's name, and that color early. They change from green to scarlet and crimson and glow in these colors until late autumn, persisting on the stems even after the leaves of other shrubs have fallen. The flowers are small, white in long, slender clusters.

There are no native deutzias. Of the fifty species that compose the genus *Deutzia,* two are from Mexico, the rest from Asia. They are well-known shrubs, that have no common name, usually with hollow twigs and mostly shreddy bark, opposite, short-stalked, toothed leaves, and generally white flowers mostly in terminal clusters. They were named for Johan Van der Deutz, a Dutch patron of botany.

One of the smallest of the deutzias and possibly the best known, is the species called bridal wreath. It is a low-growing shrub heavily ladened in May with masses of pure white flowers. One minor characteristic that adds to their beauty is that the yellow anthers are set on white filaments. Another species widely cultivated is known as the pride-of-Rochester, a double-flowered variety. It is an arching shrub with shreddy, reddish brown bark, oval leaves, and white or pinkish flowers in spirelike clusters.

Doubtless, the most important genus of the saxifrage family is *Ribes,* a Latinized version of an Arabic name for a plant with an acid juice, because it includes the currant, gooseberry, and several related shrubs grown for ornament. It is a large genus of sometimes prickly shrubs, common to the north Temperate Zone, with alternate, simple, but usually lobed, finger-fashion leaves, prevailingly greenish, yellowish, or reddish flowers, the sepals usually colored and larger than the petals that are sometimes very small or wanting, and true, juicy berries that are bristly in the gooseberry.

491. Cluster of Currants

The most common wild gooseberry east of the Mississippi River is the prickly wild gooseberry, a familiar undershrub of the northern woods. It grows to a height of four feet, with many prickles on the lower part of the stems, though the latter may sometimes be entirely smooth. The flowers are little green bells on slender one- to three-flowered peduncles, and are rich in nectar. The berries are brownish red and are covered with prickles like a bur. When mature, they are sweet and pleasant, though the prickles are sometime very sharp. They are excellent for pies, jellies and preserves.

The round-leaved gooseberry is a plant of rocky banks. It is a low shrub with rounded, three to five-lobed leaves, wedge-shaped at the base, greenish purple flowers on two to three-flowered short peduncles, and smooth, purplish berries having an agreeable flavor. Also a low shrub, the smooth gooseberry is commonly smooth, but sometimes has scattered prickles. It has slender, reclined, reddish brown, often crooked branches, alternate, simple, round leaves that are either solitary or in clusters, small, greenish white, dull purplish, bell-shaped flowers, and perfectly smooth, round, yellowish green or reddish berries. This gooseberry grows in wet woods and low grounds.

A native of Europe and the parent of the English gooseberry wherefrom many varieties have been produced, the European gooseberry is a garden shrub, but locally, an escape. It is a rigid, stocky plant, with thick branches, the fruiting ones without prickles, round, three to five-lobed, thick, very glossy, pubescent leaves, greenish, bell-shaped flowers, and large, oval, yellowish green or red berries that are minutely pubescent, and often with scattered prickles or glandular-tipped hairs.

An erect shrub, three to five feet high, the wild black currant is a plant of alluvial thickets and rich banks. It has slightly heart-shaped leaves, three to five-lobed, and toothed, greenish white or yellow, bell-shaped flowers, and shining, nearly round, and smooth black berries that are used for pies and jellies, but as they have a peculiar flavor, some people are not particularly fond of them.

The fetid or skunk currant, a plant of cold, damp woods, may be recognized by its long, prostrate, trailing stems, its deeply heart-shaped leaves, its small greenish flowers, and its pale red berries that are bristly glandular. The berries have a rather peculiar taste and are somewhat sour, though not unpleasant. When bruised, the entire plant emits a disagreeable odor, whence, its name.

The golden or buffalo currant is a native shrub found from Minnesota to Texas, and westward to Washington and California, but often cultivated throughout the East as an ornamental shrub where it has escaped to woods and waste places. It is a shrub of long, slender, upright or curving stems, growing four to eight feet high, with three to five-lobed, wedge-shaped, or heart-shaped leaves, golden yellow flowers with a pleasing, spicy odor in short, leafy-bracted racemes, and black, sometimes yellowish black berries with rather an insipid flavor. They are useful in making pies and jellies. The golden currant is a graceful and attractive plant, especially in early spring when the flowers and leaves appear together, but the flowers develop more quickly and soon the wandlike branches become covered with thick clusters of yellow flowers before the leaves have made much headway.

The swamp red currant is a low shrub of wet woods and bogs with reclining branches, three to five-lobed,

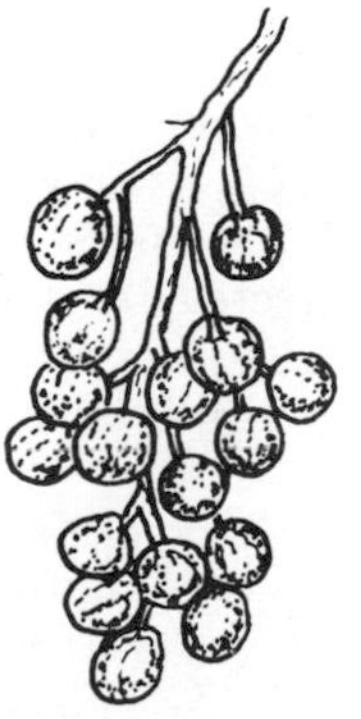
492. Leaves and Flowers of Red Currant

heart-shaped leaves, that are white, woolly or downy on the lower surface, grayish brown or purplish flowers, and smooth red berries that are sour. Not so are the berries of the red garden currant, famous for pies and jellies. This shrub, a native of Europe, and commonly planted in gardens wherefrom it has escaped to fencerows, thickets and open woods, is familiar to almost everyone. It has nearly erect branches, three to five-lobed, nearly smooth leaves, and racemes of yellowish green flowers that are followed by plump, juicy berries a quarter, to a third of an inch in diameter.

169. Screw Pine (Pandanaceae)

The screw-pine family is a group of tropical trees or shrubs with rather odd, spiral clusters of pineapplelike leaves at the tips of the main stem and branches. They are often climbers when they form prop roots or stiltlike aerial roots. The family contains three genera and about four hundred species that are widely distributed throughout the Old World tropics, especially in Malaya and in the Indian and Pacific Islands.

They have long, swordlike, stiff leaves, in tufts at the ends of the branches, that are arranged in three, spirally twisted series, with spines on the margins and back of the midrib. The flowers lack both a calyx and a corolla, but have few to many stamens, and a one to many-celled ovary. The fruit is a cluster of drupaceous fruits and is berrylike and often large.

Certain species yield fibers that are used for making baskets, fiber hats, and various other articles. Other species, of the genus *Pandanus,* a Latinized version of a Malayan name for some species, are cultivated for ornament, two species in particular being very popular as house plants and widely grown by florists. When fully grown, they have a distinct trunk, but when cultivated as pot plants, the trunk is rarely developed. Tubbed specimens in greenhouses and plants grown outdoors in the South usually develop fairly large trunks, but in the tropics, these plants may grow from thirty to sixty feet high and appear palmlike in their huge, terminal crown of leaves. One of the most popular pot or house plants has leaves that are two to three feet long, are usually arching, and are prominently white or silver-banded, and spiny-margined. Flowers are rarely produced in cultivated specimens.

170. Sedge (Cyperaceae)

The sedges look like grasses, resembling them in color, and are often spoken of as "grass," but they are not grasses, differing from them in several respects. Unlike the grasses whose stems are round and hollow, the sedges have, with few exceptions, solid stems and that are in many species, sharply triangular. Also unlike the grasses whose sheaths are usually split on the side of the stem opposite the leaf, the sheaths of the sedges are perfectly closed. Furthermore, each sedge blossom is protected by a single scale, though a perianth is sometimes present in the form of small bristles, whereas several scales enclose each grass flower.

There are three thousand species of sedges, contained in some eighty-five genera, and are found throughout the world, usually in moist places. They are quite protean in form. Some are very small, appearing only a few inches above the ground, while others are stout and erect, growing as high as a man's head with great flowering heads of many spikelets. Still others, rising leafless, look like green bayonets, tipped with cylindrical flower clusters, and still others are broad-leaved and spreading, seemingly escapes from the tropics.

They have narrow leaves, and small, inconspicuous, green, perfect or unisexual flowers arranged in spikelets, the latter solitary or grouped in spikelike or panicled clusters. The flowers may have a perianth of scales, or bristles, or hairs, or lack one altogether, with generally three stamens, though sometimes two or one, and a one-celled, one-ovuled ovary. The fruit is a tiny, dry, usually three-sided achene. The sedges often form extensive growths in swamps and along pond margins, and are mostly of little economic value.

The sedges of the genus *Cyperus,* a group of six

hundred species, are known variously as "earthnuts," "bulrushes," and "sweet rushes." They have elongated, narrow leaves that are mostly basal, growing from a tuberous rootstalk, and flattened spikelets that bear flowers without bristly perianths, the spikelets in heads or spikes. Many members of the genus have served the world since remote antiquity, including the famed papyrus, whose fibers were made into the parchment of the ancients. The nutlike tubers of certain species are edible, and the roots of a few are fragrant and aromatic, yielding a perfume.

The papyrus of the Egyptians is an aquatic sedge, with stems six to eight feet high, essentially leafless,

493. Bristle-spiked Cyperus

but clothed with sheaths, the terminal cluster of flower spkelets umbellike, with fifty to one hundred threadlike drooping rays. A typical species found in our country is the bristle-spiked cyperus, common in moist or dry soil by the wayside and with greenish straw-colored spikelets. A common sedge of sandy places is the edible cyperus, or chufa, also called earthnut or earth almond. It is a perennial two to

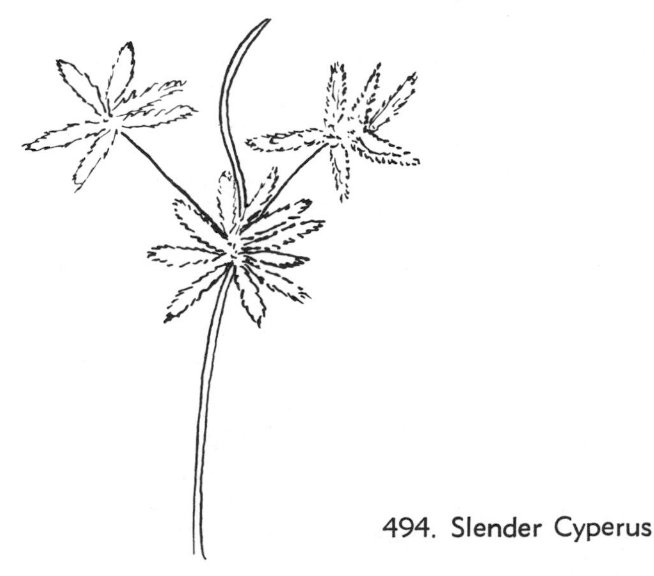

494. Slender Cyperus

three feet high, rising from an edible rootstalk, that is cultivated, especially in southern Europe, for its nutlike tubers that are said to have a sweet taste when boiled or roasted. The slender cyperus, common throughout the country, grows on hillsides. It is a small, slender plant with globose, dull greenish brown flower heads.

495. Pond Sedge

Anyone may well be excused for mistaking the pond sedge for a flowering plant of some other family, for it bears little resemblance to other members of the sedge family. It has hollow, jointed stems, one to three feet tall, with short leaves that are not sedgelike in appearance, and flowers that are borne with the leaves along the stem. They are in spikes that are composed of narrow, green spikelets, one-half to an inch long. The pond sedge, the only one of its genus *Dulichium,* grows along the borders of ponds and streams in company with other marsh plants.

The smallest species of the spike rushes are only a few inches in height, while the largest are sometimes as much as five feet tall. All the spike rushes are much alike in appearance, but as a group differ from other sedges. They have slender, round or four-angled stems that are usually soft and weak and that usually grow close together. The leaves are reduced to sheaths, and the flowers are in a small, solitary spikelet. The spike rushes thrive in the shallow water of ponds and swamp margins and generally in wet ground. In many places, the large spike rush grows by streams and ditches where its roots are often under water. The slender spike rush is common in open marshes where it may easily be recognized by its soft, dark green, hair-like stems that glisten in the bright sunshine and sway with every passing breeze. The spike rushes belong to the genus *Eleocharis,* Greek for marsh and grace in allusion to the marsh habit of some species,

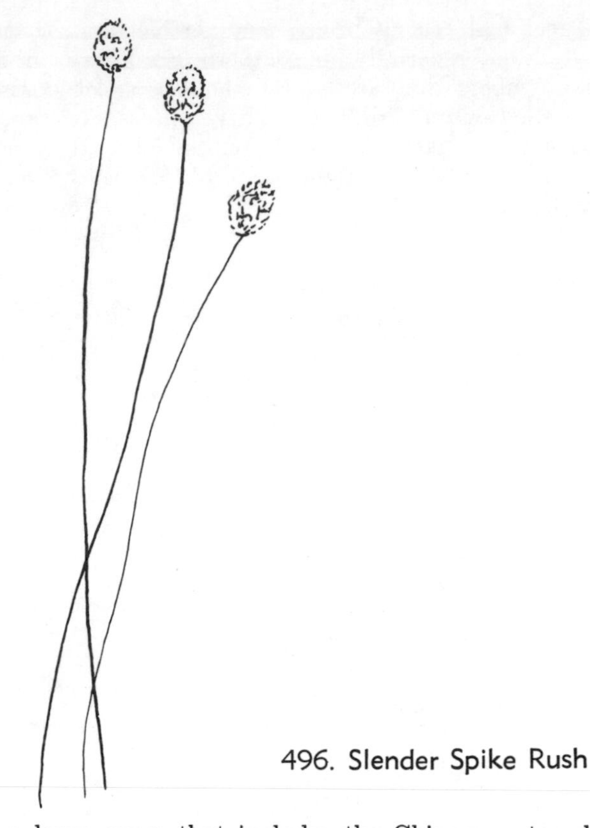
496. Slender Spike Rush

a large group that includes the Chinese water chestnut whose edible corm is rich in sugar and protein. It is widely used in China and in Chinese restaurants here.

In sandy fields and in railroad embankments even in the sand between the tracks, the sand mat will

497. Sand Mat

be found. It is a low and slender sedge, rarely a foot in height of the genus *Stenophyllus,* Greek for narrow and leaf, its tufts of dark green, threadlike stems capped with blackish green spikelets. Small grassy plants, the slender fimbristylis and the autumn fibri-

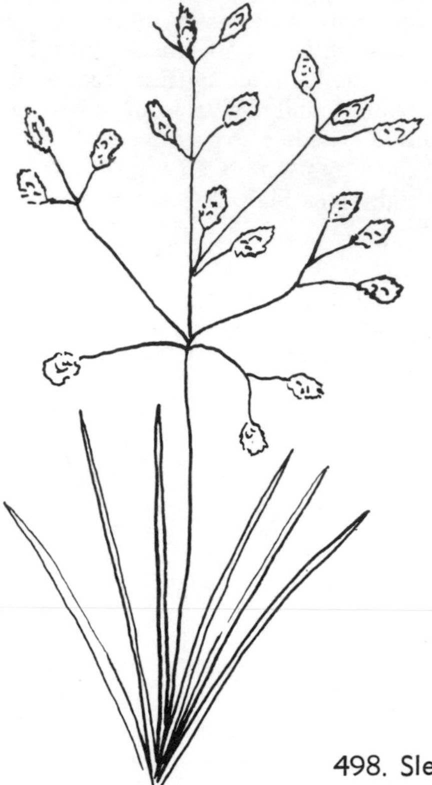
498. Slender Fimbristylis

stylis, genus *Fimbristylis,* from *fimbria* for fringe, and *stylus* for style, are usually found in moist soils. They have low and slender stems, three to sixteen inches tall, narrow leaves, and very narrow, greenish brown spikelets, in loose terminal umbels.

In early colonial days beautifully worked chair bottoms were fashioned of the chairmakers rush, easily recognized by its stiff, triangular stems, often shoulder high, one to three leaves, and one to seven oblong, brown spikelets about half an inch in length. It is found in saltwater marshes and by inland streams and is one of two hundred species of the genus *Scirpus,* Latin for bulrush, whose members are generally known as the bulrushes. They are aquatic or marsh, perennial herbs with grasslike leaves, in three rows, sheathing from the stem, the margins finely toothed, and numerous flowers in clublike spikes or terminal clusters.

A common species is the salt-marsh bulrush, a striking plant occurring most frequently near the coasts. It has stout, sharply angled stems, from one to five feet tall, and flowers in a dense, compact inflorescence that is composed of a cluster of five to twenty large, oblong, brownish spikelets. Even taller is the giant bulrush that can reach a height of nine feet and attain a diameter of an inch at its base. It

499. Chairmakers Rush

is a stout, leafless species growing along the margins of ponds and quiet streams where its smooth, round stems are more readily seen than its small terminal umbels of brownish spikelets. Common in late summer, the wool grass is a handsome plant of low meadows, its stems, that are often shoulder high, bearing a profusion of rather narrow, long, drooping leaves with margins of minute teeth. In the genus the perianth is present in the form of bristles, but in the wool grass it is long and downy, covering the spikelets of the conspicuous terminal umbel with dull gray wool, a characteristic that makes it easily recognizable. Both the wood bulrush and the leafy bulrush are leafy and graceful plants, found in swamps and low meadows. They are usually from two to six feet tall, with broad leaves and terminal umbels of many small spikelets, the flowering heads in dull tones of green and brown.

To the genus *Eriophorum,* Greek for wool or cotton and bearing, belong the cotton grasses, or as they should more properly be called the cotton sedges. They are grasslike plants, native in cool region bogs, with terminal flower heads that are furnished with long, silky hairs. They look like tufts of cotton.

The seeds of the beaked rushes of the genus *Rhynchospora,* Greek for a snout and seed, have a beaked achene. They are tufted plants of bogs and swamps with narrow and threadlike or flattened leaves, and elliptical or globular spikelets in terminal or axillary clusters, the flowers white, green, or rich brown with white stigmas and pale anthers. In the clustered beaked rush, the leaves are dark green and the clusters of dark brown, pointed spikelets are borne at

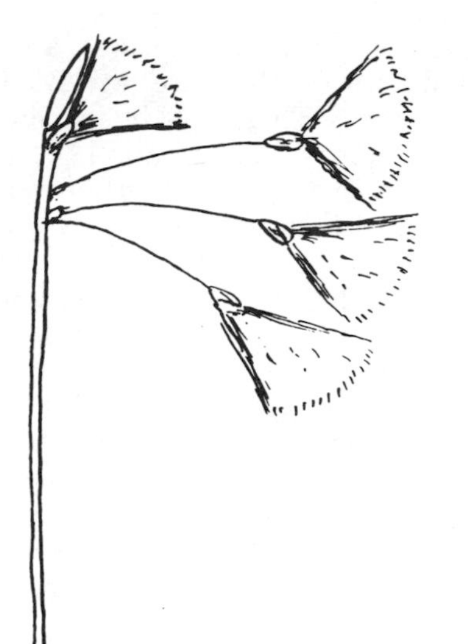

500. Slender Cotton Grass

intervals along the stem. In the white beaked rush the leaves are light in color and the flower clusters are located near the summit of the stem.

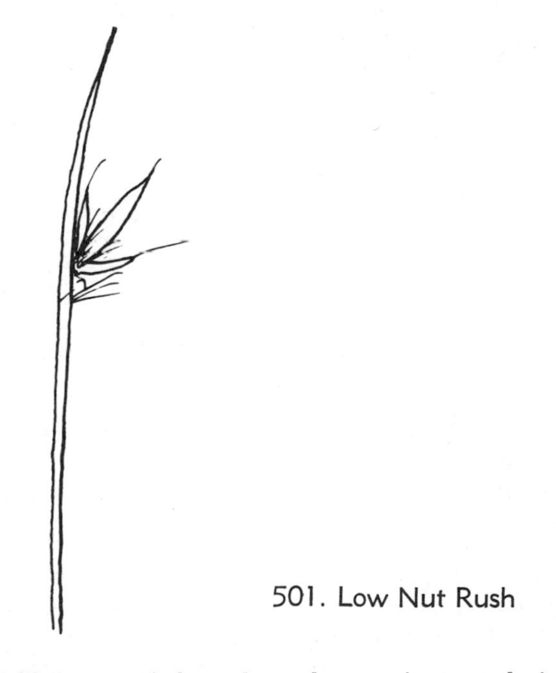

501. Low Nut Rush

Unlike most of the sedges, the staminate and pistillate flowers of the nut rushes are on different spikelets. The nut rushes of the genus *Scleria,* Greek for hardness, in allusion to the hardened fruit, are small, slender sedges, of marshes and low meadows, with narrow, much-elongated leaves and a white, bony, nutlike fruit. The low nut rush has sharp, three-angled stems and a few narrow leaves whereabove are the spikelets in four to six sessile, green or purplish clusters. The ripened, shining, white seed is very prominent and calls attention to the plant more quickly than do the flowers.

Most of the sedges belong to the genus *Carex,* a large group of a thousand species that are widely distributed in temperate regions. They grow in abundance in wet meadows, by the brooksides, and in all swampy places, some species often covering hundreds of square miles. In a locality rich with sedges, as many as fifty species may be observed in the course of a summer.

The members of the genus are tufted perennials, with solid, triangular stems, grasslike leaves, and small, green flowers crowded in flattish, spikelike clusters that are grouped at the top of the stalks. The staminate and pistillate flowers in some species occur in the same spike, in other species they occur in separate spikes, but in any case, each flower is protected by a scale or bract. The fruit is a hard achene, enclosed in a sac (perigynium), borne in the axil of a bract, or scale.

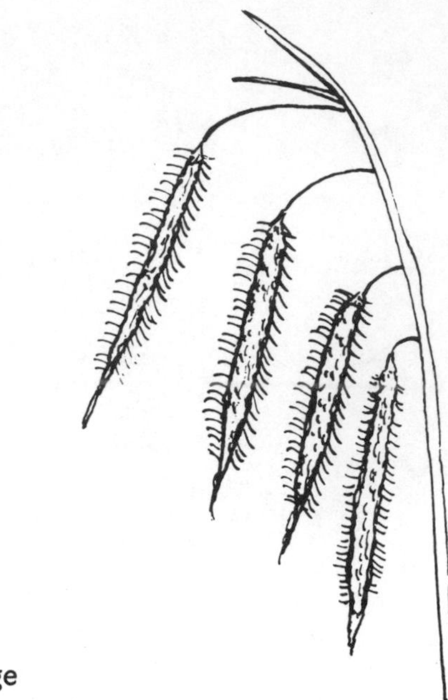

503. Fringed Sedge

502. Tufted Sedge

One hundred and fifty species of the genus are native to the eastern and central states alone, and of these, the following might be mentioned: the tufted sedge, common in wet soil and by damp waysides, with long, gracefully spreading leaves; the Pennsylvania sedge, the earliest of the sedges and one of the most abundant, found in the dry soil of hills and open woodlands where it carpets the ground with tufts of slender leaves; the bladder sedge, a common species of open marshes and a slender plant with one to three short, few-flowered, pistillate spikes whereabove the narrow staminate spike is borne on a slender stalk; the porcupine sedge, also of open marshes, with narrow, yellowish green leaves; the fringed sedge, a showy species of swamps and open woods, blooming in midsummer when its long, drooping spikes are conspicuous with their spreading, long-awned, brownish green scales; the little prickly sedge, of low grounds and easily recognized by its threadlike leaves and slender stems that form low tufts of a glistening green; and the fox sedge, a common species in low grounds, but also found by waysides, with short, green or brownish green spikes, densely crowded in clusters that form what appears to be a rough, terminal spike.

504. Fox Sedge

171. Sesame (Pedaliaceae)

The sesame, whose seeds are the benne (or benny) of commerce, is a member of a family of about fourteen genera and fifty species of annual or perennial, hairy, tropical herbs with opposite leaves and often with handsome flowers having two-lipped corollas.

The sesame of the genus *Sesamum,* the Greek version of the Arabic name for the sesame, is known in Africa as simsim and its seeds are used as food and for their oil. It is a roughly haired herb, one to two feet high, with oblong or narrow leaves, the lower often three-parted, the upper sometimes alternate, and white or pink solitary flowers occurring in the leaf axils. The corolla is two-lipped, the upper lip two-lobed, the lower three-lobed. The fruit is an oblong four-angled capsule.

The sesame is little grown in the United States and suited only to the South, but can be grown as an annual northward. The commercial production of benne is chiefly African and Indian.

172. Silk-cotton Tree (Bombacaceae)

A group of tropical trees of more than usual interest because of the famous baobab tree, the silk-cotton tree family comprises only about twenty genera and possibly a hundred and fifty species. They have palmately compound leaves, solitary flowers or flowers in panicles, the flowers with a five-toothed calyx, five petals, five or many stamens, and one style, two to five stigmas. The fruit is a capsule containing seeds, invested with copious silky hairs.

The baobab tree is a huge tree of the genus Adansonia, named for Michel Adanson, a French botanist, the trunk sometimes thirty feet in diameter, but much smaller when cultivated in southern Florida. It is a native of Africa, but often cultivated in India and South America, as well as in Florida, with such a massive trunk that when hollowed out, serves the natives as a home and provides them with their dugout canoes. It usually has about five leaflets to a leaf, white flowers, nearly six inches across, with very showy purple stamens. The flowers are solitary and hang on extraordinarily long stalks that later support the huge fruit. The latter, known as monkey's bread, is hard-shelled, white, about a foot long, with the pulp mealy and acid. As with other members of the family the flowers have a five-toothed calyx, five petals, and five stamens.

The silk-cotton tree, wherefrom the family derives its name, is a tree one hundred to one hundred and fifty feet high, with large, wide-spreading branches. At the base, the trunk flares out into immense, flank-like buttresses that may extend as much as thirty feet from the trunk. It is one of a number of species of the genus *Ceiba,* the native name for them in tropical America, with compound leaves of seven only, tapering leaflets, white or pinkish flowers, in clusters that are six to eight inches long, the flowers large and showy, and a leathery capsule. The seeds are surrounded by cottonlike fibers. These fibers, the kapok of commerce, are resilient, buoyant and water-resistant, hence, they are excellent for life preservers and mattresses. The silk-cotton tree is a native of the West Indies and is cultivated extensively in Java for its fibers. It is also planted in Florida as a shade tree.

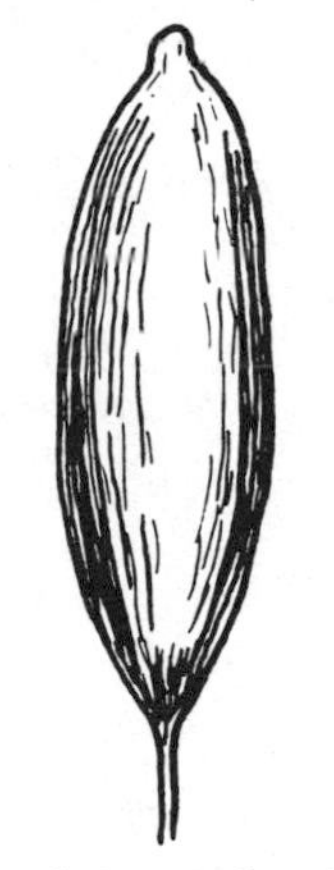

505. Fruit (Capsule) of the Silk-cotton Tree

173. Silk-tassel Tree (Garryaceae)

The silk-tassel tree family has only one genus, *Garrya,* named for Nicholas Garry, secretary of the Hudson's Bay Company, consisting of some fifteen species of evergreen shrubs, chiefly from the western part of North America, with leaves that are conspicuously silky and hairy on the lower surface and with flowers in catkins.

One of the species found growing on the dry slopes of the foothills of lower California and the adjacent states is the silk-tassel tree, also called tassel tree or quinine bush, though it does not yield quinine, sometimes cultivated for ornament. It is a hairy-branched shrub up to six feet high, with opposite, entire, leathery, elliptic or oblong leaves that are densely hairy beneath, and staminate and pistillate flowers on different plants in rather long racemes, the flowers without petals, but with four sepals and four stamens. The fruit is a nearly round, velvety berry.

174. Silver-Vine (Dilleniaceae)

The silvervine family is a small group of trees and shrubs embracing about eleven genera and three hundred species of mostly trees, shrubs or woody vines, all being tropical, except the genus *Actinidia.* They have alternate leaves, flowers with five sepals, five petals, and numerous stamens, and a pulpy fruit that in some instances, is edible.

The silvervine wherefrom the family gets its name, and that belongs to the genus *Actinidia,* Greek for ray, in allusion to the radiating styles, is a native of eastern Asia, but cultivated throughout the South. It sometimes climbs to a height of fifteen feet and has alternate, simple leaves that are often (especially on the staminate plants) splashed with silvery white or yellowish blotches. The flowers are white and fragrant and the fruit is an ovoid, multi-seeded, beaked yellow, edible berry.

Another member of the same genus and also cultivated in the South is the tara vine, a high-climbing, densely leafy, woody vine with lustrously green, broadly oval leaves, brownish white flowers, and an ellipsoid, yellowish, sweet, edible berry. The genus contains twenty-five species whereof two others, the yangtao and the Kolomikta vine, are also grown in the South. The former is often thirty feet long, its branches and twigs covered with shaggy hairs that are red when the plant is young, and has nearly round leaves, whitish flowers that ultimately become yellowish, and a nearly globular, hairy, acid berry that is edible and not unlike a gooseberry in flavor. The latter is not so high-climbing, usually not more than ten feet, with generally oblong leaves that are often white or pink-blotched, white flowers, and an oblong or ovoid, greenish yellow, sweet and edible berry.

175. Soapberry (Sapindaceae)

The soapberry family is a rather large group of one hundred and twenty genera and over one thousand species of trees and shrubs that are found almost entirely in the tropics. They have alternate (opposite in one genus), simple or compound leaves, flowers nearly regular, mostly unisexual, and usually in clusters (cymes or panicles). The individual flowers have four to five sepals and the same number of petals that, however, may be lacking, eight to ten stamens, and a two to four-celled ovary with one or more ovules in each cell. The fruit may be dry or fleshy, a drupe, nut, or sometimes winged.

Only three of the five native species are of any interest. These are the wingleaf soapberry, the wild china tree, and the Florida soapberry. The wingleaf soapberry, found in southern Florida, is not over thirty feet high, with seven to nine, elliptic or oblong, evergreen leaflets borne on a stout, interrupted winged rachis, small, inconspicuous, greenish or whitish flowers borne in a large terminal panicle, and a nearly

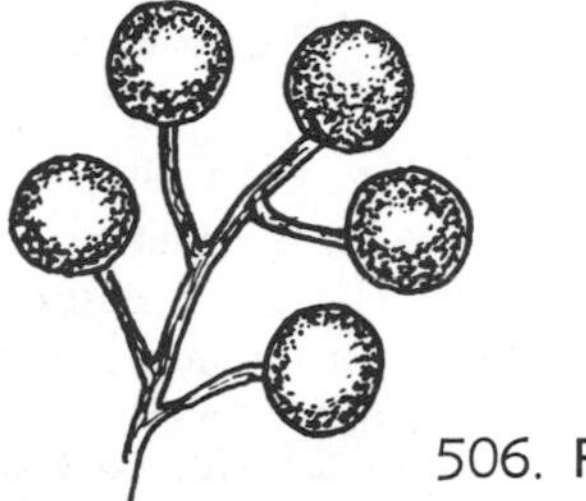

506. Fruit of Wingleaf Soapberry

round, orange brown, fleshy or leathery berry with bony black seeds, the pulp producing a good lather in water. As an early explorer of southern Florida wrote, the soapberry bears "sope berries like a musket ball that washeth as white as sope."

The wild china tree has not the limited range of the soapberry. It is a common, small tree, twenty to forty feet high, of bottomlands from Kansas to Texas, Arizona and east to Florida, with seven to thirteen leaflets that are narrowly oblong and pointed at both ends, small white flowers in dense clusters, and inverted, egg-shaped yellow berries. The pulp is reported to contain about thirty percent of saponin, a chemical substance having the property of producing a soapy lather. The wood, because of its tough and pliant character, has been used in making cotton baskets.

The Florida soapberry is a small coastal tree that may be distinguished from the wingleaf soapberry by its somewhat larger leaves and its unwinged rachises. All three species belong to the genus *Sapindus,* Latin for soap, together with *Indian,* in allusion to the Indians' use of the berries for soap.

The other two of the five species are the butterbough and the inkwood. The butterbough is a tree occasionally fifty feet in height, found in hammocks and marl soils of lower peninsular Florida and the Florida Keys with persistent leaves, white flowers, and a one-seeded, juicy, dark purple, globular berry. The inkwood is a shrub or small tree growing in hammocks and flat, pine woods of the Florida Keys, with persistent leaves, white flowers, and a black, sweet, fleshy, ovoid drupe.

An introduced species that has become naturalized and now found growing in waste places in the Southeast is the balloon vine, an extensively branching, herbaceous vine with alternate bipinnate leaves, the leaflets oval, pointed, and coarsely toothed, and small, but numerous unisexual white flowers in clusters that bear tendrils. The flowers have four sepals and petals and a three-celled ovary. The fruit is a three-valved papery, inflated capsule, containing black seeds with a white heart-shaped spot. The balloon vine is a quick-growing plant useful for trellises and low buildings, but in the North, it is treated as an annual.

Cultivated plants of the family of any interest are the akee, the longan, and the litchi. The akee is a tropical African tree grown for its fruit (the akee), which is a three-celled capsule. In each cell there is a single black seed whereto is attached the edible white aril. Only ripe arils should be used as over or underripe arils are apt to be poisonous, and when used, they are preferably fried in butter. The genus *Blighia,* whereto the akee belongs, was named after William Bligh, the captain of the *Bounty.*

The longan is an Asiatic tree also grown for its fruit that is round, about an inch in diameter, yellowish brown, the outer husk thin and shell-like, the white, juicy pulp edible. Much cultivated for its fruit is the litchi, variously called litchi, litchee, leechee, or lychee. It is a Chinese tree, round-topped, of medium size, with shining green, leathery leaflets and small greenish white flowers. Its fruit is a drupe, the fleshy aril of its seed being the edible part of the fruit that is so handsome that the tree is often considered an ornamental. When the fruit is dried it is the litchi nut of the stores. All three of these species, the akee, longan, and litchi are cultivated in Florida, the last two also in California.

Not uncommon in the subtropical gardens of Florida and California is the varnish leaf or akeake, a shrub eight to twelve feet high whose nativity is unknown, but doubtless tropical American. It grows wild in New Zealand, wherefrom comes the vernacular, akeake. It is one of fifty species of tropical and chiefly Australian shrubs and trees of the genus *Dodonaea,* named for Rembert Dodoens, Dutch physician, with alternate, oblong leaves, small flowers without petals, but with three to seven inconspicuous sepals, and eight stamens, generally in terminal clusters (racemes). The fruit is an angled capsule. The fruits, in shades of red and bronze and borne in showy clusters, are both interesting and highly ornamental.

176. Spiderwort (Commelinaceae)

About our dooryards and in our gardens in the Northeast, the dayflower often takes possession of the soil to the exclusion of other plants. Elsewhere, it may be

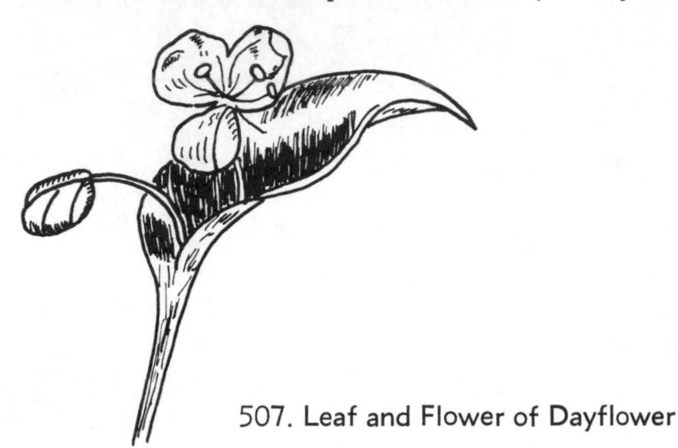

507. Leaf and Flower of Dayflower

found on riverbanks and other wet, shady places. There are several species, essentially Southern in range. Our Northeastern species is a naturalized Asiatic that has extended its range to Texas. The flowers open only in the morning and close by noon. They are rather odd from a botanical point of view. The three sepals are unequal, and so, too, are the three petals, two being rounded, showy, and blue, while the other is inconspicuous and somewhat whitish. Of the six stamens, three are fertile, and one of these is bent inward. The other three are sterile and small, with cross-shaped anthers. The flower is altogether an unusual one.

Linnaeus named the genus whereto the dayflowers belong *Commelina,* after three Dutch brothers named Commelyn. They were all botanists and two of them—commemorated in the two showy blue petals—published, while the third did not and failed to amount to much, much like the inconspicuous whitish third petal. The stem of the dayflowers is rather fleshy, smooth, and mucilaginous and the leaves are generally lance-shaped, those on the base of the stem with sheathing petioles, the floral ones heart-shaped and folded to form a hood about the flowers. In foreign countries they are used as potherbs.

Most of our garden plants seem to have come from the Old World or tropical regions, few of them have

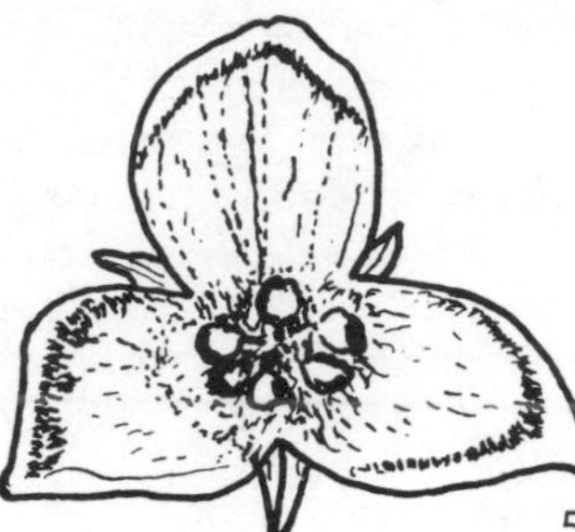

508. Flower of Spiderwort

been native transplants, but one of them is the spiderwort, a plant of rich, moist woods. As has happened so often it was the English who showed us the value of the spiderwort as a cultivated garden ornamental, for it appears that the seeds of this wild native were sent to the gardener of Charles I by some relative of his in the Virginia colony. His name was John Tradescant, for whom the plant and its kin came to be named, who planted the seeds and later found the plant worthy of a place in Hampton Court.

The spiderwort is a robust, strong-growing plant, and forms dense clumps. It has an upright, mucilaginous stem with light green leaves that are long, bladelike, sheathing the stem at the base, and regular flowers with three purplish blue petals and golden or orange anthers. Like the flowers of the dayflower, they open only for a part of the day. The plant was named spiderwort from the fact that when the stem is broken, the viscid juice that it contains is drawn out in slender strands like a spider web.

The spiderwort is a relative of the wandering Jew, a very common plant in greenhouses, widely used for hanging baskets, and useful as a house plant. All three, the dayflower, spiderwort, and wandering Jew, are members of the spiderwort family that comprises about twenty-six genera and three hundred and fifty species of succulent, annual and perennial herbs that are widely distributed and most abundant in the warmer parts of the world. The leaves are alternate, broad or narrow, without marginal teeth, and sheathing at the base. The flowers are regular or irregular, with three petals that are commonly blue and often showy, but that soon wither, three green sepals, usually six stamens, and a two- to three-celled ovary with few or many ovules. The fruit is a dry capsule.

177. Spurge (Euphorbiaceae)

The spurge family is a huge group of two hundred and fifty genera and more than four thousand interesting species of herbs, shrubs, and trees, with milky juice, extensively distributed throughout the world, being very common in the tropics and Temperate Zones of both hemispheres. The species vary greatly in their leaf, stem, and flower structure. Thus, there are some that have complete flowers—that is, flowers with both calyx and corolla—but the majority lack the corolla or even both the corolla and calyx. Some of these latter, the so-called "naked" flowers, have colored leaves and appendages that appear like the petals of flowers. Many species have ordinary leaves and stems, others are leafless, succulent and show a most remarkable resemblance to the cacti.

The members of the family generally have alternate, rarely opposite, simple, deeply lobed or compound leaves that are sometimes much reduced; usually small and inconspicuous monoecious or dioecious flowers having a various number of sepals, sometimes none, commonly no petals, one to many stamens (sometimes one thousand), and an ovary that is usually three-celled with one or two ovules in each cell. The fruit may either be a capsule or drupe. Sometimes the naked pistillate flowers are surrounded by numerous staminate flowers and sterile stamens, the whole group being in turn surrounded by a corollalike involucre, the entire structure being known as a "cyathium." In a number of species the uppermost leaves surrounding the flowers are brightly colored and are popularly known as flowers. The spurge family includes many plants that are highly prized as pot herbs and ornamentals, and others that are of economic value such as the castor oil plant, various rubber plants, and the cassava or tapioca plant whose roots are the source of tapioca and farina.

A troublesome weed of the southern states is a plant known as the hogwort. It is a species of dry, sandy fields, roadsides, and waste places with a stem one to two feet high, erect, branching, densely, softly woolled with star-shaped hairs. The leaves are also woolly, oblong lance-shaped, entire, silvery green, with a rounded or heart-shaped base. The flowers are clustered at the summit of the stem and branches. The staminate flowers are on a short raceme, the pistillate flowers are crowded below. The staminate have a five-parted calyx, five spatulate fringed petals, alternating with as many glands, and usually ten stamens. The pistillate have a seven to twelve-lobed calyx, no petals, and three styles twice or three cleft. They develop into gray or brownish, rounded, oblong seeds, with a tiny knob or caruncle at the point. The hogwort is one of a number of species of a genus that is best represented in the Southeast where over a dozen species are found. A few more occur in the central states, and one occurs on the Pacific Coast.

The silver bush is not a bush, but more of an erect herb. A southern species, occurring from Kansas to Arkansas and Texas, it has an erect stem covered with soft, silky hairs, sessile leaves that are oblong-ovate to lanceolate, entire and also covered with soft hairs, and monoecious flowers in racemes, with

509. Leaves and Flowers of Hogwort

five sepals and five petals. Both the silver bush and hogwort differ from the rest of the family by having a complete calyx and corolla.

The following three species have a calyx, but no corolla. The spurge nettle is a southern plant of dry, sandy soil with a slender, branching, bright green stem, bristly with stinging hairs, that produce, when in contact with the skin, a painful and lasting irritation. The leaves, roundish, heart-shaped in outline, but three to five-lobed, are similarly armed. The staminate and pistillate flowers are separate, the staminate ones usually in terminal clusters, the fertile ones in the axils below. The calyx of the staminate flowers is white, fuzzy, five-lobed, salver-shaped, and that of the pistillate flowers is also five-lobed, but smaller. The fruit is a three-celled, three-seeded, wrinkled, and bristly haired pod. The second of the three species, the three-seeded mercury, is a somewhat hairy plant of fields and open places. It has a stem six inches to two feet high, often purplish or brown, with long, ovate, thin, dark green leaves often turning to a coppery brown, and inconspicuous greenish flowers, the staminate on a tiny spike, the pistillate just below, and both supported by a large, leafy, cut-lobed bract. The third species, the queen's delight, is a southern species of sandy and dry soil. Its leaves, oblong-lanceolate and finely toothed, are almost sessile, and its monoecious flowers are in spikes. They have a two to three-parted calyx, two or three stamens, and develop into a three-celled, three-lobed, three-seeded capsule.

A native of Africa, the castor oil plant, of the genus *Ricinius*, the classical Latin name, has, for a long time, been cultivated in various countries for the oily seeds that are a source of the medicinal castor oil. It is a gigantic herb, treelike in the tropics, and cultivated in our gardens, wherefrom it has escaped and is now found growing in waste places, for its striking foliage and for its seeds. It is a tall, stately annual, forty-eight to one hundred and fifty inches high, with alternate, lobed leaves, the lobes toothed, and the leaves often three feet wide. The flowers are small, without petals, in dense, terminal clusters, the pistillate above, the staminate below, and the fruit is a capsule containing beautifully marked, poisonous seeds. There are several varieties. There is one with larger green leaves, another with red stems and bluish-gray leaves, and still another with red leaves, and even one with variegated leaves.

A native of tropical America and named after Joel Poinsett, a South Carolina physician, the Poinsettia is a showy plant and a Yuletide symbol. In the wild it is a shrub two to ten feet high, or even more, but as cultivated, is a winter-blooming pot plant two to four feet high. It has ovalish or elliptic leaves, usually shallowly lobed or wavy-margined, and slightly toothed. The upper leaves become brilliantly colored, vermilion bracts. The flowers are small with green and yellow attendant structures, in a floral structure typical of the family.

A related and native species found growing on slopes and in rocky soil in the middle states, the painted leaf or Mexican fire plant is a showy annual herb one to three feet high with variable leaves, the upper ones bright red.

The tapioca plant, a native of Brazil, but now grown in Florida, belongs to the genus *Manihot,* the Brazilian name for the plant, and is a woody herb or shrub, three to nine feet high, with three to seven-lobed leaves, the lobes narrow and tapering, and greenish yellow flowers, the staminate and pistillate in different clusters on the same plant and both without petals. The fruit is a capsule. The plant has a fleshy root rich in starches that are used in making the commercial tapioca. The root, however, contains hydrocyanic acid and is poisonous if eaten raw.

Another species of Brazil, the Ceara rubber tree produces a milky latex used in making certain grades of rubber. Some years ago, the Para rubber tree of the Amazon jungles was the source of most of our rubber, but today, our natural rubber comes from English and Dutch plantations in the Indo-Malayan region. The Para rubber tree grows up to sixty feet as cultured, but higher in the Amazon Valley. It has alternate, compound leaves, the three leaflets somewhat oblong, and greenish flowers that structurally are typical of the family. Its milky juice is the source of the rubber.

There are six species of gorgeously colored, tropical foliage plants, universally called crotons, but only one, the common croton, is of garden interest. In its many forms it is cultivated throughout the tropical world and in northern greenhouses. It is a shrub or small tree with smooth, rather oval or oblong leaves that, in the cultivated forms, are variously marked,

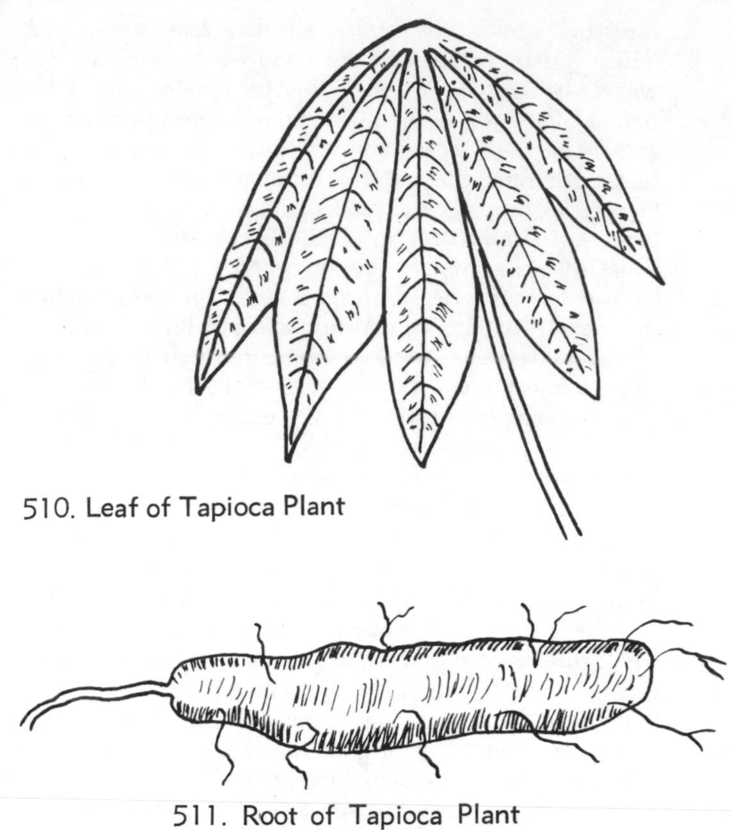

510. Leaf of Tapioca Plant

511. Root of Tapioca Plant

streaked, blotched, or banded with green, white, red, and yellow. The flowers are small and inconspicuous with the structural characters of the family, and the fruit is a roundish, white capsule, splitting into two, berrylike segments. Over a hundred named forms are in cultivation, and besides the color variations, some have finely cut leaves, others curled leaves, still others, crisp-margined leaves and a few have spirally twisted leaves.

Most of our native members of the spurge family belong to the genus *Euphorbia,* supposedly named for Euphorbus, a physician to King Juba of Numidia. It is a large group of one thousand species of wide distribution and of great diversity in habit. All have a milky juice, that is poisonous in some species, and their flowers lack sepals and petals, and in structure, are typical of the family. The fruit is a capsule that often opens explosively. Our native species are generally known as spurges.

Of all our spurges, the spotted spurge, a prostrate plant common throughout North America, seems the most hardy and adaptable. It is able to grow anywhere, and to withstand the most adverse of conditions, often appearing between the cracks of flags and paving stones in cities. It has long, branching, fibrous roots with many fine rootlets, a round, finely haired, usually dark red stem branching in all directions. The leaves are opposite, oblong, dark green, usually with a purplish brown spot near the center. The flowers are whitish or ruddy, inconspicuous, and grow at the bases of the leaves. Its poisonous juice will irritate the skin to a red rash. The flowering spurge, a hardy, perennial herb often cultivated in the garden, occurs in dry fields, old pastures, and waste places. It is an upright plant, sparsely branched, with leaves narrowly oblong to lance-shaped, varying somewhat in size, and a flower cluster with showy white appendages. The seaside spurge is a prostrate, spreading species common in the sand of the seashore, with inconspicuous flowers at the bases of the small, linear, oblong leaves.

Of all of our native spurges, the snow-on-the-mountain is perhaps the most handsome, but dangerous to handle, as its copious juice, when in contact with the skin, causes a swelling and eruption similar to that produced by poison ivy. Yet, it is widely grown as one of the most popular garden annuals. It is rather odd that a plant native to the dry plains of the West should have such a name as snow-on-the-mountain, but the white bracts of the flower cluster against a background of green leaves are somewhat reminiscent of snow on a mountain. It is a bushy, multi-branched herb, eight to fifteen inches high, with oblonglike leaves, the lower green, the upper white-margined. It is these white-margined leaves and the white bracts that crown the stem and surround the insignificant flowers that are the reason for its popularity.

A native of Europe, but an escape from gardens, now found in fields, roadsides, and in waste places, the cypress spurge is a perennial herb spreading by horizontal rootstocks with erect stems, thickly clustered with bright, light green leaves that are almost threadlike and terminated by large, flat, dome-shaped flower clusters. The insignificant flowers are indeterminate in color, but generally greenish, dull yellow, tan, or russet red with crescent-shaped glands. Altogether the clusters are quite ornamental. Also an immigrant from Europe and growing in fields, roadsides, and waste places, the sun spurge has a smooth, erect, stout stem, obovate and finely toothed leaves, and insignificant flowers also of an indeterminate color, but usually green and tan.

A common species of the family grown in greenhouses is the crown-of-thorns. A native of Madagascar, it is a creeping, spiny plant, its stems not over three feet long, generally dull purple brown, but the young stems often green. The spines are about an inch long, thick and stout, the leaves somewhat oblong and thin, rather sparse, often lacking, and the flower clusters are long-stalked, the bracts a brilliant red.

There are several other tropical American species that extend into southern Florida. These are the mill-

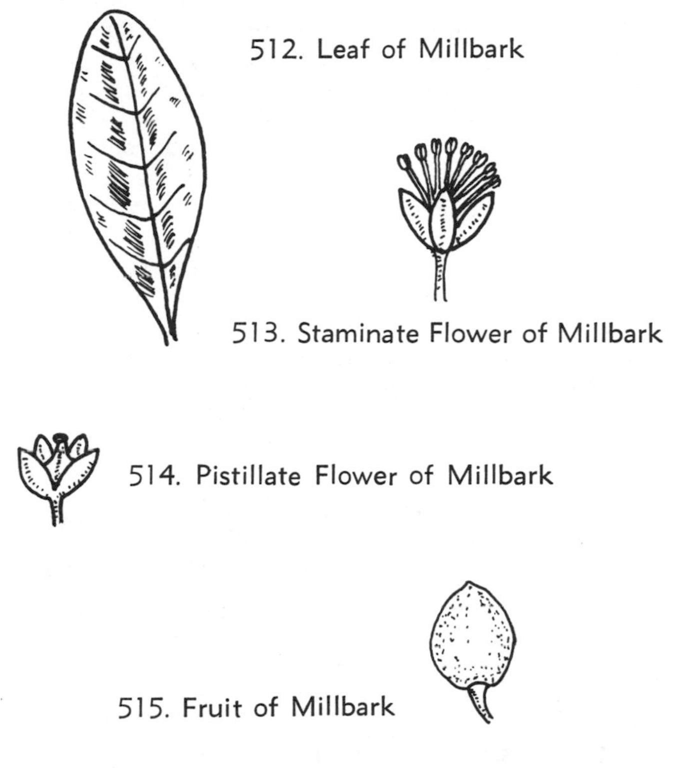

512. Leaf of Millbark

513. Staminate Flower of Millbark

514. Pistillate Flower of Millbark

515. Fruit of Millbark

bark, a small tree or shrub with alternate, simple, persistent, leathery leaves, staminate flowers in many-flowered clusters, the pistillate solitary or occasionally in clusters of two or three, and an ivory white drupe; the guiana plum, a shrub or small tree with elliptical leaves, flowers with only four sepals, and a globular, scarlet fruit; the machineel, a shrub or bushy tree, with broadly ovate leaves, yellowish green flowers, and a pear-shaped, yellow green drupe; and the oysterwood, a small tree, with persistent, leathery, elliptical leaves, the staminate flowers in clusters, the pistillate solitary, and a reddish brown to brownish black capsule.

178. Staff Tree (Celastraceae)

The bittersweet vine, in late summer or early fall, likely will have little, cottony tufts all along the stems. They are the egg masses of the two-marked tree hopper, a small brownish black insect with an enormous hornlike projection over its head. It resembles a thorn so much that it is difficult to distinguish it from such a plant structure.

The insect feeds on the sap of the bittersweet and in summer, both adults and immature forms are usually found clustered together. No matter how the vine twists and turns, they always rest with their heads towards the top so that the sap, that they suck, can flow more easily down their throats.

The bittersweet is a twining, shrubby vine found along roadsides, streams, on old stone walls, sometimes climbing trees to a height of twenty feet or more, and often planted in the garden as a covering for an arbor, trellis, or a wall, not only for its handsome foliage, but also for its brilliant autumn fruit. Its leaves grow alternately on the stem. They are light green, smooth, ovate or ovate-oblong, and finely toothed. Its flowers are small, greenish white, and

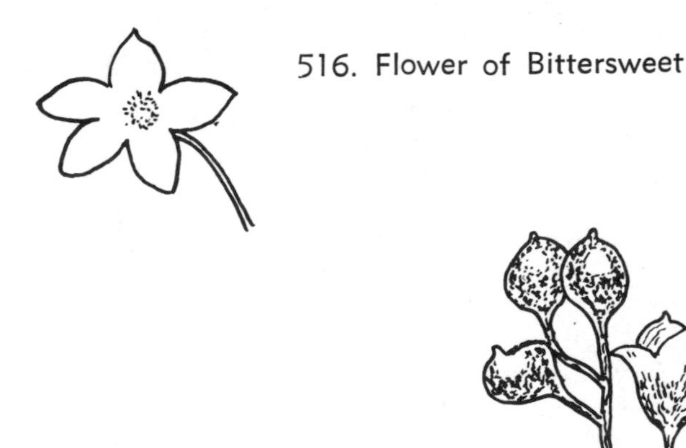

516. Flower of Bittersweet

517. Fruit of Bittersweet

grouped in a loose, spikelike terminal cluster. In the fall they are succeeded by beautiful orange, globular berries in loose clusters but that, however, strictly speaking are capsules whose orange shell divides into three parts, bends backward, and exposes the pulpy scarlet envelope of the seed within. The fruit is highly decorative and if picked and placed in a warm room before the shells open, it will expand and remain in good condition throughout the winter.

The bittersweet is a member of the staff-tree family that comprises about forty-five genera and over four hundred species of mostly erect trees, shrubs, or woody climbers that are widely distributed throughout the world, though not found in the colder regions. The leaves are alternate or opposite, membranous or leathery, and simple. The flowers are small, greenish or white, commonly perfect, either solitary, or in axillary terminal racemes or cymes, the calyx and corolla of four to five sepals and petals, the latter inserted on a conspicuous disk, the stamens of the same number as the sepals and petals, the ovary two- to five-celled, the style one, and the stigma three- to five-lobed. The fruit may either be a capsule, berry, samara, or drupe.

The staff-tree family contains two evergreen shrubs. One occurs in the mountains of Virginia and West Virginia and is known as mountain lover. The other occurs from British Columbia to California and New Mexico. The former is a low shrub with trailing, rooting branches, linear leaves, and tiny, reddish flowers. The latter is a spreading shrub with stiff branches, somewhat elliptic leaves, and white to reddish flowers.

Our deciduous shrubs belong to the genus *Euonymus*, Greek meaning good name, that is used rather ironically, as some of the plants have the reputation of poisoning cattle. There are one hundred and twenty species of the genus, about a dozen or so popular garden plants that are grown primarily for their foliage or showy fruits or both. Their leaves are opposite and stalked, their flowers are greenish, white, or yellowish with the floral characters of the family, and their fruit is a capsule.

One of the more common members of the genus is the strawberry bush found in moist woods and on wooded riverbanks. It is a tall, upright shrub, two to seven feet high, with four-angled and ash-colored twigs, thick, bright green, ovate or elliptical leaves, and solitary, greenish white flowers. The fruiting capsule is rough, warty, crimson when ripe, and the seeds are an orange scarlet. When ripe, the capsule opens disclosing the seeds, and the scarlet crimson combination is most brilliant and effective. This is also true of other members, and for this reason, they are ranked with the barberries, snowberries, and the winterberries of the holly family as decorative shrubs.

A similar species, but with a trailing habit, the trailing strawberry bush is hardly over a foot high, with its main stems lying on the ground and rooting at the nodes. Its leaves are wedge-shaped at the base and the flowers are purplish green. The fruit is like that of the aforementioned species.

Another native shrub, but sometimes a small tree, six to twelve feet high, occasionally as much as twenty-five feet high when growing in a favorable location,

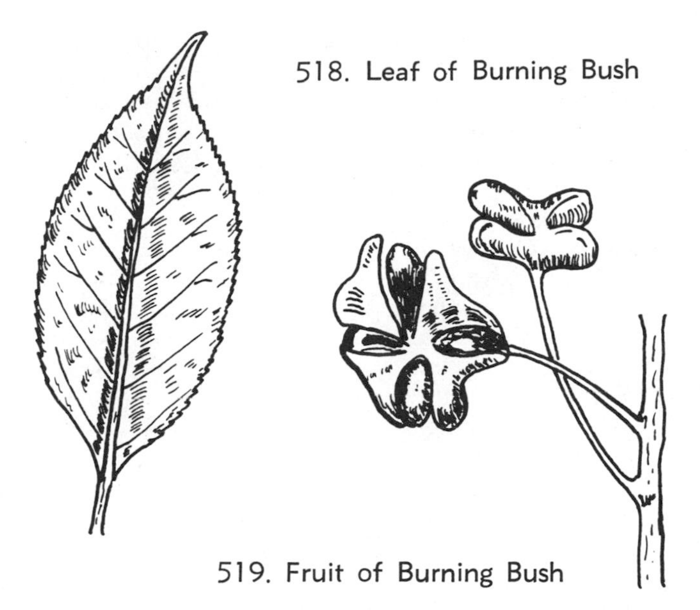
518. Leaf of Burning Bush

519. Fruit of Burning Bush

the burning bush, also known as the spindle tree or wahoo, is found in rich, moist woods, ravines, and stream bottoms. It has dark gray, bark roughened with perpendicular ridges through vertical splitting, short-stemmed, lustreless, deep bluish green leaves, and flowers with four rounded, spreading deep purple petals. The fruit is a deeply divided, small, smooth capsule with three to five lobes, light, lustreless crimson red, the seeds an orange scarlet. This species is widely cultivated.

The burning bush is well named for it retains its flame-colored fruit long after the leaves have fallen, and until the winter storms have beaten it to the ground. The Indians called the plant the wahoo and used the wood to make arrows. The leaves and fruits have been said to have purgative properties. Whether this is so or not, a European relative, the European spindle tree, introduced into America as an ornamental, has poisoned sheep and goats and children have suffered from eating its fruits.

The European spindle tree, so-called because its wood has been used in making spindles, knitting needles, and other small articles requiring a hard, close-textured wood, is similar to our burning bush, but has smaller leaves and creamy white or greenish yellow flowers that are fewer in number. The shrub is frequently an escape.

As a final note, several Asiatic evergreen shrubs of the same genus as our own burning bush are often planted as ornamentals in gardens.

179. Storax (Styracaceae)

Three native species of handsome trees of the Southeast are widely cultivated for ornament. Belonging to the genus *Halesia,* named for Stephen Hale, an early writer on botany, they are chiefly trees of the undercanopy of the forest, with alternate, stalked, toothed leaves, flowers in small, hanging clusters from the twigs of the previous season, the flowers with a four-toothed calyx, a four-lobed, bell-shaped corolla, eight to sixteen stamens, and a two to four-celled ovary with four ovules in each cell. The fruit is a somewhat oblong, rather fleshless, dry drupe.

Delightfully called the silver bell tree because of its profusion of white bell-shaped flowers in early spring, the first of the three species is sometimes eighty to ninety feet tall, with a straight trunk, short, stout, erect branches, and a narrow head. It has a brown or reddish brown, scaly bark, with shallow furrows and broad ridges, elliptical or slightly obovate, finely toothed, deep green leaves that turn a pale yellow in the fall, and two to five flowers in a cluster. The fruit is a four-winged, dry, oblong drupe. The silver bell tree is usually found growing in well-drained, rich soil in sheltered situations.

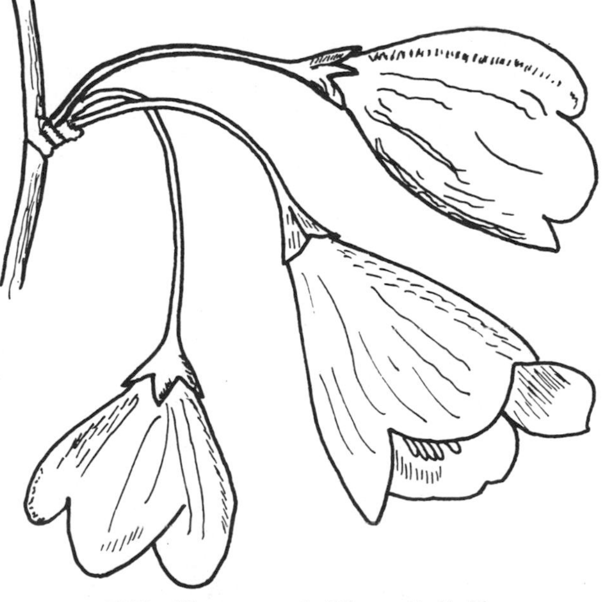
520. Flowers of Silver Bell Tree

The second of the three species, the snowdrop tree, so-called for much the same reason, that is, its profusion of small, bell-shaped, white flowers that also appear in early spring, never becomes a large tree, thirty feet being its maximum height. It is, in some respects, similar to the silver bell, but differs in that it prefers swampy ground, has larger leaves and more showy flowers, but the chief difference is in its fruit that is two-winged.

521. Fruit of Snowdrop Tree

Said to be a variety of the silver bell, the tisswood or mountain silver bell is a somewhat larger tree with larger leaves and occurs at higher elevations in the mountains of North Carolina, Georgia and Tennessee. The third species, the little silver bell, is often not more than a large bushy shrub, though it may attain the dimensions of a small tree. It grows on dry, sandy, upland sites and may be distinguished from the others by its club-shaped fruits.

The preceding species are members of the storax family that embraces six genera and one hundred species of trees and shrubs that are widely distributed in southern Europe, Asia, and the Americas. They have alternate, simple leaves, perfect, regular flowers with a calyx united, its four or five segments cleft or lobed, a generally bell-shaped corolla of four to eight petals, partly or wholly united at the base, four to eight or eight to sixteen stamens, a two- to five-celled ovary, and a fruit that is either dry or fleshy. Many of the species, because of their beautiful floral displays and the comparative ease of propagating them, are widely used as ornamentals.

Most of the species of the family are contained in the genus *Styrax,* an old Greek name for one of the species, and have the characteristics of the family. There are several native species that are known as storax or silver bells: the storax or bigleaf silver bell, the smooth storax or American silver bell, and the downy storax.

The storax or bigleaf silver bell is a southern shrub, three to twelve feet high, of moist woods and stream banks. It has elliptical or obovate, abruptly pointed, deep green leaves, showy white flowers with five narrow, mostly recurved, soft, downy lobes, and ten yellow stamens, in racemelike clusters, and an oval, dry, one-seeded drupe. Also a southern shrub, the smooth storax, or American silver bell, grows from three to nine feet high in moist thickets and on the banks of streams. The leaves are a deep green, smooth, and broadly elliptical, the flowers are white, showy, and fragrant, and the fruit is globular, leathery, and brown. The downy storax is similar to the bigleaf silver bell, but with slightly downy leaves, somewhat larger flowers, and a globular drupe.

All of these species may be planted as ornamentals, but more generally used for lawn planting are several species native to China and Japan. Also cultivated for ornament is another Asiatic shrub or tree of a different genus known as the epaulette tree. It has oblong, minutely toothed leaves, white, fragrant flowers in hanging clusters, and a bristly, ten-ribbed drupe.

180. Strawberry Shrub (Calycanthaceae)

Many plants have come to America from England and other parts of Europe, and, for that matter, from all parts of the world. A few have gone from America to the Old World. One of the earliest to be taken to England was the strawberry shrub where it has long been a favorite, as it has been in our own gardens because of its delightful fragrance.

The strawberry shrub, also known as the Carolina allspice and sweet-scented shrub, is a compact shrub, four to eight feet high, of southern mountains, rich

522. Leaf of Strawberry Shrub

523. Flower of Strawberry Shrub

524. Fruit of Strawberry Shrub

woods, and stream banks, with many slender, ascending stems, and slightly downy, brown branchlets. The leaves are elliptic or oval, sharp-pointed, toothless, dark green above, downy and grayish beneath. The flowers are dark maroon red or dark purple brown with the odor of sweet strawberries when crushed. The fruit is an ovoid, nodding, leathery receptacle containing many smooth, shining achenes. The stems and leaves, too, are spicy-aromatic when crushed.

The strawberry shrub is a member of the family of the same name that comprises only two genera, one being Asiatic, the other North American. They have opposite, entire leaves and solitary flowers on leafy, side branchlets. The flowers have sepals and petals alike, many stamens, and many one-celled ovaries. The fruit is a small, pear-shaped or egg-shaped receptacle containing many, dry, mostly one-seeded achenes.

The American genus *Calycanthus* Greek for calyx and flower, in allusion to the colored calyx, consists of three species: the one that has already been mentioned, the smooth strawberry shrub, and the sweet shrub. The smooth strawberry shrub is a similar shrub to the preceding with smooth branchlets and similar elliptical leaves, but somewhat larger. The flowers are similar, too, but a little smaller, dull green, ruddy-stained, or maroon and develop into the same kind of fruit. It is also a southern species, found in the same kind of situations, but with a smaller range. The sweet shrub is a California species and much larger than its two relatives with large, ovate leaves and larger, fragrant flowers having an intense maroon red color, but that fades to an ocher brown with age. The fruit is cuplike with a spiky crown. The shrub grows along streams.

181. Sumac (Anacardiaceae)

Everyone who rambles about outdoors should know the poison ivy and should be able to recognize it quite easily by its three leaves, or rather leaflets, for the leaves are compound. It is an erect shrub or vine with a woody stem that has a tendency to climb to considerable heights winding its way about tree trunks, over stone walls and fences, by means of aerial rootlets and spreading by running roots over the ground. The leaves are compound and consist of three ovate leaflets that are glossy or dull green, toothless or with a few coarse teeth, and that are inclined to droop. The flowers are insignificant, dull white or greenish, borne in loose, slender clusters, at the junction of the leaf stem and branch. The fruit is a small, white or cream-colored, nearly globular drupe.

The poison ivy is poisonous in all its parts, especially the leaves, the painful, itching skin eruption that is known as poison ivy poisoning being caused by an irritating nonvolatile oil. The plant should be avoided

525. Compound Leaf of Poison Ivy

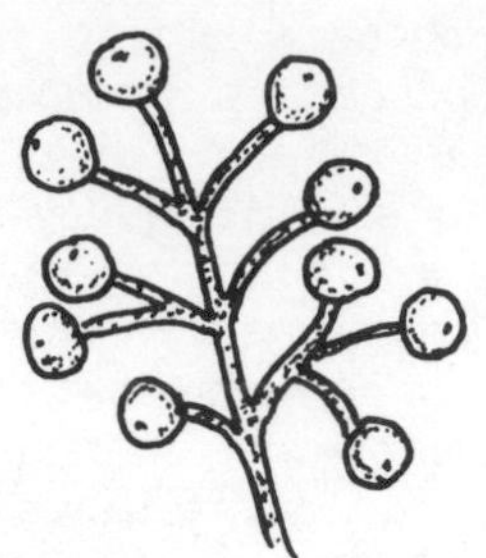
526. Fruit of Poison Ivy

at all costs, though some people seem to be immune to its toxic properties. Most people that have had poison ivy poisoning have come in contact with it, but it may also be acquired by coming in contact with the smoke from burning its leaves or stems or any part of the plant.

Equally as poisonous, if not more so, is the poison sumac, a related plant. It is a tall shrub or small tree, five to ten feet high, of low, wet ground or swamps. It has a short trunk that forks near the ground, the coarse branches spreading irregularly, and producing a round-topped head. The gray bark is smooth on young trees, rough on old ones, and the leaves are compound with seven to thirteen leaflets that are smooth, elliptical or obovate, toothless and abruptly pointed, light green above, paler beneath. The flowers are similar to those of the poison ivy and the fruit is somewhat similar too, being a small, greenish, ivory or dull white drupe about the size of a pea in long, slender, pendulous clusters and generally persisting throughout the winter. So any shrub or small tree growing in swampy places with white berries should be avoided for the poison sumac is viciously poisonous and, like the poison ivy, is poisonous in all of its parts. Some people, however, seem to be immune to its toxic properties, as with the poison ivy.

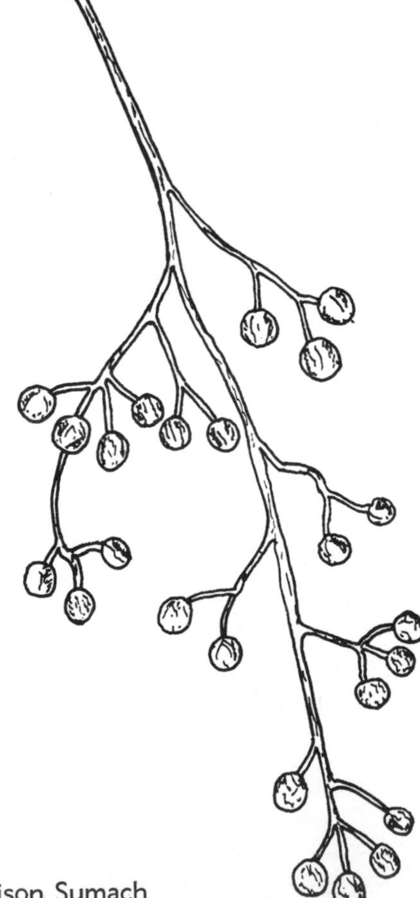
527. Fruit of Poison Sumach

Another relative of the poison ivy and the poison sumac is the poison oak, an erect, small southern shrub of thin woods with oaklike leaflets. There is also a western poison oak, a shrub with compound leaves made up of three leaflets, that are smaller than those of the poison ivy and usually irregularly lobed or toothed.

The preceding, as well as the following species of sumacs belong to the genus *Rhus,* the old Greek name of the sumac, one of seventy-three genera of the sumac family, also known as the cashew family. The family includes six hundred species of mostly tropical trees and shrubs, although a few species extend into southern Europe and temperate Asia and America.

They have a resinous bark, alternate, simple or compound leaves, perfect or imperfect flowers, the flowers with three to seven sepals and petals and as many, or twice as many stamens as petals, and mostly a one-celled ovary, and a fruit that is usually a drupe.

The sumacs number one hundred and fifty widely scattered species, and are shrubs or small trees usually with pinnately compound leaves with an odd leaflet at the end, small, greenish, perfect or imperfect flowers with five sepals, petals, and stamens, and a small drupe as the fruit, smooth and white or gray in the poisonous species, red and hairy and clustered in the nonpoisonous species.

528. Fruiting Cluster of Staghorn Sumach

Few trees are so conspicuous on the autumn landscape as the staghorn sumac when its soft, velvety, maroon, conical fruit clusters project into the air like fiery torches that burn on with gradual abatement into the middle of winter. The staghorn sumac gets its name partly from the fact that in winter, its forking leafless branches resemble the horns of a stag, and partly from the fact that the soft, velvety down that covers the growing shoots have a marked similarity to the incipient horns of the stag. Anyway, the staghorn sumac is a tall shrub or low tree of irregular contour and straggling habit and a flat-topped head, of uplands and gravelly banks. Its bark is dark brown gray, smooth on young trees, rough and scaly on older ones, and its twigs are coarse and brittle, the younger ones covered with long, soft or velvety, sepia brown hairs. The leaves have eleven to thirty-one, oblong, lance-shaped leaflets, dark green above, pale to white beneath, that lift and sway with every passing breeze. When naked in the winter, the staghorn sumac appears rather ugly and unshapely, but once the leaves have appeared, the tree acquires an entirely different almost beautiful character. The flowers are whitish green or yellowish green in dense, conical, hairy clusters, the staminate and pistillate on separate trees, the latter developing into tiny, globular acid drupes that are covered with deep red hairs and are clustered in large, compact panicles.

The fruit is very sour, as is also the fruit of the smooth sumac, but in spite of its sour taste, it is an important winter food for many species of wildlife. Several of our most important game birds rely on the sumac as a winter food, and so do some of the songbirds that winter in the North. Rabbits and the hoofed browsers also feed on the fruit as well as on the bark and twigs. If placed in water for a short time, the "berries" make a pleasing and agreeable drink known as "Indian lemonade." As a matter of fact they have been used as a substitute for lemon juice.

The smooth sumac is much similar to the staghorn sumac, but is much smaller and usually a shrub, two to ten feet high, with smooth, young twigs and is common in neglected fields, along roadsides, and borders of woods. Both the staghorn and smooth sumac make thickets on their own account, but sometimes make thickets together. They cannot be distinguished by their size alone, for the young staghorn, before it completes its growth, is often the height of its smaller relative. They can be distinguished, however, by the young twigs that are clothed in velvet in the staghorn and without hairs in the smooth, though they have a pale bluish or whitish bloom.

In the fall, the sumacs contribute their share to the autumn pageant. They gleam in scarlet and gold, often deepening to crimson and orange, and brighten waste places and neglected fields along fencerows, and up rocky, gravelly mountainsides where they "fling their magnificent beauty," "Like glowing lava streams, the sumac crawls/ Up the mountain's granite walls."

A sumac with aromatic leaves, the fragrant sumac is a straggling shrub, one to six feet high, of rocky woods. The compound leaves consist of only three leaflets, the lateral ones being asymmetrical, coarsely toothed, olive green above, and slightly downy beneath. The flowers are pale yellow or greenish yellow in small clusters, and bloom before the leaves expand, and the fruit is a red, globular drupe, downy, in small clusters.

Also a shrub, but sometimes attaining the proportions of a tree in the southern mountains, the dwarf or mountain sumac is a handsome species, one and a half to four feet high, with the twigs and branchlets softly haired, the twigs also with a balsamic or turpentine flavor to the taste. The leaves are six to twelve inches long, with nine to twenty-one leaflets that are smooth, shining, ovate lance-shaped and nearly always toothless. An odd feature of the leaves is that the central leaf stem is winged on each side between the pairs of leaflets. This species is the latest of the sumacs to blossom, the flowers sometimes not appearing until early August. They are greenish in a dense, terminal branched cluster and its panicles of bright red fruits may also be used for making Indian lemonade.

A similar, but western species, is the skunkbush, so-called because its leaves give off a fetid odor when bruised. The staghorn, smooth, fragrant, dwarf and the skunkbush sumacs, all contain tannin that is present in the greatest quantity in the leaves and flowers. There are several other western species such as the laurel sumac, having simple, evergreen and aromatic leaves and whitish fruits; the lemonade berry or mahogany, a shrubby tree of the seacoast also with simple, evergreen leaves and, of course, a far different tree from the mahogany of the tropics; and the sugar bush of inland drier hillsides.

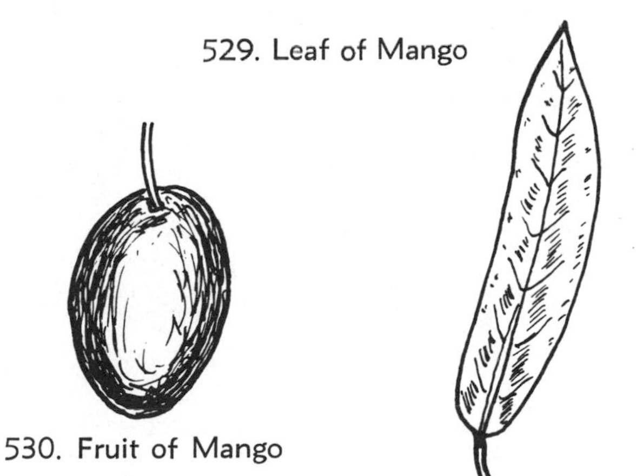
529. Leaf of Mango
530. Fruit of Mango

The sumac family does not consist entirely of sumac trees, but also includes other species as the mango, the California pepper tree, both introduced species, the pistachio and the cashew, and two native trees, the poisonwood and the smoke tree. The mango is a native of the Malay Archipelago, widely cultivated throughout the tropics and, to a limited extent, in the United States, with large and leathery leaves, green or yellow flowers, in stiff erect clusters, and an edible greenish stone fruit whose taste has been likened to that of an apricot, a peach, or a pineapple. Where the mango has been cultivated for the last four thousand years, the fruit is as important as the apple is to us.

The pepper tree, a native of Peru, has become one of the most popular shade trees in California where it has been widely planted on the streets. It is an

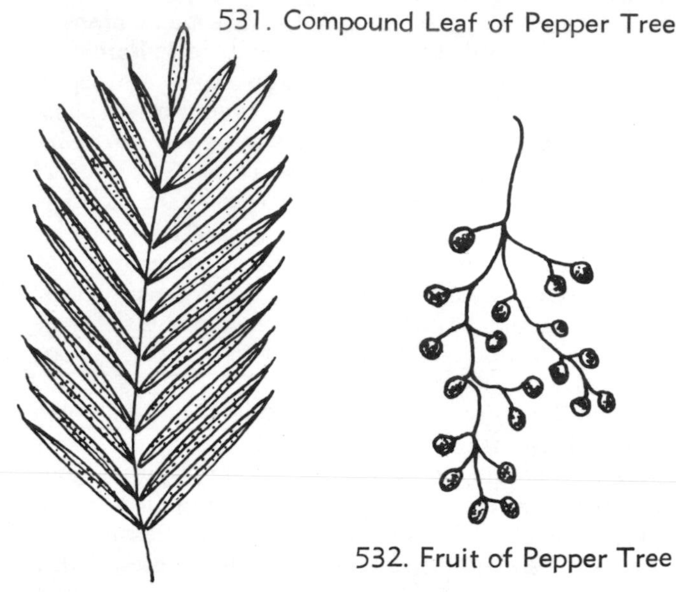

531. Compound Leaf of Pepper Tree

532. Fruit of Pepper Tree

evergreen tree, and incidentally not a true pepper, with gracefully drooping branches, numerous leaflets that are narrowly lance-shaped, yellowish white flowers in a multi-branched terminal cluster, and a rose red fruit (drupe) that persists throughout most of the winter. A somewhat similar species from Brazil, more often a shrub or a small tree, is planted as an ornamental in Florida.

The pistachio, whose fruit is the pistachio nut of commerce, is a small tree, a native of the Mediterranean region and the Orient, with compound leaves of one to five pairs of leaflets, small, inconspicuous flowers, without petals, and in lateral clusters (panicles), and a fruit that is a dry drupe, oblong or egg-shaped, red and wrinkled, the kernel of its stone rich, oily, green or yellowish green and the source of pistache. The United States has its own pistachio, a small tree of the Rio Grande region of Texas, with compound leaves divided into nine to nineteen leaflets, naked red-tinted flowers, in clusters, and dark brown, fleshy drupes.

The cashew, a tropical American species, is an evergreen tree with simple, rather leathery leaves, and many small, fragrant, yellowish pink flowers in terminal clusters. The fruit is fleshy, red or yellow, technically a part of the receptacle, and commonly called cashew apple in the tropics. It is about two and one half inches long, and neither edible, nor safe when raw. The kidney-shaped nut, which is the true fruit, contains the edible kernel, that, when roasted, is the cashew nut of commerce.

A beautiful shrub or small tree, the poisonwood, is as poisonous as the poison ivy and the poison sumach, causing a dermatitis similar to that produced by the other two. It occurs in the pinelands and hammocks of southern peninsular Florida. The leaves are somewhat ashlike, the flowers are yellow green in drooping axillary panicles, and the fruit is an orange red drupe. The pendulous clusters of the "berries" have given the tree the name of coral sumach.

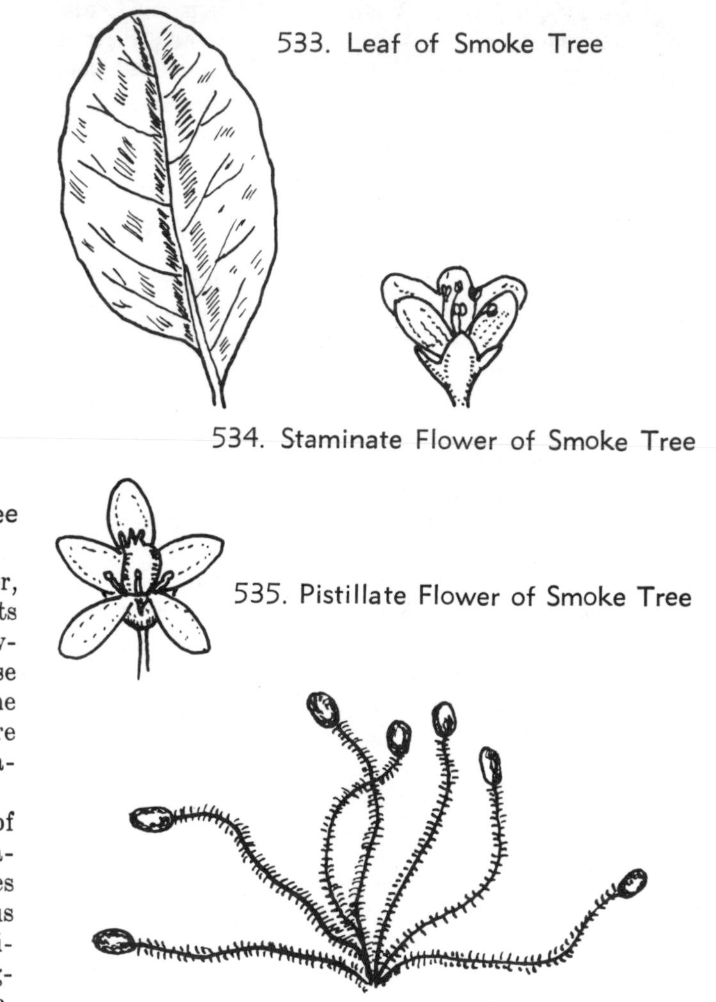

533. Leaf of Smoke Tree

534. Staminate Flower of Smoke Tree

535. Pistillate Flower of Smoke Tree

536. Fruit of Smoke Tree

Its pithy stems, its aromatic, resinous juice, and general habit stamp the smoke tree as a member of the sumach family. The American smoke tree is a beautiful shrub or small tree, sixteen to forty feet high, of southern wooded riverbanks, with large, simple leaves, four to six inches long and quite unlike the sumach leaves. They are more or less elliptical, toothless, tapering to the base, deep green above, paler beneath. The flowers are small, greenish white, the staminate and pistillate on separate trees, in large, loose, thin clusters, the flower stems becoming elongated and feathery in fruit, that is a very small, sepia brown drupe. The elongated and feathery stems are like graceful and delicate plumes, are tinted pink to green, and in the aggregate, form a great cloud of rosy haze or smoke that has given the tree its common name, and that in late summer, makes it a thing of beauty. The American smoke tree is a rather rare species and makes a handsome lawn specimen. A somewhat similar Eurasian shrub is also planted as an ornamental.

182. Sundew (Droseraceae)

Few plants could be more innocent-looking than the tiny sundew of swamps and bogs, with its raceme of buds and solitary little blossom that opens only in the sunshine, but it can be a deadly trap to fungus gnats and other small insects. These small creatures, attracted to the jewels that dot the leaves and sparkle in the sunshine, find to their dismay, that the jewels are a viscid substance, secreted by glands in the leaves, more sticky than honey and that cling to their feet like glue. Then the pretty reddish hairs that cover the leaves reach out like tentacles and clasp them in a slowly closing, tight embrace. Should an insect struggle in a vain attempt to escape, the hairs only move the faster. No torture implement of the Inquisition was more cleverly designed than the leaves of the sundew, for once an insect becomes imprisoned, the leaf rolls inward and forms a temporary stomach for a complex fluid, similar to the gastric juice in the stomach of animals, secreted to digest the hapless victim. One can watch the operation of the trap by placing a small insect on the leaves.

There are more than ninety species of sundews, most being found in Australia, but several are native to North America. They belong to the genus *Drosera,* Greek from dewy in reference to the glistening leaf hairs. All are small, perennial herbs, with a basal rosette of much-modified leaves that are covered with glistening, sticky hairs. The flowers are usually in small clusters at the end of a short stalk that extends upwards from the rosette of leaves. The flowers are regular, perfect, with five sepals and five petals, five or more stamens, and a one-celled ovary and a single style. They are followed by a small capsule.

One of the more common of our sundews is the round-leaved sundew, a very small plant with long-stemmed, round leaves that lie close to, or upon the ground. The curious leaves resemble in shape, a long-handled frying pan. Both they and the stems are covered with long, fine, red hairs. The flower stem is red, erect, and smooth and bears about four or six small flowers in a one-sided cluster that bends over, the blossoms opening one at a time and only in the sunshine. They are visited by the fungus gnats and other small insects.

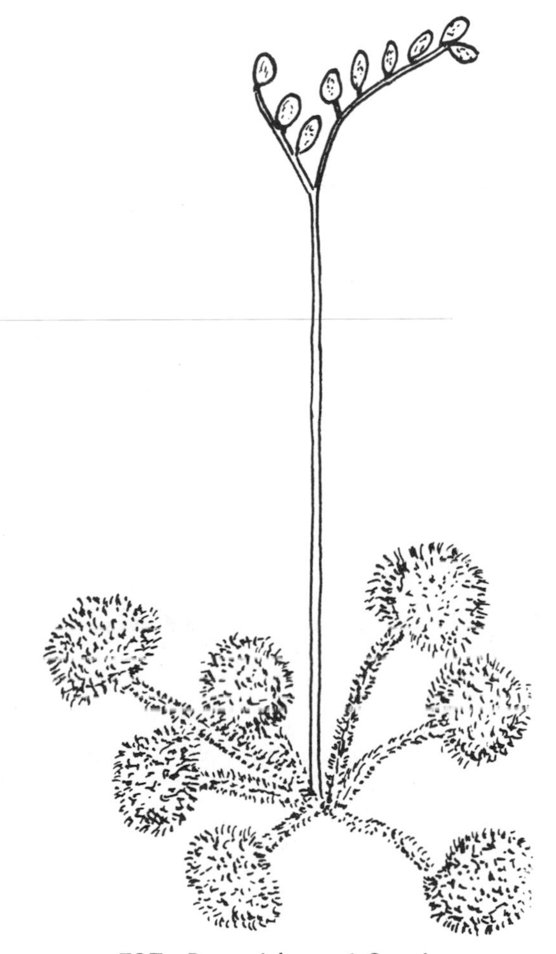

537. Round-leaved Sundew

Less common, but a similar species, the long-leaved sundew differs in having elongated, blunt-tipped leaves and naked leaf stems, the red hairs appearing only upon the little leaves. Unlike either of these two species, the thread-leaved sundew has leaves that are reduced to a mere threadlike shape without a distinct stem. The flowers of this species are pinkish purple, while those of the long-leaved sundew are white. All sundews are found in bogs and in moist sandy ground. All are also insectivorous, capturing insects whereon they feed.

And why, one may well ask, are they adapted for such a purpose? Simply because the soil wherein they live does not provide a sufficient supply of nitrogen compounds, hence, they must obtain them elsewhere. One may examine the roots of any sundew and will find that they testify to the small use they serve, namely, to absorb water for the manufacture of the sticky fluid the leaves exude. Incidentally, the sap of the sundews stains paper a madder purple.

The genus of sundews is one of four genera that comprise the sundew family, a small group of herbs whose botanical characteristics are essentially those of the sundews. Besides the genus of sundews, the only other genus of interest to us is *Dionaea,* having only one species, the famous Venus flytrap that seems to do everything, but think, when it captures its prey. It is a small, perennial plant of sandy bogs and pinelands of the Carolinas, sometimes found in the neighboring states, with a basal rosette of highly specialized leaves, and small, white flowers in a terminal cluster. The leaves are uniquely adapted for catching and digesting insects. The blade of the leaf consists of two hinged lobes or valvelike segments, whose margins are set with sharp teeth, and with three short, stiff spines in the center. When an insect chances to come in contact with these spines, the two lobes or valvelike segments immediately close, and imprison the insect, the marginal teeth interlocking and making escape difficult if not impossible. Once an insect is caught in the trap, glands along the surface of the leaf begin to secrete digestive juices that act upon the insect and prepare it for absorption into the tissues of the plant. When the insect has become completely digested, the leaf opens and the trap is ready for another victim.

One needs only to watch the behavior of the sundew and the Venus flytrap to realize that their activities in capturing insects are analogous to a simple animal reflex action—that is, the reception of a stimulus and the transmission of an impulse to a place where movement or an activity can take place and that there is only a slight difference between plants and animals when certain functions normally associated only with animals may also be found in plants as, for instance, a response to an external stimulus.

A third genus of the family, *Drosophyllum,* consists of only one species, the flycatcher of Portugal and Morocco. It is a slight, shrubby plant with narrow, glandular leaves and yellow flowers having ten stamens. Unlike the sundews and Venus flytrap, it lives in dry, rocky situations, but like the others, its leaves have many glands that secrete numerous beads of a sticky substance. An insect coming into contact with it becomes inextricably caught and soon is unable to move. Digestive juices are then secreted and before long, the insect becomes part of the plant's tissues. Those who have seen the plant covered with dead and dying insects say it is a memorable experience.

183. Sweetleaf (Symplocaceae)

The sweetleaf family contains only one genus, but the genus, *Symplocos,* Greek for connected, in allusion to the often united stamens, is a large one, of nearly three hundred species of trees and shrubs, found in most tropical and warm regions with the exception of Africa, with one species native to southeastern United States. They have alternate, simple, deciduous leaves, but evergreen in some species, usually small and inconspicuous, perfect and/or imperfect, fragrant flowers in spikes or racemes. The flowers have a five-lobed calyx, five or ten-lobed corolla, fifteen or more stamens that are often in bunches and fastened to the corolla, and a three-celled ovary. The fruit is a dry drupe. The leaves and bark of several species produce yellow dyes, while the roots of others contain substances used in the preparation of tonics. Several species are sometimes cultivated, including our own native species known as the horse sugar, or sweetleaf.

The horse sugar or sweetleaf is a shrub or small, open-headed tree, ten to thirty feet high, with a short trunk, and slim, ascending branches. It has à smooth, light gray bark with a reddish cast, thin, leathery, elliptical, leaves, lustrous dark green above, paler beneath, and fragrant white or yellowish flowers in small, close axillary clusters, and an orange or brown, nutlike drupe with one seed. It is a southern species found in rich woods and thickets. The leaves are sweet-tasting, hence, its name of sweetleaf, and as horses and cattle also relish the leaves it is also known as horse sugar. The bark of both the stems and roots is bitter and aromatic, yields a yellow dye, and has tonic medicinal properties, but is of no commercial importance.

184. Tallowwood (Olacaceae)

The tallowwood family is a group of twenty-seven genera and about one hundred and fifty species of tropical trees, shrubs, and woody vines scattered throughout the tropical forests of the world. They have simple leaves, small flowers, and a one-seeded fruit. Two species of the family are found in the lower Florida peninsula. They are the tallowwood and the gulf graytwig.

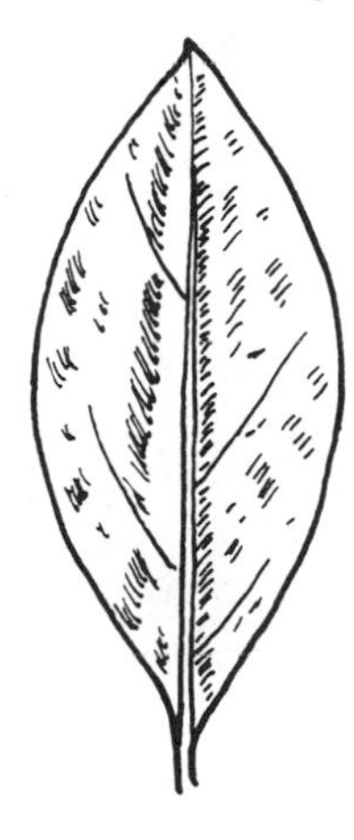

538. Leaf of Tallowwood

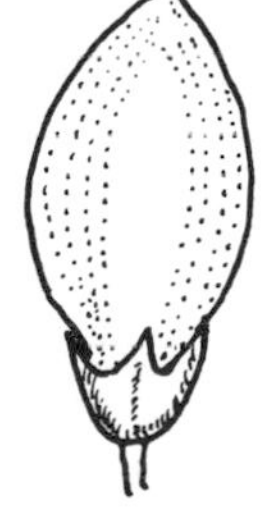

540. Fruit of Tallowwood

The tallowwood of the genus *Ximenia,* named for F. Ximenes, a Spanish missionary in Mexico, is a large shrub, with long, thorny, vinelike branches found growing on hammocks, sand dunes, and in pinelands, as well as on the sandy, gravelly, and coral-rock margins of freshwater lakes and tidewaters. It has alternate, simple, persistent, leathery, oblong to elliptical leaves, and perfect, regular, bell-shaped, fragrant, yellowish white flowers in two- to four-flowered, axillary, stalked clusters. The flowers have a four- to five-lobed calyx, four, hairy petals, eight stamens, and a one-celled ovary. The fruit is an ovoid, yellow or yellowish red drupe with an astringent pulp and a profusely pitted, reddish stone.

The gulf graytwig of the genus *Schoepfia,* is a large shrub or small, bushy tree with ascending, thornless branches growing on hammocks and on the shady or gravelly margins of tidewaters. It has alternate, simple, persistent, leathery, ovate-elliptical leaves, perfect, regular flowers borne in axillary clusters of two or three or solitary. The flowers have a cuplike calyx, four united petals, four stamens, and a one-celled ovary. The fruit is an ovoid, scarlet drupe with a thin, fleshy pulp and a hard, but brittle, stone. The graytwig is of little value, unlike the tallowwood whose bark contains appreciable quantities of a water-soluble astringent suitable for tanning leather and whose fruit is edible, being the source of food for many indigenous and migratory birds.

185. Tamarisk (Tamaricaceae)

The tamarisk family is a small group of about one hundred species of curious, salt-resistant shrubs and trees contained in only four genera. They appear to be leafless and are somewhat cedarlike, but actually they have innumerable, scalelike, tiny leaves that are pressed flat against the twigs. Their flowers are regular and perfect, very small, and stalkless, with four or five sepals, four or more stamens, and a one-celled ovary, prevailingly pink, in small spikes, the latter grouped in a terminal, branching cluster, being quite attractive in the mass. The fruit is a small capsule.

One of the four genera, *Tamarix,* is an interesting group of seventy-five species from Eurasia or Asia Minor, many being salt-tolerant plants of semi-desert places and growing in pure sand. Known as tamarisks, they are very feathery, slender plants and superficially appear leafless because of the small leaves that are set closely on the twigs, both the twigs and leaves being shed in the fall. The flowers are very small in dense racemes, that are grouped in terminal panicles, and the capsule is tiny. They are excellent plants for the seaside garden. The best-known species in cultivation is the salt cedar, also called French tamarisk. It is a shrub or small tree, fifteen to twenty-five feet high with bluish foliage and white or pinkish flowers. Some members of the genus *Myricaria,* derived from *Myrica,* the old Greek name of the tamarisk, are also of horticultural interest, particularly the German or false tamarisk. It is somewhat more hardy than the true tamarisks, wherefrom it is separated only by technical characters.

186. Tea (Theaceae)

Few plants are so highly esteemed by horticulturists as the camellias, showy-flowered trees and shrubs named after a Jesuit, George Camellus or Kamel, who travelled extensively in Asia during the seventeenth century. They have alternate, evergreen, toothed leaves and usually solitary, nearly stalkless flowers, with five to seven sepals and mostly five petals, the fruit a woody capsule. The flowers are red in the typical plant, but of other colors or even double in the many horticultural forms.

There are thirty odd species of camellias and an unknown number of varieties. Camellias are grown in gardens from southeastern Virginia through the Carolinas to Florida, westward along the Gulf coast to Texas, and along the Pacific coast to British Columbia. Some six species are grown in this area, the common camellia and the sasanqua and their many cultivated varieties probably being the most popular.

The camellias form one of the sixteen genera of the tea family, a group of one hundred and seventy-five species of trees and shrubs of the tropics and the warmer temperate parts of East Asia. They have alternate, simple, usually evergreen, leathery leaves, regular and perfect solitary flowers, though sometimes a few in a raceme or panicle, that are often large and showy. The flowers have five to seven sepals, mostly five petals, both separate or united at the base, many stamens, and a three to five-celled ovary. The fruit is usually a dry, rather woody capsule, though sometimes like a drupe.

Only a few species of the family are native to North America. These include the silky camellia, the mountain camellia and the loblolly bay—all from the southeastern states. The silky camellia is a mountain shrub, three to fifteen feet high, with tan brown, hairy twigs and ascending branches. The leaves are ovate or elliptical, pointed at the tip, finely-toothed, deep green above, paler beneath and the flowers are large, lustrous cream white, cup-shaped, with five obovate petals that are finely broken on the edge, and with many stamens, the anthers bluish purple. The fruit is a five-celled, nearly spherical capsule.

The mountain camellia is a somewhat similar shrub, but the twigs are smooth, the leaves are slightly larger and very obscurely sharp-toothed, the flowers often have six petals with orange anthers, and the fruit is distinctly angular and pointed. Both species are also known as the stewartia or stuartia, named for John Stuart, Earl of Bute, patron of botany.

The loblolly bay, also known as tan bay, is an evergreen tree, forty to seventy-five feet high, with

541. Leaf of Loblolly Bay

542. Flower of Loblolly Bay

a slender, straight trunk, and a narrow, compact head, found growing in swamps from tidewater Virginia along the coast to the delta of the Mississippi. It has red brown bark, lustrous, leathery, dark green leaves that are narrowly elliptical or lanceolate and serrate on the margins, and fragrant, showy white, solitary flowers, the petals spreading out like great wild roses often two or three inches broad. The fruit is a dry, woody, pointed capsule covered with silky hairs. The bark was once used for tanning, hence, its name of tan bark.

The loblolly bay is one of two species of the genus *Gordonia,* named for James Gordon, an English nurseryman, native to the United States, the other species Asiatic. The second species, sometimes known as the Franklinia, is a shrub or small tree, not over twenty-five feet high, with erect branches, oblongish, deciduous leaves that turn a bright crimson in the fall, flowers larger than those of the loblolly bay, and a fruit that is nearly globe-shaped. Both species are cultivated in the South, though the Franklinia can be grown farther north if protected.

The Franklinia has a rather unique and somewhat remarkable history. In 1790 William Bartram, an American naturalist, found the tree growing in groves along the Altamaha River and sent specimens to his father's garden that had been established in Philadelphia. From there, the plant was introduced into cultivation. Thus far, there was nothing unusual, for many plants are found growing in the wild and their cultivation undertaken. But now comes the strange part. Since Bartram's discovery of the tree, it has never since been found growing in the wild. Many efforts have been made to rediscover the tree, but all have proved futile and it is now believed that no other wild specimen exists. Recently it was reported, though erroneously, as rediscovered. All our specimens of the tree are now found in gardens. They are lineal descendants and the sole representatives of those found by Bartram.

Although the camellias are of horticultural interest and importance, the one commercially important member of the tea family is the tea plant, an Asiatic shrub or small tree, never over thirty feet high, and as cultivated, is usually a small shrub, with alternate, leathery, elliptic leaves (wherefrom tea is derived), that are shallowly toothed. The flowers are white, fragrant nodding, and solitary, though sometimes in two to four-flowered clusters, the flowers with five to

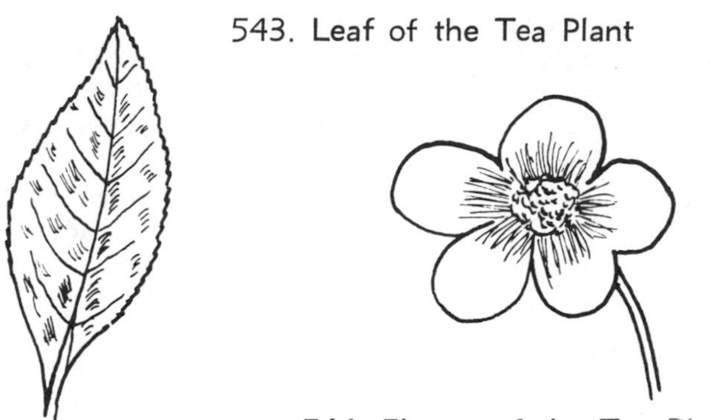

543. Leaf of the Tea Plant

544. Flower of the Tea Plant

seven sepals, five petals, and many stamens. The fruit is a woody capsule.

The origin of the tea plant remains obscured in the mass of Chinese legends, but it appears to have been cultivated for medicinal purposes by 2500 B.C. By the ninth century it had become a common Oriental beverage, but its appearance in Europe did not occur until the sixteenth century when explorers brought tea from Java to England and Holland. Soon thereafter, it was served in London coffee houses, but being expensive it was a luxury that only the wealthy could afford. The early colonists brought the tea drinking habit to America and, as is known, tea became of significant importance in the events preceding the Revolution. The tea plant is grown in the United States essentially for interest and ornament, as no tea is commercially produced here. An attempt to do so was made in the Carolinas, but it met with failure primarily because cheap labor was not available. It proved to be an interesting experiment, but nothing more.

The finest teas are made from the youngest and tenderest of the leaves. For all grades, the leaves are slightly bruised before being further treated. In making green teas, such as the oolong and souchong varieties, the bruised leaves are dried in firing machines and then packed and shipped. In making black teas, such as the orange pekoe, the bruised leaves are permitted to dry slowly so that fermentation can take place. After fermentation, the blackened leaves are treated as in the case of the green teas. Tea leaves contain volatile oils, that give the beverage its aromatic fragrance, alkaloids, that give tea its stimulating effect, and tannins, that are removed only when the leaves are boiled.

187. Teasel (Dipsacaceae)

Long a garden favorite because it is easy to grow and because it has a long blooming period, the sweet scabious is a perennial, native to southern Europe, but in our gardens it is treated as an annual. It grows two to three feet high, has a branching habit, basal leaves that are broadly lance-shaped and cut in lyre-shaped lobes, the margins coarsely toothed, and dark purple, pink, or white flowers in heads that are two inches across. The flowers are rich in nectar and are visited by various insects, and as the flowers of a cluster develop gradually, the whole head offers a standing invitation for a considerable time, hence, the insects return to the same head day after day. In southern Europe the flowering heads were used extensively at one time, and perhaps still are, for funeral wreaths, hence, the sweet scabious is known as mourning widow. In our country it is also known as mourning bride.

There are a number of scabiosas, some seventy species as a matter of fact of the genus *Scabiosa,* from the Latin for itch, in allusion to the medicinal use of some of species. They are hardy annual or perennial herbs, wth simple, opposite, oval or lance-shaped leaves, often lobed or deeply cut, and blue, purple, brownish black, reddish brown, pink, cream, or white flowers in terminal heads. The flowers have a calyx represented by bristles, a tubular corolla and four stamens. The fruit is a one-celled capsule.

The genus *Scabiosa* is one of about seven genera that comprise the teasel family, a group of one hundred and forty species of annual or perennial herbs of the Old World, but with an almost worldwide distribution. They have essentially the same characteristics as the Scabiosas, opposite or whorled leaves that are toothed or deeply cleft, small, perfect flowers in dense, bracted, involucrate heads, the individual flowers with a poorly developed calyx that is cuplike or divided into bristly segments on the rim of the receptacle, a tubular or funnel-form corolla of four or five lobes, four or two to three stamens, and a one-celled, one-ovuled ovary. The fruit is a one-seeded

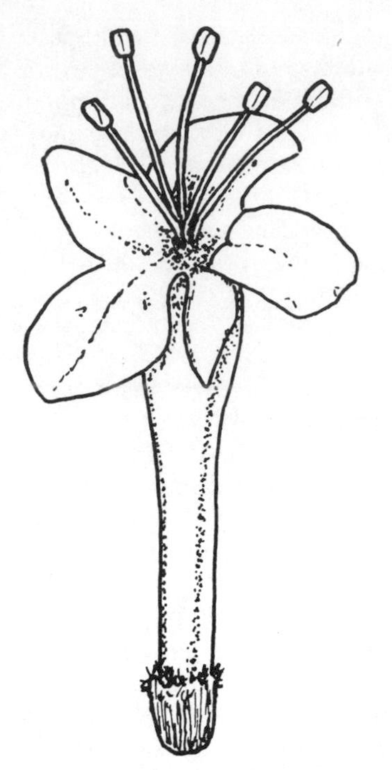

545. Flower of Teasel

achene that is often crowned by persistent bristles.

Aside from the scabiosas, the only other member of the family of any interest is the fuller's teasel of the genus *Dipsacus,* from the Greek for thirst, in allusion to the water-holding leaf bases, whose peculiar fruiting head is used to raise the nap on woolen cloth. It is a biennial European herb, four to six feet high, with opposite leaves that are nearly a foot long, and whose bases are connected and cuplike. The flowers are pale lilac, tubular and small, and crowded in a dense, cylindrical head that is provided with a number of hooked bracts that are spine-tipped (the teasel). The plant is grown commercially for these teasels.

188. Theophrasta (Theophrastaceae)

The theophrasta family is a small group of five genera and about seventy species of tropical trees and shrubs. The largest genus, *Jaquinia,* comprises about thirty-five species known as the joewoods, one being the joewood, found growing in the dry, sandy, and coral soils adjacent to saltwater of lower Florida.

The joewood is a shrubby tree, about fifteen to twenty feet high, with stout, sometimes distorted, spreading branches that form a dense, nearly globular crown, alternate, or occasionally opposite, or nearly whorled, simple, persistent, leathery, oblong, obovate, or spatula-shaped leaves that are yellow green above, and

546. Leaf of Joewood

547. Fruit (Berry) of Joewood

paler beneath. The flowers are yellow, perfect, and fragrant in terminal and axillary racemes, the individual flowers with a five-lobed, bell-shaped calyx, a five-lobed, funnel-shaped corolla, five stamens, and a one-celled ovary. The fruit is a nearly globular, orange red, multi-seeded berry.

189. Tobira (Pittosporaceae)

The tobira family is largely Australian with nine genera and over one hundred and fifty species of trees, shrubs, and vines with alternate, often leathery, thick leaves, and rather showy flowers, in various sorts of clusters. The flowers, nearly always regular, have five sepals, five petals, five stamens, the ovary with one style. The fruit is either a capsule or berry-like.

The most important of the nine genera is *Pittosporum,* Greek for resinous and seed, in allusion to the sticky seed, a group of one hundred species, several being grown for ornament in the South, especially in southern California. They have alternate, or on young twigs, apparently whorled leaves that are wavy-margined and faintly toothed or without teeth, flowers solitary or in clusters, usually terminal, but sometimes in the leaf axils, and a fruit that is a capsule with sticky seeds. The members of the genus are known as the Australian laurels, and some of them are very popular shrubs and trees in California, Florida, and along the Gulf Coast.

A few of the species that might be mentioned are the tobira, a shrub, six to eight feet high, with thick and leathery, oval-shaped leaves, greenish white, fragrant flowers in terminal clusters, and a native of China and Japan, useful for hedges in California and Florida, and often grown as a pot plant in greenhouses; the butterbush, a small tree up to twenty feet, with drooping branches, lance-shaped leaves, yellow flowers, solitary or in small clusters, and a yellow fruit; the diamond-leaf laurel, a tree sixty to eighty feet tall, with oval or diamond-shaped leaves, white flowers in terminal clusters (corymbs), and an orange yellow fruit; the Victorian box, also called cheesewood, a shrub fifteen to twenty feet high, with leathery, shining green, oval leaves with rolled margins, yellowish green flowers in dense, terminal clusters; and the tarata, a tree up to forty feet high, with wavy-margined, elliptic leaves and yellowish, fragrant flowers in terminal compound umbels.

190. Torchwood (Burseraceae)

Only two representatives of the torchwood family occur in the United States, one species in southern Florida and the other in California and Arizona. Both are of the same genus, *Bursera,* named after Joachim Burser, a German botanist and physician, one of twenty genera of the family, a group of about four hundred species of trees and shrubs with resinous bark and wood. They have alternate, pinnately compound leaves, perfect or polygamous flowers, in clustered racemes or panicles, the flowers with a four- to five-lobed calyx, four to five petals, with twice as many stamens as petals, a two- to five-celled ovary, and a drupaceous fruit. The resinous exudations from the bark are used medicinally and for compounding incense. Frankincense is the product of several Arabian species, while myrrh is obtained from a small tree found on the shores of the Red Sea.

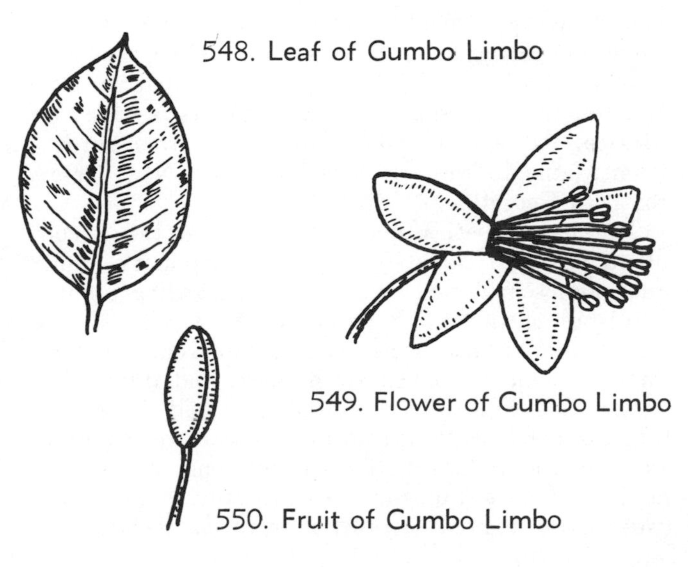

548. Leaf of Gumbo Limbo

549. Flower of Gumbo Limbo

550. Fruit of Gumbo Limbo

The genus *Bursera* contains about eighty species of tropical American trees, two occurring in the United States, one of them being the gumbo limbo found in Florida. It is a moderately large, round-headed tree, fifty to sixty feet high, with alternate, persistent or tardily deciduous leaves, odd pinnately compound with mostly five or seven leaflets, the leaflets oblong to ovate, leathery, bright green above, paler beneath. The polygamous flowers in many-flowered racemes appear with or before the leaves in early spring. The fruit looks like a green berry as it develops, but upon ripening develops into a dark red, three-angled leathery pod containing a single (sometimes two), triangular, bony nutlet. The tree resembles the willows because of its ability to sprout from a stump, or from pieces of any size set into the ground. The name gumbo limbo appears to be of Negro origin.

191. Turnera (Turneraceae)

The turnera family is a group of herbs or shrubs having flowers with five stamens, three styles, and a free ovary. There are six genera and about ninety species, most being tropical American. The genus *Turnera,* named after William Turner, an English herbalist, is a large group of tropical American plants, typifying the family. They are herbs and shrubs with alternate leaves and solitary, axillary, yellow flowers, the peduncle and petiole often coherent. The drug damiana, used as a stimulant, tonic, and aphrodisiac, is obtained from several species.

Among the ninety species are those known as the piriquetas. They are tropical and subtropical herbs with bright to deep yellow petals and spherical, green capsules of three cells in the axils of the upper leaves. Four species occur in Florida, one extending as far north as North Carolina.

192. Unicorn Plant (Martyniaceae)

The members of the unicorn-plant family are mainly tropical herbs with chiefly opposite, simple leaves, and perfect, irregular flowers, with a four- to five-cleft, or four- to five-parted calyx, a tubular corolla, four stamens, a one-celled ovary with few to many ovules, and a fleshy, drupelike fruit. The family contains three genera and about ten species, two being found in southwestern United States, the third native from Indiana to Iowa, south and west to Texas and New Mexico, but also found as an escape from gardens from Maine to western New York and New Jersey, south to Georgia.

This latter species, known primarily as the unicorn plant, but also called devil's claw, elephant's trunk, and proboscis flower, is a unique sprawling annual, multi-branched, that grows one to three feet high. Its leaves are alternate, or nearly alternate, roundish-oval, thick, soft, heart-shaped at the base, and from seven to ten inches wide. Its flowers are bell-shaped, or funnel-shaped, yellowish purple, white or pink purple, an inch and a half to two inches long, in axillary racemes of several blossoms, and its fruit is a hanging, two-beaked, curved capsule, four to six inches long,

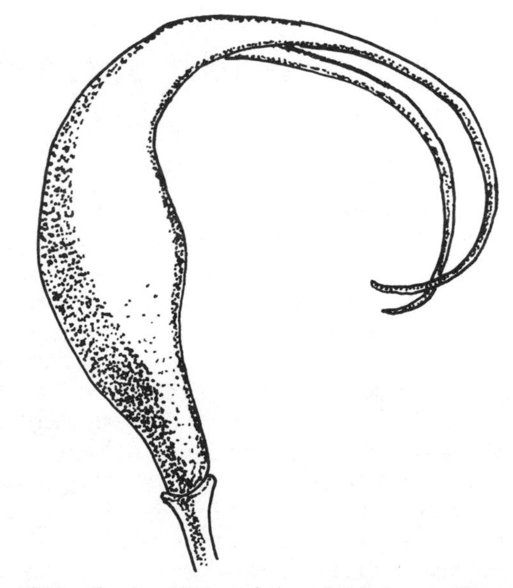

551. Fruit (Capsule) of Unicorn Plant

and woody when mature. The fruits when young are soft, tender, green and somewhat fleshy, and in that stage are used much like pickling cucumbers. The unicorn plant of the genus *Martynia,* named for John Martyn, a professor of botany at Cambridge, England in the eighteenth century, is often grown in eastern gardens just for pickling purposes. The pickling stage is in late summer and early autumn.

193. Valerian (Valerianaceae)

Once a favorite in old gardens, specially prized for the fragrance of its flowers, the common valerian or garden heliotrope may still be found in the modern garden, or may be seen growing along the roadside or on the border of a cultivated field, for it has long been an escape. A native of Europe and northern Asia, it is a strong-growing, perennial herb with thick and spreading rootstocks that are used medicinally. The stem is finely haired and the leaves are opposite and compound with eleven to twenty-one, lance-shaped leaflets that are sharply toothed, the upper ones being the exception. The flowers are numerous, small, white, pink, or lavender, and are set in compact, roundish clusters at the ends of the branches. They are succeeded by small achenes.

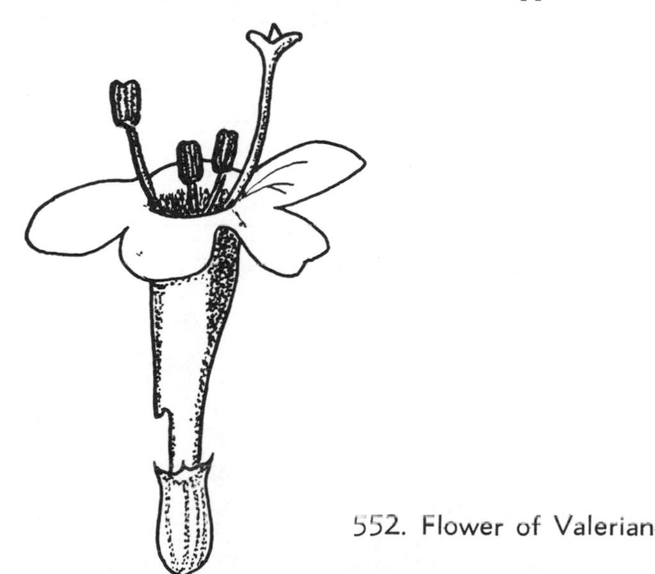

552. Flower of Valerian

Another import from Europe is the red valerian, a compact, bushy perennial plant, easily grown and long a garden favorite, with gray green foliage and many crimson flowers in dense, terminal clusters. The red valerian is also known as Jupiter's-Beard in allusion to the whorl of feathery appendages that develop from the calyx as the fruit matures; as Harriet Keeler, in her book, *Our Garden Flowers* remarks, "a very large name for a very small thing."

This is true with so many other groups of plants. A native species of valerian grows in wet or swampy ground, and, hence, is known as the swamp valerian. It is an erect, smooth plant with compound leaves of deep green leaflets, and small, dull magenta pink, paler pink or white flowers clustered in a terminal spike. Northern in range, it refuses to grow south of a line from Maine to Michigan.

The valerian family, from the Latin *valere* to be strong and powerful, in allusion to the medicinal virtue of the plants, is a family comprising nine genera and three hundred species of annual or perennial herbs (rarely shrubs) of the northern hemisphere, often with unpleasantly odorous roots. The leaves are opposite. The flowers are small, usually unisexual, sometimes irregular and attractively clustered. They have a tubular, five-parted, five-lobed calyx, a five-lobed tubular corolla, one to three stamens, and a one- to three-celled ovary, two cells being empty, the third with a single ovule. The fruit is an achene, sometimes crowned by the beautiful, radiate calyx (pappus), with its many plumose segments, or by a crown of bristles or hooks.

There are several members of the valerian family that are cultivated for their leaves that are used for salads. One of them is the corn salad, a native of Europe where it is grown as a potherb and salad. It is also cultivated in some parts of our country, but has escaped and become naturalized from Maine, south to Virginia and Arkansas. It is a smooth annual plant that grows from six to twelve inches tall in fields, waste places and occasionally, along roadsides. The stem leaves are opposite, sessile, and often toothed, the basal leaves are finger-shaped in a dense rosette, and all are tender and rather succulent. The flowers are light blue, in roundish, terminal clusters.

The leaves are best served as a salad with lettuce, and water cress, and with a salad dressing, for alone they are somewhat tasteless. In some markets in the East, the plant is commonly sold by the name of field salad.

There are several native relatives of the corn salad, but, except for the fact that they have white flowers, they are so nearly like the European corn salad and also like one another, that it is difficult to distinguish between them, their identifying characters being their fruit or seeds. Their leaves are also collected and used for salads and as a substitute for spinach. One species, called the goosefoot corn salad, grows in moist ground as does another species known as the beaked corn salad. The former occurs from western New York to Minnesota, south to Virginia and Kentucky, the latter from Massachusetts to Minnesota, and south to Florida and Texas.

194. Vervain (Verbenaceae)

When the snow lies on the ground and the sun shines brightly, the dried stalks of the blue vervain etch delicate shadows on the snow. To many people this is the plant's greatest appeal. Certainly the purple blossoms are too small to be attractive, nor are the slender stalks appealing, though they branch upward like the arms of a candelabra. The reason is, they have buds

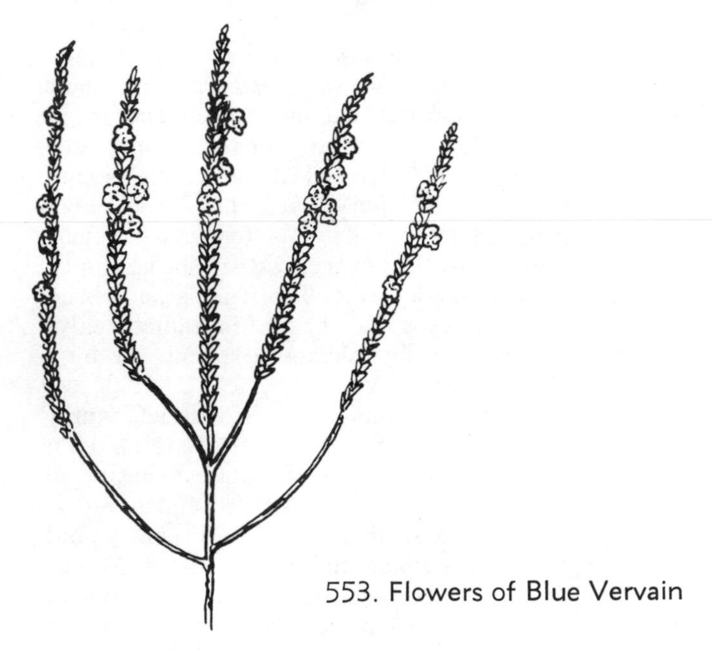

553. Flowers of Blue Vervain

at the top, flowers in the middle, and brown nutlets at the bottom, an arrangement that grates on our sense of the aesthetic. The name of the plant is misleading, for the flowers are not blue, nor do they approach any semblance of blue. Flowers, however, were not designed for man's enjoyment, but to attract insects, and so, many bumblebees, honeybees, and other insects are usually seen about the blossoms, sometimes appearing as if they were asleep on the blossoming spikes.

At the time of Shakespeare, the vervain was believed to be a charm against witches, and in ancient times as a charm to bring lost love back, for as Virgil writes:

> Bring forth water, and encircle all altars with a soft fillet:
> Burn thereon oily vervain and male frankincense,
> That I may try, by sacred magic spells,
> To make my lover madly love.

Undoubtedly the plant whereto Virgil refers was the verbena, the *herba sacra,* used according to Pliny in ancient Roman sacrifices. Anyhow, the early Christians accorded the vervain growing on Mount Calvary with healing virtues, and the Druids counted it among their sacred plants. Later, witches, regarding it as an enchanter's plant, gathered it and used it as an ingredient in their brew to help them perform mischief with their incantations. However, it also seemed to work against them, for Drayton, when speaking of the vervain, says of it, "gainst witchcraft much avayling."

There are several species of vervains that grow wild. Besides the blue vervain, in fields everywhere, there is the narrow-leaved vervain, that is like the blue vervain, but with some narrower leaves and usually only one spike at the end of a branch; the hoary vervain, a plant covered with fine, white-woolled hair, and essentially a plant of the western plains and prairies; the European vervain, an escape from gardens that has become a troublesome weed in many places; and the white vervain, a species with white flowers.

All the vervains are annual or perennial herbs with usually toothed or lobed, opposite leaves, small flowers in various shades of white, lilac, rose, and purple, in terminal spikes or rounded clusters, the calyx tubular and five-toothed, the corolla also tubular, opening salver-wise and with five lobes, stamens four in pairs, and the fruit, a small drupe containing several nutlets.

The vervains belong to a family of the same name, though it is also called the verbena family. It has much the same characteristics of the vervains themselves. The leaves are opposite or whorled, simple or compound. The flowers are usually irregular and often two-lipped, and have a four to five-lobed or toothed calyx, a tubular or funnel-shaped corolla, four stamens, two shorter than the others, and a simple and terminal style. The fruit is a drupe or berry and contains two to four nutlets. The family contains about a thousand species of herbs, shrubs, and trees, in sixty-seven genera, most occurring in south temperate and tropical countries. Only a few species are native to the United States. In addition to the vervains, there are the lemon verbena, also known as mat grass, a plant with creeping stems, thick, wedge-shaped leaves, and axillary heads of rose purple or white flowers; the fog fruit, also a creeping plant, with bluish white flowers; the beautyberry, also called the French mulberry, a southern shrub with gray, scaly twigs, leaves with lower surfaces white-woolled, small, bluish, tubular flowers in showy clusters, and small, bright purple fruits, especially beautiful in early fall; the black mangrove, a shrub or small tree in Florida, but in the tropics reaching sixty to seventy feet high, of low tidal shores, with opposite, mostly oblong to elliptical, simple, persistent leaves, dark green above, gray and woolly below, complete, white, tubular flowers in short, terminal, spikelike clusters, and a small, somewhat compressed, egg-shaped, one-seeded capsule; and the fiddlewood, a shrub or small tree, fifteen to thirty feet high of Florida shores and hammocks adjacent to tidewaters, with opposite, simple, persistent, leathery, oblong to obovate leaves, bright green above, pale green below, complete, fragrant, white, tubular flowers in axillary racemes, and a nearly globular, reddish-brown to purplish black, four-seeded drupe.

The vervain family is well represented among the garden ornamentals, notably the verbenas and the shrubby and climbing lantanas, the latter found only in the South. The verbenas, in various colors, are useful bedding plants, though they are sometimes grown in the greenhouse as pot plants for house decoration. One of the fifty species of lantanas is also grown as a pot plant for its profuse bloom, but the lantanas are generally outdoor plants, the species known as the red or yellow sage being a very ornamental shrub throughout the South.

195. Vine (Vitaceae)

A native vine that rambles over the bushes of a thicket, festoons rocky woodlands, and climbs trees with the ease of an acrobat, the Virginia creeper has also found a home in many gardens where it decorates fences and stone walls or covers the porch or the side of the house with equal abandon, making its way by means of modified flowerstalks that are now branching tendrils, each branch with an adhesive disk at the end.

> In the course of about two days, writes Darwin, after a tendril has arranged its branches so as to press upon any surface, its curved tips swell, become red, and form on their under sides little disks or cushions with which they adhere firmly.

That these disks have a very tenacious hold can be seen from the fact that one contracting tendril, with the aid of these little disks, can support the entire weight of a branch that it lifts up. According to Darwin, one tendril with five disk-bearing branches could support a weight of ten pounds, even after ten years exposure to winds and rains.

The Virginia creeper is a vigorous, tall-growing vine with deep green, compound leaves of five elliptic to oblong leaflets, pointed and toothed, and small, insignificant, yellow green or whitish green flowers in clusters opposite the leaves. It is a decorative plant when in leaf, but considerably moreso in the fall when the leaves turn a brilliant deep red and the flowers have given way to beautiful, small, cadet blue berries, the berries hanging from red stalks.

A close relative, but a native of China and Japan, the Boston ivy or Japanese ivy is also a good climber and is preeminent for covering brick and stone walls. It was brought originally to America as a conservatory plant, but then it was discovered that it could withstand the Boston climate. Being a hardy and vigorous grower, it has ever since been used in various cities as a wall cover, clinging to bricks, stone and mortar in places where, exposed to wind and rain, no other vine could survive. Moreover, its broad, waxlike, shining leaves, that change to scarlet and crimson in the fall, and blue black berries, unite to make it additionally desirable.

554. Compound Leaf of Virginia Creeper

Another native vine also used for covering walls is the pepper vine. It is a southern species, with twice-compound leaves, the leaflets broadly oval, small, greenish or yellowish flowers, and greenish blue berries, found along riverbanks from Virginia southward

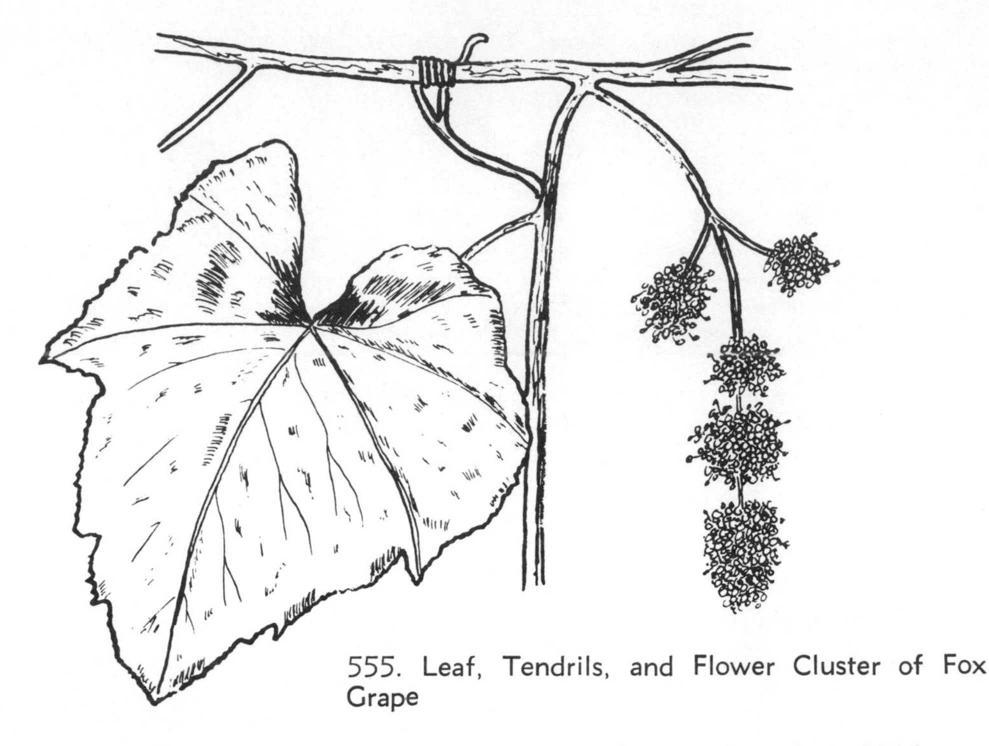

555. Leaf, Tendrils, and Flower Cluster of Fox Grape

and westward to the central states. It has fewer tendrils than the Virginia creeper, and also lacks the sucking disks of the creeper, so when used for a wall covering it should first be tied.

Also a native southern vine is the possum grape or marine ivy. It is a handsome, foliaged, strong-growing vine with simple or compound leaves of three leaflets, small, inconspicuous flowers, and black berries. It is widely grown in the South for arbors and the like.

The vines that have been mentioned belong to the vine family, a group of twelve genera and over five hundred species, many being shrubs or trees in the tropics, but prevailingly woody vines in the Temperate Zone. Except for the few native species, such as the Virginia creeper, all our other native members are the grapes of the genus *Vitis,* Latin name for grape. They are woody vines, climbing by tendrils, and though a few are somewhat decorative, most are useful as the source of all the grapes used as fruit, for jelly, and for wine-making. Usually, they have shreddy or peeling bark, forked tendrils, and alternate, often lobed, finger-fashion, toothed leaves. The flowers are small, greenish, unisexual or polygamous, the staminate and pistillate sometimes on different plants, in small clusters, the calyx entire or with very small teeth of four to five sepals, petals and stamens of the same number, the ovary two- to four-celled with one ovule in each cell, and the stigma two- to four-lobed. The fruit is the familiar grape, a true berry, with two to four, pear-shaped, grooved seeds.

The parent of the Isabella, Concord, and Catawba grapes, the northern fox grape is the common wild grape of the North, found growing in thickets from Maine to Georgia and westward to Minnesota. It is a strong-growing vine with a leaf, tendril, or flower cluster at nearly every joint. The leaves are nearly round, green above, but whitish or pale rusty beneath, the flower cluster is two to four and a half inches long, and the grapes are purplish black, thick-skinned, the pulp sweet, but musky.

Perhaps the most common grape of the Northern States, west to New England, the riverbank or frost grape is abundant along streams where it festoons the thickets, and being a high-climbing vine often takes entire possession of a tree. It is a species with smooth, greenish branches, sharply three-lobed leaves, and nearly black berries, densely covered with a bloom and rather sweet.

Found in thickets throughout eastern United States, the summer or pigeon grape is also a high-climbing vine, with broadly oval, three to five-lobed leaves, dull above, rusty beneath, flowers in panicles four and a half to six inches long, the grapes black with somewhat of a bloom and usually juicy and sweet.

Another high-climbing vine is the frost grape, also called the chicken grape. It is a rather stout species, the stem sometimes a foot or even more in diameter in old plants, with broadly, oval leaves, green on both sides and toothed, the flowers in panicles, and the grapes black, thick-skinned, slightly bloomy, and sweet after a frost.

The original of the scuppernong grape, the muscadine grape is a strong-growing vine of southern riverbanks. The stems are sometimes ninety feet long, with nearly round leaves, coarsely and triangularly toothed, green above, yellowish green beneath. The flowers are in short and dense panicles, and the grapes are dull purple, thick-skinned, the pulp decidedly musky.

The source of all the finer wine grapes is the species *vinifera,* commonly known as the wine grape. Originally wild in the Caucasus, it has been cultivated for centuries throughout much of Europe and Asia, and is now the leading grape in California, in regard to commercial wine production. It is a strong-growing vine, with stems forty-five to sixty feet long, nearly round leaves, three to five-lobed, the lobes toothed, the flowers in panicles, and the grapes slightly shaped like a football, black and with a bloom, or red or green in many varieties.

The chief use of grapes is for making wine. They are also gathered and sold as fresh fruit and are used for making jellies, as a source of grape juice, and for the production of raisins, made by gathering the grapes in trays and sun-drying them for a period varying from a week to a month.

Grapes, and the berries of other members of the vine family, such as the Virginia creeper, are a favorite food of game birds, many songbirds, and some fur bearers. In the summer, the dense foliage provides good escape and shelter cover, as well as nesting sites for songbirds. The bark of the vines is often used in nest building.

196. Violet (Violaceae)

I am invariably afield in early spring to follow the bypaths of the wildwood in search of an early flower. It may not be what some would say is an exciting or adventurous pastime but it has its compensations as those of you who do the same would know. I usually visit familiar places where I am sure to renew old acquaintances; I also seek out new and strange places for the unexpected I might find there.

Anyone who goes botanizing at this time of the year knows that many of the early wildflowers can be found merely by going to the places where they grow. As for instance, to the woodland swamp for the skunk cabbage, to the pasture for the cinquefoil, to the hillside thicket for the wood anemone. Others take a bit of searching, like the hepatica, perhaps, and the mayflower and the wild ginger. All these flowers, and many others too, are firmly fixed in our affections—yes, even the skunk cabbage in spite of its name and unseemly odor—but a perennial favorite of the author is the violet. There are a number of species—eighty or so in our country—but whenever the wild violets are mentioned it is the common violet that usually comes to mind, for it blossoms everywhere —in woods, along the roadsides, and in fields, meadows, and marshes, but to best advantage in cool, shady dells:

The violet blooms with every spring,
With every spring the breezes blow
And once again the robins sing
A song more sweet than June can know.

One thinks of the violets as low-growing, herbaceous plants, usually found in bunches, and with flowers peeping out from among the leaves, as they are in temperate regions, and is unmindful of the fact that some members of the violet family become shrubs and even small trees in the tropics. The family is a fairly large one with about fifteen genera and four hundred species of wide distribution in temperate and tropical regions.

Our violets belong to the genus *Viola,* the classical Latin name for the violet. They are low-growing plants, generally of tufted habit with both basal and stem leaves. The basal leaves are simple, heart-shaped or oval, sometimes lobed, stalked, and toothed. The stem leaves are alternate, simple, oval, usually stalked, and also toothed. Two stipules at the base of the stem leaves are usually cut into three lobes. The flowers are stalked, solitary, sometimes nodding, and

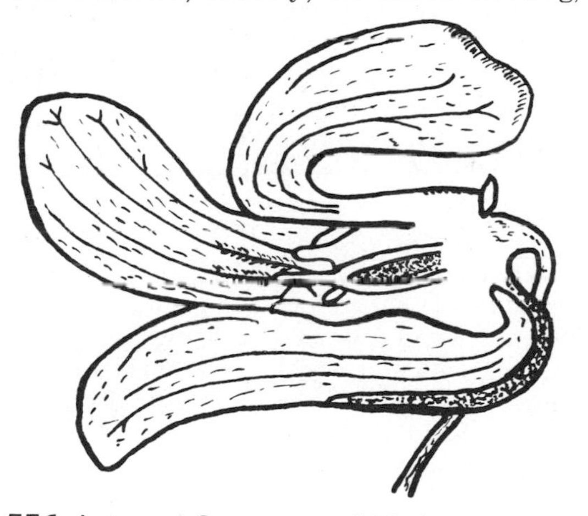

556. Internal Structure of Violet Flower

have a calyx of five sepals, a corolla of five petals, four arranged in pairs, each pair differing, the lower petal spurred, five stamens, and a one-celled ovary with many ovules. The fruit is a three-celled capsule opening elastically.

Unfortunately, violets hybridize rather easily, so that identifying them becomes a matter for a trained botanist. However, a number of them can be readily recognized by their general pattern, and, though violets are found in a variety of habitats, most species usually prefer just one kind of situation that helps to identify them.

The violets are divided into two groups—those having the leaf stalks coming directly from underground rootstalks, the so-called stemless species, and those having the leaves coming from a common stem that are known as the leafy-stemmed violets. Our common violet is a stemless species, as is the rather familiar and attractive bird's foot violet, that can readily be identified by its narrowly divided, finely cut leaves, and by its habit of growing in dry, sandy places. The early blue violet, whose leaves are also heart-shaped like those of the common violet, but deeply lobed, at times almost as much cut as the leaves of the bird's foot, is also a stemless species and common in rich woods. The arrow-leaved violet, with its long, arrow or halberd-shaped leaves and found in moist, open meadows and marshes, is another member of the group.

If one examines a violet flower closely one would find that it has an upper pair of petals, a lateral pair that is narrower than the upper pair, and a broad, lower petal that serves as a resting place for the bees and butterflies that come seeking the nectar. The lower petal extends backwards to the stem as a spur or sac that contains the nectar, and that gives the violet its characteristic shape. Cross-fertilization is effected by various species of bees and butterflies, the smaller mason bees and the mining bees of the family Halictidae being the most useful, the larger bumblebees and the butterflies generally being regarded as mere interlopers.

In many species, the flowers are sterile and incapable of producing seeds, the seeds being produced by peculiar, greenish flowers that are often mistaken for buds or seed vessels. They are located, if not actually underground, then not far above it among the leaves, and lack petals. Known as cleistogamous flowers, they never open, and without insect aid, are fertilized and ripen quantities of fertile seed.

The parent of the florist's violets is the sweet violet or English violet, a species indigenous to the three continents of Europe, Asia, and Africa. It is an escape from our gardens and is being established among our wild flowers. It figures prominently in European literature and doubtless is the one Shakespeare had in mind when he wrote:

> Violets dim,
> But sweeter than the lids of Juno's eyes
> Or Cytherea's breath.

Both Homer and Virgil mention the violet and it was a favorite of the Greeks who considered themselves complimented when called violet-crowned.

Violets are usually thought of as being blue, purple, or some such shade, but there are white violets and yellow ones also. As a matter of fact, it is believed that violets were originally white, though some botanists maintain that they were originally yellow. In the Canada violet, with its white or pale lavender blossoms, one can observe the plant in the process of changing from the white ancestral type to purple, which is generally regarded as being the highest point in the scale of chromatic evolution. The sprightly Canada violet, usually found in hilly woods, is a member of the leafy-stemmed group. A glance at this robust plant and one would suspect that the members of this group are larger plants than the stemless species.

There are several species of white violets, such as the lance-leaved violet and the sweet white violet, a sweet-scented species with heart-shaped leaves. Both prefer wet meadows, moist woodlands, and the borders of streams, and both are purple-veined, the veinings presumably serving as guidelines to the nectaries for the benefit of their insect visitors.

The word violet is so synonymous with the colors blue or purple, that it is somewhat difficult to believe that a violet can be yellow. Yet, there are two species of yellow violets, the downy yellow and the yellow round-leaved violet. The former has an erect, leafy stem covered with fine hairs, the latter a smooth stem. Both may be found in much the same kind of situation, in moist thickets or woodlands, though it is not unusual to find them in a dry place, especially the round-leaved species. They have pale golden yellow flowers with veins of madder purple, the veinings serving as pathways to the nectaries for the bees that visit them, although the common sulphur butterfly is an occasional visitor.

The most famous violet of all, and doubtless the most beloved, is the pansy. It is the oldest known English garden flower and was well established by the early seventeenth century. In his quaint style, Gerard writes of the pansy in his *Herball* that

557. Leaf and Flower of Pansy

the stalks are weake and tender, whereupon grow floures in form a figure like the Violet, and for the most part of the same bigness, of three colours, whereof it took the syrname Tricolor, that is, to say, purple, yellow, and white or blew; by reason of the beauty and braverie of which colours they are very pleasing to the eye, for smell they have little or none at all.

The word pansy comes from the French *pensee* meaning thought, indirectly a remembrance, well in keeping with this bright and cheerful and most delightful of all our flowers. As Shakespeare put it: "There is rosemary, that's for remembrance; pray you love, remember: and there is pansies, that's for thoughts."

The pansy, originally the heartsease of Europe, probably evolved from the violet known as Viola tricolor, native to the cooler parts of the continent. The pansy of the early English garden was unlike the improved pansy of today. Now there are giant pansies, three to four inches in diameter, on stalks a foot or more high, with colors that range through pure white, pure yellow deepening to orange and maroon and darkening to brown, as well as a variety of blues, and purples, and violets. There is also a rich and velvety shade that is spoken of as black, but there are no pure black flowers.

197. Walnut (Juglandaceae)

It used to be that one would go nutting in the October woods, but today it seems that our nutting is done in the supermarkets, leaving the nuts to the squirrels and other woodland creatures.

Times have changed, and the simple pleasures of yesterday have given way to more sophisticated diversions. Gone, too, is the day when one would collect and store the nuts for winter use.

As Whittier says in "Snow Bound":

> And close at hand, the basket stood
> With nuts from brown October's woods.

Presumably, Whittier's nuts were hickory nuts. Walnuts are probably the best known of our nuts, but walnut trees are scarce in New England, those growing along roadsides and rocky hillsides having come from planted trees. Occasionally, however, a native tree may be seen here and there.

The hickories are peculiarly North American trees. They will not be found elsewhere, except one species in China, and they all occur, except for a Mexican species, east of the Rocky Mountains. Europe once had numerous species and there were others in Greenland and in western North America, but the glacier removed them all from the face of the earth and the only record of them is in the Tertiary Rocks.

"Hickory" is an Indian word and

> is derived from the name of the liquor obtained by pounding the kernels. These the Indians beat into pieces with stones and putting them, shells and all, into mortars, mingling water with them, with long wooden pestells pound them so long together untill they make a kind of mylke, or oylie liquor, which they call powcohicora.*

* *Historie of Travaile into Virginia Britannia.*

They allowed this liquor to ferment and the fermented beverage was added to venison broth and the result was a rich food. "Hickory milk," strained of the shell fragments and thickened with meal, made corncakes that a king would not despise. The early settlers were

558. Fruit (Drupe) of Shagbark Hickory

taught by the Indians that an oil pressed from the nuts was excellent in cookery, and was said to equal olive oil in flavor.

The best of the hickory nuts are doubtless those of the shagbark hickory. They are large, sweet, pleasant and slightly aromatic. Many consider them to be our most important native nut. They can be bought in the stores and can also be collected in the woodlands.

One can easily recognize the shagbark hickory from its peeling bark, remarkably shredded and shaggy and loosely attached, hanging in strips commonly a foot long and several inches wide, with the tips curved out from the trunk. The shagbark is a ruggedly, picturesque, stately tree with wide spreading boughs, a pale brown gray bark, and large compound leaves, with five to seven leaflets, commonly five, the upper ones obovate, tapering towards the base, finely toothed, and dark yellowish green above, paler beneath. The flowers are in catkins, the green staminate in clusters of three, the hairy and greenish pistillate with spreading-divided stigmas single or few in a terminal cluster. The fruit is a globular, flattened dry drupe, with the

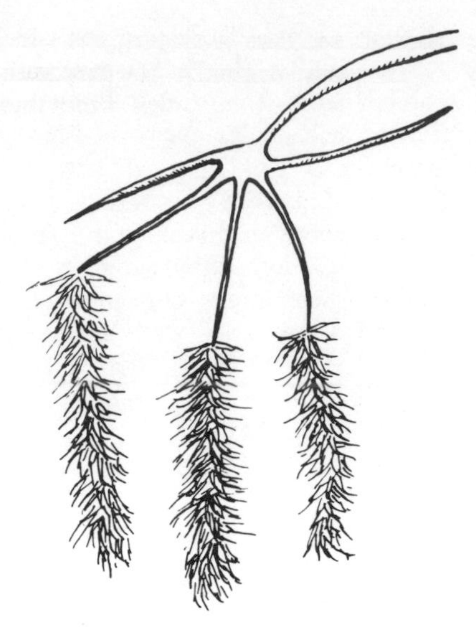

559. Staminate Catkins of Shagbark Hickory

husk separating into four sections. The nut is thin-shelled, buff white, and the kernel is very sweet.

The hickories of the genus *Carya,* Greek for the walnut, a related tree, although the genus was long called *Hicoria,* a Latinized form of the old Indian name for the tree or their nuts, are a group of valuable timber and nut trees with alternate, pinnately compound leaves, with an odd and usually larger leaflet at the end. The staminate and pistillate flowers, both without petals, occur on the same tree but in different clusters, the staminate in pendulous catkins with three to ten stamens, the pistillate with an ovary enclosed by a four-lobed involucre. The fruit is a fleshy drupe, that becomes hard and woody with age, and that separates into four valves each containing a nut.

The shagbark hickory is a tree of rich uplands, whereas the shellbark hickory, a similar species, prefers the river bottoms. It is much like the shagbark, but with coarse-flaked, much less shaggy bark and very stout, orange-colored twigs that are a distinguishing feature at any season. The leaves have somewhat larger leaflets and the terminal bud is exceptionally large, another distinguishing feature. Unlike the fruit of the shagbark, that of the shellbark is ovoid or ellipsoid, with a very thick husk, and the nut is yellowish white, pointed at both ends.

The hickory flowers are not conspicuous in either size or color, but in spring, both trees are a spectacle to behold. For, first, the buds loose their two outer scales and the silky inner ones shine in the sun like lighted tapers. They grow in size and as they loosen, a cluster of small leaves is revealed and clothed in the softest velvet. Then, the scales turn back displaying rich yellow and orange tones, and the leaves, delicate in texture and coloring, begin to unfold in imitation of the opening of a flower. When the trees are completely dressed in their leafy splendor, then the staminate catkins begin to shake out their pollen and the pistillate flowers, clustered at the tips of the leafy shoots, spread wide their stigmas to capture the golden dust. Once the flowering has passed, the spent catkins litter the ground and the summer proceeds to mature the nuts.

The mockernut is a disappointment, at least its nuts are. They are relatively large, but the kernel, though sweet, is small and is also so difficult to remove from the shell that it is not worth the trouble of getting it out. Sometimes a kernel will not even be found.

The mockernut is a medium-sized tree, sixty to seventy feet high, common on rich hillsides or bottom lands, with a light or ashen brown bark that looks more like that of an ash than a hickory, for it is broken by shallow fissures into short, confluent ridges that are corky-surfaced. Besides the character of its bark, the mockernut may be recognized by its large leaves of seven to nine leaflets, that are deep yellow green turning a clear or rusty yellow in autumn, and their aromatic, resin-scented fragrance when crushed. Its large nuts are also a distinguishing feature, as is its large terminal bud.

The pignut is a tree with an unfortunate name, given by the early settlers who apparently did not think much of the nuts, regarding them only fit for pigs. Some authors of tree books say that the kernels are astringent and bitter, others that they are of pleasant flavor. The fact is that they are extremely variable in quality. One nut can be quite pleasing to the taste, another will leave a faint bitter taste in the mouth. They are not as good eating as those of the shagbark, but generally, they can be eaten with enjoyment.

A beautiful, medium-size tree, fifty to sixty feet tall, but occasionally attaining a height of one hundred and twenty feet, the pignut is a common tree of dry woods and hillsides. Its bark is gray with shallow fissures and flattened, confluent ridges. Its leaves, with five to seven leaflets, are a deep yellow green, and its fruit is pear-shaped with a thin husk and a dull, light brown nut.

From its name, one would suspect that the nuts of the bitternut are bitter, as indeed they are, so bitter, in fact, that even the wildlife refuse them. The bitternut, or swamp hickory as it is also known, is a tall tree, fifty to seventy-five feet high, sometimes as much as a hundred feet tall, of swamps or rich woodlands, but also occurring on high uplands. Its bark is thin, light brown or brown gray and finely fissured, the ridges strongly confluent and separating into somewhat thin flakes, and it has the smallest leaflets of any of the hickories, being narrow, almost slender, and suggesting willow leaves in their contour. They are a distinguishing character, as is the fruit to some extent, being globular or obovate with four, prominent, winged sutures from the apex halfway to the base. But perhaps the most important character whereby to identify the bitternut is its flattened, tapering, yellow buds that are always present no matter what the season.

To many people, especially those living in the South, the pecan is probably the best known of the hickory tribe, being widely planted in the southern states for its nuts that are, perhaps, the best of the hickories. The pecan is a tall, slender tree, and the largest of the hickories, growing to a height of one hundred and seventy feet, reaching its highest development in the lowlands and river bottoms of the lower Mississippi valley. The bark of the pecan is smoother than that of most hickories, but moderately rough, broken into small, scaly plates, a buffish gray or a light reddish brown. The leaves have nine to fifteen leaflets. They are narrow, finely toothed, sharply pointed, often curved, and a deep yellow green. The staminate flowers are in catkins, five to six inches long, several in a cluster, while the pistillate flowers are in terminal spikes, two to five in a spike. The fruit is oblong or olive-shaped, with a thin husk splitting into four sections, the nut smooth, a lustrous tan brown, and thin-shelled, the kernel delicately flavored and sweet. By selection, many improved varieties have been developed, and in the South, many large pecan orchards may be seen.

There are other hickories, to be sure: the small-fruited hickory, with rough and scaly, but not shaggy bark, the twigs often a tan red, and whose foliage is similar to that of the pignut; the water hickory, a slender tree of wet land or swamps, with a roughly sculptured and ridged, bitter little nut; and the nutmeg hickory, a tall, straight, western species with a lustrous foliage that makes the tree the most beautiful of all the hickories.

The walnut family, whereto the hickories belong, is a small, but important group of deciduous trees often with resinous, aromatic, glandular leaves, comprising only six genera and about forty species native to the temperate or warmer regions of the northern hemisphere. Their leaves are alternate and compound, with pinnately arranged leaflets, the odd one at the tip. Their flowers are monoecious, the staminate with no petals and sometimes no sepals, and with several stamens in pendulous, spikelike, multi-flowered clusters or catkins, the pistillate sessile, with a three- to five-lobed calyx and a two- to four-celled ovary with one ovule. Their fruit is a drupe or nut, with a dehiscent or indehiscent husk.

There are about fifteen species of walnuts of the genus *Juglans*, the old Latin name of the English walnut, and, literally, means the acorn of Jupiter, found mostly in North America and Eurasia. They are generally tall trees closely related to the hickories, but do not have the shaggy bark of some of them, and with the characteristics of the family. Their fruit is a large, fleshy, non-splitting drupe containing the seed that is commonly, but incorrectly, called a nut.

In the eastern and central United States there are two species, the black walnut and the butternut. The black walnut is a tall, handsome tree, fifty to seventy-five feet high, sometimes as much as one hundred and fifty feet, with perpendicular roots, a straight trunk, and stout branches that are nearly horizontal below, but at a sharp angle above, forming a symmetrical, round-headed tree. Its bark is a medium brown or a dark brown with confluent ridges and deep furrows. Its leaves, of fifteen to twenty-three leaflets, are a bright yellow-green above, somewhat paler beneath, the leaflets ovate lance-shaped, and turn yellow in the fall. Its flowers, appearing with the leaves, are

560. Fruit (Drupe) of Walnut

greenish, the staminate are in catkins, three to six inches long, on wood of preceding year, and the pistillate on new shoots, in axillary few-flowered clusters or solitary, with prominent red stigmas. Its fruit is almost spherical, the husk, roughly dotted, is a dull green, strongly aromatic and spongy and does not split open like a hickory nut, the shell is hard, deeply sculptured, and the kernel is sweet, oily, and edible.

Originally a forest tree, common on hillsides and rich bottomlands, today the black walnut is not often seen in the dense woods, but rather along fencerows, roadsides, and woodland borders. The nut is a commercial item and is sold in the markets. The husk is sometimes used for dyeing and tanning.

Those who are familiar with our woodlands know the butternut, also known as the white walnut, with its characteristic fruit, an elongated, long husk that is clammy and sticky and covered with hairs. The kernel is sweet and delicious, but very oily. The frugal country housewife at one time pickled the nuts, placing the husks and all in vinegar, sugar, and spices. In winter they were said to be a delectable relish with meats. The husks and bark of the tree contain a rich brown dye that the colonists used in dyeing woollen cloth. During the Civil War it was used in dyeing the uniforms of the backwoods regiments. The so-called

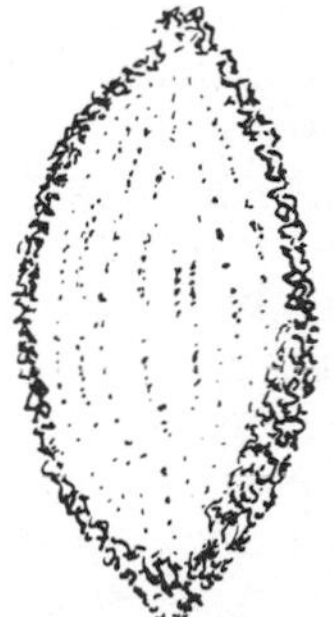

561. Fruit (Drupe) of Butternut

"butternut" jeans were a homemade, home-dyed uniform that could withstand the toughest wear.

The butternut is a round-headed, medium-sized tree, thirty to forty-five feet high, occasionally as much as ninety feet tall, and is common in rich, moist lowlands and fertile hills. The bark is a light brown gray with shallow fissures and broad flat ridges. The leaves, with seven to seventeen, lance-shaped leaflets, are a yellow green and, on long petioles, wave with every passing breeze. The staminate flowers appear in catkins, three to five inches long, the pistillate with bright red stigmas in spikes. The fruit is egg-shaped, with a sticky husk that is green throughout the summer, but becoming brown after the time of frost. The shell is thick, deeply sculptured, and the kernel is sweet and very oily.

Several other walnuts are the Texas walnut, that grows along canons and streams of the Southwest and whose small, thick-shelled nuts are much esteemed by the Mexicans and Indians; the California walnut, a beautiful tree growing along the West Coast and used as a stock whereon to graft the English walnut; and the English walnut, more correctly called the Persian walnut, for it is native to that country now known as Iran.

The English walnut is the walnut of classical literature. The Greeks knew of it and so did the Romans who curiously enough called the nut the acorn. So, when Ovid states that the people of the golden age lived upon "Acorns that had fallen/ From the towering tree of Jove," he is speaking of the walnut, not the oak.

Both the Greeks and Romans strewed walnuts at their weddings and Horace and Virgil mention the custom. And Spenser says that walnuts were used in Christmas games.

From ancient Persia the English walnut was transported to southern Europe, from there to England, and eventually to America. Parkinson describes in 1640 a kind of "French wallnuts, which are the greatest of any, within whose shell are often put a paire of fine gloves neatly foulded up together." He also knew of another variety "whose shell is so tender that it may easily be broken between one's fingers, and the nut itself is very sweete."

The English walnut, that produces the large edible nuts of commerce, is cultivated commercially in California and Oregon where it is restricted by climatic conditions and the blight disease, but there are hardy varieties that succeed reasonably well in the Middle Atlantic states and the Great Lakes region. It is a tree up to one hundred feet high, with a silvery gray bark, leaves of five to nine leaflets, and a nearly round, green, smooth fruit about two inches in diameter, the nut much sculptured. As the various hickory nuts and the walnuts are used as human food, it is not surprising that the wildlife should also make use of them for the same purpose. The squirrels and chipmunks especially relish hickory nuts, also eaten by several game birds and songbirds. The bark and leaves, too, are eaten by various mammals. Of the walnuts, only the black walnut appears to be of any importance, four species of squirrels eating the nuts, while the English walnut seems to be of interest only to the crow, though jays are known to peck through the shells. Among our butterflies only the caterpillars of the banded hairstreak and the hickory hairstreak include the walnuts and hickories among their food plants, the caterpillars of the hickory hairstreak feeding exclusively on the hickories.

198. Waterleaf (Hydrocaryaceae)

The waterleaf family comprises eighteen genera and over two hundred species of chiefly herbs that are very abundant in North America, but also found elsewhere. They have mostly alternate leaves, though they are opposite in some genera, often lobed or deeply cut, regular flowers, in terminal or lateral, one-sided clusters, the flowers perfect, with a five-parted calyx, a five-lobed or five-parted corolla, five stamens, a one-celled ovary, and a fruit that is a two-valved capsule.

A common species of the family is the Virginia waterleaf, a weak, unbranched, leafy perennial of rich woods, with a slender, smooth stem, one to two feet high, pinnately compound leaves, the leaflets ovate-oblong or ovate-lanceolate and coarsely toothed, and white, or pale lavender flowers in a few-flowered cluster. The Virginia waterleaf is one of about six species of the genus *Hydrophyllum,* Greek for water and leaf, and is also called Indian salad and Shawnee salad, from the fact that the young shoots are eaten in the spring as a salad by the Indians who appear to prize them highly.

Another common species is the Canada waterleaf, that is somewhat similar, but with compound leaves of five to seven broad lobes and white flowers, the stalk of the cluster shorter than the leafstalks. It is about eighteen inches high and also occurs in the woods.

The genus *Nemophila,* Greek for grove and love, in

allusion to the plants liking a shady place, contains several species that are of garden interest such as the fiesta flower, a plant of scrambling habit, climbing by means of prickles on the stems to a height of three to six feet, with deeply cut leaves and violet flowers; and the baby blue eyes, a garden favorite, that grows to a height of six inches, having compound leaves cut into seven to nine leaflets, and bright blue, bell-shaped flowers. These two preceding species are natives of California. An eastern species is found in moist woods from Virginia to Florida and west to Arkansas and Texas, and is called the small-flowered nemophila. It has very slender, diffusely branched stems, pinnately compound leaves of three to five, roundish or wedge-obovate, sparingly cut-lobed divisions, and small, white or blue flowers.

The genus *Phacelia,* Greek for a bundle in allusion to the flowers, is a fairly large group of one hundred and fourteen species, with simple or compound, alternate, fleshy, sometimes hairy leaves and blue, purple, or white flowers arranged in rolled, one-sided racemes, the racemes unrolling as the flowers open. One species, the silky phacelia, grows in dry soil from South Dakota to Idaho and western Nebraska. A second species, the loose-flowered phacelia, occurs in moist thickets and along streams from Ohio and Illinois south to Alabama. A third species, the small-flowered phacelia, prefers shaded banks from New York to Georgia and Kansas.

The phacelias, essentially annuals, but occasionally perennials, are important bee plants and are easily cultivated.

199. Waterlily (Nymphaeaceae)

It is a long time since the author read "Hiawatha," and he remembers little of the poem, yet a few lines remain, for whenever he sees the yellow water lily he recalls that it was his canoe that:

> Floated on the river/ Like a yellow leaf in autumn/
> Like a yellow water lily.

One will not detect any odor from the yellow water lily. How unlike its cousin, the white pond lily, whose white, golden-centered chalices perfume the air with their delicious fragrance and, needless to say, attract the bees and flowerflies and other insects that enjoy its favors. They must, however, visit the flowers in the morning, for they close in early afternoon.

One is likely to look at a rich green lily pad and fail to grasp its leathery and waterproof design. It can withstand, not only heavy rain, but the wear and tear of the wavelets that often toss it about like a raft in an angry sea. One is also apt to be unmindful that it serves as a hatchery for eggs of many kinds and as a pasturage for the larvae of insects and snails. Sometimes the bryozoans find it a convenient place to locate their colony.

There are fifty or so species of waterlilies distributed among the eight genera of the water-lily family. They are typically aquatic, perennial herbs with conspicuous floating leaves and large, showy flowers of considerable ornamental value. They occur in both the tropics and north Temperate Zone. Most of them have thick rootstalks that creep in the mud and wherefrom arise long-stalked, large and simple leaves that may either be erect or floating. The flowers are large, showy and usually solitary with three to four to six or more sepals, three or many petals, often showing gradual transitions from green sepals to colored petals and two often sterile, petallike stamens. There are three to numerous stamens and two to eight or more carpels, free or united into a many-celled ovary. The fruit is a two-valved capsule.

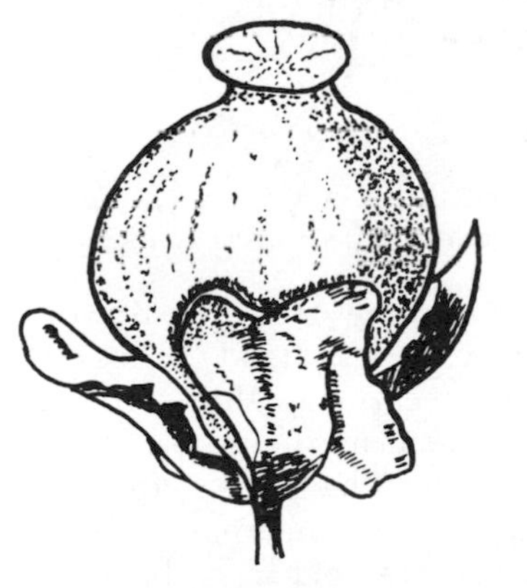

562. Fruit of Waterlily

The yellow water lily or yellow pond lily or spatterdock, as it is also known, is common in standing water, ponds, and slow streams. It has oval or egg-shaped leaves, and green and yellow, cup-shaped flowers, the six sepals unequal, concave, thick, fleshy, and green sometimes tinged with purple, the petals stamen-like, oblong, fleshy, short and yield nectar. When the flower first opens there is a triangular opening over the stigma that is so small that an entering insect must come in contact with it. The next day, the flower expands fully and the anthers beneath the stigma unfold, spread outward, and expose the pollen. Thus, cross-fertilization is ensured. The mining bees of the tribe Halictus are the chief agents in the transfer of pollen.

Unlike the yellow water lily, the white water lily is fragrant. It is found in still waters everywhere. The leaves are roundish, leathery, thick, purplish red beneath, and dark green above. The flowers are white, often five inches in diameter, opening in the morning and closing at noon, the sepals are green, tinged with reddish brown, the petals broadly lance-shaped, and the golden stamens are many, the outermost becoming petallike and white. Bees and flowerflies are the chief visitors, but beetles and butterflies attend the flowers too. In both species, the fruit is berrylike, that of the white water lily maturing beneath the surface of the water, that of the yellow pond lily above. "Resplendent in beauty, the lotus/ Lifted her golden crown above the heads of the boatmen." So wrote Longfellow in *Evangeline* of the American lotus or water chinquapin, probably the most beautiful and stately wild flower of our flora. It grows in lakes, ponds, and slow-moving streams and has large, horizontal root-

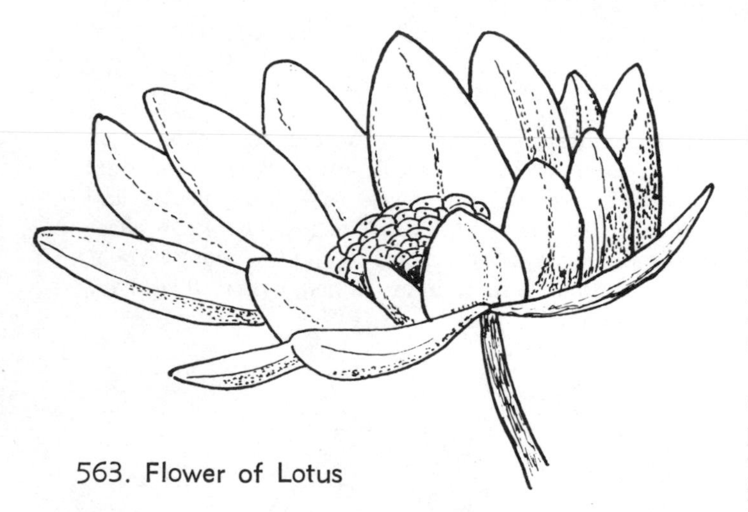

563. Flower of Lotus

stocks that are from two to five feet beneath the surface of the water and wherefrom extend the leaves. They are from one to five feet long, almost round, somewhat cup-shaped; some float on the water; others are held high above the water. The flowers are four to ten inches in diameter, with numerous stamens and petals, a pale yellow, and are held high above the water by thick stems that are sometimes six or seven feet long. The fruit is a nearly hemispherical, flat-topped receptacle containing the seeds that, when ripe, are dark brown, nearly round, about half an inch in diameter, and with a hard, sharp point at the top. They are not unlike an acorn and are often called duck acorns, probably because ducks like them, finding the meat very rich. The seeds were much prized by the Indians. Indeed, the entire plant is said to be edible, particularly the large, starchy roots that are said to taste like the sweet potato.

The American lotus is one of two species of lotus, the other being what is called the Egyptian lotus, but the real Egyptian lotus is a water lily, while the plant of Tennyson's poem, prized by the lotus-eaters, may have been a prickly African shrub whose olive-like fruit has the sweetness of honey and the taste of the date. In any event, the Egyptian lotus is an interesting plant with beautiful pink flowers about six inches in diameter and similar to the water lily. In the center of the flower and surrounded by rows of yellow stamens, there is a curious top-shaped body whose flat, upper surface appears to be studded with tiny, green acorns that are the ovaries, with the little knob at the top of each being the stigma.

Both the lotus and the water lilies are widely used as ornamentals in the water garden, the latter varying in size from our native white water lily to the enormous Victoria or royal water lily that has huge, circular leaves six feet in diameter and creamy, white flowers with a delicate, pineapple fragrance. They bloom only at night and for several nights. After the first night the color changes to a pink or red. It later develops into a prickly berry about the size of a man's head. This water lily is native to South America and is the largest in the world.

Sometimes included in the water lily family, but at other times in a family of its own, is the aquarium plant sold in pet stores under the name of Cabomba, though also known as fanwort and water shield. It is a plant of ponds with finely dissected, submerged leaves of value for supplying oxygen to the water. The flowers are white, with three sepals and three petals, and bloom on the surface. Also included in the water-lily family and sometimes with cabomba in the latter's family is the purple bonnet or water shield. It occurs in ponds and slow streams where it roots in the mud and sends up long, jelly-coated stalks that are attached to the oval, floating leaves. The flowers are purple, about one half of an inch in diameter, with three or four narrow petals.

200. Water Milfoil (Haloragaceae)

It is relatively easy to recognize the mare's tail, an aquatic plant that grows on damp shores or in shallows, because of its whorls of six to twelve stiff, stringy leaves. Above the water they are short and project stiffly out from the stem, but beneath the surface they grow long and are flaccid and drooping.

The mare's tail is one of about one hundred, widely distributed species of the water milfoil family that comprise about eight genera. They are perennial, rarely annual herbs, mainly aquatic, with alternate or whorled leaves, the submerged ones often divided and comblike, and perfect or monoecious or dioecious flowers in interrupted spikes, solitary, or clustered. The flowers have a two- to four-lobed calyx, two to four petals or none at all, one to eight stamens, and a one- to four-celled ovary. The fruit is a nutlet or drupe.

The mare's tail of the genus *Hippuris,* Greek for mare's tail, occurring in ponds and quiet streams, has a slender stem, linear or lanceolate leaves, and minute flowers that are arranged in a circlet around the stem at the base of the leaves.

The water milfoils, wherefrom the family derives its name, are contained in the genus *Myriophyllum,* Greek meaning myriad-leaved and that describes the plants. There are twenty species. They are all aquatic herbs with alternate or whorled leaves, the leaves above the water, entire, toothed, or comblike, the ones beneath the surface very finely divided that give the plants a bottlebrush appearance. The flowers are monoecious, the staminate have a two to four-lobed calyx, two to four petals, and four to eight stamens, the pistillate with a two to four-celled ovary with an ovule in each cell. The water milfoils are all rooted to the bottom, but their long, graceful branches usually reach to the surface and their flowers are borne above it.

A common species is the spiked water milfoil, commonly found in slow streams and shallow ponds. Another species is the whorled water milfoil. It occurs in both deep and shallow water, has exceptionally, finely divided leaves and purplish flowers. A third species, growing on the sandy bottoms of ponds and streams and with a more limited range than the two preceding, is the slender water milfoil, a slender species with nearly leafless flowering stems and purplish flowers. An imported species from Brazil, the "parrot's feather" that is sold by aquarium dealers, has become established around New York.

Other species of the family are the two native plants called the mermaid weeds of the genus *Proserpina,* from the Middle Latin meaning forward creeping. They are low, perennial herbs, with the stems creeping at the base, have alternate leaves, and small perfect flowers, sessile in the axils, and either solitary or three to four together. The common mermaid weed occurs on the muddy borders of ponds or in shallow water, the cut-leaved mermaid weed in sandy swamps, near the coast.

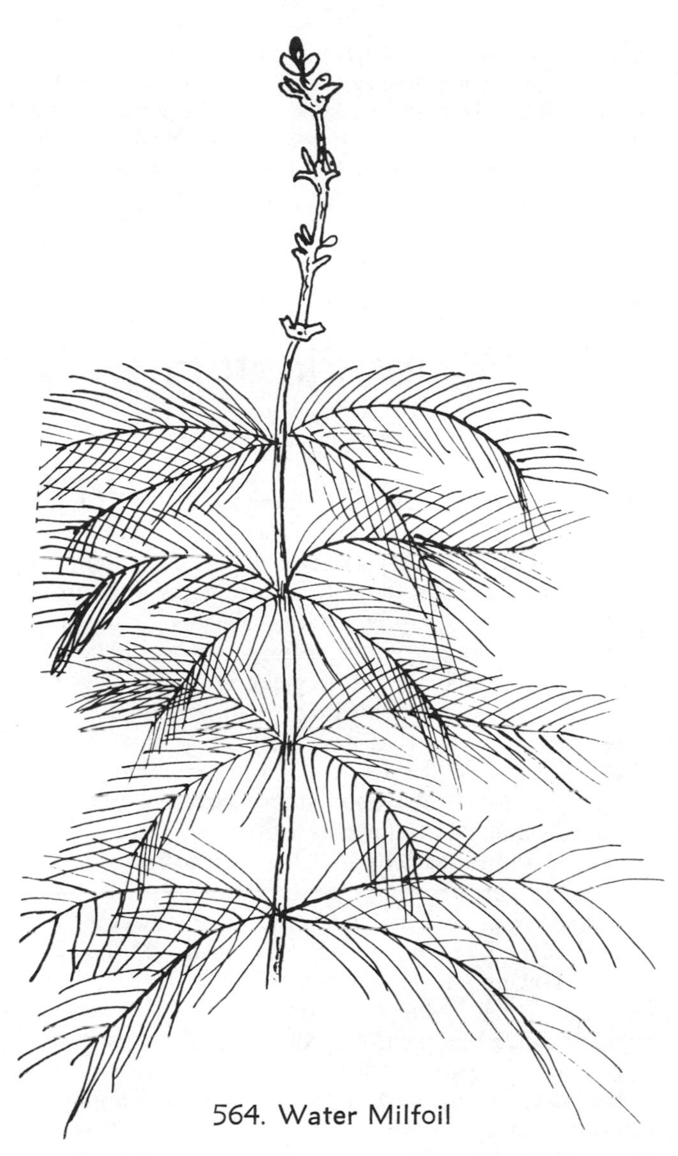

564. Water Milfoil

201. Water Nut (Trapaceae)

The water nut family has only one genus containing three species that are native to Europe, tropical Asia and Africa. They are aquatic herbs with opposite, plumelike, submerged leaves and clustered rhombic-ovate, toothed, floating leaves having inflated petioles. The flowers are perfect with a four-parted calyx, four petals, four stamens, and a two-celled ovary, with one ovule in each cell. The flowers are solitary and occur in the axils. The fruit is a one-seeded, spiny nut.

Of the three species only one, the water chestnut, has become naturalized and may be found in ponds, rivers, bays, and canals in Massachusetts, New York, and the lower Potomac. It is rooted in the mud and has stems that are often several feet long. The submerged leaves are finely segmented, the floating leaves are coarsely toothed with the upper surface glossy and the lower surface hairy. The flowers are white and are borne among the floating leaves. When the spiny nuts mature, they drop to the bottom where the seeds remain viable all winter. The seeds are edible.

202. Water Plantain (Alismaceae)

If one studies the edge of a pond, one will find that the plants that grow in the water do not grow haphazardly, but in clearly defined zones. Thus, one will find the emergent water plants near the shore: bulrushes, burweeds, reeds, marsh grasses, white-blossomed arrowheads, arrow arums, water plantains, massed blue spikes of the pickerel weed, and tall phalanxes of cattails crowding every nook and corner. These are all plants that have their roots in water, but with their stems, leaves, and flowers in the air, the submerged parts of their stems covered with the simple desmids and diatoms.

Then farther out where the water is knee-deep or a little deeper one will see the deep aquatics with floating leaves, such as the water lily, spatterdock or yellow water lily, and water smartweed with its rose-colored flowers. Here, too, are the free floating duckweeds and just below the surface the little vinelike liverwort and bladderwort. And in still deeper water the ribbonlike leaves of the pondweeds and freshwater eelgrass stream upward from the bottom. This is the zone of submerged water plants: meadows of waterweeds, stoneworts, and water crowfoots, the latter with finely dissected leaves. Seen beneath the surface, these leaves are little more than forked hairs that spread out in dainty patterns, but collapse and become a soggy mass when lifted out of water.

How these plants are adapted to live in water may be seen by observing the arrowhead for instance. By looking carefully one will find a plant with broad, arrowlike leaves, another with broad, ribbonlike leaves, and a third with narrow, grasslike leaves though they may grow only a few inches apart. One may even find one with all three kinds of leaves, but though they may grow only a few inches apart, they occur at varying depths. Hence, it would seem to follow that variation in the width of leaves must be correlated with depth of water. The arrowhead growing in the deepest water must have narrow, grasslike leaves that bring the greatest possible surface area in contact with the air dissolved in the water, and yet, that can glide harmlessly through the water; broad leaves would be torn to shreds by the water currents. Then, as the water becomes progressively more shallow, the grasslike leaves are discarded for the more broadly ribboned ones, that in turn give way to the conventional, broad, arrowlike leaves more suited to function in the air when the plant grows along the margin of the pond. Hence, the arrowhead is able to adjust itself to varying degrees of water depth, to drought or flood, and can live in and out of water with all the facility of a frog.

The different arrowheads that have just been described are not different species, but merely different forms of the same species, *Sagittaria latifolia,* to give it its scientific name. Actually, there are four forms that, though ill-defined, are in most instances, recognizable. They are *forma obtusa* with very broad obtuse leaves; *forma hastata,* leaf blades and their basal lobes oblong, lance-shaped and acute; *forma gracilis,* leaf blades and their lateral bases narrowly linear; and *forma diversifolia,* leaf blades partly arrowlike and partly lance-shaped or elliptical without basal lobes.

The typical arrowhead is a smooth species, with leaves nearly always arrow-shaped, and an erect flowerstalk with three-petaled white flowers in circles of threes. It grows in the quiet waters of streams and on the margins of ponds.

The arrowhead belongs to the genus *Sagittaria,* Latin for arrow in allusion to the arrow-shaped leaves. The genus includes thirty or more species of perennial, aquatic or marsh herbs that are distributed throughout the temperate and tropical regions of the world. They are of an erect habit with thick and tuberlike rootstalks, ribbonlike submerged leaves, roundish floating leaves, and arrow-shaped, aerial leaves, that is, those above the surface of the water. The flowers are in whorls or circles on leafless stalks and may be one of three orders: flowers with both stamens and pistil; staminate and pistillate flowers on the same plant, the staminate ones being above the pistillate ones; and staminate and pistillate flowers on different plants. The sepals are three in number, small and greenish white, while the petals are also three in number and white or spotted. Both stamens and pistils are numerous as are the ovaries that are crowded in a spherical or somewhat depressed head on a globular

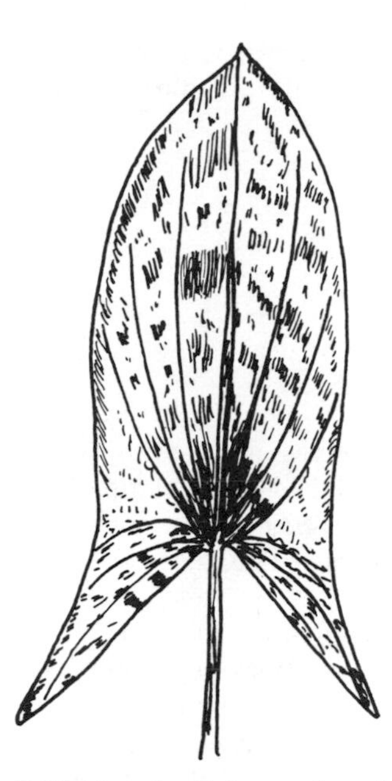

565. Leaf of Arrowhead

receptacle, in fruit forming flat, membranous, winged achenes.

There are about two dozen species of arrowheads in the United States, all alike except for botanical distinctions that identify the various species. Their pollen is distributed by various agents, not the least being the insects that frequent wet places such as the dragonflies.

The Indians found the tuberous roots of the arrowhead serviceable as food, boiling them like potatoes or roasting them in hot ashes. The Oregon Indians called it the *wapatoo* and Lewis and Clark record that they "purchased from the old squaw, for armbands and rings, a few wappatoo roots on which we subsisted. They are nearly equal in flavor to the Irish potato and afford a very good substitute for bread."

The genus *Sagittaria* is one of fourteen genera of the water plantain family whose characteristics are essentially those of the genus. It contains less than a hundred species common to freshwater habitats all over the world, but apart from the arrowheads, the only species of any interest is the water plantain of the genus *Alisma,* that is found sometimes growing in water, but more often in the mud along the edge of a pond or stream. It grows to a height of three feet, with leaves on stalks that come from the root. They are olive green and variable in shape, often narrow and grasslike if floating, but oblong or ovate if the plant grows in mud. The flowers are very small and numerous, white or pale pink, in a cluster (panicle) on a symmetrically-branched, tall stalk. They are possibly fertilized by the flowerflies and like those of the arrowheads, develop into small achenes.

566. Flowering Stem of Arrowhead

203. Water Starwort (Callitricaceae)

The water starworts are low, slender, usually tufted aquatic herbs with hairlike stems, entire, spatulate or linear leaves, minute, perfect or monoecious flowers, solitary or two or three together in the leaf axils, the flowers without sepals and petals, but with a single stamen and a four-celled ovary having an ovule in each cell. The fruit is nutlike.

There are about twenty species, all belonging to a single genus, *Callitriche,* Greek for beautiful hair, in allusion to the plants' hairlike stems, and all of wide geographic distribution. There are about eleven species found in the United States, among them being the terrestrial water starwort of damp places, the dried plant with the odor of sweet clover; the vernal water starwort, common in quiet waters; the larger water starwort, found in ponds and slow streams; and the autumnal water starwort, of lakes and cold streams, the latter a perennial, while the others are annuals.

204. Waterwort (Elatinaceae)

The waterwort family is a small group of two genera and twenty-five species of low herbs of wide geographical distribution. They have opposite or whorled, entire or toothed leaves with membranous stipules between them and small, axillary flowers that are either solitary or in clusters. The flowers have two to five sepals, the same number of petals, and as many or twice as many stamens as petals, and a two to five-celled ovary. The fruit is a capsule.

The waterworts of the genus *Elatine,* Greek for fir-like in reference to the leaves, are aquatic plants with their stems partly buried in the sand or mud of ponds and slow streams. About four species occur in the United States. The American waterwort is an erect and spreading plant of pond and stream margins with obovate leaves and small, solitary, sessile, axillary flowers. This is our common species occurring from New Hampshire south to Virginia and west to Illinois and Texas. The three other species are more localized.

The second genus of the family, *Bergia,* named in honor of Dr. P. J. Bergius, a Swedish botanist of the 18th century, is represented in the United States by a single species, the Texas bergia. It is a terrestrial species, growing in swamps and wet places, with prostrate or ascending, diffusely-branched, pubescent stems, spatulate or obovate leaves, and small, axillary or clustered flowers. A western species, its range is from southern Illinois to Kansas, and southward and westward.

205. White Mangrove (Combretaceae)

Associated with the red mangrove of the Florida coast is the white mangrove, not a true mangrove because it lacks the aerial roots that are so characteristic of the red mangrove. It is one of nearly three hundred species contained in about fifteen genera that comprise the white mangrove family, a group of evergreen trees, shrubs and vines, with astringent juice and widely distributed throughout the tropics. They have alternate or opposite, simple, entire, leathery, persistent leaves, regular, perfect or polygamous flowers with a five-lobed calyx, five petals, five to ten stamens, a one-celled ovary, and a drupaceous fruit. A few species produce valuable fruits, astringents, and dyes. Others are important in building land surfaces along tidewaters. Still others, such as the East Indian laurel and the African limba, are a source of timber for furniture. A few are planted as ornamentals in tropical countries. The East Indian almond has even become naturalized in parts of peninsular Florida.

Three species of the family occur in Florida. They are the white mangrove, the black olive tree or the oxhorn bucida, and the button mangrove or buttonwood. The white mangrove is a tree thirty to sixty feet high, with stout, spreading branches forming a narrow, round-topped head, of tidal bays, lagoons, and swales. It has opposite, oblong to broadly elliptical leaves, perfect and unisexual, white flowers in ter-

567. Leaf of White Mangrove

568. Flower of White Mangrove

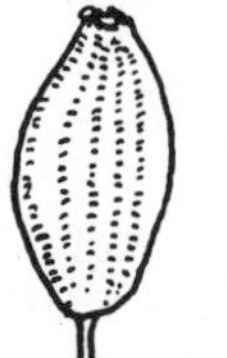

569. Fruit (Drupe) of White Mangrove

minal and axillary spikes, and an obovoid, leathery, ten-ribbed, reddish brown drupe.

The black olive tree or oxhorn bucida is a tree sometimes with a single, straight trunk, at other times with a short, prostrate stem producing several straight, upright, secondary stems, forty to fifty feet high, with stout, horizontal branches forming a broad head growing in brackish marshes. It has alternate, mostly obovate leaves, blue green above, yellow green below, with rusty hairs on the midrib, perfect, greenish white flowers in spikes, and an ovoid or conical, five-angled, light brown, leathery drupe.

The button mangrove or buttonwood is a tree, forty to sixty feet high, though sometimes a shrub, with a narrow, but shapely crown, of low, muddy, tidewater shores of lagoons and bays. It has alternate, ovate, oval, or obovate leaves, dark green above, paler below, perfect, small, flowers in dense, globular heads, and a minute, reddish, scalelike, leathery drupe. The bark is bitter and astringent and has been used in tanning leather, and in medicine as an astringent and tonic.

The genus *Terminalia,* Latin for terminal, in allusion to the leaves often being borne toward the end of the shoot, comprises over a hundred, handsome shade trees, some yielding valuable products, such as gums, resins, myroblans, tanning extracts and so on. Only one species, the Indian almond, sometimes also known as the tropical almond, is commonly cultivated, indeed, it is one of the most widely planted street trees in the tropics and used for such a purpose in Florida. It is a tree up to eighty feet high, with a smooth, brownish gray bark, alternate, rather oval, entire, leathery, glossy green leaves, and small, inconspicuous flowers that are followed by dryish, greenish or reddish drupes whose seeds are almondlike and edible. The leaves turn a handsome copper red before falling, although the tree is never entirely naked. It is a native of Malaya.

Another interesting plant of the family is the Rangoon creeper from Indo Malaya and the Phillipines, grown for ornament in southern Florida. It is a quick-growing vine, without tendrils, but with oblong, stalked, abruptly pointed, opposite leaves and showy flowers that bloom all summer in Florida. The fragrant corolla is white in the morning, but strangely turns to pink or red by evening. The flowers are in umbellike clusters that have been likened to stars falling from a bursting rocket, the flowers seeming to dangle in the air though they are attached by the long green calyx tubes. The fruit is a dry, leathery, five-angled capsule.

206. Wild Cinnamon (Canellaceae)

The wild cinnamon family is a small group of tropical trees and shrubs represented in the United States by a single species found growing in the woodlands of southern Florida. It belongs to the genus *Canella,* Latin for cinnamon, is the only one of the genus, and is variously known as the canella, cinnamon bark, whitewood, or wild cinnamon.

The wild cinnamon is a moderately large, tropical tree, reaching a height of about twenty-five feet, with a usually compact and ovoid crown. It has alternate, simple, persistent, leathery, obovate leaves that are dotted with glands, lustrous, bright green above, paler below, perfect, white or pinkish flowers in terminal or subterminal cymes. The flowers have three nearly circular sepals, five petals, and twenty stamens, their filaments united into a tube that encloses the pistil. The fruit is a globular, bright red, fleshy, two- to four-seeded berry, the seeds kidney-shaped. The orange-colored inner bark, known as canella bark, or simply canella, is highly aromatic, and is used as a condiment and in medicine as a tonic.

570. Leaf of Wild Cinnamon

571. Flower of Wild Cinnamon

572. Fruit (Berry) of Wild Cinnamon

207. Willow (Salicaceae)

Somehow the wild shrubs do not seem to have the same appeal as the trees and wild flowers, though one exception is the pussy willow that everyone knows when its buds begin to open and show their furry coats, a sure sign that spring is on the way. Later, the yellow catkins decorate the roadsides, low meadows, and riverbanks where they are sure to catch the eye. The yellow color is the color of countless pollen grains that await the wind to carry them to sticky stigmas, that may be near at hand or more distant, for the pussy willow is dioecious and the staminate and pistillate flowers occur on separate plants. Keep some of the pistillate flowers, softer and silkier in appearance than the staminate, and less yellow in tone, under observation for a few days and one will see them develop into conic-shaped capsules that contain undeveloped seeds. When the seeds are fully matured, the capsules open, the seeds are released, and, since they are furnished with long, silky down that functions as an effective parachute, are carried by the wind, perhaps to lodge eventually in some faraway place.

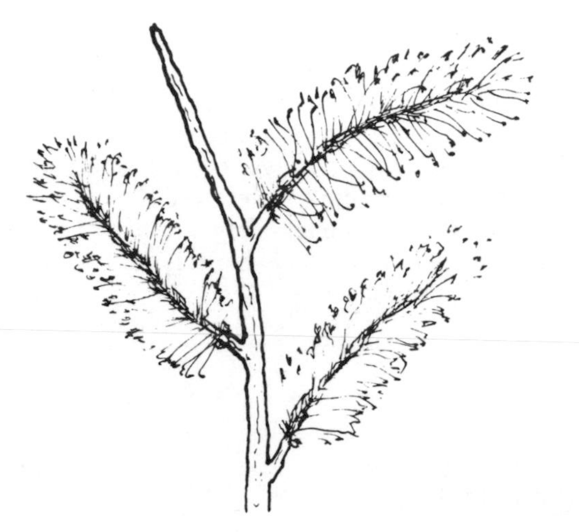

573. Sterile Catkins of Pussy Willow

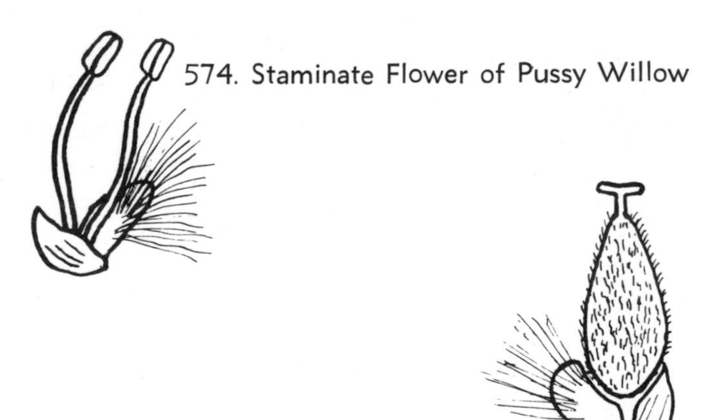

574. Staminate Flower of Pussy Willow

575. Pistillate Flower of Pussy Willow

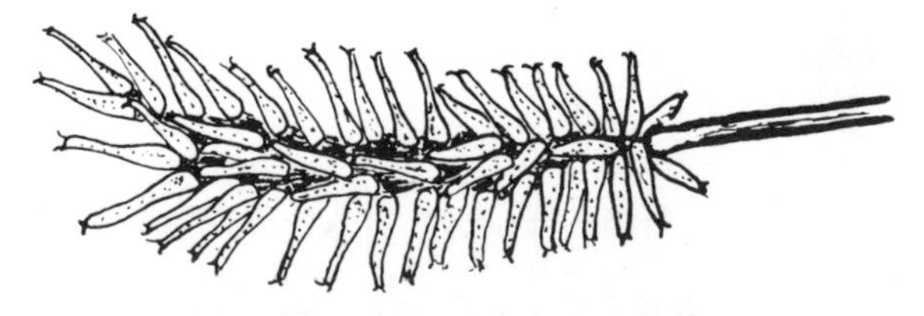

576. Capsules of Pussy Willow

The pussy willow is a large shrub, usually six to twelve feet high, though sometimes it becomes a small tree twenty-five feet in height with a trunk diameter of twelve inches. It has dark purplish red branchlets that are smooth or softly-haired and lance-shaped or elliptical leaves, pointed at both ends, usually with irregular and somewhat wavy teeth, bright green above, smooth with a white bloom beneath.

The pussy willow is one of several hundred species of trees and shrubs that comprise the willow family, though many of the willows of Arctic-alpine situations are dwarfed, carpetlike plants, like many herbaceous, woody perennials. The family consists of only two genera, the willows and poplars, widely distributed, but more abundant in the temperate and cold regions of the northern hemisphere. Many of the trees and shrubs of both genera grow rapidly, and the wood of most of them is usually light and soft.

Their leaves are alternate, simple, and deciduous, and their flowers occur in catkins, appearing with or before the leaves unfold, the staminate and pistillate on separate trees. The individual flowers are minute, each in the axil of a small bract. They lack both sepals and petals, have two to many stamens, and a

577. Staminate Flower of a Willow

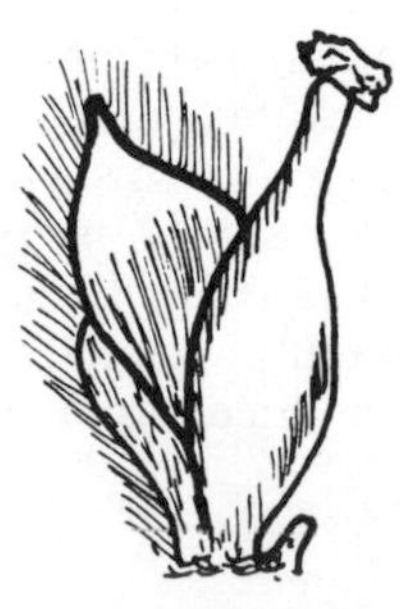

578. Pistillate Flower of a Willow

one-celled ovary with many ovules. The fruit is a one-celled capsule, and the seeds are surrounded with silky tufts. Many willows and poplars are used for ornamental purposes, some species of willows are used for basket-making, and others are useful as soil binders. Some poplars serve as windbreaks.

The willows of the genus *Salix,* the classical Latin name for a willow, differ greatly in size and habit of growth, but are otherwise very much alike in other respects, and as there are many natural and induced hybrids, the naming of a willow is rather difficult even for the expert. The willows have an abundant watery juice, furrowed, scaly bark, a soft, pliant, tough wood, slender branches, and large fibrous roots that are remarkable for their toughness, size, and the tenacity to live. For this reason, they make excellent soil binders and are often planted along streams to protect the banks against the action of the water. Their leaves are usually narrow, mostly lance-shaped, and tapering at both ends with generally serrate margins. Their catkins are erect, whereas in the closely related poplars, they are mostly drooping. The willows come from an ancient line. From impressions of their leaves in the cretaceous rocks it is possible that they are one of the oldest forms of dicotyledonous plants. They are tolerant of all soils and their only requirement is water, and they venture farther into the Arctic than any other woody plant except the birch.

579. Leaf of Black Willow

The black willow is usually found leaning over the water of streams and lakes. It is a native species, mostly a shrub, but often a tree, twenty to ninety feet high, common in the eastern states, and derives its name from the black bark of old trees. One may recognize it by its long (two to four inches), narrow, yellow green, shining leaves that taper gradually to a long point and that are finely toothed. The leaves often curve like a sickle. An interesting feature of the foliage is that each leaf bears small, green stipules that are crescent-shaped and finely toothed, and persist through the summer.

The use of the willow dates back to prehistoric man who fashioned the strong, yielding, flexible withes into ropes. Romans used it for making baskets as is done today. As the Roman, Martial wrote:

> From Britain's painted sons I came,
> And Basket is my barbarous name;
> But now I am so modish grown
> That Rome would claim me for her own.

Herodotus, however, was the first of the ancient writers to mention the willow and relates that the ancient Scythians used it as a divining rod.

The crack willow was doubtless the willow that the Romans used in making baskets, for it has served such a purpose in Europe for centuries and was the willow introduced into America before the Revolution, in the special interest of basket manufacture. Since that time, it has become well established and is found along the banks of streams and in various moist situations. It is one of the largest naturalized willows, frequently fifty to seventy feet high and sometimes as much as eighty feet. The twigs are yellow green, polished, and very brittle, hence, its name, and after a high wind, the ground beneath the tree is frequently covered with them. The tree can be identified by its leaves that are four to seven inches long, narrow, with a wedge-shaped base, a long, tapering, pointed apex, and a saw tooth margin. They are dark, shining green above, smooth, whitish, and with a bloom beneath. Just at the junction of the leaf with the petiole there are two tiny excrescences.

Widely planted in parks and gardens because of its decorative, plumy and drooping foliage, the weeping willow is one of the better known willows. A native of Europe it was introduced into England either in 1692 or 1730, and, sometime later, was brought to this country where it has spread locally along riverbanks and lake shores. It is a tree up to forty feet high, with long and pendulous branches, and leaves five to six inches long, deep green above, paler beneath, and finely toothed. Another naturalized willow is the yellow willow or golden osier, a venerable-looking tree, with a short trunk, and regular spreading top. Its preferred habitat is moist, rich soil and it can be found in such places, often beside a stream. Its bright yellow twigs give the tree its name, and also provide the means whereby it can be identified. The winter and early spring color of the tree, due to its showy twigs, is exceptionally handsome.

Besides the black willow and pussy willow there are a number of other native willows such as the shining willow, a lustrous-leaved species, mostly a shrub, but often a small tree, found on the banks of streams and swamps; the balsam willow, a shrub of swamps and lowland thickets whose leaves, when crushed, have a spicy, balsamic odor; the heart-leaved willow, a beautiful shrub sometimes attaining the size and proportions of a small tree, and growing along the watercourses, preferring locations that are frequently submerged; the bog willow, a slender shrub of cold bogs and inland swamps; the prairie willow, an upland, grayish species, with slender, white-woolled twigs; the sage willow, a low, slender, and tufted shrub of sandy plains and dry, rocky land, its leaves white-woolled on their lower surface that give the bush a sort of grayish effect; the sandbar willow, mostly a shrub, but sometimes a tree, of riverbanks where at times it forms dense thickets, the first shrub to spring up on newly formed sandbars where it holds the soil in place with its long, rigid roots; and the peach-leaved willow, a common willow of the west where it is found along the banks of streams.

The second group of plants of the willow family are the poplars, genus *Populus,* the classical Latin name of the poplar, some being called cottonwoods and aspens. There are thirty species, most of them from the North Temperate zone, a few from the warmer parts of northern Africa. They are quick-growing, soft-wooded trees, with light greenish gray bark on the younger branches and stalked, usually ovalish leaves, their other characters those of the family.

Common on alluvial soil, on riverbanks, and the borders of swamps, the balsam poplar, also known as the tacamahac, is a northern, stout tree up to ninety feet high with a perpendicular trunk, spreading branches, and dark green leaves, silvery beneath, narrowly ovate, heart-shaped at the base, pointed at the tip, and finely toothed. The spring buds are long-pointed, reddish brown, shining, resinous, and sticky with a yellowish balsamic exudation. Bees use this waxy substance, as well as that of other trees such as the fir and horse chestnut, to strengthen the edges of their cells or to fill crevices in their hive, the substance being known as propolis. A very similar species, but with heart-shaped leaves, is the Balm of Gilead whose origin is unknown. Possibly it may be a native of Asia, but more likely it is of hybrid origin and considered by some botanists to be a clone.

Rare in New England, but common in the South Atlantic states and abundant in the lower Mississippi valley, the swamp cottonwood is a tree of low wetlands, thirty to seventy-five feet high, sometimes as much as ninety feet tall, with a trunk diameter of three feet and dark gray brown bark, perpendicularly short furrowed. Its leaves are ovate, broad at the base, narrowed toward the blunt apex, somewhat coarsely and shallowly, round-toothed, olive green above, paler beneath, and its flowers are in long, drooping catkins three to four inches long. It is also called the downy poplar because the leaves retain the down on their veins more abundantly than do other poplars.

Comparatively rare and of small size in the eastern

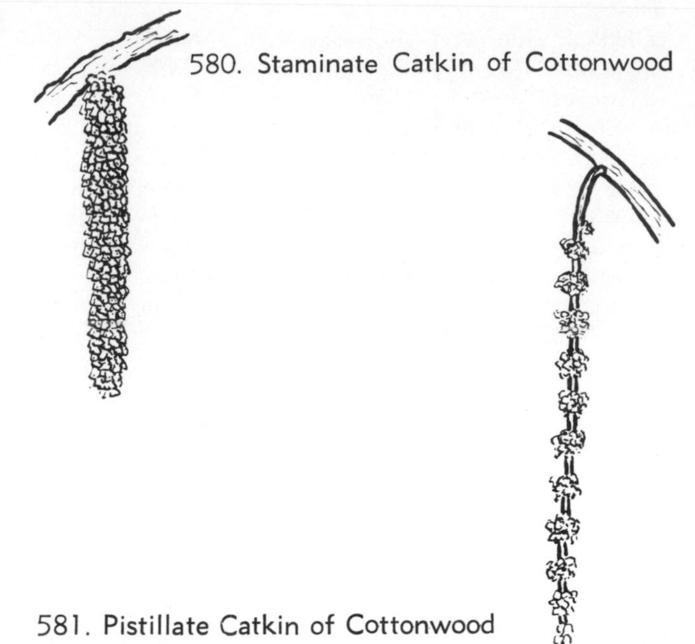

580. Staminate Catkin of Cottonwood

581. Pistillate Catkin of Cottonwood

states, the cottonwood, also known as the Carolina poplar and necklace poplar, is the largest of the poplars and most abundant along the streams between the Appalachians and Rocky Mountains where it attains a height of one hundred feet. Its bark is strongly seamed with conspicuously, confluent ridges, ashen gray when old, gray green when young, and its leaves are triangular, abruptly sharp-pointed, coarsely toothed, bright shining green above, paler beneath, on long stems. The staminate catkins are red and pendant, while the pistillate are green.

There are several western species, that include the black cottonwood, a tall and stately tree, with a broad, rounded crown supported upon heavy, upright limbs and a giant of the genus; the lance-leaved cottonwood, a slender species; and the Fremont cottonwood, a poplar similar to the Carolina poplar, but with smaller leaves and pubescent buds.

When the landscape is at the transition point between winter and spring, few people have noticed an aspen shake out its countless tassels that dance upon the south wind or have looked at one of them closely. The aspen is said to be the most widely distributed tree in North America, and though quite common, it has little value, except perhaps to the beaver, the tree being one of this animal's favorite foods. It is rarely planted, and is often an undergrowth in an oak wood where it is likely to remain unnoticed. Yet, it frequently forms a little thicket of its own on a gravelly bank by the roadside, or on the border of a swamp, though it is not a water-loving plant.

Its catkins, appearing before the leaves, are furry and show a touch of pink. With a magnifying glass or a hand lens, one can note how the scales of the staminate flowers are deeply cut into three or four, linear divisions, and are fringed with long, soft, gray hairs.

582. Staminate Flower of Aspen

583. Pistillate Flower of Aspen

There are six to twelve stamens. The pistillate flowers, found on separate trees, have a two-lobed stigma with an ovary surrounded by a broad oblique disk.

"Trembling like an aspen leaf" is an often quoted phrase. Homer was, perhaps, the first to speak of this peculiarity of the aspen:

Some wove the web,
Or twirled the spindle, sitting, with a quick
Light motion like the aspen's glancing leaves.

Other poets speak of it too, as Spenser for instance, who wrote: "His hand did quake/ And tremble like a leaf of aspen green," and Tennyson says: "willows whiten, aspens quiver."

584. Leaf of Aspen

There is no mystery as to why the aspen's leaves quiver. Look at the stem or petiole of a leaf and it will be found that it is round and slender near the twig, but flattened broadly near the leaf—that is, flattened in a plane at right angles with the blade. In other words, it has four flat surfaces exposed so that it would have to be a most cautious breeze that could pass by without disturbing the leaf's unstable equilibrium.

There are two native aspens, the American aspen and the large-toothed aspen. The American aspen is a rather small tree, thirty to forty feet high, sometimes sixty feet, with bark on old trees almost black near the base, a pale greenish brown or yellow brown, or nearly white higher on the trunk and on young stems, often roughened with horizontal bands or wartlike excrescences, and marked below the branches with large, dark scars. The leaves are broadly heart-shaped, a dull, lusterless, dark blue green with fine teeth, and that turn a bright, clear yellow in fall. The large-toothed aspen, common on the borders of streams and on hillsides, is similar in character, but much coarser throughout, and with thick leaves that have large, rounded teeth on the margins.

Besides our native poplars there are three introduced species that are in wide cultivation. They are the white or silver poplar, common in dooryards and along highways, with leaves lobed or cut finger-fashion and prominently white beneath; the black poplar, a wide-spreading tree with broadly, triangular-oval, or wedge-shaped leaves, finely blunt-toothed; and the Lombardy poplar, a familiar spirelike tree with almost vertical branches and leaves similar to those of the black poplar, and an Italian variety of the latter. Both the black poplar and the Lombardy poplar are escapes from cultivation.

Both the willows and poplars are useful to wildlife, the willows especially in the North, not only because they are more plentiful there, but because it is the region where there are more browsing and bud-eating species. The buds and small, tender portions of the twigs are a staple food of several species of grouse. The twigs, foliage and bark are eaten by rabbits and such hoofed browsers as the elk, moose, and deer. These animals also feed on the twigs and leaves of the poplars, the grouse on the catkins and buds, and the beavers and porcupines relish the bark. Various species of willows also serve as food plants for the caterpillars of several species of butterflies, such as the green comma, the hoary comma, the Compton tortoise shell, the viceroy, the Acadian hairstreak, the striped hairstreak, and Persius dusky wing. The caterpillars of several species of butterflies include various species of poplars in their dietary. They are the Compton tortoise shell, the mourning cloak, the viceroy, the banded purple or white admiral, the red-spotted purple, the tiger swallowtail, the dreamy dusky wing, and Persius dusky wing.

208. Witch Hazel (Hamamelidaceae)

It almost seems axiomatic that trees and shrubs should blossom in the spring or at least in summer, but the witch hazel is an exception and when one comes across it in the autumn woods, with yellow blossoms shining in the bright October sunshine, one can but wonder with the poet:

> Has time grown sleepy at his post
> And let the exiled summer back?
> Or is it her regretful ghost,
> Or witchcraft of the almanac?

What is equally surprising, the flowers appear at the same time as the fruit is maturing; fruit that has taken a year to develop for the fruit is from the flowers of the previous autumn. One must be careful of the fruit for when they are ripe, they explode and send the large, bony, shining black, white-tipped seeds on their way, as if they were propelled from a machine gun, to a distance as much as forty feet or more. And should one take home a twig or two with their blossoms and place them in the living room, one may suddenly be subjected to an artillery barrage.

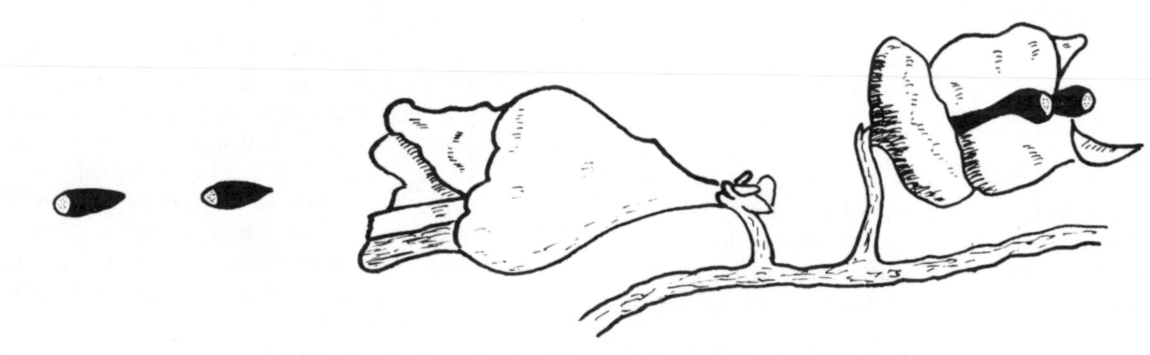

585. Exploding Fruit (Capsule) and Seeds of Witch Hazel

The witch hazel is a tall shrub, ten to twelve feet high, sometimes a slender, unsymmetrical tree, twenty or more feet high, of copses, or low, rich woods. The bark is sepia or deep brown, blotched with lighter brown, scaly or smooth. The leaves are alternate, unsymmetrical, oval or obovate, wavy-margined or coarsely serrate, and turn a spotted, dull gold yellow in autumn, sometimes hanging on the branches all winter. The flowers are a bright yellow and are in fours: four lobes to the calyx, four strap-shaped petals, and four stamens. Only the pistil varies from this fourfold plan for it has a two-celled ovary and two styles. The flowers are staminate or pistillate or perfect on the same tree and eventually develop into a two-beaked, two-celled, woody capsule that splits open explosively when ripe to shoot out its two bony seeds. The empty, gaping, brown capsules remain on the branches all winter and serve to identify the shrub at this time of the year.

Why the name of witch hazel? A witch is an evil person, more accurately an evil woman, one who is supposed to be under the influence of evil spirits or to have magic powers, but there is nothing evil about the witch hazel. It seems, however, that when the English settlers came to America, they saw in the witch hazel a certain resemblance of the shrub to an English hazel tree of elm lineage that they called *wych hazel.* And so in time our native shrub came to be known as witch hazel.

According to an old superstition, a branch of the English hazel might be used as a divining rod to locate a hidden spring of water, a buried treasure, or a precious ore. As Swift wrote:

> They tell us something strange and odd
> About a certain magic rod
> That, bending down its top divines
> Where'er the soil has hidden mines;
> Where there are none, it stands erect
> Scorning to show the least respect.

And so, the early settlers, seeing a resemblance of our witch hazel to their own hazel tree, thought that our witch hazel, too, had magic powers. Many a hole has been dug in America where the witch hazel indicated there was a hidden spring or buried treasure. But the most the witch hazel has done for us, except, perhaps, to provide us with an ornamental, is to furnish us with witch hazel extract, a soothing preparation, but of no curative properties. It is said that the Indians introduced it to the settlers. The seeds, buds, and twigs form dietary items for the pheasant, bobwhite, ruffed grouse, squirrel, the cottontail rabbit, the white-tailed deer, and the beaver.

Somewhat similar to the witch hazel, but flowering in late winter or spring is the springtime witch hazel, a small shrub of streamsides. Its twigs and leaves are woolly, the latter somewhat oblong or broadest towards the tip, coarsely toothed, and its flowers are light yellow, or reddish towards the base. Both species of witch hazels belong to the genus *Hamamelis,* including two other species, one native to Japan, the other to China. The genus is one of twenty-three genera that comprise the witch-hazel family, a group of one hundred species of trees and shrubs widely scattered in the North Temperate and tropical zones. Their leaves are deciduous or evergreen, usually alternate and simple. Their flowers are small, perfect or imperfect, with four or five sepals and the same number of petals or none at all, four or more stamens, and one pistil with a two-celled ovary, enclosing one or more ovules in each cell. The fruit is a woody capsule.

Three genera of the family occur in the United States. They are *Hamamelis, Fothergilla,* and *Liquidambar.* The genus *Fothergilla,* named for John Fothergill, an English physician, is a small genus of shrubs commonly called witch alder. A southern relative of the witch hazel is a beautiful shrub of streamsides, low wet ground, or sandy pine lands known variously as witch alder, dwarf witch alder or fothergilla. It is a low shrub, usually under three feet high, with broadly, wedge-shaped leaves, olive green above, bluish white and hairy below, fine white flowers in pomponlike, terminal spikes, and a seed capsule somewhat similar to that of the witch hazel. There are two other species, also southern in range, and all are rather ornamental.

The genus *Liquidambar,* Latin for liquid and Arabic for amber, in allusion to the fragrant resin of an Asiatic species, consists of five or six species, only one, the sweet gum, being native to the New World, the others being found in Asia. The sweet gum is a large, beautiful tree up to one hundred and twenty feet high, with a straight trunk and short, slender branches, forming a pyramidal or oblong head. The bark is gray brown, or reddish brown, furrowed and scaly on old trees, ashen gray on young trees and the leaves are a deep glossy green, starlike or cleft into five lobes, somewhat resembling those of a maple, the lobes finely toothed, fragrant when crushed, and turning deep yellow, rich red, and dark maroon in autumn. The flowers are small, inconspicuous, the staminate in terminal, hairy racemes, the pistillate in solitary, swinging balls from the axils of the upper leaves. The latter develop into a woody, globose head of beaked capsules.

The sweet gum is essentially a southern species of low, wet woodlands, but is a desirable ornamental in the milder parts of the North. The tree is fairly easy to recognize by its star-shaped leaves and the fruit balls that persist on the branches into the winter, but the distinguishing feature of the tree is the peculiar appearance of its small branches and twigs, the bark being so arranged that, with a little imagination, it has a sort of reptilian form. For this reason, the tree has been called alligatorwood.

The sap of the sweet gum is resinous and fragrant that can easily be discovered by crushing a leaf or bruising a twig. On exposure to the air it hardens into a resin or gum. Known as copal-balsam or copalm, it is sometimes used as a substitute for storax. The commercial storax, a balsam used in the manufacture of soaps, perfumes, and pharmaceutical preparations, is obtained from the resinous exudations of the bark of the Asiatic species. It is said that the resinous gum obtained from our native species is sometimes used as a chewing gum.

209. Wood Sorrel (Oxalidaceae)

Of all our woodland plants, one is not likely to find one that is as dainty as the wood sorrel. It is a stemless, perennial of cool, damp places, growing from a creeping rootstock, with compound leaves, each from the root and each composed of three, light green, heart-shaped leaflets that droop and fold together after nightfall, and delicate, pretty flowers nearly an inch broad, with five notched petals, pinkish white, striped with crimson, or white with crimson lines, and each on a slender long scape that, like the leafy stems, arise from the rootstock. The small mining bees and the flowerflies visit the blossoms to ensure cross-fertilization, but the plant also has cleistogamous flowers to make sure of its survival. The wood sorrel, or common wood sorrel, to distinguish it from other wood sorrels, has almost a worldwide distribution, being found in Europe, Asia, northern Africa as well as in America. It is sometimes known as the shamrock.

The wood sorrels are a very large and interesting group of herbaceous plants, over three hundred spe-

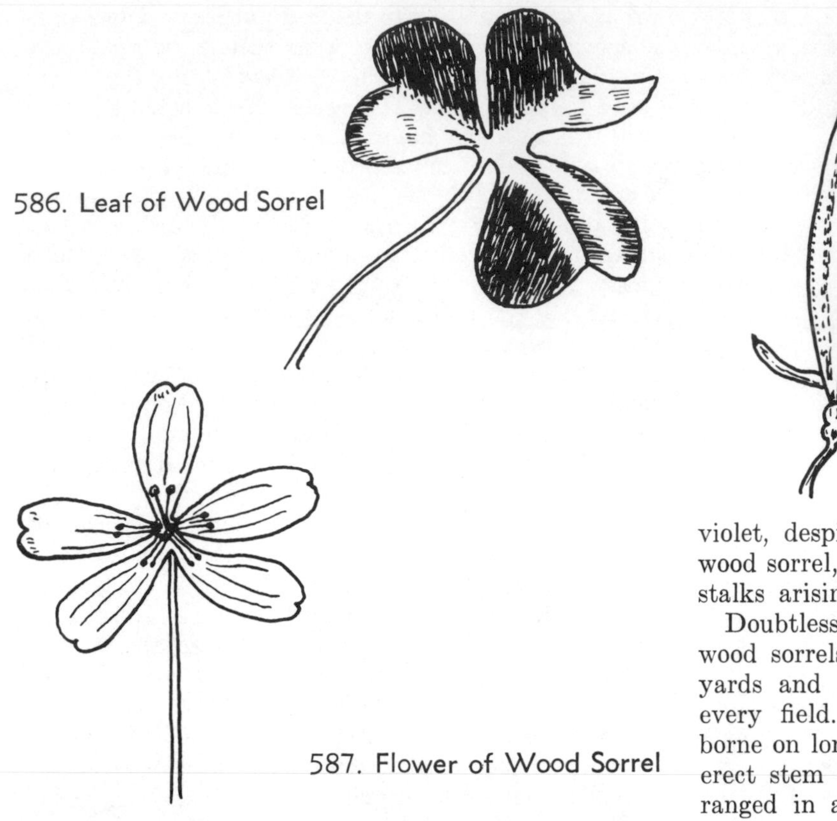

586. Leaf of Wood Sorrel

587. Flower of Wood Sorrel

588. Fruit of Wood Sorrel

cies, most occurring in the tropics and subtropics. They belong to the genus *Oxalis,* Greek for sour in reference to the sour juice that most of the species contain, the principal genus of the wood sorrel family.

All the wood sorrels have compound, cloverlike leaves. The leaflets are arranged finger-fashion and fold up at night or in dark weather. The flowers are perfect, solitary or more often in few-flowered clusters, and have five sepals and five petals, sometimes united at the base, ten stamens though sometimes five of them lack anthers, and a five-celled ovary with five styles. The fruit is typically a capsule with many seeds. Although these characteristics are for the genus *Oxalis,* they are essentially the same for the family as a whole.

Unlike the common wood sorrel that bears only one flower on a stalk, the violet wood sorrel bears three to six on a single stem. This, too, is another dainty, woodland species with leaves like those of the common wood sorrel and with flowers that are pinkish purple, lavender, or light magenta, but never violet, despite its name. And also like the common wood sorrel, it is a stemless perennial, with the flower stalks arising directly from the root.

Doubtless, the best-known representative of the wood sorrels is the yellow wood sorrel found about yards and gardens, and by every roadside and in every field. The leaves, each of three leaflets, are borne on long petioles that extend from a light green, erect stem and the flowers on pedicels that are arranged in a cluster at the summit of the peduncle arising from the leaf axil. The flowers, a rather deep lemon yellow, are scarcely one-half inch broad, with five long, ovate petals and ten yellow stamens alternately long and short, and are succeeded by hairy seed pods. The smaller bees and the flowerflies visit the blossoms, and also at times do various skipper butterflies. Some books also call the plant lady's sorrel. It originally came to us from Europe.

An evergreen tree of tropical India and China, the carambola, produces a juicy, fragrant, acid or sweet fruit used as a vegetable and as a dessert. It belongs to the genus *Averrhoa,* named for Averroes, Arabian philosopher, a small group of Asiatic trees with alternate, pinnately compound leaves, small, fragrant flowers, and a drooping, fleshy fruit. The carambola is a symmetrical tree, twenty to thirty feet high, with leaves of five to nine leaflets, purple marked flowers, and a yellowish brown, smooth-skinned, fruit, about four inches long, nearly egg-shaped, with a watery pulp, somewhat acid, and used fresh or for jellies.

210. Yam (Dioscoreaceae)

In the minds of many people there exists a certain amount of confusion between the yams and sweet potatoes that one buys at the local supermarket and many use the words interchangeably. They are not the same, of course. The sweet potato is a greenish tuber belonging to the morning glory family, while the yam is a yellowish tuber and a member of the yam family, a group of herbaceous or slightly woody, twining vines with fleshy or woody rootstalks that furnish the yams of commerce. There are some two hundred species and eight genera in the family that are mostly natives of America, though a few are found in the Old World.

They have slender stems, alternate or opposite leaves, small, greenish or whitish, dioecious or monoecious, regular flowers in spikes, racemes, or panicles. The flowers have six segments that are not easily separable as sepals and petals, six or three stamens, and a three-celled ovary. The fruit is a decidedly three-angled or winged capsule.

The yams of commerce belong to the genus *Dioscorea,* named for Dioscorides, one of the Greek fathers of botany. There are one hundred and sixty species, mostly tropical, a few cultivated for food and ornament. There is a native member of the genus called the wild yam, found in moist thickets. It has a woody, but not large rootstalk, and is a twining vine six to fifteen feet long, with ovate, entire leaves, greenish yellow flowers in drooping panicles, and a yellowish-green capsule.

Frequently cultivated for ornament is the Chinese yam or potato, also called cinnamon vine from the fragrance of its flowers. A native of China, it has tubers two to three feet long, an angled, sometimes twisted, tall-climbing stem, opposite, smooth, shining, heart-shaped or halberd-shaped leaves, and small white flowers in racemose panicles in the axils of the leaves. In these same axils, there often occur little bulblets the size of peas, that will grow if planted.

Another member of the genus is the air potato, an Indo-Malayan vine that is chiefly cultivated for its aerial tubers. It lacks the large underground tubers of most yams and has round stems that often bear in the leaf axils large tubers, eight to twelve inches long and weighing several pounds. It has alternate, heart-shaped or roundish leaves, the stalks longer than the blades, and flowers in many drooping, slender spikes.

Although the underground tubers of the yampee, also called cush-cush, are not large, they are prized for their flavor. A native of South America, it has a winged or angled stem, three to five-lobed leaves, with the staminate flowers in racemes, and the pistillate flowers in spikes. Unlike the other species that can be grown in the North, the yampee can only be cultivated in the Far South.

Sometimes grown for ornament in extreme southern Florida is the guayabo or gunda. It is a West Indian herbaceous, rather shrubby vine, arising from a large tuberous root, with alternate, roundish or heart-shaped leaves, and small, greenish, inconspicuous flowers, the staminate and pistillate on separate plants. The plant, belonging to the genus *Rajania,* named for John Ray, a famous British botanist, often produces aerial tubers that are irregular and cockscombed.

211. Yellow-eyed Grass (Xyridaceae)

The yellow-eyed-grass family consists of two genera and sixty species of perennial or annual tufted herbs, mostly of tropical distribution in both the Old World and the New. They have narrow, grasslike leaves, sheathing the base of a narrow scape, and perfect, mostly yellow, nearly or quite regular, solitary flowers, sessile in the axils of leathery, overlapping bracts (scales), forming terminal, ovoid, spherical or cylindric heads. The calyx has three sepals, the two lateral ones small, keeled, persistent, the other one larger and membranous (lacking in the South American genus *Abolboda.*) The corolla has three spreading lobes and a narrow tube. There are three fertile stamens, and three infertile ones. The ovary is one-celled, with few or many ovules, the style three-cleft. The fruit is an oblong, three-valved capsule.

Our native species have the same name as the family. The slender yellow-eyed grass is an inhabitant of sandy marshes and bogs where it lives with the sundew, cranberry, and marsh St. John's wort. In

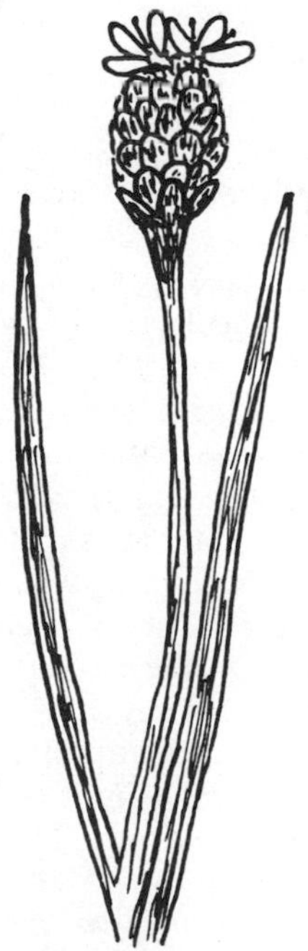

589. Leaves and Flowers of Yellow-eyed Grass

such places, its yellow dots are seen everywhere. It rises a foot or more high, with a somewhat flattened stem, bearing at the top a small, brownish, conelike head of scales wherefrom spring one, two, or three flowers showing just three, wide-open, golden petals. Sometimes the stem and grasslike leaves twist as they grow old. A large and taller species, the fringed yellow-eyed grass, of southern pine barrens, has a more flattened, stouter stem, and a head of bracts over half an inch, sometimes an inch long. Found in the sandy shores of lakes and pools from Maine to Indiana and southward the Carolina yellow-eyed grass has scapes one to two feet tall, twisted or straight. Its leaves are grasslike and quite long. The flower head measures about one half inch in length. The northern yellow-eyed grass is a dwarf and slender species occurring in the peat bogs of mountain regions. It has a straight or slightly twisted stem, narrow leaves about two inches long, and small, ovoid flower heads.

212. Yew (Taxaceae)

Spenser called the yew the "shooter eugh" because the English yeoman had his bow made of its tough wood. Shakespeare, in Richard II, has Scroop say to his fallen king: "Thy very beadsmen learn to bend their bows/ Of double-fatal yew against thy state." The Bard of Avon also refers to the yew in *Twelfth Night* when the Clown, lamenting the indifference of his loved-one, sings, "My shroud of white, stuck all with yew,/ O, prepare it." Both Spenser and Shakespeare were referring to the European yew, a tree whose history is interwoven with the development of civilization and whose wood is tough and elastic. In early English folklore, the yew was the saddest of all trees, except the cypress, and branches of the tree were used to decorate a house where a body lay awaiting burial. Even the mourners bound chaplets of yew around their heads. In England, yews have long been planted in churchyards and many superstitions have associated the yew with ghosts and fairies.

The European yew is a close-bodied, compact tree with a broad, round head and growing to a height up to sixty feet, its foliage and fruit dangerously poisonous. Our own native yew, a straggling or half-prostrate, evergreen shrub, rarely over three feet high, is common in shady, rocky woodlands, especially in evergreen woods. At first sight, the American yew looks like a hemlock seedling that has not done too well, hence, it is also called the ground hemlock, because its leaves are like those of the hemlock. On closer inspection, however, they are seen to be stronger and of a darker green, with their lower surface a decided yellow green, whereas those of the hemlock are a pale blue green. They are about an inch long and taper suddenly to a minute, pricklelike point. The staminate flowers have four to eight stamens collected into a globular head. The pistillate flowers consist of an erect ovule on a ringlike disk that enlarges as the fruit matures to become a bright red, fleshy, cup-shaped aril with the appearance of a berry. The top of the black seed within is uncovered. The berry is sweet and slightly resinous, and, unlike that of the European yew, is not poisonous. It is well not to eat too many of them at a time, and it is better not to chew or swallow the seeds as they may be poisonous, although there is no evidence that they are.

Both the American and European yews belong to the genus *Taxus,* Latin for yew, although it is also the classical word for bow. There are eight species, though some botanists consider them to be merely forms of a single species that is found over much of the North Temperate zone. They are beautiful, slow-growing, evergreen shrubs with scaly, reddish brown bark in age and spirally arranged, flattened, linear leaves, a dark glossy green on the upper surface, a more yellow green on the lower surface with two yellowish or grayish green bands. The staminate and pistillate flowers occur on separate plants and the fruit is berrylike, scarlet, red, or brown.

The yews have been cultivated since the time of the Greeks, especially the English yew. The English yew is a very slow-growing tree and has not found much favor in our country. Its shrubby, cultural varieties, such as the Irish yew, and the Japanese yew, are more desirable for evergreen planting. The Irish yew has a very handsome, columnar form with upright branches and is of a fine dark green color. The Japanese yew is probably the most horticulturally important of all the yews and the best for hedges. It is normally a tree up to forty feet high, but under cultivation, more often a bushy shrub, with leaves about an inch long and a scarlet fruit.

In addition to the American yew there are two other native species, the Pacific yew and the Florida yew. The Pacific yew is a small forest tree of the Pacific slope. It has a broad head, long horizontal, pendulous branches and a thin bark covered with purplish scales. The leaves are a warm, deep yellow green, paler beneath and a fruit similar to that of the American yew. The Florida yew is a small tree of bushy habit with foliage that is aromatic when bruised. It is a southern species of local distribution.

The yews are members of the yew family, a group of one hundred, widely distributed evergreen trees and shrubs classified into eleven genera. They generally have spreading, horizontal branches, purple, scaly bark, and spiny, linear leaves. The flowers are minute, with the staminate and pistillate commonly on separate plants, the staminate globular and scaly, the pistillate naked ovules that develop into a fleshy, berrylike fruit. The only obvious difference between the yew family and the pine family is the fruit.

Closely related to the yews are the torreyas, named for John Torreya, an American botanist. They are handsome, evergreen trees with fissured bark and branches in whorled tiers. The leaves are narrow, stiff, and almost prickle-tipped, with two white or brownish bands beneath. The staminate flowers consist of six to eight groups of four stamens each, the pistillate of stalkless ovules that develop into a drupelike, fleshy fruit.

There are six species of torreyas, including the California nutmeg, the stinking cedar, and the Japanese torreya. The California nutmeg is a tree up to sixty feet high, with gray brown bark, drooping branches, shining green leaves, and an egg-shaped fruit. The stinking cedar, also known as the Florida torreya, is a tree about twenty feet high, sometimes more, with leaves of a decidedly unpleasant odor when bruised. The Japanese torreya is a Japanese species up to seventy-five feet high, with dark, shining leaves, and a stalkless fruit that is green, but faintly streaked with purple.

There are several other related species such as the plum yews from China and Japan, small, bushlike trees with flattened needles and nutlike fruits surrounded by a thin, fleshy envelope; a group of evergreen trees and shrubs of the genus *Podocarpus,* mostly from the southern hemisphere; and phyllocladium, a plant from New Zealand with scalelike leaves and its twigs enlarged to form flattened, greenlike structures that botanists call cladodes.

Appendix

Check List of Species

Acanthus	ACANTHACEAE
Bear's breech	*Acanthus mollis*
Black-eyed Susan	*Thunbergia alata*
Sky-flower	*Thunbergia grandiflora*
Smooth ruella	*Ruella strepens*
Hairy ruella	*Ruella ciliosa*
Dense-flowered water willow	*Diantherea americana*
Loose-flowered water willow	*Diantherea ovata*

African violet	GESNERIACEAE
African violet	*Saintpaulia ionantha*
Gloxinia	*Sinningia speciosa*

Akebia	LARDIZABALACEAE
Fiveleaf akebia	*Akebia quinata*

Amaranth	AMARANTHACEAE
Green amaranth	*Amaranthus retroflexus*
Tumbleweed	*Amaranthus graecizans*
Love-lies-bleeding	*Amaranthus gangeticus*
Prince's feather	*Amaranthus caudatus*
Joseph's coat	*Amaranthus tricolor*
Globe amaranth	*Gomphrena globosa*
Celosia	*Celosia* species
Water hemp	*Acnida cannabina*
Juba's bush	*Iresine paniculata*
Froelichia	*Froelichia floridana*

Amaryllis	AMARYLLIDACEAE
Yellow star grass	*Hypoxis hirsuta*
Atamasco lily	*Zephyranthes Atamasco*
Swamp lily	*Crinum americanum*
Spider lily	*Hymenocallis occidentalis*
Century plant	*Agave americana*
False sisal	*Agave dicipiens*
Wild century plant	*Agave neglecta*
False aloe	*Agave virginica*
Desert agave	*Agave deserti*
Belladonna lily	*Amaryllis*
Trumpet daffodil	*Narcissus pseudo-narcissus*
Paper white narcissus	*Narcissus tazetta*
Poet's narcissus	*Narcissus poeticus*
Snowdrop	*Galanthus nivalis*
Early spring snowflake	*Leucojum vernum*
Late spring snowflake	*Leucojum aestivum*
Fall snowflake	*Leucojum autumnale*
Tuberose	*Polianthese tuberosa*

Annato	BIXACEAE
Annatto	*Bixa orellana*

Arrow grass	JUNCAGINACEAE
Bog arrow grass	*Scheuchzeria palustris*
Seaside arrow grass	*Triglochin maritima*
Marsh arrow grass	*Triglochin palustris*
Three-ribbed arrow grass	*Triglochin striata*

Arrowroot	MARANTACEAE
Arrowroot	*Maranta arundinaceae*
Zebra plant	*Calathea zebrina*
Powdery talia	*Thalia dealbate*

Arum	ARACEAE
Skunk cabbage	*Symplocarous foetidus*
Calla lily	*Zantedeschia ethiopica*
Jack-in-the-Pulpit	*Arisaema triphyllum*
Elephant ear	*Colocasia antiquorum*
Dasheen or taro	*Colocasia esculenta*
Dragon arum	*Arisaema Dracontium*
Arrow arum	*Peltandra virginica*
Water arum	*Calla palustris*
Golden club	*Orontium aquaticum*
Sweet flag	*Acorus Calamus*

Australian Oak	PROTEACEAE
Australian silk-oak	*Grevillea robusta*
Silver tree	*Leucadendron argenteum*
Australian honeysuckle	*Banksia ericaaefolia*
Australian honeysuckle	*Banksia integrifolia*
Sea urchin	*Hakea laurina*
Queensland nut	*Macadamia ternifolia*

Banana	MUSACEAE
Common banana	*Musa sapientum*
Plantain	*Musa paradisiaca*
Dwarf banana	*Musa cavendishi*
Abyssinian banana	*Musa ensete*
Traveller's tree	*Ravenala madagascariensis*
Bird-of-paradise	*Strelitzia reginae*
Wild plantain	*Heliconia bihai*

Barberry	BERBERIDACEAE
May apple	*Podophyllum peltatum*
Blue cohosh	*Caulophyllum thalictroides*
Twinleaf	*Jeffersonia diphylla*
European barberry	*Berberis vulgaris*
Japanese barberry	*Berberis thunbergi*
American barberry	*Berberis canadensis*
Magellan barberry	*Berberis buxifolia*
Wintergreen barberry	*Berberis julianae*
Trailing mahonia	*Mahonia aquifoloium*
Oregon grape	*Mahonia nervosa*
California barberry	*Mahonia pinnata*
Nandina	*Nandina domestica*
Large-flowered barrenwort	*Epimedium grandiflorum*
Umbrella leaf	*Diphylleia cymosa*

Batis	BATIDACEAE
Batis	*Batis maritima*

Bayberry	MYRICACEAE
Bayberry	*Myrica pennsylvanica*
Sweet gale	*Myrica gale*
Wax myrtle	*Myrica cerifera*
California wax myrtle	*Myrica california*
Sweet fern	*Comptonia peregrina*

Beech	FAGACEAE
Cork oak	*Quercus Suber*
White oak	*Quercus alba*
Post oak	*Quercus stellata*
Overcup oak	*Quercus lyrata*
Bur oak	*Quercus macrocarpa*
Swamp white oak	*Quercus bicolor*
Chestnut oak	*Quercus montana*
Basket oak	*Quercus prinus*
Yellow oak	*Quercus muhlenbergii*
Dwarf chinquapin oak	*Quercus prinoides*
Red oak	*Quercus rubra*
Scarlet oak	*Quercus coccinea*
Black oak	*Quercus velutina*
Spanish oak	*Quercus falcata*
Pin oak	*Quercus palustris*
Bear oak	*Quercus ilicifolia*
Black jack oak	*Quercus marilandica*
Shingle oak	*Quercus imbricaria*
Willow oak	*Quercus phellos*
Water oak	*Quercus nigra*
Live oak	*Quercus virginiana*
California live oak	*Quercus agrifolia*
Interior live oak	*Quercus wislizeni*
Mesa oak	*Quercus engelmanni*
Canyon oak	*Quercus chrysolepsis*
Blue oak	*Quercus douglasi*
Valley oak	*Quercus lobata*
Oregon oak	*Quercus Garryana*
California black oak	*Quercus keloggi*
Copper beech	*Fagus sylvatica*
American beech	*Fagus grandifolia*
American chestnut	*Castanea dentata*
Japanese chestnut	*Castenea crenata*
Spanish chestnut	*Castanea sativa*
Chinkapin	*Castanea pumila*
Golden chinkapin	*Castanea chrysophylla*
Tanbark oak	*Lithocarpus densiflora*

Beefwood	CASUARINACEAE
Beefwood	*Casuarina equisetifolia*

Begonia	BEGONIACEAE
Common Rex Begonia	*Begonia rex*

Bellflower	CAMPANULACEAE
Harebell	*Campanula rotundifolia*
Canterbury bells	*Campanula medium*
Peach bells	*Campanula persicifolia*
Coventry bells	*Campanula trachelium*
Carpathian bellflower	*Campanula carpatica*
Chimney bellflower	*Campanula pyramidalis*
Creeping bellflower	*Campanula rapunculoides*
Tall bellflower	*Campanula americana*
Marsh bellflower	*Campanula aparinoides*
Venus's looking-glass	*Specularia perfoliata*
Balloonflower	*Platycodon grandiflorum*

Bignonia	BIGNONIACEAE
Catalpa	*Catalpa bignonioides*
Hardy catalpa	*Catalpa speciosa*
Desert willow	*Chilopsis linearis*
Black calabash tree	*Enallagma latifolia*
West Indian calabash tree	*Crescentia cujete*
Trumpet creeper	*Bignonia radicans*
Cross vine	*Bignonia caprreolata*
Jacaranda	*Jacaranda ovalifolia*

Sausage tree — *Kigelia pinnata*
Candle tree — *Parmentiera cerifera*
Flame vine — *Pyrostegia ignea*
Yellow elder — *Stenolobium stans*
Cape honeysuckle — *Tecomaria capensis*

Birch — BETULACEAE

Paper birch — *Betula papyrifera*
Gray birch — *Betula populifolia*
European white birch — *Betula pendula*
Red birch — *Betula nigra*
Yellow birch — *Betula lutea*
Black birch — *Betula lenta*
Swamp birch — *Betula pumila*
Northern birch — *Betula borealis*
Newfoundland Dwarf birch — *Betula michauxii*
Dwarf birch — *Betula glandulosa*
Minor birch — *Betula minor*
Speckled alder — *Alnus incana*
Smooth alder — *Alnus rugosa*
Mountain alder — *Alnus viridis*
Seaside alder — *Alnus maritima*
Black alder — *Alnus glutinosa*
Mountain alder — *Alnus tenuifolia*
Red alder — *Alnus rubra*
White alder — *Alnus rhombifolia*
Hazelnut — *Corylus americana*
Beaked hazelnut — *Corylus cornuta*
Hop hornbeam — *Ostrya virginiana*
American hornbeam — *Carpinus caroliniana*

Birthwort — ARISTOLOCHIACEAE

Wild ginger — *Asarum canadense*
Virginia snakeroot — *Aristolochia serpentaria*
Dutchman's pipe — *Aristolochia durior*

Bladder nut — STAPHYLEACEAE

Bladder nut — *Staphylea trifolia*

Bladderwort — LENTIBULARIACEAE

Greater bladderwort — *Utricularia vulgaris*
Purple bladderwort — *Utricularia purpurea*
Closed bladderwort — *Utricularia cleistogama*
Common butterwort — *Pinguicula vulgaris*

Bloodwort — HAEMODORACEAE

Redroot — *Lachnanthese tinctoria*

Boldo — MONIMIACEAE

Boldo — *Peumus boldus*

Borage — BORAGINACEAE

Forget-me-not — *Myosotis scorpiodes*
Smaller forget-me-not — *Myosotis laxa*
Spring forget-me-not — *Myosotis virginica*
Virginia cowslip — *Mertensia virginica*
Viper's bugloss — *Echium vulgare*
Hound's tongue — *Cynoglossum officinale*
Wild comfrey — *Cynoglossum virginicum*
Virginia stickseed — *Lappula virginiana*
European stickseed — *Lappula Lappula*
Many-flowered stickseed — *Lappula floribunda*
Common borage — *Borago officinalis*
Common heliotrope — *Heliotropium peruvianum*

Box — BUXACEAE

Common box — *Buxus sempervirens*
Allegheny spurge — *Pachysandra procumbens*
Japanese spurge — *Pachysandra terminalis*

Brazil nut — LECHYTHIDACEAE

Brazil nut — *Bertholletia excelsa*

Broomrape — OROBANCHACEAE

One-flowered Broomrape — *Thalesia uniflora*
Branched Broomrape — *Orobanche ramosa*
Clover Broomrape — *Orobanche minor*
Louisiana Broomrape — *Orobanche ludoviciana*
Beechdrops — *Leptamnium virginianum*
Squawroot — *Cohopholis americana*

Buckeye — HIPPOCASTANACEAE

Horse chestnut — *Aesculus hippocastanum*
Ohio buckeye — *Aesculus glabra*
Sweet buckeye — *Aesculus octandra*
California buckeye — *Aesculus californica*
Red buckeye — *Aesculus pavia*
Painted buckeye — *Aesculus sylvatica*
Bottlebush buckeye — *Aesculus parviflora*

Buckthorn — RHAMNACEAE

New Jersey tea — *Ceanothus americanus*
Red-root — *Ceanothus intermedius*
Deer brush — *Ceanothus velutinus*
Snow brush — *Ceanothus cordulatus*
California lilac — *Ceanothus divaricutus*
Alder-leaved buckthorn — *Rhamnus alnifolia*
Lance-leaved buckthorn — *Rhamnus lanceolata*
Carolina buckthorn — *Rhamnus caroliniana*
Buckthorn — *Rhamnus cathartica*
Alder buckthorn — *Rhamnus frangula*
Red berry — *Rhamnus crocea*
Coffee berry — *Rhamnus californica*
Cascara sagrada — *Rhamnus purshiana*

Buckwheat — POLYGONACEAE

Field sorrel — *Rumex acetosella*
Patience dock — *Rumex patientia*

Curled dock	*Rumex crispus*
Bitter dock	*Rumex obtusifolius*
Canaigre	*Rumex hymenosepalus*
Great water dock	*Rumex Britannica*
Swamp dock	*Rumex verticillatus*
Golden dock	*Rumex persicarioides*
Common persicaria	*Polygonum pennsylvanicum*
Lady's thumb	*Polygonum persicaria*
Knotgrass	*Polygonum aviculare*
Erect knotweed	*Polygonum erectum*
Mild water pepper	*Polygonum hydropiperoides*
Water pepper	*Polygonum hydropiper*
Water smartweed	*Polygonum punctatum*
Halberd-leaved tearthumb	*Polygonum arifolium*
Arrow-leaved tearthumb	*Polygonum sagittatum*
Climbing false buckwheat	*Polygonum scandens*
Mountain fleece	*Polygonum amplexicaule*
Silver-lace vine	*Polygonum auberti*
Mexican bamboo	*Polygonum cuspidatum*
Prince's feather	*Polygonum orientale*
Common buckwheat	*Fagopyrum esculentum*
Rhubarb	*Rheum rhaponticum*
Wild buckwheat	*Eriogonum fasciculatum*
Sulphur flower	*Eriogonum umbellatum*
Coast jointweed	*Polygonella articulata*
Brunnichia	*Brunnichia cirrhosa*
Sea grape	*Coccolobis uvifera*
Pigeon plum	*Coccolobis floridana*

Burmannia	BURMANNIACEAE
Northern burmannia	*Burmannia biflora*

Bur reed	SPARGANIACEAE
Great bur reed	*Sparganium eurycarpum*
Smaller bur reed	*Sparganium simplex*
Branching bur reed	*Sparganium americanum* variety *androcladum*

Buttercup	RANUNCULACEAE
Tall buttercup	*Ranunculus acris*
Bulbous buttercup	*Ranunculus bulbosus*
Creeping buttercup	*Ranunculus repends*
Early buttercup	*Ranunculus fascicularis*
Bristly crowfoot	*Ranunculus pennsylvanicus*
Cursed crowfoot	*Ranunculus sceleratus*
Small-flowered crowfoot	*Ranunculus abortivus*
Hepatica	*Hepatica triloba*
Wood anemone	*Anemone quinquefolia*
Tall anemone	*Anemone virginiana*
Long-fruited anemone	*Anemone cylindrica*
Canada anemone	*Anemone canadensis*
Mountain anemone	*Anemone trifolia*
Rue anemone	*Anemonella thalictroides*
Meadow rue	*Thalictrum polygamum*
Early meadow rue	*Thalictrum dioicum*
Purple meadow rue	*Thalictrum dasycarpum*
Virgin's bower	*Clematis virginiana*
Leatherleaf Clematis	*Clematis viorna*
Japanese Clematis	*Clematis paniculata*
Scarlet Clematis	*Clematis texensis*
Sugarloaf clematis	*Clematis douglasi*
Golden clematis	*Clematis tangutica*
Traveler's-joy	*Clematis vitalba*
Marsh marigold	*Caltha palustris*
Goldthread	*Coptis trifolia*
Wild columbine	*Aquilegia canadensis*
Garden columbine	*Aquilegia vulgaris*
Tall larkspur	*Delphinium exaltatum*
Dwarf larkspur	*Delphinium tricorne*
Sky-blue larkspur	*Delphinium azureum*
Field larkspur	*Delphinium consolida*
Rocket larkspur	*Delphinium ajacis*
Common monkshood	*Aconitum napellus*
Wild monkshood	*Aconitum uncinatum*
Black cohosh	*Cimiicifuga racemosa*
Red baneberry	*Actaea rubra*
White baneberry	*Actaea alba*
Orangeroot	*Hydrastis canadensis*
Peony	*Paeonia* species
Globeflower	*Trollius europaeus*
Christmas rose	*Helloborus niger*
Love-in-a-mist	*Nigella damascena*
Fennel-flower	*Nigella sativa*

Cacao	STERCULIACEAE
Cacao tree	*Theobroma cacao*
Cola tree	*Cola acuminata*
Flannel bush	*Fremontia californica*

Cactus	CACTACEAE
Barbados gooseberry	*Peireskia aculeata*
Eastern prickly pear	*Opuntia compressa*
Western prickly pear	*Opuntia humifusa*
Indian fig	*Opuntia ficus-indica*
Tuna	*Opuntia megacantha*
Nopal	*Opuntia lindheimeri*
Cane cactus	*Opuntia leptocaulis*
Tree cane cactus	*Opuntia imbricata*
Jumping cholla	*Opuntia bigelovii*
Beavertail	*Opuntia basilaris*
Foxtal cactus	*Mammillaria deserti*
Fish-hook cactus	*Mammillaria tetrancistra*
Devil's pincushion	*Mammillaria robustispina*
Sunset cactus	*Mammillaria grahamii*
Arizona pincushion	*Mammillaria arizonica*
Indian strawberry cactus	*Echinocereus chrysocentrus*
Claret cup cactus	*Echinocereus triglochidiatus*
Rainbow cactus	*Echinocereus chloranthus*
Lady Finger cactus	*Echinocereus pentalophus*
Lace cactus	*Echinocereus reichenbachii*
Green-flowered pitaya	*Echinocereus viridiflorus*
Niggerhead cactus	*Echinocactus polycephalus*

California barrel cactus	*Echinocactus acanthodes*
Bisnaga	*Echinocactus visnaga*
Traveller's friend	*Echinocactus covillei*
Saguaro	*Cereus giganteus*
Pipe organ cactus	*Cereus thurberi*
Old-man's-beard	*Cereus senilis*
Night-blooming cereus	*Cereus serpentinus*
Mistletoe cactus	*Rhipsalis Cassutha*
Christmas cactus	*Zygocactus truncatus*

Caltrop	ZYGOPHYLLACEAE
Ground Burnet	*Tribulus terrestris*
Great caltrop	*Kallstroemia maxima*
Creosote bush	*Larrea divaricata*
Lignum-vitae	*Guaiacum sanctum*

Canna	CANNACEAE
Canna	*Canna hybrida*
Golden canna	*Canna flaccida*

Caper	CAPPARIDACEAE
Caper	*Capparis spinosa*
Rocky mountain bee plant	*Cleome serrulata*
Spider flower	*Cleome spinosa*

Carpetweed	AIZOACEAE
Carpetweed	*Mollugo verticillata*
Sea purslane	*Sesuvium maritimum*
Sea fig	*Mesembryanthemum chilense*
Ice plant	*Mesembryanthemum crystallinum*
Hottentot fig	*Mesembryanthemum edule*

Carrot	UMBELLIFERAE
Wild carrot	*Daucus carota*
Garden carrot	*Daucus carota sativa*
Early meadow parsnip	*Ziziz aurea*
Meadow parsnip	*Thaspium aureum*
Water parsnip	*Sium cicutaefolium*
Cow parsnip	*Heracleum lanatum*
Parsley	*Apium petroselinum*
Hemlock parsley	*Conioselinum chiense*
Fool's parsley	*Aethusa cynapium*
Cowbane	*Oxypolis rigidior*
Water hemlock	*Cicuta maculata*
Poison hemlock	*Conium maculatum*
Sweet cicely	*Osmorrhiza claytonii*
Sanicle	*Sanicula marilandica*
Water pennywort	*Hydrocotyle Americana*
Mock Bishop-weed	*Ptilimnium capillaceum*
Harbinger-of-spring	*Eirgenia bulbosa*
Sea holly	*Eryngium amethystium*
Button snakeroot	*Eryngium aquaticum*
Caraway	*Carum carvi*
Dill	*Anethum graveolens*
Coriander	*Coriandrum sativum*
Fennel	*Foeniculum vulgare*
Celery	*Apium graveolens*

Cattail	TYPHACEAE
Common cattail	*Typha latifolia*
Narrow-leaved cattail	*Typha angustifolia*

Coca	ERYTHROXYLACEAE
Coca plant	*Erythroxylon coca*

Composite	COMPOSITAE
Common dandelion	*Taraxacum officinale*
Red-seeded dandelion	*Taraxacum erythrospermum*
Fall dandelion	*Leontodon autumnalis*
Dwarf dandelion	*Krigia virginica*
Large-leaved aster	*Aster macrophyllus*
Heart-leaved aster	*Aster cordifolius*
Wavy-leaved aster	*Aster undulatus*
Spreading aster	*Aster patens*
Smooth aster	*Aster laevis*
Showy aster	*Aster spectabilis*
New England aster	*Aster novae-angliae*
Panicled white aster	*Aster paniculatus*
Heath aster	*Aster ericoides*
Many-flowered aster	*Aster multiflorus*
Golden star	*Chrysopsis mariana*
Grass-leaved golden aster	*Chrysopsis graminifolia*
Ground gold flower	*Chrysopsis falcata*
Blue-stemmed goldenrod	*Solidago caesia*
Broad-leaved goldenrod	*Solidago latifolia*
White goldenrod	*Solidago bicolor*
Bog goldenrod	*Solidago uliginosa*
Showy goldenrod	*Solidago speciosa*
Sweet goldenrod	*Solidago odora*
Rough-stemmed goldenrod	*Solidago rugosa*
Canada goldenrod	*Solidago canadensis*
Gray goldenrod	*Solidago nemoralis*
Common sunflower	*Helianthus annuus*
Tall sunflower	*Helianthus giganteus*
Ten-petaled sunflower	*Helianthus decapetalus*
Woodland sunflower	*Helianthus divaricatus*
Jerusalem artichoke	*Helianthus tuberosus*
Joe-pye weed	*Eupatorium purpureum*
Boneset	*Eupatorium perfoliatum*
Upland boneset	*Eupatorium sessilifolium*
Climbing hempweed	*Mikania scandens*
White snakeroot	*Eupatorium urticaefolium*
Tall blazing star	*Liatris scariosa*
Blazing star	*Liatris squarrosa*
Gay feather	*Liatris spicata*
Tall ironweed	*Vernonia altissima*
New York ironweed	*Vernonia noveboracensis*

Robin's plantain	*Erigeron pulchellus*
Common fleabane	*Erigeron philadelphicus*
Daisy fleabane	*Erigeron ramosus*
Canada fleabane	*Erigeron canadense*
Pussytoes	*Antennaria* species
Pearly everlasting	*Anaphalis margaritacea*
Sweet everlasting	*Gnaphalium polycephalum*
Elecampane	*Inula helenium*
Chicory	*Cichorium intybus*
Tansy	*Tanecetum vulgare*
Oxeye Daisy	*Chrysanthemum leucanthemum*
Black-eyed Susan	*Rudbeckia hirta*
Tall coneflower	*Rudbeckia laciniata*
Burdock	*Arctium Lappa*
Beggar-ticks	*Bidens frondosa*
Larger bur marigold	*Bidens laevis*
Smaller bur marigold	*Bidens cernua*
Spanish needles	*Bidens bipinnata*
Yarrow	*Achillea millefolium*
Common thistle	*Cirsium lanceolatum*
Pasture thistle	*Cirsium pumilum*
Canada thistle	*Cirsium arvense*
Sow thistle	*Sonchus oleraceus*
Field sow thistle	*Sonchus arvensis*
Spiny-leaved sow thistle	*Sonchus asper*
Common ragweed	*Ambrosia artemisiaefolia*
Great ragweed	*Ambrosia trifida*
Devil's paintbrush	*Hieracium aurantiacum*
King devil	*Hieracium pratense*
Canada hawkweed	*Hieracium canadense*
Rattlesnakeweed	*Hieracium venosum*
Lettuce	*Lactuca sativa*
Wild lettuce	*Lactua canadensis*
Hairy lettuce	*Lactua hirsuta*
Tall blue lettuce	*Lactua spicata*
Prickly lettuce	*Lactua scariola*
Rattlesnake root	*Prenanthes alba*
Lion's-foot	*Prenanthes serpentaria*
Oxeye	*Heliopsis helianthoides*
Sneezeweed	*Helenium autumnale*
Mayweed	*Anthemis cotula*
Mugwort	*Artemesia vulgaris*
Fireweed	*Erechtites hieracifolia*
Golden ragwort	*Senecio aureus*
Summer chrysanthemum	*Chrysanthemum carinatum*
Pyrethrum	*Chrysanthemum coccineum*
Costmary	*Chrysanthemum balsamita*
Corn marigold	*Chrysanthemum segetum*
Marguerite daisy	*Chrysanthemum frutescens*
Giant daisy	*Chrysanthemum uliginosum*
Shasta daisy	*Chrysanthemum maximum*
Feverfew	*Chrysanthemum parthenium*
African marigold	*Tagetes erecta*
French marigold	*Tagetes patula*
Zinnia	*Zinnia* species
Common garden dahlia	*Dahlia pinnata*
Castus dahlia	*Dahlia juarezi*
Cosmos	*Cosmos bipinnatus*
Black cosmos	*Cosmos diversifolius*
Calendula	*Calendula officinalis*
Bachelor's button	*Centaurea cyanus*
Dusty miller	*Centaurea gymnocarpa*
Sweet sultan	*Centaurea moschata*
Mountain bluet	*Centaurea montana*
Stoke's aster	*Stokesia laevis*
China aster	*Callistephus chinensis*
Annual gaillardia	*Gaillardia pulchella*
Perennial gaillardia	*Gaillardia aristata*
Arctotis	*Arctotis stoechadifolia*
Golden coreopsis	*Coreopsis tinctoria*
Golden-wave	*Coreopsis drummondi*
Tall coreopsis	*Coreopsis tripteris*
English daisy	*Bellis perennis*
Ageratum	*Ageratum houstonianum*
Strawflower	*Helichrysum bracteatum*
Tall globe thistle	*Echinops exaltatus*
Perennial globe thistle	*Echinops ritro*
Golden glow	*Rudbeckia laciniata*
Groundsel bush	*Baccharis halimifolia*
Marsh elder	*Iva frutescens*
Sea-oxeye	*Borrichia frutescens*
Santolina	*Santolina chamaecyparissus*

Coriaria	CORIARIACEAE
European species	*Coriaria myrtifolia*
New Zealand wineberry	*Coriaria sarmeniosa*
Tutu	*Coriaria ruscifolia*
Japanese ornamental	*Coriaria japonica*

Corkwood	LEITNERIACEAE
Corkwood	*Leitneria floridana*

Crowberry	EMPETRACEAE
Black crowberry	*Empetrum nigrum*
Purple crowberry	*Empetrum atropurpureum*
Broom crowberry	*Corema conradii*
Sandhill-rosemary	*Ceratiola ericoides*

Custard apple	ANNONACEAE
Papaw	*Asimina triloba*
Soursop	*Annona muricata*
Sweetsop	*Annona squamosa*
Custard apple	*Annona reticulata*
Dwarf papaw	*Asimina parviflora*
Narrow-leaf papaw	*Asimina angustifilia*
Showy papaw	*Asimina speciosa*
Pond apple	*Annona glabra*

Cyrilla	CYRILLACEAE
Leatherwood	*Cyrilla racemiflora*

Buckwheat tree	*Cliftonia monophylla*

Diapensia	DIAPENSIACEAE
Pyxie	*Pyxidanthera barbulata*
Diapensia	*Diapensis lapponica*
Galax	*Galax aphylla*
Shortia	*Shortia galacifolia*

Dipterocarpus	DIPTEROCARPACEAE
Sal tree	*Shorea robusta*
Piny resin	*Vateria indica*
Sumatra camphor-tree	*Dryobalanops aromatica*

Dogbane	APOCYNACEAE
Spreading dogbane	*Apocynum androsae-mifolium*
Indian hemp	*Apocynum cannabinum*
Climbing dogbane	*Trachrlospermum difforme*
Texas star	*Amsonia Amsonia*
Blue dogbane	*Amsonia brevifolia*
Periwinkle	*Vinca minor*
Larger periwinkle	*Vinca major*
Madagascar periwinkle	*Vinca rosea*
Common oleander	*Nerium oleander*
Sweet oleander	*Nerium indicum*

Dogwood	CORNACEAE
Flowering dogwood	*Cornus florida*
Red-osier dogwood	*Cornus stolonifera*
Round-leaved dogwood	*Cornus circinnata*
Silky dogwood	*Cornus Amomum*
Panicled dogwood	*Cornus paniculata*
Alternate-leaved dogwood	*Cornus alternifolia*
Bunchberry	*Cornus canadensis*
Tupelo	*Nyssa sylvatica*
Tupelo gum	*Nyssa aquatica*

Duckweed	LEMNACEAE
Greater duckweed	*Spirodela polyrhiza*
Lesser duckweed	*Lemna minor*
Ivy-leaved duckweed	*Lemna trisulca*
Columbia wolffia	*Wolffia columbiana*
Brazil wolffia	*Wolffia brasiliensis*

Ebony	EBENACEAE
Persimmon	*Diospyros virginiana*
Black persimmon	*Diospyros texana*
Japanese persimmon	*Diospyros kaki*

Elm	ULMACEAE
American elm	*Ulmus americana*
Slippery elm	*Ulmus fulva*
Cork elm	*Ulmus racemosa*
Wahoo elm	*Ulmus alata*
Red elm	*Ulmus serotina*
Cedar elm	*Ulmus crassifolia*
English elm	*Ulmus campestris*
Hackberry	*Celtis occidentalis*
Mississippi hackberry	*Celtis mississippiensis*
European nettle	*Celtis australis*
Planer tree	*Planera aquatica*

Evening primrose	ONAGRACEAE
Evening primrose	*Oenothera biennis*
Sundrops	*Oenothera fruticosa*
Fireweed or great willow herb	*Epilobium angustifolium*
Hairy willow herb	*Epilobium hirsutum*
Linear-leaved willow herb	*Epilobium lineare*
Downy willow herb	*Epilobium strictum*
Purple-leaved willow herb	*Epilobium coloratum*
Seedbox	*Ludwigia alternifolia*
Water purslane	*Ludwigia palustris*
Enchanter's nightshade	*Circaea lutetiana*
Fuchsia	*Fuchsia speciosa*

False mermaid	LIMNANTHACEAE
False mermaid	*Floerkea proserpinacoides*
Meadow-foam	*Limnanthes douglasi*

Figwort	SCROPHULARIACEAE
Butter-and-eggs	*Linaria vulgaria*
Blue toadflax	*Linaria canadensis*
Great mullein	*Verbascum thapsus*
Moth mullein	*Verbascum blattaria*
Turtlehead	*Chelone glabra*
Monkeyflower	*Mimulus rimgens*
American brooklime	*Veronica americana*
Marsh speedwell	*Veronica scutellata*
Thyme-leaved speedwell	*Veronica seryyllifolia*
Common speedwell	*Veronica officinalis*
Culver's root	*Veronica virginica*
Downy false foxglove	*Gerardia flava*
Fern-leaved false foxglove	*Gerardia pedicularia*
Smooth false foxglove	*Gerardia virginica*
Foxglove	*Gerardia purpurea*
Seaside gerardia	*Gerardia maritima*
Slender gerardia	*Gerardia tenuifolia*
Painted cup	*Castilleja coccinea*
Wood betony	*Pedicularis candensis*
Cow-wheat	*Melampyrum lineare*
Yellow rattle	*Rhinanthus crista-galli*
False pimpernel	*Ilysanthes dubia*
Hairy pentstemon	*Pentstemon hirsutus*
Large-flowered pent-stemon	*Pentstemon grandiflorus*
Smooth pentstemon	*Pentstemon pentstemon*
Foxglove beard-tongue	*Pentstemon digitalis*
Common snapdragon	*Antirrhinum majus*

Torenia	*Torenia fournieria*
Paulownia	*Paulownia tomentosa*

Flax	LINACEAE
Common flax	*Linum usitatissimum*
Wild yellow flax	*Linum virginianum*
Yellow flax	*Linum striatum*
Flowering flax	*Linum grandiflorum*
Perennial flax	*Linum perenne*
Blue flax	*Linum lewisii*

Flowering rush	BUTOMACEAE
Flowering rush	*Butomus umbellatus*
Water poppy	*Hydrocleis nymphoides*

Four-o'clock	NYCTAGINACEAE
Four-o'clock	*Mirabilis jalopa*
Bougainvillaea	*Bougainvillaea spectabilis*
Heart-leaved umbrella-wort	*Allionia nyctaginea*
Sand verbena	*Abronia umbellata*

Frog's bit	HYDROCHARITACEAE
Elodea	*Anarcharis canadensis*
Vallisneria	*Vallisneria spiralis*
Frog's bit	*Hydrocharis morsus-ranae*

Fumitory	FUMARIACEAE
Dutchman's breeches	*Dicentra Cucullaria*
Squirrel corn	*Dicentra candensis*
Bleeding heart	*Dicentra spectabilis*
Wild bleeding heart	*Dicentra eximia*
Pale corydalis	*Corydalis sempervirens*
Golden corydalis	*Corydalis aurea*
Climbing fumitory	*Adlumia fungosa*
Fumitory	*Fumaria officinalis*

Gentian	GENTIANACEAE
Fringed gentian	*Gentiana crinita*
Downy gentian	*Gentiana puberula*
Bottle gentian	*Gentiana andrewsii*
Five-flowered gentian	*Gentiana quinquefolia*
Yellow gentian	*Gentiana lutea*
Rose pink	*Sabbatia angularis*
Sea pink	*Sabbatia stellaris*
Large marsh pink	*Sabbatia dodecandra*
Slender marsh pink	*Sabbatia campanulata*
Yellow bartonia	*Bartonia virginiana*
Lesser centaury	*Centaurium umbellatum*
Spiked centaury	*Centaurium spicatum*
American columbo	*Frasera caroliniensis*
Pennywort	*Obolaria virginica*
Buckbean	*Menyanthes trifoliata*
Floating heart	*Nymphoides aquaticum*

Geranium	GERANIACEAE
Wild geranium	*Geranium maculatum*
Carolina cranesbill	*Geranium carolinianum*
Herb robert	*Geranium robertianum*
Show geranium	*Pelargonium domesticum*
Ivy geranium	*Pelargonium peltatum*
Rose geranium	*Pelargonium graveolens*
Fish geranium	*Pelargonium hortorum*
Nutmeg geranium	*Pelargonium odoratissimum*
Alfilaria	*Erodium circutarium*
Desert storksbill	*Erodium texanum*
White storksbill	*Erodium macrophyllum*

Ginger	ZINGIBERACEAE
Common ginger	*Zingiber officinale*
Ginger lily	*Hedychium coronarium*
Cardamon	*Elettaria cardamomum*

Ginkgo	GINKGOACEAE
Ginkgo	*Ginkgo biloba*

Ginseng	ARALIACEAE
Chinese ginseng	*Panax schinseng*
Ginseng	*Panax quinquefolium*
Dwarf ginseng	*Panax trifolium*
Spikenard	*Aralia racemosa*
Wild sarsaparilla	*Aralia nudicaulis*
Bristly sarsaparilla	*Aralia hispida*
Hercules' club	*Aralia spinosa*
English ivy	*Hedera helix*
Rice paper plant	*Tetrapanax papyriferum*

Globe daisy	GLOBULARIACEAE
Globe daisy	*Globularia* species

Goodenia	GOODENIACEAE
Scaevola	*Scaevola plumieri*

Goosefoot	CHENOPODIACEAE
Pigweed	*Chenopodium album*
Feather geranium	*Chenopodium borrys*
Mexican tea	*Chenopodium ambrosiodes*
Good-King-Henry	*Chenopodium bonus-henricus*
Strawberry blite	*Chenopodium capitatum*
Spreading orache	*Atriplex patula*
Garden orache	*Atriplex hortensis*
Spearscale	*Atriplex hastata*
Bugseed	*Corispermum hyssopifolium*
Russian thistle	*Salsola kali*
Greasewood	*Sarcobatus vermiculatus*

Samphire	*Salicornia europaea*
Slender samphire	*Salicornia bigelovii*
Woody glasswort	*Salicornia ambigua*
Tall sea blite	*Dondia americana*
Mock cypress	*Kochia scoparia*
Beet	*Beta vulgaris*
Spinach	*Spinacia oleracea*

Gourd	CUCURBITACEAE
Cucumber	*Cucumis sativa*
Melon	*Cucumis melo*
Bur gherkin	*Cucumis anguria*
Hubbard squash	*Cucurbita maxima*
Turban squash	*Cucurbita maxima* variety *turbaniformis*
Pumpkin	*Cucurbita pepo*
Scallop and summer crookneck squash	*Cucurbita pepo* variety *melopepo*
Ornamental yellow gourd	*Cucurbita pepo* variety *ovifera*
Ornamental gourd	*Lagenaria siceraria*
Watermelon	*Citrullus vulgaris*
Citron	*Citrullus vulgaris* variety *citroides*
Missouri gourd	*Cucurbita foetidissima*
Climbing wild cucumber	*Echinocystis lobata*
One-seeded bur cucumber	*Sicyos angulatus*
Creeping cucumber	*Melothria pendula*

Grass	GRAMINEAE
Corn	*Zea Mays*
Sweet corn	*Zea Mays* variety *rugosa*
Pop corn	*Zea Mays* variety *everta*
Flint corn	*Zea Mays* variety *indurata*
Dent corn	*Zea Mays* variety *indentata*
Bamboo	*Bambusa* species
Wheat (Bread)	*Triticum sativum*
Wheat (Durum)	*Triticum spelta*
Barley	*Hordeum vulgare*
Squirrel-tail grass	*Hordeum jubatum*
Rye	*Secale cereale*
Common oats	*Avena sativa*
Wild oat	*Avena fatua*
Rice	*Oryza sativa*
Wild rice	*Zizania aquatica*
Sugar Cane	*Saccharum officinarum*
Sorghum	*Holcus sorghum*
Kafir	*Holcus sorghum* variety *caffrorum*
Chicken-corn	*Holcus sorghum* variety *drummondi*
Sorgho	*Holcus sorghum* variety *saccharatus*
Broomcorn	*Holcus sorghum* variety *technicus*
Johnson grass	*Holcus halepensis*
Millet	*Panicum miliaceum*
Old witch grass	*Panicum capillare*
Forked panic grass	*Panicum dichotomum*
Round-fruited panic grass	*Panicum sphaerocarpon*
Hispid panic grass	*Panicum clandestimun*
Sea beach panic grass	*Panicum amaroides*
Kentucky bluegrass	*Poa pratensis*
Low spear grass	*Poa annua*
Wood spear grass	*Poa alsodes*
Canada blue grass	*Poa compressa*
False Redtop	*Poa triflora*
Sheep's fescue	*Festuca ovina*
Meadow fescue	*Festuca elatior*
Red fescue	*Festuca rubra*
Reed canary grass	*Phalaris arundinacea*
Canary grass	*Phalaris canariensis*
Nerved Manna grass	*Glyceria nervata*
Tall Manna grass	*Glyceria grandis*
Rattlesnake grass	*Glyceria canadensis*
Brown Bent-grass	*Agrostis canina*
Rough hairgrass	*Agrostis hyemalis*
Redtop	*Agrostis alba*
Wood muhlenbergia	*Muhlenbergia sylvatica*
Nimble will	*Muhlenbergia schreberi*
Rock muhlenbergia	*Muhlenbergia sobolifera*
Marsh muhlenbergia	*Muhlenbergia racemosa*
Meadow muhlenbergia	*Muhlenbergia mexicana*
Long-awned hairgrass	*Muhlenbergia capillaris*
Yellow foxtail	*Setaria glauca*
Green foxtail	*Setaria viridis*
Bristly foxtail	*Setaria verticillata*
Italian millet	*Setaria italica*
Palm grass	*Setaria palmifolia*
Cockspur grass	*Echinochloa crusgalli*
Couch grass	*Agropyron repens*
Bearded wheat grass	*Agropyron caninum*
Large crab grass	*Digitaria sanguinalis*
Small crab grass	*Digitaria humifusa*
Orchard grass	*Dactylis glomerata*
Sweet vernal grass	*Anthoxanthum odoratum*
Vanilla grass	*Hierochloe odorata*
Slender paspalum	*Paspalum setaceum*
Field paspalum	*Paspalum laeve*
Poverty grass	*Aristida dichotoma*
Timothy	*Phleum pratense*
Ray grass	*Lolium perenne*
Darnel	*Lolium temulentum*

Heath	ERICACEAE
Mayflower	*Epigaea repens*
Madrona	*Arbutus menziesi*
Strawberry tree	*Arbutus unedo*
Creeping snowberry	*Chiogenes hispidula*
Bearberry	*Arctostaphylos Uva-ursi*
Manzanita	*Arctostaphylos* species
Wintergreen	*Gaultheria procumbens*
Spotted wintergreen	*Chimaphila maculata*
Pipsissewa	*Chimaphila umbellata*
Shinleaf	*Pyrola elliptica*
Round-leaved pyrola	*Pyrola americana*
Small pyrola	*Pyrola secunda*
One-flowered pyrola	*Moneses uniflora*
Indian pipe	*Monotropa uniflora*
False beechdrops	*Monotropa hypopitys*

Pinxter flower	*Azalea nudiflorum*
Swamp honeysuckle	*Azalea viscosa*
Flame azalea	*Azalea calendulacea*
Tree azalea	*Azalea arborescens*
Mountain azalea	*Azalea canescens*
Great laurel	*Rhododendron maximum*
Mountain rosebay	*Rhododendron catawbiense*
Lapland rosebay	*Rhododendron Lapponicum*
Rhodora	*Rhodora canadensis*
Mountain laurel	*Kalmia latifolia*
Sheep laurel	*Kalmia angustifolia*
Swamp laurel	*Kalmia polifolia*
Sweet pepperbush	*Clethra alnifolia*
Mountain sweet pepperbush	*Clethra acuminata*
Labrador tea	*Ledum groelandicum*
Wild rosemary	*Andromeda polifolia*
Bog rosemary	*Andromeda glaucophylla*
Swamp andromeda	*Lyonia ligustrum*
Staggerbush	*Lyonia mariana*
Fetterbush	*Lyonia nitida*
Mountain fetterbush	*Andromeda floribunda*
Cassiope	*Cassiope hypnoides*
Four-angled cassiope	*Cassiope tetragona*
Swamp leucothoe	*Leucothoe racemosa*
Catesby's leucothoe	*Leucothoe Catesbaei*
Downy leucothoe	*Leucothoe axillaris*
Cassandra	*Chamaedaphne calyculata*
Heather	*Calluna vulgaris*
Allegheny menziesia	*Menziesia pilosa*
Smooth menziesia	*Menziesia glabella*
Sand myrtle	*Leiophyllum buxifolium*
Sorrel-tree	*Oxydendrum arboreum*
Dwarf huckleberry	*Gaylussacia dumosa*
Box huckleberry	*Gaylussacia brachycera*
Dangleberry	*Gaylussacia frondosa*
Black huckleberry	*Gaylussacia baccata*
Farkleberry	*Vaccinium arboreum*
Deerberry	*Vaccinium stamineum*
Dwarf blueberry	*Vaccinium pennsylvanicum*
Late blueberry	*Vaccinium vacillans*
Highbush blueberry	*Vaccinium corymbosum*
Bog bilberry	*Vaccinium uliginosum*
Dwarf bilberry	*Vaccinium caespitosum*
Large cranberry	*Vaccinium macrocarpon*
Small cranberry	*Vaccinium oxycoccus*
Mountain cranberry	*Vaccinium vitis-idaea*
Southern mountain cranberry	*Vaccinium erythrocarpum*

Holly	AQUIFOLIACEAE
European holly	*Ilex aquifolium*
American holly	*Ilex opaca*
Yaupon	*Ilex vomitoria*
Paraguay tea	*Ilex paraguayensis*
Dahoon	*Ilex cassine*
Swamp holly	*Ilex decidua*
Black alder	*Ilex verticillata*
Smooth winterberry	*Ilex laevigata*
Inkberry	*Ilex glabra*
Mountain holly	*Nemopanthus mucronata*

Honey-bush	MELIANTHACEAE
Honey-bush	*Melianthus major*

Honeysuckle	CAPRIFOLIACEAE
Tartarian honeysuckle	*Lonicera tartarica*
Fly honeysuckle	*Lonicera xylosteum*
Swamp fly honeysuckle	*Lonicera oblongifolia*
American fly honeysuckle	*Lonicera canadensis*
Mountain fly honeysuckle	*Lonicera caerulea* variety *villosa*
Twinberry	*Lonicera involucrata*
Coral honeysuckle	*Lonicera sempervirens*
Sweet wild honeysuckle	*Lonicera caprifolium*
Bush honeysuckle	*Diervilla lonicera*
Weigela	*Weigela* species
Elder	*Sambucus canadensis*
Red-berried elder	*Sambucus racemosa*
Snowberry	*Symphoricarpus racemosus*
Coral-berry	*Symphoricarpus orbiculatus*
Wolfberry	*Symphoricarpus occidentalis*
Withe-rod	*Viburnum cassinoides*
Naked withe-rod	*Viburnum nudum*
Hobblebush	*Viburnum alnifolium*
Nannyberry	*Viburnum lentago*
Black haw	*Viburnum prunifolium*
Southern black haw	*Viburnum rufidulum*
Maple-leaved arrowwood	*Viburnum acerifolium*
Downy arrowwood	*Viburnum pubescens*
Arrowwood	*Viburnum dentatum*
Cranberry tree	*Viburnum opulus*
Snowball	*Viburnum opulus sterile*
Squashberry	*Viburnum pauciflorum*
Wayfaring tree	*Viburnum lantana*
Twinflower	*Linnaea borealis* variety *americana*
Horse gentian	*Triosteum perfoliatum*

Hornwort	CERATOPHYLLACEAE
Hornwort	*Ceratophyllum demersum*

Horseradish tree	MORINGACEAE
Horseradish tree	*Moringa oleifera*

Illicium	ILLICIACEAE
Star-anise	*Illicium parviflorum*
Purple-anise	*Illicium floridanum*

Indian plum	FLACOURTIACEAE
Indian plum	*Flacourtia indica*

Kei-apple	*Dovyalis caffra*
Kitambilla	*Dovyalis hebecarpa*

Iris	IRIDACEAE
Larger blue flag	*Iris versicolor*
Slender blue flag	*Iris prismatica*
Dwarf iris	*Iris verna*
Crested dwarf iris	*Iriscristata*
Fleur-de-lis	*Iris germanica*
Saffron crocus	*Crocus sativus*
Dutch crocus	*Crocus aureus*
Scotch crocus	*Crocus biflorus*
Cloth of gold crocus	*Crocus susianus*
Common crocus	*Crocus vernus*
Gladiolus	*Gladiolus* species
Freesia	*Freesia* species
Blackberry lily	*Belamcanda chinensis*
Blue-eyed grass	*Sisyrinchium angustifolium*

Jewelweed	BALSAMINACEAE
Spotted jewelweed	*Impatiens biflora*
Pale jewelweed	*Impatiens pallida*
Common balsam	*Impatiens balsamina*
Sultana	*Impatiens sultani*

Jipijapa	CYCLANTHACEAE
Panama-hat plant	*Carludovica palmata*

Joint fir	GNETACEAE
Tumboa	*Tumboa bainesii*
Joint fir	*Ephedra* species
Mexican tea	*Ephedra* species

Katsura tree	TROCHODENDRACEAE
Katsura tree	*Cercidiphyllum japonicum*

Laurel	LAURACEAE
Sassafras	*Sassafras albidum*
Avocado	*Persea americana*
Red bay	*Persea borbonia*
Swamp bay	*Persea pubescens*
California laurel	*Umbellularia californica*
Spice bush	*Benzoin aestivale*
Pond spice	*Litsea geniculata*
Camphor tree	*Cinnamomum camphora*
Cassia-bark tree	*Cinnamomum cassia*
Cinnamon tree	*Cinnamomum zeylandicum*

Leadwort	PLUMBAGINACEAE
Sea lavender	*Limonium carolinianum*
Shrubby plumbago	*Plumbago capensis*
Common thrift	*Armeria maritima*
Prickly thrift	*Acantholimon* species

Lily	LILIACEAE
Easter lily	*Lilium longiflorum* variety *eximium*
White trumpet lily	*Lilium longiflorum*
Madonna lily	*Lilium candidum*
Upright lily	*Lilium elegans*
Japanese lily	*Lilium speciosum*
Golden-banded lily	*Lilium auratum*
Tiger lily	*Lilium tigrinum*
Meadow lily	*Lilium canadense*
Wood lily	*Lilium philadelphicum*
Turk's-cap lily	*Lilium superbum*
Tulip	*Tulipa* species
Common garden hyacinth	*Hyacinthus orientalis*
Grape hyacinth	*Muscari* species
Siberian squill	*Scilla sibirica*
Star hyacinth	*Scilla amoena*
Spanish bluebell	*Scilla hispanica*
Sea onion	*Scilla verna*
Crown imperial	*Fritillaria imperialis*
Chiondoxa	*Chiondoxa luciliae*
Lily-of-the-valley	*Convallaria majalis*
Star-of-Bethlehem	*Ornithogalum umbellatum*
Foxtail lily	*Eremus* species
Garden asparagus	*Asparagus officinalis*
Asparagus fern	*Asparagus plumosus*
Red-hot poker	*Kniphofia uvaria*
Wake-robin	*Trillium erectum*
Stemless trillium	*Trillium sessile*
Large flowering trillium	*Trillium grandiflorum*
Nodding trillium	*Trillium cernuum*
Dwarf white trillium	*Trillium nivale*
Painted trillium	*Trillium undulatum*
Yellow adder's tongue	*Erythronium americanum*
Clintonia	*Clintonia borealis*
Solomon's seal	*Polygonatum biflorum*
Great Solomon's seal	*Polygonatum commutatum*
False Solomon's seal	*Smilacina stellata*
Three-leaved false Solomon's seal	*Smilacina trifola*
False spikenard	*Smilacina racemosa*
Twisted stalk	*Streptopus amplexifolius*
Canada mayflower	*Maianthemum canadense*
Perfoliate bellwort	*Uvularia perfoliata*
Large-flowered bellwort	*Uvularia grandiflora*
Sessile-leaved bellwort	*Uvularia sessilifolia*
Mountain bellwort	*Uvularia puberula*
Indian cucumber	*Medeola virginiana*
Devil's bit	*Chamaelirium luteum*
Bunch flower	*Melanthium virginicum*
American white hellebore	*Veratrum viride*
Onion	*Allium cepa*
Wild onion	*Allium cernuum*
Meadow garlic	*Allium canadense*
Field garlic	*Allium vineale*
Wild leek	*Allium tricoccum*
Carrion flower	*Smilax herbacea*
Greenbrier	*Smilax rotundifolia*

Day lily	*Hemerocallis fulva*
Yellow day lily	*Hemerocallis flava*
Joshua tree	*Yucca brevifolia*
Desert candle	*Yucca whipplei*
Spanish dagger	*Yucca mohavensis*
Adam's needle	*Yucca filamentosa*
Chives	*Allium schoenoprasum*
Shallot	*Allium ascalonicum*

Linden	TILIACEAE
American linden	*Tilia americana*
European linden	*Tilia europaea*
Downy basswood	*Tilia Michauxii*
White basswood	*Tilia heterophylla*
Silver linden	*Tilia tomentosa*
Small-leaved linden	*Tilia cordata*
Corktree	*Entela arborescens*
Sparmania	*Sparmannis africana*
Jute	*Corchorus capsularis*
Jew's mallow	*Corchorus orinocensis*

Lizard's tail	SAURURACEAE
Lizard's tail	*Saururus cernuus*
Yerba	*Anemopsis californica*

Loasa	LOASACEAE
Prairie lily	*Mentzelia decapetala*
Blazing star	*Mentzelia laevicaulis*
Few-seeded Mentzelia	*Mentzelia oligosperma*

Lobelia	LOBELIACEAE
Cardinal flower	*Lobelia cardinalis*
Blue lobelia	*Lobelia syphilitica*
Pale spiked lobelia	*Lobelia spicata*
Water lobelia	*Lobelia dortmanna*
Indian tobacco	*Lobelia inflata*

Logania	LOGANIACEAE
Carolina jessamine	*Gelsemium sempervirens*
Indian pink	*Spigella marilandica*
Butterfly bush	*Buddleia officinalis*
Strychnine	*Strychnos nux-vomica*
Natal orange	*Strychnos spinosa*

Loosestrife	LYTHRACEAE
Long purples	*Lythrum salicaria*
Milk Willow-herb	*Lythrum alatum*
Hyssop loosestrife	*Lythrum hyssopifolia*
Swamp loosestrife	*Decodon verticillatus*
Water purslane	*Diplidis diandra*
Clammy cuphea	*Cuphea petiolata*
Toothcups	*Ammannia coccinea*
Cigar flower	*Cuphea platycentra*
Mignonette-tree	*Lawsonia inermis*
Crape myrtle	*Lagerstroemia indica*

Lopseed	PHRYMACEAE
Lopseed	*Phryma leptostachya*

Madder	RUBIACEAE
Bluets	*Houstonia coerulea*
Partridge berry	*Mitchella repens*
Cleavers	*Galium aparine*
Sweet-scented bedstraw	*Galium triflorum*
Yellow bedstraw	*Galium verum*
Rough bedstraw	*Galium asprellum*
Northern bedstraw	*Galium boreale*
Buttonbush	*Cephalanthus occidentalis*
Sweet woodruff	*Asperula odorata*
Dyer's woodruff	*Asperula tinctoria*
White bedstraw	*Galium mollugo*
Coffee	*Coffea arabica*
Quinine tree	*Cinchona officinalis*
Madder	*Rubia tinctorum*
Cape jasmine	*Gardenia jasminoides*
Fevertree	*Pinckneya pubens*
Caribbean princewood	*Exostema caribaeum*
Roughleaf velvetseed	*Guettarda scabra*
Everglades velvetseed	*Guettarda elliptica*

Madeira vine	BASELLACEAE
Madeira vine	*Boussingaultia baselloides*
Malabar Nightshade	*Basella rubra*

Magnolia	MAGNOLIACEAE
Great-flowered magnolia	*Magnolia grandiflora*
Sweet bay	*Magnolia virginiana*
Cucumber tree	*Magnolia acuminata*
Great-leaved magnolia	*Magnolia macrophylla*
Umbrella tree	*Magnolia tripetala*
Ear-leaved umbrella tree	*Magnolia fraseri*
Tulip tree	*Liriodendron tulipifera*
Yulan magnolia	*Magnolia denudata*
Purple magnolia	*Magnolia soulangeana*
Starry magnolia	*Magnolia stellata*

Mahogany	MALIACEAE
Mahogany	*Swietenia mahagoni*
Chinaberry tree	*Melia azedarach*
Cigarbox cedar	*Cedrela odorata*

Makomako	ELAEOCARPACEAE
Makomako	*Aristotelia racemosa*

Mallow	MALVACEAE
Common mallow, cheeses	*Malva rotundifolia*
High mallow	*Malva sylvestris*
Musk mallow	*Malva moschata*

Marsh mallow	*Althaea officinalis*
Hollyhock	*Althaea rosea*
Rose mallow	*Hibiscus moscheutos*
Crimson-eye rose mallow	*Hibiscus oculiroseus*
Halberd-leaved rose mallow	*Hibiscus militaris*
Rose-of-China	*Hibiscus rosa-sinensis*
Rose-of-Sharon	*Hibiscus syriacus*
Flower-of-an-Hour	*Hibiscus trionum*
Cotton rose	*Hibiscus mutabilis*
Okra	*Hibiscus esculentus*
Red false mallow	*Malvastrum coccineum*
Yellow false mallow	*Malvastrum angustum*
Desert five spot	*Malvastrum rotundifolia*
Purple poppy mallow	*Callirhoe involucrata*
Wax mallow	*Malvaviscus* species
Indian mallow	*Abutilon Abutilon*
Flowering maple	*Abutilon* species
Prickly sida	*Sida spinosa*
Paroquet bur	*Sida acuta*
Maple-leaved mallow	*Sphaeralcea acerifolia*
Globe mallow	*Sphaeralcea* species
Tree cotton	*Gossypium arboreum*
Cotton	*Gossypium* species

Malpighi	MALPIGHIACEAE
Barbados cherry	*Malpighi glabra*
Thryallis	*Thryallis glauca*
Butterfly-vine	*Stigmaphyllon ciliatum*

Mamey	GUTTIFERAE
Mangosteen	*Garcina mangostana*
Mamey	*Mammea americana*
Calaba	*Calophyllum antillanum*

Mangrove	RHIZOPHORACEAE
Red mangrove	*Rhizophora mangle*

Maple	ACERACEAE
Red maple	*Acer rubrum*
Silver maple	*Acer saccharinum*
Sugar maple	*Acer saccharum*
Southern sugar maple	*Acer floridanum*
Black sugar maple	*Acer saccharum* variety *nigrum*
Striped maple	*Acer pennsylvanicum*
Mountain maple	*Acer spicatum*
Chalk maple	*Acer leucoderme*
Dwarf maple	*Acer glabrum*
Vine maple	*Acer circinatum*
Big-leaved maple	*Acer macrophyllum*
Box elder	*Acer negundo*
Sycamore maple	*Acer pseudo-platanus*
Norway maple	*Acer platanoides*

Mayaca	MAYACACEAE
Mayaca	*Mayaca Aubleti*

Meadow beauty	MELASTOMACEAE
Meadow beauty	*Rhexia virginica*

Mezereum	THYMELAEACEAE
Leatherwood	*Dirca palustris*
Mezereum	*Daphne mezereum*
Paper tree	*Edgeworthia papyrifera*
Rice flower	*Pimelea* species

Mignonette	RESEDACEAE
Mignonette	*Reseda odorata*

Milkweed	ASCLEPIADACEAE
Common milkweed	*Asclepias syrica*
Purple milkweed	*Asclepias purpurascens*
Swamp milkweed	*Asclepias incarnata*
Poke milkweed	*Asclepias phytolaccoides*
Four-leaved milkweed	*Asclepias quadrifolia*
Whorled milkweed	*Asclepias verticillata*
Butterfly weed	*Asclepias tuberosa*

Milkwort	POLYGALACEAE
Fringed polygala	*Polygala paucifolia*
Common milkwort	*Polygala viridescens*
Cross-leaved milkwort	*Polygala cruciata*
Short-leaved milkwort	*Polygala brevifolia*
Whorled milkwort	*Polygala verticillata*
Racemed milkwort	*Polygala polygama*
Seneca snakeroot	*Polygala Seneca*

Mint	LABIATAE
Catnip	*Nepeta cataria*
Peppermint	*Mentha piperita*
Spearmint	*Mentha spicata*
Wild mint	*Mentha canadensis*
American pennyroyal	*Hedeoma pulegiodes*
False pennyroyal	*Isanthus brachiatus*
Mountain mint	*Pycnanthemum virginianum*
Horse mint	*Mentha longifolia*
Corn mint	*Mentha arvensis*
Water mint	*Mentha aquatica*
Thyme	*Thymus vulgaris*
Sage	*Salvia officinalis*
Marjorum	*Majorana hortensis*
Sweet basil	*Ocimum basilicum*
Summer savory	*Satureia hortensis*
Rosemary	*Rosmarinus officinalis*
Common lavender	*Lavendula spica*
Garden coleus	*Coleus blumei* variety *verschaffelti*
Salvia	*Salvia splendens*
Blue salvia	*Salvia patens*
Gill-over-the-ground	*Nepeta hederaceae*
Oswego tea	*Monarda didyma*

Wild bergamot	*Monarda fistulosa*
Horse balm	*Collinsonia canadensis*
Bugleweed	*Lycopus virginicus*
Mad-dog skullcap	*Scutellaria lateriflora*
Blue curls	*Trichostema dichotomum*
Seal-heal	*Prunella vulgaris*
Horehound	*Marrubium vulgare*
Motherwort	*Leonurus cardiacea*
Henbit	*Lamium amplexicaule*
Hemp nettle	*Galeopsis tetrahit*
Hedge nettle	*Stachys palustris*
False dragonhead	*Physostegia virginiana*
Wood sage	*Teucrium canadense*
Hyssop	*Hyssopus officinalis*

Mistletoe	LORANTHACEAE
European mistletoe	*Viscum album*
American mistletoe	*Phoradendron* species
Dwarf mistletoe	*Arceuthobium* species

Moonseed	MENISPERMACEAE
Carolina moonseed	*Cocculus carolinus*
Canada moonseed	*Menispermum canadense*
Cupseed	*Calycocarpum Lyoni*

Morning glory	CONVOLVULACEAE
Morning glory	*Ipomoea purpurea*
Moon-flower	*Calonyction aculeatum*
Manroot	*Ipomoea pandurata*
Cypress vine	*Quamoclit pennata*
Sweet potato	*Ipomoea batatas*
Hedge bindweed	*Convolvulus sepium*
Field bindweed	*Convolvulus arvensis*
Upright bindweed	*Convolvulus spithamaeus*
Common dodder	*Cuscuta gronovii*

Moschatel	ADOXACEAE
Musk-root	*Adoxa Moschatellina*

Mulberry	MORACEAE
Red mulberry	*Morus rubra*
Texas mulberry	*Morus microphylla*
White mulberry	*Morus alba*
Black morus	*Morus nigra*
Paper mulberry	*Broussonetia papyrifera*
Osage orange	*Maclura pomifera*
Strangler fig	*Ficus aurea*
Shortleaf fig	*Ficus laevigata*
Common fig	*Ficus carica*
Common household rubber plant	*Ficus elastica*
Peepul tree	*Ficus religiosa*
Banyan tree	*Ficus benghalensis*
Breadfruit tree	*Artocarpus communis*
Hemp	*Cannabis sativa*
Common hop	*Humulus lupulus*

Mustard	CRUCIFERAE
Sea cabbage	*Brassica oleracea*
Shepherd's purse	*Capsella bursa-pastoris*
Peppergrass	*Lepidium virginicum*
Yellow rocket	*Barbarea vulgaris*
Common whitlow-grass	*Druba verna*
Toothwort	*Dentaria diphylla*
Watercress	*Nasturtium officinale*
Spring cress	*Cardamine bulbosa*
Bitter cress	*Cardamine pratensis*
Rock cress	*Arabis hirsuta*
Black mustard	*Brassica nigra*
White mustard	*Brassica alba*
Hedge mustard	*Sisymbrium officinale*
Field mustard	*Brassica arvensis*
Radish	*Raphanus sativus*
Horseradish	*Armoracia rusticana*
Brussels sprouts	*Brassica oleracea* variety *gemmifera*
Cauliflower	*Brassica oleracea* variety *botrytis*
Broccoli	*Brassica oleracea* variety *italica*
Turnip	*Brassica rapa*
Sweet alyssum	*Lobularia maritima*
Stock	*Mathiola incana*
Rocket candytuft	*Iberis amara*
Purple candytuft	*Iberis umbellata*
Evergreen candytuft	*Iberis sempervirens*
Honesty	*Lunaria annua*
Dame's violet	*Hesperis matronalis*
Wallflower	*Cheiranthus cheiri*

Myoporum	MYOPORACEAE
Bastard sandlewood	*Myoporum sandwicense*

Myrsine	MYRSINACEAE
Marlberry	*Ardisia paniculata*
Guiana rapanea	*Rapanea guianensis*

Myrtle	MYRTACEAE
Strawberry guava	*Psidium cattleianum*
Common guava	*Psidium guajava*
Brazilian guava	*Psidium guineense*
Allspice tree	*Pimenta officinalis*
True bayberry	*Pimenta acris*
Peppermint gum	*Eucalyptus amygdalina*
Blue gum	*Eucalyptus globulus*
Red gum	*Eucalyptus rostrata*
Scarlet-flowered gum	*Eucalyptus ficifolia*
Clove tree	*Eugenia aromatica*
True myrtle	*Myrtus communis*
Simpson nakedwood	*Anamomis simpsonii*
Twinberry nakedwood	*Anamomis dicrana*
Pale lidflower	*Calyptranthes pallens*
Myrtle-of-the-river lid-flower	*Calyptranthes zuzygium*
White-stopper eugenia	*Eugenia axillaris*
Spiceberry eugenia	*Eugenia rhombea*

Boxleaf eugenia	*Eugenia buxifolia*
Redberry eugenia	*Eugenia confusa*

Naiad	NAIADACEAE
Large naias	*Naias marina*
Slender naias	*Naias flexilis*
Threadlike naias	*Naias gracillima*

Nasturtium	TROPAEOLACEAE
Dwarf nasturtium	*Tropaeolum minus*
Common nasturtium	*Tropaeolum majus*

Nepenthes	NEPENTHACEAE
Pitcher plant	*Nepenthes* species

Nettle	URTICACEAE
Tall nettle	*Urtica gracilis*
Great nettle	*Urtica dioica*
Western nettle	*Urtica holosericea*
False nettle	*Boehmeria cylindrica*
Ramie	*Boehmeria nivea*
Richweed	*Adicea pumila*
Wood nettle	*Urticastrum divaricatum*
Pellitory	*Parietaria pennsylvanica*
Baby-tears	*Helxine soleiroli*

Nightshade	SOLANACEAE
Ancestral petunia	*Petunia axillaris*
Violet petunia	*Petunia violacea*
Common garden petunia	*Petunia hybrida*
Potato	*Solanum tuberosa*
Tomato	*Lycopersicum esculentum*
Cherry pepper	*Capsicum frutescens* variety *cerasiforme*
Red cluster pepper	*Capsicum frutescens* variety *fasciculatum*
Long pepper	*Capsicum frutescens* variety *longum*
Cone pepper	*Capsicum frutescens* variety *conoides*
Eggplant	*Solanum melongena*
Tobacco	*Nicotiana tabacum*
Tree tobacco	*Nicotiana glauca*
Nicotiana	*Nicotiana alata*
Salpiglossis	*Salpiglossis sinuata*
Datura	*Datura metel*
Browallia	*Browallia* species
Matrimony vine	*Lycium* species
Butterfly flower	*Schizanthus pinnatus*
Nightshade	*Solanum dulcamara*
Black nightshade	*Solanum nigrum*
Virginia ground cherry	*Physalis virginiana*
Strawberry tomato	*Physalis pubescens*
Jimson weed	*Datura stramonium*
Purple thorn apple	*Datura tatula*
Jerusalem cherry	*Solanum pseudo-capsicum*

Nolana	NOLANACEAE
Nolana	*Nolana* species

Nutmeg	MYRISTIACACEAE
Nutmeg	*Myristica fragrans*

Ochna	OCHNACEAE
Many-flowered ochna	*Ochna multiflora*

Ocotillo	FOUQUIERIACEAE
Ocotillo	*Fouquieria splendens*

Oleaster	ELAEAGNACEAE
Silveryberry	*Elaeagnus commutata*
Oleaster	*Elaeagnus angustifolia*
Gumi	*Elaeagnus multiflora*
Buffalo berry	*Shepherdia argentea*
Canadian buffalo berry	*Shepherdia canadensis*

Olive	OLEACEAE
White ash	*Fraxinus americana*
Black ash	*Fraxinus nigra*
Red ash	*Fraxinus pennsylvanica*
Green ash	*Fraxinus pennsylvanica lanceolata*
Blue ash	*Fraxinus quadrangulata*
Biltmore ash	*Fraxinus biltmoreana*
Pumpkin ash	*Fraxinus profunda*
Water ash	*Fraxinus caroliniana*
Dwarf ash	*Fraxinus anomala*
Arizona ash	*Fraxinus velutina*
Oregon ash	*Fraxinus oregona*
Flowering ash	*Fraxinus dipetala*
European ash	*Fraxinus excelsior*
Fringe-tree	*Chionanthus virginica*
Lilac	*Syringa vulgaris*
Forsythia	*Forsythia* species
Privet	*Ligustrum vulgare*
California privet	*Ligustrum ovalifolium*
Wax privet	*Ligustrum japonicum*
Swamp privet	*Forestiera acuminata*
Olive tree	*Olea europaea*
American devilwood	*Osmanthus americanus*
Tea olive	*Osmanthus fragrans*
Spanish jasmine	*Jasminum grandiflorum*
Common white jasmine	*Jasminum officinale*
Italian jasmine	*Jasminum humile*
Winter jasmine	*Jasminum nudiflorum*

Orchid	ORCHIDACEAE
Showy orchis	*Orchis spectabilis*
Shin plasters	*Orchis rotundifolia*
Large purple-fringed orchis	*Habenaria fimbriata*

Smaller purple-fringed orchis	*Habenaria psycodes*
Purple orchis	*Habenaria peramoena*
White-fringed orchis	*Habenaria blephariglottis*
Yellow-fringed orchis	*Habenaria ciliaris*
Yellow-crested orchis	*Habenaria cristata*
Small pale green orchis	*Habenaria flava*
Ragged-fringed orchis	*Habenaria lacera*
Green round-leaved orchis	*Habenaria orbiculata*
Moccasin flower	*Cypripedium acaule*
White lady's slipper	*Cypripedium candidum*
Large yellow lady's slipper	*Cypripedium parviflorum* variety *pubescens*
Small yellow lady's slipper	*Cypripedium parviflorum*
Ram's head lady's slipper	*Cypripedium arietinum*
Showy lady's slipper	*Cypripedium regiae*
Ladies' tresses	*Spiranthes cernua*
Grass-leaved ladies' tresses	*Spiranthes praecox*
Slender ladies' tresses	*Spiranthes gracilis*
Heart-leaved twayblade	*Listera cordata*
Broad-lipped twayblade	*Listera convallarioidis*
Large twayblade	*Liparis liliifolia*
Arethusa	*Arethusa bulbosa*
Calypso	*Calypso bulbosa*
Calopogon	*Calopogon pulchellus*
Rose pogonia	*Pogonia ophioglossoides*
Nodding pogonia	*Pogonia trianthophora*
Whorled pogonia	*Pogonia verticillata*
Lesser rattlesnake plantain	*Epipactis repens*
Downy rattlesnake plantain	*Epipactis pubescens*
Early coral root	*Corallorhiza trifida*
Small-flowered coral root	*Corallorhiza odontorhiza*
Large coral root	*Corallorhiza maculata*
Adam-and-Eve	*Aplectrum hyemale*
Green adder's mouth	*Microstylis uniflora*
White adder's mouth	*Microstylis monophylla*
Crane-fly orchid	*Tipularia unifolia*
Tropical cultivated orchids	*Cattleya* species
Vanilla plant	*Vanilla fragrans*

Orpine	CRASSULACEAE
Live-forever	*Sedum telephium*
Wild stonecrop	*Sedum ternatum*
Mossy stonecrop	*Sedum acre*
Yellow stonecrop	*Sedum reflexum*
Widow's cross	*Sedum pulchellum*
Jade plant	*Crassula cordata*
River leek	*Tillaeastrum aquaticus*
Live leaf	*Bryophyllum pinnatum*
Live forever	*Echeveria* species
Ditch stonecrop	*Penthorum sedoides*
House leek	*Sempervivum tectorum*

Palm	PALMACEAE
Hog cabbage	*Pseudophoenix sargenti*
Royal palm	*Roystonea regia*
Coconut palm	*Cocos nucifera*
Date palm	*Phoenic dactylifera*
Canary Island palm	*Phoenix canariensis*
Cabbage palmetto	*Sabal palmetto*
Dwarf palmetto	*Sabal minor*
Delta palmetto	*Sabal deeringiana*
Scrub palmetto	*Sabal etonia*
Saw palmetto	*Serenoa repens*
Saw cabbage palm	*Acoelorraphe wrightii*
Silvertop palmetto	*Thrinax microcarpa*
Florida thatch palm	*Thrinax parviflora*
Silver palm	*Coccothrinax argentea*
Needle palm	*Rhapidophyllum hystrix*
California fan palm	*Washingtonia filifera*
Dwarf fan palm	*Chamaerops humilis*
Mexican blue palm	*Erythea armata*
Windmill palm	*Trachycarpus fortunei*
Raphia palm	*Raphia ruffia*
Betel palm	*Areca catechu*
Sugar palm	*Arenga saccharifera*
Oil palm	*Elaeis guineensis*
Brazilian ivory palm	*Phytelephas macrocarpa*

Papaya	CARICACEAE
Papaya	*Carica papaya*

Passionflower	PASSIFLORACEAE
Passionflower	*Passiflora incarnata*
Yellow passionflower	*Passiflora lutea*
Purple granadilla	*Passiflora edulis*
Yellow granadilla	*Passiflora laurifolia*
Giant granadilla	*Passiflora quadrangularis*

Pea	LEGUMINOSAE
Garden sweet pea	*Lathyrus odoratus*
Everlasting pea	*Lathyrus latifolius*
Garden pea	*Pisum sativum*
String bean	*Phaseolus vulgaris*
Lima bean	*Phaseolus limensis*
Scarlet runner bean	*Phaseolus coccineus*
Lentil	*Lens esculenta*
Peanut	*Arachis hypogaea*
Soybean	*Glycine max*
Red clover	*Trifolium pratense*
White clover	*Trifolium repens*
Alsike clover	*Trifolium hybridum*
Rabbit-foot clover	*Trifolium arvense*
Yellow clover	*Trifolium agrarium*
White melilot	*Melilotus alba*
Yellow melilot	*Melilotus officinalis*
Trailing bush clover	*Lespedeza procumbens*
White bush clover	*Lespedeza capitata*
Hairy bush clover	*Lespedeza hirta*
Violet bush clover	*Lespedeza violacea*
Wild indigo	*Baptisia tinctoria*
Blue false indigo	*Baptisia australis*
Rattlebox	*Crotalaria sagittalis*
Blue lupine	*Lupinus perennis*

Bluebonnet	*Lupinus subcarnosus*
Naked-flowered tick trefoil	*Desmodium nudiflora*
Pointed-leaved tick trefoil	*Desmodium grandiflora*
Prostrate tick trefoil	*Desmodium michauxii*
Canadian tick trefoil	*Desmodium canadense*
Nonesuch	*Medicago lupulina*
Alfalfa	*Medicago sativa*
Cow vetch	*Vicia cracca*
Common vetch	*Vicia sativa*
Hairy vetch	*Vicia hirsuta*
Ground nut	*Apios tuberosa*
Wild bean	*Phaseolus polystachyus*
Hog peanut	*Amphicarpaea monoica*
Beach pea	*Lathyrus maritimus*
Wild licorice	*Glycyrrhiza lepidota*
Common licorice	*Glycyrrhiza glabra*
Wild senna	*Cassia marylandica*
Partridge pea	*Cassia chamaecrista*
Wild sensitive plant	*Cassia nictitans*
Coffee senna	*Cassia occidentalis*
Kentucky coffee-tree	*Gymnocladus dioica*
Honey locust	*Gleditsia triacanthos*
Locust	*Robinia pseudo-acacia*
Water locust	*Gleditsia aquatica*
Redbud	*Cercis candensis*
Yellow wood	*Cladrastis lutea*
Woad waxen	*Genista tinctoria*
Scotch broom	*Cytisus scoparius*
Furze	*Ulex europaeus*
False indigo	*Amorpha fruticosa*
Lead plant	*Amorpha canescens*
Rose acacia	*Robinia hispida*
Huisache	*Acacia farnesiana*
Green wattle	*Acacia decurrens*
Cat's claw	*Acacia greggii*
Kangaroo thorn	*Acacia armata*
Golden wattle	*Acacia longifolia*
Mesquite	*Prosopis pubescens*
Smoke tree	*Dalea spinosa*
Ironwood	*Olneya tesota*
Palo verde	*Cercidium torrevanum*
Horse bean	*Parkinsonia aculeata*
Wild tamarind	*Lysiloma bahamensis*
Silk tree	*Albizzia julibrissin*
Chinese wisteria	*Wistaria sinensis*
Japanese wisteria	*Wistaria floribunda*
American wisteria	*Wistaria frutescens*
Kentucky wisteria	*Wistaria macrostachys*
Kudzu vine	*Pueraria thunbergiana*

Pepper	PIPERACEAE
Pepper	*Piper nigrum*
Cubeb	*Piper cubeba*
Betel	*Piper betle*

Phlox	POLEMONIACEAE
Wild blue phlox	*Phlox divaricata*
Downy phlox	*Phlox pilosa*
Wild sweet William	*Phlox maculata*
Ground pink	*Phlox subulata*
Creeping phlox	*Phlox nivalis*
Garden phlox	*Phlox paniculata*
Greek valerian	*Polemonium reptans*
Jacob's ladder	*Polemonium vanbruntiae*
Cup-and-saucer vine	*Cobaea scandens*
Prickly phlox	*Gilia californica*
Scarlet gilia	*Gilia aggregata*
Bird's-eyes	*Gilia tricolor*

Pickerelweed	PONTEDERIACEAE
Pickerelweed	*Pontederia cordata*
Mud plantain	*Heteranthera reniformis*
Water hyacinth	*Eichhornia crassipes*

Pine	PINACEAE
White pine	*Pinus strobus*
Sugar pine	*Pinus lambertiana*
Whitebark pine	*Pinus albicaulis*
Limber pine	*Pinus flexilis*
Foxtail pine	*Pinus balfouriana*
Hickory pine	*Pinus aristata*
Western white pine	*Pinus monticola*
Pitch pine	*Pinus rigida*
Western yellow pine	*Pinus ponderosa*
Long-leaved pine	*Pinus palustris*
Slash pine	*Pinus caribaea*
Digger pine	*Pinus sabiniana*
Bigcone pine	*Pinus coulteri*
Knobcone pine	*Pinus attenuata*
Red pine	*Pinus resinosa*
Jack pine	*Pinus banksiana*
Eastern yellow pine	*Pinus echinata*
Shore pine	*Pinus contorta*
Lodgepole pine	*Pinus murrayana*
Scotch pine	*Pinus sylvestris*
Austrian pine	*Pinus nigra*
Nut pine	*Pinus monophylla*
Arizona pine	*Pinus cembroides*
Rocky Mountain nut pine	*Pinus edulis*
Four-leaved pine	*Pinus quadrifolia*
White spruce	*Picea glauca*
Red spruce	*Picea rubens*
Black spruce	*Picea mariana*
Engelmann spruce	*Picea engelmanni*
Colorado blue spruce	*Picea pungens*
Sitka spruce	*Picea sitchensis*
Norway spruce	*Picea abies*
Balsam fir	*Abies balsamea*
Fraser fir	*Abies fraseri*
Alpine fir	*Abies lasiocarpa*
Giant fir	*Abies grandis*
White fir	*Abies concolor*
Silver fir	*Abies amabilis*
Noble fir	*Abies nobilis*
Red fir	*Abies magnifica*
Douglas fir	*Pseudotsuga taxifolia*
Eastern hemlock	*Tsuga candensis*
Carolina hemlock	*Tsuga caroliniana*
Western hemlock	*Tsuga heterophylla*
Mountain hemlock	*Tsuga mertensiana*

Eastern larch	*Larix laricina*
Western larch	*Larix occidentalis*
Alpine larch	*Larix lyalli*
European larch	*Larix decidua*
Bald cypress	*Taxodium distichium*
White cedar	*Chamaecyparis thyoides*
Yellow cypress	*Chamaecyparis nootkatensis*
Lawson cypress	*Chamaecyparis lawsoniana*
American arbor vitae	*Thuja occidentalis*
Giant arbor vitae	*Thuja plicata*
Big-tree	*Sequoia gigantea*
Redwood	*Sequoia sempervirens*
Red cedar	*Juniperus virginiana*
Common juniper	*Juniperus communis*
Prostrate juniper	*Juniperus communis* variety *depressa*
Creeping juniper	*Juniperus horizontalis*
Rocky Mountain red cedar	*Juniperus scopulorum*
Western juniper	*Juniperus occidentalis*
Cedar-of-Lebanon	*Cedrus libani*
Deodar	*Cedrus deodara*
Monterey cypress	*Cupressus macrocarpa*
Arizona cypress	*Cupressus arizonica*
Macnab cypress	*Cupressus macnabiana*
Italian cypress	*Cupressus sempervirens*
Monkey-puzzle	*Araucaria araucana*
Bunyabunya	*Araucaria bidwilli*
Moreton bay pine	*Cupressus cunninghami*
Norfolk island pine	*Cupressus excelsa*

Pineapple	BROMELIACEAE
Pineapple	*Ananas comosus*
Spanish moss	*Tillandsia usneoides*
Pinguin	*Bromelia pinguin*

Pink	CARYOPHYLLACEAE
Bouncing bet	*Saponaria officinalis*
Corncockle	*Agrostemma githago*
Bladder campion	*Silene latifolia*
Starry campion	*Silene stellata*
Sleepy catchfly	*Silene antirrhina*
Night-flowering catchfly	*Silene noctiflora*
Wild pink	*Silene pennsylvanica*
Sweet William catchfly	*Silene armeria*
Cushion pink	*Silene aculis*
Sea campion	*Silene maritima*
Moss campion	*Silene schafta*
Fire-pink	*Silene virginica*
Evening lychnis	*Lychnis alba*
Scarlet lychnis	*Lychnis chalcedonica*
Cuckooflower	*Lychnis flos-cuculi*
Mullein pink	*Lychnis coronaria*
Rose-of-heaven	*Lychnis coeli-rosa*
Common chickweed	*Stellaria media*
Thyme-leaved sandwort	*Arenaria serphyllifolia*
Mountain sandwort	*Arenaria groenlandica*
Pine-barren sandwort	*Arenaria caroliniana*
Sand spurry	*Spergularia rubra*
Corn spurry	*Spergularia arvensis*
Field chickweed	*Cerastium arvense*
Snow-in-summer	*Cerastium tomentosum*
Mouse-ear chickweed	*Cerastium vulgatum*
Baby's breath (perennial)	*Gypsophila paniculata*
Baby's breath (annual)	*Gypsophila elegans*
Sweet William	*Dianthus barbatus*
Clove pink	*Dianthus caryophyllus*
Deptford pink	*Dianthus armeria*
Maiden pink	*Dianthus deltoides*
China pink	*Dianthus chinensis*
Garden pink	*Dianthus plumarius*

Pipewort	ERIOCAULACEAE
Pipewort	*Eriocaulon* species

Pitcher plant	SARRACENIACEAE
Pitcher plant	*Sarracenia purpurea*
Trumpets	*Sarracenia flava*
Western pitcher plant	*Darlingtonia californica*

Plane tree	PLATANACEAE
Sycamore	*Platanus occidentalis*
Arizona sycamore	*Platanus wrightii*
California sycamore	*Platanus racemosa*
London plane tree	*Platanus acerifolia*
Oriental plane	*Platanus orientalis*

Plantain	PLANTAGINACEAE
Common plantain	*Plantago major*
Red-stem plantain	*Plantago rugelii*
Narrow-leaved plantain	*Plantago lanceolata*

Pokeweed	PHYTOLACCACEAE
Pokeweed	*Phytolacca americana*
Umbra	*Phytolacca dioica*
Rouge-plant	*Rivina humulis*

Pomegranate	PUNICACEAE
Pomegranate	*Punica granatum*

Pondweed	POTAMOGETONACEAE
Common pondweed	*Potamogeton natans*
Curled-leaved pondweed	*Potamogeton crispus*
Sago pondweed	*Potamogeton pectinatus*
Ditch grass	*Ruppia maritima*
Eel grass	*Zostera marina*
Horned pondweed	*Zannichellia palustris*

Poppy	PAPAVERACEAE
Bloodroot	*Sanguinaria canadensis*

Celandine	*Chelidonium majus*
Celandine poppy	*Stylophorum diphyllum*
Mexican prickly poppy	*Argemone mexicana*
Carolina poppy	*Argemone alba*
Prickly poppy	*Argemone intermedia*
Western prickly poppy	*Argemone platyceras*
Creamcups	*Platystemon californicus*
Matilija poppy	*Romneya coulteri*
Tree poppy	*Dendromecon rigida*
California poppy	*Eschscholtzia californica*
Iceland poppy	*Papaver nudicaule*
Oriental poppy	*Papaver orientale*
Corn poppy	*Papaver rhoeas*
Opium poppy	*Papaver somniferum*

Primrose	PRIMULACEAE
Wild or common loosestrife	*Lysimachia vulgaris*
Four-leaved loosestrife	*Lysimachia quadrifolia*
Bulb-bearing loosestrife	*Lysimachia terrestris*
Moneywort	*Lysimachia nummularia*
Star flower	*Trientalis borealis*
Fringed loosestrife	*Steironema ciliatum*
American cowslip	*Dodecatheon meadia*
Featherfoil	*Hottonia inflata*
Scarlet pimpernel	*Anagallis arvensis*
Common primrose	*Primula vulgaris*

Purslane	PORTULACACEAE
Spring beauty	*Claytonia virginica*
Purslane	*Portulaca oleracea*
Portulaca	*Portulaca grandiflora*
Pussy paws	*Spraguea umbellata*
Bitterroot	*Lewisia rediviva*
Indian lettuce	*Montia perfoliata*

Quassia	SIMAROUBACEAE
Ailanthus	*Ailanthus altissima*
Mexican alvaradoa	*Alvaradoa amorphoides*
Bitterbush	*Picramnia pentandra*
Paradise tree	*Simarouba glauca*

Rafflesia	RAFFLESIACEAE
Rafflesia	*Rafflesia arnoldi*

Riverweed	PODOSTEMACEAE
Riverweed	*Podostemum ceratophyllum*

Rockrose	CISTACEAE
Frostweed	*Helianthemum canadense*
Rockrose	*Helianthemum nummularium*
Rockrose	*Cistus* species
Beach heather	*Hudsonia tomentosa*
Golden heather	*Hudsonia ericoides*
Pinweed	*Lechea* species

Rose	ROSACEAE
Prairie rose	*Rosa setigera*
Swamp rose	*Rosa palustris*
Pasture rose	*Rosa carolina*
Meadow rose	*Rosa blanda*
Dog rose	*Rosa canina*
Sweetbrier	*Rosa eglanteria*
Pearl bush	*Exochorda racemosa*
Fire thorn	*Pyracantha* species
Japanese quince	*Chaenomeles lagenaria*
Flowering almond	*Prunus glandulosa*
Bridal wreath	*Spiraea prunifolia*
Thunberg's spirea	*Spiraea thunbergi*
Van Houtte's spirea	*Spiraea vanhouttei*
Meadowsweet	*Spiraea latifolia*
Steeplebush	*Spiraea tomentosa*
Ninebark	*Physocarpus opulifolius*
Low juneberry	*Amelanchier humulis*
Round-leaved shadbush	*Amelanchier sanguinea*
Oblong-leaved shadbush	*Amelanchier oblongifolia*
Long-rooted shadbush	*Amelanchier stolonifera*
Shadbush	*Amelanchier canadensis*
Purple-flowering raspberry	*Rubus odoratus*
Red raspberry	*Rubus strigosus*
Black raspberry	*Rubus occidentalis*
Blackberry	*Rubus allegheniensis*
Leaf-bracted blackberry	*Rubus argutus*
Thornless blackberry	*Rubus candensis*
Mountain blackberry	*Rubus villosus*
Low running blackberry	*Rubus flagellaris*
Swamp blackberry	*Rubus hispidus*
Loganberry	*Rubus loganobaccus*
Apple	*Malus pumila*
Crabapple	*Malus coronaria*
Peach	*Amygdalus persica*
Pear	*Pyrus communis*
Wild black cherry	*Prunus serotina*
Wild red cherry	*Prunus pennsylvanica*
Chokecherry	*Prunus virginiana*
Wild plum	*Prunus americana*
Canada plum	*Prunus nigra*
Chickasaw plum	*Prunus angustifolia*
Beach plum	*Prunus maritima*
American sloe	*Prunus alleghaniensis*
Sand cherry	*Prunus pumila*
Red chokeberry	*Aronia arbutifolia*
Black chokeberry	*Aronia melanocarpa*
Purple chokeberry	*Aronia atropurpurea*
American mountain ash	*Sorbus americana*
European mountain ash	*Sorbus aucuparia*
Cockspur thorn	*Crataegus crus-galli*
White thorn	*Crataegus monogyna*
Scarlet haw	*Crataegus pedicellata*
Black thorn	*Crataegus calpodendron*
Dotted haw	*Crataegus punctata*
Common cinquefoil	*Potentilla canadensis*
Marsh five-finger	*Potentilla palustris*
Silvery cinquefoil	*Potentilla argentea*

Shrubby cinquefoil	*Potentilla fruticosa*
Wild strawberry	*Fragaria virginiana*
American wood strawberry	*Fragaria americana*
Creeping dalibarda	*Dalibarda repens*
Agrimony	*Agrimonia striata*
White avens	*Geum canadense*
Purple avens	*Geum rivale*
Yellow avens	*Geum strictum*
Long-plumed avens	*Geum triflorum*
Goat's beard	*Aruncus sylvester*
Queen-of-the-prairie	*Filipendula rubra*

Rue	RUTACEAE
Prickly ash	*Zanthoxylum americanum*
Hercules-club	*Zanthoxylum clava-herculis*
Wild lime	*Zanthoxylum fagara*
Torchwood	*Amyris elemifera*
Turpentime broom	*Thamnosma montana*
Eastern hop tree	*Ptelea trifoliata*
Western hop tree	*Ptelea baldwinii*
Rue	*Rue graveolens*
Gas-plant	*Dictamnus albus*
Cork tree	*Phellodendron amurense*
Sweet orange	*Citrus sinensis*
Sour orange	*Citrus aurantium*
King orange	*Citrus nobilis*
Tangerine	*Citrus nobilis* variety *deliciosa*
Citron	*Citrus medica*
Grapefruit	*Citrus paradisi*
Lemon	*Citrus limonia*
Bergamot	*Citrus bergamia*
Lime	*Citrus aurantifolia*
Kumquat	*Fortunalla japonica*

Rush	JUNCACEAE
Bog rush	*Juncus effusus*
Grass-leaved rush	*Juncus marginatus*
Yard rush	*Juncus tenuis*
Toad rush	*Juncus bufonius*
Black rush	*Juncus roemerianus*
Common wood rush	*Luzula campestris*
Hairy wood rush	*Luzula saltuensis*

St.-John's-wort	GUTTIFERAE
St.-John's-wort	*Hypericum perforatum*
Shrubby St.-John's-wort	*Hypericum prolificum*
Great St.-John's-wort	*Hypericum ascyron*
Spotted St.-John's-wort	*Hypericum punctatum*
Canada St.-John's-wort	*Hypericum canadense*
Marsh St.-John's-wort	*Hypericum virginicum*
St.-Andrew's-cross	*Ascyrum hypericoides*
St.-Peter's-wort	*Ascyrum stans*

Sandalwood	SANTALACEAE
Sandalwood	*Santalum* species
Bastard toadflax	*Comandra umbellata*
Buffalo-nut	*Pyrularia pubera*
Nestronia	*Nestronia umbellata*
Buckleya	*Buckleya distichophylla*

Sapodilla	SAPOTACEAE
Sapodilla	*Sapota achras*
Gutta percha tree	*Palaquium gutta*
Marmalade plum	*Achras zapota*
Star apple	*Chrysophyllum cainito*
Satinleaf	*Chrysophyllum oliviforme*
Canistel	*Lucuma nervosa*
Southern buckthorn	*Bumelia lycioides*
Woolly bumelia	*Bumelia languginosa*
Tough bumelia	*Bumelia tenax*
Small bumelia	*Bumelia smallii*

Saxifrage	SAXIFRAGACEAE
Early saxifrage	*Saxifraga virginiensis*
Swamp saxifrage	*Saxifraga pennsylvanica*
Yellow mountain saxifrage	*Saxifraga aizoides*
Lettuce saxifrage	*Saxifraga micranthidifolia*
Mitrewort	*Mitella diphylla*
Naked mitrewort	*Mitella nuda*
False mitrewort	*Tiarella cordifolia*
Alumroot	*Heuchera americana*
Golden saxifrage	*Chrysosplenium americanum*
Grass of parnassus	*Parnassia caroliniana*
False goat's beard	*Astilbe biternata*
Coral bells	*Heuchera sanguinea*
Garden mock orange	*Philadelphus coronarius*
Native mock orange	*Philadelphus inodorus*
Large-flowered syringa	*Philadelphus grandiflorus*
Garden hydrangea	*Hydrangea paniculata*
Wild hydrangea	*Hydrangea arborescens*
Itea	*Itea virginica*
Bridal wreath	*Deutzia gracilis*
Pride-of-Rochester	*Deutzia scabra*
Prickly wild gooseberry	*Ribes cynosbati*
Round-leaved gooseberry	*Ribes rotundifolium*
Smooth gooseberry	*Ribes oxyacanthoides*
European gooseberry	*Ribes grossularia*
Wild black currant	*Ribes floridum*
Fetid currant	*Ribes prostratum*
Golden currant	*Ribes aureum*
Swamp red currant	*Ribes rubrum*
Red garden currant	*Ribes sativum*

Screw pine	PANDANACEAE
Screw pine	*Pandanus* species

Sedge	CYPERACEAE
Papyrus	*Cyperus papyrus*
Bristle-spiked cyperus	*Cyperus strigosus*

Edible cypress	*Cyperus esculentus*
Slender cyperus	*Cyperus filiculmis*
Pond sedge	*Dulichium arundinaceum*
Large spike rush	*Eleocharis palustris*
Slender spike rush	*Eleocharis tenuis*
Chinese water chestnut	*Eleocharis tuberosa*
Sand mat	*Stenophyllus capillaris*
Slender fimbristylis	*Fimbristylus frankii*
Autumn fimbristylis	*Fimbristylus autumnalis*
Chair-makers rush	*Scirpus americanus*
Salt-marsh bulrush	*Scirpus robustus*
Giant bulrush	*Scirpus validus*
Wool grass	*Scirpus cuperinus*
Wood bulrush	*Scirpus sylvaticus*
Leafy bulrush	*Scirpus polyphyllus*
Cotton grass	*Eriophorum* species
Clustered beaked rush	*Rhynchospora glomerata*
White beaked rush	*Rhynchospora alba*
Low nut rush	*Scleria verticillata*
Tufted sedge	*Carex stricta*
Pennsylvania sedge	*Carex pennsylvanica*
Bladder sedge	*Carex intumescens*
Porcupine sedge	*Carex hystricina*
Fringed sedge	*Carex crinita*
Little prickly sedge	*Carex scirpoides*
Fox sedge	*Carex vulpinoidea*

Sesame	PEDALIACEAE
Sesame	*Sesamum orientale*

Silk-cotton tree	BOMBACACEAE
Baobab tree	*Adansonia digitata*
Silk-cotton tree	*Ceiba pentandra*

Silk-tassel tree	GARRYACEAE
Silk-tassel tree	*Garrya elliptica*

Silvervine	DILLENIACEAE
Silvervine	*Actinidia polygama*
Tara vine	*Actinidia arguta*
Yangtao	*Actinidia chinensis*
Kolomikta vine	*Actinidia kolomikta*

Soapberry	SAPINDACEAE
Wingleaf soapberry	*Sapindus saponaria*
Wild China tree	*Sapindus drummondii*
Florida soapberry	*Sapindus marginatus*
Butterbough	*Exothea paniculata*
Inkwood	*Hypelate trifoliata*
Balloon vine	*Cardiospermum halicacabum*
Akee	*Blighia sapida*
Longan	*Euphoria longana*
Litchi	*Litchi chinensis*
Varnish-leaf	*Dodonaea viscosa*

Spiderwort	COMMELINACEAE
Dayflower	*Commelina* species
Spiderwort	*Tradescantia virginiana*
Wandering Jew	*Tradescantia fluminensis*

Spurge	EUPHORBIACEAE
Hogwort	*Croton capitatus*
Silverbush	*Argythamnia sericophylla*
Spurge nettle	*Jatropha stimulosa*
Three-seeded mercury	*Acalypha virginica*
Queen's delight	*Stillingia sylvatica*
Castor oil plant	*Ricinus communis*
Poinsettia	*Poinsettia pulcherrima*
Mexican fire plant	*Poinsettia heterophylla*
Tapioca plant	*Manihot esculenta*
Ceara rubber tree	*Jatropha glaziovii*
Para rubber tree	*Hevea brasiliensis*
Common croton	*Codiaeum variegatum pictum*
Spotted spurge	*Euphorbia maculata*
Flowering spurge	*Euphorbia corollata*
Seaside spurge	*Euphorbia polygonifolia*
Snow-on-the-mountain	*Euphorbia marginata*
Cypress spurge	*Euphorbia cyparissias*
Sun spurge	*Euphorbia helioscopia*
Crown-of-thorns	*Euphorbia splendens*
Millbark	*Drypetes diversifolia*
Guiana plum	*Drypetes lateriflora*
Manchineel	*Hippomane mancinella*
Oysterwood	*Gymnanthes lucida*

Staff tree	CELASTRACEAE
Bittersweet	*Celastrus scandens*
Mountain lover	*Pachistima canbyi*
Strawberry bush	*Euonymus americanus*
Trailing strawberry bush	*Euonymus obovatus*
Burning bush	*Euonymus atropurpureus*
European spindle tree	*Euonymus europeus*

Storax	STYRACACEAE
Silverbell tree	*Halesia carolina*
Snowdrop tree	*Halesia diptera*
Tisswood	*Halesia monticola*
Little silverbell	*Halesia parviflora*
Bigleaf silver bell	*Styrax grandifolia*
American silver bell	*Styrax americana*
Downy storax	*Styrax pulverulenta*
Epaulette tree	*Pterostyrax hispida*

Strawberry shrub	CALYCANTHACEAE
Strawberry shrub	*Calycanthus floridus*
Smooth strawberry shrub	*Calycanthus fertilis*
Sweet shrub	*Calycanthus occidentalis*

Sumac	ANACARDIACEAE
Poison ivy	*Rhus radicans*

Poison sumac	*Rhus vernix*
Poison oak	*Rhus quercifolia*
Western poison oak	*Rhus diversiloba*
Staghorn sumac	*Rhus typhina*
Smooth sumac	*Rhus glabra*
Fragrant sumac	*Rhus aromatica*
Mountain sumac	*Rhus copallina*
Skunkbush	*Rhus trilobata*
Laurel sumac	*Rhus laurina*
Lemonade berry	*Rhus integrifolia*
Sugar bush	*Rhus ovata*
Mango	*Mangifera indica*
California pepper tree	*Schinus molle*
Pistachio nut	*Pistacia vera*
Cashew	*Anacardium orientale*
Poisonwood	*Metopium toxiferum*
Smoke tree	*Cotinus americanus*
Pistachio	*Pistacia texana*
Eurasian smoke tree	*Cotinus coggygria*

Sundew	DROSERACEAE
Round-leaved sundew	*Drosera rotundifolia*
Long-leaved sundew	*Drosera longifolia*
Thread-leaved sundew	*Drosera filiformis*
Venus fly trap	*Dionaea muscipula*
Portuguese flycatcher	*Drosophyllum lusitanicum*

Sweetleaf	SYMPLOCACEAE
Horse sugar	*Symplocos tinctoria*

Tallowwood	OLACACEAE
Tallowwood	*Ximenia americana*
Gulf graytwig	*Schoepfia chrysophylloides*

Tamarisk	TAMARICACEAE
Tamarisk	*Tamarix* species
Salt cedar	*Tamarix gallica*
German tamarisk	*Myricaria germanica*

Tea	THEACEAE
Common camellia	*Camellia japonica*
Sasanqua	*Camellia sasanqua*
Silky camellia	*Stewartia malachodendron*
Mountain camellia	*Stewartia ovata*
Loblolly bay	*Gordonia lasianthus*
Franklinia	*Gordonia alatamaha*
Tea	*Thea sinensis*

Teasel	DIPSACACEAE
Sweet scabious	*Scabiosa atropurpurea*
Scabiosa	*Scabiosa* species
Fuller's teasel	*Dipsacus fullonum*

Theophrasta	THEOPHRASTACEAE
Joewood	*Jaquinia keyensis*

Tobira	PITTOSPORACEAE
Tobira	*Pittosporum tobira*
Butterbush	*Pittosporum phillyraeoides*
Diamond-leaf laurel	*Pittosporum rhombifolium*
Victorian box	*Pittosporum undulatum*
Tarata	*Pittosporum eugenioides*

Torchwood	BURSERACEAE
Gumbo limbo	*Bursera simaruba*

Turnera	TURNERACEAE
Piriquetas	*Piriqueta* species

Unicorn plant	MARTYNIACEAE
Unicorn plant	*Martynia louisianica*

Valerian	VALERIANACEAE
Common valerian	*Valeriana officinalis*
Red valerian	*Centranthus rubra*
Swamp valerian	*Valeriana uliginosa*
Corn salad	*Valerianella locustra*
Goosefoot corn salad	*Valerianella chenopodifolia*
Beaked corn salad	*Valerianella radiata*

Vervain	VERBENACEAE
Blue vervain	*Verbena hastata*
Verbena or European vervain	*Verbena officinalis*
Narrow-leaved vervain	*Verbena angustifolia*
Hoary vervain	*Verbena. stricta*
White vervain	*Verbena urticaefolia*
Lemon verbena	*Lippia lanceolata*
Fog fruit	*Lippia canescens*
Beautyberry	*Callicarpa americana*
Black mangrove	*Avicennia nitida*
Fiddlewood	*Citharexylum fruticosum*
Verbena	*Verbena* species

Vine	VITACEAE
Virginia creeper	*Parthenocissus virginica*
Boston ivy	*Parthenocissus tricuspidata*
Pepper-vine	*Ampelopsis arborea*
Possum grape	*Cissus incisa*
Northern fox grape	*Vitis labrusca*
Riverbank grape	*Vitis vulpina*
Summer grape	*Vitis aestivalis*

Frost grape	*Vitis cordifolia*
Muscadine grape	*Vitis rotundifolia*
Wine grape	*Vitis vinifera*
Violet	VIOLACEAE
Common violet	*Viola papilionacea*
Bird's foot violet	*Viola pedata*
Early blue violet	*Viola palmata*
Arrow-leaved violet	*Viola sagittata*
Sweet violet	*Viola odorata*
Canada violet	*Viola canadensis*
Lance-leaved violet	*Viola lanceolata*
Sweet white violet	*Viola blanda*
Downy yellow violet	*Viola pubescens*
Yellow round-leaved violet	*Viola rotundifolia*
Pansy	*Viola tricolor hortensis*
Walnut	JUGLANDACEAE
Shagbark hickory	*Carya ovata*
Shellbark hickory	*Carya laciniosa*
Mockernut	*Carya alba*
Pignut	*Carya glabra*
Bitternut	*Carya cordiformis*
Pecan	*Carya pecan*
Small-fruited hickory	*Carya microcarpa*
Water hickory	*Carya aquatica*
Nutmeg hickory	*Carya myristicaeformis*
Black walnut	*Juglans nigra*
Butternut	*Juglans cinerea*
Texas walnut	*Juglans rupestris*
California walnut	*Juglans californica*
English walnut	*Juglans regia*
Waterleaf	HYDROPHYLLACEAE
Virginia waterleaf	*Hydrophyllum virginianum*
Canada waterleaf	*Hydrophyllum canadense*
Fiesta-flower	*Nemophila aurita*
Baby-blue eyes	*Nemophila menziesi*
Small-flowered nemophila	*Nemophila microcalyx*
Silky phacelia	*Phacelia leucophylla*
Loose-flowered phacelia	*Phacelia bipinnatifida*
Small-flowered phacelia	*Phacelia dubia*
Water lily	NYMPHAEACEAE
Yellow water lily	*Nymphaea advena*
White water lily	*Nymphaea odorata*
American lotus	*Nelumbium pentapetalum*
Egyptian lotus	*Nelumbium indica*
Royal water lily	*Victoria regia*
Cabomba	*Cabomba caroliniana*
Purple bonnet	*Brasenia schreberi*
Water milfoil	HALORAGACEAE
Mare's tail	*Hippuris vulgaris*
Spiked water milfoil	*Myriophyllum spicatum*
Whorled water milfoil	*Myriophyllum verticillatum*
Slender water milfoil	*Myriophyllum tenellum*
Parrot's feather	*Myriophyllum proserpinacoides*
Common Mermaid-weed	*Proserpinaca palustris*
Cut-leaved mermaid-weed	*Proserpinaca pectinata*
Water nut	HYDROCARYACEAE
Water chestnut	*Trapa natans*
Water plantain	ALISMACEAE
Arrow-head	*Sagittaria latifolia*
Water plantain	*Alisma plantago-aquatica*
Water starwort	CALLITRICHACEAE
Terrestrial water starwort	*Callitriche austini*
Vernal water starwort	*Callitriche palustris*
Larger water starwort	*Callitriche heterophylla*
Autumnal water starwort	*Callitriche bifida*
Waterwort	ELATINACEAE
American waterwort	*Elatine americana*
Texas bergia	*Bergia texana*
White mangrove	COMBRETACEAE
East indian laurel	
African limba	
East indian almond	*Terminalia catappa*
White mangrove	*Laguncularia racemosa*
Black olive tree	*Bucida buceras*
Button-mangrove	*Conocarpus erecta*
Rangoon creeper	*Quisqualis indica*
Wild cinnamon	CANELLACEAE
Wild cinnamon	*Canella winterana*
Willow	SALICACEAE
Pussy willow	*Salix discolor*
Black willow	*Salix nigra*
Crack willow	*Salix fragilis*
Weeping willow	*Salix babylonica*
Shining willow	*Salix lucida*
Balsam willow	*Salix balsamifera*
Heart-leaved willow	*Salix cordata*
Bog willow	*Salix pedicellaris*
Prairie willow	*Salix humilis*
Sage willow	*Salix tristis*
Sandbar willow	*Salix longifolia*
Peach-leaved willow	*Salix amygdaloides*
Balsam poplar	*Populus balsamifera*
Balm of gilead	*Populus candicans*
Swamp cottonwood	*Populus heterophylla*

Carolina poplar	*Populus deltoides*
Black cottonwood	*Populus trichocarpa*
Lance-leaved cottonwood	*Populus acuminata*
Fremont cottonwood	*Populus fremontii*
American aspen	*Populus tremuloides*
Large-toothed aspen	*Populus grandidentata*
White poplar	*Populus alba*
Black poplar	*Populus nigra*
Lombardy poplar	*Populus nigra* variety *italica*

Witch hazel	HAMAMELIDACEAE
Witch hazel	*Hamamelis virginiana*
Springtime witch hazel	*Hamamelis vernalis*
Witch-alder	*Fothergilla gardeni*
Sweet gum	*Liquidambar styraciflua*

Wood sorrel	OXALIDACEAE
Wood sorrel	*Oxalis acetosella*
Violet wood sorrel	*Oxalis violacea*
Yellow wood sorrel	*Oxalis corniculata*
Carambola	*Averrhoa carambola*

Yam	DIOSCOREACEAE
Wild yam	*Dioscorea villosa*
Chinese yam	*Dioscorea batatas*
Air potato	*Dioscorea bulbifera*
Yampee	*Dioscorea trifida*
Guayabo	*Rajania pleioneura*

Yellow-eyed grass	XYRIDACEAE
Slender yellow-eyed grass	*Xyris flexuosa*
Fringed yellow-eyed grass	*Xyris fimbriata*
Carolina yellow-eyed grass	*Xyris caroliniana*
Northern yellow-eyed grass	*Xyris montana*

Yew	TAXACEAE
English yew	*Taxus baccata*
American yew	*Taxus canadensis*
Irish yew	*Taxus baccata* variety *fastigiata*
Pacific yew	*Taxus brevifolia*
Florida yew	*Taxus floridana*
California nutmeg	*Torreya californica*
Stinking cedar	*Torreya taxifolia*
Japanese torreya	*Torreya nucifera*
Japanese yew	*Taxus cuspidata*
Plum yews	*Cephalotaxus* species
Podocarpus	*Podocarpus* species
Phyllocladus	*Phyllocladus* species

Glossary

ACCESSORY ORGAN: A floral structure not directly concerned with reproduction.

ACHENE: A small, dry, hard, one-celled, one-seeded, indehiscent fruit.

AGGREGATE FRUIT: A cluster of fruits developed from the ovaries of a single flower.

ALTERNATE: Said of leaves on a stem that are not opposite to each other on the axis, but arranged singly at different heights.

ANGIOSPERM: A plant having the seeds borne within a pericarp.

ANNUAL: A plant that germinates, grows, flowers, fruits and dies within a single year.

ANTHER: The pollen-bearing part of a stamen.

APETALOUS: Having no petals.

APEX: The tip, point, or outermost part.

APICAL: Of, or at the apex.

AREOLE: A small space marked out upon a surface.

ARIL: An appendage growing at or about the hilum of a seed.

AWN: A bristle-shaped appendage.

BASAL: Of, or at the base.

BASE: Part of a plant organ nearest its own point of attachment.

BEAK: Any process somewhat like the beak of a bird, terminating the fruit or other parts of a plant.

BERRY: A thin-skinned, fleshy fruit with numerous scattered seeds.

BRACT: A modified leaf associated with a flower or an inflorescence.

BUD: A terminal or axillary structure on a stem consisting of a small mass of meristematic tissue, covered wholly, or in part by overlapping leaves; an unexpanded flower.

BULB: A subterranean leaf bud with fleshy scales or coats.

BULBLET: A small bulb, especially one borne upon a stem.

CALYX: A collective term for the sepals of a flower.

CAMBIUM: A layer of soft-growing tissue between the bark and the wood of trees and shrubs.

CAPSULE: A dry, dehiscent fruit composed of more than one carpel.

CARPEL: A floral organ that bears and encloses ovules; a simple pistil, or one member of a compound pistil.

CARYOPSIS: A dry, one-seeded indehiscent fruit with the seed coat and pericarp completely united, forming a single grain, as in wheat or barley.

CATKIN: A spike inflorescence bearing staminate or pistillate, apetalous (no petals) flowers.

CELL: A small mass or unit of protoplasm surrounded by a cell membrane, and in plants also by a wall, containing one or more nuclei.

CLASS: A subdivision of the plant kingdom ranking above an order.

CLASSIFICATION: A system of grouping plants in groups or divisions according to structural characteristics.

CLEFT: Split or divided.

CLEISTOGAMOUS: Fertilized or self-pollinating in the bud, without the opening of the flower.

CLONE: A group of plants, often many thousands, all originating from one seedling plant, wherefrom they have subsequently increased vegetatively, by any method other than by seeds.

CLUSTER: A group of flowers, in other words, an inflorescence.

COLUMN: The united stamens and pistils in orchids.

COMPLETE: A flower that bears sepals, petals, stamens and pistils.

COMPOUND LEAF: One divided into separate leaflets.

CONE: Typically the flower and fruit cluster of the pine and its relatives.

CONIFEROUS: Cone-bearing.

CORM: A short, often globose, upright, underground stem that stores food.

COROLLA: Collectively, the petals of a flower.

CORYMB: A simple inflorescence wherein the pedicels, growing along the peduncle, are of unequal length, those of the lowest flowers being longest, those of the upper flowers, shortest.

COTYLEDON: A food-digesting and food-storing part of an embryo; a seed leaf.

CROSS-POLLINATION: The transfer of pollen from the stamen of a flower of one plant to the stigma of a flower of another plant.

CYME: A flower cluster wherein the growing apex ceases growth early, all its meristematic tissues being used up in the formation of an apical flower, with other flowers developing farther down on the axis, the youngest flower appearing farthest from the apex.

DECIDUOUS: Refers to plants that lose their leaves regularly each year, as opposed to those that retain their leaves for more than one year.

DEHISCENT: Splitting or opening along definite seams when mature.

DICOTYLEDON: A plant having two cotyledons or seed leaves.

DRUPE: A simple, fleshy fruit wherein the inner wall of the ovary becomes hard and stony and encloses one or more seeds.

DRUPELET: A diminutive drupe.

EGG: The female germ cell of a plant.

EGG CELL: The female germ cell, or egg proper, exclusive of any envelopes derived from, or consisting of other cells.

ELLIPTICAL: Oblong with rounded ends.

ENTIRE: Without teeth or divisions.

EPIPHYTE: A plant growing attached to another plant, but not parasitic; an air plant.

EQUITANT: Astride; used of conduplicate leaves that enfold each other in two ranks, as in iris.

ESSENTIAL ORGANS: Floral structures necessary to reproduction, that is, stamens and pistils.

FAMILY: A group of related genera united because they all have a family resemblance, although quite distinct from one another.

FEATHER-FASHION: The leaflets of a compound leaf arranged like the segments of a bird's feather, that is, arising from opposite sides of a common axis.

FINGER-FASHION: The leaflets of a compound leaf arranged like the fingers of the hand, that is, all arising from approximately the same point.

FOLLICLE: A simple, dry, dehiscent fruit, producing several to numerous seeds and composed of one carpel that splits along one seam.

FRUIT: A matured ovary or cluster of matured ovaries, the seed-bearing product of a plant.

GENUS (PL. GENERA): The simplest grouping of plants, so classified because they are more like each other than like any other group.

GLUME: A chafflike bract, especially one found at the base of a grass spikelet.

GYMNOSPERM: A plant bearing naked seeds, without an ovary.

HABITAT: The site wherein a wild plant grows, such as fields, woods, bogs etc.

HAUSTORIUM: A suckerlike structure that penetrates host tissues and cells and absorbs nourishment from them.

HEAD: A dense cluster of sessile or nearly sessile flowers on a short axis or receptacle.

HEART-SHAPED: Ovate with two rounded lobes and a sinus at the base; usually employed to define such a base.

HERB: A plant with none, or with very little woody tissue above ground.

HERBACEOUS: Referring to an herb.

HESPERIDIUM: A type of berry wherein the outer fruit wall becomes leathery.

HILUM: A scar on a seed coat, marking the point of attachment of the seedstalk to the seed.

HIP: A pod containing the ripe seed of a rosebush.

IMPERFECT: A flower that bears only stamens, or only pistils.

INCOMPLETE: A flower that lacks one or more of the four kinds of floral organs.

INDEHISCENT: A fruit that does not split open along regular seams.

INFLORESCENCE: A flower cluster.

INTERNODE: The portion of a stem between two nodes.

INVOLUCRE: The circle or collection of bracts surrounding a flower cluster, of a single flower.

IRREGULAR: Showing inequality in the size, form, or union of the similar parts of a flower.

KEEL: A ridge, like the keel of a boat; the two anterior, united petals of a papilionaceous flower.

LANCE-SHAPED: Much longer than broad and tapering towards a slender, pointed tip.

LEAF: A lateral, green, expanded outgrowth of a stem.

LEAF AXIL: The upper angle between a leaf petiole and the stem wherefrom it grows.

LEAF BASE: The part of the leaf at the point of attachment to the leafstalk.

LEAF BLADE: The usually flat, expanded part of a leaf, as distinguished from its stalk.

LEAFLET :A single division of a compound leaf.

LEAFSTALK: The stalk that attaches the leaf to the stem.

LEGUME: A simple, dry, dehiscent fruit formed of a single pistil and splitting along two sutures.

LEMMA: The lower of the two bracts enclosing the flower in the grasses.

LENTICEL: A pore wherethrough the exchange of gases takes place.

LIGULE: A thin, often scarious projection from the summit of the sheath in grasses.

LIMB: The expanded portion of a corolla above the throat wherein the petals are more or less united.

LINEAR: Long and narrow with parallel margins.

LIP: Each of the upper and lower divisions of a two-lipped calyx or corolla.

LOBE: Any segment of an organ, especially if rounded.

LODICULE: A minute, thin, transparent scale at the base of a grass flower.

MEMBRANACEOUS: Thin, rather soft, and more or less translucent.

MERISTEMATIC TISSUE: A tissue whose cells are capable of frequent division, and thus, are responsible for growth.

MIDRIB: The central or main rib of a leaf.

MONOCOTYLEDON: A plant haivng only one cotyledon.

MONOECIOUS: Having staminate and pistillate flowers on the same plant.

MULTIPLE FRUIT: A cluster of matured ovaries produced by several flowers.

NECTAR: A sweet liquid produced by flowers.

NECTARY: A floral gland that secretes nectar.

NODE: The place upon a stem that normally bears a leaf or whorl of leaves.

NOMENCLATURE: A set or system of names or terms.

NUT: An indehiscent, dry, one-seeded, hard-walled fruit, produced from a compound ovary.

NUTLET: A diminutive nut.

OBLONG: Longer than broad and with nearly parallel sides.

OBOVOID: In the form of an inverted egg.

OPPOSITE: Situated on diametrically opposed sides of an axis

ORBICULAR: Circular.

ORDER: A division in classification containing a number of families.

OVAL: In the shape of the longitudinal section of an egg; egg-shaped.

OVARY: The part of the pistil that contains the ovules.

OVATE: Egg-shaped.

OVULE: The structure in the ovary of a flower that, when fertilized, becomes the seed.

PALEA OR PALET: The upper bract with which, the lemma encloses the flower.

PALMATE (LEAF): Radially lobed or divided, that is, with leaflets, or with lobes or veins of a simple leaf, radiating from one point.

PALMATELY: In a palmate manner.

PANICLE: A compound inflorescence that has several main branches with pedicellate flowers arranged along its axis.

PAPILIONACEOUS: Having a standard, wings, and keel, as in the peculiar corolla of the sweet pea.

PAPPUS: Any appendage or tuft of appendages forming a crown of various characters at the summit of the achene.

PEDICEL: The stalk of an individual flower of an inflorescence.

PEDUNCLE: The stalk of a solitary flower, or the main stalk of an inflorescence.

PEPO: An indehiscent, fleshy, many-seeded fruit with a hard rind.

PERENNIAL: Lasting year after year.

PERFECT: A flower that bears both stamens and pistils.

PERIANTH: The floral envelop; the calyx and corolla taken together.

PERICARP: The matured ovary.

PERIGYNIUM: The inflated sac that incloses the ovary in *Carex*.

PERSISTENT: Long-continuous, as a calyx upon the fruit; leaves remaining on the twigs through the winter.

PETAL: A floral leaf in the whorl between the stamens and sepals; a division of the corolla.

PETIOLE: A leafstalk.

PHOTOSYNTHESIS: The process whereby plants manufacture their own food.

PHYLLODIUM (PL. PHYLLODES): A somewhat dilated petiole having the form of and serving as a leaf blade.

PINNA: One of the primary divisions of a pinnate or compoundly pinnate leaf.

PINNATE: Compound, with the leaflets arranged on each side of a common petiole.

PISTIL: The ovule-producing part of a flower, consisting of ovary, style, and stigma.

PISTILLATE: Provided with pistils, that is, in its more proper sense, without stamens.

POD: A dry, dehiscent fruit splitting along two sutures.

POLLEN: The fecundating grains contained in the anther.

POLLEN GRAIN: The male reproductive tissue of flowering plants.

POLLINIUM (PL. POLLINIA): A coherent mass of pollen grains, often with a stalk bearing an adhesive disk.

POLYCOTYLEDON: Having more than two cotyledons.

POME: A fleshy fruit consisting of a ripened receptacle surrounding the ovary.

PSEUDOBULB: The swollen, stemlike, often grooved base of many orchids.

PUBESCENT: Covered with hairs, especially if short, soft, and downlike.

PYXIDIUM: A capsule that divides by circumscissile dehiscence, the upper portion falls off as a lid or cap.

RACEME: A simple inflorescence wherein the flowers, each with its own pedicel, are spaced along a common, more or less elongated, axis.

RACHILLA: The axis of a spikelet in grasses, specifically the floral axis as opposed to that of the spikelet.

RACHIS: The elongated axis of a spike or raceme; in a pinnately compound leaf the extension of the petiole bearing the leaflets.

RADICLE: The primary root of an embryo plant or seedling.

RECEPTACLE: The enlarged end of the stem whereto the parts of a flower are attached or the collected flowers of a head.

REGULAR: Uniform in shape or structure.

RHIZOME: A horizontal underground stem, usually rooting at the nodes and becoming erect at the apex.

RIB: A primary or prominent vein of a leaf..

ROOT: The underground part of a plant that supplies it with nourishment and anchors it in the ground.

ROOTSTALK: A rhizome.

ROSTELLUM: A little beak; in orchids the apex of the column.

ROTATE: Wheel-shaped; flat and circular in outline.

SAMARA: A dry, indehiscent, one seeded, winged fruit.

SCALE: A rudimentary leaf serving to protect a bud before expansion, such as a bud scale.

SCAPE: A peduncle rising from the ground, naked or without proper foliage.

SCHIZOCARP: A simple, dry, indehiscent fruit composed of two fused carpels that split apart at maturity, each part usually with one seed.

SEED: A complete embryo plant protected by one or more seed coats; a ripened ovule with the embryo and seed coats.

SEGMENT: One of the parts of a leaf or other like organ that is cleft or divided.

SELF-FERTILIZATION: The transfer of pollen from the stamen to the stigma of the same flower or of another flower on the same plant.

SELF-POLLINATION: Self-fertilization.

SEPAL: The outermost floral organ, usually green, that encloses the other parts of the flower in the bud; a division of the calyx.

SEPTUM: Any kind of a partition.

SESSILE: Lacking a stalk; attached by the base without a stalk or stem.

SHEATH: A tubular envelope, as the lower part of the leaf in grasses.

SHRUB: A woody perennial, smaller than a tree, usually with several stems.

SILICLE: A silique broader than it is long.

SILIQUE: A simple, dry, dehiscent fruit developed from two fused carpels that separate at maturity, leaving a persistent partition between.

SIMPLE: Of one piece; not compound.

SINUS: The cleft or recess between two lobes.

SOLITARY: Alone, single.

SPADIX: A spike with a fleshy axis.

SPATHE: A large bract or pair of bracts enclosing an inflorescence.

SPECIES: A group of individual plants more like each other than anything else and all belonging to a single genus.

SPERM CELL: The male reproductive unit.

SPIKE: An inflorescence wherein the sessile flowers are arranged on a more or less elongated common axis.

SPIKELET: A unit of a grass inflorescence with one or more flowers and their bracts.

SPINE: A sharp, woody or rigid outgrowth from the stem.

SPUR: A hollow, saclike or tubular extension of some part of a blossom, usually nectariferous.

STAMEN: The pollen-producing organ of a flower.

STAMINATE: Having stamens only; applied to a flower with stamens but no pistils.

STAMINODE: A sterile stamen, or any structure without anther corresponding to a stamen.

STEM: The main ascending axis of a plant.

STERILE: Unproductive, as a flower without a pistil or stamen, or a stamen without an anther.

STIGMA: The part of the pistil, usually the apex, that receives pollen and whereon pollen grains germinate.

STIPULE: An appendage at the base of a petiole or on each side of its insertion.

STOMA: An orifice in the epidermis of a leaf communicating with internal air cavities.

STROBILE: An inflorescence marked by imbricated bracts or scales.

STYLE: The usually attenuated portion of the pistil connecting the stigma with the ovary.

SYNCARP: A collective fruit.

TAXONOMY: The science that deals with the classification of plants or more specifically with the laws governing such classification.

TENDRIL: A slender prolongation of the stem or leaf for use in climbing.

TERETE: Having a circular, transverse section.

THALLUS: A simple plant body not differentiated into roots, stems, and leaves.

THROAT: The orifice of a united corolla or calyx.

THYRSE: A contracted cylindrical or ovoid, and usually compact panicle; a form of mixed inflorescence wherein the main axis is racemose and the secondary or later axes are cymose.

TRANSPIRATION: The normal escape of water vapor through the pores of a leaf.

TREE: A woody plant, of some considerable size, with a single stem or trunk.

TUBER: A thickened and short, subterranean branch having numerous buds or eyes.

TWIG: A young shoot or branch; a portion of a stem.

UMBEL: An inflorescence wherein the stems of the flowers are of approximately the same length and grow from the same point.

UTRICLE: A small bladdery one-seeded fruit; any small bladderlike body.

VEIN: A thread of fibro-vascular tissue in a leaf or other organ.

VISCID: Glutinous; sticky.

WHORL: An arrangement of leaves, etc, in a circle around the stem.

WING: The lateral petal of a papilionaceous corolla.

WOOLLY: Clothed with long and tortuous or matted hairs.

Bibliography

Anderson, A. W. *How We Got Our Flowers*. New York: Dover, 1966.

Armstrong, M. *Fieldbook of Western Wildflowers*. New York: Putnams, 1915.

Bailey, L. H. *How Plants Get Their Names*. New York: Dover, 1963.

———. *The Cultivated Conifers of North America*. New York: Macmillan, 1933.

———. *The Standard Encyclopedia of Horticulture*. New York: Macmillan, 1925.

Baker, M. F. *Florida Wildflowers*. New York: Macmillan, 1952.

Benson, L. *Plant Classification*. Boston: Heath, 1957.

Berry, J. B. *Western Forest Trees*. Gloucester, Mass: Peter Smith, 1964.

Bowers, C. G. *Rhododendrons and Azaleas*. New York: Macmillan, 1960.

Britton, N. L., and Brown, A. *An Illustrated Flora of the Northern United States and Canada*. New York: Dover, 1970.

Britton, N. L., and Rose, J. N. *The Cactaceae*. New York: Dover, 1937.

Brown, H. B. *Cotton*. New York: McGraw-Hill, 1927.

Cheney, R. H. *Coffee*. New York: New York University Press, 1925.

Clarkson, R. E. *The Golden Age of Herbs and Herbalists*. New York: Dover, 1972.

Clements, E. S. *Flowers of Prairie and Woodland*. New York: Wilson, 1947.

Clements, E. S., and Clements, F. E. *Flower Families and Ancestors*. New York: Wilson, 1928.

———. *Rocky Mountain Flowers*. New York: Wilson, 1945.

Coit, J. E. *The Citrus Fruits*. New York: Macmillan, 1915.

Copeland, E. B. *The Coconut*. New York: Macmillan, 1921.

Dana, W. S. *How To Know The Wildflowers*. New York: Dover, 1963.

Dunsterville, G. C. K. *An Introduction to the World of Orchids*. New York: Doubleday, 1964.

Fairchild, D. *Exploring for Plants*. New York: Macmillan, 1930.

Fernald, M. L., and Kinsey, A. C. *Edible Wild Plants of Eastern North America*. Cornwall-on-Hudson, N.Y.: Idlewild Press, 1943.

Francis, M. E. *The Book of Grasses*. New York: Doubleday, 1912.

Free, M. *All About the Perennial Garden*. New York: Doubleday, 1955.

Gerard, J. *Leaves from Gerard's Herbal*. Edited by M. Woodward. New York: Dover, 1969.

Gray, A. *Manual of Botany*. New York: American Book Company, 1950.

Greene, E. F., and Blomquist, H. L. *Flowers of the South*. Chapel Hill, N.C.: University of North Carolina, 1953.

Grieve, M. *A Modern Herbal*. New York: Dover, 1971.

Grimm, W. C. *Recognizing Native Shrubs*. Harrisburg, Pa.: Stackpole, 1966 a.

———. *The Book of Trees*. Harrisburg, Pa.: Stackpole, 1966 b.

Hall, C. J. *Cocoa*. New York: Macmillan, 1914.

Harding, Mrs. E. *Lilacs in My Garden*. New York: Macmillan, 1936.

———. *Peonies in the Garden*. Philadelphia: Lippincott, 1923.

Harlow, W. M. *Trees of the Eastern and Central United States and Canada*. New York: Dover, 1942.

Harlow, W. M., and Harrar, E. S. *Textbook of Dendrology*. New York: McGraw-Hill, 1958.

Harrar, E. S., and Harrar, J. G. *Guide to Southern Trees*. New York: Dover, 1962.

Hitchcock, A. S. *Manual of the Grasses of the United States*. New York: Dover, 1971.

Hotchkiss, N. *Common Marsh Underwater and Floating-leaved Plants of the United States*. New York: Dover, 1972.

Hottes, A. C. *Climbers and Ground Covers*. New York: De La Mare, 1947.

Houghton, A. D. *The Cactus Book*. New York: Macmillan, 1930.

House, H. D. *Wild Flowers*. New York: Macmillan, 1961.

Hume, H. H. *Azaleas and Camellias*. New York: Macmillan, 1931.

Hunt, T. F. *The Cereals of America*. New York: Judd, 1919.

Hylander, C. J. *The World of Plant Life*. New York: Macmillan, 1956.

Hylander, C. J., and Johnston, E. F. *The Macmillan Flower Book*. New York: Macmillan, 1954.

Jacob, H. *Coffee, the Epic of a Commodity*. New York: Viking, 1935.

Jenkins, D. H. *Annuals for Every Garden*. New York: Barrows, 1945.

Keeler, H. L. *Our Garden Flowers*. New York: Scribners, 1910.

———. *Our Native Trees*. New York: Scribners, 1912.

———. *Our Northern Shrubs*. New York: Dover, 1969.

Knapp, A. *Cocoa and Chocolate*. London: Chapman and Hall, 1920.

Lemmon, R. S., and Johnson, C. C. *Wildflowers of North America*. Garden City, N.Y.: Hanover House, 1961.

Macself, A. J. *Gladioli*. New York: Scribners, 1925.

Martin, A. C., Zim, H. S., and Nelson, A. L. *American Wildlife and Plants*. New York: Dover, 1951.

Mathews, F. S. *Fieldbook of American Trees and Shrubs*. New York: Putnams, 1915.

———. *Fieldbook of American Wildflowers*. Revised edition by Norman Taylor. New York: Putnams, 1955.

McCurrach, J. C. *Palms of the World*. New York: Horticultural, 1960.

McKelvey, S. D. *The Lilac*. New York: Macmillan, 1928.

McKenney, M., and Peterson, R. T. *A Field Guide to the Wildflowers of the Northeastern and Central States*. Boston: Houghton Mifflin, 1968.

McLean, F., Clark, W., and Fischer, E. *The Gladiolus Book*. New York: Doubleday, 1927.

McMinn, H. E., and Maino, E. *Pacific Coast Trees*. Berkeley, Cal.: University of California Press, 1935.

Medsger, O. P. *Edible Wild Plants*. New York: Macmillan, 1944.

Menninger, E. A. *Fantastic Trees*. New York: Viking, 1967.

Moldenke, H. N. *American Wildflowers*. New York: Macmillan, 1949.

Moore, H. E. *African Violets, Gloxinias and Their Relatives*. New York: Macmillan, 1957.

Morris, F., and Eames, E. *Our Wild Orchids*. New York: Scribners, 1929.

Muenscher, W. C. *Aquatic Plants of the United States*. Ithaca, N.Y.: Comstock, 1944 a.

———. *Poisonous Plants of the United States*. New York: Macmillan, 1944 b.

———. *Rice: Garden Spice and Wild Pot-herbs*. Ithaca, N.Y.: Comstock, 1955.

Nicolas, J. H. *The Rose Manual*. New York: Doubleday, 1934.

Niklitschek, A. *Water Lilies and Water Plants*. New York: Scribners, 1933.

Northern, H., and Northern, R. *Ingenious Kingdom*. Englewood Cliffs, N.J.: Prentice-Hall, 1970.

Parsons, M. E. *The Wildflowers of California*. New York: Dover, 1966.

Peattie, D. C. *A Natural History of Western Trees*. Boston: Houghton Mifflin, 1953.

———. *Natural History of Trees of Eastern and Central North America*. Boston: Houghton Mifflin, 1950.

Percival, J. *The Wheat Plant*. New York: Dutton, 1921.

Percival, M. S. *Floral Biology*. New York: Pergamon Press, 1965.

Peterson, M. G. *How To Know The Wild Fruits*. New York: Dover, 1973.

Petrides, G. A. *A Field Guide to the Trees and Shrubs*. Boston: Houghton Mifflin, 1958.

Phillips, G. A. *Delphiniums*. New York: Macmillan, 1933.

Platt, R. *The Great American Forest*. Englewood Cliffs, N.J.: Prentice-Hall, 1965.

Pohl, R. W. *How To Know The Grasses*. Dubuque, Iowa: Brown, 1954.

Preston, R. J. *Rocky Mountain Trees*. New York: Dover, 1968.

Rickett, H. W. *Wildflowers of America*. New York: Crown, 1963.

Robbins, W., and Ramaley, F. *Plants Useful to Man*. Philadelphia: Blakiston, 1937.

Rockwell, F. F., and Grayson, E. C. *The Complete Book of Annuals*. New York: Doubleday, 1955.

———. *The Complete Book of Bulbs*. New York: Doubleday, 1953.

Rogers, J. E. *The Tree Book*. New York: Doubleday, 1916.

Rogers, W. E. *Tree Flowers of Forest, Park, and Street*. New York: Dover, 1965.

Rohde, E. S. *The Old English Herbals*. New York: Dover, 1971.

Salisbury, F. B. *The Biology of Flowering*. New York: Natural History Press, 1971.

Sargent, C. S. *Manual of the Trees of North America*. New York: Dover, 1961.

Spencer, E. R. *All About Weeds*. New York: Dover, 1957.

Stanford, E. E. *General and Economic Botany*. New York: Appleton-Century, 1937.

Stout, A. B. *Daylilies*. New York: Macmillan, 1934.

Stuart, W. *The Potato, Its Culture, Uses, History, and Classification*. Philadelphia: Lippincott, 1923.

Sturtevant, E. L. *Sturtevant's Edible Plants of the World*. Edited by U. P. Hedrick. New York: Dover, 1972.

Sudworth, G. B. *Forest Trees of the Pacific Slope*. New York: Dover, 1967.

Taylor, N. *Taylor's Encyclopedia of Gardening*. Edited by N. Taylor. Boston: Houghton Mifflin, 1948.

Thornber, J. J., and Bonker, F. *The Fantastic Clan*. New York: Macmillan, 1932.

Ukers, W. H. *Romance of Tea*. New York: Knopf, 1936.

Weldon, L. W., Blackburn, R. D., and Harrison, D. S. *Common Aquatic Weeds*. New York: Dover, 1969.

Wherry, E. T. *Wildflower Guide: Northeastern and Midland United States*. New York: Doubleday, 1948.

Wilder, L. B. *The Fragrant Garden*. New York: Dover, 1974.

Wilkie, D. *Gentians*. New York: Scribners, 1936.

Zucker, L. *Flowering Shrubs*. Princeton, N.J.: Princeton University Press, 1966.

Index